Developments in Marine Geology

Volume 6

Geology of the China Seas

Developments in Marine Geology

Volume 6

Geology of the China Seas

By

Pinxian Wang, Qianyu Li & Chun-Feng Li
State Key Laboratory of Marine Geology,
Tongji University, Shanghai

ELSEVIER   AMSTERDAM • BOSTON • HEIDELBERG • LONDON • NEW YORK • OXFORD
PARIS • SAN DIEGO • SAN FRANCISCO • SINGAPORE • SYDNEY • TOKYO

Elsevier
The Boulevard, Langford Lane, Kidlington, Oxford OX5 1GB, UK
Radarweg 29, PO Box 211, 1000 AE Amsterdam, The Netherlands

British Library Cataloguing in Publication Data
A catalogue record for this book is available from the British Library

Library of Congress Cataloging-in-Publication Data
A catalog record for this book is available from the Library of Congress

ISBN: 978-0-444-59388-7
ISSN: 1572-5480

For information on all Elsevier publications
visit our website at store.elsevier.com

Contents

1. Introduction   1

2. General Outline of the China Seas   11

  2.1 Introduction: Marginal Seas Between Asia and Pacific   11
  2.2 Bathymetry and Geomorphology   12
    2.2.1 East China Sea   16
    2.2.2 South China Sea   20
    2.2.3 Sensitivity to Eustatic Changes   22
  2.3 River Deltas and Catchments   23
    2.3.1 Rivers Emptying into the China Seas   23
    2.3.2 Yellow River (Huanghe)   26
    2.3.3 Yangtze River (Changjiang)   29
    2.3.4 Pearl River (Zhujiang)   31
    2.3.5 Red River (Sông Hồng)   33
    2.3.6 Mekong River   34
    2.3.7 Small Mountainous Rivers   37
  2.4 Oceanography and Climatology   37
    2.4.1 Monsoon   37
    2.4.2 Surface Circulation   42
    2.4.3 Temperature and Salinity   49
    2.4.4 Ocean Connection and Deepwater Circulation   57
    2.4.5 Deepwater Circulation   62
  2.5 Oceanographic Summary   64
    2.5.1 Monsoon and seasonality   65
    2.5.2 The role of "mixing mill"   66
    2.5.3 Regional impact   67
    References   67

3. Tectonic Framework and Magmatism   73

  3.1 Introduction   73
  3.2 Outline of Present-Day Tectonic Framework   74
  3.3 Mesozoic Tectonic Background   79
  3.4 Cenozoic Continental Margin Rifting and Subsidence   89
    3.4.1 Deep Crustal Structures   90
    3.4.2 Cenozoic Rifting and Subsidence Processes   102
  3.5 The SCS and Okinawa Trough   115
    3.5.1 SCS basin   116
    3.5.2 Okinawa Trough   137
  3.6 Magmatisms and Neotectonics   140
    3.6.1 Regional Magmatisms Revealed by Magnetic Data   142
    3.6.2 Magmatic Episodes   145

3.6.3 Regional Seismicity   153
3.6.4 Hydrothermal Activities   156
3.6.5 Post-Miocene Continental Margin Neotectonics   161
3.7 Tectonic Summary   164
References   165

4. Sedimentology   183

4.1 Introduction   183
4.2 Modern Sediment Distribution   184
4.2.1 Sediment Distribution Patterns   184
4.2.2 East China Sea   193
4.2.3 South China Sea   201
4.3 Depositional History   210
4.3.1 East China Sea Shelf   210
4.3.2 South China Sea Shelf   215
4.3.3 Deep Sea   226
4.4 Terrigenous Sediments   233
4.4.1 Deltaic Sands and Offshore Muds   233
4.4.2 Relict Sands and Sand Ridges   259
4.5 Biogenic Sediments   268
4.5.1 Carbonate Buildups and Coral Reefs   268
4.5.2 Calcareous Components   280
4.5.3 Siliceous Components   293
4.5.4 Organic Carbon   299
4.6 Volcanic Sediments   310
4.6.1 Okinawa Trough Volcanic Deposition   310
4.6.2 South China Sea Volcanic Deposition   312
4.7 Sedimentologic Summary   316
4.7.1 Origin of Sediments   316
4.7.2 Sediment Transport Dynamics   319
References   324

5. Basins and Stratigraphy   341

5.1 Introduction   341
5.2 An Overview of Lithostratigraphy   342
5.2.1 Basins and Pre-Cenozoic Basements   342
5.2.2 Cenozoic Stratigraphy   347
5.2.3 Biostratigraphic Framework   368
5.3 Regional Stratigraphy and Sequence Stratigraphy   381
5.3.1 East China Sea Sector   381
5.3.2 Northern South China Sea Sector   403
5.3.3 Southern South China Sea Sector   420
5.4 Isotopic and Astronomical Stratigraphy   438
5.4.1 Pliocene–Pleistocene Isotopic Records   440
5.4.2 Neogene Isotopic Records   443

5.5 Stratigraphic Summary 447
    5.5.1 Lithostratigraphy Patterns 447
    5.5.2 Stratigraphy and Evolution of the China Seas 449
    References 451

6. Paleoceanography and Sea-Level Changes 469

6.1 Introduction 469
6.2 Paleoceanography: Long-Term Trends 470
    6.2.1 East China Sea: Okinawa Trough 470
    6.2.2 South China Sea: Ocean Connection and Basin Evolution 472
    6.2.3 South China Sea: Upper Waters and Monsoon Climate 478
6.3 Late Quaternary Paleoceanography and Sea-Level Changes 494
    6.3.1 East China Sea 494
    6.3.2 South China Sea 515
6.4 Summary 551
    References 553

7. Hydrocarbon and Mineral Resources 571

7.1 Introduction 571
7.2 Hydrocarbon Resources 572
    7.2.1 Bohai Basin 572
    7.2.2 North Yellow Sea Basin 581
    7.2.3 South Yellow Sea Basin 583
    7.2.4 ECS Basin 585
    7.2.5 Northern SCS 594
    7.2.6 Southern SCS 609
7.3 Mineral Resources 617
    7.3.1 Coastal Placers 617
    7.3.2 Phosphorite 619
    7.3.3 Sulfide Deposits 620
    7.3.4 Ferromanganese Nodules/Crusts 620
7.4 Other Resources 622
    7.4.1 Gas Hydrate Potentials 622
    7.4.2 CO_2 gas Fields 626
7.5 Summary 628
    References 631

8. History of the China Seas 643

8.1 Introduction 643
8.2 Occurrences of Pre-Oligocene Marine Deposits 644
    8.2.1 Late Mesozoic Marine Deposits 644
    8.2.2 Paleogene Marine Deposits 647
8.3 Formation of Modern China Seas 652
    8.3.1 Formation of the Marginal Basins 652
    8.3.2 China Seas in Glacial Cycles 654

8.4 Debates on Cenozoic Marine Transgressions in China   657
    References   663

9. Concluding Remarks   667

9.1 Relatively Young Age of Formation   667
9.2 Effect of Topographic Reversal of East Asia   667
9.3 Restricted Connection with the Open Ocean   668
9.4 Extensive Continental Shelves   669
    References   670

Index   673

Chapter 1

Introduction

The China Seas, that is, East and South China Seas (SCSs), have recently become a focus of attention for the global community because of many new and ongoing scientific endeavors and rapid development of economy in the surrounding countries. Not only has the fast growth of hydrocarbon production attracted exploration activities of oil companies worldwide, but also the successive discoveries of gas hydrates and hydrothermal vents have also stimulated geological investigations in the region. The China Seas play a substantial role in environmental change and conservation for sustainable development as they are surrounded by the most densely populated areas in the world but have been threatened by all kinds of marine hazards, ranging from harmful algal bloom, hypoxia, and submarine earthquakes to monsoon flooding. On the other hand, the well-preserved thick sediment piles in the China Seas offer a unique opportunity to study the long-term climate variability, particularly the East Asian monsoon, and to fill up the gap in the poorly preserved geological records of the western Pacific.

The geological work on the China Seas can be traced back to the 1930s when Tingying Ma published his study on coral growth rate in the SCS (Ma, 1936; see Yang & Oldroyd, 2003, for comments). However, it was not until after World War II did Chinese scientists start to follow his step to explore the coral reefs in the late 1940s. China conducted the first survey of its entire coastal seas in the late 1950s (Zeng, 1994), resulting in some pioneering works in the early 1960s, such as the work on shelf sediment distribution by Qin (1963). In 1960, the first marine geology team was established in Tianjin for oil exploration in the Bohai Gulf and later in the Yellow Sea and the East China Sea (ECS). Even during the decade-long turmoil of the so-called Cultural Revolution, Chinese geologists and geophysicists continued their offshore exploration efforts and subsequently discovered oil and gas in the China Seas. The first oil-producing well was drilled in the Bohai Gulf, northern SCS, and the ECS, respectively in 1967, 1977, and 1982 (Xiao, 2000). In 2011, the offshore hydrocarbon output reached one-fifth of the total domestic production in China. The Chinese economic reform since 1978 has drastically increased the industrial and academic investments in marine science and technology.

Developments in Marine Geology, Vol. 6. http://dx.doi.org/10.1016/B978-0-444-59388-7.00001-9

Therefore, the China Seas have seen a boom in geological investigations over the last decades. The unprecedented development in marine geology of the China Seas can be demonstrated simply by the number of publications. Since 1980, at least 121 monographs and atlases on geology of the China Sea have been published in China (Table 1.1), written mostly in Chinese and only 9 of them in English.

Along with Chinese scientists across the Taiwan Strait, international scientists, especially those from the countries surrounding the China Seas, have contributed greatly as well. Japanese scientists, for example, have been studying the Okinawa Trough for about 30 years, and their discoveries from deep-water investigations are the most remarkable. Over 200 dives of the human-operated vehicles (HOV) and remotely operated vehicles (ROV) between 1984 and 2001 (Glasby & Notsu, 2003) had led to the discoveries of active hydrothermal fields with sulfide deposits in the mid-Okinawa Trough (Halbach et al., 1989) and sediment-hosted CO_2 lakes in the southern Okinawa Trough (Inagaki et al., 2006). Since the 1980s, Korean geologists studied many parts of the Yellow Sea and the adjacent ECS (Chough, 1983; Chough, Lee, & Yoon, 2000). For decades, many SE Asian countries have participated in offshore hydrocarbon exploration and marine geological researches in the SCS, as summarized by Wang and Li (2009).

The scientific attraction of the China Seas has resulted in rapid growth of international research projects and joint cruises. The first joint project between the United States and China was on the sediment dynamics of the Yangtze estuary and the shallow ECS, with successful cruises organized in 1980 and 1981 (Jin & Milliman, 1983). Another United States–China joint project was the integrated geophysical survey of the northern SCS margin, with a cruise from October to December 1985 (Nissen et al., 1995). Numerous international cruises on geological subjects have been implemented since 1990 using German and French research vessels, and the first cruise of the Ocean Drilling Program (ODP) in the China Seas took place in 1999 (Table 1.2).

TABLE 1.1 Monographs and Atlases on Geology of the China Seas Published in China from 1990 to 2012, Based on Incomplete Statistics

| Years | China Seas | East China Sea | | South China Sea | Sum |
		Yellow Sea and Bohai	East China Sea Proper		
1980–1989	9	4	7	11	31
1990–1999	14	6	8	22	50
2000–2012	17	4	10	9	40
Total	40	14	25	42	121

TABLE 1.2 Major International Geological Cruises to the China Seas Since 1990 Are Shown with their Main Themes

Cruise	Time	Theme	Reference
R/V Sonne, Germany			
SO72a	Oct. 25–Nov. 18, 1990	Sedimentation	Wong (1993)
SO95	Apr. 12–Jun. 05, 1994	Paleomonsoon	Sarnthein et al. (1994)
SO114	Nov. 20–Dec. 12, 1996	Pinatubo ash	Wiesner, Kuhnt, and Shipboard Scientific Party (1997)
SO115	Dec. 13–Jan. 16, 1997	Sunda Shelf	Stattegger, Kuhnt, Wong, and Scientific Party (1997)
SO132	Jun. 17–Jul. 09, 1998	Sedimentation	Wiesner, Kuhnt, and Shipboard Scientific Party (1998)
SO140	Apr. 03–May 04, 1999	Sedimentation	Wiesner, Stattegger, Kuhnt, and Shipboard Scientific Party (1999)
SO177	Jun. 02–Jul. 02, 2004	Gas hydrates	Suess (2005)
SO220	Apr. 14–May 16, 2012		
SO221	May 17–June 07, 2012		
R/V JOIDES Resolution, the United States (Ocean Drilling Program)			
ODP 184	Feb. 11–Apr. 12, 1999	Asian Monsoon	Wang, Prell, and Blum (2000)
ODP 195, Site 1202	Apr. 28–May 1, 2001	Okinawa Trough	Wei (2006)
IODP 349	Jan. 26–Mar. 30, 2014	Tectonics	Li, Lin, and Kulhanek (2013)
R/V Marion Dufresne, France			
MD106 (IMAGES III)	Apr. 16–Jun. 30, 1997	"IPHIS"	Chen, Beaufort, and the Shipboard Scientific Party of the IMAGES III/MD 106-IPHIOS Cruise (Leg II) (1998)
MD122 (IMAGES VII)	Apr. 30–Jun. 18, 2001	"WEPAMA"	Bassinot (2002)

Continued

TABLE 1.2 Major International Geological Cruises to the China Seas Since 1990 Are Shown with their Main Themes—Cont'd

Cruise	Time	Theme	Reference
MD147 (IMAGES XII)	May 15–Jun. 08, 2005	"Marco Polo 1"	Laj, Wang, and Balut (2005)
MD190 (CIRCEA)	Apr. 29–May 21, 2012	Monsoon, ocean, climate	
R/V L' Atlanta, France			
"Donghai"	Apr. 22–17 May, 1996	Sedimentation	Berné, Liu, Guéguen, et al. (1996)

The ever-closer international collaboration has efficiently promoted the development of geological science in the China Seas. An example is the joint cruise with the R/V Sonne between Germany and China in 1994 (Sarnthein et al., 1994), which brought the SCS into the global focus of paleomonsoon studies and secured the success of the subsequent ocean drilling leg in 1999. In 2004, scientists onboard the same vessel discovered cold-seep carbonates in the northern SCS (Suess, 2005), further stimulating gas hydrate exploration in the region. Of particular significance was ODP Leg 184 targeted on the evolution and variability of the East China monsoon (Wang et al., 2000), a groundbreaking endeavor in the deep-sea research of the China Seas.

The last decades also witnessed closer scientific cooperation and exchanges inside Asia. Most prominent is the series of meetings called "the international conference on Asian Marine Geology." Initiated by Chinese and Japanese marine scientists in 1988, the conference has been held seven times in five Asian countries (Table 1.3) to promote collaboration between Asian marine geologists. Over the decades, there have been a number of international research projects on geology in the China Seas and beyond, such as the UNESCO/IOC projects on late Quaternary paleogeographic mapping in 1990s (Wang et al., 1997) and on SCS fluvial sediments (*FluSed*) (2008–2014) and the IGCP Project 475 on Deltas in the Monsoon Asia-Pacific Region (DeltaMAP) since 2003 (Chen, Watanabe, & Wolanski, 2007), among others.

In addition to those volumes in Chinese listed in Table 1.1, more and more papers in English on various subjects of China Sea geology have appeared recently, with the majority as journal articles dealing with specific and regional topics. The present volume intends to synthesize the rich data

TABLE 1.3 International Conferences on Asian Marine Geology

No.	Dates	Venue	Publications
1	Sep. 07–10, 1988	Shanghai, China	Wang, Lao, and He (1990)
2	Aug. 19–22, 1991	Tokyo, Japan	Tokuyama, Taira, and Kuramoto (1995)
3	Oct. 17–21, 1995	Cheju Is., Korea	
4	Oct. 10–14, 1999	Qingdao, China	
5	Jan. 13–18, 2004	Bangkok, Thailand	
6	Aug. 29–Sep 1, 2008	Kochi, Japan	
7	Oct. 11–14, 2011	Goa, India	

scattered in journals and books, particularly those generated by Chinese marine and petroleum geologists, and to provide an updated overview on geology of the seas. In Chapter 2, we will briefly review the general features of the China Seas and their environments, including the bathymetry and geomorphology, river deltas and catchments, and oceanography and climatology of the region. Chapter 3 will discuss the tectonic history of the basin formation, the widely spread magmatism, and neotectonics. Chapter 4 describes the sedimentation patterns in the modern China Seas, followed by those in the late Quaternary glacial cycles. The major sedimentary basins and their stratigraphic features are introduced in Chapter 5, emphasizing on marine depositional sequences. Chapter 6 is devoted to paleoceanography and sea-level changes, largely on the basis of results from the ocean drilling and sediment cores. In Chapter 7, we outline the production and potential of hydrocarbon and mineral resources in the China Seas, including the occurrences of gas hydrate and hydrothermal sulfides. Chapter 8 is an attempt to synthesize the geological history of the China Seas. The book concludes with some remarks on marginal seas in Chapter 9. This is a teamwork product, with each chapter assigned to one of authors. Chapters 2, 6, and 9 and this chapter were drafted by Wang; Chapters 3 and 7, by Q. Li; Chapters 4 and 5, by C.-F. Li; and Chapter 8, by Wang and Q. Li.

A trouble in reading literature from this region is the lack of uniform use of geographic names. For example, Chinese Pinyin is frequently used by Chinese authors but may differ from the usage elsewhere. To ease the readers, we list synonyms of geographic names in the China Seas in Table 1.4.

TABLE 1.4 Geographic and Basin Names Frequently Referred to in this Volume Using Chinese "Pinyin" Are Listed to Show their Synonyms, with Names in Brackets Used Intermittently

Name Used in this Volume	Synonym	Location
Strait		
Bashi Strait	Luzon Strait	Taiwan to Luzon (the Philippines)
Gulf		
Bohai Gulf	Bohai Sea	Northernmost ECS
Beibuwan or Beibu Gulf	Gulf of Tonkin	Vietnam to China
River		
Yellow River	Huanghe	Northern China to Bohai Gulf
Yangtze River	Changjiang	Central China to southern ECS
Pearl River	Zhujiang	Southern China to northern SCS
Red River	Song Hong, Honghe	Indochina peninsula to SW SCS
Reef/Island		
Dongsha	Pratas Islands	Northeastern SCS
Nansha	Spratly Islands and surroundings	Southern SCS
Yongshu Reef	Fiery Cross or Northwest Investigator	Southern SCS
Meiji Reef	Mischief Reef	Southern SCS
Huanglu Reef	Royal Charlotte Reef	Southern SCS
Sanjiao Reef	Livock Reef	Southern SCS
Xian-e Reef	Alicia Anne Reef	Southern SCS
Xinyi Reef	First Thomas Shoal	Southern SCS
Zhubi Reef	Subi Reef	Southern SCS
Xisha	Paracel Islands	Northern SCS
Yongxing Island	Woody Island	Northern SCS
Shi Island	Rocky Island	Northern SCS

TABLE 1.4 Geographic and Basin Names Frequently Referred to in this Volume Using Chinese "Pinyin" Are Listed to Show their Synonyms, with Names in Brackets Used Intermittently—Cont'd

Name Used in this Volume	Synonym	Location
Chenhang Island	Duncan Island	Northern SCS
Dongdao Island	Lincoln Island	Northern SCS
Zhongsha	Macclesfield Bank	Northern SCS
Huangyandao	Scarborough Shoal	Northern SCS
Liyue	Reed Bank	Southern SCS
Nantong Reef	Louisa Reef	Southern SCS
Trench/Trough		
Xisha Trought	Palawan–Borneo Trough	Southern SCS
Basin		
Taixinan (Tainan) Basin	SW Taiwan Basin	Northeastern SCS
Pearl River Mouth Basin	Zhujiangkou Basin	Northern SCS
Qiongdongnan Basin	SE Hainan Basin	Northern SCS
Beibuwan Basin	Beibu Gulf Basin	Northwestern SCS
Yinggehai Basin	Song Hong Basin	Northwestern SCS
Zhongjiannan Basin	Nha Trang Basin	Northwestern SCS
Wan'an Basin	Nam Con Son Basin	W South China Sea
Zengmu Basin	E Natuna–Sarawak Basins	Southern SCS

ACKNOWLEDGMENTS

This work was supported by the National Natural Science Foundation of China (91128000, 91028007, and 91228203). The authors express their deep gratitude to Dr. Yulong Zhao who actively participated in the preparation of this book and greatly contributed to its final completion. Helve Chamley is acknowledged for initiation of the work. This work was greatly benefited from fruitful discussions with Chiyu Huang, James T. Liu, Jinhai Zhao, Jianyi Jia, Huanjiang Chen, and Jialin Wang. We thank Weilin Zhu, Gongcheng Zhang, and Huaiyang Zhou for providing valuable data and advices. The cover figure and some text figures were provided by Wei Huang. Jian Wang, Taoran Song, Weinan Liu, Liang Dong, Xixi Dong, Xiao Li, Zhenghua Liu, Anlin Ma, Anhui Meng, and Hao Zhang are thanked for their technical assistance in figure preparation. We thank all publishers and copyright holders for granting permissions to use published figures.

REFERENCES

Bassinot, F. (2002). *IPF les rapports des campagnes à la mer. WEPAMA Cruise MD 122/ IMAGES VII on board RV "Marion Dufresne"*, Leg 1, 301 pp.

Berné, S., Liu, Z., Guéguen, B., et al. (1996). *Donghai Cruise. Preliminary report: Shangai 22 April, 17 May 1996*. Plouzané: IFREMER.

Chen, M. T., Beaufort, L., & the Shipboard Scientific Party of the IMAGES III/MD 106-IPHIOS Cruise (Leg II). (1998). Exploring quaternary variability of the East Asian monsoon, Kuroshio Current, and Western Pacific Warm Pool systems: High-resolution investigations of paleoceanography from the IMAGES III (MD 106)—IPHIS cruise. *TAO Taipei, 9*(1), 129–142.

Chen, Z., Watanabe, M., & Wolanski, E. (2007). Sedimentological and ecohydrological processes of Asian deltas: The Yangtze and the Mekong. *Estuarine, Coastal and Shelf Science, 71*, 1–2.

Chough, S. K. (1983). *Marine geology of Korean seas*. Springer, 157 pp.

Chough, S. K., Lee, H. J., & Yoon, S. H. (2000). *Marine geology of Korean seas* (2nd ed.). Elsevier, 313 pp.

Glasby, G. P., & Notsu, K. (2003). Submarine hydrothermal mineralization in the Okinawa trough, SW of Japan: An overview. *Ore Geology Reviews, 23*, 299–339.

Halbach, P., Nakamura, K.-I., Wahsner, M., Lange, J., Sakai, H., Käselitz, L., et al. (1989). Probable modern analogue of Kuroko-type massive sulphide deposits in the Okinawa trough back-arc basin. *Nature, 338*, 496–499.

Inagaki, F., Kuypers, M. M. M., Tsunogai, U., Ishibashi, J.-I., Nakamura, K.-I., Treude, T., et al. (2006). Microbial community in a sediment- hosted CO_2 lake of the southern Okinawa Trough hydrothermal system. *Proceedings of the National Academy of Sciences of the United States of America, 103*(38), 14164–14169.

Jin, Q., & Milliman, J. D. (1983). In: *Proceedings of international symposium on sedimentation on the continental shelf with special reference to the East China sea. April 12-16, 1983*. Hangzhou, China: China Ocean Press, 952 pp.

Laj, C., Wang, P., & Balut, Y. (2005). *IPEV les rapports de campagnes à la mer. MD147/MARCO POLO-IMAGES XII à bord du "Marion Dufresne"*, 59 pp.

Li, C. F., Lin, J., & Kulhanek, D. K. (2013). South China Sea tectonics: Opening of the South China Sea and its implications for southeast Asian tectonics, climates, and deep mantle processes since the late Mesozoic. *Integrated Ocean Drilling Program Expedition, 349*. http://dx.doi.org/10.2204/iodp.sp.349.2013, Scientific Prospectus.

Ma, T. Y. H. (1936). Reef corals and the growth rate of corals along the Sea of china. *Geological Review, 1295–300* (in Chinese).

Nissen, S. S., Hayes, D. E., Buhl, P., Diebold, J., Yao, B., Zeng, W., et al. (1995). Deep penetration seismic soundings across the norhtern margin of the South China Sea. *Journal of Geophysical Research, 100*, 22407–22433.

Qin, Y. (1963). A preliminary study on the topography and bottom sediment types of the continental shelf of China Sea. *Oceanologia et Limnologia Sinica, 5*(1), 71–86 (in Chinese, with Russian abstract).

Sarnthein, M., Pflaumann, U., Wang, P., & Wong, H. K. (Eds.), (1994). *Preliminary report on SONNE-95 Cruise "Monitor Monsoon" to the South China Sea. Berichte-reports*. In *Geol.-Palaont. Inst. Univ. Kiel*, Vol. 48, 225 pp.

Stattegger, K., Kuhnt, W., Wong, H. K., & Scientific Party. (1997). Cruise report SONNE 115 SUNDAFLUT. Berichte-report. In *Institüt für Geowissenschaften, Univ. Kiel: Vol. 86*, 211 pp.

Suess, E. (2005). *RV SONNE cruise report SO 177, Sino–German cooperative project, South China Sea continental margin: Geological methane budget and environmental effects of methane emissions and gashydrates: IFM-GEOMAR Reports*, 133 pp.

Tokuyama, H., Taira, A., & Kuramoto, S. (1995). Asian marine geology. *Marine Geology, 127*, 117.

Wang, P., Bradshaw, M., Ganzei, S. S., Tsukawaki, S., Bin Hassan, K., Hantoro, W. S., et al. (1997). West Pacific marginal sea during last glacial maximum: Amplification of environmental signals and its impact on monsoon climate. In P. Wang & W. Berggren (Eds.), *Proceedings of the 30th international geological congress 13* (pp. 65–85). The Netherlands: VRP Publishers.

Wang, P., Lao, Q., & He, Q. (Eds.), (1990). *Proceedings of the first international conference on asian marine geology, Sept. 7-10, 1988.* Beijing: China Ocean Press, 395 pp.

Wang, P., & Li, Q. (Eds.), (2009). *The South China Sea: Paleoceanography and sedimentology:* Springer Publishing, 506 pp.

Wang, P., Prell, W. L., & Blum, P. (Eds.), (2000). *Proceedings of the ocean drilling program, initial reports: Volume 184.* Texas A&M University, College Station, TX: Ocean Drilling Program [CD-ROM].

Wei, K.-Y. (2006). Leg 195 synthesis: Site 1202—Late quaternary sedimentation and paleoceanography in the southern Okinawa trough. In M. Shinohara, M. H. Salisbury, & C. Richter (Eds.), *Proceedings of the Ocean Drilling Program Scientific Results: Vol. 195.* http://www-odp.tamu.edu/publications/195_SR/VOLUME/SYNTH/SYNTH3.PDF.

Wiesner, M. G., Kuhnt, W., & Shipboard Scientific Party. (1997). *Cruise report, R/V Sonne SO-114, Manila—Kota Kinabalu.* Institute for Biogeochemistry and Marine Chemistry, University of Hamburg, IBMC Library Ref No IIA 942, 55 pp.

Wiesner, M. G., Kuhnt, W., & Shipboard Scientific Party. (1998). *Cruise report, R/V Sonne SO-132, Singapore—Manila.* Institute for Biogeochemistry and Marine Chemistry, University of Hamburg, IBMC Library Ref No IIA 943, 120 pp.

Wiesner, M. G., Stattegger, K., Kuhnt, W., & Shipboard Scientific Party. (1999). *Cruise report, R/V Sonne SO-140, Singapore-Nha Trang-Manila. Berichte-reports. Reihe Institut für Geowissenschaften Universität Kiel: Vol. 7,* 157 pp.

Wong, H. K. (Ed.), (1993). *Quaternary sedimentation processes in the South China Sea,* R/V Sonne, Cruise SO 72A, BMBF, Germany, final report.

Xiao, H. (Ed.), (2000). *Chronicles of marine geological survey in peoples republic of China.* China Ocean Press, 259 pp (in Chinese).

Yang, J.-Y., & Oldroyd, D. (2003). A Chinese palaeontologist, Ma ting Ying (1899–1979): From coral growth-rings to global tectonics. *Episodes, 26*(1), 19–25.

Zeng, C.-K. (1994). Introduction. In D. Zhou, Y.-B. Liang, & C.-K. Zeng (Eds.), *Oceanology of China seas I.* Kluwer Academic Publishers.

General Outline
of the China Seas

2.1 INTRODUCTION: MARGINAL SEAS BETWEEN ASIA AND PACIFIC

A series of marginal seas separate Asia from the Pacific, the largest continent from the largest ocean, by straddling the world's largest subduction zone of the western Pacific. From north to south, four large marginal seas are the sea of Okhotsk, sea of Japan, East China Sea (ECS), and South China Sea (SCS) (Figure 2.1A). These seas cover approximately $7 \times 10^6 \, \text{km}^2$ in area and $7.6 \times 10^6 \, \text{km}^3$ in volume, accounting for 2% and 0.5% of the global ocean, respectively. However, the role of these marginal seas in affecting the global climate changes goes far beyond their size, because they lie in the site of the most active material and energy exchanges between continent and ocean.

Table 2.1 compares the morphological features of the four major seas, including their size and the extent of isolation from the open ocean. This is an area of the most active sea–land interactions, covering all the lithosphere, hydrosphere, and atmosphere. Apart from the atmospheric transmission of vapor and dust, both river and oceanic currents enter the marginal seas where they are able to mix (Figure 2.1B and C). The hydrologic system in the marginal seas is thus highly sensitive to changes in both the continent and ocean and, in particular, to sea-level fluctuations during glacial–interglacial cycles. Even a minor drop of the sea level can cause major changes in water circulation and consequently the nature of regional land–sea interactions (Wang, 1999, 2004).

The ECS and SCS comprise the southern half of these four western Pacific marginal seas. Compared to their northern counterparts, the China Seas are distinguished by two major features that are critical for sedimentologic and paleoceanographic studies. The first is their lower latitudes in position and hence higher water temperature resulting in generally good carbonate preservation, in contrast to extremely poor carbonate preservation in the sea of Okhotsk and sea of Japan (Figure 2.2). The second feature is the high supply of terrigenous sediments by five large rivers into the China Seas, in contrast to

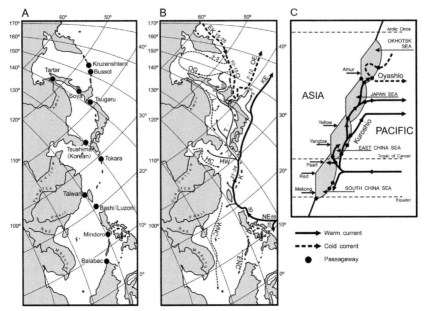

FIGURE 2.1 The system of western Pacific marginal seas between the Asian continent and the Pacific Ocean (Okhotsk, Japan, and East China and South China Seas) and the western boundary currents. (A) Main passageways (black dots) and rivers emptying into the marginal seas. (B) Western boundary currents flowing through the marginal seas. Numbers give the water transport in Sv (1 Sv = 10^6 km^3/s). EK, East Kamchatka Current; NNC, East Nansha Coast Current; HC, Huanghai (Yellow Sea) Coastal Current; HW, Huanghai Warm Current; K, Kuroshio Current; KE, Kuroshio Extension; NE, North Equatorial Current; O, Oyashio Current; OE, Oyashio extension; OG, Okhotsk Gyre; S, Soya Current; T, Tsushima Current. (C) A simplified sketch showing the hydrographic system in the marginal seas (Wang, 2004).

no large rivers that empty into the sea of Japan and only one Amur River into the sea of Okhotsk. The river-derived high sedimentation rates are responsible for the occurrences of the two widest continental shelves on earth, that is, the ECS shelf and the Sunda Shelf of the southern SCS (Figure 2.3).

With these advantageous geologic features, the China Seas are endowed with a special attraction for Earth scientists worldwide. Unlike the Atlantic, the western Pacific lacks well-preserved paleoclimate archives as its bottom is bathed with more corrosive waters. Therefore, the well-preserved hemipelagic sediments in the China Seas with high depositional rates provide some indispensable geologic records for paleoclimate study in the western Pacific region.

2.2 BATHYMETRY AND GEOMORPHOLOGY

The China Seas extend from 41°N to the equator with a total area of about 4.7×10^6 km^2 (Figure 2.4). This binary basin consists of, in the northeast,

TABLE 2.1 Morphological Features of the Marginal Seas Between the Asian Continent and the Pacific Ocean

Sea Basin	Area (10^3 km^2)	Volume (10^5 km^3)	Shallow Sea Area (<200 m) (%)	Aver. Depth (m)	Max Depth (m)	Sill Depth (m)	Passage Width (km)
Sea of Okhotsk	1590	1365	41.2	777	3374	~2000	455
Sea of Japan	980	1713	26.3	1361	4049	130	136
East China Sea	1170	303	75.6	370	2719	>2000	981
South China Sea	3500	4242	52.4	1212	5377	2400	950

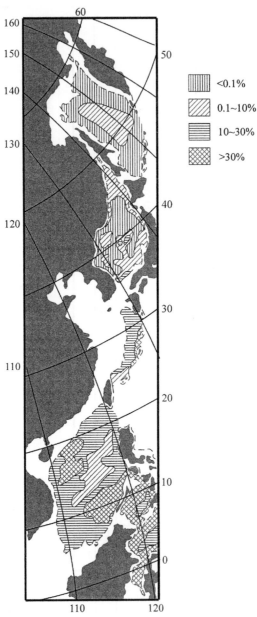

FIGURE 2.2 Distribution of carbonate (%) in surface sediments of the four western Pacific marginal seas deeper than 200 m (Wang, 1999).

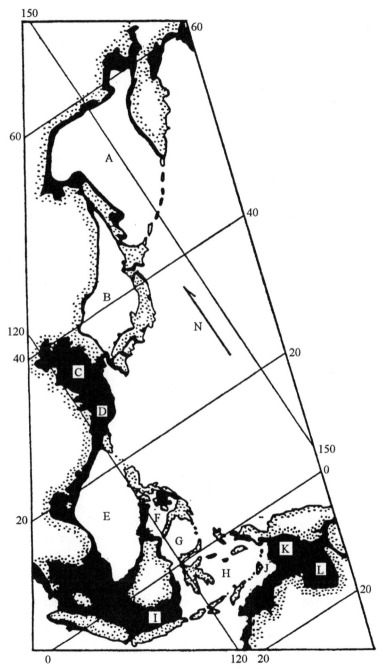

FIGURE 2.3   Emerged shelves (solid areas) in the western Pacific marginal seas at the Last Glacial Maximum (LGM). (A) The sea of Okhotsk, (B) Sea of Japan, (C) Yellow Sea and Bohai Gulf, (D) East China Sea, (E) South China Sea, (F) Sulu Sea, (G) Celebes Sea, (H) Banda Sea, (I) Java Sea, (J) Timor Sea, (K) Arafura Sea, and (L) Gulf of Carpentaria (Wang, 1999).

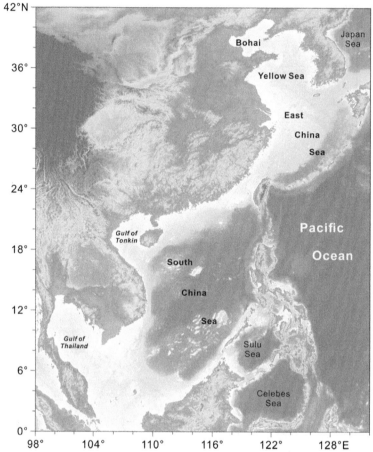

FIGURE 2.4 The China Seas.

the ECS dominated by a wide continental shelf and, in the southwest, the SCS centered at its deepwater basin with an oceanic crust. Although the two seas are well separated, they are connected through the shallow Taiwan Strait.

2.2.1 East China Sea

The ECS, as referred to in this book, includes the Yellow Sea and its innermost bay in the north called the Bohai Sea or Bohai Gulf (Figure 2.5). The ECS is bounded by the Korean Peninsula, by Kyushu and the Ryukyu Islands on the east, by the island of Taiwan on the south, and by mainland China and the Asian continent on the west and north.

The ECS proper has an area of about 770,000 km², and its floor is divided into two contrasting parts: the continental shelf with shelf break at 140–160 m and the Okinawa Trough with an average water depth of 370 m. The shelf occupies ~2/3 of the ECS proper by area and features with a large delta of the Yangtze River (Changjiang) that segregates to sand ridges in the middle and outer shelf (Figure 2.6). The Okinawa Trough is bordered by the Ryukyu Islands on the east and makes up the rest ~1/3 of the ECS area. The trough is

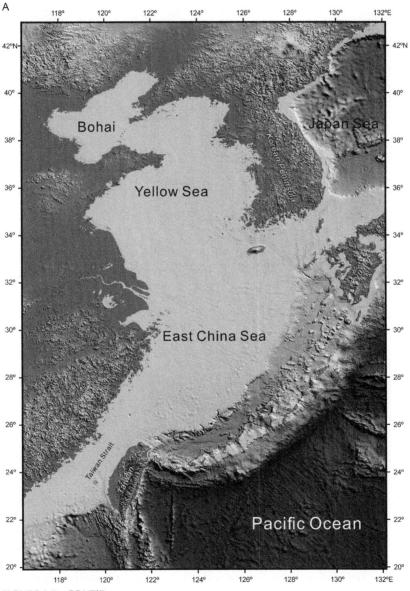

FIGURE 2.5—CONT'D

B

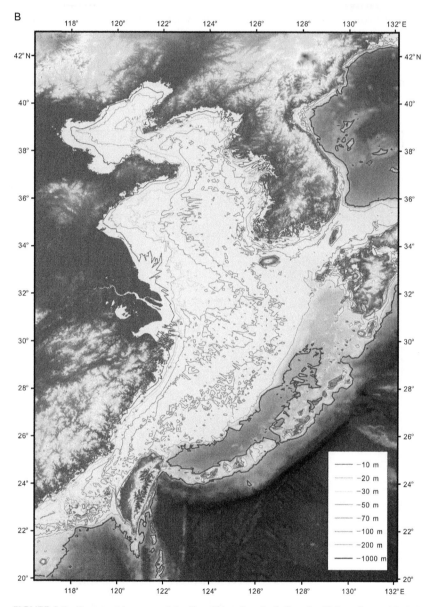

FIGURE 2.5 Topographic maps of the East China Sea (including the Yellow Sea and Bohai Gulf). (A) Geomorphological sketch; (B) bathymetry. *By Wei Huang.*

shallower in the NE part, deepening toward the SW end where the maximum depth exceeds 2700 m.

The Yellow Sea is the northern extension of the ECS shelf, with a total area of about 400,000 km². It is a semiclosed basin bordered by the Korean

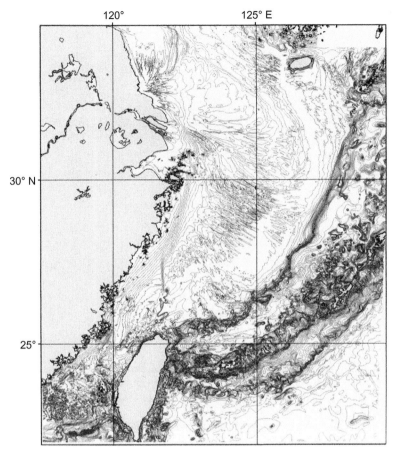

FIGURE 2.6 Geomorphological features of the ECS. Isobaths interval: 5 m in water depth <100 m, 10 m between 100 and 200 m, 100 m beyond 200 m (Li, 2008).

Peninsula on the east and mainland China on the west, and its southern limit is the line running from Chiju Island, South Korea, to the mouth of the Yangtze. The Yellow Sea is a shallow shelf with a water depth of 44 m in average, and its deepest site is located in its southeast close to the Chiju Island. The name of the Yellow Sea originates from the sediment particles from the Yellow River that turn the surface of the water yellow in color, but it is called the West Sea by some Korean authors. Traditionally, the Yellow Sea is divided into a larger southern Yellow Sea and a smaller northern Yellow Sea, by a line from the easternmost tip of China's Shandong Peninsula to the western end of North Korea's South Hwanghae Province. For the northern Yellow Sea, a name of Korean Bay has also been in use (Figure 2.7).

The Bohai Gulf is often called the Bohai Sea in China where it has been considered as one of four seas off the Chinese coast together with SCS,

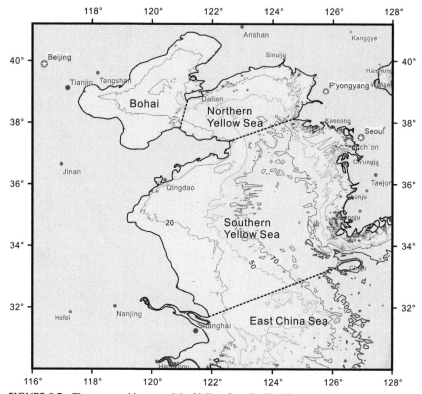

FIGURE 2.7 The topographic map of the Yellow Sea. *By Wei Huang.*

ECS, and Yellow Sea. In reality, the Bohai is a shallow and relatively small gulf basin of the Yellow Sea, with a total area of about 77,000 km^2 and an average depth of only 18 m. The Bohai Gulf is an inner basin of China, bounded by the Liáodōng Peninsula on the northeast and Shandong Peninsula on the south. Several rivers drain the Bohai from its SW and NE borders, and the Bohai Strait in the SE connects it with the Yellow Sea at a maximal water depth of 86 m. In and surrounding the Strait, a series of sand ridges are developed (Figure 2.8).

2.2.2   South China Sea

The SCS is the largest marginal sea in the western Pacific, embracing an area of about 3.5×10^6 km^2 and extending from the Tropic of Cancer to the equator, across over 20° of latitude. It is bordered on the NW by the Asian continent and on the SE by a series of islands, from Taiwan, Luzon, and Palawan on the east to Borneo and Sumatra on the south. The deep basin, the continental slope, and the continental shelf, respectively, cover about 15%, 38%, and 47% of the total SCS area with an average water depth of 1140 m. The major

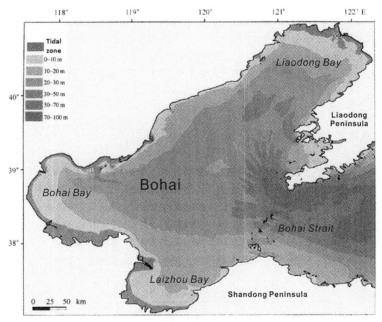

FIGURE 2.8   Topography of the Bohai Gulf (Wang, 2012).

feature of the SCS topography is the rhomboid deep basin with an oceanic crust, extending from NE to SW (Figure 2.9). The deep basin has an average water depth of ~4700 m, with a maximum of 5559 m reported from its eastern margin yet to be confirmed by survey.

The central deep basin is surrounded by continental and island slopes, topographically dissected and often studded with coral reefs (Figure 2.10). The northern slope with Dongsha islands is separated by the Xisha Trough from the western slope with Xisha (Paracel Islands) and Zhongsha reefs (Macclesfield Bank), while the southern slope is occupied by the Nansha terrace, the largest reef area in the SCS. The Nansha Islands are scattered on a carbonate platform known as "Dangerous Grounds," covering a broad area of ~570,000 km^2. The eastern slope is narrow and steep, bordered by deepwater Luzon Trough and Manila Trench (Figure 2.9).

The continental shelf is better developed on the northern and southern sides of the SCS, and both these shelves are narrower in the east and broader in the west. On the northern shelf, there are a number of submarine deltas developed off the Pearl and other river mouths, and fringing reefs are growing in the NW along the coasts of Leizhou Peninsula and Hainan Island. Exceeding ~300 km in width, the Sunda Shelf in the southwestern SCS is one of the largest shelves in the world. A number of submerged river valleys observed on the bottom of the Sunda Shelf are relict of a large river network, the paleo-Sunda rivers, from the Last Glacial Maximum (see Figure 6.52).

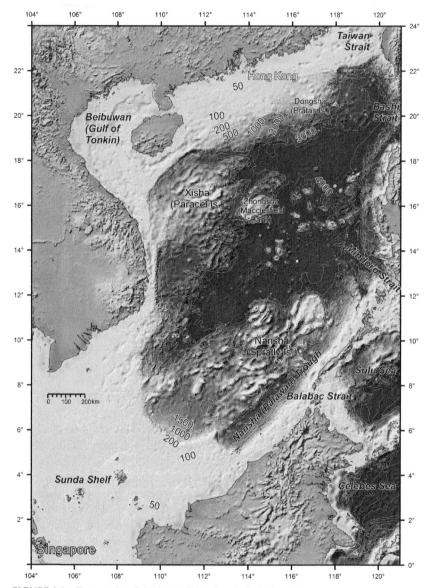

FIGURE 2.9 Topography of the South China Sea. Isobaths in meters.

2.2.3   Sensitivity to Eustatic Changes

One of the most remarkable features of the China Seas is their high sensitivity
to sea-level fluctuations in the glacial cycles for two reasons. First is their
extensive continental shelves, as the ECS shelves, including the entire Yellow
Sea and the Bohai Gulf, and the Sunda Shelf of the SCS are among the largest
continental shelves of the modern world (Figures 2.5 and 2.9). Second is the

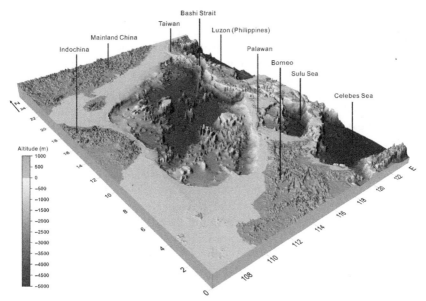

FIGURE 2.10   Three-dimensional diagram shows the main geomorphological features of the SCS. *From Wang and Li (2009)*.

limited connection with the open ocean, particularly for the SCS, as its water exchange with the Pacific is restricted to the Bashi Strait between Taiwan and Luzon islands with a sill depth of ~2400 m (Table 2.1; Figure 2.9).

As a result, even a small drop of the sea level can cause major changes in sea–land interactions and hence the regional environmental responses. During the Last Glacial Maximum, for example, the entire ECS was emerged with only the Okinawa Trough covered by seawater. In the south, the exposed Sunda Shelf was drained by the paleo-Sunda river that radically changed the oceanographic and sedimentologic settings in the SCS. The modern SCS is open on its northeastern and southwestern sides, but was largely isolated from the open ocean during the Last Glaciation. Its southern connection to the Indian Ocean is at present limited to the uppermost 30–40 m, but was completely closed at lowered sea-level stand in glacial times, when the Taiwan Strait was also exposed. Deeper passages of the semiclosed SCS that maintained its connection to other sea basins throughout the glacial cycles are the Bashi Strait with a sill depth of ~2400 m connecting to the Philippine Sea and the Mindoro and Balabac Straits connecting to the Sulu Sea, with sill depths of ~420 m and ~100 m, respectively (Figure 2.9).

2.3   RIVER DELTAS AND CATCHMENTS

2.3.1   Rivers Emptying into the China Seas

As the major contributor of terrigenous sediments on Earth, Asia provides about 50% of the total sediment load transported to the global ocean by rivers

(Lu & Chen, 2008). Among the rivers emptying into the China Seas, five major ones play a dominant role: the Yangtze (Changjiang), Yellow (Huanghe), Pearl (Zhujiang), Red (Sông Hồng), and Mekong Rivers (Table 2.2; Figure 2.11) (Wang et al., 2011). Together, these five rivers supply ~1900 Mt (10^6 t) of sediment annually, accounting for ~10% of the global river sediment flux (Milliman & Farnsworth, 2011). The voluminous river-delivered sediment was fully responsible for the development of mega deltas and extensive continental shelves that feature the China Seas.

The high sediment supply to the China Seas has been largely preceded by topography and climate of the region and geologic features of the drainage basins but enhanced considerably by human activities over the last couple of millennia. The sources of the Yangtze, Yellow, and Mekong Rivers are all located in the Tibetan Plateau at elevations exceeding 4000–5000 m, resulting in a strong topographic gradient of their courses. Under the prevailing monsoonal climate, the watersheds of the Yangtze, Pearl, Red, and Mekong, except the Yellow in the north, enjoy plentiful rainwater resources, with annual precipitation over 1000 mm on average (Table 2.2). A combination of vertical contrast in topography and monsoon precipitation favors erosion and transport of terrigenous material in the four large drainage regions. However, the Yellow River has semiarid climate conditions in its watershed and drains the Loess Plateau in its middle course. The easily erodible loess landscape has contributed to high suspended-sediment concentrations in the Yellow River water (~25 g/l). Therefore, although the water discharge of the Yellow River is nearly $20\times$ lesser than the Yangtze, its sediment discharge is more than twice as much as the Yangtze provides (Table 2.2).

As a distinct geographic feature caused by the high sediment supply, the development of mage deltas in the shallow China Seas depends on some boundary conditions such as sediment supply level, accommodation space, coastal energy, and density difference of waters to shape their morphology and size (Syvitski & Saito, 2007). The two types of deltas have been distinguished: the river-dominated type or "male deltas," such as the Yellow, Red, and Mekong deltas, and the tide-dominated type or "female deltas," including the Yangtze and Pearl Deltas (Duc, Nhuan, & Ngoi, 2012; Wang, Yu, & Huang, 2012).

As nearly half of a billion people in the world live on or near deltas, often in megacities, the anthropogenic shrinkage of mega deltas in areas since the recent decades has become an alarming issue (Saito, Chaimanee, Jarupongsakul, & Syvitski, 2007; Syvitski et al., 2009). The mega-delta subsidence in recent decades is caused by a decrease of sediment supply and a relative sea-level rise (i.e., land subsidence) and by excess groundwater extraction. Together with the destruction of coastal ecosystems, such as mangrove deforestation, these activities have resulted in the present severe coastal erosion of the mega deltas of Asia (Saito et al., 2007). Over a thousand years, the five major rivers mentioned earlier in the text have annually delivered

TABLE 2.2 Features of the Five Major Rivers Discharging into the China Seas: Catchment Area, River Length, Maximal Elevation, Annual Precipitation of the Area, Water Discharge, Sediment Discharge, and Delta Area

River	Area (10^3 km^2)	Length (km)	Max Elev. (m)	Ann. Prec. (mm)	Water Disch. (km^3/Year)	Sed. Disch. (Mt/Year)	Delta Area (10^3 km^2)
Yellow	790	5400	4300	300	50	1080	36
Yangtze	1800	6300	5100	1190	900	420	66
Pearl	440	2100	1800	1600–2300	300	70	41
Red	150	1100	2200	1120	120	130	14
Mekong	810	4400	5400	1300	500	160	93

Mt, million ton.
Based on Wang (2012) and Wang et al. (2011).

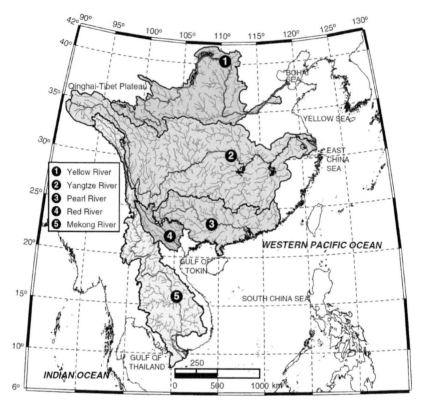

FIGURE 2.11 Five major rivers discharging into the China Seas and their catchment areas (Wang et al., 2011).

~ 1900 or $\sim 2000 \times 10^6$ t of sediment to the China Seas, but the present sediment flux has reduced to $\sim 600 \times 10^6$ t only, mainly due to dam construction and other human interventions (Wang et al., 2011). The societal concern about environment problems in river catchments and deltas has stimulated research activities devoted to rivers and deltas, in particular in the Asian region, which have greatly improved our knowledge of fluvial processes and delta development. Numerous international projects and meetings on Asian rivers have generated a large number of publications, and at least 12 special issues published in scientific journals over the past 12 years have contributed to river-related topics (Table 2.3), from which many significant results are to be cited in the following sections.

2.3.2 Yellow River (Huanghe)

The Yellow River is the second-longest river in China after the Yangtze and the sixth-longest in the world with an estimated length of 5400 km, draining

TABLE 2.3 Special Issues on Asian Large Rivers Published Within the Last 12 Years

Year	Topic	Journal	Volume (Issue)
2001	Yangtze River, China	Geomorphology	41 (2/3)
2007	Sedimentologic and biological processes of catchment to estuary: Example from the Yangtze and Mekong Rivers	Estuarine, Coastal, and Shelf Science	71 (1/2)
	Large monsoon rivers of Asia	Geomorphology	85 (3/4)
2008	Large Asian rivers and their interactions with estuaries and coasts	Quaternary International	186
2009	Short- and long-term processes, landforms, and responses in large rivers	Geomorphology	113
	Larger Asian rivers: Climate change, river flow, and sediment flux	Quaternary International	208
2010	Harmonizing river catchment and estuary	Estuarine, Coastal, and Shelf Science	86 (3)
	Larger Asian rivers: Climate change, river flow, and watershed management	Quaternary International	226
2011	The mega deltas of Asia: Interlinkage of land and sea and human development	Earth Surface Processes and Landforms	36
	Larger Asian rivers: Climate, hydrology, and ecosystems	Quaternary International	244
2012	Geomorphology of large rivers	Geomorphology	147/148
	Larger Asian rivers: Climate, water discharge, water, and sediment quality	Quaternary International	282

a total basin area of 790×10^3 km^2. Originating from the Tibetan Plateau, the river is commonly divided into three segments with very different geomorphologies. In its upper reaches, the elevation of the Yellow River drops by 3500 m, on an average grade of 0.10%. The middle stream of the Yellow River passes through the Loess Plateau, where the large amount of mud and sand discharges into the river and makes the Yellow River the most sediment-laden river in the world. In the lower reaches, the river is confined to a levee-lined course before emptying into the Bohai Gulf (Figure 2.12).

The extremely high sediment load of the Yellow River can be learned from silt concentration measured at some stations between Longmen and Lijin

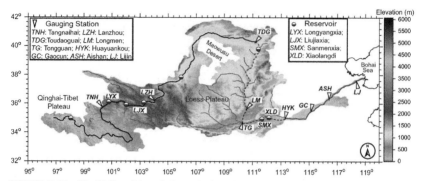

FIGURE 2.12   The Yellow River drainage basin, with locations of major gauging stations and reservoirs. The mainstream is divided into three segments: the upper reaches (bold line), the middle reaches (dashed line), and the lower reaches (bold line) (Wang et al., 2010).

(see LM and LJ in Figure 2.12), which varied from 222 to 933 kg/m^3 during 1966–1977 (Yu, 2002). The high sediment load is of course related to the Loess Plateau, but rapid erosion of the plateau has been greatly facilitated by human intervention. Prior to deforestation and farming of the Loess Plateau about 2500 years ago, a relatively small amount of sediment was exported into the river. Since then, however, the increase of sediment load following the erosion of the Loess Plateau has been 3- to 10-fold by estimation (Xu, 2003). A consequence is rapid sedimentation within the lower reaches of the river, leading to increasing channel migration and annual flooding. About 2500 years ago, the Chinese began constructing a complex dike system along the lower reaches of the river. On average, about 500–1000 Mt of sediment was deposited annually along the lower reaches of the river, mostly as channel deposits, thereby elevating the river bed, which in turn necessitated the continual building of ever higher dikes. In some places, the elevated river now sits 10 m or more above the flat surrounding floodplain, one result being that no tributaries flow into the lower Yellow River (Figure 2.12; Milliman & Farnsworth, 2011; Ren, 1985).

In consequence, the Yellow River has become restless and changed its lower course frequently due to floods and dike breaches. According to historical records, dike breaches had occurred more than 1500×, causing 26× of major shifts of river course in the last 3000 years. The last shift of its course took place in 1855, when the Yellow River moved its mouth from the south to the north side of the Shandong Peninsula, a lateral distance of more than 300 km, and afterward, the river again flowed into the Bohai Gulf instead of the Yellow Sea, as before the twelfth century. The Yellow River has accordingly generated two separate deltas: the old delta with an area of about 150,000 km^2 at the Yellow Sea and the modern one of 5400 km^2 at the Bohai Gulf (Yu, 2002).

Currently, the Yellow River faces the most serious water problem. Draining through much of arid northern China, the annual runoff of Yellow River has declined to <20 mm/year over the past 30 years, making it as perhaps the driest major river in the world. The last decades have also seen the increased construction of dams and reservoirs to store water for irrigation and to mitigate disastrous floods. As of the early 1990s, more than 2600 reservoirs of various sizes were built throughout the Yellow River watershed. In addition, a considerable quantity of river water (and thus sediment) has been removed from the lower reaches of the river for agricultural and industrial utilization. All these together have caused a dramatic decrease in both water and sediment discharge from the Yellow River. Historically, it delivered more than 1000 Mt/year of sediment to the Bohai Gulf or the Yellow Sea during the period 1128–1855; now, the annual sediment load during 1987–2008 decreased to only 40 Mt/year after the constructions of a series of large reservoirs (Figure 2.13; Milliman & Farnsworth, 2011; Wang et al., 2011).

2.3.3   Yangtze River (Changjiang)

The Yangtze is Asia's largest river in terms of basin area (1,800,000 km²), length (6300 km), and water discharge (900 km³/year). Its catchment can be divided into three reaches from west to east. The upstream includes the Jinsha Jiang segment where the river originates from the Tibetan Plateau with river valleys deeply incised into rocky canyons and the Chuan Jiang segment of the Sichuan Basin (Figure 2.14). The Yangtze midstream starts at the end of the Three Gorges where the slope decreases dramatically and the channel becomes much broader. In its downstream, the Yangtze enters a flood plain, and the channel can be wider than >15 km and as shallow as ~6 m in the estuarine section (Chen, Li, Shen, & Wang, 2001).

The unusual length and topographic gradient of the river course have made sediment transport by the Yangtze very complicated. About half of the water discharge is derived from the river upstream, but most of the sediment discharge is provided by its upper basin, although a considerable portion of this upstream-derived sediment is trapped in a series of lakes along the midstream.

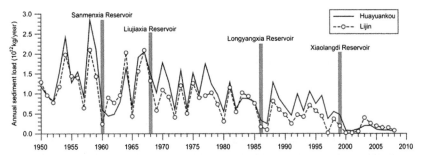

FIGURE 2.13   Annual sediment load recorded at Huayuankou and Lijin stations along the downstream of the Yellow River (see Figure 2.12 for locations: HYK and LJ) from 1950 to 2008. The gray bars indicate the starting time of major reservoirs (Wang et al., 2011).

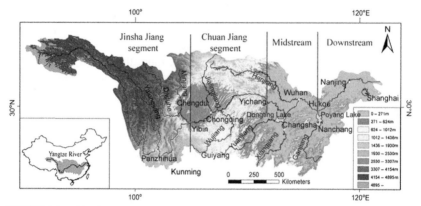

FIGURE 2.14 The Yangtze River drainage basin with tributaries. The mainstream is divided into four segments: the Jinsha Jiang and Chuan Jiang segments of the upstream, the midstream, and downstream (He, Zheng, Huang, Jia, & Li, 2012).

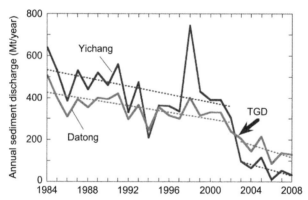

FIGURE 2.15 Comparison of annual sediment discharges in the Yangtze upstream and downstream, 1984–2008. Yichang: at the end of the upstream, 70 km downward from the Three Gorges Dam. Datong: downstream, ~600 km landward of the Yangtze estuary. TGD—the closing of the Three Gorges Dam in 2003 (Milliman & Farnsworth, 2011).

Climatically, the Yangtze River effectively divides the wet South China from the dry North China. Due to climatic and geologic differences, the Yangtze's southern tributaries have higher runoff but lower sediment yields than the northern tributaries.

Over the past 50 years, annual Yangtze sediment discharge has declined substantially, from ~500 Mt/year through much of the 1950s and 1960s, to less than 400 Mt/year in the mid and late 1990s, to less than 150 Mt/year in the recent years (Figure 2.15). The stepwise decline of sediment discharge is a typical case to demonstrate how human activities can modify the sediment discharge from rivers. Since 1950, the Yangtze River has been strongly fragmented by more than 50,000 dams, most of which are medium and small in size constructed in the tributaries. The large reservoirs before 2003, with a total storage capacity of about 68 km^3, trap most of the sediment derived from the major tributaries in the upper and middle reaches.

Most remarkable change occurred at the closure of the world's largest dam, the 189 m high Three Gorges Dam (TGD) in 2003, forming a 600 km long reservoir with a holding capacity of nearly 40 km^3. In the 20 years prior to 2003, annual average sediment discharge at Yichang, about 70 km downstream from the TGD, gradually declined from ~500 to ~400 Mt/year, or about 90 Mt/year greater than at Datong in the lower stream. With the closing of the TGD, however, sediment discharge past Yichang dropped precipitously to <100 Mt/year in 2003 and 2004 and an average of 35 Mt/year since 2005. Sediment discharge at Datong, while also declining, has become about 100 Mt/year greater than at Yichang (Figure 2.15), indicative of channel and bank erosion downstream of the TGD (Milliman & Farnsworth, 2011; Yang et al., 2006).

The observed changes in sediment discharge have a much broader influence. For example, mean particle size in the suspended sediment at Yichang has decreased by roughly a factor of 3 after the construction of the TGD, whereas 300 km downstream at Hankou, its twofold increase may indicate sediment deposition behind the TGD and the corresponding downstream erosion. Another almost immediate result from the closing of the TGD has been a noticeable erosion of the Yangtze's submarine delta (Yang, Milliman, Li, & Xu, 2011), and accelerated shoreline erosion seems to follow suit. Moreover, the decrease of suspended load and the changes in dissolved matter at the Yangtze River mouth will have a series of impacts on the biogeochemistry of the ECS (Chen, 2000; Koshikawa et al., 2007).

2.3.4   Pearl River (Zhujiang)

The Pearl River is the second largest river in China in terms of total water discharge (Table 2.2) and contributes water flow, sediment load, and nutrient materials to the northern SCS. It consists of three stream systems, the Xijiang (West River), Beijiang (North River), and Dongjiang (East River), among which the 1600 km long Xijiang covers ~77% of the total river length and ~87% of the Pearl River Basin (Wang et al., 2011; Figure 2.16). After merging into a common delta, the Pearl River Delta, the three rivers are considered as the Pearl's tributaries. By straddling the Tropic of Cancer, the Pearl River Basin embraces a region of subtropical to tropical monsoon climate. The high precipitation rates in the watershed, ranging from 1200 to 2200 mm/year, support the water discharge of the river.

With an area of almost 17,200 km^2, the Pearl River Delta is one of the world's most complicated river networks, as measured by the density of channel length per unit at 0.68–1.07 km/km^2. The estuary passes through this

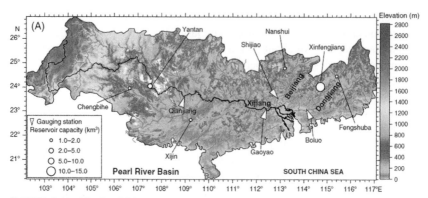

FIGURE 2.16 The Pearl River drainage basin, with locations of major gauging stations and reservoirs (Wang et al., 2011).

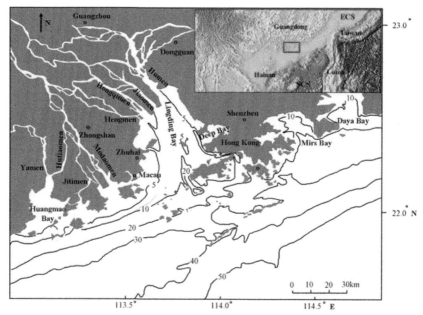

FIGURE 2.17 Pearl River estuary and adjacent waters showing eight major river inlets: Humen, Jiaomen, Hongqimen, Hengmen, Modaomen, Jitimen, Hutiaomen, and Yamen. Contour lines represent water depths in meters (Ji, Sheng, Tang, Liu, & Yang, 2011).

complex river network and flows into the SCS through eight inlets with a total annual runoff of about 300 km³/year. Water depths in the Pearl River estuary vary from 2 to 5 m on the western side to 15 m over the eastern side (Figure 2.17).

The annual mean sediment discharge of the Pearl River averaged ~70 Mt/year (Zhang, Wei, Zheng, Zhu, & Zhang, 2012), but fell to 45 Mt/year in 2008, or

about 60% of its original flux. This is again a consequence of the human intervention. Reservoirs have decreased the sediment load from the watershed, and excavation of sand in the lower channel and land-use change in the river basin have
also influenced sediment delivery from the Pearl River to the SCS (Wang et al.,
2011; Zhang et al., 2012). From 1986 to 2003, an estimated $>8.7 \times 10^8$ m^3 of sand
was excavated, resulting in average downcutting depths of 0.59–1.73, 0.34–4.43,
and 1.77–6.48 m in the main channels of the Xijiang, Beijiang, and Dongjiang
Rivers, respectively, causing a 1.59–3.12 m decrease of water levels in the
upstream of the Pearl River Delta (Luo, Zeng, Ji, & Wang, 2007).

2.3.5   Red River (Sông Hồng)

The Red River system has a total watershed area of $\sim150 \times 10^3$ km^2. Geopo-
litically, the Red River system has 50.3% of its area situated in Vietnam,
48.8% in China, and 0.9% in Laos. Originating from the mountains of Yunnan
Province in China, the river takes in the locally abundant red laterite soils,
which give the river its characteristic red color and hence the name. In its
middle reaches, the river is confined to a straight and narrow fault-controlled
valley. The lower reach is 255 km in length and comprises a triangular Red
River Delta (14,000 km^2 in area) where the river branches into a number of
distributaries and discharges into the Gulf of Tonkin, in the northwestern
SCS (Maren, 2007; Figure 2.18).

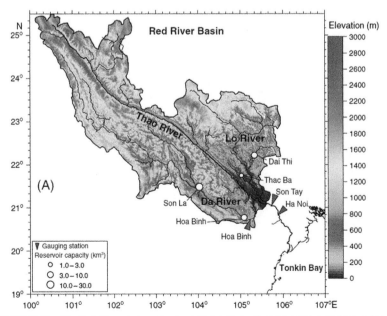

FIGURE 2.18   The Red River drainage basin, with locations of major gauging stations and reservoirs (Wang et al., 2011).

The subtropical monsoon climate supplies abundant rain (1600 mm/year) to the Red River system, with 85–95% falling during the summer season (Le, Garnier, Billen, Thery, & Chau, 2007). A high chemical weathering rate in the Red's watershed has contributed large amounts of dissolved load to the SCS (Moon, Huh, Qin, & Pho, 2007).The sediment load from the Red River was previously estimated as ~130 Mt/year (e.g., Milliman & Syvitski, 1992). More recently, a modeling approach based on 1997–2004 data has estimated a mean annual suspended load of 40 Mt/year under present conditions (Le et al., 2007). This 4× decrease of sediment load is obviously related to dam constructions. For example, the Hòa Bình reservoir with a storage capacity of 9.5 km^3 and height of 128 m, built during 1985–1989, has subsequently trapped a significant share of sediment. After 1989, the sediment load at gauging station Sơn Tây in the lower reach of the Red River (Figure 2.18) decreased to 50 Mt/year (1989–2008) from 115 Mt/year (1960–1988) mostly in response to the Hòa Bình Dam (Wang et al., 2011).

2.3.6 Mekong River

The Mekong is another international river draining through six countries in SE Asia: China, Myanmar, Laos, Thailand, Cambodia, and Vietnam (Figure 2.19B). Being the largest river in the region, the Mekong River is

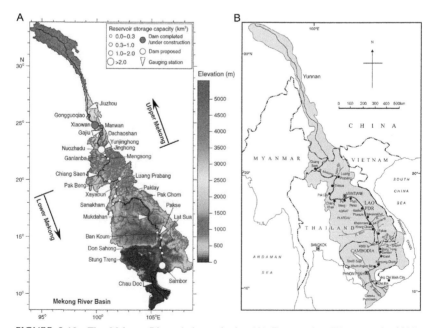

FIGURE 2.19 The Mekong River drainage basin. (A) Topography (Wang et al., 2011); (B) geography (Gupta & Liew, 2007).

~4400 km in length with a total drainage area of 810×10^3 km^2 and features two contrasting parts. The upper Mekong is in the Tibetan Plateau, sharing the "Three Rivers area" with the Yangtze and Yellow Rivers. The upper basin, where the river is called the Lancang and confined within steep and narrow gorges, makes up ~24% of the total Mekong area and contributes 18% of the freshwater discharging to the SCS (Kummu & Varis, 2007). Out from the Lancang, it drops 4500 m in altitude before running into the lower basin where the river becomes wide and gentle and flows for a further 2600 km before entering the SCS via a complex delta system in Vietnam (Figure 2.19A).

The sediment load from the Mekong River to the sea was estimated at 160 Mt/year by Milliman and Syvitski (1992). About 50% of sediment in the river is derived from the steep and unstable slopes of the upper Mekong Basin, as estimated at station Yunjinghong near the end of the upper basin (Figure 2.18A) at 89 Mt/year (Fu, He, & Lu, 2008). A large portion of the sediment yielded from the upper Mekong is deposited in the middle reaches where the slope becomes mild, but sediment yield from the lower basin offsets the sediment loss and keeps relatively stable sediment flux from the Mekong River to the SCS (Wang et al., 2011).

In its lower reach, the Mekong River first forms the border between Thailand and Laos and then flows through Cambodia before entering its delta in Vietnam (Figure 2.19B). Geomorphologically, the river remains incised into bedrock until ~18°N. Further south, the river wanders within Quaternary alluvium until it switches to a bedrock-constrained multichannel and alluvial to deltaic again further downstream (Meshkova & Carling, 2012). Bank erosion occurs mostly in the lower reach, but the erosion process is relatively slow, with an average annual erosion rate only of 0.1% of the channel width at the middle section along the Thailand–Laos boundary (Kummu, Lu, Rasphone, Sarkkula, & Koponen, 2008).

The Mekong River delta is among the world's largest deltas, measuring about 62×10^3 km^2 in plane area with over 80% in Vietnam and the remainder in Cambodia, where the delta apex lies at more than 200 km from the river mouth. From its apex, the delta fans out to the southeast and, along with the Cà Mau Peninsula, separates the SCS from the Gulf of Thailand. In Vietnam, the Mekong River bifurcates to form two main channels, the Bassac River and the Mekong River, and flows through the lowermost Mekong delta into the SCS (Figure 2.20) (Nguyen et al., 2000; Tamura et al., 2010).

As all other mega deltas in Asia, the Mekong delta is currently under pressure of the dam constructions and other human activities (Chen, Saito, & Goodbred, 2005; Syvitski et al., 2009). The delta has been vulnerable to flood historically, and some unusually large floods in the recent years had particularly hit it hard. As shown by numerical modeling, the flood levels in the delta depend on the combined impacts of high river flows, storm surges, and sea-level rise and the siltation of the Mekong estuary resulting from the

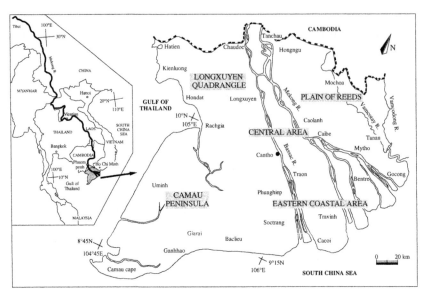

FIGURE 2.20   The Mekong River delta (Nguyen, Ta, & Tateishi, 2000).

construction of dams (Hoa, Nhan, Wolanski, Cong, & Shigeko, 2007). At present, the Mekong River is moderately fragmented and river flow is slightly regulated by dams and reservoirs largely along its upper course. In the lower Mekong Basin, 11 reservoirs of various sizes with a total storage capacity of 4.1 km^3 are proposed to be constructed along the main stream. In the upper Mekong Basin, there are eight reservoirs completed with a total storage capacity of 24 km^3 (Wang et al., 2011). The first large reservoir on the main upstream of the Mekong River, the Manwan reservoir, was built in 1993 by China and has since trapped most sediment from upstream, up to about 300 Mt/year from 1993 to 2003 (Fu et al., 2008). However, there appears to have been little impact on the sediment load in the lower Mekong Basin and the sediment seaward flux. At station Mukdahan in Laos, the sediment load after completion of the Manwan reservoir has instead increased by ~50%, possibly because intensive land use and intensification of human disturbance in the lower Mekong Basin surpass the dam influence during the last 20 years (Wang et al., 2011).

There are other environmental problems in the Mekong delta. For example, the area of mangrove forests in Trà Vinh province in Vietnam has decreased by 50% since 1965 largely due to shrimp farming (Thu & Populus, 2007). Dam constructions, however, remain as the most significant impact factor. With the completion of the proposed reservoirs in the lower Mekong, the total reservoir storage capacity along the river will increase to 28 km^3, which, it will be sure, causes a drastic reduction in the sediment discharge level and subsequently affects the biogeochemical cycle in the southern SCS (Wang et al., 2011).

2.3.7   Small Mountainous Rivers

Along with these five large rivers, smaller rivers also play important roles in the transfer of terrigenous sediment to the China Seas. This is particularly true for the SCS where small mountainous rivers from surrounding islands contribute large amounts of sediments to the marine basin, with Taiwan as the most typical example.

Taiwan is generally recognized as having the highest sediment production in the world because of its high relief, steep gradients, frequent tectonic activity, highly erodible rocks, heavy rainfall, and frequent typhoons (Milliman & Syvitski, 1992), and more than 30% of the total sediment from Taiwan is discharged as hyperpycnal concentrations (Kao & Milliman, 2008). Based on historical measurements and estimates, modern Taiwan rivers collectively discharge >300 Mt of sediment to the surrounding seas each year, a figure comparable to the discharge from large rivers (Figure 2.21; Liu et al., 2008). Over much of the Holocene, the rivers of western Taiwan alone have discharged at least 100 Mt/year on average, while those in southern Taiwan may contribute further 50 Mt/year (Liu et al., 2008). A larger portion of the sediment shed from Taiwan has been eventually resting in the SCS and ECS.

Similar observations have been reported also for other small rivers in islands of the southern SCS margin and along the Vietnam coast, and their impacts on sediment deposition and biogeochemistry in the SCS should not be overlooked.

2.4   OCEANOGRAPHY AND CLIMATOLOGY

The oceanography of the China Seas is largely controlled by two major factors: the underlying morphology of the enclosed marginal sea basin and the overlying atmospheric circulation epitomized by the East Asian monsoon. Recent progress in modern oceanography has accumulated valuable information for geologists to reconstruct past oceanographic evolution in the region. Here, we provide an update of the current knowledge on oceanography of the China Seas starting from the Asian monsoon.

2.4.1   Monsoon

Monsoon-related seasonal changes are the most prominent feature of the China Seas' oceanography. Not only the surface circulation in the SCS has a seasonally reversal characteristic driven by the East Asian monsoon, but also the ECS circulation has seasonal patterns constrained by the East Asian monsoon through its effects on the Kuroshio Current and on the Yangtze outflow. The Asian monsoon is the largest monsoon system in the modern world, and its two subsystems, the Indian or South Asian monsoon and the East Asian monsoon, are closely related and interact with each other, but have

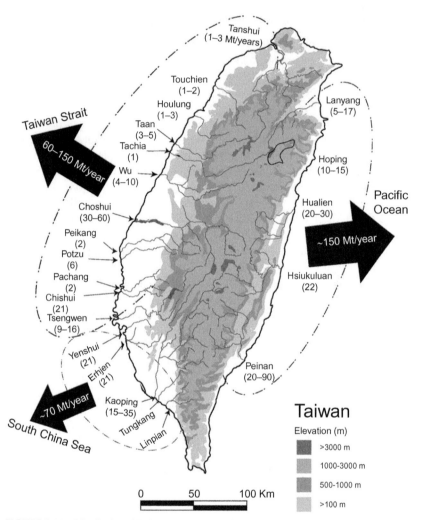

FIGURE 2.21 Distribution of Taiwan mountainous rivers and their annual sediment loads to the surrounding seas (Liu et al., 2008).

different structures dictated by different sea–land patterns. For East Asia, the ocean is to the east and the Tibetan Plateau to the west, while South Asia has the ocean to the south and the Tibetan Plateau to the north, resulting in three major distinctions between the two monsoon subsystems: (1) For the South Asian subsystem, the winter monsoon is largely blocked by Tibet and hence is much weaker than its summer counterpart, whereas East Asia has the most intensive winter monsoon in the world. (2) The South Asian summer monsoon is tropical in nature, while the East Asian monsoon has a subtropical in addition to the tropical component (Chen, 1992). (3) Humidity is sourced by the East Asian monsoon primarily from the western Pacific warm pool (WPWP) but is largely from the Indian Ocean by the South Asian monsoon.

Precipitation and wind fields of the Asian monsoon show contrasting patterns between winter and summer (Figure 2.22). In winter, three branches of monsoon flow southward, respectively, over the Arabian Sea, the Bay of Bengal (Indian monsoons), and the SCS (East Asian monsoon), forming strong cross-equatorial northerlies there (Figure 2.22; Wang et al., 2003). Only the northerly over the SCS can be traced backward to its origin over the cold Siberia, from where the world's coldest winter air mass streams southward along the East Asian coastal zone, generating the strongest winter monsoon. As the circulation of the Asian winter monsoon encompasses a larger meridional domain, the

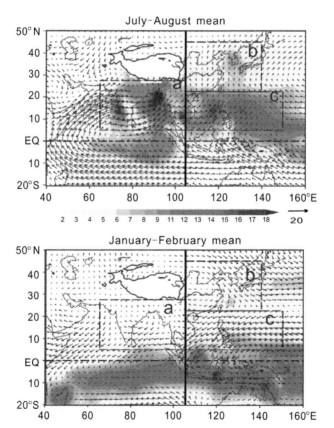

FIGURE 2.22   Asian monsoon system is shown in climatological mean precipitation rates (shaded in mm/day) and 925 hPa wind vectors (arrows) during July–August (upper panel) and January–February (lower panel). Dotted squares show three major summer precipitation areas: (a) Indian tropical monsoon, (b) western North Pacific tropical monsoon, and (c) East Asian subtropical monsoon. *Modified from Wang, Clemens, and Liu (2003).*

winter climate in the China Seas is predominated by processes of tropical–extratropical exchanges (Chang, Wang, & Hendon, 2006).

According to the new "global monsoon" concept, the monsoon systems of different continents constitute a global-scale seasonally varying atmospheric overturning circulation (Trenberth, Stepaniak, & Caron, 2000), which can be easily recognized in Figure 2.22. The East Asian summer monsoon, however, comprises not only a tropical component connected with cross-equatorial flows from the south but also an eastern component originated from the subtropical easterlies in the western periphery of the western Pacific subtropical high. Along with the western North Pacific monsoon trough or the ITCZ (the intertropical convergence zone), an East Asian subtropical front or "Meiyu (plum rain) front" forms during summer. As a result, two major summer monsoon areas are distinguished in East Asia: the East Asian subtropical summer monsoon domain covering eastern China, Korea, Japan, the ECS, sea of Japan, and the adjacent Pacific (Figure 2.22B) and the western North Pacific tropical summer monsoon domain including the SCS, Philippines, and the Philippine Sea (Figure 2.22C). The Indian and western North Pacific summer monsoons, including the SCS monsoon, are of tropical nature in which the low-level winds reverse from winter easterlies to summer westerlies, but the East Asian summer monsoon is a subtropical structure in which the low-level winds reverse primarily from winter northerlies to summer southerlies (Wang & Lin, 2002). This definition does not fully agree with the conventional view of Chinese meteorologists, who usually embraces the SCS in the East Asian monsoon region together with the ECS. If the SCS region is included, then the East Asian summer monsoon domain becomes a hybrid type of tropical and subtropical monsoon (Ding & Chan, 2005). In this book, we use the broader sense of the East Asian monsoon in discussing its influence on the entire China Seas.

Between the two parts of the China Seas, the monsoon climate differs not only in the tropical versus subtropical nature but also in the annual cycle. In the SCS, the northerly winter monsoon lasts nearly 6 months from November to April, but in the ECS, it lasts about 8 months, from September to April. Comparatively, the southerly summer monsoon is weaker in strength and shorter in duration for the entire China Sea, lasting nearly 4 months (mid-May to mid-September) in the SCS and only 2 months (July to August) in the ECS (Chu & Wang, 2003; Lee & Chao, 2003). The mean surface wind stress over the SCS averages nearly 0.2 N/m^2, reaching 0.3 N/m^2 in mid-December before dropping to nearly 0.1 N/m^2 in June.

The low-level wind patterns of the China Seas are affected by topographic features of the surrounding land. As seen from Figure 2.23, winter winds in the ECS intensify toward the south and follow the contour directions of the basin, which are north to northwest in the northern reaches but northeast in the southern reaches (Sun, 2006). In the SCS, the northeasterly monsoon prevails in winter, and two wind speed maxima exceeding 10 m s^{-1} occur,

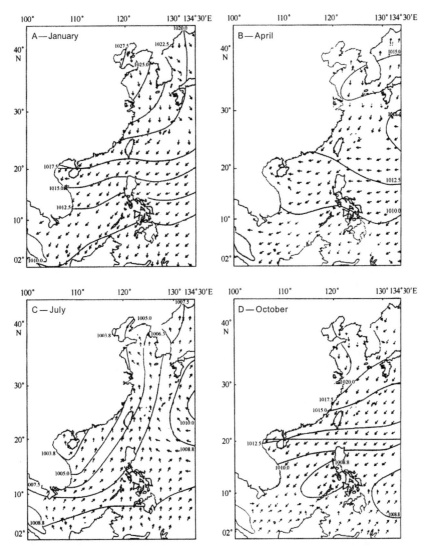

FIGURE 2.23   Seasonal variations of the surface pressure (isolines, in h/Pa) and wind field (arrows) in the China Seas (Sun, 2006).

respectively, in the Taiwan and Bashi Straits and around 11°N off the coast of southern Vietnam near the southern tip of Annam Cordillera (Figure 2.24A), both due to orographic forcing by mountains. The northeasterly winds are blocked by the Indochina coastal mountain range, giving rise to a wind jet off-shore, similar to the southwesterly jet in summer (Figure 2.24B) (Liu et al., 2004; Xie et al., 2003).

The SCS has recently become a hot spot worldwide in modern and paleo-monsoon studies because of its crucial role in the development of Asian

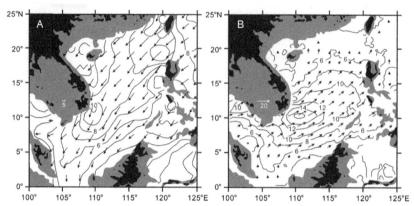

FIGURE 2.24 QuikSCAT surface winds over the SCS. (A) Winter wind velocity (vector) and its magnitude (contours at 2 m/s intervals) averaged for December–February 2000–2002. (B) Summer wind stress vectors and their magnitude (contours in 10^{-2} per Nm^2) averaged for June–August 2000–2002. Land topography with elevations >500 m is shaded in black and 0–500 m in gray. Note the dominant NE–SW direction for the maximal wind velocity (arrows). *Modified from Xie, Xie, Wang, and Liu (2003) and Liu, Jiang, Xie, and Liu (2004).*

monsoon and its magnificent sediment record of monsoon evolution. Since the 1990s, a number of paleomonsoon cruises to the SCS were organized, leading to numerous research projects mostly on variability and evolution of the East Asian monsoon over the Quaternary. Meanwhile, the modern monsoon studies have revealed complicated relations between monsoon intensity and spatial variations in precipitation. For example, the "South China Sea Monsoon Experiment (SCSMEX)" from 1996 to 2001 found that a strong (weak) monsoon over the SCS often leads to lesser (more) precipitation over the middle and lower reaches of the Yangtze River Basin and more (lesser) precipitation in North China (Ding, Li, & Liu, 2004). However, many of the paleoreconstructions underestimated the complicity of the monsoon system and frequently suffer from a simplistic approach to monsoon proxies and paleomonsoon interpretation (see Wang et al., 2005, for a review). All these call for a closer collaboration between the modern climatology and paleoclimatology and for awareness to underscore the importance of adequate modern monsoon knowledge before making paleomonsoon interpretations.

2.4.2 Surface Circulation

2.4.2.1 East China Sea

Although the earliest scheme of the ECS circulation pattern was first described more than 70 years ago by Uda based on in-field measurements during the 1930s, it has depicted nearly all major components of the circulation we know today (Figure 2.25). In general, the circulation system in the ECS comprises the Kuroshio Current from the ocean and the coastal currents

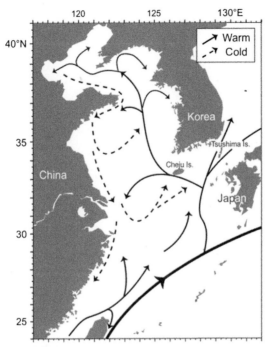

FIGURE 2.25   The circulation pattern of the East China Sea in June generated by Uda in 1934. *Reproduced from Isobe (2008).*

associated with river discharges. Many revised versions of the circulation pattern in the ECS published subsequently (e.g., Guan, 1994; Isobe, 2008; Lee & Chao, 2003; Niino & Emery, 1961; Su, 1998) all offer this basic concept.

The warm Kuroshio, or "Black Tide," enters the ECS from east of Taiwan, flows northeastward along the edge of the continental shelf along approximately the 200 m isobath, and leaves the ECS through the Tokara Strait southwest of Kyushu. South of about 28°N, the Kuroshio runs into the continental shelf and is uplifted, resulting in onshore intrusion of its oceanic waters (Hsueh, 2000). Two currents in the ECS are associated with the Kuroshio: The Tsushima Warm Current west of Kyushu transports warm and saline Kuroshio waters into the Yellow Sea, and the Taiwan Warm Current into the ECS shelf. In the Yellow Sea, the winter circulation is dominated by the northward-flowing Yellow Sea Warm Current and the southward-flowing coastal currents, the China and Korea coastal currents (Figure 2.26; Guan, 1994).

The recent progress in oceanography has challenged the classical view of circulation pattern in the ECS. For example, the Yellow Sea Warm Current was often considered as a branch of the Tsushima Current, which in turn

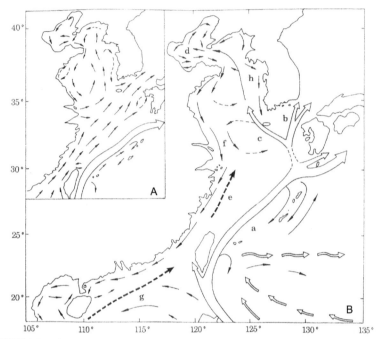

FIGURE 2.26 Schematic view of the major current systems in the East China Sea. (A) Winter; (B) summer. a, Kuroshio; b, Tsushima Warm Current; c, Yellow Sea Warm Current; d, Bohai circulation; e, Taiwan Warm Current; f, China coastal current; g, South China Sea Warm Current; h, Korea coastal current (Guan, 1994).

was branched from the Kuroshio (Figures 2.25 and 2.26). Recent studies, however, found that the Yellow Sea Warm Current is developed mainly in winter as a compensating current for the coastal currents (Xu, Wu, Lin, & Ma, 2009). It is also debated whether the Tsushima Current is a branch of the Kuroshio. Fang, Zhao, and Zhu (1991) proposed connectivity between the Taiwan and Tsushima Warm Currents and termed the shelf currents as the "Taiwan–Tsushima–Tsugaru Warm Current System" to reflect their idea of its extension to the North Pacific through the Japan Sea. Following suit, Isobe (2008) suggested a supporting role of the Taiwan–Tsushima Warm Current by the onshore Kuroshio intrusion across the ECS shelf break (gray area in Figure 2.27; see Section 2.4.4.1 for further discussion).

Another issue is the discharges from large rivers, especially the Yangtze and Yellow Rivers. Satellite data show a significant turbid water plume extending in the southeast direction from the Yellow Sea coasts of China to the shelf edge south of Chiju in winter (Figure 2.28). This is the way how the suspended sediments from the old Yellow River mouth are transported into the Okinawa Trough. As shown by satellite images, the turbid plume grows in fall, reaches its maximum expansion and intensity in winter–spring,

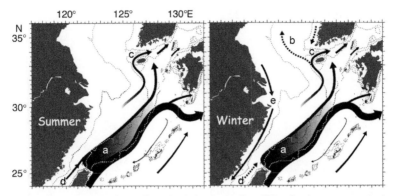

FIGURE 2.27   Schematics for surface circulation in the East China Sea in summer and winter. Broken arrows represent surface currents that are possibly sporadic events in winter. Dotted lines denote depth contour lines of 50, 100, and 200 m. A, Shelf edge; B, Yellow Sea Warm Current; C, Chiju Warm Current; D, Taiwan Strait Warm Current; E, China coastal current (Isobe, 2008).

and subsides in the late spring. In summer, the plume becomes coastally trapped. In comparison, the Yangtze River diluted water in summer only transports a small amount of the suspended sediment from Yangtze to the outer shelf south of Chiju, without entering the Yellow Sea (Figure 2.28; Yuan et al., 2008). Instead, the suspended load from the Yangtze is distributed along the China coastal current, mostly transported in winter.

According to the model results, the ECS shelf circulation in winter is induced primarily by the northerly monsoonal winds, and the Kuroshio-forced circulation over the shelf is very weak. Under the northerly wind forcing, the currents over the ECS shelf flow southward in an opposite direction to the Yellow Sea Warm Current. The resulting divergence, therefore, forces the Yellow Sea Coastal Current to move offshore to form the East China Sea Current (Figure 2.28), and this cross-shelf current plays an important role in material exchange between the continent and ocean (Yuan & Hsueh, 2010).

2.4.2.2 South China Sea

The first picture of the SCS circulation by Wyrtki (1961) on the basis of hydrographic and ship-drift data depicts basically a cyclonic gyre in winter and an anticyclonic gyre in summer, driven by the seasonally reversing monsoon. Over the years, the Wyrtki model of SCS circulation has been confirmed by dynamical studies (e.g., Gan, Li, Curchitser, & Haidvogel, 2006; Lu et al., 2001; Shaw & Chao, 1994) and observations (e.g., Qu, 2000; Xu, Qiu, & Chen, 1982). These distinct features are in fact reflections of the upper-layer circulation in response to the seasonal changes of the monsoon wind stress curl, with additional influence from the Kuroshio in its northern part (Qu, 2000). In winter, there is a basin-wide cyclonic gyre

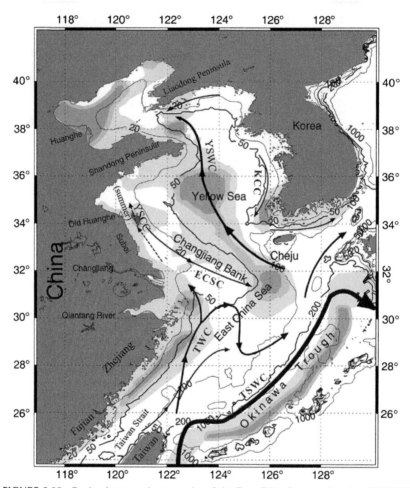

FIGURE 2.28 Regional map and topography of the East China Sea area based on ETOPO5 dataset. Contour unit is meter. Shaded areas show the distribution of mud patches at the ocean bottom, with darker colors indicating finer-grain sediments. YSWC, Yellow Sea Warm Current; ECSC, East China Sea Current; YSCC, Yellow Sea Coastal Current; KCC, Korea Coastal Current; TSWC, Tsushima Warm Current; TWC, Taiwan Warm Current. The dash arrow along the Subei coast marks the YSCC in summer (Yuan, Zhu, Li, & Hu, 2008).

(Figure 2.29A), while in summer, the circulation splits into a weakened cyclonic gyre ~12°N and a strong anticyclonic gyre in the south (Figure 2.29B).

Over the past 20 years, the surface circulation in the SCS has been reviewed by such authors as Huang and Wang (1994), Hu, Kawamura, Hong, and Qi (2000), Su (2004), and Sun (2006). Its seasonal features can be demonstrated with the distribution of climatological mean January and July dynamic heights (dyn cm) at 100 m water depth (Figure 2.30), which shows a large cyclonic gyre in winter (Figure 2.30B), dominant north of a line roughly

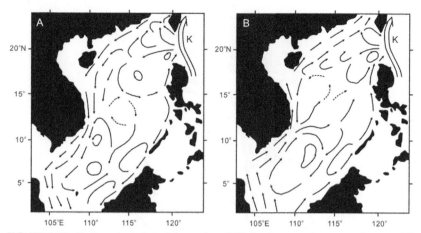

FIGURE 2.29   Surface circulation of the modern SCS showing opposite patterns between (A) winter and (B) summer. K, Kuroshio Current. *Modified from Fang, Fang, Fang, and Wang (1998).*

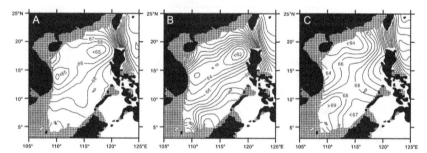

FIGURE 2.30   Distribution of the mean dynamic heights (contour in dyn cm) at −100 m in the SCS is shown for three periods: (A) annual mean, (B) January monthly mean, and (C) July monthly mean. Region with water depth shallower than 100 m is stippled. *Modified from Qu (2000).*

from 10°N, 110°E to 15°N, 120°E, and two weakly developed gyres in summer (Figure 2.30C). A large cyclonic eddy, the west Luzon eddy, is present west of Luzon in winter and another strong cyclonic eddy, the east Vietnam eddy, occurring off central Vietnam in both winter and summer (Qu, 2000; Su, 2004). The much stronger winter gyre helps to separate the SCS circulation along the previously mentioned line into the northern and southern parts with different oceanographic features.

Therefore, the SCS surface circulation can be summarized into a schematic diagram with seasonally alternating basin-scale gyres (Figure 2.31) and associated strong western boundary currents. In winter, a southward jet flows along the entire western boundary, while in summer, a northward jet

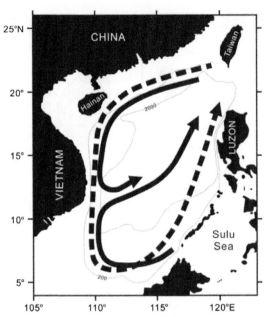

FIGURE 2.31 Sketched surface circulation of the SCS with seasonally alternating, basin-scale cyclonic gyres: winter (dashed line); summer (solid line). The summer pattern involves a cyclonic gyre in the northern and an anticyclonic gyre in the southern parts (Wang & Li, 2009).

flows along the western boundary in the southern SCS, apparently veering eastward off central Vietnam near 12°N (Wang, Chen, & Su, 2006). This cross-basin eastward jet, or "summer southeast Vietnam offshore current" of Fang, Fang, Shi, Huang, and Xie (2002), has a maximum velocity of around 0.8 m/s and plays a significant role in the SCS oceanography by dividing it into the northern and southern parts. The eastward jet is accompanied by a dipole structure with an anticyclonic cell south of the jet and a cyclonic cell north of it (Liu et al., 2002; Wang et al., 2006).

Along with the major gyres, other specific features of the SCS surface circulation have been observed. The South China Sea Warm Current, for example, is a northeastward counterwind current in winter occurring off the Guangdong coast in the NE part of the SCS. After its discovery in the 1960s, the SCS Warm Current has been confirmed by observations and numerical modeling as likely connecting with the Taiwan Strait Warm Current and the Taiwan Warm Current. All these three warm currents lie along the seaward side of the downwind China coastal current and flow toward the NE throughout the year (Guan & Fang, 2006; Hu et al., 2000). Although their origin remains a subject of future studies, they are probably associated with the Kuroshio invasion into the SCS and together have been playing a significant role in water exchanges between the Pacific and the China Seas.

2.4.3   Temperature and Salinity

2.4.3.1   East China Sea

The distribution of sea surface temperature (SST) and sea surface salinity (SSS) in the ECS is largely stipulated by the contrast between the Kuroshio and river discharges, mainly from the Yangtze and Yellow Rivers. In winter, the lowest temperature occurs in the Bohai Gulf ($-1.5 \sim 3.6$ °C) and the northern coastal area of the Yellow Sea ($0 \sim 2$ °C), the only areas in the China Seas with winter sea ice (Shi & Wang, 2012). The winter SST in the Yellow Sea ranges from 0 to 13 °C, showing a general trend to increase southward. Another feature is the intrusion of the Yellow Sea Warm Current during winter as a warm tongue into the central area of the Yellow Sea, proximately between 34°N and 35°N, 122°E and 124°E (Wang, Liu, & Meng, 2012; Zhang, Wang, Lü, Cui, & Yuan, 2008). While in the ECS proper, the winter SST isotherms clearly follow the courses of warm currents, especially the Kuroshio, Tsushima, and Taiwan Warm Currents (Figure 2.32A; Sun, 2006).

The summer SST gradient in the ECS is much gentler, with $24 \sim 27$ °C in the Bohai and Yellow Seas and fairly close to 26–29 °C in the ECS proper (Figure 2.32B; Sun, 2006), although the water structure in the Yellow Sea is thoroughly reorganized. Instead of the warm current in winter, the Yellow Sea Cold Water Mass now occupies the central Yellow Sea in summer. From early summer to fall, <10°C cold water under the seasonal thermocline can be easily mapped by bottom temperature in the central area of the Yellow Sea (Figure 2.33B), ascribable to summer stratification in the area. Above the depth of 20 m, the top layer water becomes warmer and less saline, but the water temperature below the depth of 30 m remains cool due to the shielding effect of the thermocline from the heat exchange (Zhang et al., 2008). In general, the water column in the ECS, including the Yellow Sea and Bohai Gulf, is stratified in summer because of river discharge but strongly mixed in winter because of wind disturbance (Figure 2.34).

As a result of the exchanges between the saline oceanic water and freshened coastal water caused by river discharges, the SSS pattern in the ECS exhibits significant seasonal variations. This can be demonstrated with the Yangtze River diluted water, a mix of river runoff and shelf water with low salinity (Kim et al., 2009). In summer, it extends northeastward, reaching the Chiju Island area, while in winter, it is restricted in the coastal area and running mostly southward due to a weaker Yangtze runoff and strong monsoon winds (Figure 2.32C and D; Sun, 2006).

2.4.3.2   South China Sea

In the SCS, the seasonal reversal in circulation leads to seasonal contrast in SST and SSS patterns. In summer, surface water temperature in the SCS is relatively uniform, with only minor variations between 28.5 and 29.5 °C

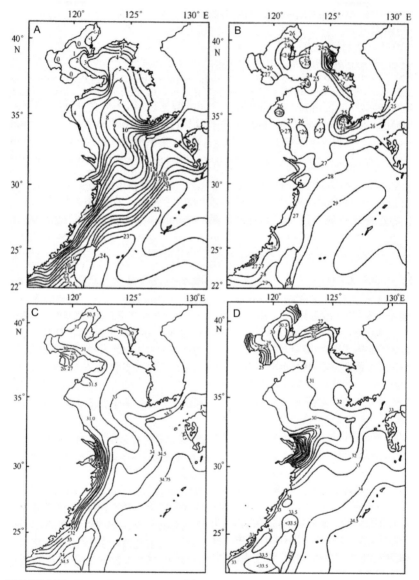

FIGURE 2.32 Sea surface temperature (°C) and sea surface salinity (‰) in the modern East China Sea show distinct seasonality patterns: (A) SST in winter; (B) SST in summer; (C) SSS in winter; (D) SSS in summer (Sun, 2006).

(Figure 2.35B). In winter, an intense western boundary current flows southward along the continental slope, separating the Sunda Shelf to the west and the deep basin to the east (dashed line in Figure 2.31). This current transports cold water from the north and causes a distinct cold tongue south of Vietnam,

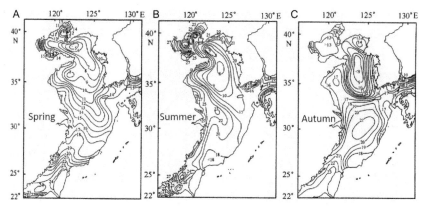

FIGURE 2.33 Bottom temperature (°C) in the modern East China Sea. (A) May; (B) August; (C) November (Sun, 2006).

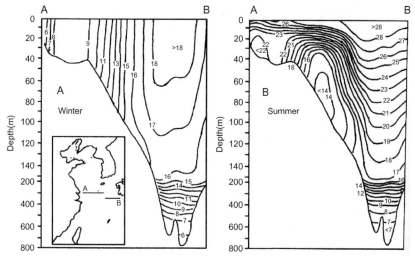

FIGURE 2.34 Vertical distributions of temperature (°C) along sections A and B in the East China Sea (Sun, 2006).

leading to a steeper temperature gradient in the winter SCS (Figure 2.35A). Compared to the neighboring Pacific and Indian Oceans, the winter SST in the SCS is considerably lower and formulates a conspicuous gap in the Indo-Pacific Warm Pool (Liu et al., 2004).

Seasonal SSS changes in the SCS are shown in Figure 2.35C and D. In winter, salinity is higher in the north due to the Kuroshio influence, with a high-salinity tongue in the top 100 m 12°N propagating to the southwest. The winter salinity decreases southward because of increasing precipitation,

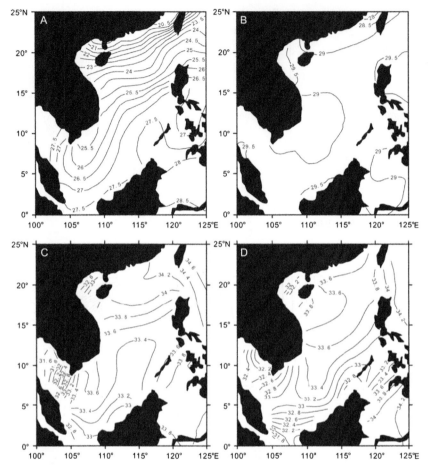

FIGURE 2.35 Sea surface temperature (°C) and sea surface salinity (‰) in the modern SCS showing distinct seasonality patterns. (A) SST in winter; (B) SST in summer; (C) SSS in winter; (D) SSS in summer (Wang, Du, & Shi, 2002).

causing isohalines running parallel to latitudes (Figure 2.35C). In summer, salinity varies less in the north but very large in the south, with higher values occurring off the Vietnam coast due to summer upwelling and a low-salinity tongue outside the Mekong estuary (Figure 2.35D; Wang et al., 2002). Clearly, freshwater input is a significant factor for SSS variations. For example, waters near the Pearl River mouth is highly stratified especially during the wet season, with a strong halocline inside the estuary and a plume formed in the surface layer just outside the estuary. Soon after forming, the plume turns toward the west during the dry season, but turns to the east in the wet season due to the southwesterly monsoon (Dong, Su, Wong, Cao, & Chen, 2004). Once the surface plume reaches beyond the inner shelf, it can easily

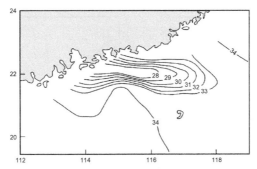

FIGURE 2.36   Sea surface salinity distribution outside the Pearl River estuary (in ‰) (Su, 2004).

extend further east across the shelf, possibly driven by the SCS Warm Current and the favorable southwesterly summer monsoon winds (Figure 2.36; Su, 2004).

Oceanographic data show an upper-layer thermohaline front in winter, running across the SCS basin from the South Vietnam coast to Luzon island, roughly along the 25.5 isotherm (Figure 2.35A). As seen from a comparison of the upper-layer (0–300 m deep) T–S diagrams between the northern and southern SCS (Figure 2.37), the front separates two kinds of water masses: high-temperature, low-salinity water in the south and low-temperature, high-salinity water in the north. Both the temperature and salinity ranges with depth are larger in the south of the front than in the north, and the salinity minimum occurs at a shallower depth in the north of the front than in the south. Thus, the thermohaline variability is larger in the south than in the north of the cross-basin current (Figure 2.37) (Chu & Wang, 2003).

Vertically, the uppermost layer of water in the SCS is well mixed to a seasonally varying depth range due to monsoon influence. In winter, the mixed layer is thickest (~80 m) in the NW part, becoming thinner in the south (~40 m; Figure 2.38A). The mixed layer pattern in summer is reversed, becoming thicker in the south (~50 m) and thinner in the northwest (~30 m; Figure 2.38C). In spring, the mixed layer decreased to about 30 m because of the weakened wind flow (Figure 2.38B), while in fall, it is relatively thick (~50 m) over the entire SCS basin, except a slightly thinner layer (~40 m) off the Vietnam coast (Figure 2.38D) (Shi et al., 2001). Below the mixed layer, a seasonal thermocline exists throughout the year in the deep basin with distinct variations in its thickness and average depth range (Table 2.4), and the thermocline slope in the deep sea in winter basically mirrors the summer pattern due to monsoon influence, with a west–east difference of over 50 m. It is therefore obvious that the upper water of the SCS piles up in the northwestern part in winter and in the southeastern part in summer (Liu et al., 2000; Shaw, 1996).

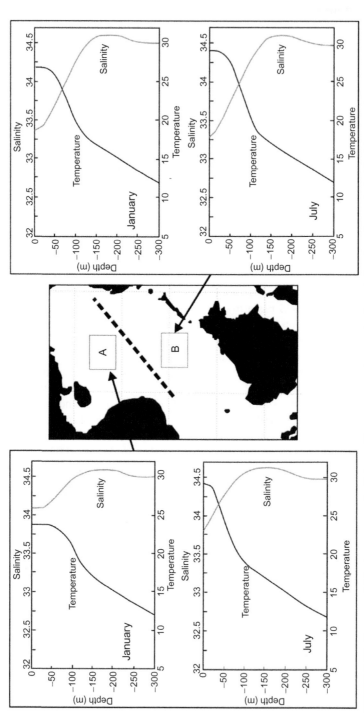

FIGURE 2.37 January and July temperature and salinity profiles are compared across the thermohaline front (heavy dashed line) between the northern (A) and southern (B) SCS. *Modified from Chu and Wang (2003).*

FIGURE 2.38   The mixed layer depth in the SCS in (A) January, (B) April, (C) July, and (D) October (Shi, Du, Wang, & Gan, 2001).

TABLE 2.4 Temporal Variations of the Seasonal Thermocline in the South China Sea

Season	Thickness (m)	Average Depth (m)
Spring	~150	~75
Summer	100–120	75–85
Fall	100–120	75–85
Winter	<100	~100

Based on Liu, Yang, and Wang (2000).

Another important feature of the SCS oceanography is the development of upwelling. Coastal upwelling occurs off the southern coast of Vietnam in summer and off the Luzon island in winter, corresponding to the two eddies in Figure 2.30. The winter Luzon cold eddy and the summer Vietnam cold eddy can be recognized from the upper-layer temperature distribution

(Yang & Liu, 1998) or from the relative low subsurface temperature and relative high-salinity data at −100 m (Figure 2.39; Sun, 2006).

Along with the two major upwelling regions, shelf break upwelling along the edge of the Sunda Shelf may also occur from October to December (Chao et al., 1996). Moreover, frequent tropical cyclones with 10 typhoons averaging per annum traverse the SCS, also causing upwelling of cold subsurface water into the mixed layer (Wong et al., 2007).

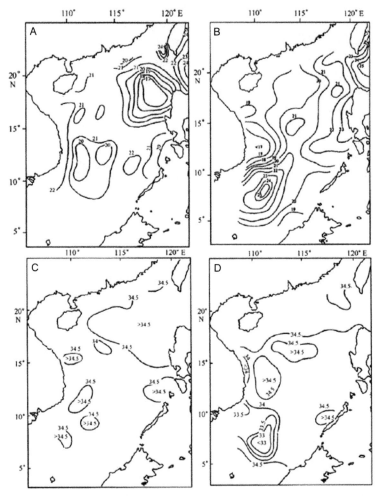

FIGURE 2.39 Temperature (°C) and salinity (‰) at 100 m depth in the modern South China Sea show distinct seasonality patterns: (A) temperature in winter; (B) temperature in summer; (C) salinity in winter; (D) salinity in summer (Sun, 2006).

2.4.4   Ocean Connection and Deepwater Circulation

2.4.4.1   Kuroshio Intrusion

The Kuroshio provides the main inflow of the oceanic water to the China Seas. It originates from a bifurcation of the North Equatorial Current off the east coast of the Luzon islands at about 15°N and flows northward. A small part of the Kuroshio water passes through the Bashi Strait, where it exchanges with the SCS, while the main Kuroshio enters the ECS east of Taiwan before exiting through the Tokara Strait south of Kyushu Island (Figure 2.28), with an estimated mean transport of ~21.5 Sv at the entrance (Lee et al., 2001). The exchanges between the Kuroshio and shelf waters in the northern and southern ECS are different. South of 28°N, the subsurface water of the Kuroshio is uplifted by the east–west running continental shelf break, generating a surface-layer circulation featuring a flow of cool and less saline shelf water toward the Kuroshio. North of 28°N, the Kuroshio is directed toward the coast of Kyushu and is forced to turn eastward, and the separation of the flow on the western side gives rise to a northward current that supplies tropical heat and salt to the Yellow Sea and the sea of Japan (Hsueh, 2000).

This notion on the Kuroshio is supported by recent observations. According to 16-year (1993–2008) satellite data, the shoreward intrusion of the Kuroshio mainstream chiefly occurs near the southwest of Kyushu where an estimated transport of 1.8 Sv of the Kuroshio veers toward the Tsushima Strait throughout the year. The surface shoreward intrusion in other parts of the Kuroshio is generally weak and most of the transport that deviates shoreward from the core (around the 200 m isobath) of the Kuroshio recirculates within the stream (Liu & Gan, 2012). Furthermore, the previously discussed Taiwan–Tsushima Warm Current has to be supported by onshore Kuroshio intrusion across the ECS shelf break (Figure 2.27), as the northward flow through the Taiwan Strait is much too small to be accounted as the main southern source to the Tsushima Warm Current (Isobe, 2008; Teague et al., 2003).

A more controversial issue is how the Kuroshio water invades the SCS through the Bashi Strait. It remains debatable whether the Kuroshio intrudes as a direct current or a loop around inside the SCS (e.g., Metzger & Hurlburt, 2001; Yuan, 2002) or the Kuroshio water flows through the Strait only via submesoscale processes such as eddies (Su, 2004). Regardless of all the controversies, the Kuroshio water does enter the SCS one way or another and exchanges with the SCS water through the Bashi Strait, the only deepwater connection between the SCS and the Pacific.

2.4.4.2   Water Exchange with Oceans

Water exchanges of the ECS and SCS with the open ocean are different, although the Kuroshio intrusion provides the main vehicle for such exchanges in the two seas. The SCS is comparatively more complicated, as it is connected to not only the Pacific but also the Indian Ocean and participates in the Indo-Pacific throughflow. Only the surface waters in the SCS exchange freely with those in the neighboring seas, while deeper waters flowing into the SCS are primarily from the western Philippine Sea through the Bashi Strait. The tropical water, originating from the high-salinity pool in the subtropical North Pacific, occurs as a salinity maximum at around 150 m in the SCS. The North Pacific Intermediate Water, flowing from the subpolar North Pacific, occurs as a salinity minimum centered around 500 m. Deep water in the western Philippine Sea overflows the sill of the Bashi Strait at a depth of 2400 m and fills the deep SCS. The water transport through the Bashi Strait influences the circulation and heat budget of the SCS and determines the nature of its deep water. The annual water budget of the SCS is summarized in Table 2.5.

The Bashi Strait transport has a sandwiched vertical structure, with the Philippine Sea water entering the SCS at the surface and in deeper parts and with the net Bashi Strait transport out of the SCS at intermediate depths. The net transport of the western Pacific water into the SCS is estimated at the order of 4 or 4.2–5.0 Sv (Qu, 2000; Su, 2004). As a result, the SCS plays the role of a "mixing mill" that mixes surface and deep waters from the western Pacific and returns them through the Bashi Strait at intermediate depths (Yuan, 2002).

Water transport through the Bashi Strait can be illustrated using field experiment results. In October 2005, a sandwiched vertical structure was observed, with the net volume of westward transport of 9 Sv in the upper layer

TABLE 2.5 Characteristics of the Annual Water Budget of the SCS Show Variations in Input and Output Volumes

Input	Flux (km³/Year)	Output	Flux (km³/Year)
Surface inflow from the western Philippine Sea	2.7×10^{-1}	Surface outflow to the western Philippine Sea	-2.4×10^{-1}
Deepwater inflow from the western Philippine Sea	3.7×10^{-2}	Net outflow at intermediate depths to the western Philippine Sea	-5.8×10^{-2}
Surface inflow from Sulu Sea	1.8×10^{-2}	Net surface outflow to Java Sea	-1.8×10^{-2}
Riverine input	1.7×10^{-3}	Surface outflow to East China Sea	-1.1×10^{-2}

Modified from Wong et al. (2007).

(<500 m) and 2 Sv in the deep layer (>1500 m) and the net volume of east-ward transport of 5 Sv in the intermediate layer (500–1500 m) (Figure 2.40A; Tian et al., 2006). In July 2007, however, only a two-layer vertical structure was observed, with eastward transport at both the upper and intermediate depths, and the westward transport staying in the deeper layer (Figure 2.40B; Yang et al., 2010). Clearly, the discrepancies between the two experiments underscore the need for many more *in situ* observations toward a better understanding of the spatiotemporal variations in water exchanges at the Bashi Strait.

Unlike the ECS, the SCS is connected to both the Pacific and Indian Oceans and exchanges the surface water between them. On the Pacific side, through the Bashi Strait, there is a small (~1 Sv) net outflow of surface water (0–350 m depth) from the SCS in the wet season, but a net inflow (~3 Sv) in the dry season. The differences are mainly formulated by compatible inflow and outflow of Sunda Shelf water on the Indian Ocean side. Below 350 m, the SCS water flows out, but the western Philippine Sea water again flows into the SCS below 2400 m, making the deep SCS waters homogenous with the same property as the Philippine Sea (Chen & Huang, 1996). A rigorous exchange of water between the SCS and the western Philippine Sea through the Bashi Strait has resulted in a very short residence time of the deep water in the SCS, and the corresponding basin-wide upwelling rate is as high as about 10–90 m/year, subsequently giving rise to a thin mixed layer and a shallow top of the nutricline in the SCS (Wong et al., 2007).

As a result, significant difference in the vertical structure of waters occurs between the two sides of the Bashi Strait. To the west of the strait (~122°E), the temperature and salinity contours above 600 m shoal up by about 50–200 m in the SCS than in the Philippine Sea, with a continuous shoaling of the salinity contours to about 1000 m (Figure 2.41). Thus, at the same depth above 600 m, the SCS water is cooler than the western Philippine Sea water, and the thickness of the mixed layer in the SCS is also only about half of that in the western Philippine Sea. Such apparent "upwelling" hydrographic features are found in the upper 600 m of water over the entire deep SCS Basin. Below about 600 m, however, the SCS water becomes slightly warmer than the Philippine Sea water (Su, 2004).

Waters from the two sides of the Basin Strait also differ in biogeochemistry. The subsurface water of the SCS west of Luzon is more enriched in nutrients and depleted in oxygen relative to the water east of the Strait, but the situation reverses in the intermediate water (Gong et al., 1992). Comparison between dissolved oxygen concentrations at different depth intervals on both sides of the Strait clearly reveals this feature. Specifically, the oxygen content on the Pacific side is relatively higher above 600–700 m, lower between 700 and 1500 m, and then higher again below 1500 m (Figure 2.42) (Qu, 2002).

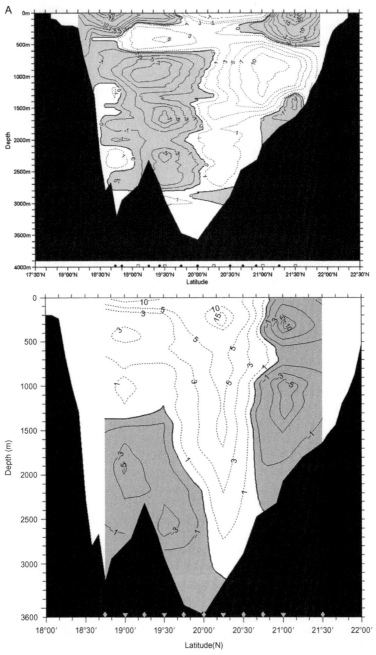

FIGURE 2.40 Flow velocity (cm/s) across the Bashi Strait. (A) Subinertial flow velocity in October 2005 (Tian et al., 2006); (B) subtidal flow velocity in July 2007 (Yang et al., 2010). Dashed lines and positive values denote the eastward flow; solid lines and negative values denote the westward flow (also shaded gray). The seabed topography is shaded black.

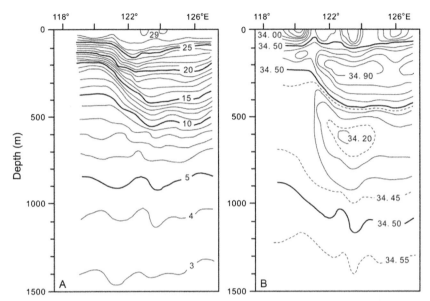

FIGURE 2.41   Profiles show variations in (A) temperature (°C) and (B) salinity (‰) across the Bashi Strait (at ~122°E) along 19.5°N in the summer of 1965. *From Su (2004).*

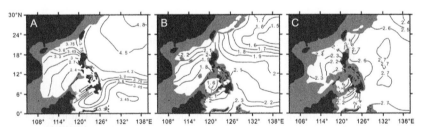

FIGURE 2.42   Variations of dissolved oxygen concentration (ml/l) at (A) 140 m, (B) 1000 m, and (C) 2000 m water depths on both sides of the Bashi Strait. Note a relative high level in intermediate waters of the SCS compared to other marginal sea basins (Qu, 2002).

On a broader aspect, water of the Pacific origin enters the SCS through the Bashi Strait, and from there, part of the water continues southward into the Java Sea and returns to the Pacific through Makassar Strait. This circulation has been coined "the SCS throughflow" (Figure 2.43), acting hypothetically as a heat and freshwater conveyor in regulating the SST pattern in the Indonesian maritime continent and its adjoining western Pacific and eastern Indian Oceans (Qu, Song, & Yamagata, 2005, 2006, 2009). Consequently, the connection of the southern SCS with the Indian Ocean through such passages as the Karimata Strait must have played a significant role in the

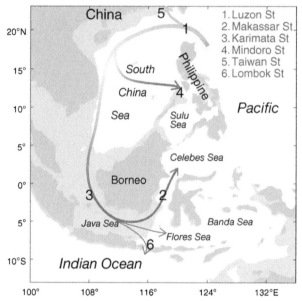

FIGURE 2.43 A schematic diagram of the South China Sea throughflow. Water entering the South China Sea through the Bashi (Luzon) Strait is lower in temperature and higher in salinity than water leaving it through the Karimata, Mindoro, and Taiwan Straits (Qu et al., 2006).

throughflow between the Indian and Pacific Oceans despite their shallow water and should no longer be ignored (Susanto et al., 2010).

2.4.5   Deepwater Circulation

Since the ECS is dominated by continental shelf with only the southern part of the Okinawa Trough deeper than 1000 m, researches on the deepwater circulation in the China Sea are virtually restricted to the SCS part. As discussed in the preceding text, all waters below ∼2000 m in the SCS are of Pacific origin and have the same hydrographic properties as the western Pacific water at about 2000 m. The incoming Pacific water crosses the Bashi Strait, sinks, spreads out, and fills the SCS deep basin. As a result, the deepwater temperature in the SCS is above 2.1 °C even at 4000 m, almost 0.8°C warmer than in the western Pacific at the same depth (Chen et al., 2001). Across the Bashi Strait, a persistent density difference exists between the Pacific and the SCS (Figure 2.44). Water on the Pacific side is well stratified, with potential density increasing from about 36.48 kg/m³ at 1000 m to 37.00 kg/m³ at 3500 m. But no obvious deepwater stratification occurs on the SCS side, where water density is vertically uniform with a density range of only about 0.02 kg/m³ below 2000 m (Figure 2.44) (Qu et al., 2006).

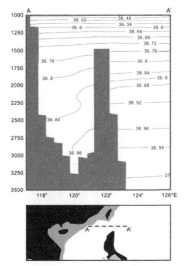

FIGURE 2.44   Profile A–A′ of potential density (kg/m^3) along 20°N, across the Bashi Strait. Lower panel shows the location and nearby bottom topography, with <200 m area shaded (Qu et al., 2006).

Due to the lack of observations, deepwater circulation in the SCS has been projected mainly from indirect data, all revealing a cyclonic pattern. For example, the water density field based on the synthetic salinity data (Figure 2.45A–C) shows that, upon entering the SCS through the Bashi Strait, waters of Pacific origin first turn northwestward and then southwestward along the continental margins off southeast China and east Vietnam, forming a cyclonic deep-layer circulation. The intrusion of deep Pacific water is also evident when using oxygen distribution as a passive tracer (Figure 2.45D). As it spreads over the SCS, water from the Pacific gradually gets mixed with ambient waters, loses its Pacific characteristics, and returns in less dense layers. Therefore, density distributions within the sea basin suggest a cyclonic deep boundary current system, as might be expected for an overflow-driven abyssal circulation (Qu et al., 2006).

This mode of the deep SCS circulation is confirmed recently by an analysis of updated monthly climatology of observed temperature and salinity using the US Navy Generalized Digital Environment Model (GDEM Version 3.0). The cyclonic circulation lies from about 2400 m to the bottom, and the boundary current transport of the cyclonic circulation is around 3.0 Sv (Figure 2.46). The quantitative portrayal of the deep SCS circulation suggests that the cyclonic circulation is mainly forced by the Luzon overflow, with bottom topography playing an important role (Wang et al., 2011).

In summary, the abyssal basin of the SCS is filled constantly by the Pacific water flowing down the sill of the Bashi Strait, and the deep SCS water

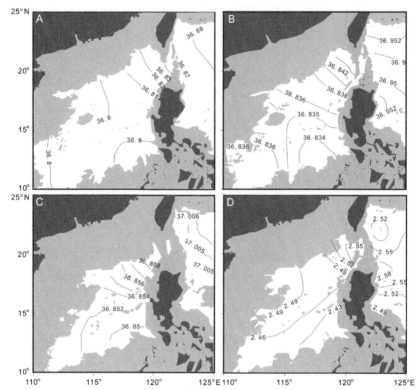

FIGURE 2.45 Potential density in kg/m^3 is calculated from the synthetic salinity depths at (A) 2000 m, (B) 2500 m, and (C) 3500 m, while (D) shows dissolved oxygen concentration (ml/l) along 36.84 density surface, lying at depths roughly between 2000 and 3000 m (Qu et al. 2006).

upwells into the intermediate SCS water between 350 and 1350 m, which is then exported out of the SCS again through the Bashi Strait (Su, 2004). The deep SCS water is estimated to have a fast flushing time of 40–50 years (Chen et al., 2001) or even less than 30 years (Qu et al., 2006). Therefore, the intermediate, deep, and bottom waters are essentially the same age. Because of a short residence time, the amount of particulate matter decomposition in the water column is sufficient to produce only a small maximum of chemical properties in the vertical profiles (Chen et al., 2001).

2.5 OCEANOGRAPHIC SUMMARY

The paramount features of the China Seas' oceanography are shaped by land–sea interactions. The exchanges of material and energy between Asia and the Pacific largely determine the oceanography of the China Seas, and in turn, the China Seas exert profound influence on both the Asian land and the Pacific

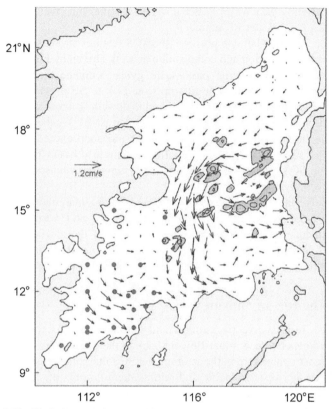

FIGURE 2.46   Vertical-averaged geostrophic current (cm/s) from 2400 m to the bottom. The light shading indicates water depths shallower than 2400 m. The dark shading indicates sea mountains shallower than 3600 m. The solid dots denote stations where oxygen exceeds 2.15 ml/l at 3000 m layer (Wang et al., 2011).

water. Basically, the characteristics of the China Seas' oceanography can be considered from the following three aspects.

2.5.1   Monsoon and seasonality

The China Seas are distinguished by prominent seasonal contrast almost in all aspects of oceanography, as the region is prevailed by East Asian monsoons. The water column in the ECS shelf, including the Yellow Sea and Bohai Gulf, is stratified in summer with freshened river runoff floating at the top but well mixed in winter due to intensive northern winds (Figure 2.34). Meanwhile, in summer, the Yangtze River diluted water extends northeastward into the Chiju Island area, while in winter, the Yangtze runoff is restricted in the coastal district and running mostly southward (Figure 2.32C and D). The sharp seasonality is characteristic also in the middle part of the Yellow Sea, where the Yellow Sea Warm Water invades in winter but gives its way to the Yellow Sea Cold Water in summer.

Despite its lower-latitude position, the SCS features strong seasonal variations in water circulation and composition as well. Driven by East Asian monsoons, seasonally alternating basin-scale gyres dominate the SCS surface circulation, and the thermocline tilts up from NW to SE in winter but to the opposite direction in summer. Two coastal upwelling areas are associated with monsoons: west of Luzon in winter and east of Vietnam in summer. Unlike the Indian monsoon, the local winter monsoon exceeds the summer monsoon in strength and in time duration on a yearly basis. Therefore, it is unwise for SCS studies to indiscriminately copy the monsoon proxies or climate cyclicity from other regions.

Inversely, the China Seas serve as a dynamic marginal sea basin in modulating the East Asian monsoon. Not only the SST variations exert a direct influence on the monsoon intensity, but also the glacial exposure of the extensive shelves and the reorganization of water currents in the geologic past must have caused significant changes in the monsoon system.

2.5.2   The role of "mixing mill"

The China Seas receive river runoff and ocean water, mix them up in the basins, and generate new water flowing back to the ocean. This "mixing mill" effect is most remarkable in the voluminous SCS that mixes surface and deep waters from the western Pacific and returns them through the Bashi Strait at intermediate depths (Yuan, 2002). The SCS also forms part of a throughflow for water exchanges between the Indian and Pacific Oceans (Qu et al., 2009), although mostly restricted to its upper-layer water. For the deeper water such as the North Pacific Intermediate Water, the SCS functions as a "cul de pas" with its inflow and outflow though the same Bashi Strait (You, Chern, & Yang, 2005), yet the outflowing water has significantly different physical and biogeochemical properties after mixing in the SCS Basin.

The China Seas are also one of the most significant regions in generating internal waves in the world's oceans. As observed at the Bashi and Tokara Straits and on the continental shelf and slope of the ECS, the tidal flow over such topographic features as a sill or shelf break in stratified water often produces nonlinear internal waves (e.g., Liu, Wai, Lozovatsky, & Fernando, 2009; Niwa & Hibiya, 2004; Tian et al., 2003). Recent observations in the northeastern SCS reveal that nonlinear internal waves emanate nearly daily from the Bashi Strait during spring tide and rapidly propagate westward to the shallow region west of 118°E. The amplitude of these waves reach 140 m or more, forming the largest free propagating nonlinear internal waves observed in the internal basin. Their propagation speed at about 2.9 m/s is also faster than any internal waves previously recorded in the world's oceans. The most likely source region of these giant internal waves is the near 2000 m deep ridge in the northern Bashi Strait (Liu et al., 2006). As shown by recent observations, these internal waves energize the top 1500 m of the water column and move large amounts of nutrient up. As a result, diapycnal diffusivity in the SCS and the Bashi Strait is elevated by two orders of magnitude over that of the smooth bathymetry in the North Pacific, exhibiting higher values at depths greater than about 1000 m (Tian, Yang, & Zhao, 2009). All these new findings are of great importance for our understanding of the mixing processes in the China Seas.

2.5.3   Regional impact

As a specific feature of the North Pacific, the western boundary currents are running through a series of marginal seas (Figure 2.1C; Wang, 2004). Therefore, the entire system of the oceanic circulation in the western Pacific is highly sensitive to changes in the China Seas. The Kuroshio, for example, brings high temperature and salinity to the ECS, the sea of Japan, and the sea of Okhotsk and finally meets with the Oyashio in the NW Pacific. As the ECS is the first marginal sea on its long journey to the north, any changes in the ECS on the Kuroshio will come to impact the subsequent marginal basins and the NW Pacific at large. As a result, variations of ocean currents in the ECS can lead to changes in heat and salinity transport from low to high latitude in the NW Pacific region.

In the south, the SCS also receives a part of the Kuroshio water and constitutes a key section of the Indonesian throughflow (Figure 2.43), consequently affecting the water exchanges between the Pacific and Indian Oceans. However, the regional influence of the China Seas' oceanography is not restricted to the ocean, but changes in water properties such as the local SST have been directly influencing the monsoon circulation as well. All these highlight the importance of modern variations and geologic evolution of the China Seas for our understanding of environmental changes in East Asia and the western Pacific.

REFERENCES

Chang, C.-P., Wang, Z., & Hendon, H. (2006). The Asian winter monsoon. In B. Wang (Ed.), *The Asian monsoon* (pp. 89–127). Springer.

Chao, S. V., Shaw, P. T., & Wu, S. Y. (1996). Deep water ventilation in the South China Sea. *Deep Sea Research Part II, 43,* 445–466.

Chen, L. (1992). Features of the East Asian monsoon. In M. Murakami & Y. Ding (Eds.), *Studies of Asian monsoon in Japan and China* (pp. 220–235). Ibaraki, Japan: Meteorological Research Institute.

Chen, C. T. A., & Huang, M. H. (1996). A mid-depth front separating the South China Sea water and the Philippine Sea water. *Journal of Oceanography, 52,* 17–52.

Chen, C. T. A. (2000). The Three Gorges Dam: Reducing the upwelling and thus productivity in the East China Sea. *Geophysical Research Letters, 27,* 381–383.

Chen, Z., Li, J., Shen, H., & Wang, Z. (2001). Yangtze River of China: Historical analysis of discharge variability and sediment flux. *Geomorphology, 41,* 77–91.

Chen, Z., Saito, Y., & Goodbred, S. L. Jr., (2005). *Mega-deltas of Asia-Geological evolution and human impact*. Beijing, China: China Ocean Press, 268 pp.

Chu, P. C., & Wang, G. (2003). Seasonal variability of thermohaline front in the central South China Sea. *Journal of Oceanography, 59*, 65–78.

Ding, Y., & Chan, J. C. L. (2005). The East Asian summer monsoon: An overview. *Meteorology and Atmospheric Physics, 89*, 117–142.

Ding, Y., Li, C., & Liu, Y. (2004). Overview of the South China Sea monsoon experiment. *Advances in Atmospheric Sciences, 21*, 343–360.

Dong, L., Su, J., Wong, L. A., Cao, Z., & Chen, J. C. (2004). Seasonal variations and dynamics of the Pearl River plume. *Continental Shelf Research, 24*, 1761–1777.

Duc, D. M., Nhuan, M. T., & Ngoi, C. V. (2012). An analysis of coastal erosion in the tropical rapid accretion delta of the Red River, Vietnam. *Journal of Asian Earth Sciences, 43*, 98–109.

Fang, G. H., Fang, W. D., Fang, Y., & Wang, K. (1998). A survey of studies on the South China Sea upper ocean circulation. *Acta Oceanographica Taiwanica, 37*, 1–16.

Fang, W., Fang, G., Shi, P., Huang, Q., & Xie, Q. (2002). Seasonal structures of upper layer circulation in the southern South China Sea from in situ observations. *Journal of Geophysical Research, 107*, C11. http://dx.doi.org/10.1029/2002JC001343.

Fang, G., Zhao, B., & Zhu, Y. (1991). Water volume transport through the Taiwan Strait and the continental shelf of the East China Sea measured with current meters. In K. Takano (Ed.), *Oceanography of Asian Marginal Seas* (pp. 345–348). Amsterdam: Elsevier.

Fu, K. D., He, D. M., & Lu, X. X. (2008). Sedimentation in the Manwan reservoir in the Upper Mekong and its downstream impacts. *Quaternary International, 186*, 91–99.

Gan, J., Li, H., Curchitser, E. N., & Haidvogel, D. B. (2006). Modeling South China Sea circulation: Response to seasonal forcing regimes. *Journal of Geophysical Research, 111*. http://dx.doi.org/10.1029/2005JC003298, C06034.

Gong, G. C., Liu, K. K., Liu, C. T., & Pai, S. C. (1992). The chemical hydrography of the South China Sea west of Luzon and a comparison with the West Philippine Sea. *TAO Taipei, 13*, 587–602.

Guan, B. (1994). Patterns and structures of the currents in Bohai, Huanghai and East China Seas. In D. Zhou, Y. B. Liang, & C. K. Zeng (Eds.), *Oceanology of China Seas: Vol. 1.* (pp. 17–26). Dordrecht, The Netherlands: Kluwer Academic Publishers.

Guan, B., & Fang, G. (2006). Winter counter-wind currents off the southeastern China coast: A review. *Journal of Oceanography, 62*, 1–24.

Gupta, A., & Liew, S. C. (2007). The Mekong from satellite imagery: A quick look at a large river. *Geomorphology, 85*, 259–274.

He, M., Zheng, H., Huang, X., Jia, J., & Li, L. (2012). Yangtze River sediments from source to sink traced with clay mineralogy. *Journal of Asian Earth Sciences, 69*, 60–69.

Hoa, L. T. V., Nhan, N. H., Wolanski, E., Cong, T. T., & Shigeko, H. (2007). The combined impact on the flooding in Vietnam's Mekong River delta of local man-made structures, sea level rise, and dams upstream in the river catchment. *Estuarine, Coastal and Shelf Science, 71*, 110–116.

Hsueh, Y. (2000). The Kuroshio in the East China Sea. *Journal of Marine Systems, 24*, 131–139.

Hu, J., Kawamura, H., Hong, H., & Qi, Y. (2000). A review on the currents in the South China Sea: Seasonal circulation, South China Sea warm current and Kuroshio intrusion. *Journal of Oceanography, 56*, 607–624.

Huang, Q., & Wang, W. (1994). Current characteristics of the South China Sea. In D. Zhou, Y. B. Liang, & C. K. Zeng (Eds.), *Oceanology of China Seas: Vol. 1.* (pp. 39–47). Dordrecht, The Netherlands: Kluwer Academic Publishers.

Isobe, A. (2008). Recent advances in ocean-circulation research on the Yellow Sea and East China Sea shelves. *Journal of Oceanography, 64*, 569–584.

Ji, X., Sheng, J., Tang, L., Liu, D., & Yang, X. (2011). Process study of circulation in the Pearl River Estuary and adjacent coastal waters in the wet season using a triply-nested circulation model. *Ocean Modelling, 38*, 138–160.

Kao, S.-J., & Milliman, J. D. (2008). Water and sediment discharge from small mountainous rivers, Taiwan: The roles of lithology, episodic events and human activities. *Journal of Geology, 116*, 431–448.

Kim, H.-C., Yamaguchi, H., Yoo, S., Zhu, J., Okamura, K., Kiyomoto, Y., et al. (2009). Distribution of Changjiang diluted water detected by satellite chlorophyll-a and its interannual variation during 1998–2007. *Journal of Oceanography, 65*, 129–135.

Koshikawa, M. K., Takamatsu, T., Takada, J., Zhu, M., Xu, B., Chen, Z., et al. (2007). Distributions of dissolved and particulate elements in the Yangtze estuary in 1997–2002: Background data before the closure of the Three Gorges Dam. *Estuarine, Coastal and Shelf Science, 71*, 26–36.

Kummu, M., Lu, X. X., Rasphone, A., Sarkkula, J., & Koponen, J. (2008). Riverbank changes along the Mekong River: Remote sensing detection in the Vientiane-Nong Khai area. *Quaternary International, 186*, 100–112.

Kummu, M., & Varis, O. (2007). Sediment-related impacts due to upstream reservoir trapping, the lower Mekong River. *Geomorphology, 85*, 275–293.

Le, T. P. Q., Garnier, J., Billen, G., Thery, S., & Chau, V. M. (2007). The changing flow regime and sediment load of the Red River, Vietnam. *Journal of Hydrology, 334*, 199–214.

Lee, H.-J., & Chao, S.-Y. (2003). A climatological description of circulation in and around the East China Sea. *Deep-Sea Research II, 50*, 1065–1084.

Lee, T. N., Johns, W. E., Liu, C. T., Zhang, D., Zantopp, R., & Yang, Y. (2001). Mean transport and seasonal cycle of the Kuroshio east of Taiwan with comparison to the Florida Current. *Journal of Geophysical Research, C106*, 22143–22158.

Li, J. (Ed.). (2008). *Regional geology of the East China Sea* (31 p.). Beijing: China Ocean Press.

Liu, K.-K., Chao, S.-Y., Shaw, P.-T., Gong, G.-C., Chen, C.-C., & Tang, T. Y. (2002). Monsoon-forced chlorophyll distribution and primary production in the South China Sea: Observations and a numerical study. *Deep-Sea Research I, 49*, 1387–1412.

Liu, Z., & Gan, J. (2012). Variability of the Kuroshio in the East China Sea derived from satellite altimetry data. *Deep-Sea Research I, 59*, 25–36.

Liu, Q., Jiang, X., Xie, S.-P., & Liu, W. T. (2004). A gap in the Indo-Pacific warm pool over the South China Sea in boreal winter: Seasonal development and interannual variability. *Journal of Geophysical Research, 109*. http://dx.doi.org/10.1029/2003JC002179, C07012.

Liu, J. P., Liu, C. S., Xu, K. H., Milliman, J. D., Chiu, J. K., Kao, S. J., et al. (2008). Flux and fate of small mountainous rivers derived sediments into the Taiwan Strait. *Marine Geology, 256*, 65–76.

Liu, C.-T., Pinkel, R., Hsu, M.-K., Klymak, J. M., Chen, H.-W., & Villanov, C. (2006). Nonlinear internal waves from the Luzon Strait. *Eos, Transactions American Geophysical Union, 87*, 449–451.

Liu, Z., Wai, H., Lozovatsky, I. D., & Fernando, H. J. S. (2009). Late summer stratification, internal waves, and turbulence in the Yellow Sea. *Journal of Marine Systems, 77*, 459–472.

Liu, Q., Yang, H., & Wang, Q. (2000). Dynamic characteristics of seasonal thermocline in the deep sea region of the South China Sea. *Chinese Journal of Oceanology and Limnology, 18*(2), 104–109 (in Chinese).

Lu, X. X., & Chen, X. (2008). Large Asian rivers and their interactions with estuaries and coasts. *Quaternary International, 186*, 1–3.

Lu, Z., Yang, H., & Liu, Q. (2001). Regional Dynamics of Seasonal Variability in the South China Sea. *Journal of Physical Oceanography, 31*, 272–284.

Luo, X.-L., Zeng, E. Y., Ji, R.-Y., & Wang, C.-P. (2007). Effects of in-channel sand excavation on the hydrology of the Pearl River Delta, China. *Journal of Hydrology, 343*, 230–239.

Maren, D. S. V. (2007). Water and sediment dynamics in the Red River mouth and adjacent coastal zone. *Journal of Asian Earth Science, 29,* 508–522.

Meshkova, L. V., & Carling, P. A. (2012). The geomorphological characteristics of the Mekong River in northern Cambodia: A mixed bedrock–alluvial multi-channel network. *Geomorphology, 147–148,* 2–17.

Metzger, E. J., & Hurlburt, H. E. (2001). The nondeterministic nature of Kuroshio penetration and eddy shedding in the South China Sea. *Journal of Physical Oceanography, 31,* 1712–1732.

Milliman, J. D., & Farnsworth, K. L. (2011). *River discharge to the coastal ocean. A global synthesis.* Cambridge: Cambridge University Press, 384pp.

Milliman, J. D., & Syvitski, J. P. M. (1992). Geomorphic/tectonic control of sediment discharge to the ocean: The importance of small mountainous rivers. *Journal of Geology, 100,* 525–544.

Moon, S., Huh, Y., Qin, J., & Pho, N. V. (2007). Chemical weathering in the Hong (Red) River basin: Rates of silicate weathering and their controlling factors. *Geochimica et Cosmochimica Acta, 71,* 1411–1430.

Nguyen, V. L., Ta, T. K. O., & Tateishi, M. (2000). Late Holocene depositional environments and coastal evolution of the Mekong river delta, southern Vietnam. *Journal of Asian Earth Sciences, 18,* 427–439.

Niino, H., & Emery, K. O. (1961). Sediments of shallow portions of East China Sea and South China Sea. *Geological Society of America Bulletin, 72,* 731–762.

Niwa, Y., & Hibiya, T. (2004). Three-dimensional numerical simulation of M2 internal tides in the East China Sea. *Journal of Geophysical Research, 109.* http://dx.doi.org/10.1029/2003JC001923, C04027.

Qu, T. (2000). Upper layer circulation in the South China Sea. *Journal of Physical Oceanography, 30,* 1450–1460.

Qu, T. (2002). Evidence for water exchange between the South China Sea and the Pacific Ocean through the Luzon Strait. *Acta Oceanologica Sinica, 21*(2), 175–185.

Qu, T., Du, Y., Meyers, G., Ishida, A., & Wang, D. (2005). Connecting the tropical Pacific with Indian Ocean through South China Sea. *Geophysical Research Letters, 32,* L24609. http://dx.doi.org/10.1029/2005GL024698.

Qu, T., Girton, J. B., & Whitehead, J. A. (2006). Deepwater overflow through Luzon Strait. *Journal of Geophysical Research, 111,* C01002. http://dx.doi.org/10.1029/2005JC003139.

Qu, T., Song, Y. T., & Yamagata, T. (2009). An introduction to the South China Sea throughflow: Its dynamics, variability, and application for climate. *Dynamics of Atmospheres and Oceans, 47,* 3–14.

Ren, M. (Editor in Chief). (1985). *Modern sedimentation in coastal and nearshore zone of China* (466 p.). Beijing: Ocean Press & Springer-Verlag.

Saito, Y., Chaimanee, N., Jarupongsakul, T., & Syvitski, J. P. M. (2007). Shrinking mega-deltas in Asia: Sea-level rise and sediment reduction impacts from case study of the Chao Phraya Delta. *Inprint Newsletter of the IGBP/IHDP Land Ocean Interaction in the Coastal Zone, 2007*(2), 3–9.

Shaw, P. T. (1996). Winter upwelling off Luzon in the Northeastern South China Sea. *Journal of Geophysical Research, 101,* 16435–16448.

Shaw, P. T., & Chao, S. Y. (1994). Surface circulation in the South China Sea. *Deep-Sea Research I, 41,* 1663–1683.

Shi, P., Du, Y., Wang, D. X., & Gan, Z. J. (2001). Annual cycle of mixed layer in South China Sea. *Tropical Oceanology, 20,* 10–17 (in Chinese).

Shi, W., & Wang, M. (2012). Sea ice properties in the Bohai Sea measured by MODIS-Aqua: 2. Study of sea ice seasonal and interannual variability. *Journal of Marine Systems, 95,* 41–49.

Su, J. (1998). Circulation dynamics of the China Seas North of 18°N. In A. R. Robinson & K. H. Brink (Eds.), *The sea, vol. 11* (pp. 483–505). New York, NY: Wiley.

Su, J. (2004). Overview of the South China Sea circulation and its influence on the coastal physical oceanography outside the Pearl River Estuary. *Continental Shelf Research, 24,* 1745–1760.

Sun, X. P. (2006). *Regional oceanography off China Coasts.* Beijing: China Ocean Press (in Chinese), 376pp.

Susanto, R. D., Fang, G., Soesilo, I., Zheng, Q., Qiao, F., Wei, Z., et al. (2010). New surveys of a branch of the Indonesian throughflow. *Eos, Transactions American Geophysical Union, 91*(30), 261. http://dx.doi.org/10.1029/2010EO300002.

Syvitski, J. P. M., Kettner, A. J., Overeem, I., Hutton, E. W. H., Hannon, M. T., Brakenridge, G. R., et al. (2009). Sinking deltas due to human activities. *Nature Geosciences, 2,* 681–685.

Syvitski, J. P. M., & Saito, Y. (2007). Morphodynamics of deltas under the influence of humans. *Global and Planetary Changes, 57,* 261–282.

Tamura, T., Horaguchi, K., Saito, Y., Nguyen, V. L., Tateishi, M., Ta, T. K. O., et al. (2010). Monsoon-influenced variations in morphology and sediment of a mesotidal beach on the Mekong River delta coast. *Geomorphology, 116,* 11–23.

Teague, W. J., Jacobs, G. A., Ko, D. S., Tang, T. Y., Chang, K. I., & Suk, M. S. (2003). Connectivity of the Taiwan, Cheju, and Korea straits. *Continental Shelf Research, 13,* 63–77.

Thu, P. M., & Populus, J. (2007). Status and changes of mangrove forest in Mekong Delta: Case study in Tra Vinh, Vietnam. *Estuarine, Coastal and Shelf Science, 71,* 98–109.

Tian, J. W., Yang, Q., & Zhao, W. (2009). Enhanced diapycnal mixing in the South China Sea. *Journal of Physical Oceanography, 39,* 3191–3203.

Tian, J. W., Zhou, L., Zhang, X., Liang, X., Zheng, Q., & Zhao, W. (2003). Estimates of M2 internal tide energy fluxes along the margin of Northwestern Pacific using TOPEX/POSEIDON altimeter data. *Geophysical Research Letters, 30,* 1889. http://dx.doi.org/10.1029/2003GL018008.

Tian, J. W., Yang, Q., Liang, X., Xie, L., Hu, D., Wang, F., et al. (2006). Observation of Luzon Strait transport. *Geophysical Research Letters, 33,* L19607. http://dx.doi.org/10.1029/2006GL026272.

Trenberth, K. E., Stepaniak, D. P., & Caron, J. M. (2000). The global monsoon as seen through the divergent atmospheric circulation. *Journal of Climate, 13,* 3969–3993.

Wang, P. (1999). Response of Western Pacific marginal seas to glacial cycles: Paleoceanographic and sedimentological features. *Marine Geology, 156,* 5–39.

Wang, P. (2004). Cenozoic deformation and the history of sea-land interactions in Asia. In P. Clift, P. Wang, D. Hayes, & W. Kuhnt (Eds.), *AGU Geophysical Monograph: Vol. 149. Continent-Ocean interactions in the East Asian Marginal Seas* (pp. 1–22).

Wang, P., Clemens, S., Beaufort, L., Braconnot, P., Ganssen, G., Jian, Z., et al. (2005). Evolution and variability of the Asian monsoon system: State of the art and outstanding issues. *Quaternary Science Reviews, 24,* 595–629.

Wang, Y. (Ed.), (2012). *Regional oceanography of china seas—Marine geomorphology* (p. 676). Beijing, China: China Ocean Press (in Chinese).

Wang, H., Bi, N., Saito, Y., Wang, Y., Sun, X., Zhang, J., et al. (2010). Recent changes in sediment delivery by the Huanghe (Yellow River) to the sea: Causes and environmental implications in its estuary. *Journal of Hydrology, 391,* 302–313.

Wang, G., Chen, D., & Su, J. (2006). Generation and life cycle of the dipole in the South China Sea summer circulation. *Journal of Geophysical Research, 111.* http://dx.doi.org/10.1029/2005JC003314, C06002.

Wong, G. T. F., Tseng, C. M., Wen, L. S., & Chung, S. W. (2007b). Nutrient dynamics and nitrate anomaly at the SEATS station. *Deep Sea Research Part II, 54,* 1528–1545.

Wang, B., Clemens, S. C., & Liu, P. (2003). Contrasting the Indian and East Asian monsoons: Implications on geological timescales. *Marine Geology, 201,* 5–21.

Wang, D., Du, Y., & Shi, P. (Eds.), (2002). *Climatological atlas of physical oceanography in the upper layer of the South China Sea* (p. 168). Beijing, China: Meteorological Press (in Chinese).

Wang, P., & Li, Q. (Eds.), (2009). *The South China Sea. Paleoceanography and sedimentology.* Springer, 506pp.

Wang, B., & Lin, H. (2002). Rainy season of the Asian–Pacific summer monsoon. *Journal of Climate*, *15*, 386–398.

Wang, F., Liu, C., & Meng, Q. (2012). Effect of the Yellow Sea warm current fronts on the westward shift of the Yellow Sea warm tongue in winter. *Continental Shelf Research*, *45*, 98–107.

Wang, H., Saito, Y., Zhang, Y., Bi, N., Sun, X., & Yang, Z. (2011). Recent changes of sediment flux to the western Pacific Ocean from major rivers in East and Southeast Asia. *Earth-Science Reviews*, *108*, 80–100.

Wang, Z., Yu, G. A., & Huang, H. Q. (2012). Gender of deltas and parasitizing rivers. *International Journal of Sediment Research*, *27*, 1–19.

Wong, G. T. F., Ku, T. L., Mulholland, M., Tseng, C. M., & Wang, D. P. (2007a). The South East Asian Time-series Study (SEATS) and the biogeochemistry of the South China Sea-An overview. *Deep Sea Research Part II*, *54*, 1434–1447.

Wyrtki, K. (1961). *Scientific results of marine investigations of the South China Sea and Gulf of Thailand 1959–1961. Naga reports 2.* (pp. 164–169). San Diego, CA: Scripps Institution of Oceanography, University of California.

Xie, S.-P., Xie, Q., Wang, D., & Liu, W. T. (2003). Summer upwelling in the South China Sea and its role in regional climate variations. *Journal of Geophysical Research*, *108*, C3261. http://dx.doi.org/10.1029/2003JC001867.

Xu, J. (2003). Sedimentation rates in the lower Yellow River over the past 2300 years as influenced by human activities and climate change. *Hydrological Processes*, *17*, 3359–3371.

Xu, X. Z., Qiu, Z., & Chen, H. C. (1982). The general descriptions of the horizontal circulation in the South China Sea. In *Proceedings of the 1980 symposium on hydrometeorology, Chinese society of oceanology and limnology* (pp. 137–145). Beijing, China: Science Press (in Chinese).

Xu, L. L., Wu, D. X., Lin, X. P., & Ma, C. (2009). The study of the Yellow Sea Warm Current and its seasonal variability. *Journal of Hydrodynamics*, *21*, 159–165.

Yang, H. J., & Liu, Q. Y. (1998). The seasonal features of temperature distributions in the upper layer of the South China Sea. *Oceanol Limn Sinica*, *29*, 501–507 (in Chinese).

Yang, S. L., Milliman, J. D., Li, P., & Xu, K. (2011). 50,000 dams later: Erosion of the Yangtze River and its delta. *Global and Planetary Change*, *75*, 14–20.

Yang, Q. X., Tian, J. W., & Zhao, W. (2010). Observation of Luzon Strait transportin summer 2007. *Deep-Sea Research I*, *57*(2010), 670–676.

Yang, Z. S., Wang, H., Saito, Y., Milliman, J. D., Xu, K., Qiao, S., et al. (2006). Dam impacts on the Changjiang (Yangtze) River sediment discharge to the sea: The past 55 years and after the Three Gorges Dam. *Water Resources Research*, *42*. http://dx.doi.org/10.1029/2005WR003970, W04407.

You, Y. Z., Chern, C.-S., & Yang, Y. (2005). The South China Sea, a cul-de-sac of North Pacific Intermediate Water. *Journal of Oceanography*, *61*, 509–527.

Yu, L. (2002). The Huanghe (Yellow) River: A review of its development, characteristics, and future management issues. *Continental Shelf Research*, *22*, 389–403.

Yuan, D. L. (2002). A numerical study of the South China Sea deep circulation and its relation to the Luzon Strait transport. *Acta Oceanologica Sinica*, *21*, 187–202.

Yuan, D., & Hsueh, Y. (2010). Dynamics of the cross-shelf circulation in the Yellow and East China Seas in winter. *Deep-Sea Research II*, *57*, 1745–1761.

Yuan, D., Zhu, J., Li, C., & Hu, D. (2008). Cross-shelf circulation in the Yellow and East China Seas indicated by MODIS satellite observations. *Journal of Marine Systems*, *70*, 134–149.

Zhang, S. W., Wang, Q. Y., Lü, Y., Cui, H., & Yuan, Y. L. (2008). Observation of the seasonal evolution of the Yellow Sea Cold Water Mass in 1996–1998. *Continental Shelf Research*, *28*, 442–457.

Zhang, W., Wei, X., Zheng, J., Zhu, Y., & Zhang, Y. (2012). Estimating suspended sediment loads in the Pearl River Delta region using sediment rating curves. *Continental Shelf Research*, *38*, 35–46.

Chapter 3

Tectonic Framework and Magmatism

3.1 INTRODUCTION

Tectonics of the China Seas has been synthesized by numerous Chinese authors (e.g., Cai, 2005; Chen & Wen, 2010; Gong, 1997; Gong, Li, & Xie, 1997; Jia & Gu, 2002; Jin, 1989, 1992; Li, 2012; Liu, 1992, 1994; Liu & Li, 2001; Liu, Huang, & Yang, 1988; Liu, Zhao, & Fan, 2002; Qin, Zhao, & Chen, 1988; Wang, Prell, et al., 2000; Xu, Liu, Zhang, et al., 1997; Yao, et al., 1994; Yu & Li, 1992; Zhai, Chen, & Zhang, 2001; Zhang, 1986, 2008; Zhang, Chen, & Zhang, 2005; Zhou, Liang, & Zeng, 1994; Zhu, 2007). These works are thorough and multifaceted, covering topics of tectonics, geophysics, and hydrocarbon and mineral resources.

The first round of regional marine investigations was made mostly by scientists from the United States. Wageman, Hilde, and Emery (1970) are among the first to carry out reflection seismic surveys in the East China Sea (ECS). Some of the earliest efforts to understanding the regional crustal structures in the ECS were made by Ludwig et al. (1973) and Leyden, Ewing, and Murauchi (1973), who applied sonobuoy refraction measurements to reveal upper crustal velocity distributions along 2D profiles. In the South China Sea (SCS), early pioneering geophysical studies include heat flow measurement (Karig, 1971, 1973), magnetic anomaly identification (Ben-Avraham & Uyeda, 1973; Bowin, Lu, Lee, & Schouten, 1978), and crustal seismic velocity estimation (Ludwig, Kumar, & Houtz, 1979). Taylor and Hayes (1980, 1983) summarized regional tectonic history of the SCS based on these early investigations.

With major breakthroughs in hydrocarbon explorations in the late 1970s and early 1980s, extensive geologic studies in the China Seas surged in the 1980s. Zhang (1986) first synthesized regional tectonics of the China Seas. Other earliest landmark contributions are regional geologic syntheses of the Yellow Sea (Institute of Oceanology, Chinese Academy of Sciences, 1982; Qin, 1990b), Bohai Sea (Qin, 1990), ECS (Jin, 1992; Qin et al., 1988), and SCS (Jin, 1989; Liu et al., 1988).

At the meantime, international collaborations between Guangzhou Marine Geological Survey and Lamont–Doherty Geological Observatory led to two phases of joint geophysical surveys in the SCS (Yao, et al., 1994). The first comprehensive geophysical atlas of the China Seas was compiled (Liu, 1992). Several memoirs on offshore hydrocarbon explorations were also published, reflecting much deepened understanding with new acquisitions of large volumes of data (e.g., Gong, 1997; Gong et al., 1997; Xu et al., 1997). Hydrothermal activities and heat flow in the Okinawa Trough and ECS were also studied preliminarily (Li, 1995; Yu & Li, 1992).

ODP Leg 184 in the SCS marked the start of a second round of extensive research (Wang, Prell, et al., 2000). Several comprehensive research projects were set up to study geodynamics and petroleum geology (e.g., Gao & Li, 2002; Li & Gao, 2003, 2004). Numerous deep-sounding surveys with ocean-bottom seismometers were carried out. Many new geologic and geophysical memoirs and atlases were compiled (Chen & Wen, 2010; Gong, Li, & Xie, 2004; Li, Ding, Gao, Wu, & Zhang, 2011; Li, Ding, Wu, Zhang, & Dong, 2012; Liu et al., 2002; Ren et al., 2013; Zhang, 2008; Zhu, 2007).

Currently, the China Seas are at the forefront of the most active sea-going research activities. Early 2011, the South China Sea Deep (SCSD), a comprehensive 8-year research program, was funded by the National Science Foundation of China (NSFC) (Wang, 2012). With a total budget of ~30 M in U.S. dollars, this program has funded, among many other projects, coincident refraction/reflection surveys and the first passive-source OBS array and deep-tow magnetic survey in the China Seas. Furthermore, a new IODP Expedition (# 349) was launched at the end of January 2014 to retrieve basaltic rocks from oceanic crust and study the opening history of the SCS and its bearings on climatic and deep mantle processes (Li, Lin, & Kulhanek, 2013). These landmark research activities are anticipated to significantly elevate our understanding of the marginal basins.

It is noted that many of the early synthetic work were published in Chinese and presented their discussions either on a particular region or in a region by region style. In this chapter, we try to synthesize tectonics of the entire study area from an evolutionary perspective, that is, following first the main chronological thread of regional tectonic events and then focusing on local geodynamic features whenever further discussions on them are deemed necessary.

3.2 OUTLINE OF PRESENT-DAY TECTONIC FRAMEWORK

The China Seas occupy a large area composed of distinctly different tectonic blocks (Figure 3.1). To illustrate these differences, we will start from the middle Triassic when the North China Craton collided with the Yangtze Craton along the Qinling–Dabie–Sulu orogenic belt (Figure 3.1). The welding of these two blocks forced the closure of a northeastern branch of the

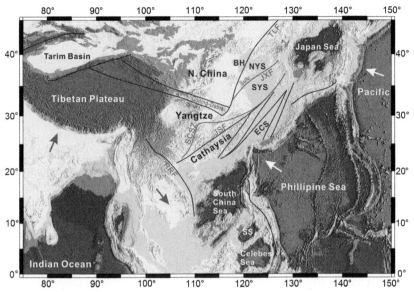

FIGURE 3.1   Regional tectonophysical units surrounding the China Seas. BH, Bohai Sea; NYS, Northern Yellow Sea; SYS, Southern Yellow Sea; ECS, East China Sea; SS, Sulu Sea; TLF, Tan-Lu fault; JXF, Jiaxiang Fault; JSF, Jiangshao Fault; RRF, Red River Fault; SCSZ, South China suture zone. Black lines are regional faults. SCSZ and JXF delineate block boundaries that are often reactivated by faulting. Arrows mark the directions of plate movement. *Modified from Li (2004).*

Paleo-Tethys and formed a unified continental basement, upon which subsequent development of China Seas occurred.

Orogenic belts and large-scale faults often serve as the tectonic boundaries between different seas. The Cenozoic evolution of the Bohai Basin has been strongly controlled by the Tan-Lu fault (Allen, MacDonald, Zhao, Vincent, & Brouet-Menzies, 1997, 1998), which originated partly from the North China and Yangtze collision in the middle Triassic and has remained active since then. The North Yellow Sea is located between the Tan-Lu fault and the Sulu orogeny. The Sulu orogen is bounded to the south by the Jiaxiang Fault. Both the Bohai and the North Yellow Seas are located within the North China block (Figure 3.1). In contrast, the South Yellow Sea to the south of the Sulu orogen is located in the Yangtze Block, which was largely covered by the Paleo-Tethys before its collision with the North China block.

East of the Tan-Lu fault, crustal faults are mostly in NE strikes, of which the Jiaxiang Fault and the Jiangshao Fault form the northern and southern boundaries of the lower Yangtze Block, respectively. NE-striking crustal faults developed in the ECS continental shelf form the boundaries of the ECS Basin and of depressions within the basin (Li, Zhou, Ge & Mao, 2007; Li, Zhou, et al., 2009). Based on the present-day seismicity and crosscutting relationships between faults and strata interpreted from reflection seismic data, these NE-striking regional crustal faults on the ECS continental shelf are not active faults since they all ceased their activities at least 2 Ma (Li, Zhou, Ge & Mao, 2007; Li, Zhou, et al., 2009). These faults were induced by the subduction of a Paleo-Pacific Plate beneath the Eurasian Plate in the late Mesozoic, and later in the early Cenozoic, they controlled or influenced formations of basins and rifts in the area.

Whether these regional faults and the Yangtze Block had extended east to the Korean Peninsula is still a focus of discussion. Even for those who support the eastward extension of the Sulu belt and the Yangtze Block to the Korean Peninsula, they are far from reaching an agreement upon the ways and styles of tectonic connections. The Jiangshao Fault appears to be connected with the Ogcheon Belt in Korea (Figure 3.2); however, Paleozoic and early Mesozoic tectonic belts may have been strongly reformed by later magmatism and tectonic movements. The apparent geologic difference between the Korean Peninsula and the Yangtze Block may be largely caused by late magmatism and shall not be taken as evidences supporting the argument that the faults and the Yangtze Block did not extend to the Korean Peninsula during the late Mesozoic (Li, Chen, & Zhou, 2009).

The boundary between the South Yellow Sea and the ECS is the Jiangshao Fault, which developed roughly along the ancient boundary between the Yangtze and the Cathaysian block in the offshore area (Figure 3.1). The welding of the Yangtze and the Cathaysian blocks, both with basement rocks older than 2.0 Ga (Chen & Jahn, 1998; Qiu, Gao, McNaughton, Groves, & Ling, 2000; Wang, Fan, Guo, Peng, & Li, 2003), likely formed along the South China suture zone in the Neoproterozoic at about 930–1000 Ma (Charvet, Shu, Shi, Guo, & Faure, 1996; Chen, Foland, Xing, Xu, & Zhou, 1991; Li & McCulloch, 1996; Shi & Li, 2012), although much younger ages were also proposed (e.g., Haynes, 1988; Hsü et al., 1988; Wu, 2003). The Cathaysia itself is characteristically segmented by a group of northeast-trending faults (Figure 3.1), reflecting mostly late Mesozoic and early Cenozoic rifting and volcanic tectonics. The Jiangshao Fault is an important tectonic boundary and has experienced multiple stages of evolution (Shu & Xu, 2002; ZBGMR, 1989). Before the middle Triassic, this fault was a major paleogeographic boundary, but it later became a belt of extensive magmatism, rifting, and strike-slip faulting (Gilder et al., 1996; Shu et al., 2009).

Faulting and volcanisms have in general taken a southeastward younging trend within the Cathaysia, due most likely to the deepening of the subducted Paleo-Pacific Plate and eastward retreat of the Paleo-Pacific subduction zone (Li & Li, 2007; Shi & Li, 2012; Zhou & Li, 2000). The early Cenozoic opening of the ECS Basin, located within the Cathaysian block, is one of a series of consequences of this prolonged regional extension since the late Mesozoic (Figure 3.2). Today, this regional rifting activity continues east of the ECS Basin in the Okinawa Trough, though the rifting of this back-arc basin is

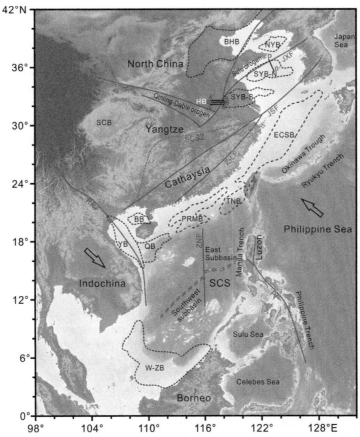

FIGURE 3.2 Tectonophysical outlines of the China Seas and major Cenozoic sedimentary basins. BHB, Bohai Basin; NYB, Northern Yellow Sea Basin; SYB-N, northern depression of the Southern Yellow Sea Basin; SYB-S, southern depression of the Southern Yellow Sea Basin; ECSB, East China Sea basin; TNB, Taixinan Basin; PRMB, Pearl River Mouth Basin; BB, Beibuwan Basin; QB, Qiongdongnan Basin; YB, Yinggehai Basin; W-ZB, Wan'an–Zengmu Basin; HB, Hefei Basin; SCS, South China Sea; TLF, Tan-Lu fault; JXF, Jiaxiang Fault; JSF, Jiangshao Fault; RRF, Red River Fault; Z-LF, Zhenghe–Lianhuashan Fault; ZNF, Zhongnan Fault; PF, Philippine Fault; SCSZ, South China suture zone; SCB, Sichuan Basin. AA′, BB′, CC′, and DD′ are four seismic sections to be shown in Figures 3.4 and 3.11. Arrows mark the directions of plate movement.

due to the subduction of the western Philippine Sea Plate, started about 10 Ma, rather than to the direct subduction of the Paleo-Pacific Plate as in the late Mesozoic and early Cenozoic (Hsu, Sibuet, & Shyu, 2001; Park, Tokuyama, Shinohara, Suyehiro, & Taira, 1998; Shinjo, 1999). The western Philippine Sea Plate is subducting northwestward beneath the Eurasian Plate at a rate of about 7 cm/year (Seno, Stein, & Gripp, 1993) and forms the NE-trending Ryukyu Trench–Ryukyu arc–Okinawa Trough system. Although this active subduction and back-arc rifting process has started only recently since the late Cenozoic (Kizaki, 1986; Lee & Lawver, 1995) owing to the northward and clockwise movement of the Philippine Sea Plate (Hall, 2002; Hall, Ali, Anderson, & Baker, 1995), it is generally argued from the distribution of Cathaysian magmatism that the Mesozoic Paleo-Pacific subduction also followed roughly along the present-day Ryukyu Trench and the northern SCS continental margin.

Some of the depressions and uplifts in ECS Basin can be traced down to the south of Taiwan around the northeastern SCS, though geographically, this regional basin feature is interrupted partly by the uplift of Taiwan caused by the late Neogene collision between the Luzon Arc and the Eurasian continental margin. Indeed, the Pearl River Mouth Basin on the northern SCS continental margin developed in the same early Cenozoic extensional regime as did the ECS Basin.

Culminating these extension and rifting episodes and the formation of sedimentary basins, the SCS is the only basin of the China Seas that has witnessed the final continental breakup and subsequent seafloor spreading. The SCS is uniquely located at the junction of the Eurasian Plate, the western Pacific Plate, and the Indo-Australian Plate (Figure 3.1), and therefore, a variety of different tectonic forces may have contributed to its opening. These mechanisms include India–Eurasia continental collision and a consequent tectonic extrusion process mainly along the Red River Fault zone (Briais, Patriat, & Tapponnier, 1993; Flower, Russo, Tamaki, & Hoang, 2001; Schärer et al., 1990; Tapponnier, Peltzer, Le Dain, Armijo, & Cobbold, 1982), extension related to subduction of the Pacific Plate along the western Pacific margin (Hall, 2002; Taylor & Hayes, 1980) or to the subduction under Sabah/Borneo (Holloway, 1982), and extension related to an upwelling mantle plume (e.g., Fan & Menzies, 1992). Considering the long and persistent rifting and extension all along the East China margins, however, the SCS basin does not appear to be an exception from other basins of the China Seas in its early rifting mechanism —they are all attributable initially to the retreat and roll back of the subducting Pacific Plate (Shi & Li, 2012), though their later evolutional paths were affected unavoidably by other tectonic activities.

Seafloor spreading in the SCS started at about 32 Ma and stopped at about 16 Ma (Briais et al., 1993; Taylor & Hayes, 1980, 1983). The Luzon Arc started to form by the eastward subduction of the SCS plate about 15 Ma (Briais et al., 1993; Yang et al., 1996), and the arc migrated progressively to the northwest (Seno et al., 1993). The Luzon Arc collided either with the Eurasian continental margin (Chemenda, Yang, Stephan, Konstantinovskaya, & Ivanov, 2001; Clift, Schouten, & Draut, 2003; Ho, 1986; Suppe, 1984; Teng, 1990) or with the preexisting Ryukyu arc (Hsu & Sibuet, 1995), forming one of the youngest and most active orogen in the world, the Taiwan island. The Luzon Arc first began to strike mid-Taiwan along the Longitudinal Valley in an oblique angle, and the collision then progressively propagated to the south (Davis, Suppe, & Dahlen, 1983; Suppe, 1984), causing present-day active collision in southern Taiwan (Yu, Chen, & Kuo, 1997) and tectonic escape in southwestern Taiwan (Lacombe, Mouthereau, Angelier, & Deffontaines, 2001).

Despite its short evolutionary history, the SCS has nearly undergone a complete Wilson cycle from continental breakup and seafloor spreading to subduction, and, therefore, it is also well suited for studying various plate boundary activities, such as continental rifting (e.g., Hayes & Nissen, 2005), oceanic subduction (the Manila Trench; e.g., Li, Zhou, Li, Chen, et al., 2007), strike-slip faulting (the Red River Fault; e.g., Clift & Sun, 2006), and active orogenic processes (Taiwan; e.g., Huang, Xia, Yuan, & Chen, 2001) (Figure 3.2). The exact timings of these events are still debatable, and even more so is the opening sequence of different subbasins of the SCS. However small the size of the SCS may appear, it can be subdivided into at least three subbasins (i.e., East, Southwest, and Northwest Subbasins) showing distinctly different magnetic anomalies, indicating a rather complex opening history.

3.3 MESOZOIC TECTONIC BACKGROUND

Mesozoic tectonics formed the unified continental basement upon which developed the China Seas and the Cenozoic basins. Two major tectonic events transformed East Asia in the Mesozoic. The first is the early Mesozoic closure of the Paleo-Tethys and a series of subsequent or nearly contemporaneous collisions between Indochina and Gondwana-derived continents, including collisions between Indochina and Sibumasu leading to the closure of the southern branch of the Paleo-Tethys; between Indochina and Southeast China; and between North China and South China (Yangtze and Cathaysia combined) terminating the northern branch of the Paleo-Tethys (Figure 3.3). This major event is traditionally coined as the Indosinian orogeny, a term originally conceived in Vietnam almost a century ago (Deprat, 1914; Fromaget, 1927), but later also adopted to describe an episode of regional folding, uplift, orogeny, and magmatism in Southeast China during the middle Triassic. In fact, this adoption of terminology without involving a genetic link has caused misunderstanding and confusion (Carter & Clift, 2008; Lepvrier et al., 2004; Lepvrier, Van Vuong, Maluski, Thi, & Van, 2008). Early Mesozoic deformations in Southeast China were associated not only with Tethyan tectonics and the Indosinian orogeny but also in a large part caused by, or are at least contemporary with, the northwestward subduction of the Paleo-Pacific Plate (Carter & Clift, 2008; Li & Li, 2007; Li, Li, Li, & Wang, 2006; Shi & Li, 2012) (Figure 3.3). Today, the Indochina is bordered to the northeast by the Red River Fault (Figures 3.1 and 3.2), which developed roughly along the Triassic or older suture zone between the Indochina and South China (Lepvrier et al., 2004).

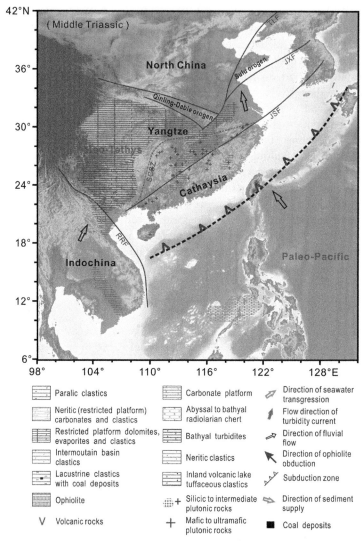

FIGURE 3.3 Middle Triassic distribution of sediments and tectonic pattern showing the subduction of the Paleo-Pacific Plate, the retreat of the Paleo-Tethys, and the coastal uplifting. Arrows mark the directions of plate movement at that time. NB, northern branch of the Paleo-Tethys; SB, southern branch of the Paleo-Tethys. *Modified from Shi and Li (2012).*

This second event, the Mesozoic Andes-type subduction of the Paleo-Pacific Plate, induced a middle Triassic orogen in Southeast China (e.g., Hamilton, 1979; Hilde, Uyeda, & Kroenke, 1977; Holloway, 1982; Isozaki, 1997; Jahn, Chen, & Yen, 1976; Li & Li, 2007; Xiao & Zheng, 2004; Yang, Feng, Fan, & Zhu, 2003; Zhou & Li, 2000). But since as early as the late Triassic, the eastern Asian continental margin has been predominately an extensional regime (Chen, Lee, & Shinjo, 2008; Gilder et al., 1996; Li et al., 2006; Zhou, Sun, Shen, Shu, & Niu, 2006). The causes for this extension and associated widespread magmatism are still debated but in general fall into two main groups of hypothesis: a flat subduction model of the Paleo-Pacific Plate and subsequent slab rollback or foundering (Li & Li, 2007; Li, Zhou, Li, Hao, & Geng, 2007; Zhou & Li, 2000) and an eastward postorogenic extension after the middle Triassic orogeny, followed by a Cretaceous subduction along the Southeast China coast (Chen et al., 2008; Zhou et al., 2006). Despite their apparent differences, both models share in common that Southeast China was characterized by the Triassic tectonic convergence followed by the Jurassic and Cretaceous tectonic extension and rifting.

The closure of the northern branch of the Paleo-Tethys between North China and South China blocks caused the buildup of the Qinling–Dabie–Sulu orogeny (Figure 3.2), which remains today to be the largest coesite- and diamond-bearing ultrahigh-pressure metamorphic belt in the world. Multidisciplinary studies indicate that the continental subduction of the Yangtze Block beneath the North China block occurred from ~240 to 220 Ma, and soon after high- and ultrahigh-pressure metamorphosed slab was exhumed rapidly in ~220–200 Ma (e.g., Xu et al., 2003). A variety of different exhumation mechanisms have been proposed, and it seems evident that the orogen experienced multiple stages of exhumation (Dong & Huang, 2000; Li, Li, Hou, Yang, & Wang, 2005; Wang & Cong, 1996), incorporating potentially different mechanisms such as slab break-off, delamination, thrusting, erosion, and magmatism.

The closure of the northern branch of the Paleo-Tethys also had a far-reaching impact on the development of large fault systems and sedimentary basins. The Tan-Lu fault extends from the Dabie orogen northeastward all the way to the Siberia for nearly 2500 km and is certainly among the longest continental faults (Figure 3.3). Xu, Zhu, Tong, Cui, and Lin (1987) initially suggested a model of a postorogenic transcurrent fault with a large offset of 1000–1500 km, which, however, cannot be easily reconciled with a sudden termination of the lateral movement near the southern edge of the Dabie orogen. Okay, Sengör, and Satir (1993) suggested that the Tan-Lu fault evolved gradually from a transform fault, and later, its sinistral movement was transferred to thrusting and exhumation of the Dabie orogen. Yin and Nie (1993) proposed that the Tan-Lu fault developed from the lithospheric indentation of the northeastern part of the South China block into the North China block. Conversely, Li (1994) suggested that during the continent–continent collision, the upper crust of the South China block in the lower Yangtze region detached from the lower crust and thrust over the North China block, whereas the lower part of the Yangtze lithosphere subducted beneath the North China block along a subsurface suture located over 400 km away to the south of the Sulu orogen. Tan-Lu fault could have also initiated as part of a subduction zone along the eastern margin of the North China block, which acted as a rectangular promontory penetrated South China approximately normal to the Tan-Lu fault (Zhang, 1997; Zhang, Cai, & Zhu, 2006). Despite these various models of the fault, few have doubted its genetic link to the North China—South China collision.

The Tan-Lu fault is undoubtedly a major lithospheric boundary between the North China and lower Yangtze Blocks. It shows a very broad fault zone up to 25 km wide with mostly two main branches that are primarily northwestward dipping. Reflection seismic, gravimetric, and magnetic data all indicate that a significant uplift and exhumation occurred in a large area east of the Tan-Lu fault, including the Sulu orogeny (Li, Wang, et al., 2012). Seismic sections in the southern segment of the fault reveal an average of ~6 km of vertical offset that has occurred across the Tan-Lu fault since the early Jurassic (Figure 3.4). The Hefei Basin developed immediately to the west of the Tan-Lu fault, above the Paleozoic–Mesozoic unconformity caused by the Triassic North China–lower Yangtze collision. The direct outcome of this persistent differential uplift and erosion after the North China–lower Yangtze collision is that much of the Mesozoic (the Jurassic and upper Triassic in particular) sediments are absent east of the Tan-Lu fault (Figure 3.5).

It is interesting to note that this type of differential vertical movement continues to happen in Southeast China. The Zhenghe–Lianhuashan Fault was a major belt of metamorphism and intermittent magmatism during the Mesozoic (GDBGMR, 1988). From topography, it is also evident that recent differential vertical movement occurred along the fault, with its eastern wall moved preferentially upward (Figure 3.2).

The closure of the southern branch of the Paleo-Tethys between the Indochina and South China (Yangtze and Cathaysia together) probably terminated early Triassic carbonate sequences of Tethyan domain in the northern SCS continental margin (Shi & Li, 2012). In the early Triassic, much of the South China was then under the influence of the Paleo-Tethys, and the deepest water depths were near the projected South China suture zone (Figure 3.2). The southeast coastal area was largely under erosion and was a source for clastic materials. However, further south near the present-day northern SCS margin, probably an area of marine deposition existed, since the Paleo-Tethys can reach there through the narrow channel between the Indochina and South China (Shi & Li, 2012). Thick Mesozoic and Paleozoic strata identified from seismic sections in northern SCS margin are indicative of a long period of deposition, and seismic facies resembling that of carbonate sequences has been identified in the deep section (Shi & Li, 2012).

On the southern SCS margin, abyssal to bathyal radiolarian chert deposition persisted throughout much of the Mesozoic (Fontaine, 1979; Kiessling & Flügel, 2000; Suzuki, Takemura, Yumul, David, & Asiedu, 2000; Yeh & Cheng, 1996; Zamoras & Matsuoka, 2004; Zamoras et al., 2008) (Figure 3.7). These southern blocks (particularly northeast Palawan) were in contact with the Eurasian continental margin back then and received

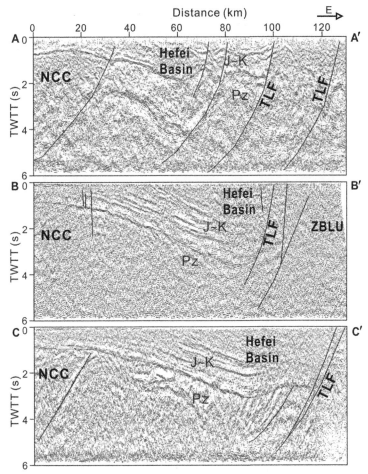

FIGURE 3.4   Reflection seismic transects crossing the Tan-Lu faults. Data provided by China Petrochemical Corporation. The horizon marks the Paleozoic–Mesozoic unconformity. Subvertical lines are faults. TLF, Tan-Lu fault; ZBLU, Zhangbaling uplift; K, Cretaceous; J, Jurassic; Pz, Paleozoic; NCC, North China Craton; TWTT, two-way travel time. For these sections, the maximum estimated depth could reach 15 km. See Figure 3.2 for locations of these profiles. *From Li, Wang, et al. (2012).*

thick chert deposits caused by the Paleo-Pacific subduction and tectonic accretion of bathyal marine sediments. It has been widely suggested that a late Mesozoic Paleo-Pacific subduction zone of Andes type once existed along the Southeast China continental margin (Hamilton, 1979; Hayes et al., 1995; Hilde et al., 1977; Holloway, 1982; Jahn et al., 1976; Taylor & Hayes, 1983; Xiao & Zheng, 2004; Yang et al., 2003; Zhou & Li, 2000). Many studies suggest that the northeast SCS area was once situated in a transition zone between the Tethys and Paleo-Pacific (Chen, Xu, & Xu, 1998; Li, 2004;

Geology of the China Seas

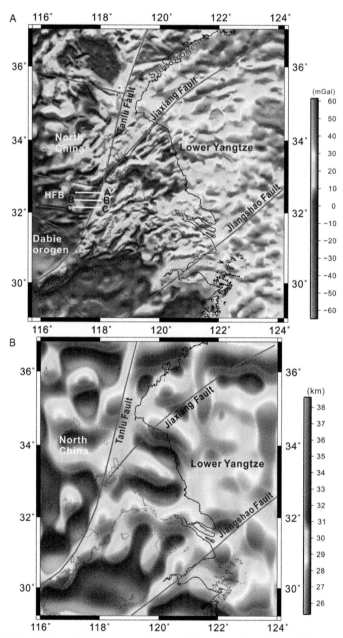

FIGURE 3.5 (A) Regional Bouguer gravity map; (B) estimated preliminary Moho topography from the Bouguer gravity anomalies shown in (A) using the Parker–Oldenburg method. AA′, BB′, and CC′ are three seismic sections shown in Figure 3.4. *From Li, Wang, et al. (2012).*

Liu & Quan, 1996; Sun et al., 1989; Xia, Huang, & Huang, 2004; Zamoras & Matsuoka, 2001; Zhou, Chen, Sun, & Xu, 2005). Paleontological studies in the area indicate that some late Mesozoic macrofossils show closer affinity with Tethyan biota, whereas some others show characteristics more typical of the Pacific biotic province (Chen et al., 1998; Hutchison, 1989; Kudrass, Hiedicke, Cepek, Kreuzer, & Müller, 1986; Liu & Quan, 1996). Therefore, during the early Triassic, the SCS was an area of strong interaction between the Paleo-Tethys and Paleo-Pacific. This interaction was hampered in the middle Triassic by subsequent or nearly contemporaneous collisions between Indochina and Sibumasu and between Indochina and Southeast China (Carter & Clift, 2008; Lepvrier et al., 2004, 2008, 2011; Metcalfe, 2009).

In Southeast China, the middle Triassic also witnessed a westward regression of the Paleo-Tethys and simultaneous expansion and westward migration of the coastal orogeny (Figure 3.6), due likely to the flat subduction of the Paleo-Pacific Plate (Li & Li, 2007). The Cathaysia then turned into a largely exposed land, and the waning Paleo-Tethys had restricted water supplies and became overly evaporated with deposition of restricted platform orthodolomites, evaporites, and clastics (Shi & Li, 2012). The Mesozoic Southeast China orogeny and plateau traditionally were put into affiliation with the so-called Indosinian orogeny. However, the concept of Indosinian orogeny should be confined within Vietnam and Indochina, and the extension of this terminology to Southeast China disguises the likely true cause of this Mesozoic orogeny, which is the northwestward subduction of the Paleo-Pacific Plate (Carter & Clift, 2008).

Li and Li (2007) demonstrated a sweeping orogeny that migrated from coastal South China ~1300 km into the continental interior and suggested a flat subduction model of the Paleo-Pacific. Shi and Li (2012) further validated the northwestward migration of the orogeny during the Triassic and the early Jurassic and observed that the orogeny gradually uplifted the coastal area and pushed the coastal and marine depocenter to migrate northwestward. Eventually, the large area that once received marine deposition in the early Triassic became an area of either nondeposition or lacustrine and/or intermountain basin deposition in the early Jurassic. These observations and evidences support that the orogeny was triggered by the northwestward subduction of the Paleo-Pacific Plate. The location of the Mesozoic subduction can be traced all the way from southwest Japan, through Ryukyu Islands and Taiwan, to northeast Palawan, as correlative Mesozoic radiolarian cherts representing subduction accretion have all been found in these places (Fontaine, 1979; Kiessling & Flügel, 2000; Shen, Ujiie, & Sashida, 1996; Suzuki et al., 2000; Yeh & Cheng, 1996; Zamoras & Matsuoka, 2004; Zamoras et al., 2008).

The major phase of transition from the Paleo-Tethys of primarily carbonate deposition to the Paleo-Pacific regime of mainly clastic deposition occurred

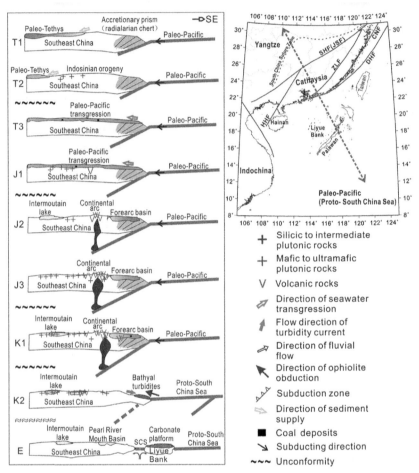

FIGURE 3.6 Sketched sections showing major tectonic events in Southeast China and adjacent seas (Shi & Li, 2012). SCS, South China Sea.

roughly at the end of the middle Triassic (ca. 220 Ma). Soon afterward, the Mesozoic orogen then experienced denudation and peneplanation, postorogenic extension/rifting, and magmatism. Postorogenic faulting, rifting, and magmatism were synchronized both temporally and spatially and migrated progressively eastward (Shi & Li, 2012), caused by possibly slab rollback, foundering, and/or gravitational collapse (e.g., Li & Li, 2007; Zhou & Li, 2000). As an outcome of complicated interplay between orogenic buildup and migration, late- or postorogenic extension, and Paleo-Pacific subduction, a transient Paleo-Pacific transgression also occurred during the late Triassic and early Jurassic (Figure 3.7). The Mesozoic evolution of Southeast China constitutes nearly a complete cycle of regional orogenic buildup and destruction.

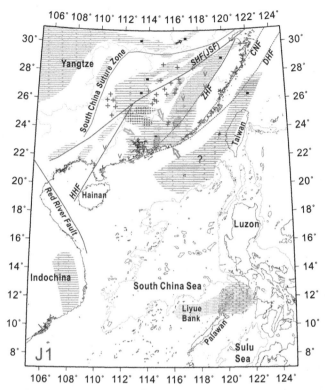

FIGURE 3.7   Early Jurassic distribution of sediments, tectonic pattern, and magmatism (Shi & Li, 2012). *Notes*: see Figure 3.3 for legend and annotations. SHF (JSF), Shihang Fault (Jiangshan–Shaoxing Fault); ZLF, Zhenghe–Lianhuashan Fault; CNF, Changle-Nan'ao Fault; DHF, Donghai Fault; HHF, Hepu–Hetai Fault.

With continued eastward younging in magmatism and faulting, an arc–forearc basin–trench system was well developed in the middle Jurassic and early Cretaceous along the present-day Southeast Asian continental margin. Further landward, the maximum volcanism in the late Jurassic formed a continental arc, which turned into a basin and range province in the Cretaceous due to prolonged extension and magmatism. At the end of the early Cretaceous, sharp paleogeographic changes occurred on the two conjugate margins of the SCS. The northern continental margin became largely an area of denudation, and for the first time since the early Triassic, abyssal to bathyal radiolarian cherts vanished from northeast Palawan (Shi & Li, 2012). This significant change in tectonic setting marks the transition from an active to a passive continental margin that happened ca. 100 Ma near the present-day northern SCS margin, probably due to an eastward retreat in the Paleo-Pacific subduction zone and pervasive continental margin extension and/or the collision of the China-Indochina margin with the West Philippine block (Shi & Li,

2012; Zhou et al., 2006). The sharp regional changes in tectonic settings from the early to late Cretaceous have also been well documented based on geochemical studies on volcanic rocks (Charvet et al., 1990; Charvet, Lapierre, & Yu, 1994; Lapierre, Jahn, Charvet, & Yu, 1997). K-rich orogenic suites generated in an active continental margin in the early Cretaceous changed to bimodal volcanism associated with a postcollisional crustal extension and rifting in the late Cretaceous, and the major tectonic event between the early and late Cretaceous is proposed to correlate with the collision of the China-Indochina margin with the West Philippine block (Charvet et al., 1990, 1994; Lapierre et al., 1997).

Extensive continental margin extension continued into the early Cenozoic and was responsible for the ultimate rifting and late Oligocene and early Miocene seafloor spreading in the SCS. Due to the prolonged erosion and exhumation in most part of the northern SCS margin from the late Cretaceous well into the early Paleogene, the unconformity between the early and late Cretaceous is, in most cases, really represented by the unconformity between the Cenozoic and Mesozoic sequences. The top of the Mesozoic is marked by an unconformity representing a stratigraphic hiatus from the upper Cretaceous to lower Oligocene (Lee, Tang, Ting, & Hsu, 1993; Lin, Watts, & Hesselbow, 2003; Tzeng, 1994).

Geophysical data revealed two large areas of well-preserved Mesozoic sedimentation on the two conjugate SCS margins. Before breakup, these two regions should have been adjacent to each other, forming a large Mesozoic basin. Mesozoic strata on the two conjugate margins are correlated and can be found all the way to the continent–ocean boundary (COB; Figure 3.8; Shi & Li, 2012).

Seismic data in northern SCS continental margin show four major unconformities (Mesozoic/Cenozoic, Jurassic/Cretaceous, Triassic–Jurassic/Jurassic, and Paleozoic/Triassic–Jurassic) that can all be tied to regional tectonic events revealed from paleolithofacies (Shi & Li, 2012). Roughly confined by these unconformities, three major phases of marine deposition on the northern SCS margin are identified. The first phase was tied to the Paleozoic and early Triassic sequences within a mostly Tethyan affiliation, but this phase was terminated by the Triassic orogeny and continent–continent collisions in a wide scale. The second phase was the late Triassic to early Jurassic Paleo-Pacific transgression with coastal to shallow marine clastic deposits, but this phase was then dismantled by a major southeastward regression synchronized with faulting and magmatism. The third phase of marine deposition, lasting from the middle Jurassic to the early Cretaceous, was limited to the forearc area, including the accretionary wedge and trench, of the Paleo-Pacific subduction zone. Depositions from this phase, together with those from the previous two phases, were once again dismembered by pervasive late Mesozoic and Paleogene continental margin extension and rifting as well as local uplifting and erosion. Clearly, the sedimentary environments differed

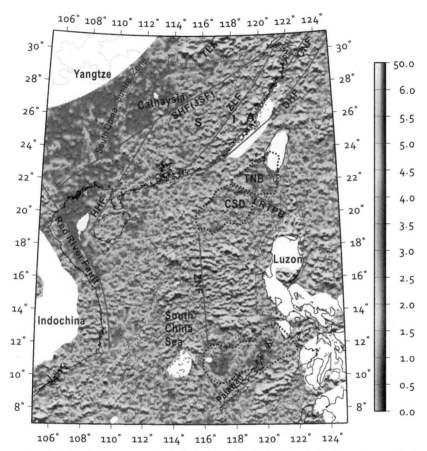

FIGURE 3.8   Regional map of 3D analytic signal amplitudes (ASA) calculated from the total field magnetic anomalies (Shi & Li, 2012). *Notes*: S, I, and A indicate areas of dominantly S-type, I-type, and A-type igneous rocks, respectively. TNB, Taixinan Basin; CSD, Chaoshan Depression; LRTPB, Luzon–Ryukyu Transform Plate Boundary; ZNF, Zhongnan Fault. The thin contour in the SCS basin is the 3000 m isobath. Dashed lines in the two conjugate margins of the SCS confine two areas with well-preserved Mesozoic sediments. See Figure 3.7 for further annotations.

markedly from phase to phase, and originally well-developed Mesozoic marine depositions experienced subsequent tectonic compression, uplifting, erosion, rifting, and magmatism. Mesozoic sedimentary basins as seen today are polycyclic superimposed basins.

3.4   CENOZOIC CONTINENTAL MARGIN RIFTING AND SUBSIDENCE

Continental margin rifting and subsidence started to prevail in the early Cenozoic throughout the China Seas and formed a series of sedimentary basins

(Figure 3.2). Early rifting of these basins started mostly around the Cenozoic/ Mesozoic boundary (Ren, Tamaki, Li, & Zhang, 2002), signifying a rather regional cause for their early extensions. Locally, the developments of these basins tend to be intimately linked to large-scale faults: the Bohai Basin (also known as the North China basin) to the Tan-Lu fault, the South Yellow Sea Basin to the Jiaxiang Fault, the ECS Basin to the NE-striking faults in the Cathaysian block, and the Yinggehai Basin to the Red River Fault (Figures 3.1 and 3.2). It is crucial to understand first the deep crustal struc-tures of these faults and basins using various geophysical techniques.

3.4.1 Deep Crustal Structures

3.4.1.1 Bohai and Yellow Sea

We have shown reflection seismic sections (Figure 3.4) and the gravity anomalies and Moho depths around the Tan-Lu fault (Figure 3.5). A third dimension to these geophysical data is magnetic anomalies, which offer a very wide spectrum of information of the Earth's interior, including seafloor spreading and plate movement, lithosphere evolution and magmatism, charac-teristics of core rotation, interactions between the Earth and outer space, mantle serpentinization and fluid flow, and geothermal field evolution.

Figure 3.9 shows a total field magnetic map with reduction to the pole and upward continuation by 10 km (Li, Chen, et al., 2009). Major tectonic struc-tures such as faults, orogens, and basement uplifts appear more readily identi-fiable from the magnetic map after the data reduction than from the original total field magnetic map. Upward continuation is a routine technique in magnetic data processing to suppress shallow anomalies and enhance deep geologic information. After these processing, the Tan-Lu fault shows evident along-strike segmentation, and to the north of the Sulu orogen, it forms a distinct boundary between mainly negative anomalies in the North Yellow Sea and positive anomalies in the Bohai Sea. This sharp contrast in magnetic anomalies appears to be rather puzzling at first sight since the two blocks here across the Tan-Lu fault are all considered as being parts of the North China Craton. Differential magmatism and tectonic activities across the Tan-Lu fault must have been dramatic enough to trigger this contrast.

On seismic sections, the Tan-Lu fault coexists with a rift zone within the Bohai Basin (Hsiao, Graham, & Tilander, 2004). The Bohai Basin formed by rifting and rapid tectonic subsidence and volcanism during the Eocene and Oligocene, with up to 7000 m of Cenozoic strata accumulated in the faulted subbasins bounded by normal faults and locally by strike-slip faults (Chang, 1991; Ye, Shedlock, Hellinger, & Sclater, 1985). This was followed by Paleo-gene postrift thermal subsidence (Figure 3.10). The Tan-Lu fault was certainly active during the Cenozoic rifting of the Bohai Basin and has actually been active ever since its formation in the early Mesozoic. There are different views on how the basin-bounding strike-slip faults (such as the Tan-Lu fault) played roles in the formation of the Bohai Basin. In the composite pull-apart model,

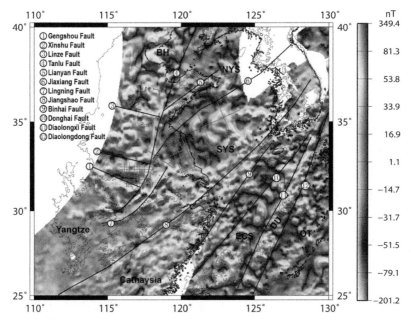

FIGURE 3.9   Magnetic anomaly map of East China and adjacent seas after the reduction to the pole and upward continuation by 10 km (Li, Chen, et al., 2009). Numbered lines are interpreted regional crustal faults. Thin gray lines mark locations of some of multichannel seismic sections in the area. Overlapped on the map are topographic contours of 100, 0, and −100 m, respectively.

the Bohai Basin was initially formed by east–west extension and was later pulled apart by the dextral movement of basin-bounding strike-slip faults (e.g., Allen et al., 1997, 1998; Nilsen & Sylvester, 1995). However, Hsiao et al. (2004) suggested that the deformational styles and the magnitude of fault activity do not align with the pull-apart model and they prefer a continental rifting model in which the Tan-Lu fault otherwise played a minor role.

There are two relatively independent structural deformation systems in the Bohai Basin, extensional and right-lateral strike-slip systems. Based on fault activities, Qi and Yang (2010) argued that the extensional rifting mainly developed from the middle Paleocene to the late Oligocene, whereas the strike-slip systems were mainly active from the Oligocene to the Miocene. Strike-slip deformation intensified as extensional deformation weakened.

The southern Yellow Sea and the lower Yangtze Block together show circular negative magnetic anomalies surrounding positive anomalies (Figure 3.9). In many cases, negative magnetic anomalies correspond with basement lows while positive anomalies with basement highs. The same conclusion can also be made in the ECS (Hsu et al., 2001). Regional crustal faults can be readily identified from Figure 3.9. West of the Tan-Lu fault and north of the Dabie orogen, crustal faults strike mainly in NW orientation and are intercepted in high angle by the Tan-Lu fault. East of the Tan-Lu fault, crustal faults are mostly in NE

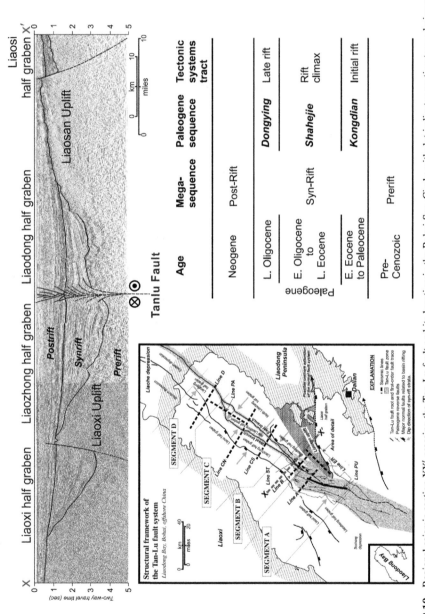

FIGURE 3.10 Regional seismic section XX′ crossing the Tan-Lu fault and its location in the Bohai Sea. Circle with dot indicates motion toward viewer. Circle with x indicates motion away from viewer. The lower-right panel shows Paleogene stratigraphic framework. *Modified from Hsiao et al. (2004).*

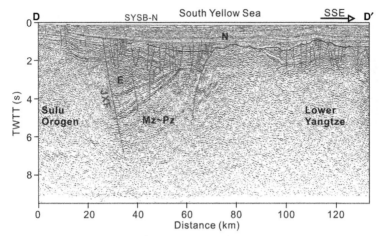

FIGURE 3.11   Seismic section DD′ from the South Yellow Sea. See Figure 3.2 for location. Data are from China National Offshore Oil Company. Transects from t1–t1′ to t5–t5′ are located offshore to the south of the Sulu orogen. The interpreted horizons mark the Neogene–Paleogene unconformity, the Eocene–Oligocene unconformity, and the Mesozoic–Cenozoic unconformity, respectively. Subvertical lines are faults. JXF, Jiaxiang Fault; N, Neogene; E, Paleogene; Mz, Mesozoic; Pz, Paleozoic; TWTT, two-way travel time. The maximum estimated depth of this section can be over 30 km. *Modified from Li, Wang, et al. (2012).*

strikes, of which the Jiaxiang Fault and the Jiangshao Fault form the northern and southern boundaries of the lower Yangtze Block, respectively.

The Jiaxiang Fault is commonly taken to be the southern boundary of the Sulu orogen. This fault is a prominent geophysical boundary (Figure 3.9). To the southwest, the Jiaxiang Fault appears to be merged with the Tan-Lu fault (Figure 3.2). From reflection seismic profiling, the main branch of the Jiaxiang Fault dips primarily to the southeast but shows a large lateral variability in faulting structures (Figure 3.11) (Li, Wang, et al., 2012). The Jiaxiang Fault often controls the northern borders of Paleogene half-grabens, which explain the gravity lows associated with the fault (Figure 3.5). Much of the Mesozoic (the Jurassic and upper Triassic in particular) sediments are absent in this area due to persistent uplifting and erosion after the North China–Yangtze collision and the preferential uplifting east of the Tan-Lu fault. Consequently, the Mesozoic behavior of the Jiaxiang Fault and how it was related to early Mesozoic collisional structures remain ambiguous (Li, Wang, et al., 2012).

3.4.1.2   East China Sea

Within the ECS Basin, the NE-trending Xihu Depression is evolved from part of a Paleogene continental margin megarift. Together with the Diaobei Depression (also known as Jilong Depression; see Hsu et al., 2001; Kong,

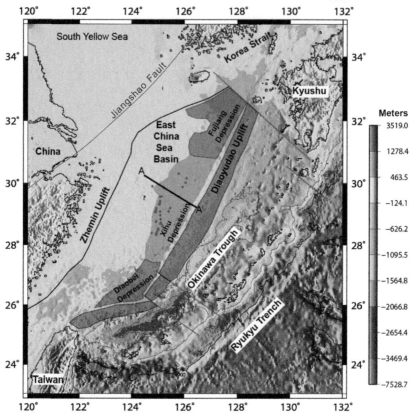

FIGURE 3.12 Relief map showing the position of the Xihu Depression and major tectonic units in the ECS region (Li, Zhou, et al., 2009). Isobaths shown are 0, 1000, 2000, and 3000 m, respectively. Bold red lines are several young NW-striking faults, and dots show well locations. AA' is the seismic line to be shown in Figure 3.13.

1998) to the south and the Fujiang Depression to the north, these three depressions form a huge continental margin rifting system subparallel to the Diaoyudao Uplift (also named as Taiwan–Sinzi belt; see Hsu et al., 2001; Kong, 1998) and the Okinawa Trough (Figure 3.12). The Xihu Depression today has a maximum depositional depth up to 15 km and ranges from about 90 to 150 km in width. NE-striking crustal faults developed in the ECS continental shelf form the boundaries of the ECS Basin and of depressions within the basin (Li, Zhou, Ge & Mao, 2007; Li, Zhou, et al., 2009). Based on the present-day seismicity and crosscutting relationships between faults and strata interpreted from reflection seismic data (Figure 3.13), it is inferred that NE-striking regional crustal faults on the ECS continental shelf are not active faults since they all ceased their activities at least 2 Ma (Li, Zhou, Ge & Mao, 2007; Li, Zhou, et al., 2009). These faults were induced by the

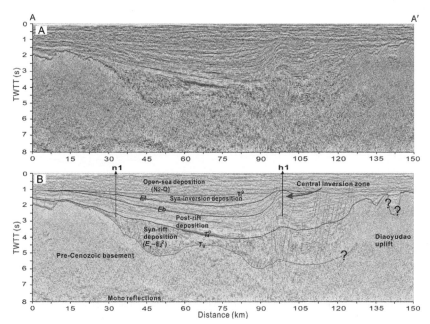

FIGURE 3.13 (A) Seismic section AA′ passing through wells h1 and n1 in the Xihu Depression (Li, Zhou, et al., 2009). See Figure 3.12 for location. (B) Interpretations of (A). T_g, basement unconformity; T_4^0, breakup unconformity; T_1^0, inversion unconformity; E_t, top of eroded layers; E_b, base of eroded layers; TWTT, two-way travel time.

subduction of a Paleo-Pacific Plate beneath the Eurasian Plate in the late Mesozoic, and later in the early Cenozoic, they controlled or influenced formations of basins and rifts in the area. The Diaoyudao Uplift to the east of the ECS Basin has well-defined positive magnetic anomalies extending from the northeastern part of Taiwan all the way to the east of the Korean Peninsula.

The Xihu Depression has experienced three stages of evolution. The first is an initial rifting stage with fast subsidence rates from the early Paleocene ($\sim$65.0 Ma) to the middle Eocene ($\sim$40.4 Ma). Synrift sediments and breakup unconformities can be clearly identified on seismic sections (Figure 3.13). The second is a thermal subsidence stage from the middle Eocene ($\sim$40.4 Ma) to the end of Miocene ($\sim$5.33 Ma). Since the Pliocene, an accelerated subsidence occurred throughout the ECS continental shelf (Yang et al., 2004; Zhong et al., 2001), and a thick open-sea sequence (up to 1300 m in thickness) was deposited (Figure 3.13).

The density model from gravity modeling shows a thin crustal layer (density $= 2.86$ g/cm^3) of around 6.5 km in thickness at depocenter (Figure 3.14). This implies that the original crust was severely thinned to the magnitude of a normal oceanic crustal thickness. Nevertheless, due to thick Cenozoic

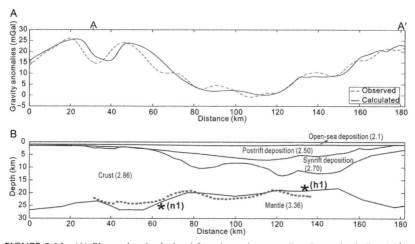

FIGURE 3.14 (A) Observed and calculated free-air gravity anomalies along seismic line AA' in Figure 3.13; (B) crustal and upper mantle density model from 2.5D gravity modeling and inversion (Li, Zhou, et al., 2009). Stars show inverted Moho depths from backstripping analyses of wells n1 and h1. Square dots represent Moho depths interpreted from reflection seismic data (Figure 3.13). Close matches are observed in Moho depths estimated from reflection seismic data, gravity modeling, and backstripping analyses, respectively. Densities in g/cm^3 are shown in parentheses.

sediments, the Xihu Depocenter now has an average Moho depth of about 20 km, whereas the Moho depths are at about 23–27 km outside of the depression.

3.4.1.3 South China Sea

In the SCS region, regional magnetic anomalies can also be used to define large-scale faults and tectonic zones (Figure 3.15). However, we will defer the discussion of magnetic anomalies to the next section of marginal sea formation but focus on two regional seismic profiles T1 and T2 shown in Figure 3.15. Profile T2 traverses the northeasternmost SCS basin, while profile T1 goes through the northern continental margin southward to the central basin. From these representative profiles, we gain a basic understanding of the crustal structures.

Structurally, the northern continental margin of the SCS is very complicated, but in general, the depressions and uplifts are in NE orientations, subparallel to the present-day continental shelf (Figure 3.2). The Taixinan Basin is a dominant structure along the northeasternmost SCS continental margin (Figure 3.2). Geometrically, it appears to connect with the ECS Basin, but their basin structures and evolutional history differ. The Taixinan Basin can be subdivided into three NE-striking structures, namely, North depression, Central uplift, and South depression, respectively (Lin et al., 2003). In the North depression, mostly northwest-dipping active normal faults cut through almost the entire Cenozoic strata, showing primarily a weak extensional

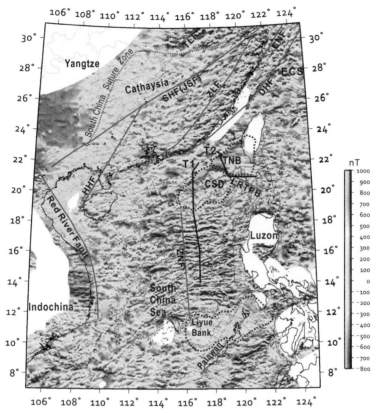

FIGURE 3.15 Regional total field magnetic anomaly map of the South China Sea and adjacent region. *Notes*: Black contours are coastal lines and contours in the ocean basins are 3000 m isobaths. T1 and T2 are two regional seismic profiles to be shown in Figures 3.18 and 3.16, respectively. See Figure 3.7 for further annotations. *From Shi and Li (2012)*.

environment and southeastward increase of thermal subsidence (Figure 3.16) (Li, Zhou, Li, Hao, & Geng, 2007). Some of the larger faults can be traced down to the basement and control the basement structures. They often behave as growth faults bounding half-grabens. In contrast, deformation and faulting are weak in the Cenozoic sedimentary layers in the South depression. The lower Miocene and upper Oligocene sequences onlap onto the Central uplift and flatten and spread out to the southeast. The Central uplift was a longstanding high during the early Miocene and late Oligocene and served both as a source of sediments due to its erosion and as a barrier between two distinct sedimentary environments (Li, Zhou, Li, Hao, et al., 2007). The Central uplift is likely composed of upper Jurassic and lower Cretaceous sedimentary rocks. After the late Miocene, the Central uplift was no longer a topographic barrier between the North and South depressions (Figure 3.16).

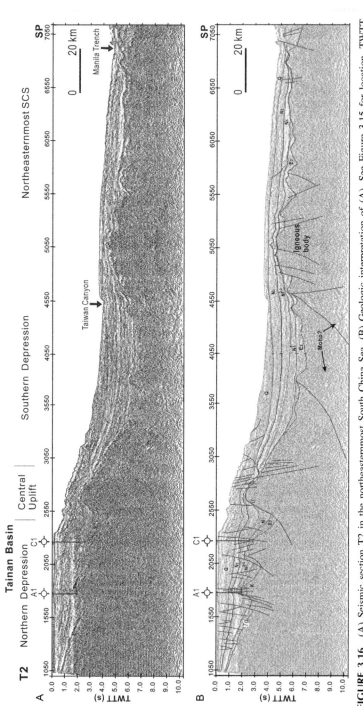

FIGURE 3.16 (A) Seismic section T2 in the northeasternmost South China Sea. (B) Geologic interpretation of (A). See Figure 3.15 for location. TWTT, two-way travel time; SP, shot point; Moho, Mohorovicic discontinuity. A1 and C1 are well names. E3, late Oligocene; N_1^1, early Miocene; N_1^{2-3}, middle to late Miocene; N_2, Pliocene; Q, Quaternary (Li, Zhou, Li, Hao, et al., 2007).

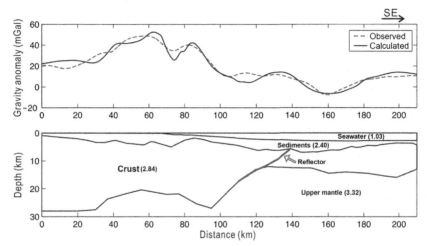

FIGURE 3.17 Gravitational modeling/inversion of a segment of profile T2 in the northeastern-most South China Sea with a four-layer depth model with densities (g/cm^3) shown in parentheses (Li, Zhou, Li, Hao, et al., 2007). See Figure 3.15 for location.

A prominent landward-dipping reflector is found within the basement of the South depression. There are large contrasts in seismic facies, Moho reflections, and crustal thickness across this reflector (Figures 3.16 and 3.17). To the north-west, the crustal thickness is up to 28 km, while to the southeast, the crust is less than 10 km in thickness beneath the South depression. The South depression could be a necking zone in the early rifting, and it could be partly floored with a silver of oceanic crust (Li, Zhou, Li, Hao, et al., 2007), though most parts of the northeasternmost SCS is floored with extended continental crust.

Lower crustal layers of high seismic velocities at around 7.2 km/s have been reported beneath the northern SCS continental margin (Kido, Suyehiro, & Kinoshita, 2001; Nissen, Hayes, Buhl, et al., 1995; Nissen, Hayes, & Yao, 1995; Yao, et al., 1994). While it has been hypothesized that these layers might be underplated upper mantle materials during conti-nental rifting (Kido et al., 2001), Nissen, Hayes, Buhl, et al. (1995) argued that high-velocity lower crustal materials beneath the SCS continental margin are due in a larger part to variability in prerift crustal structure. Gravity mod-eling shows that the high gravity anomalies on the continental margin may result from a shallow Moho (Figure 3.17), but lower crustal underplating can also equally account for the observed high gravity anomalies in the North depression of the Taixinan Basin.

The northern continent–ocean transition zone (COT) of the SCS is heavily faulted and accompanies widespread recent magmatism and extrusions (Li, Zhou, Hao, et al., 2008; Li, Zhou, Li, Hao, et al., 2007; Tsai, Hsu, Yeh, Lee, & Xia, 2004; Wang, Chen, Lee, & Xia, 2006; Yan, Deng, Liu, Zhang, & Jiang, 2006). Deep reflection seismic surveys have identified many landward-dipping reflectors, which can be interpreted as crustal faults (Hayes et al., 1995; Li, Zhou, Hao, et al., 2008; Li, Zhou, Li, Hao, et al., 2007). Seismic profile T1 shows a thorough-going crustal fault at about 300 km, which bounds a basement uplift to its south and a half-graben to its north (Figure 3.18). This listric fault and associated graben are important features along the northern SCS continental margin, and this fault in particular represents the northern edge of the COT (Li, Zhou, Li, Chen, & Geng, 2008). The angular unconformity and the fold structure in the half-graben indicate that the seafloor spreading in the SCS accompanied an episode of compression in the northern continental margin (Li et al., 2010).

Figure 3.18 shows part of the Pearl River Mouth Basin (PRMB), and between PRMB and the COB is a basement uplift called Dongsha–Penghu uplift. The Pearl River Mouth Basin has several major subbasins separated by local uplifts, again forming horst and graben structures. One of the subbasins, the Baiyun Sag located to the west of the Dongsha Islands, shows anomalously deep basement and thick Cenozoic sedimentary strata. The Dongsha–Penghu uplift, in particular, is accompanied with a magnetic anomaly (Figure 3.15) that is believed to be caused by Mesozoic prerifting magmatic bodies intruded into Mesozoic sedimentary sequences (Li, Zhou, Hao, et al., 2008). This episode of magmatism appears to be the last magmatic event before the initial rifting in the northern continental margin. The timing of early rifting and subsidence of the Pearl River Mouth Basin is similar to that of the ECS Basin but precedes the subsidence of the Taixinan Basin.

The Qiongdongnan Basin forms the western extension of the rifting that affected the Pearl River Mouth Basin, and to the south is juxtaposed with the Xisha Trough, which is a failed rift in the early opening of the SCS (Figure 3.2). The basement structure appears to be dominated by a series of tilted fault blocks with no evidence for significant volcanism related to rifting. The contrast with the Yinggehai Basin structure is marked because the strain is partitioned across many smaller faults, rather than one major structure (Clift & Sun, 2006). The Yinggehai Basin is controlled primarily by the strike-slip movement along the Red River Fault, with up to 17 km of Cenozoic stratigraphic succession (Gong et al., 1997). The Yinggehai strike-slip faults crosscut those in the extensional Qiongdongnan Basin; the seismic data indicated a cessation of major extensional faulting in the Qiongdongnan Basin prior to 24 Ma, yet strike-slip faulting in the Yinggehai continued at least until a phase of structural inversion prior to 13.8 Ma (Clift & Sun, 2006).

Based on the Parker–Oldenburg algorithm (Oldenburg, 1974; Parker, 1973), Moho depths in the SCS area (Figure 3.19) are estimated from a simple Bouguer gravity data set (Li et al., 2010; Li, Zhou, Li, Chen, et al., 2007). It is found that the Moho depths in the central basin of the SCS are mostly less than 15 km below sea level, and this depth range is consistent to previous

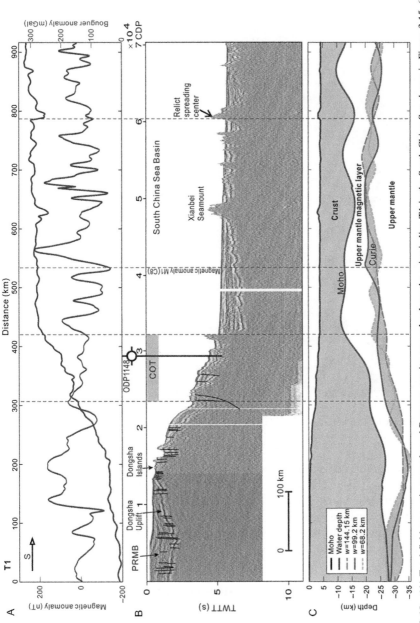

FIGURE 3.18 (A) Total field magnetic anomaly and Bouguer gravity anomaly along the seismic line T1 in the South China Sea shown in Figure 3.15. (B) The seismic profile T1 with fault interpretations. (C) Moho depths and Curie depths estimated with different window sizes along seismic line T1. TWTT, two-way travel time; COT, continent–ocean transition zone; PRMB, Pearl River Mouth Basin; w, the width of moving windows (Li et al., 2010).

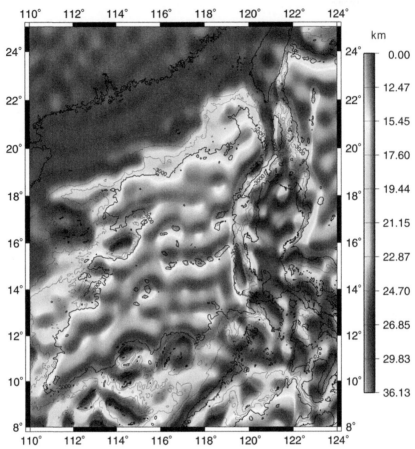

FIGURE 3.19 The Moho topography in the SCS area estimated from Bouguer gravity anomalies using the Parker–Oldenburg algorithm (Li et al., 2010).

estimates from gravity and/or seismic data (Chen & Jaw, 1996; Liu, Wang, Yuan, & Su, 1983; Nakamura, McIntosh, & Chen, 1998; Shih, 2001; Xia et al., 1997; Xu & Jiang, 1989). It is also implied that roughly the uppermost 10 km of the mantle beneath the central SCS basin is also magnetized, since Curie depths in most parts of the central basin are apparently much deeper than the average Moho depth (Figure 3.18).

3.4.2 Cenozoic Rifting and Subsidence Processes

With the preceding discussion of Mesozoic tectonics and deep structures, we gain a general view of regional tectonic units, their affiliations, and inter-relationships. In this section, we focus on rifting and subsidence processes of the major continental margin basins. With only a few exceptions, their development and evolution follow the typical two-stage model. The first stage is characterized by localized initial rifting and rapid subsidence. Mostly near-source synrift sediments are accumulated within half-grabens bounded by extensional faults on one side and onlapping unconformity on the other. Before rifting, uplifting often occur in some places in response to asthenospheric upwelling and isostasy. The second stage of postrift thermal subsidence occurs normally in a broader scale and deeper water environment. Sedimentary inputs can be carried to the depocenter from a long distance. In most cases, an unconformity often called breakup unconformity can develop between the synrift and postrift strata, reflecting a significant change in tectonic regime and sedimentary environment. About 10 to 5 Ma, except for the SCS basin and the Okinawa Trough, basin-scale thermal subsidences stopped all within the China Seas, and they were followed by rapid subsidence all along the East China continental margin. This recent accelerated subsidence certainly cannot be explained by either rifting or thermal subsidence, but must be associated with a regional or global tectonic or environmental event.

Although, for most of the basins, this two-stage model holds, the timing of initial rifting, the duration of initial rifting, and the onset timing of thermal subsidence appear to vary from basin to basin, and we will examine these similarities and differences.

3.4.2.1   Bohai and Yellow Sea Basins

Figure 3.10 shows that the Bohai Basin had a long synrifting phase started at the early Paleogene and received the deposition of primarily lacustrine, fluvial, and alluvial sediments. This phase was accompanied with block faulting, episodic rapid subsidence, and widespread calc-alkaline basaltic volcanism (Hu, He, & Wang, 2001; Ye et al., 1985). Postrifting subsidence started from the early Neogene. Between the synrifting and the postrifting megasequences is a distinct boundary, represented on seismic reflection profiles by a high-amplitude basin-wide reflector and an angular unconformity near the basin margins (Hsiao et al., 2004). This indicates widespread denudation and, therefore, possibly regional uplift during the transition between initial rifting and postrifting thermal subsidence (Ye et al., 1985). Starting from the late Miocene (12 Ma), though not synchronized everywhere, the subsidence rate increased in the Bohai Basin, and the frequent occurrence of destructive earthquakes and observations of relatively high heat flow suggest that the Quaternary tectonic activity differs from the simple thermal subsidence pattern (Hu et al., 2001; Ye et al., 1985). Since this intensified subsidence occurred not only within the Bohai Basin but also in other basins all along the China continental margin, we prefer a regional rather than local mechanism for this recent rejuvenation of sedimentation.

The Northern Yellow Sea Basin shares many similarities with the Bohai Basin in sedimentary and subsidence characters; it experienced a major phase of Paleogene rifting and initial subsidence and Neogene thermal subsidence (Cao, Wang, Zhan, & Li, 2007). Rift faulting was relatively weak and thin Paleogene strata were deposited only in a small area. Uplift and denudation occurred in the late Cretaceous before the initial Paleogene rifting and Neogene thermal subsidence.

The South Yellow Sea Basin is divided further into a north subbasin called the Northern South Yellow Sea Basin (SYSB-N) and a south subbasin named the Southern South Yellow Sea Basin (SYSB-S) (Figure 3.2). The subsidence history of the SYSB-N can be grouped into a main Paleogene rifting and subsidence phase and a secondary Neogene subsidence phase, separated by an uplift and erosion phase during the late Ologocene and early Miocene (Lee, 2010; Yi, Yi, Batten, Yun, & Park, 2003). In some places, the main subsidence phase itself consists of a rapid initial subsidence in the early Paleogene and a slow thermal subsidence phases in the late Paleogene (Figure 3.11). Initial subsidence occurred in very localized half-grabens controlled largely by the Jiaxiang Fault and received mainly nonmarine (fluvial–alluvial and lacustrine) clastic sediments with a thickness of several kilometers (Lee, 2010). Fluvial facies prevailed in the initial rifting stage but gradually changed to lacustrine facies with the deepening of water. Figure 3.11 shows that the synrift sediments were faulted and tilted before the thermal subsidence. Tectonic inversion and denudation occurred from the late Eocene/Oligocene to early Miocene in isolated zones within the basin and erosion continued into the late Miocene on the basin margins (Yi et al., 2003) (Figure 3.11). Thermal subsidence resumed in the depocenter after this compressional event, and since the middle Miocene, deposition has been dominated by mainly unconsolidated sands (KIGAM, 2006; Yi et al., 2003). Paleogene strata were strongly faulted and often titled. Neogene thermal subsidence was minor before rapid regional broad-sea subsidence started. Overall, the thermal subsidence phase appears to be insignificant in the Yellow Sea Basin. This is expected since rifting was not as pervasive here as in the Bohai or the ECS Basin, and the asthenospheric upwelling was not dramatic enough to trigger later large thermal subsidence. Subsidence patterns also vary dramatically within the basin.

3.4.2.2 East China Sea

The ECS continental shelf is characterized by NE-striking uplifts and depressions, which became active gradually eastward. With Cenozoic sediments up to 15 km in thickness, the Xihu Depression evolved from a deep Paleogene continental margin rift (Figure 3.12). Backstripping analyses show that the Xihu Depression overall experienced three stages of tectonic subsidence, an initial rifting stage from the early Paleocene to the late middle Eocene (∼40 Ma), during which thick synrift sediments were deposited at the depocenter; a thermal subsidence stage from the late middle Eocene (∼40 Ma) to

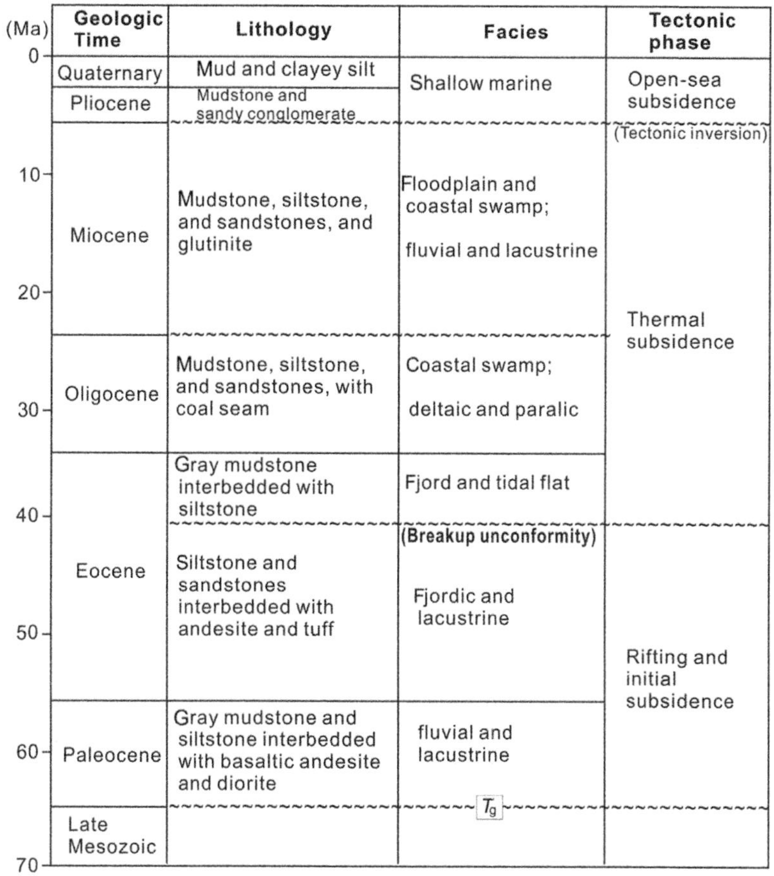

(Ma)	Geologic Time	Lithology	Facies	Tectonic phase
0	Quaternary	Mud and clayey silt	Shallow marine	Open-sea subsidence
	Pliocene	Mudstone and sandy conglomerate		
				(Tectonic inversion)
10	Miocene	Mudstone, siltstone, and sandstones, and glutinite	Floodplain and coastal swamp; fluvial and lacustrine	Thermal subsidence
20				
30	Oligocene	Mudstone, siltstone, and sandstones, with coal seam	Coastal swamp; deltaic and paralic	
40		Gray mudstone interbedded with siltstone	Fjord and tidal flat	
			(Breakup unconformity)	
	Eocene	Siltstone and sandstones interbedded with andesite and tuff	Fjordic and lacustrine	
50				Rifting and initial subsidence
60	Paleocene	Gray mudstone and siltstone interbedded with basaltic andesite and diorite	fluvial and lacustrine	
			T_g	
70	Late Mesozoic			

FIGURE 3.20   Schematic stratigraphic sequences and tectonic events in the Xihu Depression, East China Sea basin (Li, Zhou, Ge & Mao, 2007; Li, Zhou, et al., 2009).

the late Miocene (~5.5 Ma); and an accelerated open-sea subsidence since the Pliocene (Figures 3.20 and 3.21) (Li, Zhou, Ge, & Mao, 2009). The synrift and postrift sedimentary packages are clearly separated by a breakup unconformity. The rapid open-sea subsidence is quite regional and is apparently not related to the thermal subsidence (Figure 3.13). Tectonic subsidence analyses of 40 industrial wells reveal nonuniform stretching—preferential crustal extension along the depocenter but preferential mantle extension at the western flank. Uniform extension occurred only somewhere between the central inversion zone and the western margin (Li, Zhou, Ge, & Mao, 2009). The fact that the mantle stretching factors keep relatively stable while the crustal stretching factors drop considerably from the center to the flank of the depression suggests that the lithospheric mantle extension is more laterally spread out than crustal extension.

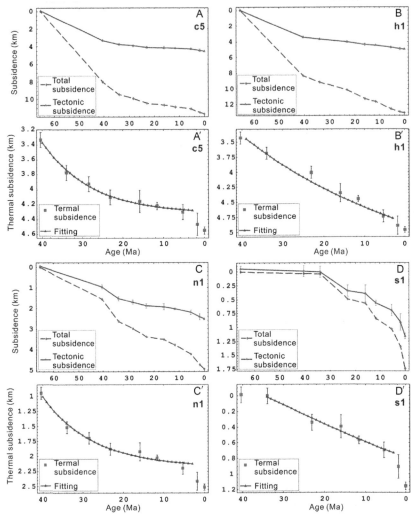

FIGURE 3.21 Total subsidence and tectonic subsidence curves of four wells c5 (A), h1 (B), n1 (C), and s1 (D) in the Xihu Depression, East China Sea basin (Li, Zhou, et al., 2009).

Similar to the South Yellow Sea Basin, tectonic inversion and denudation also occurred in the Xihu Depression from the late Oligocene to early Miocene, but mainly near the basin flanks. This event was strongly coupled with intensified magmatism, both within the basin and along its eastern margin (Li, Zhou, Ge, & Mao, 2009; Su, Li, & Ge, 2010). Another dramatic tectonic event occurred along the depocenter at the end of Miocene due to deep-rooted tectonic inversion, leading to a maximum thickness of erosion of ~1700 m (Li, Zhou, Ge, & Mao, 2007).

3.4.2.3   South China Sea

Taixinan Basin

Uplifting and denudation persisted in the Taixinan Basin from the late Cretaceous to the early Oligocene, before rifting and subsidence started in the late Oligocene (Lee et al., 1993; Lin et al., 2003; Tzeng, 1994). Unlike other basins, such as the ECS Basin and the Pearl River Mouth Basin, synrift sediments were not well developed (Lee et al., 1993; Li, Zhou, Li, Hao, et al., 2007; Lin et al., 2003). Postrift sediments in the North depression are Oligocene transgressive sandstones and thick younger marine shale (Lin et al., 2003). The Oligocene sequence is generally thin and limited to basin centers (Tzeng, 1994). The basement rocks drilled are lower-to-middle Cretaceous sandstones with interbedded shales of shallow marine to continental environments (Lin et al., 2003; Wang, Wang, et al., 2000; Xia & Huang, 2000). Apparently, the Taixinan Basin was more closely affiliated with the rifting event eventually led up to the opening of the SCS and later thermal subsidence, not directly with rifting events in the ECS Basin to the north. Seismic facies become chaotic, less continuous, and moderately deformed in the proximity of the Central uplift (Figure 3.16) of the Taixinan Basin, due to reef buildups, rock-type variations, and rock deformation near the present and past continental shelf breaks. High-angle normal faults are found above the southeast flank of the Central uplift. These are active faults cutting through the uppermost sedimentary layers and deforming the seabed into stepwise landforms, indicating frequent shallow gravity collapses at this locality. It is also noted that the slope instabilities are accompanied by ocean bottom-simulating reflectors (BSRs). Three stages of vertical differential movement between the Central uplift and the graben basement can be identified (Li, Zhou, Li, Hao, et al., 2007). The first stage corresponds to a rapid differential movement across the fault during the Oligocene, during which most of the sediments were derived from the northwest. At the start of the second stage, although the relative rate of uplift of the Central uplift was diminished, erosion occurred as it became subaerial. Source materials were carried to the graben from both the continent and the Central uplift, and the lower Miocene sequence was formed. In the third stage, the Central uplift ceased its relative uplift and started to subside, probably during the late Miocene (Lee et al., 1993), and terrestrial sedimentary sequence was deposited onlapping the Central uplift until it was buried. These features indicate significant sedimentary environmental change near the Miocene–Pliocene boundary, which was concurrent with the end of erosion on the Central uplift and with the onset of material influx from the southeast (Chen & Jaw, 1996). Pliocene–Quaternary strata can be traced continuously across the Central uplift of the Taixinan Basin.

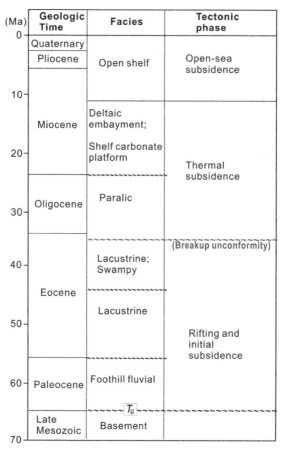

FIGURE 3.22 Generalized stratigraphic column with formation names, thicknesses, generalized lithology, and facies of the Pearl River Mouth Basin. *From Su, Whitem, and McKenzie (1989) and Chen and Li (1987).*

The Pearl River Mouth Basin

The Pearl River Mouth Basin (PREM) was formed by lithospheric extension, widespread crustal thinning, and normal faulting. Most initial rifting of the basin took place from the Paleocene to the late Eocene (~35 Ma), followed by the postrift thermal subsidence (Figure 3.22) (Su et al., 1989). The geometry of faulting suggests that considerable amounts of local footwall uplift and erosion occurred during the rifting period. This episode resulted in the formation of a series of NE–SW-trending horsts and graben (He, 1988). Sediment facies change progressively with time from terrestrial to marine (Wang, 1982; Xu, Chen, et al., 1987; Xu, Zhu, et al., 1987). The typical sedimentary sequence consists of red clastics of Paleocene to early Eocene age, overlain by dark argillaceous lacustrine shales of middle to early Eocene age and by early Oligocene coal-bearing

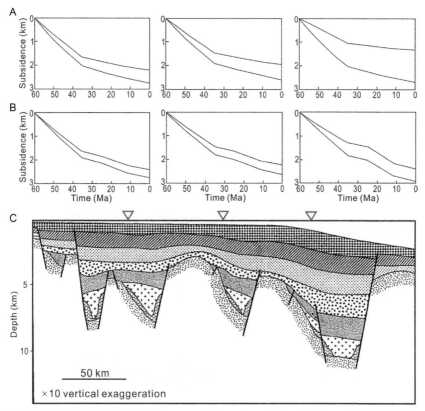

FIGURE 3.23   Corrected subsidence curves at the centers of three fault-bounded blocks in the Pearl River Mouth Basin. (A) Theoretical subsidence curves assume single phase of stretching that started at 60 Ma lasting 15 Ma; (B) as for (A) with additional phase of stretching that started at 25 Ma lasting 14 Ma; (C) profile modified from Chen and Li (1987). *From Su et al. (1989)*

littoral plain and swamp deposits. These are in turn overlain by early to middle Miocene littoral clastics and shallow bank carbonates and then by late Miocene to Quaternary neritic and bathyal clastics (Figure 3.22) (Su et al., 1989; Wu, 1988). All of the observed subsidence curves show a rapid increase in subsidence, irrelevant to sea-level changes, during the Miocene, which can be best explained by having a second minor phase of stretching between 25 and 11 Ma (Figure 3.23) (Su et al., 1989). The amount of this extension, however, did not significantly perturb the rate of postrift subsidence. Sedimentation in the central PRMB, and especially on the slope, was more rapid after 12 Ma. This three-stage subsidence pattern is very similar to what we have seen in the ECS. Clift and Lin (2001) showed that crustal extension exceeded that in the mantle lithosphere under the South China Shelf but that the two varied in phase, suggesting depth-dependent extension rather than a lithospheric-scale detachment.

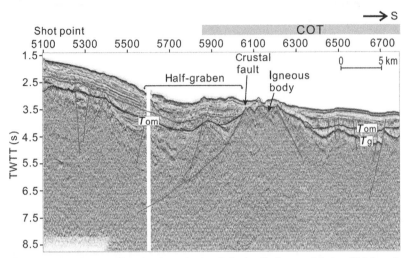

FIGURE 3.24 Seismic section showing a landward-dipping listric crustal fault, which bounds a basement uplift (igneous body?) to its south and a half-graben to its north. This section is a zoomed-in view of the seismic section shown in Figure 3.18, roughly between CDP# 20000 and 40000. The location of the seismic profile is marked by a black line T1 in Figure 3.15. Horizon T_{om} is the Oligocene–Miocene boundary coincidental with a zone of tectonic slump and erosion (Li, Jin, et al., 2004; Wang, Prell, et al., 2000). T_g, basement unconformity; TWTT, two-way travel time; COT, continent–ocean transition zone.

Near the continental break of the northern SCS continental margin, one particular sequence boundary really stands out, which, confirmed by ODP Leg 184, is the Oligocene–Miocene boundary. It has strong seismic reflections and coincides with a zone of tectonic slump and erosion (Li, Jian, & Li, 2004; Wang, Prell, et al., 2000). Li, Jian, et al. (2004) interpreted that this unconformity corresponds probably to changes in the rotation of different land blocks and to a ridge jump in the seafloor spreading ridge of the SCS. This unconformity is apparently diachronous, and to the north, it can be underlain directly by Mesozoic metasedimentary rocks in the northern continental margin. On the COT, this unconformity appears to coincide with a breakup unconformity in the half-graben, where it is likely underlain by strongly deformed pre-Miocene synrifting sediments (Figure 3.24).

Beibuwan Basin

Different from the PREM and Qiongdongnan Basin, synrift sediments in the Beibuwan Basin are much thicker, up to 6 km, than postrift sediments, being less than 2 km in thickness (Li, Lin, & Zhang, 1998; Xie, Müller, Li, Gong, & Steinberger, 2006). Wells from the Beibuwan Basin, a continental rift basin to the northwest of the PREM, show different subsidence histories to those in the main part of the South China Shelf, and extension there appears to be more uniform with depth and predates that on the South China Shelf (Clift & Lin,

2001; Su et al., 1989). The end of rifting of the Beibuwan Basin is also older (~43 Ma) than the estimate for the wells on the shelf (Clift & Lin, 2001). This progressive eastward younging in rifting activities is also similar to what we have seen in the ECS. The strain rate inversion analysis indicates that there are two episodes of rifting (Xie et al., 2006). Early rifting with relatively high stretching factors of 1.2–1.5 occurred in Paleocene and Eocene time. A second late rift event occurred in the Oligocene, but it is associated with a smaller stretching factor of less than 1.1.

Yinggehai and Qiongdongnan Basin

Controlled largely by the Red River Fault, the subsidence of the Yinggehai Basin is quite different in timing and magnitude to the PRMB. The Yinggehai Basin started to open after 45 Ma, especially after 34 Ma. It is shown that the depocenter trended northwest before about 36 Ma and then jumped southward and became nearly N–S trending and migrated toward the southeast up to 21 Ma (Figure 3.25); thereafter, the depocenter trended northwest again (Sun, Zhou, Zhong, Zeng, & Wu, 2003). This process took place in four main stages: (1) slow rifting stage with a NW-trending depocenter before 36 Ma, (2) rifting stage formed by sinistral slip of the Indochina block accompanied by rapid clockwise rotation between 36 and 21 Ma, (3) rifting thermal subsidence stage affected by sinistral slip of the Indochina block (21–5 Ma), and (4) dextral strike-slip (5–0 Ma). The shift from rifting to postrifting subsiding stage occurred at about 21 Ma (Sun et al., 2003). The infill of the Yinggehai Basin varies from alluvial, fluvial, and lacustrine (before 30 Ma) to neritic (30–10 Ma) and finally semi-abyssal (10–5 Ma) sediments (Sun et al., 2003). From about 30 Ma or younger, the northwest of the Yinggehai Basin began to inverse and a nearly NS-trending uplift formed (Zhong et al., 2004), and inversion structures migrated southeastward from the late Oligocene to middle Miocene (30–15.5 Ma), before rapidly subsiding again after ~5 Ma (Clift & Sun, 2006; Zhang, 1999).

One-dimensional tectonic subsidence analysis of the southern and central Yinggehai Basin shows long-term subsidence, with the fastest episodes of subsidence prior to 30 Ma and again after 5.5 Ma, suggesting at least two stages of extension (Clift & Sun, 2006). The period of 16–5.5 Ma is the slowest period and coincides with and postdates the folding and inversion seen in the seismic profiles (Figure 3.25), thus dividing the basin history into three distinct periods (Figure 3.26).

Xie et al. (2006) identified a rapid increase of anomalous postrift subsidence toward the deepwater areas, ranging from 900 to 1200 m in the deepwater part of the Qiongdongnan and Pearl River Mouth Basins, to 700–300 m in the continental shelf area of the basins, and to less than 250 m in the Beibuwan Basin further north. This rapid subsidence event, preceded by minor uplift, is interpreted as a thermal cooling episode induced by a late magmatic event

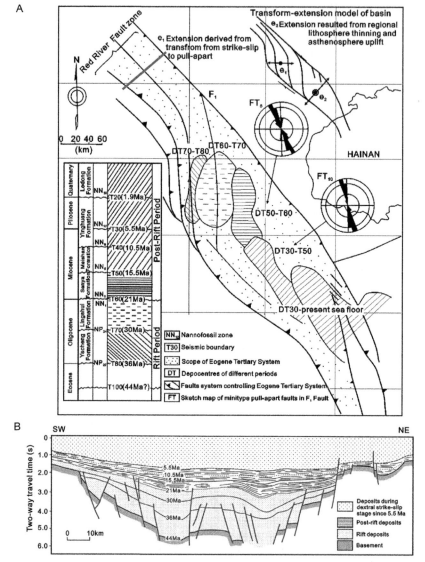

FIGURE 3.25 (A) Tectonic framework of the Yinggehai Basin and the migration of depocenters through time. The bold line is the location of the section shown in (B). (B) Interpreted seismic section in the Yinggehai Basin (Xie et al., 2006). *After Li et al. (1998) and Xie et al. (2006).*

(Xie et al., 2006). In the Yinggehai Basin, anomalous postrift subsidence is about 300–500 m in the northwestern part, substantially less than that in the middle and southeastern parts, where it ranges from 900 to 1200 m. The distinct postrift subsidence anomaly in the Yinggehai Basin may originate from a combination of the dextral movement of marginal faults since the late Miocene and the effect of dynamic topography (Xie et al., 2006).

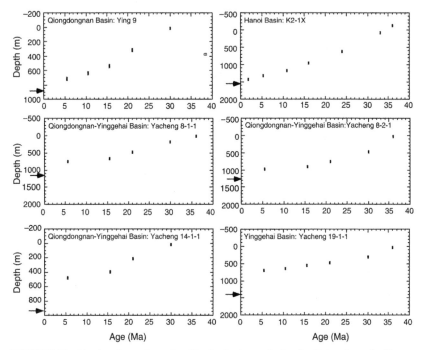

FIGURE 3.26   Results of one-dimensional backstripping analysis of well data from the Yingge-hai region showing the sediment-unloaded depth to basement through Cenozoic time (Clift & Sun, 2006). Vertical solid bars indicate uncertainty in the water depths of deposition. Arrows at left-hand axes indicate modern unloaded depth to basement.

Like in the Taixinan Basin, the rift phase in the Qiongdongnan Basin is associated with the initial Eocene and Oligocene opening of the SCS (Ru & Pigott, 1986), with postrift subsidence starting in the early Miocene (21 Ma). Synrift tectonic subsidence is largely less than 2500 m in the Qiong-dongnan Basin. In the southern parts of the Qiongdongnan and Pearl River Mouth Basins, although significant postrift subsidence is observed, synrift subsidence appears to be small or absent because of the lack of faulting in the upper crust (Xie et al., 2006). This is also the case in the Taixinan Basin, reflecting strong lower crustal necking in a narrow zone within the COT before the opening of the SCS basin. Large-scale, southward-prograding and southward-aggrading foresets are a characteristic feature of the Plio–Pleistocene in the Qiongdongnan Basin (Clift & Sun, 2006; Xie et al., 2006).

Cuu Long Basin and Wan'an (Nam Con Son) Basin

There are two large basins, the Cuu Long Basin and the Wan'an (Nam Con Son) Basin, located offshore southern Vietnam and southwest of the SCS (Figure 3.27). Initial rifting began in the Eocene–early Oligocene, followed by the uplift and rotation of the crustal blocks in the late Oligocene at the

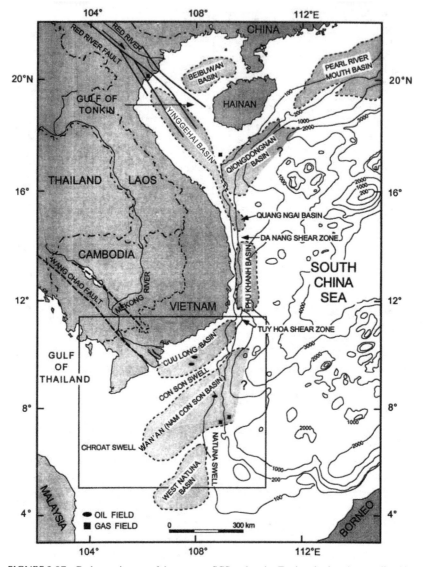

FIGURE 3.27 Bathymetric map of the western SCS and major Tertiary basins. Area outlined by box includes the Cuu Long Basin and the Wan'an (Nam Con Son) Basin (Lee, Lee, & Watkins, 2001). Contour interval is in meters.

possible onset of seafloor spreading in the southwestern SCS, which marked the transition from rifting to regional subsidence in the Cuu Long Basin (Lee et al., 2001). Postrift thermal subsidence continued afterward in the Cuu Long Basin. However, the Wan'an (Nam Con Son) Basin to the south experienced continued faulting started in the early Miocene and lasted until

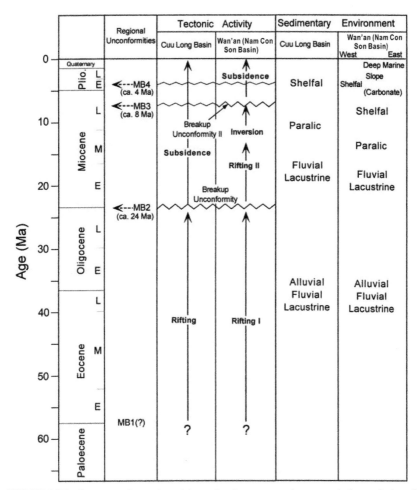

FIGURE 3.28   Summary of tectonic activity and sedimentary environments during evolution of the Cuu Long and Wan'an (Nam Con Son) Basins (Lee et al., 2001).

the late Miocene (Figures 3.28 and 3.29). Since the Wan'an (Nam Con Son) Basin is located immediately at the southwest extension of the Southwest Subbasin of the SCS, its new phase of faulting that are not observed in the Cuu Long Basin can be explained by the Miocene spreading of the Southwest Subbasin of the SCS.

3.5   THE SCS AND OKINAWA TROUGH

The continental margin basins discussed in the preceding text in this chapter already ceased their active rifting and subsidence, and sediments overfilled them to form today a broad continental margin (Figures 3.1 and 3.2).

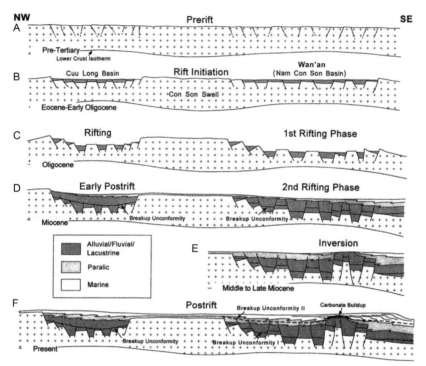

FIGURE 3.29 Schematic illustration of the geologic evolution of the Cuu Long and Wan'an (Nam Con Son) Basins (Lee et al., 2001). (A) Proto-southern Vietnam margin was characterized by zones of weakness. (B) The zones of weakness provided the locations of rift initiation in the Paleogene. (C) In the Oligocene, rifting and extension created the Cuu Long and Wan'an (Nam Con Son) Basins on either side of the Con Son swell; this initial rifting phase was characterized by rapid subsidence and infilling. (C) Faulting continued in the Wan'an (Nam Con Son) Basin in the Miocene, while only postrift subsidence occurred in the Cuu Long Basin. (E) Inversion in the middle to late Miocene terminated the Miocene faulting in the Wan'an (Nam Con Son) Basin. (F) Slowed subsidence and decreased sediment influx resulted in the shelf-slope transition in the eastern Wan'an (Nam Con Son) Basin.

Conversely, the SCS and Okinawa Trough are places with deepwater sedimentation today. The SCS is floored with oceanic crust and Okinawa Trough is still under active rifting. Therefore, these two tectonic units deserve our special treatment.

3.5.1 SCS basin

From the late Mesozoic until the end of Eocene, most of the western Pacific marginal seas were nonexistent, and East Asia was under the direct impact of the Pacific Plate (Figure 3.2). With the cessation of the Paleo-Pacific subduction and associated magmatism and the gradual closure of the Tethys and the buildup of the Tibetan Plateau, East Asia reversed its topography from being west tilting to east tilting, an event exerted a profound impact on regional environments and on continent–ocean interactions (Wang, 2004).

As one of the classical representatives of western Pacific marginal seas, the SCS basin is the ultimate product of the prolonged rifting since the late Mesozoic (Shi & Li, 2012). It went through continental margin rifting and final breakup, and seafloor spreading lasted from ~32 to ~16 Ma (Briais et al., 1993; Li & Song, 2012; Taylor & Hayes, 1980, 1983), though this exact timing is being constantly studied and questioned. Its location at the junction of the Eurasian, western Pacific, and Indo–Australian Plates makes it sensitive to all major tectonic events in these plates and complicates our understanding of its opening mechanism. Early studies attributed the SCS basin to the back-arc spreading behind the Luzon Arc by the subduction of the Philippine Sea Plate, but before long, it was realized that a back-arc model cannot fit the basin configuration; the Luzon Arc is actually formed by the eastward subduction of the SCS plate, and the extinct spreading center and magnetic anomalies are nearly perpendicular to the Manila Trench (Figures 3.30 and 3.31). Hypotheses linking the SCS opening to the subduction of the Indo-Australian Plate can also be seen occasionally. Now the SCS basin is generally considered as an Atlantic type of oceanic basin from continental rifting. Comparing with the Atlantic, the SCS basin is of course much smaller and younger. Therefore, its subsidence along the margins is less dramatic and its conjugate continental margins can be easily accessed and compared. These attributes make it an ideal natural laboratory for studying continental breakup and basin formation (Wang, 2012).

3.5.1.1   The COB and COT

From the regional topographic map (Figure 3.30), it is seen that the COT broadens significantly from the east to the west. Accurate outlining of the COB and understanding of its tectonic affiliation are key to determining the incipient ages of seafloor spreading.

Li and Song (2012) noticed that the COB correlates very well to a transition zone in free-air gravity anomalies (Figure 3.32). Areas floored with oceanic crust in the central basin correspond to mostly positive gravity anomalies, which are rimmed and surrounded by grossly negative gravity anomalies. The transition zone in between corresponds roughly with the COB, which can be verified on reflection seismic profiles. The positive free-air gravity anomalies over the oceanic crust are closely related to mantle uplift and high density of the oceanic crust, whereas in the transition zone, extension and subsidence of continental crust accompanied with draping of low-density sediments lead to overall negative free-air gravity anomalies. As can be seen from the total field magnetic anomaly map (Figure 3.31), the COB so determined encloses very well the area with clear seafloor spreading magnetic anomalies in the SCS.

There are speculations that mantle could be exhumed before the onset of seafloor spreading along the COB of the SCS. This could be the case if the

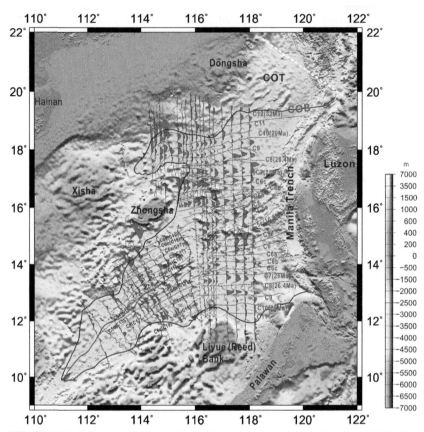

FIGURE 3.30 Bathymetry, key magnetic profiles, and interpreted ages and their spatial distributions (Li & Song, 2012; Song & Li, 2012). The thin dashed lines are sketches of interpreted negative
magnetic anomalies in the central basin. Their preliminary ages are modeled based on the geomagnetic polarity timescales CK95 (Cande & Kent, 1995). Two possible but different seafloor spreading
models are shown in the Southwest Subbasin. The solid line in the Southwest Subbasin shows the
track of the negative free-air gravity anomaly along the relict spreading center (Figure 3.33).
COT, continent–ocean transition zone; COB, continent–ocean boundary.

extensional rates and the mantle temperature were low and if depth-dependent
extension caused crustal breakup first before the breakup of continental mantle.
In these scenarios, the COB could move further seaward to the exact breakup
point of the entire continental lithosphere. However, these speculations cannot
be confirmed at this stage by geophysical data alone. The COT of the SCS
shows significant lateral variations in deformational structures, and future ocean
drilling expeditions are desired to ground-truth these interpretations.

The COT is here defined as the heavily extended and faulted narrow
zone between the less attenuated continental lithosphere and the oceanic
lithosphere. Therefore, we do not limit the COT only to those areas where

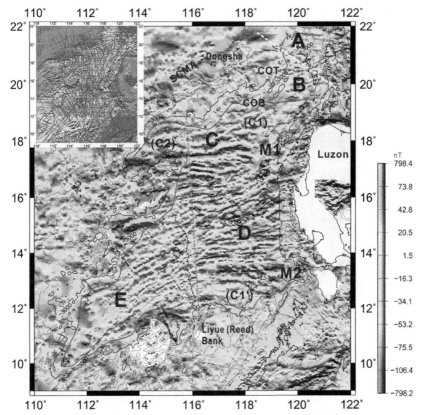

FIGURE 3.31   Total field magnetic anomaly map of the South China Sea (SCS) showing magnetic zonation (Li & Song, 2012). A, B, C (C1, C1', C2), D, and E are different magnetic zones. M1 and M2 are major magnetic boundaries in the East Subbasin, and their ages are estimated to be C8 (~26 Ma). The bold solid line marks the continent–ocean boundary (COB). The light dashed line marks the Zhongnan Fault. The black dashed line marks the Zhongnan ridge. The white bold dashed line marks the Luzon–Ryukyu Transform Boundary (LRTPB, Hsu et al., 1998; Hsu, Yeh, Doo, & Tsai, 2004; Li, Zhou, Li, Hao, et al., 2007; Li, Zhou, Hao, et al., 2008; Sibuet et al., 2002). The white thin dashed lines are sketches of negative magnetic anomalies in the central basin. SCMA, offshore South China magnetic anomaly. The upper-left inlet is the regional topography overlapped with magnetic tracks in red lines.

continental lithospheric mantle is possibly exhumed. Exhumation of mantle certainly is an indicator of extreme extension, but extreme extension can also occur in the lower crust and mantle first (Huismans & Beaumont, 2011).

The COT of the SCS is heavily faulted, attenuated, and magmatically intruded continental lithosphere (Li et al., 2010; Wang et al., 2006; Zhu et al., 2012). A large listric normal fault located immediately to the south of the Dongsha Rise can be considered as one of the north limits of the transition zone (Li et al., 2010), but the deformation styles change markedly along the margin

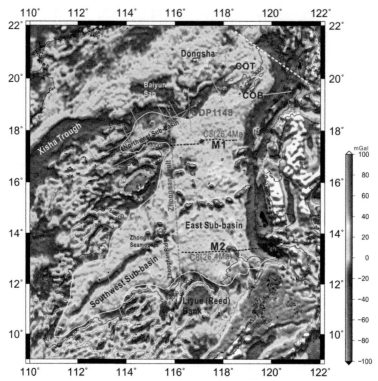

FIGURE 3.32 Free-air gravity anomaly map of the SCS (based on the 1 min grid) (Li & Song, 2012). The light solid lines are reflection seismic sections. The two bold black dashed lines mark the two magnetic anomalies M1 and M2 interpreted from Figure 3.31. The white solid line marks the interpreted COB. The solid pentacle in the Zhongnan Seamounts marks the sampling site of alkaline basalts (K–Ar age: 3.49 Ma).

(Figure 3.32). The Dongsha Rise has well-developed Mesozoic rocks and has remained almost intact in the Cenozoic rifting, but the Baiyun Sag adjacent to the Dongsha Rise suffered from extreme thinning locally. The COT is extremely wide to the southwest, with numerous troughs and faulted blocks, similar to typical nonvolcanic rifted margins found in other places. The COT is rather narrow to the east, with intensified magmatism and possible high-velocity lower crustal materials from magmatic underplating. Therefore, the eastern COT appears to be more similar to volcanically rifted margin.

3.5.1.2 Major Subbasins

Magnetic anomaly features differ markedly between different tectonic units of the SCS basin. Based on magnetic attributes like magnitudes, strikes, and frequencies, the SCS basin can be divided into five zones representing different crustal affiliations or oceanic crusts (Li, Zhou, Li, et al., 2008) (Figure 3.31).

Zones A and B are in the northeasternmost part of the SCS basin, intersected by the Luzon–Ryukyu Transform Plate Boundary (LRTPB) (Hsu et al., 1998, 2004; Sibuet et al., 2002). The northeasternmost part of the SCS basin appears to be the conflux of many different geologic processes, for example, Mesozoic northwestward subduction, early Cenozoic rifting, and late Cenozoic Manila subduction. The northern SCS passive margin may have reached the area northeast of Taiwan about 10 Ma (Clift et al., 2003; Hall, 2002; Teng, 1990, 1996). However, it was also argued that prior to 15 Ma, the Ryukyu subduction zone extended south to the northeasternmost SCS and a sector of the northeasternmost SCS basin comprised oceanic crust, older than that of the present-day SCS crust (Sibuet & Hsu, 1997; Sibuet, Hsu, & Debayle, 2004; Sibuet et al., 2002). What would be the nature of this older oceanic crust if there is? There are various hypotheses regarding a proto-SCS oceanic basin and its existence is supported by the wide occurrence of Jurassic, Cretaceous, and/or early Cenozoic marine sediments in the present southern SCS and beyond. A large part of this proto-SCS may have been subducted into, or uplifted as part of, island arcs to the south. Part of this proto-SCS oceanic crust may be preserved in the northeasternmost part of the SCS basin but these questions have not been fully sorted out. Deschamps, Monié, Lallemand, Hsu, and Yeh (2000) interpreted the Huatung Basin to the east of the Manila subduction zone as an early Cretaceous oceanic block, but this interpretation is disputed (Sibuet et al., 2002). One unsolved problem is whether and how the Huatung Basin was affiliated with the southeast-verging proto-SCS (Deschamps et al., 2000; Hall, 2002; Sibuet et al., 2002) and the northwesterly subducting Paleo-Pacific Plate.

Some have considered that the magnetic anomalies from seafloor spreading can be traced all the way northward to 21.5°N, east of Dongsha Islands, and thereby argued that the initial opening of the SCS may not be later than 37 Ma (Hsu et al., 2004). Li, Zhou, Li, Hao, et al. (2007) and Li, Zhou, Li, et al. (2008) documented the sharp contrasts between the northeasternmost SCS and the central SCS and suggested that most part of the northeasternmost SCS is underlain by a heavily attenuated and magmatically intruded block with Mesozoic strata in the basement. This area can be characterized as a COT, though extremely thinned crust beneath the South depression of the Taixinan Basin is on the order of the thickness of a typical oceanic crust (Figures 3.16 and 3.17).

Zone C is between the COB and a conspicuous negative magnetic anomaly named M1 (Li, Zhou, Li, et al., 2008). It can be further subdivided along the Zhongnan Fault into zones C1 and C2, as differences in magnetic amplitudes and strikes across the Zhongnan Fault are also apparent (Figure 3.31). Zone C2 corresponds to the traditionally defined Northwest Subbasin. The Northwest Subbasin is previously known as a small residual oceanic basin died soon after its initial opening, and it has a relict spreading center, with which magnetic anomalies are nearly symmetrical. By contrast, there is not a relict ridge in zone C1 (Briais et al., 1993; Taylor & Hayes, 1980, 1983). If so, the Zhongnan transform fault must be introduced in the northern SCS basin (Figures 3.30 and 3.31), in order to reconcile the differential spreading and movement caused by persistent spreading to its east in zone C1 after the spreading ceased to the west in zone C2 (Li & Song, 2012).

The magnetic anomaly M1 possibly is caused by an episode of strong magmatism and corresponds to an important tectonic event in the seafloor spreading of the SCS. M1 was previously identified as C8, dated as ~27.5 Ma (Briais et al., 1993; Taylor & Hayes, 1980, 1983). Li and Song (2012) reexamined the age of M1 (C8) and found it to be ~26.4 Ma based on the geomagnetic polarity timescale CK95 (Cande & Kent, 1995) and ~26.0 Ma based on the new geomagnetic polarity timescale of Gradstein et al. (2004). These ages attribute magnetic anomaly M1 to a magnetic reversal event in the middle late Oligocene.

It is argued that the tectonic event associated with C8 may not be caused by a ridge jump, but is more likely related to a change in spreading rates or magmatic magnitudes along the spreading center (Li & Song, 2012). This event could be linked with a major tectonostratigraphic event around 25 Ma revealed by the ODP Site 1148, forming an unconformity/slump zone with a 2–3 Ma of sedimentary hiatus and coincidental strong seismic reflections (Li, Jian, et al., 2004; Wang, Prell, et al., 2000). From seismic reflection profiles (Li et al., 2010), it is seen that oceanic basement older than C8 is overall without large undulations and is with thick overlying sediments, while that younger than C8 fluctuates heavily and has much thinner burials. In the ECS, there also occurred strong regional compression, uplifting and erosion, and pervasive magmatic activities at the Oligocene–Miocene transition (~23 Ma) (Su et al., 2010). Therefore, this Oligocene–Miocene transitional tectonic event occurred probably throughout East China and caused regional instability.

To the south of the relict spreading center, the counterpart of M1 is M2. Between M1 and M2 is the magnetic zone D, an area traditionally known as the East Subbasin, with large amplitudes of magnetic anomalies. Zone E (Southwest Subbasin) located to the west of the Zhongnan Fault differs markedly in magnetic strengths and strikes from Zone D (East Subbasin).

There are concurrent major changes in sedimentary and geochemical records across the Oligocene–Miocene boundary from Ocean Drilling Program (ODP) Site 1148 in the northern SCS (Li, Huang, Jiang, & Wan, 2011; Wang, Prell, et al., 2000). However, debates remain on how this change occurred. Clift, Lee, Clark, and Blusztajn (2002) proposed that the provenance changed from South China coast to the South China block inland in the late Oligocene, and this statement seems to be supported by clay mineralogy and geochemistry (Miao et al., 2008; Pang et al., 2007; Shao, 2008), palynology (Wu, Qin, & Mao, 2003), and paleoceanography (Zhao, 2005). In contrast, Li et al. (2003) argued that the sediments at ODP Site 1148 were derived from Indochina landmass and Sunda Shelf (Borneo Island) in the Oligocene, but from the northern sources in the Miocene, based on significant variations of sediment Nd isotopes. Li, Huang, et al. (2011) recently suggested that terrigenous materials at ODP Site 1148 mainly originated from the southern sources (presumptively from Palawan) during the early spreading of the SCS, because most of the terrigenous detritus from South China were captured by continental margin depressions before reaching the site. But these southern blocks gradually moved further away after 25–23 Ma, and South China gradually became the main source of the northern detritus afterward (Li, Huang, et al., 2011). The Oligocene–Miocene sedimentary discontinuity of ODP Site 1148 could result from strong tectonic activities, global sea-level rise, poor material supply, and the ocean currents together under the background of the expansion of the SCS (Li, Huang, et al., 2011).

3.5.1.3   Opening Sequences

The Zhongnan Fault may have played a central role in the multistage opening models of the SCS (Li, Zhou, Li, Hao, et al., 2007; Ru & Pigott, 1986; Taylor & Hayes, 1980, 1983; Yao, et al., 1994). As mentioned in the preceding text, in order to reconcile the markedly different magnetic anomalies and differential movement between the Northwest Subbasin (magnetic zone C2) and East Subbasin (magnetic zone C1), a transform fault should be introduced (Li & Song, 2012). The Zhongnan Fault demonstrates itself clearly on seismic profiles (Li, Zhou, Li, et al., 2008). It also shows up as a zone of low free-air gravity anomalies (Figure 3.32) and as a low depression on the topography map (Figure 3.30). This magnetic contrast either may support a model of episodic rifting of the SCS (Ru & Pigott, 1986) or can be attributed to different crustal types with the Southwest and East subbasins evolving independently. Pautot et al. (1986) suggested that the East Subbasin developed within an older, preexisting oceanic crust, whereas the Southwest Subbasin resulted from continental rifting that led to seafloor spreading. In the Southwest Subbasin, Li, Ding, et al. (2012) proposed that spreading propagated from NE to SW and showed a transition from steady seafloor spreading, to initial seafloor spreading, to continental rifting in the southwest end.

Up till now, all age controls on the oceanic crust of the SCS basin are from magnetic anomaly correlations or from empirical relationships between ages and basin depths and/or heat flow (Barckhausen & Roeser, 2004; Briais et al., 1993; Hsu et al., 2004; Ru & Pigott, 1986; Taylor & Hayes, 1980, 1983; Yao, et al., 1994). Table 3.1 lists the large uncertainties in various age estimates of the SCS from magnetic anomalies, heat flow, and bathymetry.

This large margin of error in estimated ages leads to different Cenozoic opening models of the SCS. It remains uncertain whether the SCS basin experienced primarily a single episode or multiple episodes of extension and seafloor spreading and, if multiple, in what sequence the three subbasins evolved

TABLE 3.1 Various Age Estimates of the South China Sea

Authors	Ages (Ma)	Area of Study	Year of Publication	Data
Taylor and Hayes	32–17	East Subbasin	1980, 1983	Magnetic anomaly
Briais et al.	32–16	Central SCS basin	1993	Magnetic anomaly
Yao et al.	42–35	Southwest Subbasin	1994	Magnetic anomaly
Barckhausen and Roeser	31–20.5	Central SCS basin	2004	Magnetic anomaly
Hsu et al.	37–15	Central SCS basin and northeastern SCS	2004	Magnetic anomaly
Ru and Pigott	~55 35–36 ~32	Southwest Subbasin Northwest Subbasin East Subbasin	1986	Heat flow and bathymetry

(e.g., Briais et al., 1993; Hayes & Nissen, 2005; Li, Zhou, Hao, et al., 2008; Li, Zhou, Li, Hao, et al., 2007; Pautot et al., 1986; Ru & Pigott, 1986; Taylor & Hayes, 1980; Yao, et al., 1994). It has been suggested, for example, that the opening of the East and Northwest Subbasins predated, or at least synchronized with, that of the Southwest Subbasin (Figure 3.33) (Briais et al., 1993; Hall, 2002; Hall & Morley, 2004; Hayes & Nissen, 2005; Honza, 1995; Lee & Lawver, 1995; Schluter, Hinz, & Block, 1996; Sun et al., 2008; Taylor & Hayes, 1983; Tongkul, 1994; Zhou, Ru, & Chen, 1995), a model contrasting with some others in which an earlier opening in the Southwest Subbasin is preferred (e.g., Li, Zhou, Li, Hao, et al., 2007; Ru & Pigott, 1986; Yao, et al., 1994). The latter group of models considers the sharp contrasts between the East and the Southwest Subbasins and the important roles of transform faults such as the Zhongnan Fault, which the first group often ignores.

3.5.1.4 A Discussion of the Opening Mechanism

What regional geologic processes finally triggered the continental breakup in the SCS? Hypotheses presented so far differ markedly and include (1) India–Eurasia collision and a consequent tectonic extrusion process mainly along the Red River Fault (Figure 3.34A) (Briais et al., 1993; Flower et al., 2001; Lallemand & Jolivet, 1986; Leloup et al., 2001; Schärer et al., 1990; Tapponnier et al., 1982), (2) slab pull and subduction of the proto-SCS under Sabah/Borneo (Figure 3.34B) (Hall, 2002; Holloway,

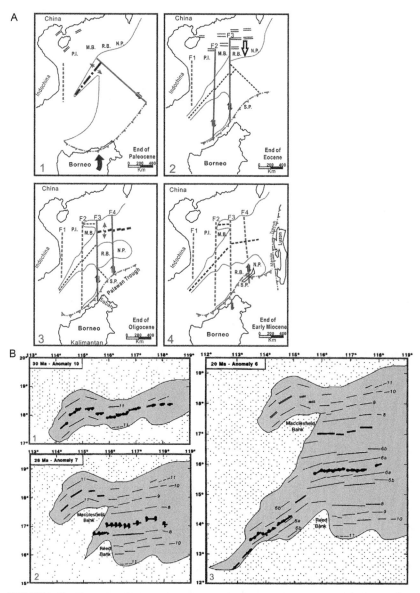

FIGURE 3.33   Examples of two groups of contrasting tectonic models for the opening phases of the South China Sea. (A) Multiphase episodic rifting model in which the Southwest Subbasin is the first to open from continental rifting. NP, Northwest Palawan; SP, South Palawan; MB, Macclesfield Bank; RB, Reed Bank. (B) Southwestward continuous propagating model in which the Southwest Subbasin is coeval with the central East Subbasin. *Panel (A) after Ru and Pigott (1986) and panel (B) after Briais et al. (1993).*

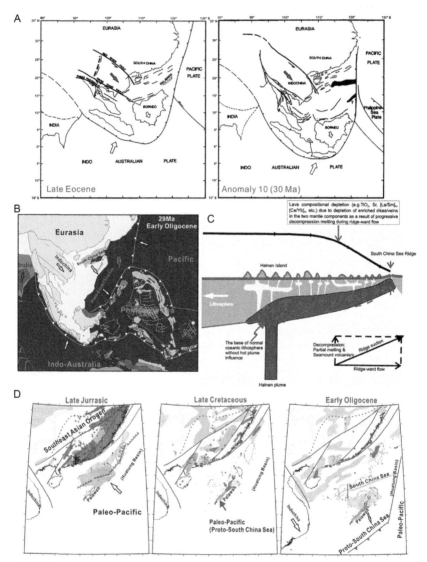

FIGURE 3.34 Different hypothetical models on the driving mechanisms of the opening of the South China Sea. (A) Opening induced by India–Eurasia continental collision and consequent tectonic extrusion (Briais et al., 1993; Flower et al., 2001; Leloup et al., 2001; Tapponnier et al., 1982, 1990); (B) opening induced by slab pull and subduction of the proto-SCS (Hall, 2002; Holloway, 1982; Taylor & Hayes, 1980, 1983); (C) opening induced by an upwelling mantle plume (e.g., Fan & Menzies, 1992; Xu, Wei, Qiu, Zhang, & Huang, 2012); (D) opening induced by regional extension related to subduction and retreat of the Pacific Plate (Shi & Li, 2012; Taylor & Hayes, 1980, 1983). See Figure 3.3 for further annotations.

TABLE 3.2 Differences Between the East and Southwest Subbasins

Attributes	East Subbasin	Southwest Subbasin
Water depths	Shallower on average	Deeper on average
Basement depths	Close (deepen to the margins)	Close (deepen to the west)
Heat flow	Lower	Higher
Magnetic strikes	East–west	Northeast–southwest
Magnetic amplitudes	Stronger	Weaker
Magnetic spectra	Preferentially low in high-wave number components	Preferentially high in high-wave number components
Free-air gravity anomalies	Higher	Lower
Curie depths	Mostly deeper	Mostly shallower
Seismicity	Stronger	Weaker

1982; Taylor & Hayes, 1980, 1983), (3) extension related to an upwelling mantle plume (Figure 3.34C) (e.g., Fan & Menzies, 1992; Xu et al., 2012), and (4) regional extension related to subduction and retreat of the Pacific Plate along the western Pacific margin (Figure 3.34D) (Shi & Li, 2012; Taylor & Hayes, 1980, 1983). In addition to these end-member models, hybrid models have been proposed, stating, for example, that the Oligocene opening was related to extrusion, whereas the Miocene opening was driven by subduction toward Borneo (Cullen et al., 2010).

Any tectonic reconstruction model of the opening of the SCS must explain the formation of distinct magnetic zones (Figure 3.31) and the observed differences in magnetic character between the various subbasins (Table 3.2). The original SCS basin before subduction along the Manila Trough could be twice the size of what we see today (Sibuet et al., 2002), so the geodynamic model should also be able to explain the formation of a large ocean basin.

The motion on the Ailao Shan–Red River Fault (RRF) has been estimated to be several hundreds of kilometers from 35 to 15 Ma (e.g., Leloup et al., 2001). Recent studies from Vietnam show that the Red River Fault changed from transpression to transtension some 33 Ma, therefore corresponding to the timing of initial SCS opening in the earliest Oligocene (Jolivet et al., 2011). However, it is difficult to imagine that the extruding Indochina block alone could trigger the opening of a large marginal basin located primarily to the east of the extruding block. It has been argued by others that only a minor amount of extension associated with the SCS spreading center was transferred to the RRF (Clift et al., 2008; Morley, 2002; Rangin et al., 1995). Regional rifting in East Asia occurred long before the India–Eurasia collision (Figure 3.34D) and was associated mainly with the subduction of the Paleo-Pacific Plate (e.g., Shi & Li, 2012; Taylor & Hayes, 1980, 1983).

Hypotheses have also been proposed regarding the existence of a proto-SCS oceanic basin (Haile, 1973; Madon et al., 2000) that was once connected to the Pacific Plate and began to close from ~44 Ma (e.g., Hall, 1996, 2002) in order to accommodate the opening of the SCS (Figure 3.34B). Supported by the wide occurrence of Mesozoic and/or early Tertiary marine sediments, a large part of this proto-SCS may have been subducted into, or uplifted as part of, island arcs formed to the south in Borneo/Sabah and Palawan (Hall, 2002; Hutchison, 1996, 2004), where remnants of the proto-SCS oceanic crust are believed to be present (Hutchison, 2005) and are one possible origin of ophiolites of South Palawan (Cullen, 2010; Pubellier et al., 2004; Rangin et al., 1990; Schluter et al., 1996; Tu et al., 1992). Slab pull force from this subducting proto-SCS plate might have contributed to the opening of the SCS.

Low-velocity mantle structures indicated by seismic tomography around the Hainan Island could be related to a deep-rooted mantle plume (Lebedev & Nolet, 2003; Lei et al., 2009; Montelli et al., 2004; Zhao, 2007), which could have been responsible for the widespread Cenozoic alkali basalts and the final opening of the SCS (Xu et al., 2012; Yan & Shi, 2007). However, lack of strong evidences for massive synrifting volcanism put the triggering mechanism by a mantle plume questionable. Based on available data (Chung & Sun, 1992; Flower, Zhang, Chen, Tu, & Xie, 1992; Tu et al., 1992), older SCS seafloor shows more distinct geochemical signature than the younger products of postspreading volcanism. Thus, it is possible to find contrasting geochemical signatures for each stage of basin development and evolution. More geochemical studies are needed to test the mantle plume hypothesis using directly sampled basement basalts within the SCS. Systematic sampling of basement rocks will be required to validate the episodes of basin development. The age and geochemical signatures of the oceanic basalts of the SCS will help understand its relationship with volcanism in the surrounding margins and the relationship of the basin opening with regional tectonic and magmatic events. This information can be integrated with geophysical data to understand the opening mechanism and geodynamic evolution of the SCS.

3.5.1.5 Manila Subduction Zone

The active eastward subduction of the SCS block along the Manila Trench formed a classical trench (Manila Trench)–accretionary prism (Kaoping Slope and Hengchun Ridge)–forearc basin (Luzon Trough)–volcanic arc (Luzon

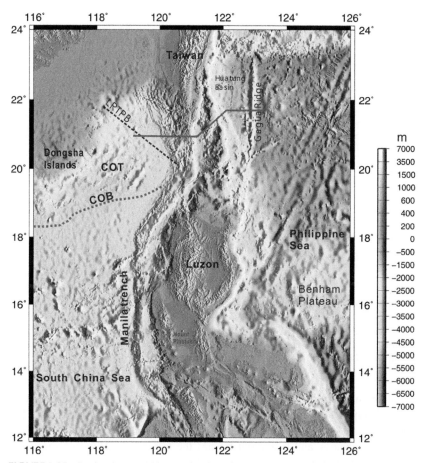

FIGURE 3.35   Regional topographic map showing major tectonic regimes of the Manila subduction zone. The solid line is the seismic section 973GMGS shown in Figure 3.36.

Arc) system (Figures 3.21, 3.35 and 3.36). With the continued northwestward subduction of the Paleo-Pacific and Pacific Plate, volcanic arcs and newly formed marginal seas in the Pacific migrate with this plate motion, resulting terrane amalgamations in sharp contacts. The Luzon Arc is believed to be formed within the Philippine Sea Plate with the eastward subduction of the SCS block. The incipient timing and triggering mechanism of this subduction process are in wide speculation. It is assumed that the subduction started soon after the cessation of seafloor spreading in the SCS about 15 Ma (Briais et al., 1993; Yang et al., 1996). The initial subduction probably developed along one of the preexisting transform faults, which should be widespread in the western Pacific. However, it is rather puzzling at the moment how could the younger and lighter SCS lithosphere subducted beneath the older and denser Philippine Sea lithosphere.

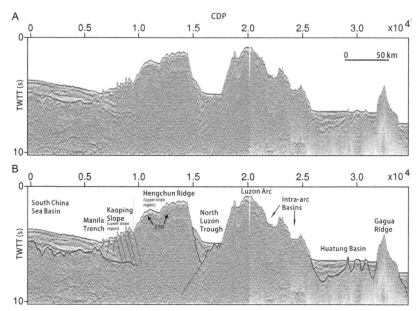

FIGURE 3.36 (A) The migrated seismic section along the 973GMGS line across the Manila Trench and the Luzon Arc; (B) preliminary geologic interpretations of faults and sedimentary basement. TWTT, two-way travel time; CDP, common depth point; BSR, bottom simulation reflector (Li, Zhou, Li, Chen, et al., 2007; Li, Zhou, Li, Hao, et al., 2007).

Sediments deposited in the SCS basin proximal to the Manila Trench can be divided into three sequences (Li, Zhou, Li, Chen, et al., 2007; Li, Zhou, Li, Hao, et al., 2007). The Miocene sequence shows weaker seismic amplitudes, varies in thickness, and tapers out toward the trench. The Pliocene sequence keeps a fairly constant thickness and good lateral continuity in seismic reflectivity, but to the east, it is also truncated considerably by a basement high. Near the trench, the Quaternary sequence shows wavy seismic reflections and thickens gradually toward the trench axis.

The accretionary prism to the south of Taiwan is composed of two main parts: the old Hengchun Ridge to the east and the young Kaoping Slope to the west (Figure 3.36). The Kaoping Slope is composed of late Neogene to Quaternary sediments received from Taiwan, while the Hengchun Ridge consists mainly of Miocene turbidites derived from the Eurasian continent (Huang et al., 1997). The Hengchun Ridge (also known as the upper slope domain; Reed, Lundberg, Liu, & Kuo, 1992) lacks coherent seismic reflections but shows well-developed BSR (Figure 3.36). The Kaoping Slope as the lower slope domain (Reed et al., 1992) shows seismic reflection characteristics different to the Hengchun Ridge and can be subdivided into two parts (Figure 3.36). Within the eastern subpart develop a series of east-dipping thrusts, severe folds, and mud diapirs on the top. On the contrary, folded but continuous seismic reflections can still be identified within the western subpart of the Kaoping Slope. In addition, a group of west-dipping, rather than east-dipping, faults characterize the toe of the western subpart. This lateral variation and zoning in structural styles indicate that, from the lower slope to the upper slope, the degree of deformation increases with ages. Beneath the Kaoping Slope, the top of the subducted SCS crust is clearly visible with undulating but strong reflections.

The Hengchun Ridge appears to be from an early stage of accretion, and it is suggested that it represents the ancient accretionary prism before the Manila Trench jumped to the west of Taiwan about 2–5 Ma (Huang, Yuan, Lin, Wang, & Chang, 2000; Li, Zhou, Li, Chen, et al., 2007; Li, Zhou, Li, Hao, et al., 2007). This jump in subduction trench was contemporary to the initiation of the arc–continent collision. The construction of the Manila accretionary prism and its eastward progressive deformation indicate that the subduction of SCS has experienced multiple phases of increased subductive activity (Li, Zhou, Li, Chen, et al., 2007; Li, Zhou, Li, Hao, et al., 2007).

There is certainly a significant lateral variation in the nature of subducted materials along the strike of the Manila subduction zone (Figure 3.35), and this has strongly affected the slab geometry and deformation patterns. From the north to the south, we observe subduction of continental lithosphere near Taiwan, transitional lithosphere (heavily attenuated continental lithosphere) in the northeasternmost SCS, oceanic lithosphere, and the relict spreading ridge of the SCS. The buoyancy of the subducting transitional lithosphere to the north is probably responsible for the concave curve in the subduction zone and relatively shallow earthquakes around the North Luzon Arc (Bautista, Bautista, Oike, Wu, & Punongbayan, 2001; Yang et al., 1996).

Along the Manila Trench axis, the water depth increases gradually to the south (Figure 3.35). This can be explained by increased subduction rate and slab angle to the south and, in a lesser degree, by the decrease in sediment input from the northern SCS continental margin and Taiwan. Based on the multibeam morphotectonic analysis of the Manila Trench accretionary wedge, Li, Jin, et al. (2004) studied the indentation tectonics and oblique subduction along Manila Trench and found that the seamount subduction does not lead to the erosion of the accretionary wedge (Figure 3.37). The indentation structures provide good indicators for analyzing regional stress field and relative plate motion of convergent margin. It is found that the oblique subduction actually is a NWW-trending obduction of Luzon microplate moving along with the NWW-trending movement of the Philippine Sea Plate (Li, Jin, et al., 2004). The subduction of seamounts, found numerous in the central South China Basin and particularly along the relict spreading center, may have been closely linked to earthquake activities in the Manila subduction zone. However, earthquake activities of this subduction zone are the least understood and remain to be important future research targets.

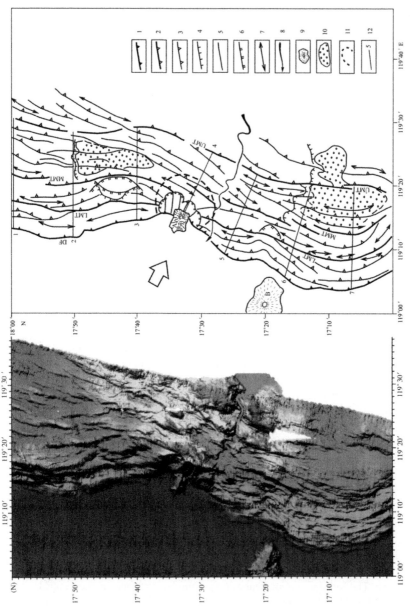

FIGURE 3.37 (Left) Multibeam shaded relief image of the accretionary wedge of middle Manila Trench. (Right) Structural interpretation of the accretionary wedge of middle Manila Trench. 1, Deformation front (DF); 2, main thrusts; 3, compressive faults; 4, extensional faults; 5, faults with unknown characters; 6, fault cliffs; 7, folds; 8, submarine canyon; 9, submarine seamounts; 10, sedimentary basins on the slope; 11, local uplift due to seamount subduction; 12, profiles and their numbers for volume statistics of the wedge. A and B are seamounts, LMT is Lower Main Thrust, MMT is Middle Main Thrust, and UMT is Upper Main Thrust (Li, Jin, & Gao, 2002; Li, Jin, et al., 2004).

From analysis of geomorphological, geochronological, geochemical, and geophysical features, Yang et al. (1996) identified a double island arc structure between Taiwan and Luzon, that is, an older western chain largely composed of Miocene to Pliocene volcanic rocks and a young active eastern chain with mostly Quaternary islets. The magmas erupted in the eastern chain were derived from more enriched mantle sources than the magmas erupted in the western chain (Yang et al., 1996).The two volcanic chains are separated by 50 km just north of Luzon (18°N) and converge near 20°N.

Yang et al. (1996) interpreted from seismological data that the double arc is caused by a slab tearing and an abrupt increase in the dip angle of the subducted Eurasian slab roughly along the COB zone (in Figure 3.35). Yang et al. (1996) further attributed the double arc to the subduction of the extinct midocean ridge of the SCS. According to their interpretation, the western chain was the volcanic front before the ridge reached the Manila Trench, but when the ridge reached the subduction zone at 5–4 Ma, its buoyancy temporarily interrupted the subduction, thus causing a time gap in magmatic activity. When subduction started again, the dip angle became shallower in response to the extra buoyancy of the downgoing ridge, leading to an eastward shift of the volcanic front (the younger eastern chain) (Yang et al., 1996). However, Bautista et al. (2001) observed noticeable inconsistencies in the aforementioned model, in particular with the spatial distribution of volcanoes and with seismicity data, and presented a new model in which the collision and subsequent partial subduction of a buoyant block (extended continental crust) in the northeasternmost SCS cause the sharp bend in the trench line and the double arc. From an observed gap in strain energy release and the abrupt change in dip from steep to shallow south of 18°N, Bautista et al. (2001) proposed that a slab tear occurred along the relict spreading center, rather than along the COB (Figure 3.38).

3.5.1.6   The Taiwan Orogen

The late Cenozoic collision between the Luzon Arc and the Eurasian continental margin marks one of the most magnificent tectonic events in the western Pacific (Figure 3.35). The oblique collision lifted Taiwan island, making it among the most active and the youngest orogens in the world. It is estimated that the Taiwan orogen started to build up several million years ago (Biq, 1972; Chai, 1972; Teng, 1990) at a relatively fast uplifting rate (>1 cm/a). The present-day northwestward movement rate of the western Philippine Sea Plate is estimated at about 7–8 cm/a (Seno et al., 1993; Yu et al., 1997). Due to the obliqueness in the convergence, incipient collision propagates southward at a rate of 5.5–12 cm/a (Davis et al., 1983; Lundberg, Reed, Liu, & Lieske, 1997; Suppe, 1984), causing the present-day active collision in southern Taiwan (Yu et al., 1997) and tectonic escape in SW Taiwan in an area corresponding to the Manila accretionary prism (Lacombe et al., 2001).

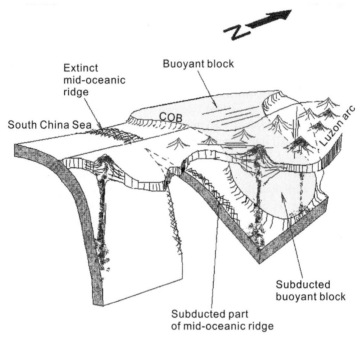

FIGURE 3.38 A proposed model of the subducted slab of the Eurasian Plate beneath Luzon (Bautista et al., 2001).

By moving southward along the convergent belt, we could slice as if progressively older evolutionary processes of the collision based on the concept of time–space equivalent (Malavieille et al., 2002; Suppe, 1984; Teng, 1990). East–west geologic sections offshore southern Taiwan could therefore help reveal tectonic scenarios back to older time when the Luzon Arc started to collide with the Eurasian continental margin (Huang et al., 1995, 1997, 2000, 2001) and therefore give significant new insights on the early orogenic process. Meanwhile, through close onshore–offshore geologic comparisons, the future evolutions of the present-day geologic units offshore southern Taiwan can be projected.

Taiwan island is now composed of distinct northeast-striking tectonic units separated by major faults (Figures 3.39 and 3.40). These units are, from the west to east, the coastal plain, Western Foothills, and Hsuehshan Range (passive Asian continental margin deposits now deformed into a fold-and-thrust belt) and western central range–Hengchun Peninsula (precollision accretionary prism), eastern central range (underthrust Eurasian continent), and the coastal range (accreted Luzon Arc–forearc). The Lishan–Laonung fault located between the western central range and the Hsuehshan Range–Western Foothills represents the boundary between the Eurasian and the Philippine Sea Plate (Huang et al., 1997). In comparison, the Longitudinal

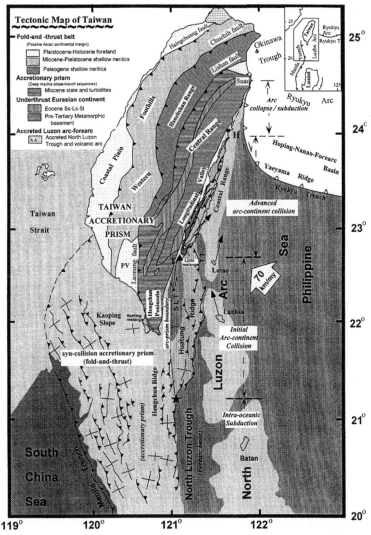

FIGURE 3.39   Tectonic framework of the Taiwan arc–continent collision illustrating the four geodynamic processes, from south to north: intraoceanic subduction, initial arc–continent collision, advanced arc–continent collision, and arc collapse/subduction. TT, Taitung Trough; SLT, Southern Longitudinal Trough. The arrows mark the declination measured in Lutao, Lanhsu, and the coastal range (Huang et al., 2001). *Data from Yang, Wang, Tsai, and Hsu (1983) and Lee, Kissel, Barrier, Laj, and Chi (1991).*

Valley between eastern central range and the coastal range marks the Luzon–Eurasia collision suture (Figures 3.39 and 3.40). The Kenting Mélange along the frontal accretionary prism in the Hengchun Peninsula and the Lichi Mélange in the 200 km coastal range east of the Longitudinal Valley stand for part

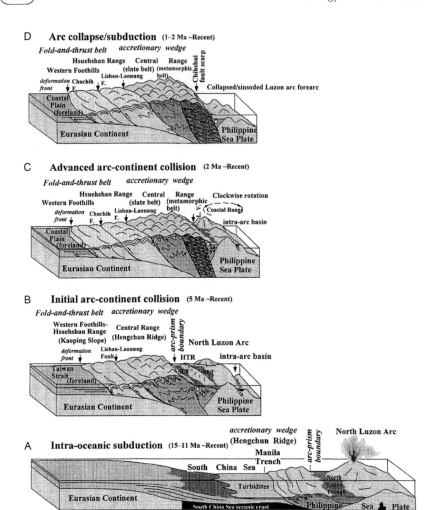

FIGURE 3.40 Evolution of Taiwan orogeny and related geodynamic processes. (A) Intraoceanic subduction; (B) initial arc–continent collision; (C) advanced arc–continent collision; and (D) arc collapse/subduction (Huang et al., 2001).

of the Manila subduction and the Luzon–Eurasia collision complex, respectively (Figures 3.39 and 3.40) (Huang et al., 2001).

In western coastal plain of Taiwan, shallow marine sequences of the northern SCS are uplifted and exposed, and they provide unique opportunities to studying the rifting, sedimentation, and paleoclimate and paleoceanography of the SCS (Huang et al., 2012). Eocene sequences in western Taiwan are all deposited in rifting basins and are covered unconformably by the late Oligocene–Neogene postrifting strata. The breakup unconformity between synrifting and postrifting sequences exists throughout the island and represents a depositional hiatus from the early Oligocene to late Eocene, equivalent to an unconformity in the Pearl River Mouth Basin (Huang et al., 2012). If this breakup unconformity was caused by the initial seafloor spreading in the SCS, then it may suggest that the initial opening was between 33 and 39 Ma in the northeast (Huang et al., 2012). Surprisingly, neither obvious stratigraphic gap nor slumping features are found in the Oligocene–Miocene transition in Taiwan, contrasting highly with what has been documented from the deepwater ODP Site 1148 (Huang et al., 2012). This might suggest that the significant event in the Oligocene–Miocene transition was not recorded everywhere in the China Seas.

To the northeast of Taiwan extends the NE-trending Ryukyu Trench–Ryukyu arc–Okinawa Trough system (Figure 3.1), where the Philippine Sea Plate is subducting in a NW direction along the Ryukyu Trench and the Okinawa Trough is undergoing active back-arc rifting (Kizaki, 1986; Miki, 1995; Shinjo, 1999; Sibuet & Hsu, 1997; Wang et al., 1999). Although this active subduction and back-arc rifting process has started only recently since the late Cenozoic (Kizaki, 1986; Lee & Lawver, 1995) owing to the northward and clockwise movement of the Philippine Sea Plate (Hall, 2002; Hall et al., 1995), it is generally argued from the distribution of Cathaysian magmatism that the Mesozoic Paleo-Pacific subduction might have also followed roughly along the present-day Ryukyu Trench and northern SCS continental margin.

The Philippine Sea Plate, bounded by active subduction zones, has itself a long and complicated multiphase history (Hall, 2002; Hall et al., 1995). The Huatung Basin located east of Taiwan was proposed as a trapped fragment of the Cretaceous "proto-SCS" or the "New Guinea Basin" within the Philippine Sea Plate (Deschamps et al., 2000). The Gagua Ridge between the Huatung Basin and the Philippine Sea Plate acted as a transverse plate boundary accommodating the northwest shear motion of the Philippine Sea Plate (Deschamps, Lallemand, & Collot, 1998; Deschamps et al., 2000; Sibuet & Hsu, 1997; Sibuet et al., 2002).

3.5.2   Okinawa Trough

The Okinawa Trough is a back-arc rifting basin behind the Ryukyu arc, triggered by the northwest subduction of the Philippine Sea Plate. Its maximum water depth approaches 2300 m in the south and progressively decreases to 200 m in the north (Figure 3.12). Refraction data reveal that the crustal thickness increases from 18 in the southern Okinawa Trough to 30 km in the northern Okinawa Trough (Hirata et al., 1991; Sibuet, Hsu, Shyu, & Liu, 1995). Seismic reflections reveal typical graben structures in the trough (Figures 3.41 and 3.42), and en echelon volcanic ridges along the central axis of the trough (Gungor et al., 2012; Park et al., 1998; Sibuet et al., 1987).

The Okinawa Trough is structurally segmented along its strike. Different segments may have different timing and mechanisms of rifting. The middle and

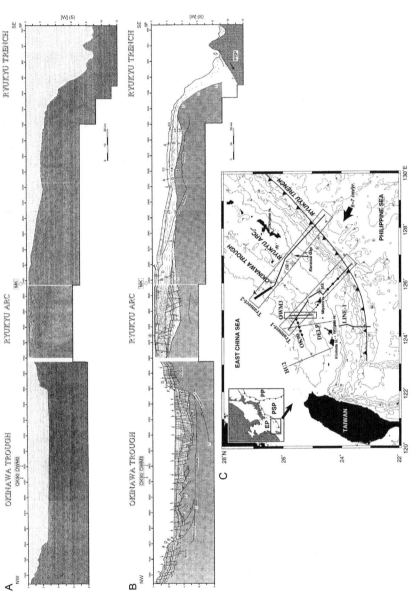

FIGURE 3.41 (A) Seismic profile of Transect-1. (B). Interpretation of seismic profile of Transect-1(a). Numbers within each seismic unit in the interpretation indicate P-wave velocity determined from OBS refraction experiment. Vertical exaggeration is about 10:1 for (A) and (B). (C) Bathymetry and physiographic features of the southern Ryukyu Island arc system and the position of the multichannel seismic profile Transect-1. Black dots indicate the positions of the OBS. Data are shown by double-circled symbols. MK and OK-1 are two offshore petroleum exploration wells. Unit and interval of water depth contours are 1 km and 1 km, respectively. EP, Eurasia Plate; PSP, Philippine Sea Plate; PP, Pacific Plate. *After Park et al. (1998).*

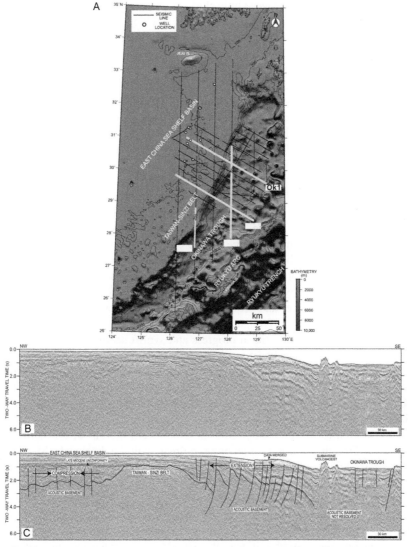

FIGURE 3.42   (A) Locations of seismic reflection data (thin and bold lines) and well locations (white dots) in the northern ECS Basin and in the western margin of the northern Okinawa Trough. Contours are bathymetry in meters. (B) Seismic reflection profile (Ok1) and (C) its interpretation showing compression. Extension in the western margin of the northern Okinawa Trough is characterized by deeply rooted, NW-dipping growth faults. The late Miocene unconformity that eroded the Diaoyudao Uplift (Taiwan–Sinzi belt) forms a conformable surface in western margin of the northern Okinawa Trough. Vertical exaggeration is approximately 10×. *After Gungor et al. (2012).*

northern Ryukyu arc experienced little clockwise rotation. The middle and northern Okinawa Trough may be related to the opening of the Japan Sea and associated with a major dextral strike-slip fault along the east side of the Diaoyudao Uplift (the Taiwan–Sinzi belt; Le Pichon & Mazzotti, 1997). The exact timing of initial opening is still controversial and most consider it to be the late Miocene, between 10 and 6 Ma (Gungor et al., 2012; Lee, Shor, Bibee, Lu, & Hilde, 1980; Letouzey & Kimura, 1986; Miki, 1995; Park et al., 1998; Sibuet et al., 1987). The key observation is a strong Miocene erosional event, marked by a sharp angular unconformity in the Okinawa Trough. The presence of the late Miocene–early Pliocene horizon ($\sim$5 Ma) and possibly older synrift strata in the trough indicates that the onset of the rifting is older than 5 Ma (Gungor et al., 2012).

The southern part of the trough opened later, in the early Pleistocene, by the clockwise rotation of the southern Ryukyu arc (Figure 3.43) (Miki, 1995; Park et al., 1998; Shinjo, 1999). From seismic data in the southern part, Park et al. (1998) identified three stages of tectonic evolution: (1) During stage 1 from the late Miocene to earliest Pleistocene, prerift deposits accumulated over a wide region from the ECS continental shelf to the forearc region; (2) stage 2 is defined by a series of tectonic processes involving crustal doming, erosion, subsidence, and sedimentation, in association with initial rifting of the southern Okinawa Trough during most of early Pleistocene time; and (3) the back-arc rifting is still in progress and synrift sedimentation has been under way since the late Pleistocene (stage 3).

The prominent late Miocene unconformity ($\sim$5 Ma) marks a significant change in the ECS. This event caused tectonic inversion and erosion in the Xihu Depression (Figures 3.13 and 3.41) and truncated the Diaoyudao Uplift (Gungor et al., 2012; Li, Chen, & Zhou, 2009; Li, Zhou, Ge, & Mao, 2007). In contrast, sedimentary strata in the western margin of the northern Okinawa Trough are cut by numerous normal faults and the late Miocene horizon forms a conformable surface (Gungor et al., 2012). Based on these observations, Gungor et al. (2012) suggested that the compressional tectonic movement that affected the ECS Basin and the Diaoyudao Uplift did not propagate into the Okinawa Trough, and the Diaoyudao Uplift acted as a buttress against the tectonic forces. However, we argue exactly the opposite; based on our own studies, it was the late Miocene extension and mantle upwelling in the Okinawa Trough that triggered the compression in the ECS Basin.

3.6 MAGMATISMS AND NEOTECTONICS

Magmatisms occur at all major stages of evolution of the China Seas. Being located at Eurasian continental margin controlled largely by regional faults, China Seas are also places of dominant neotectonics. We give a brief review of the widely spread magmatisms and neotectonics including earthquakes.

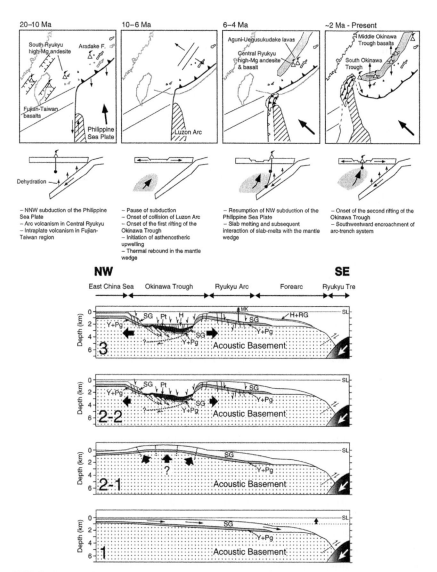

FIGURE 3.43   (Upper) Proposed evolutional model of the Okinawa Trough (Shinjo, 1999). The profiles represent sections across the middle Okinawa Trough–central Ryukyus. The vertical dimension in all sketches has been exaggerated to show details of the mantle wedge. (Lower) Tectonic evolutionary stages in the southern Ryukyu arc system since the Neogene (Park et al., 1998). (1) Stage 1 (late Miocene to earliest Pleistocene): prerift sedimentation. (2–1 and 2–2) Stage 2 (early Pleistocene): initial back-arc rifting. (3) Stage 3 (late Pleistocene to Holocene): the back-arc rifting still in progress. The geologic age of acoustic basement is pre-Cenozoic. Pg, Paleogene deposits; Y, Yaeyama Group; SG, Shimajiri Group; Pt, early Pleistocene deposits; RG, Ryukyu Group; H, Holocene sediments; SL, sea level.

3.6.1 Regional Magmatisms Revealed by Magnetic Data

Magnetic anomalies reflect primarily distribution of igneous rocks due to their much stronger susceptibility than other types of rocks. One simple magnetic attribute called 3D analytic signal amplitude (ASA) can quantify very effectively the localities and magnitudes of magmatism (Li, Zhou, Li, Chen, et al., 2008; Zhang & Li, 2011). 3D ASAs are simply defined as (Nabighian, 1984; Ofoegbu & Mohan, 1990; Roest, Verhoef, & Pilkington, 1992)

$$|A(x, y)| = \sqrt{(\partial M/\partial x)^2 + (\partial M/\partial y)^2 + (\partial M/\partial z)^2}, \qquad (3.1)$$

where M is the magnetic field anomaly. Essentially equivalent to the total magnetic gradient, ASA are independent of inclinations and declinations of source magnetizations and the Earth's magnetic field if the magnetic contacts are nearly vertical (Agarwal & Shaw, 1996; Li, 2006; Nabighian, 1972; Salem, Ravat, Gamey, & Ushijima, 2002). Unlike original magnetic anomalies, ASA are always positive, with peaks tending to be located directly above the magnetic sources and/or their contacts. ASA are advantageous over reduction to the pole in that they work extremely well at very low latitudes and, by utilizing first-order derivatives, lead to better characterization of magnetic (and thereby magmatic) boundaries at higher resolutions (Li, Zhou, Li, Chen, et al., 2008; Zhang & Li, 2011). This makes ASA well suited for studying magmatism and allows more straightforward geologic interpretation.

Figure 3.44 shows a map of ASA with regional fault interpretations around the northern China Seas (Li, Wang, et al., 2012; Zhang & Li, 2011). The Mesozoic suture zone, the Qinling–Dabie–Sulu orogenic belt, shows high magnetic signals that are mostly likely associated with postorogenic magmatism. The western part of the South Yellow Sea and part of the North Yellow Sea appear to be magmatically quiet. In contrast, strong magnetic anomalies offshore west of the Korean Peninsula have quite small wavelengths, suggesting younger and more active magmatism, instead of Mesozoic magmatic activities (Li, Chen, et al., 2009) (Figure 3.9). One representative of this recent magmatic activity is the formation of the young megavolcano, the Jeju Island. A large contrast in magnetic anomalies across the Tan-Lu fault is also observed, which could only be explained by significant differential magmatisms (Figure 3.9). The Bohai Basin shows similar patterns in magmatism to other parts of the North China Craton west of the Tan-Lu fault, in which magmatic sources appear to be deep-seated. The ECS shows similar magnetic strengths as in the Cathaysian block, and their strong magnetic anomalies are interpreted to be caused mostly by late Mesozoic Yanshanian event. Therefore, the Mesozoic basement of the ECS was more strongly transformed by magmatism than that of the Yellow Sea. The high magnetic anomaly associated with the Diaoyudao Uplift extends all the way from offshore northeastern Taiwan to the Japan Sea and suggests that the uplift might be a buried volcanic arc, which was active during the rifting of the Xihu Depression of the ECS Basin (Li, Chen, et al., 2009; Li, Zhou, et al., 2009).

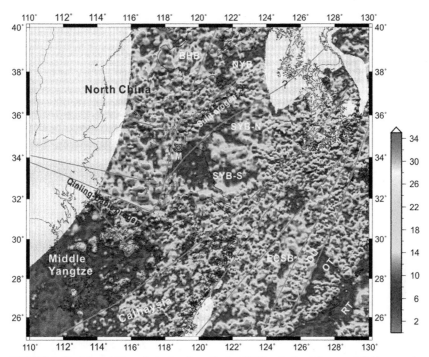

FIGURE 3.44   Map of 3D analytic signal amplitudes (ASA) calculated from total field magnetic anomalies (Li, Wang, et al., 2012). Overlapped on the map are topographic contours of 100 and 0 m, as well as major rivers and lakes. M indicates the position of the Main Hole of the Chinese Continental Scientific Drilling (CCSD) (e.g., QS Liu et al., 2009; Liu et al., 2010). DU, Diaoyudao Uplift; OT, Okinawa Trough; RT, Ryukyu Trench.

Zhang and Li (2011) estimated depth distributions of magmatic sources from ASA (Figure 3.45). The Bohai area shows high ASAs and large lateral extensions, but the detected magmatic sources are mostly deeply buried at depths up to 10 km or deeper. This indicates strong magmatic activities in the lower to middle crust of the craton from tectonic reactivation. Due to tectonic uplifting and magmatic intrusions, the Sulu–Dabie orogen has mostly shallowly buried magmatic bodies with depths smaller than 5 km. The western part of the South Yellow Sea shows predominantly deeply buried magmatic sources, but unlike the Bohai Sea, the magmatic activities in the Yellow Sea appear to be very weak and the calculated ASAs there are very small. The Cathaysian block and its offshore extension northeastward are dominated by relatively isolated and shallow magmatic bodies (mostly <5 km) that are mainly caused by the subduction of the Paleo-Pacific Plate in the late Mesozoic. Depths to magmatic sources in the ECS correlate well with basement uplifts and depressions caused by Cenozoic continental margin rifting. The Ryukyu Archipelago corresponds with a very quiet magnetic zone and shows deeply buried magmatic sources, and these observations support that the Ryukyu Archipelago itself is not the volcanic arc from the subduction

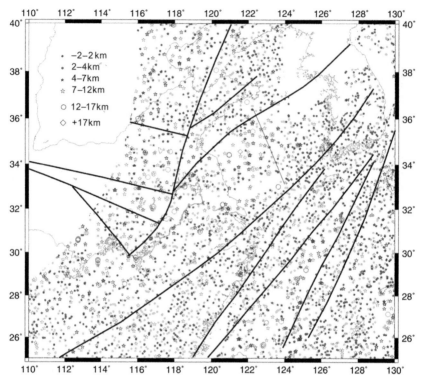

FIGURE 3.45 Depths to the top of magnetic sources estimated from 3D analytic signal amplitudes (ASA) (Zhang & Li, 2011). See Figure 3.9 for further annotations. Bold lines are regional faults. The thin line is the position of a seismic line.

of the Philippine Sea Plate but is composed of accretionary materials and/or continental blocks separated from the Eurasian Plate by opening of the Okinawa Trough. The true volcanic arc is represented by a volcanic belt slightly to the west of the Ryukyu Archipelago.

In Southeast China and the SCS, ASA increase eastward from the inland to the coast (Figure 3.8) (Shi & Li, 2012). ASA are particularly large in the zone with widespread early Cretaceous I-type and late Cretaceous A-type granites and rhyolites (Qiu, Shen, & Wang, 2002; Shu & Xu, 2002). There seems to be an eastward increase in rock susceptibilities in the Cathaysia that may have a genetic link to regional crustal and magmatic evolution (Shi & Li, 2012). Magnetic susceptibility of igneous rocks can be a proxy for geochemical composition, rock type, and tectonic setting (Aydin, Ferré, & Aslan, 2007; Ellwood & Wenner, 1981; Gregorová, Hrouda, & Kohút, 2003; Ishihara, Robb, Anhaeusser, & Imai, 2002). S (sedimental)-type volcanic–intrusive complexes found mainly in the inland area are expected to correspond with low susceptibilities, because an S type is equated with ilmenite-bearing granitoids that have their origins within thick continental crust and bear limited mantle sources and ferromagnetic minerals (Chappell & White, 1974; Gregorová et al., 2003; Ishihara, 1977; Takahashi, Aramaki, & Ishihara, 1980). An I (igneous)-type is often equated with magnetite-series granitoids and volcanic rocks (Gregorová et al., 2003; Ishihara, 1977; Takahashi et al., 1980), and the slightly higher ASA associated with I-type volcanic–intrusive complexes can therefore be reasonably explained by their higher magnetite contents due to stronger crust–mantle interactions (Figure 3.8). The A (anorogenic)-type granites and rhyolites distributed roughly along the coast may correspond to the largest magnetic susceptibilities in the Cathaysia, suggesting potentially high portions of mantle ingredients and local lithospheric thinning and asthenospheric upwelling (Hong, Guo, Li, Kang, & Xu, 1987; Hong, Wang, Han, & Jin, 1996; Jiang, 1991). It is known that A-type suites are always high in $Fe/(Fe+Mg)$ and contain Fe-rich silicates such as annite and fayalite (Clemens, Holloway, & White, 1986), which can contribute to high magnetic susceptibilities.

ASA could also reflect the magnitudes of magmatism. This appears to be the case based on regional geologic surveys that found progressively eastward younging and intensification in volcanism (FJBGMR, 1985; GDBGMR, 1988; GXBGMR, 1985; ZBGMR, 1989; Zhou & Li, 2000). Therefore, overall eastward increase in ASA reflects a combined effect of both changes in rock types and magmatic intensities (Shi & Li, 2012).

3.6.2   Magmatic Episodes

The most dramatic magmatism in East China occurred after the middle Triassic orogeny, caused by postorogen and postexhumation extension and the subduction of the Paleo-Pacific Plate. The climax of this magmatic event, traditionally known as the Yanshanian event, lasted from the late Jurassic to the early Cretaceous ($\sim$160 to 100 Ma) and was particularly evident within the Cathaysian block to the east of the Jiangshao Fault and in a small area east of the Dabie orogen between the middle and the lower Yangtze (e.g., Xu et al., 2003) (Figure 3.46). Because of the stronger late Mesozoic magnetism, the Cathaysia shows much higher magnetic anomalies than the Yangtze to the west (Figures 3.8 and 3.15). The most intense continental magmatism occurred along the entire coastal area of Southeast China in the late Jurassic, forming a massive continental arc (Figure 3.46). Most igneous rocks are silicic to intermediate, but sporadic mafic to ultramafic rocks can be found inland. The pyroclastic rocks can reach well over 1000 m in thickness and are distributed preferably along major fault zones (Shi & Li, 2012). In the early Cretaceous, I-type volcanic–intrusive complexes developed mainly along the coastal region. Strong volcanism and both mafic and silicic plutonism accompanied fault activities. Basalt and rhyolite interbedded with paralic clastics are found from drillings on the two conjugate margins of the SCS (Shi & Li, 2012). This suggests a transient early Cretaceous episode of magmatism around the SCS, reflecting potentially the continued eastward migration of magmatism (e.g., Charvet et al., 1994).

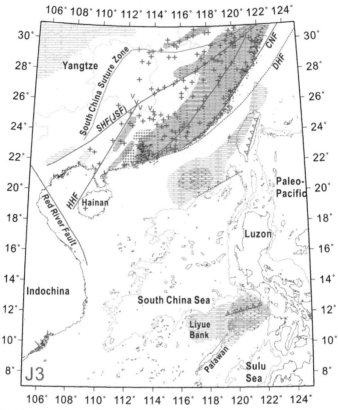

FIGURE 3.46 Late Jurassic distribution of sediments, tectonic pattern, and magmatism (Shi & Li, 2012). The Yanshanian event is marked by the development of a large continental arc in Southeast China continental margin. *Notes*: See Figure 3.3 for legend and annotations. SHF (JSF), Shihang Fault (Jiangshan-Shaoxing Fault); ZLF, Zhenghe–Lianhuashan Fault; CNF, Changle-Nan'ao Fault; DHF, Donghai Fault; HHF, Hepu–Hetai Fault.

In the late Cretaceous, the coastal fault spatially controlled the distribution of the A-type granites and rhyolites (Qiu et al., 2002).

Yan et al. (2006) summarized that the Cenozoic magmatism over the Pearl River Mouth Basin can be divided into three stages, Paleocene–Eocene (pre-spreading), Oligocene–middle Miocene (synspreading), and late Miocene–Quaternary (postspreading). The first stage produced intermediate–acidic extrusives, such as andesite, dacite, rhyolite, and tuff, with K–Ar ages of 57–49 Ma, which formed domes in the basin. The prespreading magmatism predominantly occurs on the northern margin of the SCS and in South China coastal areas and shows a bimodal affinity (Xu et al., 2012). This bimodal magmatism may have resulted from double-layered convection in a magma chamber under an extensional environment associated with lithosphere thinning (Chung, Cheng, Jahn, O'Reilly, & Zhu, 1997; Chung, Sun, Tu, Chen, & Lee, 1994).

The second-stage synspreading magmatism occurred along fissures or fault intersections within extensionally faulted depressions. Volcanism contemporaneous with seafloor spreading in the SCS was very weak in the margins and adjacent areas (Yan et al., 2006). Cored rocks were basalt and intermediate eruptives. Li and Song (2012) argued that the Oligocene–Miocene transition (~23 Ma) was accompanied with strong magmatism in the region and this event was recorded by a magnetic anomaly (C8, Figure 3.31) in the SCS. This stage of magmatism is also observed in the Xihu Depression, with magmatic sills emplaced mostly along the Oligocene–Miocene boundary (Su et al., 2010) (Figure 3.47). This stage of magmatism is also concurrent to strong faulting and erosion in the depression. From Figure 3.47 one can observe that faults played an essential role in the emplacement of volcanic structures, as some of the faults clearly developed

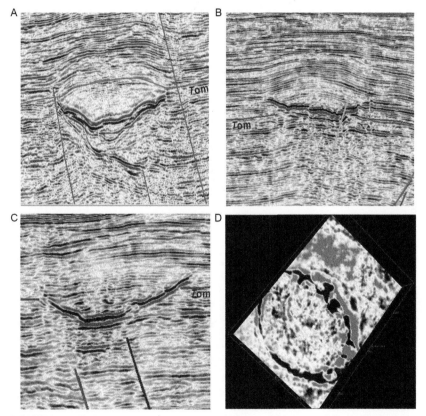

FIGURE 3.47 Widespread volcanism during the Oligocene–Miocene transition (~23 Ma) in the Xihu Depression, ECS. (A–C) Are vertical seismic sections, and (D) is a time slice of a 3D seismic volume. The strong reflections are caused by igneous complexes. Green horizon Tom, Oligocene–Miocene unconformity. *From Su et al. (2010).*

only beneath the emplacements and acted as the magma conduits. In a map view, these structures often appear as circular domes.

Postspreading magmatism is strong along the lower slope of the northern margin of the SCS and in the central basin (Lüdmann & Wong, 1999; Lüdmann et al., 2001; Yan, Zhou, & Liu, 2001). They can be easily identified from geophysical data (Figure 3.48) and high-resolution bathymetric map

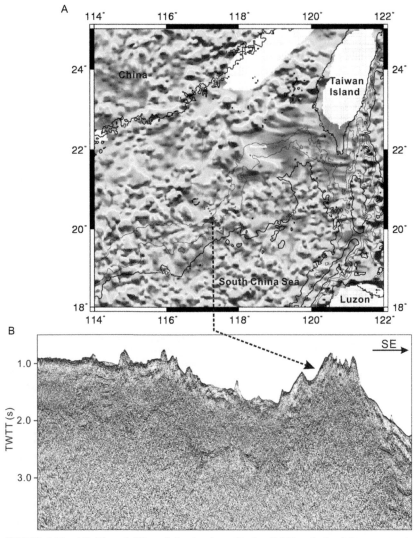

FIGURE 3.48 (A) Map of 3D analytic signal amplitudes (ASA) calculated from magnetic anomalies in the northern continental slope of the South China Sea. ASA help pinpoint exact locations of igneous bodies. The dashed line shows the location of the seismic section in (B). (B) The seismic section showing extensive magmatic extrusions. *Panel (A) from Shi and Li (2012).*

(Figure 3.49). Most known major seamounts in the central SCS basin are believed to be emplaced after the seafloor spreading stopped at about 15 Ma. K–Ar and Ar–Ar ages of dredged samples from seamounts are all younger than 15 Ma (Han, 2011; Kudrass et al., 1986; Wang, Wu, Liang, & Yin, 1985; Yan, Shi, Wang, Bu, & Xiao, 2008), the estimated cessation age

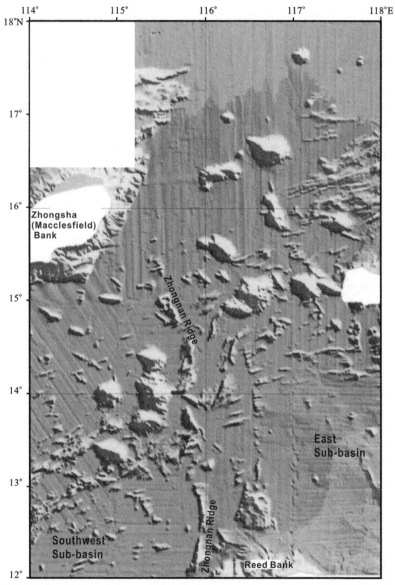

FIGURE 3.49 High-resolution multibeam bathymetric map of the South China Sea showing many seamounts from postspreading magmatism (Li, Jin, & Gao, 2002; Li, Ding, et al., 2011).

of seafloor spreading in the SCS. Most of these seamounts are along, or to the north of, the relict spreading center (Figures 3.27 and 3.49). These are mainly alkali basalts with subordinate tholeiites and display OIB-type geochemical characteristics (Xu et al., 2012). Li et al. (2010) found that these recent volcanic activities have disturbed the geothermal field and therefore heat flow measurements cannot be used to infer the oceanic crustal ages here. The exact mechanism of the widespread postspreading magmatism is still under extensive investigation.

Postspreading magmatism is also widespread in Hainan Island, Leizhou Peninsula, Thailand, and Vietnam (Figure 3.50). These are mainly alkali basalts with subordinate tholeiites and display OIB-type geochemical characteristics (Flower et al., 1992; Tu et al., 1992). Geochemical data reflect endogenous mantle processes related to disaggregation of the South China margin rather than a northward extension of the Southern Hemisphere Dupal isotope anomaly (Tu et al., 1992). The Dupal isotope anomaly and presence of high-magnesium olivine phenocrysts suggest their possible derivation from

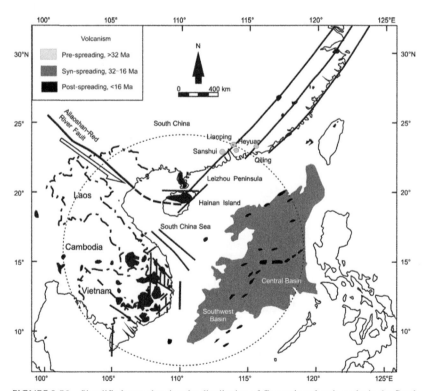

FIGURE 3.50 Simplified map showing the distribution of Cenozoic volcanic rocks in the South China Sea and its surroundings (Chung et al., 1997; Lee et al., 1998; Xu et al., 2002, 2012; Yan et al., 2006). The dashed circle outlines the area affected by the possible "Hainan plume" (Yan and Shi, 2007).

a mantle plume in Hainan (Xu et al., 2012). Xu et al. (2012) further hypothesized that the temporal and spatial distribution of these Cenozoic volcanisms may be accounted for either by stress reorganization before and after SCS spreading or ridge suction of plume flow during opening of the SCS, and the volcanic rocks within the SCS basin may not be typical mid-ocean ridge basalts (MORB). Whether the mantle plume exists in Hainan and how it may affect the opening of the SCS are still open questions.

In the northern South Yellow Sea Basin, Lee, Kwon, Yoon, Kim, and Yoo (2006) identified various igneous and related features such as stocks, laccoliths, sills, dikes, volcanic edifices, and hydrothermal vent systems, which are closely linked to basement faulting (Figures 3.51 and 3.52). On multichannel seismic reflection data, the stocks are characterized by a seismically dead zone with upturned host rocks and uplifted overburden; the sills are imaged as concordant, high-amplitude reflections with a distinct lateral extent; the dikes are characterized by steeply inclined, crosscutting reflections; the hydrothermal vent systems appear as a seismic chimney (Lee et al., 2006). These types of igneous structures can also be easily found in other parts of the China Seas and provide a unique opportunity to address their implications for hydrocarbon systems in the rift basin.

Further east, the Okinawa Trough is bounded by two major magmatic belts, the Diaoyudao Uplift to the west and the Ryukyu arc to the east. Both the Diaoyudao Uplift and the Ryukyu arc reveal themselves on the regional magnetic anomaly map as strong magnetic anomaly belts (Li, Chen, et al., 2009), but the latter is much narrower (Figures 3.9 and 3.36). We interpret the Diaoyudao Uplift as the early Cenozoic volcanic arc associated with the back-arc opening of the Xihu Depression of the ECS Basin. Today, this relict arc still shows slightly uplifted Curie-point depths (Li, Chen, et al., 2009). It probably experienced multiple phases of volcanisms, but later in the late Miocene, strong uplifting and denudation occurred, causing its flat top before rapid subsiding and burial afterward (Gungor et al., 2012). The Ryukyu arc is the active volcanic arc from the subduction of the Philippine Sea Plate and Curie points there are much shallower than surrounding areas (Li, Chen, et al., 2009). There are also localized but very high (>300 nT) magnetic anomalies within the Okinawa Trough (Figures 3.9 and 3.36), which may be related to volcanic ridges (Hirata et al., 1991) or dike intrusions or emplacement of early oceanic crust (Davagnier, Marsset, Sibuet, Letouzey, & Foucher, 1987). In the southernmost Okinawa Trough, a cluster of active submarine volcanoes, consisting of pre-back-arc rifting volcanic rocks, forms a belt (Chung, Wang, Shinjo, Lee, & Chen, 2000), and submarine volcanoes are also identified in the northern Okinawa Trough (Figure 3.34) (Gungor et al., 2012).

Of course, the most active and devastating volcanic activities are caused by the subduction of oceanic lithosphere. The two active volcanic arcs in our study area are the Ryukyu arc and the Luzon Arc. In southern Manila Trench, the subducted slab is the steepest and the subduction rate appears to

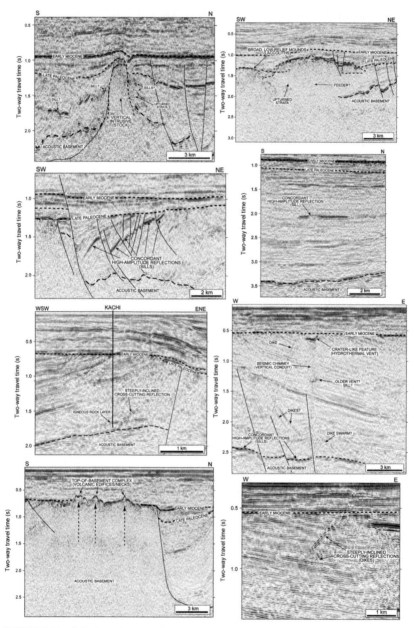

FIGURE 3.51 Seismic sections showing interpretations of various structures and traps associated with the igneous complexes observed in the northern South Yellow Sea Basin (Lee et al., 2006).

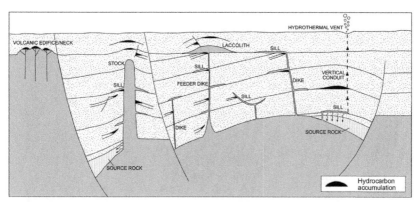

FIGURE 3.52 Schematic cross section illustrating various structures and traps associated with the igneous complexes observed in the northern South Yellow Sea Basin (Lee et al., 2006).

be the fastest of all Manila subduction zone (Figure 3.35). In addition, the Benham Plateau in the Philippine Sea Plate is moving against the Luzon, causing indentation tectonics that further pushes the southern Luzon westward (Figure 3.35). These joint forces caused a very active chain of volcanoes in Luzon, among which is the active stratovolcano Mount Pinatubo. Its recent eruption on 15 June 1991 produced one of the few largest terrestrial eruptions of the twentieth century.

3.6.3 Regional Seismicity

The eastern Eurasian continental margin is hallmarked by a series of subduction zones, which trigger most of the deep and large earthquakes with magnitudes larger than 5.0 (Figure 3.53). The largest interplate earthquakes triggering the most devastating tsunamis of this century (the 2004 M_w 9.1–9.3 Sumatra–Andaman earthquake and the 2011 M_w 9.0 Japan earthquake) all occurred along the subductions in Southeast Asia.

Essentially, no large earthquakes ($M > 5$) occurred in East China continental shelf and in the central SCS basin from early 2002 to late 2008 (Figure 3.53). When we look at a longer time window, however, large earthquakes do occur in these areas. Figure 3.54 shows all earthquakes of magnitudes larger than 3.0 compiled from 01 January 1970 to 28 February 2008 by China Earthquake Networks Center (http://www.csndmc.ac.cn/). Strong seismicity has been observed in the Bohai Basin, the Northern Yellow Sea Basin, and the Southern Yellow Sea Basin. Although numerous faults are identified from geophysical data in these regions (Figure 3.54), the occurrence of earthquakes does not appear to cluster along these faults, indicating their weak activity of these faults at the moment. Seismicity in the ECS Basin is rather weak, contrasting to other parts of China Seas shown in Figure 3.54. Deep earthquakes occur only at the subduction zones and the intraplate

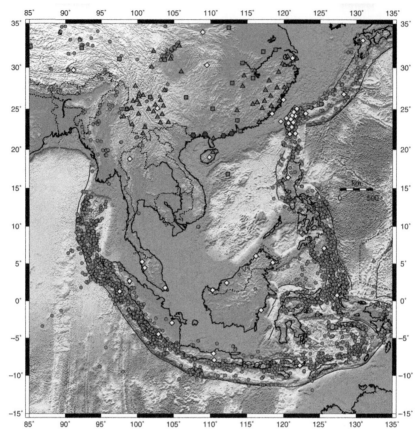

FIGURE 3.53 Seismic stations and earthquakes (gray dots) of magnitudes larger than 5.0 recorded from early 2002 to late 2008 in Southeast Asia (provided by Mei Xue at Tongji University). squares and triangles and diamonds are seismic stations. *Data from IRIS (Incorporated Research Institutions for Seismology).*

earthquakes occurred in northern China Seas are almost exclusively confined within the crust, with hypocenter depths no larger than 35 km (Figure 3.54). The largest offshore intraplate earthquakes are located in the Southern Yellow Sea Basin. For example, an $M=6.1$ earthquake (09 November 1996) occurred in an area very close to Shanghai and was felt in the coastal areas.

Li, Wang, et al. (2012) showed two long geophysical profiles with projected earthquake hypocenters ($M \geq 3.0$) that are within 2° range of the two profiles (Figure 3.54), in order to examine the relationships between earthquakes and the Curie isotherm and Moho. They found that almost all intraplate earthquakes in East China are confined within the magnetic layer, which almost coincides with the crust. Most large earthquakes with magnitudes over 4.0 tend to occur in the middle crust where the temperatures are around 300 °C (Figure 3.54)

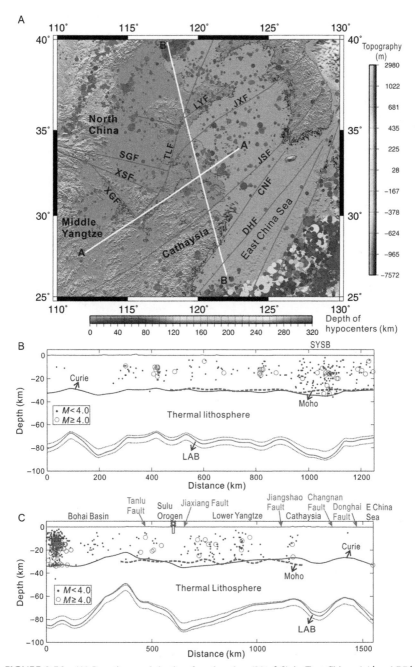

FIGURE 3.54 (A) Locations and depths of earthquakes ($M \geq 3.0$) in East China. AA′ and BB′ are the two geophysical profiles. TLF, Tan-Lu fault; LYF, Lianyan fault; JXF, Jiaxiang Fault; JSF, Jiangshao Fault; CNF, Changnan fault; DGH, Donghai fault; SGF, Shougeng fault; XSF, Xinshu fault; XGF, Xiangguang fault. (B) The integrated geologic model AA′. (C) The integrated geologic model BB′ shown in these two profiles are the bathymetry, estimated Moho and Curie-point depths, projected earthquake hypocenters ($M \geq 3.0$) within a 2° interval along the transects, and the thermal lithosphere–asthenosphere boundary (LAB). Dashed thin lines are the upper and lower bounds of the LAB (Li, Wang, et al., 2012).

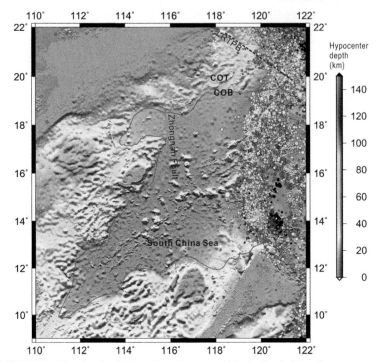

FIGURE 3.55   Map showing 7801 earthquakes occurred in the South China Sea area.

(Li, Wang, et al., 2012). The middle crust, normally considered as the transition zone between brittle upper crust and ductile lower crust, tends to have the largest strain accumulation and release in East China.

A total of 7801 earthquakes occurred in the SCS area is shown in Figure 3.55. Like what we have already seen from Figure 3.53, most earthquakes here are associated with the Manila subduction zone. Among these earthquakes, the deepest earthquake was 323.6 km beneath the surface and the largest one had a magnitude of 7.0. Most of the deep earthquakes occurred in the southern segment of the Manila Trench, confined to the south of the relict spreading center. This part of the trench is also the deepest, indicating fastest subduction rates. We confirm that the subducting SCS slab is torn apart along the relict spreading center, with the southern part subducting in a higher angle.

3.6.4   Hydrothermal Activities

The Okinawa Trough is an active back-arc spreading basin in a continental margin setting with strong volcanisms and high heat flow. Unsurprisingly, hydrothermal systems are widespread in the Okinawa Trough. The middle

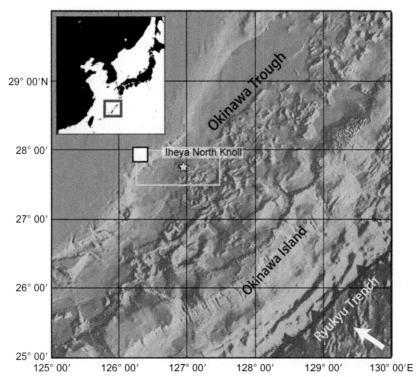

FIGURE 3.56   Location map and bathymetry of the Iheya North Knoll area of the Okinawa Trough (Tsuji et al., 2012). The yellow star (gray star in print version) shows the location of the hydrothermal field.

Okinawa Trough is the area where hydrothermal fields were first discovered in the ECS in 1984 and 1986 (Kimura et al., 1988; Uyeda, 1987). The heat flow within and around this area is anomalously high, implying unusually active hydrothermal flow and mineral deposition (Ludwig & Valencia, 1993).

Integrated Ocean Drilling Program (IODP) Expedition 331 provided an opportunity to drill into an active hydrothermal system and associated deposits at the Iheya North Knoll of the Okinawa Trough (Expedition 331 Scientists, 2010) (Figure 3.56). This expedition documented processes including formation of brines and vapor-rich fluids by phase separation and segregation, uptake of Mg and Na by alteration minerals in exchange for Ca, leaching of K at high temperature and uptake at low temperature, anhydrite precipitation, potential microbial oxidation of organic matter and anaerobic oxidation of methane utilizing sulfate, and methanogenesis (Takai, Mottl, Nielsen, & the IODP Expedition 331 Scientists, 2012). Strongly linked with the geologic setting and the thick terrigenous sediments, the chemistry of hydrothermal fluids collected from active sulfide chimneys in the Okinawa Trough is characterized by higher concentrations of CO_2, CH4, NH4, I, and K and higher alkalinity than those in typical sediment-free mid-ocean ridge hydrothermal fluids (Gamo, Sakai, Kim, Shitashima, & Ishibashi, 1991; Kawagucci et al., 2011; Konno et al., 2006; Sakai, Gamo, Kim, Shitashima, et al., 1990; Sakai, Gamo, Kim, Tsutsumi, et al., 1990 Takai & Nakamura, 2010). The relatively shallow water depth of many Okinawa Trough hydrothermal systems serves to induce subcritical phase separation (Suzuki et al., 2008) and subsequent phase segregation, as the boiling temperature of seawater decreases steeply with decreasing pressure at $\sim$100 bar (Takai et al., 2012). Phase separation and segregation sometimes produce hydrothermal fluids of quite different chemical composition at different vent sites in the same hydrothermal field, even though they are derived from the same source fluid (Kawagucci et al., 2011).

Tsuji et al. (2012) acquired seismic reflection data around the Iheya North Knoll hydrothermal in the mid-Okinawa Trough field and identified widespread porous volcaniclastic pumiceous deposits and intrusions from silicic arc volcanism (Figure 3.57). They found that the structural culmination around the knoll of the highly permeable volcaniclastic sequences that integrate multiple flow paths focuses fluid flow in the region of the hydrothermal vents, and the high concentrations of CH4 and NH4 in the fluids are likely derived from the interaction of migrating fluids with trough-fill sediments.

Carbon dioxide-rich fluids derived from hydrothermal activities associated with magmatic sources were observed to emerge from the seafloor in the hydrothermal fields, mid-Okinawa Trough (Inagaki et al., 2006; Sakai, Gamo, Kim, Shitashima, et al., 1990; Sakai, Gamo, Kim, Tsutsumi, et al., 1990). Upon contact with seawater at 3.8 °C, gas hydrates immediately form on the surface of the bubbles and these hydrates coalesce to form pipes standing on the sediments. Deep liquid CO_2 lakes could be maintained in place by a surface pavement and a subpavement cap of CO_2 hydrate ($CO_2 \cdot 6H_2O$) that traps the low-density liquid CO_2 in place (Figure 3.58) (Inagaki et al., 2006; Nealson, 2006).

In the SCS, there are two major types of hydrothermal activities. The most active hydrothermal activity is probably associated with the Manila subduction zone, where compression and dewatering of subducted sediments have driven upwelling volatiles along numerous faults and produced widespread mud volcanoes and shale diapirs in the accretionary wedge off southern Taiwan (Chi, Reed, Liu, & Lundberg, 1998; Chow, Lee, Liu, Lee, & Watkins, 2001; Chow, Lee, Sun, Liu, & Lundberg, 2000). Although still to be confirmed in the central SCS basin, the second type of hydrothermal activity is speculated to be linked to seafloor volcanisms. Seafloor spreading stopped around 15 Ma, but massive postspreading volcanisms, as young as 3.49 Ma (Jin, 1989), occurred along the relict spreading center and in the northern part of the central basin. Hydrothermal activities must have accompanied these volcanisms when they were active and soon after their extrusions.

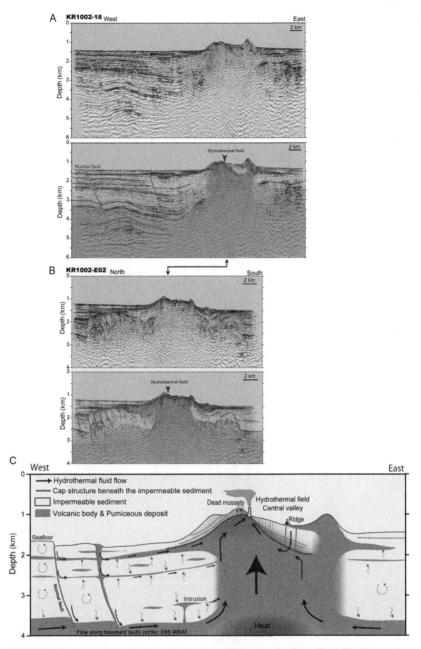

FIGURE 3.57 Prestack depth-migrated seismic profiles across the Iheya North Knoll hydrothermal field with and without geologic interpretation. (A) E–W line KR1002-18 and (B) N–S line KR1002-E2. The two profiles intersect at the Iheya North Knoll hydrothermal field (arrow). Gray shaded areas indicate volcanic bodies or deposits with high permeability. Gray lines indicate geologic boundaries, some of which are pumiceous deposits. The normal faults indicated by blue lines. (C) Model of the hydrothermal fluid system around the Iheya North Knoll based on seismic profile KR1002-18 in (A). Fluids flow laterally and updip along permeable pumiceous layers within the sedimentary sequence and laterally along basement faults within the crust. Near the knoll, fluids migrate upward due to heating and are then trapped below the hydrogeologic barrier (bold line) and flow beneath it to vents in the hydrothermal field. *From Tsuji et al. (2012).*

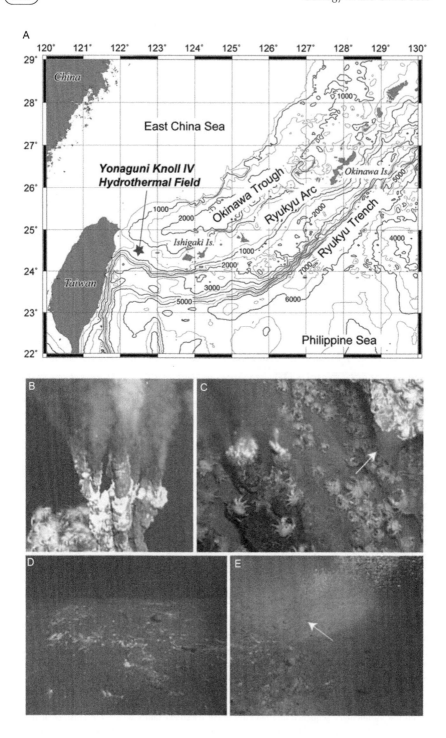

3.6.5   Post-Miocene Continental Margin Neotectonics

In Section 3.4, we have observed anomalous rapid subsidence of the Chinese continental margin after the Miocene, while faulting activities are rather weak comparing to early rifting phase and thermal subsidence phase. Nevertheless, because of active regional tectonic events in western Pacific subduction zones and the continued buildup of the Tibetan Plateau and because of close association of the continental margin basins with regional faults (Figure 3.2), post-Miocene continental margin tectonic activities, often behaving quite differently in stress fields from early tectonic regimes, are worthy of some discussion. They can create new faults and unconformities, cause depocenter migration, and modify preexisting hydrocarbon systems (Gong, Huang, & Chen, 2011). Gong et al. (2011) summarized that the dextral movements along the Tan-Lu fault are responsible for neotectonics in the Bohai Sea and the South Yellow Sea Basin (Figure 3.59), Pliocene east–west strike-slip movements in the ECS Basin are consistent with NW–SE extension of the Okinawa Trough (Figure 3.60), neotectonic deformation in the Pearl River Mouth Basin appears to be mostly related to arc–continent collision in Taiwan, and gravity slides near the shelf edge of the Qiongdongnan Basin and diapirs in the Yinggehai Basin are primarily related to sedimentary loading and the Red River Fault (Figure 3.61).

Lüdmann and Wong (1999) also correlated magmatic–tectonic events in northern SCS continental margin to two main collision phases (5–3 and 3–0 Ma) between Taiwan and the continental margin of East China. The collisional events transformed the compression into strike-slip movements, which may have remobilized many of the rift and drift faults, providing pathways for magma ascent (Lüdmann & Wong, 1999).

Many interesting features exist roughly along the transition zone between the Southwest and East Subbasins of the SCS (Figures 3.30 and 3.49); from the east to the west, they are identified as the Zhongnan Fault, the Zhongnan ridge, and the Zhongnan Seamounts. The Zhongnan ridge is subparallel to the Zhongnan Fault and appears as a low rise in a zigzag shape (Figures 3.49 and 3.58). On the magnetic anomaly map (Figure 3.31), the Zhongnan Fault seems to be correlated with a magnetic weak belt. On the gravity anomaly map, the Zhongnan ridge appears clearly as a gravity high belt (Li & Song, 2012). Seismic data show that the Zhongnan ridge has a V-shaped crest (Li, Zhou, Li, et al., 2008) (Figure 3.62), like that on the Gagua Ridge that are interpreted

FIGURE 3.58   Overview of the Yonaguni Knoll IV hydrothermal field at the southern Okinawa Trough (Inagaki et al., 2006). (A) Location map of the Yonaguni Knoll IV. (B) "Lion chimney," one of the most active black smoker vents in this field. (C) "Crystal chimney," one of the vapor-rich clear smoker vents adjacent to the Tiger black smoker. The emission of liquid CO_2 droplets was observed in close proximity. (D) White patchy area, "CO_2 hydrate zone," 50 m southward from the Tiger chimney. (E) Continuous emission of liquid CO_2 droplets from the "CO_2 lake" overlying hydrates. White arrows in C and E indicate the emission of liquid CO_2 droplets.

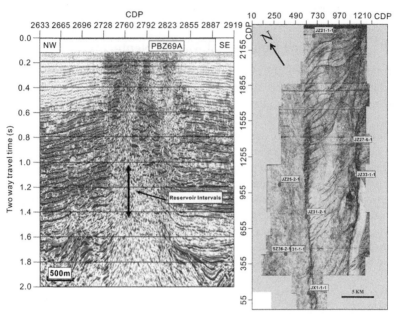

FIGURE 3.59 (Left) Shallow seismic profile in the Bohai Basin (Gong et al., 2011). Gas chimneys at shallow depths indicate gas escape along recent faults. The reservoir is poorly imaged due to shallow gas effects. Depth in milliseconds. (Right) A 3D coherence time slice from the Bohai Basin shows densely spaced, short-segmented Quaternary faults along the right-lateral Tan-Lu fault (Gong et al., 2011).

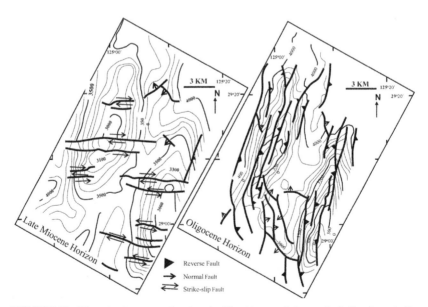

FIGURE 3.60 Time-structure maps showing significant temporal changes in fault patterns in the Xihu Depression, ECS Basin (Gong et al., 2011). Pliocene faults are mostly E–W trending shallow strike-slip structures and are underlain by deeper, older, reverse faults trending NE–SW.

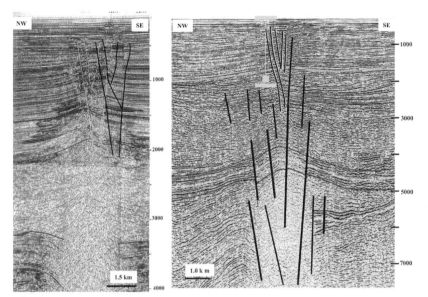

FIGURE 3.61   Left: Poststack seismic profile through a shale diapir in the Yinggehai Basin (Gong et al., 2011). The deep-rooted zone of disturbance is capped by extensional faults with bright spots (gas). Some faults extend to the seafloor with gas seepages. Depth in milliseconds. Right: Prestack depth-migrated seismic profile through the same shale diaper (Gong et al., 2011). The steeply dipping faults are interpreted to be conduits for upward gas migration. Depth in meters.

as being formed from a compressed fault zone (Deschamps et al., 1998; Li, Zhou, Li, Chen, et al., 2007). Also considering the deeply dipping strong reflections within the Zhongnan ridge, we interpret that the ridge formed by compression along a structurally fractured or weak zone. Cenozoic sediments are almost absent at the crest of the ridge. We suspect that hydrothermal fluids, deep from the mantle, may have accompanied with fracturing here, and the ridge remains to be an important structure for investigating seawater circulation within the oceanic crust and *in situ* serpentinization and magnetization in the lower crust and upper mantle caused possibly by infiltration of seawater through large transform faults. The uppermost mantle can be magnetized by *in situ* serpentinization caused by infiltration of seawater through large transform faults (Li & Lee, 2006). In the SCS, Curie isotherm is recently found to lie underneath the Mohorovicic discontinuity (Li et al., 2010), hence raising the question whether the uppermost mantle here is also magnetized from serpentinization. Serpentinization has been widely discovered as a product of water–rock interactions in deep sea and is thus worthy of future study in the SCS. It is of interest to study whether the serpentinization process had ever taken place during seafloor spreading and whether serpentinization exists or not in the deep lithosphere of the SCS (Wang, 2012).

A

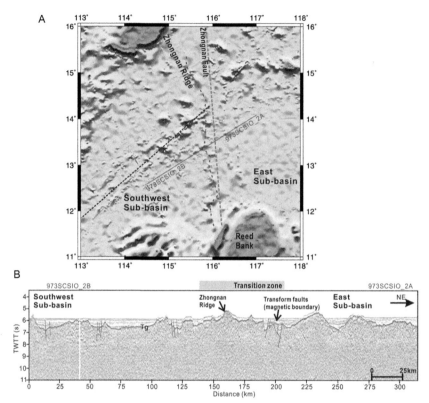

FIGURE 3.62 (A) Topographic map showing the Zhongnan ridge. The red dashed line marks the Zhongnan ridge. The bold dashed line marks the Zhongnan Fault. The thin dashed line marks the relict spreading center of the Southwest Subbasin. The solid line is the position of the seismic section in (B). (B) Seismic section crossing the Zhongnan Fault and the Zhongnan ridge. *Panel (B) from Li, Zhou, Li, Chen, and Geng (2008).*

3.7 TECTONIC SUMMARY

This chapter summarizes the tectonic framework and evolutionary history of the China Seas since the early Mesozoic. We discuss the geology of the Mesozoic basement, upon which the Cenozoic rifting basins and marine depositions develop. Mesozoic tectonics left many weak zones along collisional and subduction boundaries, which are later reactivated to form regional faults such as the Tan-Lu fault, the Jiaxiang Fault, the Red River Fault and the coastal faults in the Cathaysian blocks. Formation of Cenozoic rifting basins in the China Seas was largely controlled by these faults.

Pervasive tectonic inheritance is observed. Cenozoic continental margin rifting in the Southeast China Seas inherited the regional extension induced by the collapse of the Cathaysian Plateau and the eastward rollback and/or retreat of the Paleo-Pacific Plate, whose subduction beneath the Eurasia occurred in the late Permian and early Triassic. The ultimate product of this regional extension is the SCS oceanic lithosphere, formed from around 32 to 16 Ma. While the SCS underwent almost an entire cycle from incipient rifting to seafloor spreading and to cessation of drifting and ultimate subduction, active rifting in the Okinawa Trough did not start until the late Miocene and continues today.

Although under a large uniform extensional stress field during the Cenozoic, continental margin rifting basins in the China Seas developed in slightly different episodes. In general, basins or depressions further inland often started and stopped rifting earlier than those further seaward. Thermal subsidence appears to be more dominant for basins close to the oceanic lithosphere of the SCS, and consequently, synrifting structures in these basins appear to be less characteristic.

Mesozoic magmatism was dominated by the subduction-related continental arc in the Cathaysian block. Cenozoic magmatism is related mostly to rifting, for example, in the Okinawa Trough, and seafloor spreading in the SCS, but synrifting volcanisms are not strong in the SCS. Postrifting volcanisms of various ages exist in the central SCS basin, forming large seamounts. It is still debated how magmatism played roles in the opening of the SCS, as with other opening mechanisms proposed.

Active tectonics along subduction zones and back-arc basins trigger numerous earthquakes, volcanoes, and strong hydrothermal activities. The ongoing collision between the Luzon Arc and the Eurasian continental margin makes the Taiwan orogen one of the most rapidly growing orogens in the world.

REFERENCES

Agarwal, B. N. P., & Shaw, R. K. (1996). Comment on "An analytic signal approach to the interpretation of total field magnetic anomalies" by Shuang Qin. *Geophysical Prospecting*, *44*, 911–914.

Allen, M. B., MacDonald, D. I. M., Zhao, X., Vincent, S. J., & Brouet-Menzies, C. (1997). Early Cenozoic two-phase extension and late Cenozoic thermal subsidence and inversion of the Bohai basin, northern China. *Marine and Petroleum Geology*, *14*(7–8), 951–972.

Allen, M. B., MacDonald, D. I. M., Zhao, X., Vincent, S. J., & Brouet-Menzies, C. (1998). Transtensional deformation in the evolution of the Bohai basin, northern China. In R. E. Holdsworth, R. A. Strachan, & J. F. Dewey (Eds.), *Continental transpression and transtensional tectonics*: *Vol. 135*. (pp. 215–229). London: Special Publication of Geological Society.

Aydin, A., Ferré, E. C., & Aslan, Z. (2007). The magnetic susceptibility of granitic rocks as a proxy for geochemical composition: Example from the Saruhan granitoids, NE Turkey. *Tectonophysics*, *441*, 85–95.

Barckhausen, U., & Roeser, H. A. (2004). Seafloor spreading anomalies in the South China Sea revisited. In P. Clift, P. Wang, W. Kuhnt, & D. Hayes (Eds.), *AGU Geophysical Monograph Series*: *Vol. 149*. *Continent-ocean interactions within East Asian Marginal Seas*. (pp. 121–125).

Bautista, B. C., Bautista, M. L. P., Oike, K., Wu, F. T., & Punongbayan, R. S. (2001). A new insight on the geometry of subducting slabs in northern Luzon, Philippines. *Tectonophysics*, *339*, 279–310.

Ben-Avraham, Z. B., & Uyeda, Z. S. (1973). The evolution of the China basin and the Mesozoic paleogeography of Borneo. *Earth and Planetary Science Letters, 18,* 365–376.

Biq, C. C. (1972). Transcurrent buckling, transform faulting and transpression, their relevance in eastern Taiwan kinematics. *Petroleum Geology of Taiwan, 20,* 1–39.

Bowin, C., Lu, R. S., Lee, C. S., & Schouten, H. (1978). Plate convergence and accretion in Taiwan-Luzon Region. *American Association of Petroleum Geologists Bulletin, 62,* 1645–1672.

Briais, A., Patriat, P., & Tapponnier, P. (1993). Updated interpretation of magnetic anomalies and seafloor spreading stages in the South China Sea: Implications for the tertiary tectonics of Southeast Asia. *Journal of Geophysical Research, 98,* 6299–6328.

Cai, Q. (2005). *Oil and gas geology in China Seas.* Beijing: China Ocean Press 406 pp (in Chinese).

Cande, S. V., & Kent, D. V. (1995). Revised calibration of the geomagnetic polarity time-scale for the late Cretaceous and Cenozoic. *Journal of Geophysical Research, 100,* 6093–6095.

Cao, D.-Y., Wang, X.-G., Zhan, W.-F., & Li, W.-Y. (2007). Subsidence history of the eastern depression in the North Yellow Sea Basin. *Journal of China University of Mining and Technology, 17*(1), 90–95.

Carter, A., & Clift, P. D. (2008). Was the Indosinian orogeny a Triassic mountain building or thermotectonic reactivation event? *Comptes Rendus Geoscience, 340,* 83–93.

Chai, B. H. (1972). Structure and tectonic evolution of Taiwan. *American Journal of Science, 272,* 389–432.

Chang, C. Y. (1991). Geological characteristics and distribution patterns of hydrocarbon deposits in the Bohai Bay basin, east China. *Marine and Petroleum Geology, 8,* 98–106.

Chappell, B. W., & White, A. J. R. (1974). Two contrasting granite types in the Lachlan Fold Belt, southeastern Australia. *Pacific Geology, 8,* 173–174.

Charvet, J., Faure, M., Xu, J. W., Zhu, G., Tong, W. X., & Lin, S. F. (1990). La zone tectonique de Changle-Nanao, Chine du sud-est. *Comptes Rendus de l'Académie des Sciences Paris, 2-310,* 1271–1278.

Charvet, J., Lapierre, H., & Yu, Y. (1994). Geodynamic significance of the Mesozoic volcanism of southeastern China. *Journal of Southeast Asian Earth Sciences, 9,* 387–396.

Charvet, J., Shu, L. S., Shi, Y. S., Guo, L. Z., & Faure, M. (1996). The building of south China: Collision of Yangzi and Cathaysia blocks, problems and tentative answers. *Journal of Southeast Asian Earth Sciences, 13,* 223–235.

Chemenda, A. I., Yang, R.-K., Stephan, J.-F., Konstantinovskaya, E. A., & Ivanov, G. M. (2001). New results from physical modeling of arc–continent collision in Taiwan: Evolutionary model. *Tectonophysics, 333,* 159–178.

Chen, J., Foland, K. A., Xing, F., Xu, X., & Zhou, T. (1991). Magmatism along the southeast margin of the Yangtze block: Precambrian collision of the Yangtze and Cathaysia blocks of China. *Geology, 19,* 815–818.

Chen, J. F., & Jahn, B.-M. (1998). Crustal evolution of southeastern China: Nd and Sr isotopic evidence. *Tectonophysics, 284,* 101–133.

Chen, A. T., & Jaw, Y. S. (1996). Velocity structure near the northern Manila Trench: An OBS reflection study. *Terrestrial, Atmospheric and Oceanic Sciences, 7,* 277–297.

Chen, C.-H., Lee, C.-Y., & Shinjo, R. (2008). Was there Jurassic paleo-Pacific subduction in South China? Constraints from $^{40}Ar/^{39}Ar$ dating, elemental and Sr–Nd–Pb isotopic geochemistry of the Mesozoic basalts. *Lithos, 106,* 83–92.

Chen, S., & Li, Z. (1987). Major oil accumulation characteristics and exploration direction in the Pearl River Mouth basin. In *Proceedings of the international symposium on petroleum geology of northern continental shelf area of the South Gina Sea, compiled by Guangdong Petroleum Society and China Oil, Guangzhou* (pp. 12–23).

Chen, J., & Wen, N. (2010). *Geophysical atlas of South China Sea*. Beijing: Science Press 137 pp (in Chinese).

Chen, J., Xu, K., & Xu, R. (1998). The Triassic and Jurassic biogeography of South China. *Acta Palaeontologica Sinica*, *37*(1), 97–107 (in Chinese).

Chi, W. C., Reed, D. L., Liu, C. S., & Lundberg, N. (1998). Distribution of the bottom-simulating reflector in the offshore Taiwan collision zone. *Terrestrial, Atmospheric and Oceanic Sciences*, *9*, 779–794.

Chow, J., Lee, J. S., Liu, C. S., Lee, B. D., & Watkins, J. S. (2001). A submarine canyon as the cause of a mud volcano—Liuchieuyu Island in Taiwan. *Marine Geology*, *176*, 55–63.

Chow, J., Lee, J. S., Sun, R., Liu, C. S., & Lundberg, N. (2000). Characteristics of the bottom simulating reflectors near mud diapirs: Offshore southwestern Taiwan. *Geo-Marine Letters*, *20*, 3–9.

Chung, S.-L., & Sun, S.-S. (1992). A new genetic model for the East Taiwan ophiolite and its implications for Dupal domains in the northern Hemisphere. *Earth and Planetary Science Letters*, *109*, 113–145.

Chung, S. L., Sun, S., Tu, K., Chen, C. H., & Lee, C. (1994). Late Cenozoic basaltic volcanism around the Taiwan Strait, SE China: Product of lithosphere-asthenosphere interaction during continental extension. *Chemical Geology*, *112*(1–2), 1–20.

Chung, S.-L., Cheng, H., Jahn, B.-M., O'Reilly, S. Y., & Zhu, B. (1997). Major and trace element, and Sr-Nd isotope constraints on the origin of Paleogene volcanism in South China prior to the South China Sea opening. *Lithos*, *40*, 203–220.

Chung, S.-L., Wang, S.-L., Shinjo, R., Lee, C.-S., & Chen, C.-H. (2000). Initiation of arc magmatism in an embryonic continental rifting zone of the southernmost part of Okinawa Trough. *Terra Nova*, *12*, 225–230.

Clemens, J. D., Holloway, J. R., & White, A. J. R. (1986). Origin of an A-type granite: Experimental constraints. *American Mineralogist*, *71*, 317–324.

Clift, P., Lee, J. I., Clark, M., & Blusztajn, J. (2002). Erosional response of South China to arc rifting and monsoonal strengthening; a record from the South China Sea. *Marine Geology*, *184*(3–4), 207–226.

Clift, P., & Lin, J. (2001). Preferential mantle lithospheric extension under the South China margin. *Marine and Petroleum Geology*, *18*, 929–945.

Clift, P., Schouten, H., & Draut, A. E. (2003). A general model of arc-continent collision and subduction polarity reversal from Taiwan and the Irish Caledonides. In R. D. Larter & P. T. Leat (Eds.), *Intra-Oceanic subduction systems; tectonic and magmatic processes*: Vol. 219. (pp. 81–98). London: Special Publications of Geological Society.

Clift, P., & Sun, Z. (2006). The sedimentary and tectonic evolution of the Yinggehai - Song Hong Basin and the southern Hainan margin, South China Sea; implications for Tibetan uplift and monsoon intensification. *Journal of Geophysical Research*, *111*. http://dx.doi.org/10.1029/2005JB004048.

Cullen, A., Reemst, P., Henstra, G., Gozzard, S., & Ray, A. (2010). Rifting of the South China Sea: new perspectives. *Petroleum Geoscience*, *16*(3), 273–282.

Cullen, A. B. (2010). Transverse segmentation of the Baram-Balabac Basin, NW Borneo: refining the model of Borneo's tectonic evolution. *Petroleum Geoscience*, *16*(1), 3–29.

Davagnier, M., Marsset, B., Sibuet, J.-C., Letouzey, J., & Foucher, J.-P. (1987). Mechnismes actuels d'extension dans le bassin d'Okinawa. *Bulletin de la Societe Geologique de France*, *8*, 525–531.

Davis, D., Suppe, J., & Dahlen, F. A. (1983). Mechanisms of fold-and-thrust belts and accretionary wedges. *Journal of Geophysical Research*, *88*, 1153–1172.

Deprat, J. (1914). Étude des plissements et des zones d'écrasement de la moyenne et de la basse rivière Noire. *Memoire du Service Geologique de l'Indochine*, *3–4*, 1–59.

Deschamps, A. E., Lallemand, S. E., & Collot, J.-Y. (1998). A detailed study of the Gagua Ridge: A fracture zone uplifted during a plate reorganisation in the Mid-Eocene. *Marine Geophysical Researches, 20*, 403–423.

Deschamps, A., Monié, P., Lallemand, S. E., Hsu, S.-K., & Yeh, K. Y. (2000). Evidence for early Cretaceous oceanic crust trapped in the Philippine Sea plate. *Earth and Planetary Science Letters, 179*, 503–516.

Dong, S.-W., & Huang, D. (2000). Comparison of the Dabie and Sulu orogenic belts on both sides of the Tan-lu fault zone and initiation and development of the Tan-Lu fault zone. In X. Wang, et al. (Ed.). *On Tan-Lu Fault Zone* (pp. 1–374). Beijing, China: Geological Publishing House.

Ellwood, B. B., & Wenner, D. B. (1981). Correlation of magnetic susceptibility with data in late orogenic granites of the southern Appalachian Piedmont. *Earth and Planetary Science Letters, 54*, 200–202.

Expedition 331 Scientists. (2010). Deep hot biosphere. *IODP Expedition, 331*. http://dx.doi.org/10.2204/iodp.pr.331.2010, Preliminary Report.

Fan, W. M., & Menzies, M. A. (1992). Destruction of aged lower lithosphere and accretion of asthenosphere mantle beneath eastern China. *Geotectonica et Metallogenia, 16*, 171–180.

Flower, M. F. J., Russo, R. M., Tamaki, K., & Hoang, N. (2001). Mantle contamination and the Izu-Bonin-Mariana (IBM) 'high-tide mark': Evidence for mantle extrusion caused by Tethyan closure. *Tectonophysics, 333*, 9–34.

Flower, M. F. J., Zhang, M., Chen, C.-Y., Tu, K., & Xie, G. (1992). Magmatism in the South China Basin: 2. Post-spreading Quaternary basalts from Hainan Island, south China. *Chemical Geology, 97*(1–2), 65–87.

Fontaine, H. (1979). Note on the geology of the Calamian Islands, North Palawan, Philippines. *ESCAP-CCOP Newsletter, 6*, 3–30.

Fromaget, J. (1927). Études géologiques sur le Nord de 1'Indochine centrale. *Bulletin du Service Geologique de l'Indochine, 16*, 1–368.

Gamo, T., Sakai, H., Kim, E.-S., Shitashima, K., & Ishibashi, J. (1991). High alkalinity due to sulfate reduction in the CLAM hydrothermal field, Okinawa Trough. *Earth and Planetary Science Letters, 107*(2), 328–338.

Gao, S., & Li, J. B. (2002). *Formation and evolution of Chinese marginal seas*. Beijing: Ocean Publishing House 175 pp.

GDBGMR (1988). *Memoir of regional geology of guangdong province*. Beijing, China: Geological Publishing House 941 pp (in Chinese).

Gilder, S. A., Gill, J., Coe, R. S., Zhao, X., Liu, Z., Wang, G., et al. (1996). Isotopic and paleomagnetic constraints on the Mesozoic tectonic evolution of south China. *Journal of Geophysical Research, 101*, 16137–16154.

Gong, Z. (1997). *The Chinese offshore oil and gas fields*. Beijing: Petroleum Industry Press 223 pp (in Chinese).

Gong, Z. S., Huang, L. F., & Chen, P. H. (2011). Neotectonic controls on petroleum accumulations, offshore China. *Journal of Petroleum Geology, 34*(1), 5–28.

Gong, Z. S., Li, S. T., & Xie, T. J. (1997). *Continental margin basin analysis and hydrocarbon accumulation of the Northern South China Sea*. Beijing, China: Science Press 510 pp (in Chinese).

Gong, Z. S., Li, S. T., & Xie, T. J. (2004). *Dynamic research of oil and gas accumulation in Northern marginal basins of South China Sea*. Beijing: Science Press 339 pp.

Gradstein, F., Ogg, J., Smith, A., et al. (2004). *A Geologic Time Scale 2004*. Cambridge: Cambridge University Press.

Gregorová, D., Hrouda, F., & Kohút, M. (2003). Magnetic susceptibility and geochemistry of Variscan West Carpathian granites: Implications for tectonic setting. *Physics and Chemistry of the Earth, 28*, 729–734.

Gungor, A., Lee, G. H., Kim, H.-J., Han, H.-C., Kang, M.-H., Kim, J., et al. (2012). Structural characteristics of the northern Okinawa Trough and adjacent areas from regional seismic reflection data: Geologic and tectonic implications. *Tectonophysics, 522–523*, 198–207.

GXBGMR. (1985). *Memoir of regional geology of guangdong province.* Beijing, China: Geological Publishing House 774 pp (in Chinese).

Haile, N. S. (1973). The recognition of former subduction zones in Southeast Asia. In D. H. Tarling & S. K. Runcorn (Eds.), *Implications of Continental Drift to the Earth Sciences, Vol. 2.* (pp. 885–892). London: Academic Press.

Hall, R. (1996). Reconstructing Cenozoic SE Asia. In R. Hall & D. J. Blundell (Eds.), *Tectonic Evolution of Southeast Asia. Geological Society - Special Publications, 106*(1), 153–184.

Hall, R. (2002). Cenozoic geological and plate tectonic evolution of SE Asia and the SW Pacific: Computer-based reconstructions, model and animations. *Journal of Asian Earth Sciences, 20*, 353–431.

Hall, R., Ali, J. R., Anderson, C. D., & Baker, S. J. (1995). Origin and motion history of the Philippine Sea Plate. *Tectonophysics, 251*, 229–250.

Hall, R., & Morley, C. K. (2004). Sundaland Basins. In P. D. Clift, W. Kuhnt, P. Wang, & D. Hayes (Eds.), *American Geophysical Union Monograph: Vol. 149. Continent-ocean interactions with East Asian marginal seas* (pp. 55–87).

Hamilton, W. (1979). Tectonics of the Indonesian region. *United States Geological Survey. Professional Paper, 1078* 345 pp.

Han, X. Q. (2011). Ocean ridge basalt of Southwestern subbasin in the South China Sea: Rock geochemistry and geochronology constraints of South China Sea. In *"South China Sea evolution study of the major research plan" 2011 annual start meeting (in Chinese). Shanghai, S1-O-12* (pp. 26–27).

Hayes, D. E., & Nissen, S. S. (2005). The South China Sea margins: Implications for rifting contrasts. *Earth and Planetary Sciences Letters, 237*, 601–616.

Hayes, D. E., Nissen, S., Buhl, P., Diebold, J., Yao, B., Zeng, W., et al. (1995). Through-going crustal faults along the northern margin of the South China Sea and their role in crustal extension. *Journal of Geophysical Research, 100*, 22435–22446.

Haynes, S. J. (1988). Structural reconnaissance of the Jiangnan geoanticline: A suspected terrane of compressional tectonic character. In D. G. Howell & T. J. Wiley (Eds.), *Proceeding of the 4th international tectonostratigraphic terrane conference* (pp. 31–33). Menlo Park, CA: U.S.G.S.

He, L. (1988). Formation and evolution of South China Sea and their relation to hydrocarbon potential. *Marine Geology and Quaternary Geology, 8*, 15–28 (in Chinese with English abstract).

Hilde, T. W. C., Uyeda, S., & Kroenke, L. (1977). Evolution of the Western Pacific and its margin. *Tectonophysics, 38*, 145–152.

Hirata, N., Kinoshita, H., Katao, H., Baba, H., Kaiho, Y., Koreawa, S., et al. (1991). Report on DELP 1988 cruise in the Okinawa Trough part 3. Crustal structure of the southern Okinawa Trough. *Bulletin of Earthquake Research Institute, 66*, 37–70.

Ho, C. S. (1986). A synthesis of the geological evolution of Taiwan. *Tectonophysics, 25*, 1–16.

Holloway, N. H. (1982). North Palawan Block, Philippines -its relation to the Asian mainland and role in evolution of South China Sea. *American Association of Petroleum Geologists Bulletin, 66*, 1355–1383.

Hong, D., Guo, W., Li, G., Kang, W., & Xu, H. (1987). *Petrology of the Miarolitic Granite Belt in the Southeast Coast of Fujian province and their Petrogenesis.* (pp. 1–132). Beijing, China: Science and Technology Publishing House (in Chinese).

Hong, D., Wang, S., Han, B., & Jin, M. (1996). Post-orogenic alkaline granites from China and comparisons with anorogenic alkaline granites elsewhere. *Journal of Southeast Asian Earth Sciences, 13*, 13–27.

Honza, E. (1995). Spreading mode of backarc basins in the western Pacific. *Tectonophysics, 251*, 139–152.

Hsiao, L.-Y., Graham, S. A., & Tilander, N. (2004). Seismic reflection imaging of a major strike-slip fault zone in a rift system: Paleogene structure and evolution of the Tan-Lu fault system, Liaodong Bay, Bohai, offshore China. *American Association of Petroleum Geologists Bulletin, 88*, 71–97.

Hsu, S.-K., Liu, C.-S., Shyu, C.-T., Liu, S.-Y., Sibuet, J.-C., Lallemand, S., et al. (1998). New gravity and magnetic anomaly maps in the Taiwan-Luzon region and their preliminary interpretation. *Terrestrial, Atmospheric and Oceanic Sciences, 9*, 509–532.

Hsu, S.-K., & Sibuet, J.-C. (1995). Is Taiwan the result of arc-continent or arc-arc collision? *Earth and Planetary Science Letters, 136*, 315–324.

Hsu, S.-K., Sibuet, J.-C., & Shyu, C.-T. (2001). Magnetic inversion in the East China Sea and Okinawa Trough: Tectonic implications. *Tectonophysics, 333*, 111–122.

Hsü, K. J., Sun, S., Li, J. L., Chen, H. H., Peng, H. P., & Sengor, A. M. C. (1988). Mesozoic overthrust tectonics in south China. *Geology, 16*, 418–421.

Hsu, S.-K., Yeh, Y.-C., Doo, W.-P., & Tsai, C.-H. (2004). New bathymetry and magnetic lineations identifications in the northernmost South China Sea and their tectonic implications. *Marine Geophysical Research, 25*, 29–44.

Hu, S.-B., He, L.-J., & Wang, J.-Y. (2001). Compilation of heat flow data in the China continental area (3rd edition). *Chinese Journal of Geophysics, 44*(5), 611–626.

Huang, C.-Y., Wu, W. Y., Chang, C. P., Tsao, S., Yuan, P. B., Lind, C.-W., et al. (1997). Tectonic evolution of accretionary prism in the arc-continent collision terrane of Taiwan. *Tectonophysics, 281*, 31–51.

Huang, C.-Y., Xia, K.-Y., Yuan, P. B., & Chen, P.-G. (2001). Structural evolution from Paleogene extension to Latest Miocene-Recent arc-continent collision offshore Taiwan: Comparison with on land geology. *Journal of Asian Earth Sciences, 19*, 619–639.

Huang, C.-Y., Yuan, P. B., Lin, C. W., Wang, T. K., & Chang, C.-P. (2000). Geodynamics processes of Taiwan arc-continent collision and comparison with analogs in Timor, Papua New Guinea, Urals and Corsica. *Tectonophysics, 325*, 1–21.

Huang, C.-Y., Yuan, P. B., Song, S. R., Lin, C. W., Wang, C., Chen, M.-T., et al. (1995). Tectonics of short-lived intra-arc basins in the arc-continent collision terrane of the Coastal Range, Eastern Taiwan. *Tectonics, 14*, 19–38.

Huang, C. Y., Yen, Y., Zhao, Q. H., et al. (2012). Cenozoic stratigraphy of Taiwan: Window into rifting, stratigraphy and paleoceanography of South China Sea. *Chinese Science Bulletin, 57*, 3130–3149.

Huismans, R., & Beaumont, C. (2011). Depth-dependent extension, two-stage breakup and cratonic underplating at rifted margins. *Nature, 473*, 74–78.

Hutchison, C. S. (1989). *Geological evolution of South-East Asia.* (pp. 1–355). Oxford: Clarendon Press.

Hutchison, C. S. (1996). The "Rajang accretionary prism" and "Lupar Line" problem of Borneo. In R. Hall & D. J. Blundell (Eds.), *Tectonic evolution of southeast Asia. Geological Society - Special Publications, 106*(1), 247–261.

Hutchison, C. S. (2004). Marginal basin evolution: the southern South China Sea. *Marine and Petroleum Geology*, *21*(9), 1129–1148.

Inagaki, F., Kuypers, M. M. M., Tsunogai, U., Ishibashi, J., Nakamura, K., Treude, T., et al. (2006). Microbial community in a sediment-hosted CO_2 lake of the southern Okinawa Trough hydrothermal system. *Proceedings of the National Academy of Sciences of the United States of America*, *103*(38), 14164–14169.

Institute of Oceanology, Chinese Academy of Sciences. (1982). *Geology of the Yellow Sea and East China Sea*. Beijing: Science Publishing House 219 pp.

Ishihara, S. (1977). The magnetite-series and ilmenite-series granitic rocks. *Mining Geology (Japan)*, *27*, 293–305.

Ishihara, S., Robb, L. J., Anhaeusser, C. R., & Imai, A. (2002). Granitoid series in terms of magnetic susceptibility: A case study from the Barberton region, South Africa. *Gondwana Research*, *5*, 581–589.

Isozaki, Y. (1997). Contrasting two types of orogen in Permo-Triassic Japan: Accretionary versus collisional. *Island Arc*, *6*, 2–24.

Jahn, B. M., Chen, P. Y., & Yen, T. P. (1976). Rb–Sr ages of granitic rocks in southeastern China and their tectonic significance. *Geological Society of America Bulletin*, *86*, 763–776.

Jia, J., & Gu, H. (2002). *Oil-bearing systems and petroleum assessment of the Xihu Sag in the East China Sea*. Beijing: Geological Publishing House 204 pp (in Chinese).

Jiang, X. (1991). A tentative discussion on the geochemical characters and tectonic setting of alkali-rich granites in the coastal area of East Zhejiang. *Acta Petrologica et Mineralogica*, *10*(2), 144–153.

Jin, Q. (1989). *Geology and hydrocarbon resources of the South China Sea*. Beijing, China: Geological Press 417 pp (in Chinese).

Jin, X. (1992). *Marine geology of the East China Sea*. Beijing: China Ocean Press 524 pp.

Jolivet, L., Beyssac, O., Goffe, B., Avigad, D., Lepvrier, C., Maluski, H., et al. (2011). Oligo-Miocene midcrustal subhorizontal shear zone in Indochina. *Tectonics*, *20*, 46–57.

Karig, D. E. (1971). Structural history of the Mariana island arc system. *Geological Society of America Bulletin*, *82*, 323–344.

Karig, D. E. (1973). Plate convergence between the Philippines and the Ryukyu Islands. *Marine Geology*, *14*, 153–168.

Kawagucci, S., Chiba, H., Ishibash, J., Yamanaka, T., Toki, T., Muramatsu, Y., et al. (2011). Hydrothermal fluid geochemistry at the Iheya North field in the mid-Okinawa Trough: Implication for origin of methane in subseafloor fluid circulation systems. *Geochemical Journal*, *45*(2), 109–124.

Kido, Y., Suyehiro, K., & Kinoshita, H. (2001). Rifting to spreading process along the northern margin of the South China Sea. *Marine Geophysical Research*, *22*, 1–15.

Kiessling, W., & Flügel, E. (2000). Late Paleozoic and Late Triassic limestone from the North Palawan Block (Philippines): Microfacies and paleogeographical implications. *Facies*, *43*, 39–78.

KIGAM. (2006). *Joint study on sedimentary basins between Korea and China: Korea institute of geoscience and mineral report GAA2003002-2006(4)*. (in Koreanwith English abstract).

Kimura, M., Uyeda, S., Kato, Y., Tanaka, T., Yamamoto, M., Gamo, T., et al. (1988). Active hydrothermal mounds in the Okinawa Trough backarc basin, Japan. *Tectonophysics*, *145*, 318–324.

Kizaki, K. (1986). Geology and tectonics of the Ryukyu islands. *Tectonophysics*, *125*, 193–207.

Kong, F. (1998). *Continental margin deformation analysis and reconstruction—Evolution of the East China Sea Basin and adjacent plate interaction*. (p. 263). University of Texas at Austin, Ph.D. Dissertation.

Konno, U., Tsunogai, U., Nakagawa, F., Nakaseama, M., Ishibashi, J., Nunoura, T., et al. (2006). Liquid CO2 venting on the seafloor: Yonaguni Knoll IV hydrothermal system, Okinawa Trough. *Geophysical Research Letters, 33*(16), L16607.

Kudrass, H. R., Hiedicke, M., Cepek, P., Kreuzer, H., & Müller, P. (1986). Mesozoic and Cenozoic rocks dredged from the South China Sea (Reed Bank area) and Sulu Sea and their significance for plate-tectonic reconstructions. *Marine and Petroleum Geology, 3,* 19–30.

Lacombe, O., Mouthereau, F., Angelier, J., & Deffontaines, B. (2001). Structural, geodetic and seismological evidence for tectonic escape in SW Taiwan. *Tectonophysics, 333,* 323–345.

Lallemand, S., & Jolivet, L. (1986). Japan Sea: a pull apart basin? *Earth and Planetary Science Letters, 76*(3–4), 375–389.

Lapierre, H., Jahn, B. M., Charvet, J., & Yu, Y. W. (1997). Mesozoic felsic arc magmatism and continental olivine tholeiites in Zhejiang Province and their relationship with the tectonic activity in southeastern China. *Tectonophysics, 274,* 321–338.

Lebedev, S., & Nolet, G. (2003). Upper mantle beneath southeast Asia from S velocity tomography. *Journal of Geophysical Research, 108,* L2048.

Lee, E. Y. (2010). Subsidence history of the Gunsan Basin (Cretaceous-Cenozoic) in the Yellow Sea, offshore Korea. *Austrian Journal of Earth Sciences, 103,* 111–120.

Lee, T. Q., Kissel, C., Barrier, E., Laj, C., & Chi, W. (1991). Paleomagnetic evidence for a diachronic clockwise rotation of the Coastal Range, eastern Taiwan. *Earth and Planetary Science Letters, 104,* 245–257.

Lee, G. H., Kwon, Y. I., Yoon, C. S., Kim, H. J., & Yoo, H. S. (2006). Igneous complexes in the eastern Northern South Yellow Sea Basin and their implications for hydrocarbon systems. *Marine and Petroleum Geology, 23,* 631–645.

Lee, T.-Y., & Lawver, L. A. (1995). Cenozoic plate reconstruction of Southeast Asia. *Tectonophysics, 251,* 85–138.

Lee, G. H., Lee, K., & Watkins, J. S. (2001). Geologic evolution of the Cuu Long and Nam Con Son basins, offshore southern Vietnam, South China Sea. *American Association of Petroleum Geologists Bulletin, 85,* 1055–1082.

Lee, T.-Y., Lo, C.-H., Chung, S.-L., Chen, C.-Y., Wang, P.-L., Lin, W.-P., et al. (1998). $^{40}Ar/^{39}Ar$ dating result of Neogene basalts in Vietnam and its tectonic implication. In M. Flower, S.-L. Chung, C.-H. Lo, & T.-Y. Lee (Eds.), *American Geophysical Union Monograph*: Vol. 27. *Mantle dynamics and plate interactions in East Asia* (pp. 317–330).

Lee, C., Shor, G. G., Bibee, L. D., Lu, R. S., & Hilde, T. W. C. (1980). Okinawa Trough, origin of a back-arc basin. *Marine Geology, 35,* 219–241.

Lee, T.-Y., Tang, C.-H., Ting, J.-S., & Hsu, Y.-Y. (1993). Sequence stratigraphy of the Tainan Basin, offshore southwestern Taiwan. *Petroleum Geology of Taiwan, 28,* 119–158.

Lei, J. S., Zhao, D. P., Steinberger, B., Wu, B., Shen, F., & Li, X. (2009). New seismic constraints on the upper mantle structure of the Hainan plume. *Physics of the Earth and Planetary Interiors, 173,* 33–50.

Leloup, P. H., Arnaud, N., Lacassin, R., Kienast, J. R., Harrison, T. M., Trong, T. T. P., et al. (2001). New constraints on the structure, thermochronology, and timing of the Ailao Shan-Red River shear zone, SE Asia. *Journal of Geophysical Research: Solid Earth, 106*(B4), 6683–6732.

Le Pichon, X., & Mazzotti, S. (1997). A new model for the early opening of the Okinawa basin. In *Proceedings of Chinese Taipei ODP consortium 1997 annual meeting and long-range plan for Chinese Taipei ocean drilling program workshop, Taipei, Taiwan, 6 November 1997* 34 pp.

Lepvrier, C., Faure, M., Van, V. N., Vu, T. V., Lin, W., Trong, T. T., et al. (2011). North-directed Triassic nappes in Northeastern Vietnam (East Bac Bo). *Journal of Asian Earth Sciences, 41,* 56–68.

Lepvrier, C., Maluski, H., Tich, V. V., Leyreloup, A., Thi, P. T., & Vuong, N. V. (2004). The Early Triassic Indosinian orogeny in Vietnam (Truong Son Belt and Kontum Massif): Implications for the geodynamic evolution of Indochina. *Tectonophysics, 393*, 87–118.

Lepvrier, C., Van Vuong, N., Maluski, H., Thi, P. T., & Van, V. T. (2008). Indosinian tectonics in Vietnam. *Comptes Rendus Geosciences, 340*, 94–111.

Letouzey, J., & Kimura, M. (1986). The Okinawa Trough: Genesis of a back-arc basin developing along a continental margin. *Tectonophysics, 125*, 209–230.

Leyden, R., Ewing, M., & Murauchi, S. (1973). Sonobuoy refraction measurement in the East China Sea. *American Association of Petroleum Geologists Bulletin, 57*(12), 2396–2403.

Li, Z.-X. (1994). Collision between the North and South China blocks: A crustal-detachment model for suturing in the region east of the Tanlu fault. *Geology, 22*, 739–742.

Li, N. (1995). *Geothermics of the Okinawa Trough*. Qingdao, China: Qingdao Publishing House 138 pp.

Li, C.-F. (2004). Comparative geological study between the western Pacific marginal seas and the paleo-Tethyan marginal seas. In J. Li & S. Gao (Eds.), *Serial studies of the formation and evolution of the Chinese marginal seas: Vol. 3. Basin evolution and resources of the Chinese Marginal Seas* (pp. 46–53). Beijing, China: Ocean Press Color plate III, IV (in Chinese).

Li, X. (2006). Understanding 3D analytic signal amplitude. *Geophysics, 71*, L13–L16.

Li, Z.-X., & Lee, C.-T. A. (2006). Geochemical investigation of serpentinized oceanic lithospheric mantle in the Feather River Ophiolite, California: implications for the recycling rate of water by subduction. *Chemical Geology, 235*, 161–185.

Li, J. B. (Ed.), (2012). *Regional oceanography of China Seas—Marine geology*. Beijing: China Ocean Press 547 pp (in Chinese).

Li, C.-F., Chen, B., & Zhou, Z. (2009). Deep crustal structures of eastern China and adjacent seas revealed by magnetic data. *Science in China (Series D), 52*, 984–993.

Li, J. B., Ding, W. W., Gao, J. Y., Wu, Z. Y., & Zhang, J. (2011). Cenozoic evolution model of the seafloor spreading in South China Sea: New constraints from high resolution geophysical data. *Chinese Journal of Geophysics, 54*(12), 3004–3015.

Li, J. B., Ding, W. W., Wu, Z. Y., Zhang, J., & Dong, C. Z. (2012). The propagation of seafloor spreading in the southwestern subbasin, South China Sea. *Chinese Science Bulletin, 57*, 3182–3191.

Li, J. B., & Gao, S. (Eds.), (2003). *Lithospheric structure and dynamics of Chinese marginal seas*. Beijing: Ocean Publishing House 238 pp.

Li, J. B., & Gao, S. (Eds.), (2004). *Basin evolution and resources of Chinese marginal seas*. Beijing: Ocean Publishing House 509 pp.

Li, A.-C., Huang, J., Jiang, H.-Y., & Wan, S.-M. (2011). Sedimentary evolution in the northern slope of the South China Sea since the Oligocene and its response to tectonics. *Chinese Journal of Geophysics, 54*(6), 1084–1096.

Li, Q., Jian, Z., & Li, B. (2004). Oligocene-Miocene planktonic foraminifer biostratigraphy, site 1148, northern South China Sea. In W. L. Prell, P. Wang, P. Blum, D. K. Rea, & S. C. Clemens (Eds.), *Proceedings of the ocean drilling program, scientific results: Vol. 184.* (pp. 1–26) (Online).

Li, J. B., Jin, X., & Gao, J. (2002). Morpho-tectonic study on late-stage spreading of the Eastern Subbasin of South China Sea. *Science in China (Series D), 45*, 978–989.

Li, J. B., Jin, X., Ruan, A., Wu, S., Wu, Z., & Liu, J. (2004). Indentation tectonics in the accretionary wedge of middle Manila Trench. *Chinese Science Bulletin, 49*(12), 1279–1288.

Li, Z. X., & Li, X. H. (2007). Formation of the 1300 km-wide intra-continental orogen and post-orogenic magmatic province in Mesozoic South China: A flat-slab subduction model. *Geology, 35*, 179–182.

Li, S. G., Li, Q. L., Hou, Z. H., Yang, W., & Wang, Y. (2005). Cooling history and exhumation mechanism of the ultrahigh-pressure metamorphic rocks in the Dabie mountains, central China. *Acta Petrologica Sinica, 21*(4), 1117–1124.

Li, X. H., Li, Z. X., Li, W. X., & Wang, Y. J. (2006). Initiation of the Indosinian Orogeny in South China: Evidence for a Permian magmatic arc on the Hainan Island. *Journal of Geology, 114*, 341–353.

Li, C.-F., Lin, J., & Kulhanek, D. K. (2013). South China Sea tectonics: opening of the South China Sea and its implications for southeast Asian tectonics, climates, and deep mantle processes since the late Mesozoic. *IODP Expedition, 349*, http://dx.doi.org/10.2204/iodp. sp.349.2013, Scientific Prospectus.

Li, S. T., Lin, C. S., & Zhang, Q. M. (1998). Dynamic process of episodic rifting in continental marginal basin and tectonic events since 10Ma in South China Sea. *Chinese Science Bulletin, 43*(8), 797–810.

Li, X. H., & McCulloch, M. (1996). Secular variation in the Nd isotopic composition of Neoproterozoic sediments from the southern margin of the Yangtze block: Evidence for a Proterozoic continental collision in southeast China. *Precambrian Research, 76*, 67–76.

Li, C.-F., Shi, X., Zhou, Z., Li, J., Geng, J., & Chen, B. (2010). Depths to the magnetic layer bottom in the South China Sea area and their tectonic implications. *Geophysical Journal International, 182*, 1229–1247.

Li, C.-F., & Song, T. (2012). Magnetic recording the Cenozoic oceanic crustal accretion and evolution of the South China Sea Basin. *Chinese Science Bulletin, 57*, 3165–3181.

Li, C.-F., Wang, J., Zhou, Z., Geng, J., Chen, B., Yang, F., et al. (2012). 3D geophysical characterization of the Sulu–Dabie orogen and its environs. *Physics of the Earth and Planetary Interiors, 192–193*, 35–53.

Li, X. H., Wei, G. J., Shao, L., Liu, Y., Liang, X., Jian, Z., et al. (2003). Geochemical and Nd isotopic variations in sediments of the SCS: A response to Cenozoic tectonic in SE Asia. *Earth and Planetary Science Letters, 211*(3–4), 207–220.

Li, C.-F., Zhou, Z., Ge, H., & Mao, Y. (2007). Correlations between erosions and relative uplifts from the central inversion zone of the Xihu Depression, East China Sea Basin. *Terrestrial, Atmospheric and Oceanic Sciences, 18*, 757–776.

Li, C.-F., Zhou, Z., Ge, H., & Mao, Y. (2009). Rifting process of the Xihu Depression, East China Sea Basin. *Tectonophysics, 472*, 135–147.

Li, C.-F., Zhou, Z., Hao, H., Chen, H., Wang, J., Chen, B., et al. (2008). Late Mesozoic tectonic structure and evolution along the present-day northeast South China Sea continental margin. *Journal of Asian Earth Sciences, 31*, 546–561.

Li, C.-F., Zhou, Z., Li, J., Chen, B., & Geng, J. (2008). Magnetic zoning and seismic structure of the South China Sea ocean basin. *Marine Geophysical Research, 29*, 223–238.

Li, C.-F., Zhou, Z., Li, J., Chen, H., Geng, J., & Li, H. (2007). Precollisional tectonics and terrain amalgamation offshore southern Taiwan: Characterizations from reflection seismic and potential field data. *Science in China Series D: Earth Sciences, 50*, 897–908.

Li, C.-F., Zhou, Z., Li, J., Hao, H., & Geng, J. (2007). Structures of the northeasternmost South China Sea continental margin and ocean basin: Geophysical constraints and tectonic implications. *Marine Geophysical Research, 28*, 59–79.

Lin, A. T., Watts, A. B., & Hesselbow, S. P. (2003). Cenozoic stratigraphy and subsidence history of the South China Sea margin in the Taiwan region. *Basin Research, 15*, 453–478.

Liu, G. (Ed.), (1992). *Geophysical series maps in China Seas and adjacent regions.* Beijing: Science Press (in Chinese).

Liu, Y. (1994). *Neotectonics and crustal stability of the South China Sea.* Beijing: Science Press 284 pp (in Chinese).

Liu, Z., Huang, C., & Yang, S. (1988). *Geological tectonics and continental margin extension in the South China Sea*. Beijing: Science Press 398 pp (in Chinese).

Liu, S., & Li, S. (2001). *Petroleum prospecting in the East China Sea*. Beijing: Geological Publishing House 278 pp (in Chinese).

Liu, Q.-S., Liu, Q., Yang, T., Zeng, Q., Zheng, J., Luo, Y., et al. (2009). Magnetic study of the UHP eclogites from the Chinese Continental Scientific Drilling (CCSD) Project. *Journal of Geophysical Research*, *114*, B02106. http://dx.doi.org/10.1029/2008JB005917.

Liu, B., & Quan, Q. (Eds.), (1996). *Historical geology* (pp. 1–277). Beijing, China: Geological Press (in Chinese).

Liu, Z., Wang, Q., Yuan, H., & Su, D. (1983). Contours of Bouguer gravity anomalies and depths of Mohorovicic discontinuity in the South China Sea. *Journal of Tropical Oceanography*, *2*(2), 167–172 (in Chinese).

Liu, Q.-S., Zeng, Q., Zheng, J., Yang, T., Qiu, N., Liu, Z., et al. (2010). Magnetic properties of serpentinized garnet peridotites from the CCSD main hole in the Sulu ultrahigh-pressure metamorphic belt, eastern China. *Journal of Geophysical Research*, *115*, B06104. http://dx.doi.org/10.1029/2009JB000814.

Liu, Z., Zhao, H., & Fan, S. (2002). *Geology of the South China Sea*. Science Press 502 pp (in Chinese).

Lüdmann, T., & Wong, H. K. (1999). Neotectonic regime at the passive continental margin of the northern South China Sea. *Tectonophysics*, *311*, 113–138.

Lüdmann, T., Wong, H. K., & Wang, P. (2001). Plio-Quaternary sedimentation processes and neotectonics of the northern continental margin of the South China Sea. *Marine Geology*, *172*, 331–356.

Ludwig, W. J., Kumar, N., & Houtz, R. E. (1979). Profiler-sonobuoy measurements in the South China Sea Basin. *Journal of Geophysical Research*, *84*, 3505–3518.

Ludwig, W. J., Murauchi, S., Den, N., Buhl, P., Hotta, H., Ewing, M., et al. (1973). Structure of East China Sea–West Philippine Sea margin off Southern Kyushu, Japan. *Journal of Geophysical Research*, *78*, 2526–2536.

Ludwig, N. A., & Valencia, M. J. (1993). Oil and Mineral Resources of the East China Sea: Prospects in Relation to Maritime Boundaries. *GeoJournal*, *30*(4), 381–387.

Lundberg, N., Reed, D. L., Liu, C. S., & Lieske, J., Jr. (1997). Forearc-basin closure and arc accretion in the submarine suture zone south of Taiwan. *Tectonophysics*, *274*, 5–23.

Madon, M. B. H., Meng, L. K., & Anuar, A. (2000). Sabah Basin. In K. M. Leong (Ed.), *The Petroleum Geology and Resources of Malaysia* (pp. 499–542)) Kuala Lumpur, Malaysia (Petroliam Nasional Berhad).

Malavieille, J., Lallemand, S. E., Dominguez, S., Deschamps, A., Lu, C.-Y., Liu, C.-S., et al. (2002). Arc-continent collision in Taiwan: New marine observations and tectonic evolution. In T. B. Byrne & C.-S. Liu (Eds.), *Geology and geophysics of an arc-continent collision*: *Vol. 358*. (pp. 189–213). Taiwan, Republic of China: Geological Society of America Special Paper.

Metcalfe, I. (2009). Late Palaeozoic and Mesozoic tectonic and palaeogeographical evolution of SE Asia. In E. Buffetaut, G. Cuny, J. Le Loeuff, & V. Suteethorn (Eds.), *Late palaeozoic and mesozoic ecosystems in SE Asia*: *Vol. 315*. (pp. 7–23). London: Geological Society.

Miao, W. L., Shao, L., Pang, X., Lei, Y., Qiao, P., Li, A., et al. (2008). REE geochemical characteristics in the northern South China Sea since the Oligocene. *Marine Geology and Quaternary Geology*, *28*(2), 71–78 (in Chinese).

Miki, M. (1995). Two-phase opening model for the Okinawa Trough inferred from paleomagnetic study of the Ryukyu Arc. *Journal of Geophysical Research*, *100*, 8169–8184.

Montelli, R., Nolet, G., Dahlen, F. A., Masters, G., Engdahl, E. R., & Hung, S.-H. (2004). Finite-frequency tomography reveals a variety of plumes in the mantle. *Science*, *303*, 338–343.

Morley, C. K. (2002). A tectonic model for the Tertiary evolution of strike-slip faults and rift basins in SE Asia. *Tectonophysics, 347*(4), 189–215.

Nabighian, M. N. (1972). The analytic signal of two-dimensional magnetic bodies with polygonal cross-section—Its properties and use for automated anomaly interpretation. *Geophysics, 37*, 507–517.

Nabighian, M. N. (1984). Toward a three-dimensional automatic interpretation of potential field data via generalized Hilbert transforms—Fundamental relations. *Geophysics, 49*, 780–786.

Nakamura, Y., McIntosh, K., & Chen, A. T. (1998). Preliminary results of a large offset seismic survey west of Hengchun Peninsula, southern Taiwan. *Terrestrial, Atmospheric and Oceanic Sciences, 9*, 395–408.

Nealson, K. (2006). Lakes of liquid CO2 in the deep sea. *Proceedings of the National Academy of Sciences of the United States of America, 103*, 13903–13904.

Nilsen, T. H., & Sylvester, A. G. (1995). Strike-slip basins. In C. J. Busby & R. V. Ingersoll (Eds.), *Tectonics of sedimentary basins* (pp. 425–458). Malden, MA: Blackwell Science.

Nissen, S. S., Hayes, D. E., Buhl, P., Diebold, J., Yao, B., Zeng, W., et al. (1995). Deep penetration seismic soundings across the northern margin of the South China Sea. *Journal of Geophysical Research, 100*, 22407–22433.

Nissen, S. S., Hayes, D. E., & Yao, B. (1995). Gravity, heat flow, and seismic constraints on the processes of crustal extension: Northern margin of the South China Sea. *Journal of Geophysical Research, 100*, 22447–22483.

Ofoegbu, C. O., & Mohan, N. L. (1990). Interpretation of aeromagnetic anomalies over part of southeastern Nigeria using three dimensional Hilbert transformation. *Pure and Applied Geophysics, 134*, 13–29.

Okay, A. I., Sengör, A. M. C., & Satir, M. (1993). Tectonics of an ultrahigh-pressure metamorphic terrane: The Dabie Shan/Tongbai Shan orogen, China. *Tectonics, 12*, 1320–1334.

Oldenburg, D. W. (1974). The inversion and interpretation of gravity anomalies. *Geophysics, 39*, 526–536.

Pang, X., Chen, C., Shao, L., Wang, C., Zhu, M., He, M., et al. (2007). Baiyun movement, a great tectonic event on the Oligocene-Miocene boundary in the northern SCS and its implications. *Geological Review, 53*(2), 145–151 (in Chinese).

Park, J.-O., Tokuyama, H., Shinohara, M., Suyehiro, K., & Taira, A. (1998). Seismic record of tectonic evolution and backarc rifting in the southern Ryukyu island arc system. *Tectonophysics, 294*, 21–42.

Parker, R. L. (1973). The rapid calculation of potential anomalies. *Geophysical Journal of the Royal Astronomical Society, 31*, 447–455.

Pautot, G., Rangin, C., Briais, A., Tapponnier, P., Beuzart, P., Lericolais, G., et al. (1986). Spreading direction in the central South China Sea. *Nature, 321*, 150–154.

Pubellier, M., Monnier, C., Maury, R., & Tamayo, R. (2004). Plate kinematics, origin and tectonic emplacement of supra-subduction ophiolites in SE Asia. *Tectonophysics, 392*(1–4), 9–36.

Qi, J., & Yang, Q. (2010). Cenozoic structural deformation and dynamic processes of the Bohai Bay basin province, China. *Marine and Petroleum Geology, 27*, 757–771.

Qin, Y. (1990a). *Geology of the Bohai Sea*. Beijing: China Ocean Press 354 pp.

Qin, Y. (1990b). *Geology of the Yellow Sea*. Beijing: China Ocean Press 289 pp.

Qin, Y., Zhao, Y., & Chen, L. (1988). *Geology of the East China Sea*. Beijing: Science Press 263 pp.

Qiu, Y., Gao, S., McNaughton, N. J., Groves, D. I., & Ling, W. (2000). First evidence of >3.2 Ga continental crust in the Yangtze craton of south China and its implications for Archean crustal evolution and Phanerozoic tectonics. *Geology, 28*, 11–14.

Qiu, J., Shen, W., & Wang, D. (2002). Late Mesozoic A-type granites in the coastal areas of Zhejiang and Fujian. In D. Wang, X. Zhou, & J. Qiu (Eds.), *Petrogenesis of granitic volcanic-intrusive complexes and crustal evolution in Southeast China* (pp. 1–295). Beijing, China: Science Press.

Rangin, C., Jolivet, L., & Pubellier, M. (1990). A simple model for the tectonic evolution of Southeast Asia and Indonesia regions for the past 43 m.y. *Bulletin de la Societe Geologique de France, 6*((6), 889–905.

Rangin, C., Klein, M., Roques, D., Le Pichon, X., & Trong, L. V. (1995). The Red River fault system in the Tonkin Gulf, Vietnam. *Tectonophysics, 243*(3–4), 209–222.

Reed, D. L., Lundberg, N., Liu, C. S., & Kuo, B. Y. (1992). Structural relations along the margins of the offshore Taiwan accretionary wedge: Implication for accretion and crustal kinematics. *Acta Geologica Taiwanica, 30*, 105–122.

Ren, J.-S., Niu, B.-G., Wang, J., He, Z.-J., Jin, X.-C., Xie, L.-Z., et al. (2013). 1:5 Million International Geological Map of Asia. *Acta Geoscientica Sinica, 34*(1), 24–30.

Ren, J., Tamaki, K., Li, S., & Zhang, J. (2002). Late Mesozoic and Cenozoic rifting and its dynamic setting in Eastern China and adjacent areas. *Tectonophysics, 344*, 175–205.

Roest, W. R., Verhoef, J., & Pilkington, M. (1992). Magnetic interpretation using the 3-D analytic signal. *Geophysics, 57*, 116–125.

Ru, K., & Pigott, J. D. (1986). Episodic rifting and subsidence in the South China Sea. *AAPG Bulletin, 70*, 1136–1155.

Sakai, H., Gamo, T., Kim, E.-S., Shitashima, K., Yanagisawa, F., Tsutsumi, M., et al. (1990). Unique chemistry of the hydrothermal solution in the mid-Okinawa Trough backarc basin. *Geophysical Research Letters, 17*, 2133–2136.

Sakai, H., Gamo, T., Kim, E.-S., Tsutsumi, M., Tanaka, T., Ishibashi, J., et al. (1990). Venting of carbon dioxide-rich fluid and hydrate formation in mid-Okinawa trough Backarc Basin. *Science, 248*, 1093–1096.

Salem, A., Ravat, D., Gamey, T. J., & Ushijima, K. (2002). Analytic signal approach and its applicability in environmental magnetic applications. *Journal of Applied Geophysics, 49*, 231–244.

Schärer, U., Tapponnier, P., Lacassin, R., Leloup, P. H., Zhong, D., & Ji, S. (1990). Intraplate tectonics in Asia: A precise age for large-scale Miocene movement along the AilaoShan-Red River shear zone, China. *Earth and Planetary Science Letters, 97*, 65–77.

Schluter, H. U., Hinz, K., & Block, M. (1996). Tectono-stratigraphic terranes and detachment faulting of the South China Sea and Sulu Sea. *Marine Geology, 130*, 39–78.

Seno, T., Stein, S. A., & Gripp, A. E. (1993). A model for the motion of the Philippine Sea plate consistent with NUVEL-1 and geological data. *Journal of Geophysical Research, 98*, 17941–17948.

Shao, L. (2008). Sedimentary filling of the Pearl River Mouth Basin and its response to the evolution of the Pearl River. *Acta Sedimentologica Sinica, 26*(2), 179–185 (in Chinese).

Shen, G., Ujiie, H., & Sashida, K. (1996). Off-scraped Permian-Jurassic bedded chert thrust on Jurassic-early Cretaceous accretionary prism: Radiolarian evidence from Ie Island, central Ryukyu Island Arc. *The Island Arc, 5*, 156–165.

Shi, H., & Li, C.-F. (2012). Mesozoic and early Cenozoic tectonic convergence-to-rifting transition prior to opening of the South China Sea. *International Geology Review, 54*(15), 1801–1828.

Shih, M.-H. (2001). *Moho depth beneath the Taiwan Strait from gravity data.* (pp. 1–72). Taiwan, Republic of China: National Central University, Master Thesis.

Shinjo, R. (1999). Geochemistry of high Mg andesites and the tectonic evolution of the Okinawa Trough—Ryukyu arc system. *Chemical Geology, 157*, 69–88.

Shu, L. S., & Xu, M. J. (2002). Geological background of southeast China. In D. Wang & X. Zhou (Eds.). *Petrogenesis of granitic volcanic-intrusive complexes and crustal evolution in Southeast China* (pp. 1–295). Beijing, China: Science Press (in Chinese).

Shu, L. S., Zhou, X. M., Deng, P., Wang, B., Jiang, S. Y., Yu, J. H., et al. (2009). Mesozoic tectonic evolution of the Southeast China Block: New insights from basin analysis. *Journal of Asian Earth Sciences, 34*, 376–391.

Sibuet, J.-C., & Hsu, S.-K. (1997). Geodynamics of the Taiwan arc-arc collision. *Tectonophysics, 274*, 221–251.

Sibuet, J.-C., Hsu, S.-K., & Debayle, E. (2004). Geodynamic context of the Taiwan orogen. In P. Clift, P. Wang, W. Kuhnt, & D. Hayes (Eds.), *Geophysical Monograph Series: Vol. 149. Ocean-continent interactions within East Asian marginal seas* (pp. 127–158). Washington, D.C: American Geophysical Union.

Sibuet, J.-C., Hsu, S.-K., Le Pichon, X., Le Formal, J.-P., Reed, D., Moore, G., et al. (2002). East Asia plate tectonics since 15Ma: Constraints from the Taiwan region. *Tectonophysics, 344*, 103–134.

Sibuet, J.-C., Hsu, S.-K., Shyu, C.-T., & Liu, C.-S. (1995). Structural and kinematic evolution of the Okinawa trough backarc basin. In B. Taylor (Ed.), *Backarc basins: Tectonics and magmatism* (pp. 343–378). New York, NY: Plenum.

Sibuet, J. C., Letouzey, J., Barbier, F., Charvet, J., Foucher, J. P., Hilde, T. W. C., et al. (1987). Back arc extension in the Okinawa Trough. *Journal of Geophysical Research, 92*, 14041–14063.

Song, T., & Li, C.-F. (2012). The opening ages and mode of the South China Sea estimated from high-density magnetic tracks. *Progress in Geophysics, 27*, 1432–1442 (in Chinese).

Su, C., Li, C.-F., & Ge, H. (2010). Characteristics and causes of anomalous reflections on the Oligocene and Miocene transition in the Xihu Depression, East China Sea Basin. *Journal of Marine Science, 4*, 14–21 (in Chinese).

Su, D., Whitem, N., & McKenzie, D. (1989). Extension and subsidence of the Pearl River Mouth basin, northern South China Sea. *Basin Research, 2*, 205–222.

Sun, S., Li, J., Chen, H., Peng, H., Hsu, K. J., & Shelton, J. W. (1989). Mesozoic and Cenozoic sedimentary history of South China. *Bulletin of the American Association of Petroleum Geologists, 73*, 1247–1269.

Sun, Z., Zhong, Z., Keep, M., Zhou, D., Cai, D., Li, X., et al. (2008). 3D analogue modeling of the South China Sea: A discussion on breakup pattern. *Journal of Asian Earth Sciences, 34*, 544–556.

Sun, Z., Zhou, D., Zhong, Z. H., Zeng, Z. X., & Wu, S. M. (2003). Experimental evidence for the dynamics of the formation of the Yinggehai basin, NW South China Sea. *Tectonophysics, 372*, 41–58.

Suppe, J. (1984). Kinematics of arc-continent collision, flipping of subduction, and back-arc spreading near Taiwan. *Memoir of the Geological Society of China, 6*, 21–34.

Suzuki, R., Ishibashi, J., Nakaseama, M., Konno, U., Tsunogai, U., Gena, K., et al. (2008). Diverse range of mineralization induced by phase separation of hydrothermal fluid: Case study of the Yonaguni Knoll IV hydrothermal field in the Okinawa Trough back-arc basin. *Resource Geology, 58*(3), 267–288.

Suzuki, S., Takemura, S., Yumul, G. P., , Jr.David, S. D., & Asiedu, D. K. (2000). Composition and provenance of the Upper Cretaceous to Eocene sandstones in central Palawan, Philippines: Constraints on the tectonic evolution of Palawan. *Island Arc, 9*, 611–626.

Takahashi, M., Aramaki, S., & Ishihara, S. (1980). Magnetite series/ilmenite series vs. I-type/S-type granitoids. In I. Ishihara & S. Takenouchi (Eds.), *Granitic magmatism and related mineralization. Mining geology* (pp. 13–28). Tokyo, Japan: Society of Resource Geologists of Japan.

Takai, K., Mottl, M. J., Nielsen, S. H. H., & the IODP Expedition 331 Scientists. (2012). IODP expedition 331: Strong and expansive subseafloor hydrothermal activities in the Okinawa trough. *Scientific Drilling, 13*, 19–27.

Takai, K., & Nakamura, K. (2010). In L. Barton, M. Mendl, & A. Loy (Eds.), *Geomicrobiology: Molecular and environmental perspective* (pp. 251–283). New York, NY: Springer.

Tapponnier, P., Lacassin, R., Leloup, P. H., et al. (1990). The Ailao Shan/Red River metamorphic belt: Tertiary left-lateral shear between Indochina and South China. *Nature, 343,* 431–437.

Tapponnier, P., Peltzer, G., Le Dain, A. Y., Armijo, R., & Cobbold, P. (1982). Propagating extrusion tectonics in Asia: New insights from simple experiments with plasticine. *Geology, 7,* 611–616.

Taylor, B., & Hayes, D. E. (1980). The tectonic evolution of the South China Sea. In D. E. Hayes (Ed.), *Geophysical Monograph Series: Vol. 23. The tectonic and geologic evolution of South Eastern Asian Seas and Islands, I* (pp. 89–104). Washington, D.C: American Geophysical Union.

Taylor, B., & Hayes, D. E. (1983). Origin and history of the South China Sea basin. In D. E. Hayes (Ed.), *Geophysical Monograph Series: Vol. 27. The tectonic and geologic evolution of South Eastern Asian Seas and Islands, II* (pp. 23–56). Washington, D.C: American Geophysical Union.

Teng, L. S. (1990). Geotectonic evolution of the late-Cenozoic arc-continent collision in Taiwan. *Tectonophysics, 183,* 57–76.

Teng, L. S. (1996). Extensional collapse of the northern Taiwan mountain belt. *Geology, 24,* 949–952.

Tongkul, F. (1994). The geology of northern Sabah, Malaysia: Its relationship to the opening of the South China Sea Basin. *Tectonophysics, 235,* 131–137.

Tsai, C.-H., Hsu, S.-K., Yeh, Y.-C., Lee, C.-S., & Xia, K. (2004). Crustal thinning of the northern continental margin of the South China Sea. *Marine Geophysical Researches, 25,* 63–78.

Tsuji, T., Takai, K., Oiwane, H., Nakamura, Y., Masaki, Y., Kumagai, H., et al. (2012). Hydrothermal fluid flow system around the Iheya North Knoll in the mid-Okinawa trough based on seismic reflection data. *Journal of Volcanology and Geothermal Research, 213–214,* 41–50.

Tu, K., Flower, M. F. J., Carlson, R. W., Xie, G., Chen, C.-Y., & Zhang, M. (1992). Magmatism in the South China Basin: 1. Isotopic and trace-element evidence for an endogenous Dupal mantle component. *Chemical Geology, 97*(1–2), 47–63.

Tzeng, J. (1994). *Tertiary seismic stratigraphic analysis of the Tainan Basin.* Taipei, Taiwan: National Taiwan University M.S. Thesis (in Chinese).

Uyeda, S. (1987). Active hydrothermal mounds in the Okinawa Back-arc Trough. *Eos, Transactions American Geophysical Union, 68*(36), 737.

Wageman, J. M., Hilde, T. W. C., & Emery, K. O. (1970). Structural framework of East China Sea and Yellow Sea. *Bulletin of American Association of Petroleum Geologists, 54,* 1611–1643.

Wang, S. (1982). Basic geological structural features of the basin at the mouth of Pearl River. *Acta Petrologica Sinica, S1,* 1–13 (in Chinese).

Wang, P. (2004). Cenozoic deformation and the history of sea-land interactions in Asia. In P. D. Clift, W. Kuhnt, P. Wang, & D. Hayes (Eds.), *Geophysical Monograph Series: Vol. 149. Continent-ocean interactions in the East Asian Marginal Seas* (pp. 1–22). Washington, D.C: American Geophysical Union.

Wang, P. (2012). Tracing the life history of a marginal sea—On "The South China Sea Deep" Research Program. *Chinese Science Bulletin, 57*(24), 3093–3114.

Wang, T. K., Chen, M.-K., Lee, C.-S., & Xia, K. (2006). Seismic imaging of the transitional crust across the northeastern margin of the South China Sea. *Tectonophysics, 412,* 237–245.

Wang, K.-L., Chung, S.-L., Chen, C.-H., Shinjo, R., Yang, T. F., & Chen, C.-H. (1999). Post-collisional magmatism around northern Taiwan and its relation with opening of the Okinawa Trough. *Tectonophysics, 308,* 363–376.

Wang, Q., & Cong, B. (1996). Tectonic implication of UHP rocks from the Dabie mountains. *Science in China Series D: Earth Sciences, 39*, 311–318.

Wang, Y. J., Fan, W., Guo, F., Peng, T., & Li, C. (2003). Geochemistry of Mesozoic mafic rocks adjacent to the Chenzhou-Linwu fault, south China: Implications for the lithospheric boundary between the Yangtze and Cathaysia blocks. *International Geological Review, 45*, 263–286.

Wang, P., Prell, W. L., Blum, P., Arnold, E. M., Buehring, C. J., Chen, M.-P., et al. (2000). *Proceedings of ocean drilling program, initial report, 184.* College Station, TX: Ocean Drilling Program.

Wang, K.-D., Wang, J.-P., Xu, G.-Q., Zhong, S.-L., Zhang, Y.-Y., & Yang, H.-R. (2000). The discovery and division of the Mesozoic strata in the southwest of Donghai Shelf Basin. *Journal of Stratigraphy, 24*(2), 129–131 (in Chinese).

Wang, X. J., Wu, M., Liang, D., & Yin, A. (1985). Some geochemical characteristics of basalts in the South China Sea. *Chinese Journal of Geochemistry, 4*, 380–390.

Wu, J. (1988). Cenozoic basins of the South China Sea. *Episodes, 11*, 91–96.

Wu, H. (2003). Tectonopalaeogeographic analysis of the geologic problems related to ophiolitic belt in northeastern Jiangxi province. *Journal of Palaeogeography, 5*, 328–342 (in Chinese).

Wu, G. X., Qin, J. G., & Mao, S. Z. (2003). Deep-water Oligocene pollen record from South China Sea. *Chinese Science Bulletin, 48*(17), 1868–1871 (in Chinese).

Xia, K., & Huang, C. (2000). The discovery of Meso-Tethys sedimentary basins in the South China Sea and their oil and gas perspective. *Earth Science Frontiers, 7*(3), 227–238 (in Chinese).

Xia, K., Huang, C., & Huang, Z. (2004). Upper Triassic-Cretaceous sediment distribution and hydrocarbon potential in South China Sea and its adjacent areas. *China Offshore Oil and Gas, 16*(2), 73–83 (in Chinese).

Xia, K., et al. (1997). Geophysical field and crustal structures of the South China Sea. In Z. Gong & S. Li (Eds.). *Continental margin basin analysis and hydrocarbon accumulation of the Northern South China Sea* (pp. 1–16). Beijing, China: Science Press (in Chinese).

Xiao, G., & Zheng, J. (2004). New opinions about "residual Tethys" in northern South China Sea slope and southern East China Sea. *Geoscience, 18*(1), 103–108 (in Chinese).

Xie, X., Müller, R. D., Li, S., Gong, Z., & Steinberger, B. (2006). Origin of anomalous subsidence along the Northern South China Sea margin and its relationship to dynamic topography. *Marine and Petroleum Geology, 23*, 745–765.

Xu, S., Chen, J., Zheng, J., Li, X., & Sang, J. (1987). Tertiary depositional characteristics in the east area of Pearl River Mouth Basin. In Guangdong Petroleum Society and China Oil, Guangzhou (Ed.), *Proceedings of the international symposium on petroleum geology of northern continental shelf area of the South China Sea* (pp. 469–491).

Xu, D., & Jiang, J. (1989). Moho pattern and deep structure in the North center of South China Sea. *Donghai Marine Science, 7*(1), 48–56 (in Chinese).

Xu, D., Liu, X., & Zhang, X. (1997). *China offshore geology.* Beijing: Geological Publishing House 310 pp (in Chinese).

Xu, Y. G., Sun, M., Yan, W., Liu, Y., Huang, X. L., & Chen, X. M. (2002). Xenolith evidence for polybaric melting and stratification of the upper mantle beneath South China. *Journal of Asian Earth Sciences, 20*, 937–954.

Xu, Y. G., Wei, J. X., Qiu, H. N., Zhang, H. H., & Huang, X. L. (2012). Opening and evolution of the South China Sea constrained by studies on volcanic rocks: Preliminary results and a research design. *Chinese Science Bulletin, 57*, 3150–3164.

Xu, Z., Zhang, Z., Liu, F., Yang, J., Li, H., Yang, T., et al. (2003). Exhumation structure and mechanism of the Sulu ultrahigh-pressure metamorphic belt, central China. *Acta Geologica Sinica, 77*(4), 433–450.

Xu, J., Zhu, G., Tong, W. X., Cui, K. R., & Lin, Q. (1987). Formation and evolution of the Tancheng–Lujiang wrench fault system: A major shear system to the northwest of the Pacific Ocean. *Tectonophysics, 134*, 273–310.

Yan, P., Deng, H., Liu, H., Zhang, Z., & Jiang, Y. (2006). The temporal and spatial distribution of volcanism in the South China Sea region. *Journal of Asian Earth Sciences, 27*, 647–659.

Yan, Q. S., & Shi, X. F. (2007). Hainan mantle plume and the formation and evolution of the South China Sea. *Journal of China University of Geosciences, 13*, 311–322 (in Chinese).

Yan, Q. S., Shi, X. F., Wang, K. S., Bu, W., & Xiao, L. (2008). Major, trace elements and Sr-Nd-Pb isotope study of Cenozoic alkali basalts of the South China Sea. *Science in China Series D: Earth Sciences, 51*, 550–566.

Yan, P., Zhou, D., & Liu, Z. (2001). A crustal structure profile across the northern continental margin of the South China Sea. *Tectonophysics, 338*, 1–21.

Yang, J., Feng, X., Fan, Y., & Zhu, S. (2003). An analysis of middle-late Mesozoic tectonics, paleogeography and petroleum potential in the northeastern South China Sea. *China Offshore Oil and Gas (Geology), 17*(2), 89–103 (in Chinese).

Yang, S., Hu, S., Cai, D., Feng, X., Chen, L., & Gao, L. (2004). Present-day heat flow, thermal history and tectonic subsidence of the East China Sea Basin. *Marine and Petroleum Geology, 21*, 1095–1105.

Yang, T. F., Lee, T., Chen, C.-H., Cheng, S.-N., Knittel, U., Punongbayan, R. S., et al. (1996). A double island arc between Taiwan and Luzon: Consequence of ridge subduction. *Tectonophysics, 258*, 85–101.

Yang, K. M., Wang, Y., Tsai, Y. B., & Hsu, V. (1983). Paleomagnetic studies of the Coastal Range, Lutao and Lanhsu in eastern Taiwan and their tectonic implications. *Bulletin of the Institute of Earth Sciences, Academia Sinica, 3*, 173–189.

Yao, B., Zeng, W., Hayes, D. E., & Spangler, S. (1994). *The geological memoir of South China Sea surveyed jointly by China and USA*. Wuhan, China: China Univ. Geosci. Press 204 pp (in Chinese).

Ye, H., Shedlock, K. M., Hellinger, S. J., & Sclater, J. G. (1985). The North China basin: An example of a Cenozoic rifted intraplate basin. *Tectonics, 4*, 153–169.

Yeh, K.-Y., & Cheng, Y.-N. (1996). Jurassic radiolarians from the northwest coast of Busuanga Island, North Palawan Block, Philippines. *Micropaleontology, 42*, 93–124.

Yi, S., Yi, S., Batten, D. J., Yun, H., & Park, S.-J. (2003). Cretaceous and Cenozoic on-marine deposits of the Northern South Yellow Sea Basin, offshore western Korea: Palynostratigraphy and palaeoenvironments. *Palaeogeography, Palaeoclimatology, Palaeoecology, 191*, 15–44.

Yin, A., & Nie, S. (1993). An indentation model for the North and South China collision and the development of the Tan-Lu and Honam fault system, eastern Asia. *Tectonics, 12*, 801–803.

Yu, S.-B., Chen, H.-Y., & Kuo, L.-C. (1997). Velocity field of GPS stations in the Taiwan area. *Tectonophysics, 274*, 41–59.

Yu, P., & Li, N. (1992). *Heat flow in crust of the East China Sea*. Beijing: China Ocean Press (in Chinese).

Zamoras, L. R., & Matsuoka, A. (2001). Malampaya Sound Group: A Jurassic-early Cretaceous accretionary complex in Busuanga Island, North Palawan Block. *Journal of Geological Society of Japan, 107*, 316–336.

Zamoras, L. R., & Matsuoka, A. (2004). Accretion and postaccretion tectonics of the Calamian Islands, North Palawan Block (Philippines). *Island Arc, 13*, 506–519.

Zamoras, L. R., Montes, M. G. A., Queaño, K. L., Marquez, E. J., Dimalanta, C. B., Gabo, J. A. S., et al. (2008). Buruanga peninsula and antique range: Two contrasting terranes in Northwest Panay, Philippines featuring an arc–continent collision zone. *Island Arc, 17*, 443–457.

ZBGMR. (1989). *Memoir of regional geology of Zhejiang province.* (pp. 1–608). Beijing, China: Geological Publishing House.

Zhai, S., Chen, L., & Zhang, H. (2001). *Igneous process and subseafloor hydrothermal activity in the Okinawa trough.* Beijing: China Ocean Press 240 pp (in Chinese).

Zhang, W. (1986). *Marine and continental tectonics in China Seas and adjacent regions.* Beijing: Science Press 575 pp (in Chinese).

Zhang, K. J. (1997). North and South China collision along the eastern and southern North China margins. *Tectonophysics, 270,* 145–156.

Zhang, Q. M. (1999). Evolution of Ying-Qiong Basin and its tectonic-thermal system. *Natural Gas Industry, 19*(1), 12–17 (in Chinese).

Zhang, X. (2008). *Tectonic geology in China Seas.* Beijing: China Ocean Press 404 pp (in Chinese).

Zhang, K.-J., Cai, J.-X., & Zhu, J.-X. (2006). North China and South China collision: Insights from analogue modeling. *Journal of Geodynamics, 42,* 38–51.

Zhang, H., Chen, B., & Zhang, H. (2005). *Geology and mineral resources in Chinese offshore areas.* Beijing: China Ocean Press 384 pp (in Chinese).

Zhang, S., & Li, C.-F. (2011). Magmatic activities in eastern China and adjacent seas revealed from 3D analytic signals of magnetic data. *Geophysical and Geochemical Exploration, 35* (3), 290–297 (in Chinese).

Zhao, Q. H. (2005). Late Cainozoic ostracod faunas and paleoenvironmental changes at ODP site 1148, South China Sea. *Marine Micropaleontology, 54*(1–2), 27–47.

Zhao, D. (2007). Seismic images under 60 hotspots: Search for mantle plumes. *Gondwana Research, 12,* 335–355.

Zhong, D., Ding, L., Ji, J., Zhang, J., Liu, F., Liu, J., et al. (2001). Coupling of the lithospheric convergence of west China and dispersion of east China in Cenozoic: Link with paleoenvironmental changes. *Quaternary Sciences, 21,* 303–312 (in Chinese).

Zhong, Z., Wang, L., Xia, B., Dong, W., Sun, Z., & Shi, Y. (2004). The dynamics of Yinggehai basin formation and its tectonic significance. *Acta Geologica Sinica, 78*(3), 302–309.

Zhou, D., Chen, H., Sun, Z., & Xu, H. (2005). Three Mesozoic sea basins in eastern and southern South China Sea and their relation to Tethys and Paleo-Pacific domains. *Journal of Tropical Oceanography, 24*(2), 16–25 (in Chinese).

Zhou, X. M., & Li, W. X. (2000). Origin of late Mesozoic igneous rocks in southeastern China: Implications for lithosphere subduction and underplating of mafic magmas. *Tectonophysics, 326,* 269–287.

Zhou, D., Liang, Y.-B., & Zeng, C.-K. (Eds.), (1994). *Part IV marine geology: Vol. 2. In Oceanology of China Seas* (pp. 345–429). Dordrecht, Germany: Kluwer.

Zhou, D., Ru, K., & Chen, H. Z. (1995). Kinematics of Cenozoic extension on the South China Sea continental margin and its implications for the tectonic evolution of the region. *Tectonophysics, 251,* 161–177.

Zhou, X., Sun, T., Shen, W., Shu, L., & Niu, Y. (2006). Petrogenesis of Mesozoic granitoids and volcanic rocks in South China: A response to tectonic evolution. *Episodes, 29,* 26–33.

Zhu, W. (2007). *Gas geology in marginal basins of the Northern South China Sea.* Beijing: Petroleum Industry Press 391 pp (in Chinese).

Zhu, J., Qiu, X., Kopp, H., Xu, H., Sun, Z., Ruan, A., et al. (2012). Shallow anatomy of a continent–ocean transition zone in the northern South China Sea from multichannel seismic data. *Tectonophysics, 554–557,* 18–29.

Chapter 4

Sedimentology

4.1 INTRODUCTION

Marine geology in China started from sedimentology in the 1950s, mainly on deltas and coastal sediments. It was not until the end of the "cultural revolution" in the late 1970s, however, when China's earth scientists were able to embark on systematic researches of shallow shelf depositions as well as deltas. From then on, many institutional and interinstitutional projects, sometimes with international collaborations on marine geology survey in general and on marine sedimentology in particular, were carried out, resulting in the publication of numerous sediment atlases and monographs since the 1980s. Most of these publications are in Chinese, including such landmark works as "Geology of the Bohai Sea," "Geology of the Yellow Sea," and "Geology of the East China Sea" (Qin, 1985, 1989; Qin, Zhao, & Chen, 1988), all with an emphasis on marine sedimentology. Subsequent offshore oil and gas exploration and deep-sea research into the South China Sea and the Okinawa Trough have stimulated further development of marine sedimentology in China (He, 2006). A major shift from the old-fashioned descriptive stage occurred after the 1990s when systematic and genetic approach toward "source to sink" processes became the main trend in Chinese sedimentologic studies.

Lying in the dynamic land–sea interaction center of the marginal western Pacific, the China Seas accumulate voluminous sediments weathered from neighboring landmasses. Clay and sand discharged from rivers constitute the main part of the sediments deposited on the seafloor. Due to their vast geographic coverage, the China Seas feature distinctive regional patterns of sedimentation, as outlined in this chapter.

In the following, we will start from "Modern Sediment Distribution" to discuss major sedimentologic features in two large regional sectors, the East China Sea (ECS, including the Bohai Sea, Yellow Sea, and Okinawa Trough) and the South China Sea (SCS). This will be followed by "Depositional History" of various shelf and deep-sea areas by emphasizing on their late Quaternary deposition. "Terrigenous Sediments" includes discussions on fluvial sediments (modern or relict) that form the major river deltas, mud fields, and sand ridges on shelf. In "Biogenic Sediments," distributions of coral reefs,

183

calcareous and siliceous biogenic components, and organic carbon will be presented, followed by "Volcanic Sediments" in the Okinawa Trough and the SCS deep sea. Finally, we conclude the chapter with a summary of two points: origins of sediments and their transport dynamics.

4.2 MODERN SEDIMENT DISTRIBUTION

4.2.1 Sediment Distribution Patterns

4.2.1.1 Major Sedimentologic Features

The China Seas showcase the following three major features in sedimentology, which are beneficial to studies of the sedimentologic records for environmental changes in the region:

1. Sedimentologic patterns of the China Seas have largely resulted from the land–sea interaction between the Pacific and East Asia superimposed by sea-level fluctuations over glacial cycles, as detailed in Chapter 7. The eastern Asian continent and the surrounding island arcs provided sediments in particulate and dissolved forms, which were subsequently transported by coastal currents and oceanic currents represented by the Kuroshio and its branches. The last glaciation exposed the broad shelves and washed sediments into the deeper waters, while the postglacial transgression redistributed the sediments and generated the current picture of sediment distribution characterized by Holocene muds over some proximal parts of the shelf and partially reworked relict (palimpsest) sediments on middle and outer shelf.

2. As marginal basins, the China Seas capture the majority of land-delivered sediments, giving rise to very high sedimentation rates inside the basins compared to very low values in the open Pacific. This explains the absence from the western Pacific of large deep-sea fans in contrast to their occurrence in the Indian (Bengal and Indus) and Atlantic (Amazon). Related is the existence of wide continental shelves, especially the Sunda Shelf and the East China Sea shelf, as a large-scale coastline migration over glacial cycles had caused drastic changes in regional sedimentology.

3. The lower-latitude SCS is also characterized by intensive ventilation in its bottom water (with resident time of only 30–40 yrs) due to the inflood of North Pacific Intermediate and Deep Water, allowing better carbonate preservation in SCS deep waters than all other marginal seas in the western Pacific.

4.2.1.2 Sediment Types and Distribution

The most common sediment types in the China Seas are sand, silt, and silty clay on the shelf and increasing biogenic components in the deep sea (Figure 4.1). At present, the five largest rivers in the region—Yellow,

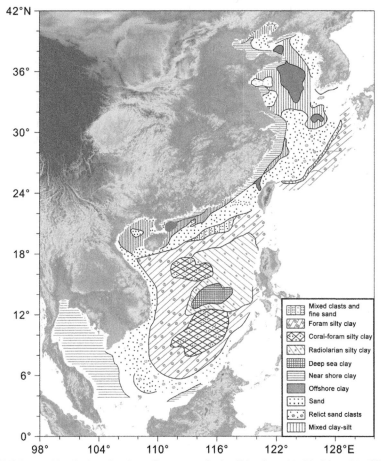

FIGURE 4.1   Distribution of major sediment types in the China Seas. *Modified from Liu (1996).*

Yangtze, Pearl, Red, and Mekong—together discharge terrigenous sand, silt, and clay to the sea at a rate of $\sim 600 \times 10^{9}$ kg/yr (Wang et al., 2011). About half of this sediment accumulates as submarine deltas at river mouths, while a significant part of the finer sediment flows further offshore or alongshore under the action of currents.

Along the coastal zone of the China Seas, deltas represent the most active deposition center of fluvial sediments. In between these deltas are thin covers of alluvial and aeolian deposits developed at Quaternary times. Water and sediment discharge from rivers, coastal environments, and basement topography, among other factors, determine the shape, size, and volume of delta buildup.

Silty clay or mud sourced from the suspended load of rivers is largely distributed at and near river mouths. Silty clay also covers the central Yellow

Sea and several offshore areas (Figure 4.1). Unlike the nearshore mud, the off-shore mud areas are probably formed by concentrating of modern far-field resuspended material and by winnowing of older mud as a result of cyclonic circulation or eddies (Liu, 1996). Sand occurs widely as sheets or ridges from middle to outer shelf, especially NE Yellow Sea, ECS shelf, northern SCS shelf, and Sunda Shelf. Sand ridges are well developed in the northwest Bohai Strait, offshore west Korea, offshore Jiangsu, middle to outer ECS shelf, Taiwan Strait, and Qiongzhou (Hainan) Strait, all driven by tidal currents or waves (Liu & Xia, 2004; Figure 4.1). The majority of sand on the middle and outer shelf is probably relict. Fine sand mixed with coarse clasts also occurs along the shelf break and the upper slope of the western Okinawa Trough and extending from shelf break even down to ~1000 m along the northern SCS slope, and they appear to be a recent deposit propelled by currents. In the SCS, coral debris and associated bioskeletons are distributed surrounding Hainan and southern Taiwan (fringing reefs) and Dongsha, Xisha, Zhongsha, and Nansha Islands (atolls).

Silty clay rich with calcareous microorganisms mainly foraminifera (up to 20–60%) and calcareous nannoplankton covers the Okinawa Trough and the outer shelf to <3000 m water depths in the SCS. Silty or sandy turbidites with either very rare or extremely abundant bioskeletons occur especially in the southern Okinawa Trough, offshore SW Taiwan, and along the eastern SCS deep sea. Silty clay rich with siliceous organisms such as radiolarians (up to 30–40%) spreads at 3000–4000 m of the SCS, while the central deep basin of the SCS below 4000 m is covered by brown abyssal clay with radiolarians (mostly <10%; Figure 4.1).

Therefore, the ESC except part of the Okinawa Trough is dominated by terrigenous sediments, while the SCS is dominated by terrigenous and bio-genic sediments. Clearly, different geographic and bathymetric settings between these two sectors determine the general patterns of their sediment distribution. Other factors such as sediment load of rivers, current/tidal dynamics, and regional climate also play an important role in influencing the regional sediment distribution (Chen, 2008; He, 2006; Li, 2012). Overall, the distribution of sediment types in the greater ECS shelf is closely related not only to the discharge of the Yellow and Yangtze Rivers but also to the action of tidal current fields, as implied also by modeling results (Chen & Zhu, 2012).

From the marine geologic point of view, the SCS is an older sea with a deep basin, the ECS is younger, and the Yellow Sea and Bohai Sea are the youngest. Before their opening, however, marine deposition started earlier in the southern ECS during the Paleocene and then in the SCS during the Eocene and extended further northward in the outer ECS since the Miocene. The Okinawa Trough received marine sedimentation mainly during the Plio-cene–Pleistocene. Marine sediment achieved its widest distribution cover to include the far north (the Bohai Sea) only since the Quaternary.

4.2.1.3   Mineral Compositions

About 97 mineral types have been identified in sediments from the China Seas, including 77 detrital and ~20 authigenic minerals and volcanic compounds (Chen, 2008). Light minerals are dominated by quartz and feldspar, while heavy minerals are mainly composed of hornblende, epidote, schistose minerals (muscovite, biotite, green mica, and chlorite), and ilmenite.

As listed in Table 4.1, the abundance of minerals varies from sea to sea depending on their sources, making regional mineral assemblages characteristic. For example, the Bohai Sea assemblage is characterized by mineral calcite, garnet, and zircon; the Yellow Sea by high proportion of schistose minerals; the ECS by high dolomite; the Okinawa Trough by pyroxene (hypersthene) and volcanic glass; the SCS shelf by ilmenite, leucoxene, tourmaline, and zircon; and the SCS deep basin by high biogenic calcite and opal, magnetite, and volcanic debris.

Clay minerals are dominated by illite, chlorite, kaolinite, and smectite (Table 4.3). Others such as halloysite, vermiculite, and palygorskite–sepiolite are rare to scarce. From north to south, illite and smectite show a trend of decrease in their abundance, while chlorite and kaolinite increase. Compared to other shelf areas, the Bohai Sea has relatively high smectite due to the supply of smectite-rich sediment from the Yellow River. Similarly, a high content of kaolinite in the northern SCS is due to the kaolinite-rich sediment from the Pearl River. All these convey the same information as the detrital minerals that the respective sediment load of the big rivers has contributed mostly to variations in regional mineral distribution in seas. For the sediments of the Yangtze and Yellow Rivers, for example, different mineral components are well revealed in XRD spectral analyses (Figure 4.2). After many years of research, the distribution patterns of clay mineral assemblages in many parts of the China Seas and the associated rivers have been established, as shown in Table 4.2 and Figure 4.3.

Authigenic minerals in the China Seas include glauconite, pyrite, goethite, rhodochrosite, siderite, barite, aragonite, calcite, gypsum, cellophane, and zeolite, with the first two being most widely distributed and most abundant. Glauconite presents mainly in three forms: granular, foliate, and bioshaped, often with a dark color in ECS and SCS sediments but a light green color in Yellow Sea sediments. Although the three glauconite types are composed of similar proportions of SiO_2 (40–50%) and Fe_2O_3 (20–30%) between them, the bioshaped glauconite often has relatively high CaO compared to other two types. Pyrite presents in various crystal aggregates from granular, oolitic to framboidal or berry-shaped, and all the types have relatively stable chemical compositions, Fe (45–48%) and S (50–54%). Authigenic pyrite is enriched in two areas of the Yellow Sea: NW (up to 63%) and central South Yellow Sea (up to 43%; Figure 4.1).

4.2.1.4   Chemical Compositions

A systematic survey on the concentration of chemical elements in shallow shelf sediments of the China Seas and major rivers was carried out by Zhao

TABLE 4.1 Average Percentage Abundance of Major Minerals in the Sediments of the China Seas

Minerals	Bohai	Yellow Sea	ECS Shelf	Okinawa Trough	Tokin Gulf	SCS Shelf	SCS Basin
All heavy minerals	5.7	3.0	6.9	2.9	0.7	0.6	1.1
Hornblende	31.8	27.9	33.6	20.2	13.1	15.3	23.37
Epidote	25.8	14.0	13.5	9.8	16.5	13.7	4.3
Ilmenite	15.3	4.7	9.7	4.4	19.2	18.93	2.97
Garnet	7.5	4.7	2.7	1.4	1.0	0.8	<0.1
Dolomite	<0.1	6.3	10.2	8.6	1.4	2.0	
Pyroxene	<0.1	0.5	3.7	23.8	0.5	0.7	8.9
Magnetite	1.5	1.9	1.7	4.9	<0.1	<0.1	18.14
Tourmaline	<0.1	0.4	0.3	0.2	8.1	5.04	
Zircon	4.1	0.2	0.7	0.2	2.0	2.5	<0.1
Metamorphic	<0.1	<0.1	0.4	<0.1	0.8	0.7	
Leucoxene	2.2	<0.1	1.1	<0.1	<0.1	14.07	
Rutile+anatase	<0.1	<0.1	<0.1	<0.1	2.0	<0.1	
Sillimanite	<0.1	<0.1	<0.1		1.33	<0.1	
Ferromanganese nodules				<0.1			19.62
All light minerals	94.3	97.0	93.1	97.1	99.3	99.4	98.9
Quartz	33.4	44.9	42.1	34.5	74.0	70.5	
Plagioclase	50.4	22.6	30.7	21.3	13.6	14.1	
K-feldspar	13.2	16.1	13.6	5.9	5.6	7.1	
Mineral calcite	2.5	1.4	1.5	0.5	0.2	0.3	
Biogenetic calcite	<0.1	<0.1	<12	<8	<7	<8	>20
Volcanic mud				<2			>15
Volcanic glass				33.3			>10

Modified from Chen (2008).

TABLE 4.2 Mean Abundance of Clay Minerals in the Sediments of Large Rivers and Seas of China as Compared to World Ocean Records

River or sea	Illite %	Chlorite %	Kaolinite %	Smectite %	Provenance
Yellow River	62	12	10	16	Loess
Yangtze River	65	11	14	<10	Central–W China
Pearl River	50	17	30	3	S China
Bohai Sea	56.6	13.2	10.2	19.4	Yellow River
Yellow Sea (N)	61.6	15.3	8.9	14.3	Yellow River
Yellow Sea (S)	56.8	16.3	9.6	17.7	Mixed
ESC shelf	59.7	19.6	8.9	11.8	Yangtze
Okinawa Trough	61.4	25.1	7.2	6.4	Mixed
Tonkin Gulf	39.3	11.7	22.1	26.7	Red+mixed
SCS northern shelf	49.7	20.7	10–30	12.2	Pearl
SCS Basin	53–67	15–20	7.5–21	4.8–16	Mixed
World ocean	40.2	12.8	14.0	34.2	

Modified from Chen (2008).

and Yan (1994). As expected, Si and Al, often compounded as SiO_2 and Al_2O_3, respectively, are the most common elements in the quartz-/feldspar-dominated sand and silt sediments of the China Seas (Table 4.3). In the greater East China Sea, Al, K, and Zr are relatively high in the Bohai sector, organic matter is relatively high in the Yellow Sea sector, and Sr and Ca are relatively high in the ECS proper. Among the elemental compounds, detrital $CaCO_3$ is high in the Bohai and northern Yellow Sea, while biogenic $CaCO_3$ is more abundant in the deeper ECS, SCS, and Okinawa Trough.

Distributions of the biogenic elements are regulated by the origin of matter, water dynamics, and sediment grain size, among others. In general, the finer the grain sizes, the higher the contents of TOC, N, P, and organic matter and the lower the contents of S and Si (Song, 2010). For example, the highest TOC contents were found in muddy sediments around the inner shelf of the southern Yellow Sea and of the East China Sea, while most regions with sandy sediments have lowest TOC contents. Relatively higher Si content in the southern Taiwan Strait than the nearby regions appears to have been caused by silica enrichment in sediment particulates and lesser river input.

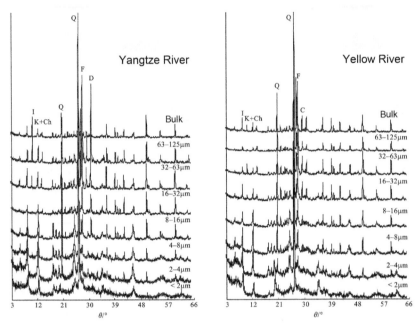

FIGURE 4.2 The XRD spectra of various grain size fractions of sediments from the Yangtze River (left) and the Yellow River (right) (Yang, Wang, & Qiao, 2009). I, illite; K, kaolinite; Ch, chlorite; Q, quartz; F, feldspar; C, calcite; D, dolomite.

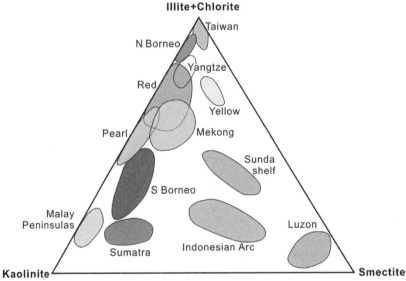

FIGURE 4.3 Positioning the main sediment sources to the China Seas in the ternary diagram of clay minerals illite + chlorite, kaolinite, and smectite. *Based on Liu, Berné, et al. (2007), Liu, Tuo, et al. (2008), Liu, Huang, et al. (2009), Liu, Zhao, et al. (2009), and Liu et al. (2012).*

TABLE 4.3 Average Concentrations of Characteristic Elements in the Sediments of Shallow China Seas and Rivers

	Bohai Sea	Yellow Sea	ECS Shelf	Okinawa Trough	SCS Shelf	Yellow River	Yangtze River	Pearl River
Major elements ($\times 10^{-2}$)								
Si	29.10	29.30	29.22	19.21	29.23	29.30	28.79	28.90
Al	6.36	6.21	5.31	5.93	5.24	4.87	6.51	6.80
Fe	3.02	3.16	3.15	3.00	2.87	2.20	6.85	4.52
Mg	1.14	1.15	1.07	1.15	1.06	0.84	1.33	0.90
Na	1.79	1.72	1.33	1.93	1.20	1.63	0.91	0.35
K	2.30	2.15	1.82	1.83	1.67	1.61	1.83	1.50
Ca	2.48	2.69	4.15	11.44	4.33	3.29	2.86	1.66
Ti	0.33	0.35	0.35	0.31	0.34	0.36	0.55	0.65
C	1.32	1.40	1.82	3.94	1.91	1.13	1.63	0.90
Cl	0.35	0.36	0.30	0.57	0.37	0.03	0.03	–
Rare and trace elements ($\times 10^{-6}$)								
Mn	560	570	520	2020	490	450	810	820
P	527	522	516	630	425	600	650	680
N	556	624	615	1130	620	500	600	–
S	328	514	472	730	652	100	170	–
B	58	69	51	62	58	52	63	60

Continued

TABLE 4.3 Average Concentrations of Characteristic Elements in the Sediments of Shallow China Seas and Rivers—Cont'd

	Bohai Sea	Yellow Sea	ECS Shelf	Okinawa Trough	SCS Shelf	Yellow River	Yangtze River	Pearl River
F	474	507	470	680	472	410	560	520
Ba	504	512	396	412	288	540	512	340
Sr	202	194	269	590	265	220	150	100
Cu	22	18	14	26	13	13	35	38
Pb	20	22	21	27	19	15	27	30
Zn	64	67	66	82	61	40	78	85
Hg	36	24	25	130	27	15	80	93
V	70	76	71	78	61	60	97	105
Co	11	13	12	13	9	9	17	18
Ni	26	26	25	36	20	20	33	35
Cr	57	64	61	58	53	60	82	86
Zr	246	229	180	120	235	354	246	270
Li	36	39	36	32	44	23	43	26
Nd	28	30	29	22	27	30	34	36
Th	10.5	12.4	10.7	7.0	11.9	13	12.4	15

Modified from Zhao and Yan (1994).

A general trend is that the concentrations of TOC, TON, TOP, and Si decreased rapidly below the 20 cm surface layer, suggesting that the decomposition of organic matter mainly occurred in the surface and subsurface layers (Song, 2010, and references).

Therefore, these results reveal features of "continent affinity" and "grain size control" in elemental concentration and distribution. Their continent affinity is reflected by a good linkage between elemental concentrations in rivers and in shallow seas where the rivers discharge into, such as the Yellow River–Bohai and Yellow Sea link, the Yangtze River–ECS link, and the Pearl River–SCS shelf link (Table 4.3). Similar linkages also exist for elements between the Red River and Tonkin Gulf, between the Mekong and Sunda Shelf, and between other rivers and their corresponding shelf areas. For grain size control, the concentration values of many elements generally vary with changes in grain size. As shown in Figure 4.4, for example, the ratios between the mean concentration of elements and the mean abundance of the total sediment, or mean (s)/mean (t), in the five main sediment types from the Yellow Sea shelf show values ranging between 0.6 and 1.2, indicating a close relationship between element distribution and sediment types. This is particularly obvious for SiO_2, K_2O, and Na_2O, all with values of mean (s)/mean (t) ratio close to 1. However, very high concentrations of such elements as Mn, C, S, Sr, and Hg in the Okinawa Trough are caused by a high level of hypothermal and biological activities there (Table 4.3).

4.2.2   East China Sea

4.2.2.1   General Characteristics

As summarized by Qin (1994), the modern sediment distribution in the ECS is characterized by three major features:

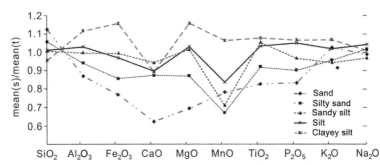

FIGURE 4.4   The ratio between the mean abundance of elemental components and sediment types in the Yellow Sea reveals a better relationship for the detrital elements (SiO_2, K_2O, Na_2O) than for the partially detrital and authigenic or biogenic elements (CaO, MgO, MnO, Fe_2O_3). *Modified from Li (2012).*

1. Terrigenous clay or mud is widespread in the Bohai Sea and Yellow Sea, especially in the western and southern parts. The clay is primarily from the (re-)suspended load of the Yellow River, forming several large mud fields in the area.
2. Terrigenous sand is widely distributed in the ECS shelf and eastern Yellow Sea. The sand is mainly fine to medium in grain size and moderately sorted, with the majority from the Yangtze River and rivers in western Korea. Sand ridges are common, especially on the ECS shelf, driven by the regional hydrodynamics.
3. Hemipelagic clay with 20% or more biogenic components characterizes the sediment in the Okinawa Trough. The main part of clay is a mixture from several sources: the Yellow River mainly in the northern trough, the Yangtze River mainly in the middle through, and Taiwan rivers mainly in the southern trough.

4.2.2.2 Continental Shelf

Except the Okinawa Trough, the entire greater East China Sea (including the Bohai Sea, Yellow Sea, and ECS s.s.) functions as a wide shelf with a total area of $101.7 \times 10^6 \text{ km}^2$. About 1/3 to 1/4 of the total shelf area is covered by sediment with 50% to >90% fine sand (Figure 4.5; He, 2006; Li, 2008, 2012; Xu, Liu, Zhang, Li, & Chen, 1997). The main sandy areas in the region are ESC outer shelf, SW and E–NE Yellow Sea, and N Bohai Strait, forming some unique tidal sand ridge systems. While most of the outer shelf sand is relict from last glacial deposition at sea-level lowstands (Qin, 1994), the sandy deposits along the NE Yellow Sea shelf appear to have been formed during the last postglacial sea-level rise (Chough, Lee, & Yoon, 2000). Similarly, although containing reworked and stained particles and microfossils, the Yangtze Shoal in the northwest ECS has been considered as a typical active offshore tidal sand sheet rather than a paleo-Yangtze submarine delta (Liu, 1997). Coarse sand and gravels are scattered as pockets or small hills or as individuals in sandy and clayey sediments often associating with dynamic hydrologic environments such as along underwater troughs.

Silt, including clayey silt, is second in abundance, and silty areas cover a large part of the Yellow Sea, northern ECS, and northern and southwestern Bohai Sea and occur as patches or narrow belts in the western to southwestern ECS (Figure 4.5).

In the Bohai Sea, silty clay covers the modern Yellow River estuary and southern and central parts of the sea. The clay near the river mouth inherits the properties of the suspended load from the Yellow River: bright yellowish to brown in color, semifluid, and rich in $CaCO_3$, although all these change proportionally at increasing distance away from the river mouth (Qin, 1985).

The Yellow Sea is featured by deposition of fine sand and silt on the shelf and mud in the central part. Apart from the central mud sheet, the distribution

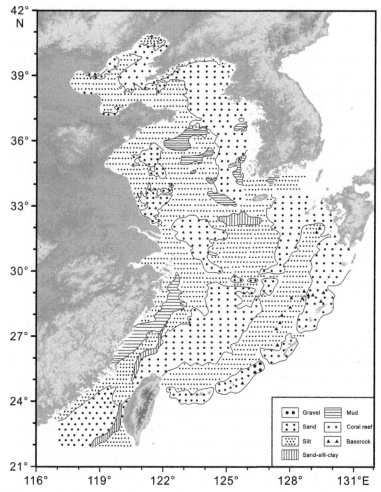

FIGURE 4.5 Distribution of sediments in the greater East China Sea including the Okinawa Trough. *Based on He (2006) and Li (2012).*

of sand and silt often shows an elongated pattern and is rarely continuous except for several sand ridge areas (Figure 4.5). Four mud zones have been distinguished, respectively, from the central, the northern, the southeastern, and the southwestern Yellow Sea, the last of which is also called the old Huanghe Delta mud (Yang, Jung, Lim, & Li, 2003, and references). Most of the mud deposits are found to have sourced largely from the resuspended load of Chinese rivers particularly the Yellow River (Yang et al., 2003; Yang & Youn, 2007). In the SE Yellow Sea mud zone and smaller mud patches along the coast of southern Korean Peninsula, mineralogical and geo-chemical features show a strong imprint of Korean rivers (Chough et al., 2000;

Chough, Lee, Chun, & Shinn, 2004; Lim, Jung, Choi, Yang, & Ahn, 2006; Yang, Li, & Yokoyama, 2006). However, a discrepancy between the accumulation rate of these mud deposits (40–150 Mt/yr) and the total river discharge in the southern Korean region (6–20 Mt/yr) indicates additional input of the Yellow–Yangtze mud by the northeastward flowing Yellow Sea Warm Current (Lim, Choi, Jung, Rho, & Ahn, 2007).

A systematic survey of suspended matter in the water column and surface sediments of the southern Yellow Sea carried out by Cai et al. (2003) indicates the existence of two high-concentration tongues in spring for all sections: a distinct one from off the Old Yellow River mouth spreading southeastward to most part of the southern Yellow Sea and a weaker one around the Shandong Peninsula extending southward. In autumn, however, these two tongues no longer exist, but a new tongue forms off the Yangtze River mouth and extends northward over 6° latitude and eastward to Chiju Island area. Depleted values of −25‰ to −26‰ δ^{13}Corg indicative of terrestrial origin occur in two areas: around the Shandong Peninsula and to the NE of the Yangtze River mouth, although they all increase to −22‰ to −23‰ due to marine degradation. The flux of settling particulate matter is highest in the boundary area between the Yellow Sea and ECS where a wide nepheloid layer is present (Guo et al., 2010).

The mud patch to the southwest of Chiju Island, or Chiju Island Mud, probably receives sediments from this nepheloid layer, which likely sources not only from the Yangtze River but also from the Old Yellow River mouth. In the nearshore area of the ECS, a silty clay belt extends southwestward from the Yangtze River mouth along the shore mainly within 50–60 m isobath and terminates near the southern Taiwan Strait. Like the clay belt surrounding the Shandong Peninsula, this ECS clay belt has been driven also by the prevalent alongshore current, albeit the clay source here comes mainly from the Yangtze River (Liu, Xu, et al., 2007; Liu, Zhu, Li, Li, & Li, 2007, and references).

It is evident, therefore, that the Yellow River is largely accountable for the shelf deposition in the Bohai and Yellow Seas and the Yangtze River for the ECS proper. The suspended and resuspended loads of these two largest Chinese rivers, as clearly shown in satellite images (Figure 4.6), are the main source of the silt and clay deposition on and along shelf. Sandy sediments are largely from the Yangtze and rivers surrounding the Bohai Sea and Yellow Sea, although the majority has been trapped at deltas and only a small portion is able to travel out to the sea by storms. However, the sand of ancient deltas formed at glacial lowstands has been reworked by currents and tides, as occurring on the ECS outer shelf. The strong tidal current dynamics in the region directly contribute to the formation of the spectacular sand ridge systems. Numerical simulation of regional tidal elevations and currents demonstrates that sand ridges form mainly in the areas with strong rectilinear tidal currents, sand sheets form mainly in the areas dominated by strong rotatory tidal currents, and clayey sediments, that is, mud patches, form mainly in the areas with weak tidal currents (Zhu & Chang, 2000).

Apart from the Yellow and Yangtze Rivers, many smaller rivers surrounding the sea also contribute to the ECS deposition, especially at estuaries and nearshore regions. To establish their role in offshore deposition quantitatively or proportionally is difficult, however, because many mineralogical and geochemical signatures are similar or intermediate between them and the two larger mainland rivers, the Yellow and Yangtze (e.g., Yang et al., 2003).

Based on the characteristic minerals, nine mineral provinces for the greater ECS have been identified (Chen, 2008; Qin, 1994), as summarized in Table 4.4 and Figure 4.7. In general, detrital minerals characterize the ECS provinces, and biogenic and volcanic minerals differentiate the mineral province in the Okinawa Trough. Many of the provinces are distributed as alongshore belts sometimes across traditional geographic units. For example, mineral provinces 1 and 2 extend from the Bohai Sea to the Yellow Sea, and province 8 widens from the southern Yellow Sea to the entire middle and outer ECS shelf. Two provinces are relatively smaller in size: province 3 from offshore Qingdao, West Yellow Sea with high epidote (26%) and common presence of aegirine, riebeckite, and calcareous nodules, and province 5 from the Old Yellow River mouth with abundant schistose minerals and dolomite, further indicating a good correspondence of these mineral provinces to sediment supply and hydroclimate conditions.

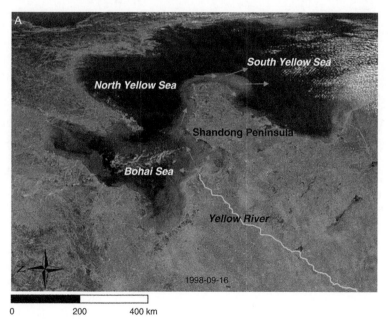

FIGURE 4.6—CONT'D

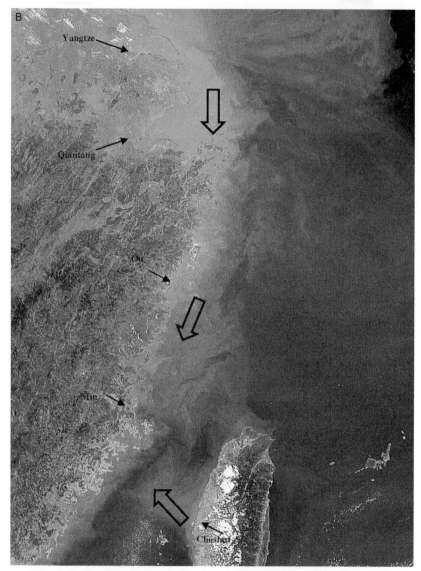

FIGURE 4.6 Satellite images of (A) the Bohai and Yellow Seas as viewed from northwest (Yang & Liu, 2007) and (B) the East China Sea as viewed from south (Xu, Lim, et al., 2009; Xu, Milliman, et al., 2009), showing the flow path of the suspended load of the Yellow River and Yangtze River, respectively.

TABLE 4.4 Characteristics of Mineral Provinces of the China Seas

Mineral Provinces	Characteristics (mean %)
1. S Bohai Sea–NW Yellow Sea	Mineral calcite 9%, schistose minerals 37%, high smectite
2. N Bohai Sea–N Yellow Sea	K-feldspar 14%, mineral calcite 0%, schistose minerals 1%
3. W Yellow Sea, offshore Qingdao	Epidote 26%, aegirine and riebeckite present, calcareous nodules
4. SE Yellow Sea	Hornblende >50%, zircon >5%
5. Old Yellow River mouth	Schistose minerals 43%, dolomite 10%
6. SE Yellow Sea	Transitional
7. W East China Sea	Dolomite 18%, heavy minerals decreasing southward
8. S Yellow Sea–central ECS	Hornblende 36%, metamorphic minerals 0.4%, schistose minerals decreasing southward
9. Okinawa Trough	Pyroxene 24%, volcanic glass 33% decreasing to the south
10. South China Sea northern shelf	Quartz 70%, ilmenite 19%, tourmaline 5%, kaolinite 10–30%, schistose minerals 15%
11. Beibuwan (Gulf of Tonkin)	High in quartz, ilmenite, tourmaline, and smectite; sillimanite 1.3%, rutile+anatase 2%, higher in the north
12. N–central deep SCS	Quartz, feldspar, and mica decreasing basinward, volcanic debris, illite, and smectite higher in the east
13. S deep SCS, Nansha Islands	Biogenic calcite 45–70%
14. SE Sunda Shelf–SE SCS margin	Biogenic calcite ~50%, clay+quartz ~50%
15. NE Sunda Shelf	High in quartz, mica, and epidote
16. Central–W Sunda Shelf	High in quartz, mica, and ilmenite

Based on Chen (2008).

4.2.2.3 Okinawa Trough

In the Okinawa Trough, sand is mainly distributed along the western and northwestern upper slope, silt in the southern part and along the west, and silty clay in the middle and northern parts (Figure 4.5; Li & Chang, 2009; Xu et al., 1997). In areas south of 26°N, the silty deposit is reddish brown to yellowish brown in the surface before changing to dark gray color further downsection. $CaCO_3$ is low, often <10%. In the middle and northern trough, the silty clay

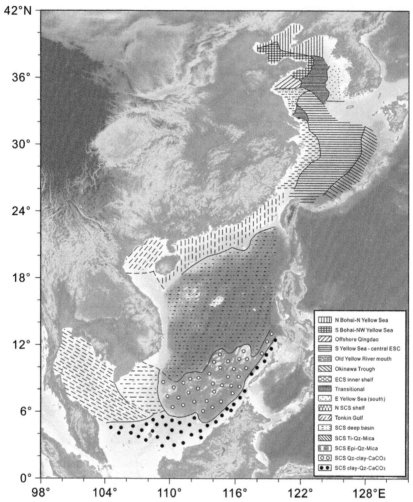

FIGURE 4.7   Mineral provinces of the China Seas. Refer to Table 4.4 for details. *Modified from Qin (1994) and Chen (2008).*

contains relatively higher $CaCO_3$, with foraminifera increasing northward from 15% to 50% or more. Volcanic debris is frequent, reaching 15–20% at some localities. Accompanying these are a high percentage of pyroxene and elements Mn, C, S, Sr, and Hg (Table 4.4), all because of a high thermal regime and frequent volcanic activities in the region.

The silt-dominated southern trough is characterized by frequent turbidites and a high sedimentation rate. Recent turbidite layers observed in many sediment cores have been dated ca. 2002, 1986, 1966, 1959, 1947, and 1922, which can be correlated with the history of major submarine earthquakes in and around the region (Huh, Su, Wang, Lee, & Lin, 2006). Clearly, the major detrital components in southern Okinawa Trough were derived primarily via the Lanyang River from Northern Taiwan, as a result of high erosional gradients, heavy summer rain falls, and frequent earthquakes (Hsu et al., 2004; Huh, Su, Liang, & Ling, 2004). A significant part of Taiwan-derived material may have been transported to the middle trough or further north by the powerful Kuroshio Current (Huh et al., 2006).

In the middle and northwestern trough, the proportion of mainland-originated sediments increases. For example, mineralogical and chemical analyses of samples from a sediment trap SST-1 (29°21.6′N, 128°13.5′E; water depth 1100 m) indicate that a Yangtze River influence was relatively high during winter, while the Yellow River- and/or aeolian-derived sediment enhanced in spring and late summer when the total flux also intensified (Katayama & Watanabe, 2003), especially in times of strong monsoons (Bian, Jiang, & Song, 2010).

4.2.3   South China Sea

4.2.3.1   General Characteristics

Due to its diverse topographic settings and great water depth, the SCS sediment shows the following four general features (Liu, Huang, et al., 2009; Liu, Zhao, Colin, Siringan, & Wu, 2009; Su & Wang, 1994):

1. Shallow shelf terrigenous clastic sediments consist chiefly of modern terrigenous clayey silt, silty clay, and bioclasts. Clay content is often higher than 80% and sand less than 15%. Bioclasts are mostly fragments of bivalve shells, gastropods, foraminifera, ostracods, and sponge spicules.
2. Hemipelagic slope ooze includes biogenic and terrigenous sediments above the carbonate compensation depth (CCD), including clayey silt, silty clay, and calcareous ooze. The ooze is gray or yellowish gray with planktonic foraminifera and other calcareous skeletons. The carbonate content is generally higher than 30% or even up to 62%, often decreasing with increasing water depth due to dissolution.
3. Carbonate reef sediments are mainly biogenic sand and gravel, mostly restricted to shallow settings (0–400 m) of the Dongsha, Xisha, and Nansha Islands. Fragments of coral and associating organisms such as bivalves, algae, and foraminifera constitute the major components of carbonate reef sediments.
4. Abyssal basin clay represents a mixed biogenic and terrigenous ooze below the CCD. Biogenic tests are much less than those in shelf and slope sediments, here mainly containing radiolarians and lesser amounts of diatoms and agglutinated foraminifera. Turbidite appears to form the dominant type of "abyssal clay" in most part of the SCS deep sea.

4.2.3.2   Continental Shelf

Unlike the ESC with a huge N–S-extending shelf, the SCS features a rhombus-shaped central deep basin surrounded by a shelf widening in the north and southwest and narrowing in the east and west. The SCS shelf occupies about half of the total sea area of 350×10^6 km². Although terrigenous sand, silt, and clay are present as the main sediment types on the ECS shelf, the SCS shelf sediments are characterized by relatively higher biological contents, higher $CaCO_3$, coarser grain size, higher depositional rates, and frequent coral fragments and volcanic debris in local areas (He, 2006; Li, 2012; Liu, Zhao, Fan, & Chen, 2002; Su & Wang, 1994; Xu et al., 1997). Su, Fan, and Chen (1989) published a sedimentary atlas outlining sediment types and their distribution features in the northern and central SCS.

Sandy sediments are widely distributed along the SCS shelf (Figure 4.8). Fine sand is the most dominant, often becoming finer and mixed with more biodebris at deeper settings and coarser and mixed with rock pebbles or clasts of bivalves in shallower areas where tidal wave activity is high. Submarine

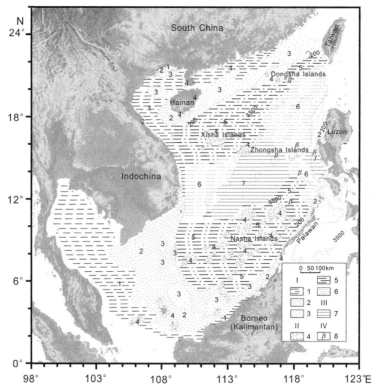

FIGURE 4.8 Distribution of sediments in the South China Sea. I Terrigenous type includes (1) nearshore modern terrigenous mud, (2) nearshore modern terrigenous sand and silt, and (3) neritic (paleolittoral) relict sand. II Biogenic type includes (4) neritic coral sand and gravel, (5) hemipelagic and abyssal calcareous ooze, and (6) abyssal siliceous ooze. III Mixed type is represented by (7) abyssal clay. IV Volcanic–biogenic–terrigenous type includes (8) volcanic material (about 5% of the sediment). *Modified from Su and Wang (1994).*

sand ridges are distributed mainly in areas immediately outside from both ends of the Qiongzhou (Hainan) Strait, produced by the vigorous cross strait current. A large area in the central to southern Taiwan Strait is also covered by sand mostly sourced from Taiwan rivers (Liu, Liu, et al., 2008). Like those on the ECS outer shelf, most sand deposits on the middle to outer shelf of the SCS, including the huge sand sheet on the very low-gradient outer Sunda Shelf and the large sand patch in the central Tonkin Gulf, are relict. For example, next to the Red River mouth, sand dominates along the shoreline between 0 and 15 m water depths, and silt from 15 to 25–30 m. Further offshore, the surface sediments are mainly sandy silt and sand of older ages (Duc et al., 2007).

Compared to the large sand cover, the distribution of silt is patchy on shelf and often forms a transitional zone between sandy and muddy deposits. Bioskeletons increase to 20–50% or more in silty and clayey sediments on the middle to outer shelf. Unlike the clayey sediment on the outer shelf, however, the large silty clay patch offshore from the eastern Leizhou Peninsula (Figure 4.1) is related to reduced current flow or eddy circulation like the mud areas on the Yellow Sea shelf (Figure 4.5). The inner and middle Sunda Shelf to ~100 m water depth, especially in the Gulf of Thailand and along the southern margin, is covered by mud and silt in the innermost part and older silt and fine sand in the outer part (Hanebuth, Voris, Yokoyama, Okuno, & Saito, 2011), which contain >80% quartz and a small amount of feldspar (Su & Wang, 1994; Xu et al., 1997). Compared to light minerals, heavy minerals are low, mostly <1.5%, in SCS sediments except in the southern shelf where heavy minerals often exceed 10% (Li, 2012).

The terrigenous sediment on SCS shelf was mainly sourced from the Pearl River (to the northern shelf), the Red River (to the Tonkin Gulf), the Mekong (to the Sunda Shelf), and rivers in the southwest Taiwan (to the northeast shelf–slope and the Taiwan Strait). Due to the prevailing alongshore current, the fine sediment load from all the mainland rivers is mostly moved southward or southwestward and deposited alongshore, as clearly revealed also in satellite images (Figure 4.9).

Provenance studies of SCS sediments have been increasingly focusing on clay minerals because other mineralogical and elemental characters are similar between the major rivers due to their common share of a source region in Tibet Plateau (e.g., Boulay et al., 2005; Liu, Chen, Chen, & Yan, 2010; Liu, Colin, et al., 2004, 2007, 2010; Liu, Huang, et al., 2009; Liu, Saito, et al., 2010; Liu, Tuo, et al., 2008; Liu, Zhao, Li, & Colin, 2007; Liu, Zhao, et al., 2009; Liu et al., 2011, 2012; Wan, Li, Xu, & Yin, 2008; Wan et al., 2010). Among clay minerals, illite is dominant, 37–82% (mean 58%), followed by chlorite 9–29% (mean 16%), smectite 0–40% (mean 14%), and kaolinite 0–24% (mean 12%; Li, 2012; Liu, Berné, et al., 2007; Liu, Huang, et al., 2009; Liu, Zhao, et al., 2009; Liu et al., 2011; Su & Wang, 1994, and references). Illite is enriched (70% or more) in the north to northeast and decreases southward. Chlorite is enriched (>20%) offshore from Luzon

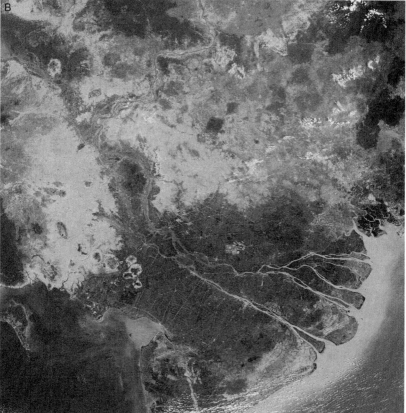

FIGURE 4.9 Satellite images of (A) the northern South China Sea (http://www.eosnap.com/public/media/2011/01/china/20101227-china1-full.jpg) and (B) the Mekong Delta in the southwestern SCS (http://www.worldalldetails.com/Pictureview/3820-The_Mekong_Delta_Vietnam_Asia_Satellite_view.html), showing the alongshore deposition features and the flow path of the sediment load of the Pearl River, Red River, and Mekong.

and decreases westward and southward. Smectite is enriched (>20%) offshore from Luzon and SE Vietnam and decreases to <5% along the northern margin and <10% offshore from Palawan and Kalimantan. Kaolinite is enriched (>15%) in the northwest and southwest and decreases eastward and northeastward. A compilation of available data indicates that clay minerals in the SCS are mainly sourced from the Pearl (illite and kaolinite), the Mekong (chlorite and kaolinite), Taiwan (illite and chlorite), Luzon (smectite), and Malay Peninsula and Sumatra (kaolinite) (Figure 4.10A; Liu, Colin, et al., 2010; Liu et al., 2012). In the smectite-poor sediments from Malay Peninsula and North Borneo, their differences in the abundance of kaolinite and illite+ chlorite may imply tectonic or climatic influences (Figure 4.10B).

Based on the characteristic minerals, Chen (2008, and references) identified five mineral provinces for the SCS shelf and two provinces for the deep sea, as summarized in Table 4.4 and Figure 4.7. In general, the northern shelf is characterized by high quartz (70%) and ilmenite (19%), and the Gulf of Tonkin by the presence of sillimanite, rutile, and anatase. Although quartz and mica are also as high as on the northern shelf, the northeast Sunda Shelf features epidote, and the central to western Sunda Shelf has ilmenite, but the province from the southeastern Sunda Shelf to the southeastern SCS margin has high biogenic calcite ($\sim$50%) and clay+quartz ($\sim$50%) (Table 4.4).

4.2.3.3   Continental Slope and Deep Basin

Sediment Distribution and Mineral Composition

The SCS deep sea is covered by two main sediment types: calcareous ooze over the slope and abyssal clay over the central basin below the carbonate compensation depth (CCD; Figure 4.8; Liu, Huang, et al., 2009; Liu, Zhao, et al., 2009; Su & Wang, 1994). The calcareous ooze over the slope refers to all biogenic and terrigenous sediments above the CCD, including clayey silt, silty clay, and calcareous ooze. The ooze is gray or yellowish gray with nannofossils and planktonic foraminifera as well as other calcareous skeletons. The carbonate abundance is generally higher than 30% or even up to 62% on the upper and middle slope but decreases with increasing water depth due to dissolution, as confirmed in deposits from many cores (e.g., Wang & Li, 2009; Wang, Prell, & Blum, 2000; Wang, 1999). The abyssal clay is a mixed biogenic and terrigenous ooze from below the CCD. In the abyssal clay, biogenic skeletons are much less than those in shelf and slope sediments and mainly contain radiolarians (mostly <10%) and lesser amounts of diatoms and agglutinated foraminifera. Volcanic substances are common locally, especially in the northeastern region. Coral debris is common in areas surrounding coral reefs on atolls and terranes, especially in the southern region.

As those on shelf, mineral components in SCS slope and deep-sea sediments show distribution patterns closely relating to grain size and their sources. Light minerals are predominant, 90–99%, while heavy minerals are

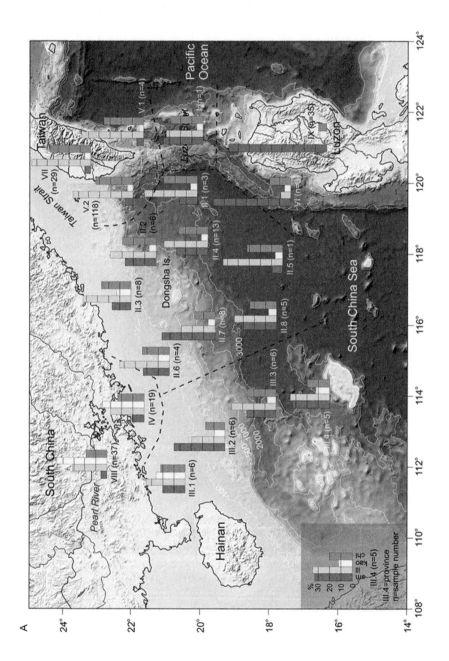

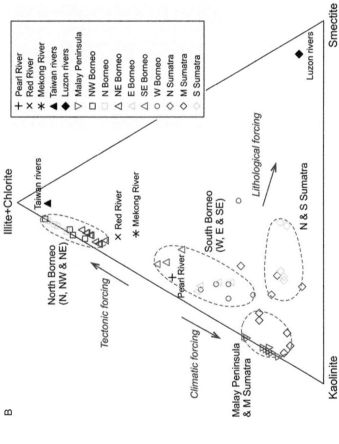

FIGURE 4.10 (A) Generalized relative abundance of clay minerals in the northern SCS and adjacent rivers: smectite, illite, kaolinite, and chlorite (Liu, Colin, et al., 2010). (B) Ternary plots of the clay minerals illite+chlorite, kaolinite, and smectite for various localities and rivers from surrounding the SCS (Liu et al., 2012).

low, averaging at ~1%, although a maximum of ~13% from some localities has been reported. For example, the central and northeastern parts of the SCS deep sea are characterized by relatively high percentages of magnetite, ferromanganese nodules, pyroxene, and volcanic glass (Yang, Lin, Zhang, Lin, & Ji, 2002). In addition to the mineral provinces identified by Chen (2008) (Table 4.4), Yang et al. (2002) recognized four mineral provinces (I–IV) for the central and northeastern SCS, and Li, Wang, Liao, Chen, and Huo (2008) distinguished three provinces (NW, W, and SW) for the west and southwestern SCS. Noteworthy is that authigenic ferromanganese nodules are characteristic in province III of Yang et al. (2002) from the northeast, while volcanic debris is common in province IV from the central and eastern SCS. These mineral distribution patterns are confirmed by geochemical analyses by Liu et al. (2011) who attributed Mn and MgO enrichment in the northeast and CaO enrichment in the northern slope region to the differences in their prevailing hydrology and sources.

Among clay minerals in sediments from the SCS slope and deep sea, illite may attend ~50–60%, chlorite ~12% or more, smectite 10–20%, and kaolinite >8%. It is no doubt that their sources are similar to those for their counterparts in shelf sediments, with the Pearl mainly supplying illite and kaolinite, the Mekong mainly chlorite and kaolinite, Taiwan mainly illite and chlorite, Luzon mainly smectite, and Malay Peninsula and Sumatra mainly kaolinite (Figure 4.10; Li, 2012; Liu et al., 2012).

Sand Dunes, Sediment Waves, and Drifts

In the northern SCS, subaqueous sand dunes are distributed mainly on the upper slope, sediment drifts on the middle slope, and sediment waves on the middle to lower slope. The upper slope sand dunes are dominated by fine to medium quartz sand, while the middle to lower slope sediment drifts and sediment waves contain abundant hemipelagic matters. Several holes at ODP Site 1144 penetrated the large sediment drift field from east of the Dongsha Islands and revealed such depositional features as a high sedimentation rate of 500 m/Ma and 10–20% biogenic carbonate (Wang et al., 2000).

Due to the mesoscale internal waves produced by inflood of the North Pacific Water (Chapter 2), large-scale sand dunes and sediment waves have developed in many parts of the northern SCS. For example, Reeder, Ma, and Yang (2011) described very large subaqueous sand dunes along a 40 km long transect southeast of 21.93°N, 117.53°E in water depths of 160–600 m (Figure 4.11). Reaching >16 m in height and >350 m in crest-to-crest wavelength, the sand dunes form on the upper continental slope, away from the influence of shallow-water tidal forcing, deep basin bottom currents, and topographically amplified canyon flows. It is envisioned that ~30 internal waves each month drive approximately 60 extremely energetic, episodic resuspension events, which dissipate extremely large amounts of energy on the upper continental slope to assist the formation of these sand dunes (Figure 4.11).

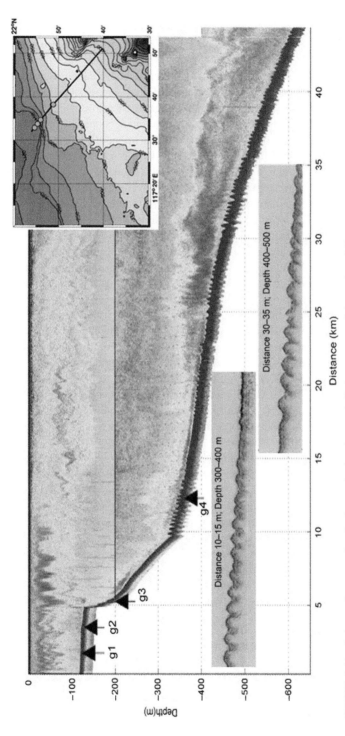

FIGURE 4.11 A time series composite of echosounder data from a shelf–slope transect of the northern SCS (inset) showing the water column structure of the internal solitary waves, superimposed by two subbottom profiles acquired by the shipborne 3.5 kHz chirp sonar system showing the nonuniformly spaced structure of the sand dunes. Bottom grab samples (g1–g4) were analyzed to determine the properties of the sand dunes. *Modified from Reeder et al. (2011).*

Much larger fields of sediment waves and drifts occur in the northeastern middle to lower slope (e.g., Gong et al., 2012; Li et al., 2013). A summary of earlier works was given by Liu, Huang, et al. (2009) and references) on the sediment drift body in the northeastern SCS, where ODP Site 1144 recovered cores over 500 m spanning the early Pleistocene to Holocene. Unlike the upper slope sand dunes, biogenic $CaCO_3$ in these sediment fields increases to ~20%. Li et al. (2013) attributed the sediment drifts in the northeastern middle slope between 800 and 1800 m to contour currents and unidirectionally migrating channels. Further downslope, to the east of the Dongsha Islands, Kuang, Zhong, Wang, and Guo (2014) recognized four large sediment wave fields (Figure 4.12). The wave fields are distributed on the western levees and adjacent overbank areas of related channels and the area immediately outside the channel mouths. Each sediment wave field consists of several to tens of rows of sediment waves, with a size up to 2.8–7.2 km in wavelength and 30–60 m in wave height (Figure 4.12B). They are mostly asymmetrical in cross sections with a thicker upslope flank and a thinner downslope flank, indicating upslope migration. Piston cores reveal that sandy turbidites developed with normal size grading dominate the channels, while massive muds with thin sandy or silty turbidite interbeds prevail in the interchannel regions on which the sediment waves situated. The sediment waves on the channel levees and related overbank areas were presumably built by the turbidity currents overspilled from the channels, while those located outside the channel mouths were likely generated by unconfined sheetlike turbidity currents from the channel mouths (Kuang et al., 2014). Another probable mechanism for these sediment wave fields could be the bottom contour currents induced by the intrusion of North Pacific Deep Water, which move westward and upward when dissipated by bottom topography and subsequently help to produce the upslope-migrating sediment waves (Gong et al., 2012).

Slump and turbidite deposition caused by slope failure and gravity flow from canyons are also widespread in other parts of the deepwater SCS. While the northeastern margin is plagued by canyons, which are typically ~1 km wide and up to 20 km long, steep-sided, and asymmetrical with a higher slope angle, the northwestern margin shows increased evidence of slumping and slope failure (Dickinson, Ware, Cosham, & Murphy, 2012). Mapping of the base Pleistocene shows the slumping has resulted in removal of Quaternary sediment along the intercanyon ridges as well as mass transport deposits along the lower slope. Several international cruises in the recent years have revealed a turbidite nature for most part of the deep basin (e.g., Wetzel & Unverricht, 2013).

4.3 DEPOSITIONAL HISTORY

4.3.1 East China Sea Shelf

Deposition on the greater ECS during the Quaternary is featured by neritic and deltaic deposits at high sea level and deltaic or alluvial deposits or

FIGURE 4.12—CONT'D

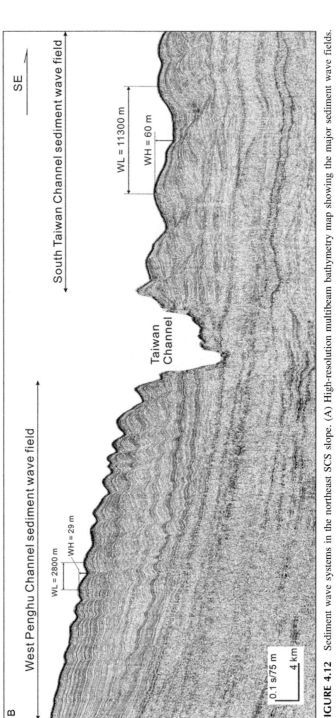

FIGURE 4.12 Sediment wave systems in the northeast SCS slope. (A) High-resolution multibeam bathymetry map showing the major sediment wave fields. (B) Seismic profile cross the sediment waves showing the asymmetrical structure and upslope migration. *From Kuang et al. (2014).*

nondeposition at low sea level. Available data indicate that marine deposition started from the early Pleistocene in the ECS proper and Yellow Sea and from the middle Pleistocene in the Bohai Sea. Marine deposition reaching a maximum over 100 km inland from the west Bohai Sea occurred during the late Pleistocene, in marine isotope stage (MIS) 5 (Li, 2012; Yang & Lin, 1993).

Core BC1 (119°54′E, 39°9′N, water depth 27 m) from the central Bohai Sea (Qin, 1985) and core QC2 (122°16′E, 34°18′N, water depth 50 m) from the southern Yellow Sea (Qin, 1989; Yang & Lin, 1993) represent the best-studied cores so far in the region. As summarized in Figure 4.13, sand, silt, and clay of alluvial and fluvial–deltaic origins are characteristic in the cores.

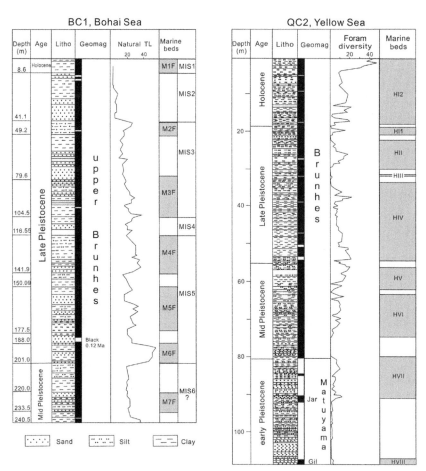

FIGURE 4.13   Major sedimentary features of (left) core BC1 from the central Bohai Sea and (right) core QC2 from the South Yellow Sea. Note that the age for the lower parts of the two cores remains to be resolved with additional solid evidence. *Modified from (Qin, 1985) and Yang and Lin (1993).*

Based on microfossils, particularly foraminifers and ostracods, seven shallow-marine beds (M1F–M7F) in core BC1 and 8 marine beds (HI–HVIII) in core QC2 have been recognized. Deposition of these beds in shallow neritic environments is demonstrated by an *Ammonia*-dominated benthic foraminiferal assemblage without or with very rare planktonic species. Radiocarbon dating confirms the Holocene age for the youngest marine beds (M1F and HI). Geomagnetic results indicate that the oldest marine bed M7F in core BC1 lies within the Brunhes chron probably having an age of 0.15–0.20 Ma in the later part of the middle Pleistocene, while HVIII in core QC2 is at the upper Olduvai chron with an age of ~1.76 Ma in the early Pleistocene. An early consensus is that the marine beds in BC1 represent marine transgressions at MIS 3 (by M2F and M3F), MIS 5 (by M4F, M5F and M6F), and MIS 6 (by M7F), respectively, which can be correlated with H1 to HV in core QC2 (Figure 4.14). Toward deltaic or deeper-water settings, however, the threefold MIS 5 interval could have merged to only one marine bed because either the other two beds are missing or all the three beds are conjoined together (Li, 2012; Yang & Lin, 1993; Tang, 1996). Therefore, actuate dating is critical for positioning these marine deposits. For example, the existence of MIS 3 transgressions in the Bohai Sea has been challenged recently by optically stimulated luminescence (OSL) dating, which shows that the marine beds previously assigned to MIS 3 based on radiocarbon dating are now proved to have

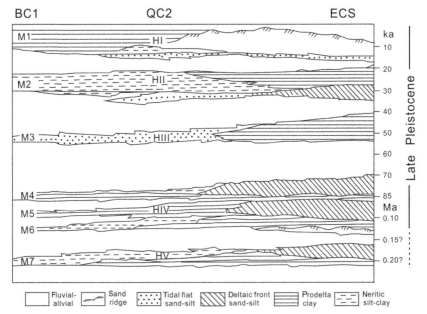

FIGURE 4.14   A sketch showing the general features of late Pleistocene marine deposition in the greater East China Sea. Note vertical scale changes. *Modified from Yang and Lin (1993).*

registered an older OSL age in MIS 5 (e.g., Yi, Lai, et al., 2012; Yi, Yu, et al., 2012).

In the western South Yellow Sea, a detailed study of cores and seismic profiles reveals the presence of a large delta active during MIS 3 that was subsequently affected by channel incision (Figure 4.15; Liu, Chen, et al., 2010; Liu, Saito, et al., 2010). The delta succession and incised-channel fills have high concentration of smectite, suggesting the paleo-Yellow River as the main contributor. Comprising two seismic facies, the delta sequence shows a coarsening-upward trend. The incised-channel system consists of two main channels with multiple tributaries in a dendritic pattern. The incised-channel fills begin with fluvial and then estuarine sediments, which are later truncated by a transgressive ravinement surface and capped by transgressive deposits (Figure 4.15).

A series of deltaic–estuarine–prodeltaic–shoreface facies developed underneath the sand ridges on the ECS (Figure 4.16; Berne et al., 2002). Normal to chaotic sedimentary structures including frequent channel downcut and refill are all present, indicating repeated deltaic to shallow-marine environments over the Quaternary glacial–interglacial cycles. Like the modern situation, sand ridge systems are also found in previous transgressive phases and at sea-level highstands, such as seismic unit U60 representing MIS 7. Core DZQ4 (29°24.75′N, 125°21.85′E, water depth 89 m, core length 51.65 m) confirms a sediment succession from MIS 6 to MIS 1, although unit U120 (8–12 m) representing the MIS 5 high sea level is relatively condensed. In contrast, the last glacial deposition during MIS 2–MIS 4, which consists of units U125, U130, U140a, and U140b (part) with a total thickness of ~26 m, is more extensive. Therefore, on the outer shelf of the ECS, the formation of sand ridges seems to have been associated more frequently with high current activities at interglacial high sea level, while muddy and silty deposition was more common at glacial lowstands when river mouths moved offshore and current activities were low (Figure 4.16).

4.3.2 South China Sea Shelf

4.3.2.1 Northern South China Sea shelf

As on the ECS, a large paleoriver–deltaic network exists on the northern SCS shelf, offshore from the present Pearl River Mouth, as mapped by core and seismic data (Figure 4.17). Buried at 5–25 m below the present seafloor (BPSF), the paleoriver channels may reach a maximum width of 6 km and a maximum incision depth of <20 m. The fan-shaped paleodeltas stretch out southeasterly in parallel to the main river channel and southwesterly and northeasterly by side expansion, and three major phases have been recognized (Bao, 1995). Phase I paleodelta lies at 30–62 m water depths and covers ~450 km². Phase II comprises two paleodeltas from 80 to 100 m, with the

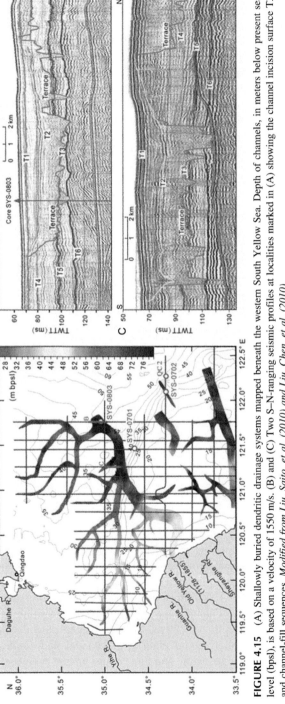

FIGURE 4.15 (A) Shallowly buried dendritic drainage systems mapped beneath the western South Yellow Sea. Depth of channels, in meters below present sea level (bpsl), is based on a velocity of 1550 m/s. (B) and (C) Two S–N-ranging seismic profiles at localities marked in (A) showing the channel incision surface T3 and channel-fill sequences. *Modified from Liu, Saito, et al. (2010) and Liu, Chen, et al. (2010).*

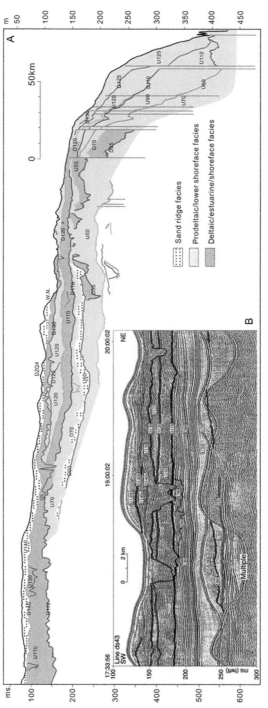

FIGURE 4.16 (A) Simplified shelf–slope seismic profile of the ECS showing the distribution of Pleistocene depositional units from U50 to U140, representing MIS 9 to MIS 2–1. (B) A detailed profile across 29°00′N /125°13′E reveals the internal structures of the seismic units, including chaotic facies in U60, U90, and U140 and main channel horizon D130 (Berne et al., 2002). (A) *Based on Berne et al. (2002).*

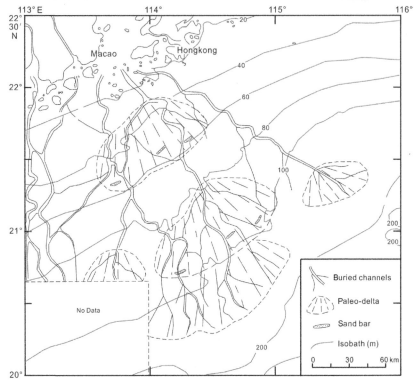

FIGURE 4.17 The buried paleoriver–paleodelta system on the northern SCS shelf offshore from the Pearl River Mouth based on seismic and sonar data and cores. *Based on Bao (1995).*

large one reaching a size of ∼590 km². Elongate sand bars developed along the front margin of the phase I and II paleodeltas. Phase III paleodeltas from 100 to 160 m water depths are largest in size, and the main one covers an area of ∼605 km² and connects to the phase II large paleodelta in the north. The phase III paleodelta front margin is marked by a slump zone with steep elevation of 10–25 m (Bao, 1995). In addition, there are smaller paleodeltas developed alongside the phase II and III main paleodeltas (Figure 4.17). However, when and how this paleoriver–deltaic network formed and functioned and what impact it had caused on the regional deposition require further studies.

Numerous cores have been drilled on the northern SCS shelf and slope for geologic survey or petroleum exploration (e.g., Yang & Kou, 1996; Qiu, Huang, & Zhong, 1999; Li, 2012; Liu, Zhao, et al., 2002, and references). Quaternary deposits are dominated by fluvial, deltaic, to shallow-marine facies, which can be grouped into four major units: Q1 (early to middle Pleistocene), Q2 (middle Pleistocene), Q3 (late Pleistocene), and Q4 (last deglacial to Holocene) (Figure 4.18). Subunits are also dividable based on lithologic and seismic characteristics. Stratigraphically, older units are dominated by

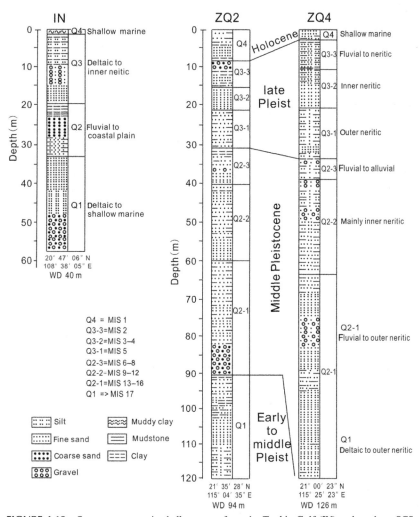

FIGURE 4.18 Quaternary strata in shallow cores from the Tonkin Gulf (IN) and northern SCS shelf (ZQ2 and ZQ4), showing major sediment facies (Li, 2012, and references). The age of units Q1–Q4 was determined by ^{14}C and TL dating and geomagnetic chronostratigraphy.

alluvial to fluvial facies and younger units by deltaic to shallow-marine facies. Although their thickness differs from place to place, early to middle Pleistocene Q1 can reach over 100 m, middle Pleistocene Q2 ranges between 50 and 100 m, and late Pleistocene Q3 and Holocene Q4 each rarely excess 40 m. While older deposit units generally become thicker seaward, Holocene Q4 often turns thicker landward, indicating the effects of progressive transgression onshore and stronger current on outer shelf in the Holocene. In the middle Pleistocene when most part of the northern SCS margin was prevailed by alluvial to deltaic deposition, the area surrounding the eastern Leizhou Peninsula was experiencing the youngest volcanic eruption, forming the Shimaoshan volcanic rock, which characterizes the upper part of late Pleistocene unit Q2.

The evolution of fluvial–deltaic–marine depositional facies on the northern SCS shelf is best illustrated in seismic profiles, such as the one shown in Figure 4.19. Clearly, the middle to late Pleistocene succession from the outer shelf is dominated by regressive units (A, B, D, and E), while the transgressive unit C is relatively thin and discontinuous. By correlation, seismic unit A is equivalent to late Pleistocene Q3 in cores, and units B to E represent different parts of middle Pleistocene Q2. The present shelf break represents a delta front zone shaped by previous lowstand deposits during the middle Pleistocene (D) and (B) (Figure 4.19; Lüdmann et al., 2001).

4.3.2.2 Sunda Shelf

Like those on many other shelves, the Quaternary deposit sequence on the Sunda Shelf also features regressive deltaic packages (Figure 4.20), as demonstrated by Hanebuth et al. (2011) in their review on sedimentation of the Sunda Shelf over the last glacial cycle. The regressive units are overall sigmoidal-promoting, extremely thick, and wide succeeding. At the Last Glacial Maximum, a huge paleoriver network developed on the emerged paleo-Sunda Shelf, mainly comprising the "Molengraaff" (North Sunda), paleo-Mekong, and paleo-Chao Phraya Rivers. Their large valleys are still prominent morphologic features on the modern seafloor with widths of up to 5 km (Hanebuth et al., 2011). This huge paleoriver network has played an important role in the supply of water and sediment at low sea level and in channel fill and sediment accumulation at subsequent postglacial transgression, producing locally complex deposits. Rapid flooding of the Sunda Shelf by a sudden sea-level rise of 16 m occurred within 300 years during meltwater pulse 1A, or 14.6 to 14.3 thousand years ago (Hanebuth, Stattegger, & Grootes, 2000).

As summarized in Table 4.5 and Figure 4.21, a total of 14 depositional units along the Sunda Shelf–slope transect were identified by Hanebuth et al. (2011) for the last glacial cycle (MIS 5–MIS 1). Units 6, 8, and 9 in Table 4.5 and Figure 4.21 are equivalent to unit a in Figure 4.20A and the blue unit in Figure 4.20B, indicating the dominance of regressive deposition.

For the paleo-Mekong drainage area, an abrupt transition from fluvial to marine deposition of incised valley-fill sediments has been observed in many cores retrieved offshore from the Mekong mouth. Valley-fill sediments mainly consist of fluvial mud, whereas postglacial sedimentation is characterized by shallow-marine carbonate sands. Sudden Holocene flooding started at ~14–12 kyr BP at outer shelf sites and completed at ~7 kyr BP at land-based sites (Figure 4.22; Tjallingii et al., 2010).

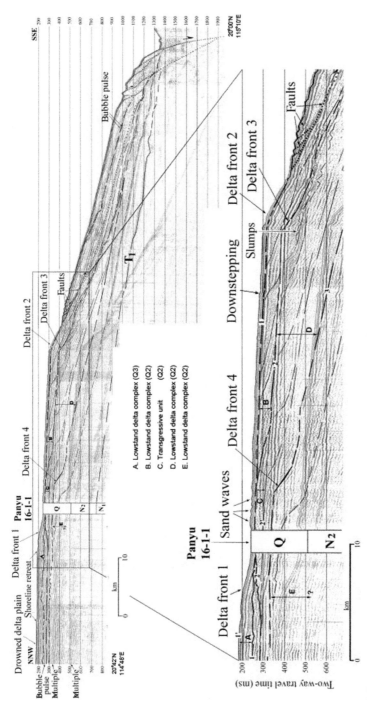

FIGURE 4.19 Seismic profile of the northern SCS outer shelf and slope showing deltaic to marine deposit units A to E, which are equivalent to late Pleistocene Q3 (A) and middle Pleistocene Q2 (B–E) by correlation to the results of well Panyu 16-1-1. Note that Holocene and early Pleistocene units are not labeled. *Modified from Lüdmann et al. (2001).*

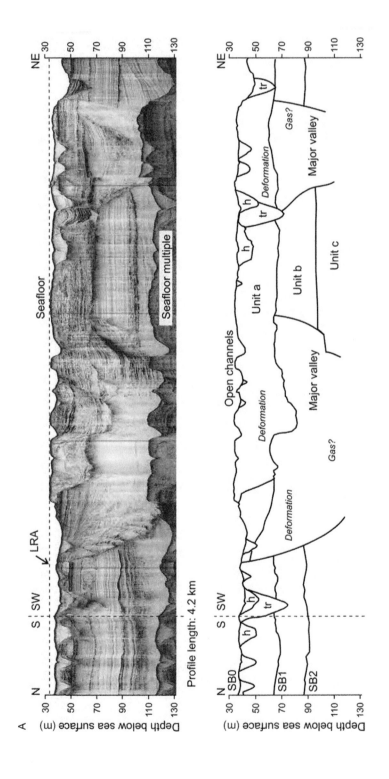

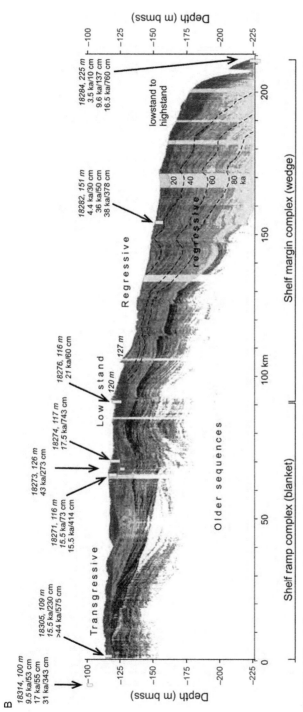

FIGURE 4.20 Sediment-acoustic (parasound) profiles from the Sunda Shelf (Hanebuth et al., 2011). (A) The inner shelf profile located at 2°N/105°E shows complex structures, including channels, and three regressive units (a = MIS 2–4, b = MIS 6, and c = MIS 8) over the last ~330 kyr. Note the patchy distribution of transgressive (Tr) and Holocene (h) deposits. (B) The outer shelf profile from 5.5°N/111°E to 7°N/112°E is marked by thick regressive sediment sequences during the last glacial period (blue), overlain by relatively thin lowstand and transgressive packages of the Holocene (brown). Also shown are sediment ages (cal ^{14}C BP) at different depths of shallow cores (yellow boxes).

TABLE 4.5 A Summary of Stratigraphic Units Along the Sunda Shelf–Slope Transect Showing Their Depositional and Lithologic Characteristics (Hanebuth et al., 2011)

Unit	Stratigraphic and Regional Position	Depositional Character	Thickness	Time of Formation	Nomenclature
1	Youngest deposits, in inner-marginal (coastal) position	Beach sands, muddy swamp deposits, lagoonal and shallow-marine muds	≤20 m	Latest transgression to highstand (8–0 kyr BP)	Highstand system, coastal wedge, Holocene sand prism
2	Modern seafloor on open shelf (off unit 1)	Marine carbonate muds	<1 m	After main parts of the shelf are flooded (13–0 kyr BP)	Condensed section
3	On central and inner shelf, mainly in channel structures	Facies succession from terrestrial, estuarine to marine–marine conditions	Several meters or less	In the course of rapid transgression (18–12 kyr BP)	Transgressive system, incised-valley fills
4	In uppermost part of unit 7, across the entire inner and central shelf	Gray stiff clays with orange flames and concretions	1 to several meters	LGM and (probably) older	Paleosols, lateritic gleysol
5	Above unit 9; locally at border between central and outer shelf	Sand-barrier and tidal flat complex	Few meters	LGM (21–19 kyr BP)	Lowstand coastal complex
6	Seabed, on top of and off unit 9	Hemipelagic sediments, increased rates during lower sea level	Tens of meters	Continuously	Open-shelf succession
7	Surrounded by units 12, 8, 4, 3, on inner and central shelf	Coarse-grained channel fills, local terrestrial sand sheets and mud fans	Up to tens of meters	During glacial exposure (~110/90–20 kyr BP)	Continental sequence

8	Local sediment bodies resting on unit 12	Shallow-marine sediments	Lens-like, ~10 m	Late regression (MIS 3; 45 ka) and older	Forced regressive system
9	On top of former shelf-margin complex, on the outer shelf/upper slope	Shallow-marine sediments, massive progradation of the shelf edge	Wedge-like, up to 80 m	Late regression and lowstand (~90–20 kyr BP)	Regressive to lowstand system, shelf-margin complex
10	Intercalated into unit 6 and unit 9, continental slope	Occasional slide deposits	Up to tens of meters	Low sea level (MIS 3, 2)	Ocean basin fill
11	Seabed, open-marine/deep-sea basin	Open-marine hemipelagic sediments	Tens of meters	Continuously	Ocean basin fill
12	Between transgressive unit 13 and terrestrial unit 7	Aggradation of terrestrial, coastal and shallow-marine sediments (dominating the stratigraphic record across the shelf)	Up to 30 m, persistent over long distances	Early regression (~115–80 kyr BP)	Regressive system, shelf-ramp complex
13	Beneath unit 7, in inner-marginal (coastal) position	Coastal facies, cp. unit 1	Few meters?	Penultimate (Eemian, MIS 5) sea-level highstand (~124–118 kyr BP)	Highstand system, coastal wedge
14	Beneath unit 12, on inner to central shelf	Channel cut-and-fill structures, terrestrial, estuarine, marginal-marine, cp. unit 3	Few meters?	Penultimate transgression (~127–124 kyr BP)	Transgressive system, Incised-valley fills

Refer to Figure 4.21 for their distribution.

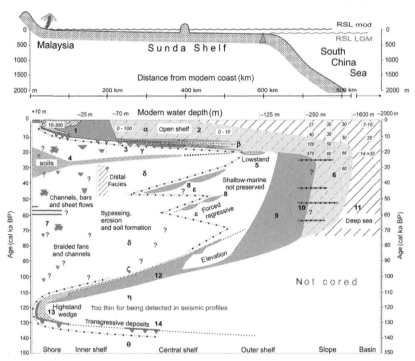

FIGURE 4.21 Interpreted environmental dynamics and major depositional facies over the last 140 kyr on the Sunda Shelf (Hanebuth et al., 2011). Depositional units (detailed in Table 4.5): 1 modern coastal highstand complex (yellow, sandy; green, muddy), 2 Holocene highstand condensed section (hatched gray), 3 deglacial transgressive unit (green), 4 late Pleistocene exposure soils (gray), 5 LGM lowstand unit (orange), 6 slope deposition (hatch gray), 7 late Pleistocene alluvial deposits (orange, gray), 8 successive force-regressive units (blue), 9 shelf-margin sediment wedge (blue), 10 slope mass movement complex (orange), 11 deep-sea sedimentation (hatched), 12 MIS 5 regressive unit (blue), 13 penultimate (Eemian) coastal highstand complex (yellow, sandy; green, muddy), and 14 penultimate transgressive unit (green). Symbols α to θ are various surfaces or hiatuses. Numbers in small boxes are sedimentation rates [cm/kyr]. Levels with significant organic matter are marked with red stripes and previous sea-level positions with dotted gray lines.

4.3.3   Deep Sea

4.3.3.1   South China Sea

ODP Leg 184 Records

A total of 17 holes at 6 deepwater sites were drilled, and about 5000 m of cores were recovered from the northern SCS slope (Sites 1144, 1145, 1146, 1147, and 1148) and the southern SCS slope (Site 1143) during Ocean Drilling Program (ODP) Leg 184 in the spring of 1999. The sediments recovered represent the simple mixing of nannofossil ooze and detrital clays derived from Asia, via the Pearl and Red Rivers on the northern margin and the

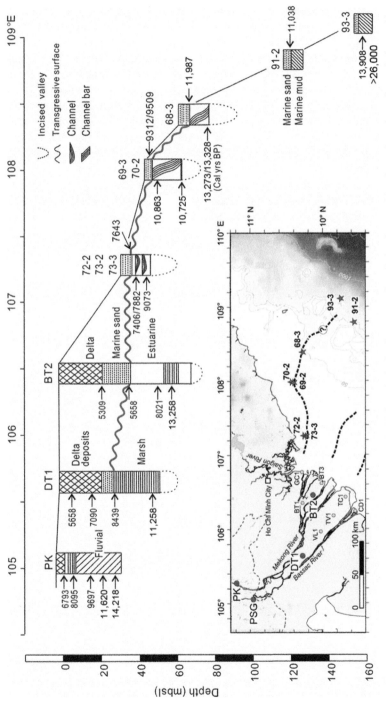

FIGURE 4.22 Schematic west–east profile in cores along the Mekong incised-valley system (inset), showing a sudden change in major sediment facies marking the regional transgressive surface, ~12 kyr BP to ~7 kyr BP (Tjallingii, Stattegger, Wetzel, & Phung Van, 2010).

Mekong and Molengraaff Rivers and directly from Borneo on the southern margin. These sediments represent deepwater deposition throughout the Neogene and back into the early Oligocene (Site 1148) and provide evidence for the evolution of the SCS Basin (Figure 4.23; Wang et al., 2000).

The lack of coarse clastic material at sites on the northern slope suggests that the shallower slope basins have been acting as efficient sediment traps. The presence of a 33-Ma-old deepwater, bathyal facies implies an older onset of the initial seafloor spreading probably in the Eocene. The slumping, brittle faulting, and mineralization observed in lithologic units VI and VII at Site 1148 mark the most intensive late Oligocene tectonic deformation in the SCS region.

Increases in carbonate accumulation in the late Miocene are recorded at several sites, indicating maximum marine carbonate production (Figure 4.23). The Pliocene and Pleistocene sediments, however, are characterized by increase of terrigenous input largely due to frequent sea-level fluctuations associated with increased magnitude of global glacial cycles.

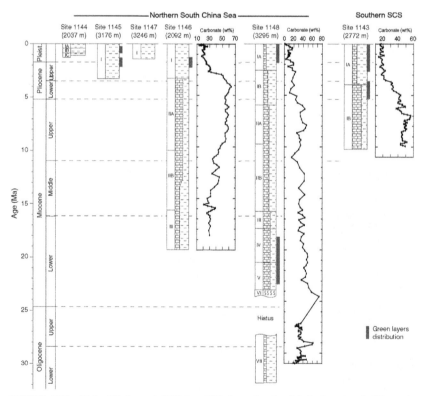

FIGURE 4.23   Major lithology at ODP Leg 184 sites, showing lithologic units, CaCO$_3$ variations, and distribution of green layers. *Modified from Wang et al. (2000).*

Green clay layers are common, either as discrete layers as thick as 3 cm or more commonly as disrupted layers, patches, or mottles. Most of the green layers are confined to the Pliocene–Pleistocene (Figure 4.23) except for a lower Miocene set recovered at Site 1148. Their common association with burrows and patches caused by burrowing implies a link to the presence of organic matter. Certainly, their green color is suggestive of reducing conditions associated with organic matter alteration. Probably, they reflect only the local variations in sediment composition and burial.

Depositional Mass

Sediment mass accumulation since the Oligocene in deepwater SCS has been estimated based on data from 94 seismic profiles, 121 boreholes, 34 industry wells, and 6 ODP 184 sites (Huang & Wang, 2006, 2007). The results indicate that the total deposit mass in the SCS since the beginning of seafloor spreading in the Oligocene is about 1.44×10^{10} Mt. The major part of this sediment is of terrigenous origin, and most of it was deposited during the Oligocene and Quaternary. For example, the Oligocene accumulation rate reached a maximum of ~ 22 g cm^{-2} yr^{-1}, followed by ~ 15 g cm^{-2} yr^{-1} in the Quaternary, while other periods registered a lower rate between 8.6 and 11.5 g cm^{-2} yr^{-1} (Figure 4.24). Today, silt discharge from all rivers into the SCS is nearly 400×10^6 t per annum, which is equivalent to ~ 18 g cm^{-2} kyr^{-1} for the SCS with an area of 3.34×10^6 km^2. Therefore, the accumulation rate in the present-day SCS is very high, largely due to intensive denudation in the modern drainage regions.

The deposit mass over the last 20 kyr can be used to characterize deposition during the Quaternary glacial–interglacial cycles. As the shallow shelf

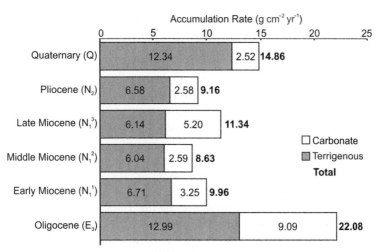

FIGURE 4.24 Estimated accumulation rate for deposition in various periods of the SCS (Huang & Wang, 2006).

sediment has been subject to intensive modifications by frequent sea-level changes, this exercise has been confined to areas deeper than 100 m water depths based primarily on data from 136 sediment cores (Figure 4.25; Huang & Wang, 2007).

From the sediment thickness results, seven deposition areas (I–VII) can be recognized (Figure 4.25). The deposit mass in area V from the southern SCS has a maximum thickness of >1000 m, or ~600 m for MIS 2 and ~400 m for MIS 1. Other areas with thicker MIS 1–2 sediments include area I from the northeast and area IV from the southwest. In contrast, the central deep-sea area VII and along the eastern margin are covered by a thin sediment layer of <100 cm from MIS 1 and MIS 2 combined (Figure 4.25). The total deposit mass for all areas is 1.84×10^6 Mt for MIS 1 and 2.56×10^6 Mt for MIS 2, with the majority being of terrigenous, or 77% for MIS 1 and 83% for MIS 2. As seen in Figure 4.26, area IV from the offshore Mekong mouth registered highest terrigenous accumulation rates of $15\text{–}20 \text{ g cm}^{-2} \text{ kyr}^{-1}$ during MIS 1 and MIS 2, followed by $11\text{–}18 \text{ g cm}^{-2} \text{ kyr}^{-1}$ in area V from offshore Borneo affected by the paleo-Rajang and paleo-Baram Rivers and by $10\text{–}16 \text{ g cm}^{-2} \text{ kyr}^{-1}$ in area I between the offshore Pearl River Mouth and Taiwan Island in the northeast.

Biogenic components comprise about one-fifth of the total deposit in the late Quaternary SCS. While carbonate has been strongly affected by dissolution, opal appears to have mainly responded to productivity variations over glacial cycles. The average opal accumulation rate is $0.28 \text{ g cm}^{-2} \text{ kyr}^{-1}$ for MIS 1, increasing to $0.44 \text{ g cm}^{-2} \text{ kyr}^{-1}$ MIS 2, probably indicative of high productivity in the glacial SCS (Figure 4.26; Huang & Wang, 2006, 2007).

4.3.3.2 Okinawa Trough

The depositional and paleoceanographic characteristics of the Okinawa Trough in the late Quaternary are revealed in many short sediment cores (Li & Chang, 2009), some of which are illustrated in Figure 4.27. The silt- and clay-dominated deposit sequence preserves signals of global climate events as well as regional volcanism. In the northern part, the Yellow River probably contributed directly to the clay deposition in the northern trough through the 4–5 km wide Goto submarine canyon at sea-level lowstands (Oiwane et al., 2011). During glacial stages, lowering of the sea level caused progradation of the Yellow River mouth approximately 900 km seaward and resulted in direct discharge into the canyon head. The large sedimentary discharge gave rise to efficient cutting and widening of the canyon. However, the occurrence of carbonate-cemented strata prevented axial incision, which has resulted in a flat canyon floor (Oiwane et al., 2011).

In core DGKS9604 (28°16.64′N, 127°01.43′E, water depth 766 m) from the middle trough, large variations in $^{87}Sr/^{86}Sr$ ratios and ε_{Nd} values occurred between 14.0 and 7.1 kyr BP, indicating a significant change in sediment

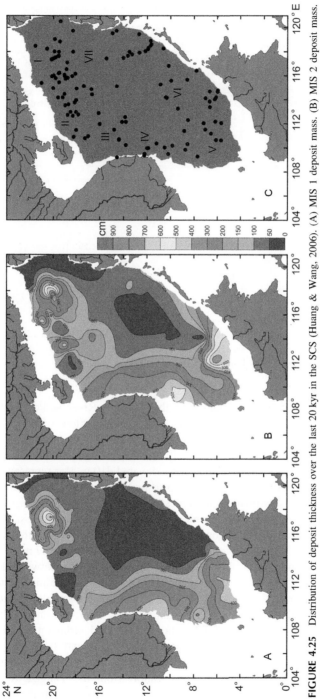

FIGURE 4.25 Distribution of deposit thickness over the last 20 kyr in the SCS (Huang & Wang, 2006). (A) MIS 1 deposit mass. (B) MIS 2 deposit mass. (C) The seven deposition areas and the location of 136 sediment cores used for the modeling exercise.

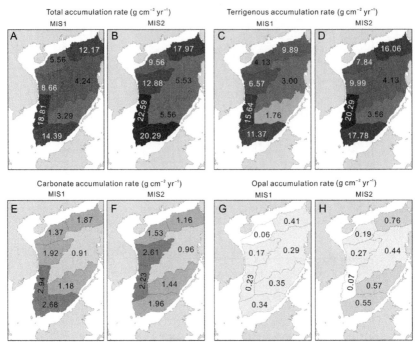

FIGURE 4.26 The average accumulation rates of various components in the seven off-shelf areas of the SCS during MIS 1 and MIS 2 (Huang & Wang, 2007).

provenances from the dominance of the Yangtze River and/or continental shelf to that of Taiwan Island due to the reentry flow of the Kuroshio (Figure 4.28; Dou et al., 2012).

In the south, as revealed by drilling at ODP Site 1202 (24°48.24′N, 122°30.01′E, water depth 1274 m), the recovered 410 m core represents a sediment sequence of the last ~70 kyr, corresponding to a sedimentation rate of about 4.2 m kyr^{-1}, which is the highest ever reported from the region (Wei, Mii, & Huang, 2005). Hemipelagic silty clay with insignificant silt or sand layers occurs from the core top down to 133 m (MIS 1–3), but the silt–sand layers increase gradually from <10% between 133 and 167 m to >50% between 223 and 279 m, followed by decreases to <10% between 310 and 337 m and to <3% between 337 and 407 m (MIS 4). Containing slate fragments, quartz grains, and shallow-marine fossils, the silt–sand layers represent fine-grained turbidite from gravity flows (Huang, Chiu, & Zhao, 2005). The provenance data from Site 1202 indicate increased sediment supply from northwestern Taiwan between 28 and 19.5 kyr BP and from East China sources between 19.5 and 11.2 kyr BP. The change in provenance at 19.5 kyr BP reflects increased fluvial runoff from the Yangtze River and strong sediment reworking from the East China Sea shelf in the course of

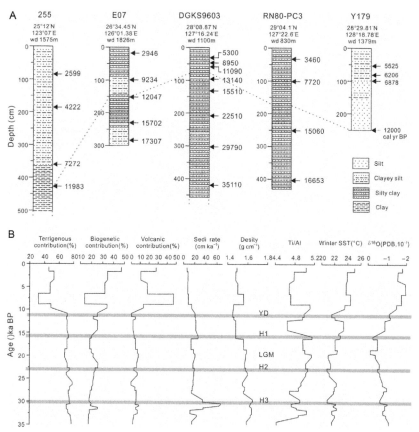

FIGURE 4.27 (A) Sediment cores from the Okinawa Trough showing main sediment types and ^{14}C ages. (B) Variations of various parameters and climate events (horizontal bars) in core DGKS9603. *Panels (A) modified from Liu, Zhu, et al. (2007) and Liu, Xu, et al. (2007) and (B) modified from Meng, Liu, Du, and Shi (2009).*

increased humidity and postglacial sea-level rise, particularly after 15.1 kyr BP (Diekmann et al., 2008). The influence of the Kuroshio on the northward transport of sediments increased especially since 14 kyr BP (Dou et al., 2010, 2012).

4.4   TERRIGENOUS SEDIMENTS

4.4.1   Deltaic Sands and Offshore Muds

Numerous works have been published on the sedimentologic characteristics and evolution of major river deltas in China, namely, the Yellow River Delta (Chen, 1991; Chen & Xue, 1997), the Yangtze River Delta (Huang, Tang, & Yang, 1996; Li & Wang, 1998; Yan & Xu, 1987; Yun, 2004; Zheng, 1999),

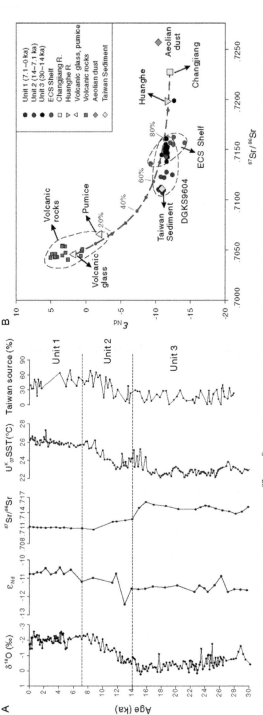

FIGURE 4.28 (A) Comparisons of ε_{Nd} values and $^{87}Sr/^{86}Sr$ ratios with other paleoceanographic proxies from core DGKS9604. (B) Discrimination plot of ε_{Nd} versus $^{87}Sr/^{86}Sr$ in core DGKS9604 and in the sediments of potential sources showing provenance changes at 14 kyr and 7.1 kyr BP. *From Dou et al. (2012).*

and the Pearl River Delta (Long, 1997; Zhao, 1990). A synthesis of these three large deltas was summarized in a monograph by Ren (1994).

The development of large river deltas in the region is characterized by an early Holocene sea-level jump and delta initiation (Hori & Saito, 2007). Sedimentary facies changes from estuarine sand and mud to shelf or prodelta mud at most river mouths indicate rapid marine influence during 9000–8500 cal BP. The sea-level jump was probably caused by the final collapse of the Laurentide Ice Sheet, resulting in catastrophic drainage of glacial lakes and rapidly landward movement of the depocenter of an estuarine system. The subsequent deceleration in sea-level rise promoted concurrent delta initiation in the region.

4.4.1.1 Yellow River Delta (Huanghe Delta)

With a delta area of 5400 km^2, the modern Yellow River Delta in the southern coast of the Bohai (Chapter 2) is characterized by a high concentration of sediment, huge sediment discharge, thin Holocene deltaic sediments, a lateral delta lobe shift, and a steep longitudinal profile in its lower reaches (Saito, Yang, & Hori, 2001, and references). Mainly consisting of silty clay of loess origin, the total sediment discharge reaches about 1080 Mt/yr for historical average over 2 kyr, but today, it declines to only about 40 Mt/yr. Most of the fluvially derived sediment appears to remain trapped within the modern deltaic system, and the rest as suspended load continues westward, northward, and eastward following the prevalent seasonal currents (Qin, 1994). Different from other river systems, the Yellow River disperses its sediments in a very unique way into the ocean, where they are influenced uniquely by formation of the negatively (hyperpycnal) buoyant sediment plumes, seasonal resuspension, and alongshore current transportation. The dominant negatively hyperpycnal plumes travel very short distance near the modern Yellow River mouth. Those fine sediments are resuspended, mainly in winter seasons, and are continually transported long distance along margins in a semi-hyperpycnal flow (Liu, Milliman, Gao, & Cheng, 2004, and references).

The Holocene delta of the Yellow River has been divided into 10 superlobes (Xue, 1993; Saito et al., 2001, and references), including a North Jiangsu superlobe (9) for deposition between 1128 and 1855 at the old delta locality in the south of the Shandong Peninsula (Figure 4.29; Liu et al., 2013). These superlobes record the recent history of the restless river. In the modern superlobe area (10), the Holocene section consists in an ascending order of pretransgression river and lake deposits, transgressive tidal flat and estuarine deposits, shelf deposits, and prograding delta and river deposits (Figure 4.29; Xue et al., 1995). The delta deposits are the thickest and are typically composed of two imbricating superlobes: AD 1855–present and AD 11–AD 1048 superlobes, representing the two major periods of recent delta progradation in the area (Saito et al., 2000).

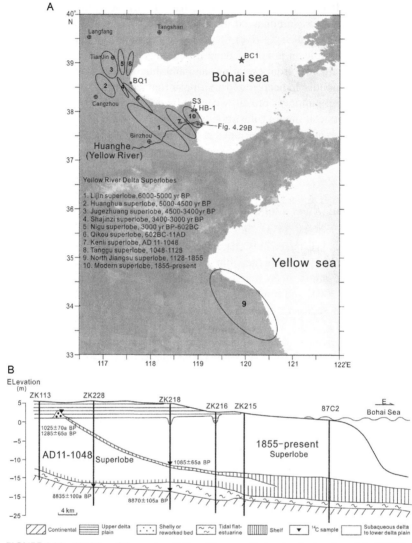

FIGURE 4.29 (A) Map of the modern Yellow River mouth showing the location of the 10 Holocene delta superlobes (Qiao et al., 2011; Saito et al., 2001). (B) Sketched profile along the river mouth showing the modern delta superlobe overlapping the AD 11-1048 superlobe (Xue, Zhu, & Lin, 1995).

The late Quaternary deltaic sequence is characterized by three intervals of marine transgressive facies, T1, T2 and T3, corresponding to depositions at MIS 1, 3, and 5, respectively. These three marine intervals have been observed in many sediment cores from the coastal area of the Bohai and Yellow Seas (e.g., Liu et al., 2013; Liu, Xue, et al., 2009; Yi, Lai, et al.,

2012; Yi, Yu, et al., 2012). For example, from about 70 km northwest of the present Yellow River mouth at 4.5 m water depth, core HB-1 encountered T1 (depositional units DU1) and T2 (DU5 to DU7), but did not penetrate T3, which is interpreted to occur as seismic unit SU7 below the core depth (Figure 4.30; Liu, Xue, et al., 2009). As a characteristic of the Yellow River deposition, silt with abundant mica is dominant in most part of the marine intervals. However, the scarcity of mica in DU 3, DU 4, (lower) DU 6, and DU 7 demonstrates that the Yellow River did not discharge into the Bohai Sea, or at least did not influence the study area, during early MIS 3 and from MIS 2 to the early Holocene (25–8.5 cal kyr BP) (Figure 4.30). It is noteworthy that, among the late Quaternary seismic units, the marine units SU5 and SU7 both feature a relatively thick sequence. Older marine beds of mainly middle Pleistocene age have been reported in cores S3 and DY from the present delta area (Yang, Li, & Cai, 2006; Zhuang et al., 1999), as well as in core BC1 from the central Bohai Sea (Figure 4.13), probably indicating the earliest marine influence in the region (Yi, Lai, et al., 2012; Yi, Yu, et al., 2012).

4.4.1.2   Yellow River Distal Subaqueous Delta

In addition to the 10 delta superlobes, a Yellow River-derived distal subaqueous delta lobe, or Shandong mud wedge, exists around the eastern tip of the Shandong Peninsula in the Yellow Sea across about 3° latitude, from 38° to 35°N (Figure 4.31; Liu, Milliman, & Gao, 2002; Liu, Milliman, et al., 2004; Yang & Liu, 2007). The mud wedge is featured by a 20–40 m thick complex sigmoidal-oblique clinoform overlying prominent pre-Holocene and transgressive surfaces. The silty clay-dominated mud wedge has been formed since the middle Holocene highstand, deposited mainly by the resuspended Yellow River sediments carried down by the coastal current (Figure 4.31).

Apart from the Yellow River, sediments from the small local streams and coastal erosion along the northern coast of the Shandong Peninsula may have contributed to the mud wedge (Figure 4.31A). While the alongshore current plays a key role in the southward growth of the delta, the upwelling and countercurrent in the deep layer of the boundary areas in the east Shandong Peninsula and North Yellow Sea may supply predeposited fine sediment landward to the mud wedge. Meanwhile, the strong Yellow Sea Warm Current and tidal currents also may play a role of eroding fine sediments in the northeastern Yellow Sea and transporting them southwestward to the Shandong mud wedge (Figure 4.31B; Liu, Milliman, et al., 2004; Liu, Xue, et al., 2009; Yang & Liu, 2007).

4.4.1.3   Yangtze River Delta (Changjiang Delta)

Covering an area of ~66,000 km^2, the Yangtze River Delta is characterized by a large water discharge, a large interseasonal water-level change,

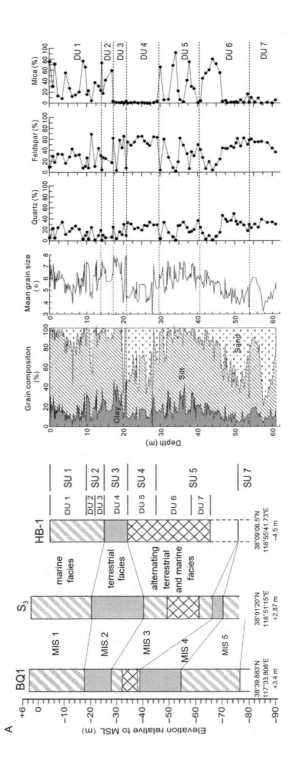

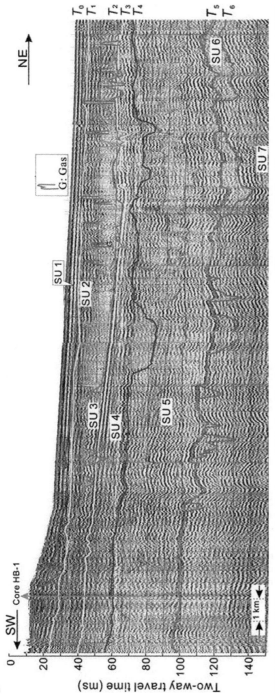

FIGURE 4.30 (A) Correlation of the late Quaternary deltaic deposition between cores on- and offshore modern Yellow River mouth, with depositional (DU1–DU7) and seismic units (SU1–SU7) corresponding to MIS 1 to MIS 5, respectively. (B) Downcore variations of sediment grain size and major mineral components in core HB-1. (C) Seismic profile through HB1 showing complex structures including channels and gas chimneys, close to reflector surfaces bounding the lowstand units SU3 and SU6. *Modified from Liu, Xue, et al. (2009).* See Figure 4.29 for locations of cores.

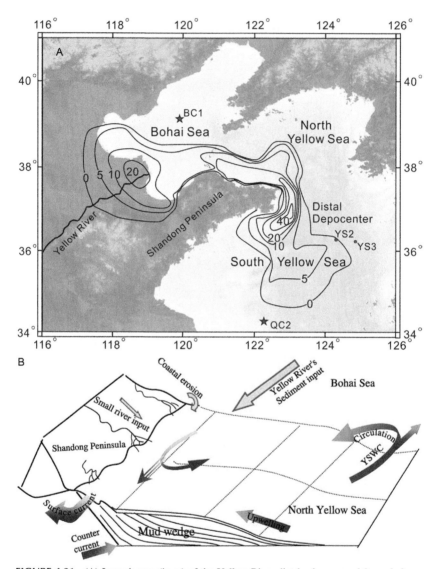

FIGURE 4.31 (A) Isopach map (in m) of the Yellow River distal subaqueous delta and along-shore mud wedge (Liu, Xue, et al., 2009). (B) A sketch showing sedimentary processes affecting the morphology of the alongshore mud wedge (Liu, Milliman, et al., 2004). YSWC = Yellow Sea Warm Current.

a tide-dominated coastal environment, continuous seaward progradation with isolated river mouth sand bodies, thick Holocene sediments, and a deep-incised valley formed during the last glacial period (Figure 4.32; Li, Chen, Zhang, Yang, & Fan, 2000; Li et al., 1979, 2002; Saito et al., 2001). Despite its length and size next to the Nile and the Amazon as the

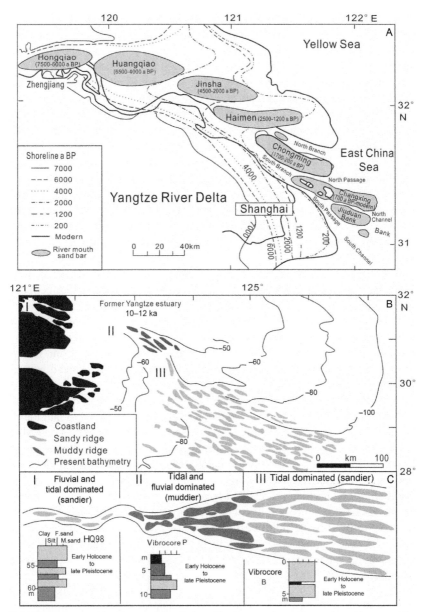

FIGURE 4.32   (A) The Yangtze River Delta with changes of coastline and sand bars, marking the delta progradation since ~7000 yr BP. (B) Tidal sandy ridge system outside the Yangtze River mouth. (C) Idealized estuarine depositional features for the Yangtze based on early Holocene to late Pleistocene sediments from three boreholes (Chen, Saito, et al., 2003; Chen, Shiau, et al., 2003; Hori et al., 2001a). *Panel (A) based on Li, Guo, Xu, Wang, and Li (1979), Li, Li, and Cheng (1983), Li and Li (1983), and Saito et al. (2001).*

third largest river on earth, the Yangtze discharges terrigenous sediments of ~420 Mt/yr, only about half of that achieved by the Yellow River. The sediment is mainly composed of sand and silt, with relatively low abundance of clay compared to other large rivers in the region. For the total suspended sediments from the Yangtze, about 44% was deposited in the submarine delta, >27% was transported southerly into the Hangzhou Bay, about 20% was delivered offshore to the sea, and only 9% was supplied and exchanged with the northern Jiangsu waters to the north of the estuary (Wan, Li, & Shen, 2009). Due to the tidal-dominated hydrology, sandbars and sand ridges are well developed inside and outside the river mouth, becoming muddier in the vicinity of the estuary owing to the mixed hydrodynamic interaction between the runoff flow and tidal currents (Figure 4.32; Hori et al., 2001a,2001b). Farther offshore, surficial sediments of the ridges were reworked, becoming sandier as tidal currents gradually dominated during the early Holocene prior to the recent delta initiation (Chen, Saito, et al., 2003; Hori et al., 2001a).

The deltaic sediments are characterized by very thinly interbedded to thinly interlaminated sand and mud (sand–mud couplets) and bidirectional ripple laminations, as a result of strong tidal influence. Sediment accumulation rate increases from ~1.1 m/kyr in prodelta to 3.5 m/kyr in the delta front to lower intertidal to subtidal flat. For the last 5000 years, average delta progradation rate was about 50 km/kyr but increased abruptly from 38 to 80 km/kyr after about 2000 yr BP, presumably due to the widespread human influence (Hori et al., 2001b).

A synthesis of data from 14 cores in the Yangtze delta region with intensive OSL and ^{14}C dating and microfossil analysis has revealed two late Pleistocene transgressive intervals, respectively, corresponding to MIS 3 and MIS 5 (Figure 4.33; Zhao et al., 2008). During MIS 5, fluvial gravelly sediment deposition prevailed with limited marine influence, indicating high-relief topography and local rivers developed with large freshwater and sediment discharge. Marine microfaunas may not have survived due to dominant freshwater process or strong erosion has scoured away the marine records of MIS 5. Tectonic subsidence is indicated to occur at a rate of 1 m/kyr since MIS 5 and lower the regional coastal area to be invaded by seawater during MIS 3, leading to muddy and sandy deposition along with widespread coastal facies (Zhao et al., 2008).

Core PD (31°37′29″N, 121°23′38″E) penetrated 318.7 m into the present Yangtze delta and recovered a Quaternary sequence of 240 m dominated by fluvial facies and a Pliocene sequence of 63 m dominated by lacustrine facies (Figure 4.34). Geochemical proxies including REE fractionation parameters and elemental ratios Cr/Th, Nb/Co, and Th/Co suggest that the Pliocene and Quaternary sediments have remarkably different provenances: mainly from proximal and more silicic sources in the Pliocene and from distal and more basic provenances in the Quaternary (Yang, Li, & Yokoyama, 2006). The

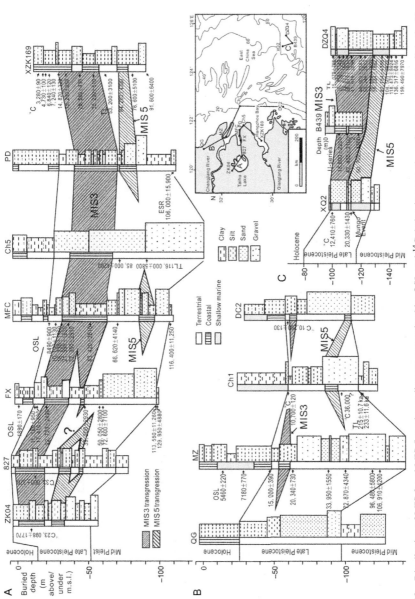

FIGURE 4.33 Distribution of MIS 3 and MIS 5 marine sediments constrained by OSL and ^{14}C dating in cores along three transects (A–C) in the Yangtze River Delta area. *Modified from Zhao, Wang, Chen, and Chen (2008).*

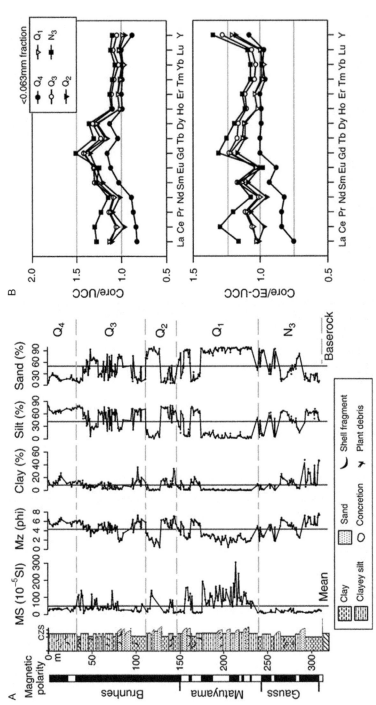

FIGURE 4.34 (A) Lithology, magnetostratigraphy, and sediment grain size compositions of core PD (Yang, Li, & Yokoyama, 2006). C, clay; Z, silt; S, sand; MS, magnetic susceptibility ($\times 10^{-5}$ SI unit); Mz, mean grain size (phi); Q4, Holocene; Q3, upper Pleistocene; Q2, middle Pleistocene; Q1, lower Pleistocene; N3, Pliocene. (B) REE fractionation patterns of the fine-grained sediments (<0.063 mm fraction) normalized to upper continental crust (UCC) and East China upper continental crust (ECUC), showing distinction of the N3 from Q1–Q4 sediments (Yang, Li, & Yokoyama, 2006).

Th–U–Pb ages of monazite grains dated at <25, 50–200, 200–400, 400–550, 800–1000, and 1800–2000 Ma are consistent with the main tectonic and magmatic events in the Yangtze Craton. The younger monazite grains with ages less than 25 Ma, signifying the Himalayan Movement, are present in all of the Quaternary strata but absent from the Pliocene sediments. Together, these data indicate that during the Pliocene, the "paleo-Changjiang" or its eastern equivalent was a locally small river draining today's lower Yangtze valley, and during the early Pleistocene, it evolved into a large river streaming down from the eastern Tibetan Plateau like the Yangtze today (Yang, Li, & Yokoyama, 2006). This view is upheld by subsequent studies of other deep cores in the Yangtze lower reach and deltaic basins, which tend to constrain the running-through time of the Yangtze on a narrow period of the late Pliocene to early Pleistocene (Fan, Wang, & Wu, 2012). However, marine influence on the Yangtze delta region was weak during the early and middle Pleistocene, and only since the late Pleistocene did marine influence become intensified, as indicated by foraminifera found in cored sediments (Yang, Li, & Cai, 2006).

4.4.1.4 Yangtze River Distal Subaqueous Delta

High-resolution seismic profiling and coring in the inner shelf of the East China Sea had revealed an elongate (~1000 km) subaqueous deltaic deposit stretching from the modern Yangtze River mouth south toward the Taiwan Strait (Figure 4.35; Li et al., 2007, 2009; Liu, Colin, & Trentesaux, 2006; Xu, Milliman, et al., 2009; Xu et al., 2012). This alongshore-distributed clinoform overlies a transgressive sand layer, thins offshore from ~40 m thickness at ~30 m water depth to <1–2 m at ~80 m isobaths, and has ~100 km across-shelf distribution. As revealed by XRD and grain size analyses, the clay-dominated mud wedge is sourced primarily from the Yangtze River, with increasing influence of Taiwan-derived silt and very fine sand near the Taiwan Strait (Figure 4.35). It is anticipated that, while Yangtze clays are dispersed southward by the coastal current mainly in winter–spring months, silts and sands from Taiwanese rivers (especially the Choshui) are transported northward by the Taiwan Warm Current mainly during summer and autumn (Xu, Lim, et al., 2009; Xu, Milliman, et al., 2009). Together with strong tides, waves, winter storms, and offshore upwelling, these sediment dispersal processes have jointly contributed to the formation of the alongshore mud wedge.

Among the four seismic units identified for the deposition sequence, the upper two units III (11–2 kyr BP) and IV (since 2 kyr BP) represent the Holocene mud wedge, while the underlying units I and II represent the last lowstand valley fills and deglacial transgressive deposits, respectively (Xu et al., 2012). Incised valleys, up to 15 m deep, are filled by flat-lying or inclined strata in unit I. Unit II is thin (<3 m) and located far from the

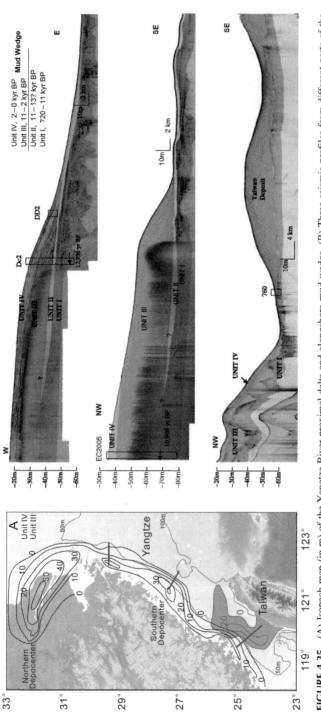

FIGURE 4.35 (A) Isopach map (in m) of the Yangtze River proximal delta and alongshore mud wedge. (B) Three seismic profiles from different parts of the mud wedge (red lines in A) showing the location of cores and variations of lithologic units from north to south. Note that units III (brown) and IV (green) form the main body of the mud wedge. *Panel (B) modified from Xu et al. (2012).*

modern Yangtze delta between 30 and 26°N in water depths between 40 and 90 m. The elongated muddier units III and IV are widely distributed, but unit III extends eastward farther than unit IV (Xu et al., 2012).

4.4.1.5   Offshore mud Fields in the Yellow and East China Seas

The Yellow Sea and the northern ECS are covered by five mud fields of different sizes (Table 4.6; Figures 4.5 and 4.36). Among them, the central Yellow Sea mud field (CYSM) with an area of 150×10^3 km^2 is the largest. Apart from the southwestern Yellow Sea mud off the northern Jiangsu coast (SWYSM) at the old Huanghe delta, these mud fields form largely in areas where the eddy circulation is prevalent (Liu, 1996; Xu et al., 1997) or tidal currents are weak (Zhu & Chang, 2000). Fine sediments either winnowed

TABLE 4.6 Mud Areas in the Yellow and East China Seas

	Area (10^3 km^2)	Deposition Rate (mm yr^{-1})	Sediment Budget (10^7 t yr^{-1})	Provenance	Current
North Yellow Sea mud (NYSM)	6.2	<2	0.6–2	Yellow River	Eddy
Central Yellow Sea mud (CYSM)	140–150	~0.2–2.7	4–5 (30?)	Yellow River	YSCC + YSWC
Southeast Yellow Sea mud (SEYSM) (Heuksan mud bank)	~8	1–17	4.9–8.7 (0.5?)	Mixed (Korean rivers)	KCC + YSWC
Southwest Yellow Sea mud (SWYSM) (old Huanghe delta mud)	~30?			Old Yellow	YSCC
Southwest Chiju Island Mud (SWCIM)	15	2–5	2–6	Yellow River	CWC (YSWC)

YSCC, Yellow Sea Cold Current; YSWC, Yellow Sea Warm Current; KCC, Korean Coastal Current; CWC, Chiju Warm Current.
Based on Yang et al. (2003).

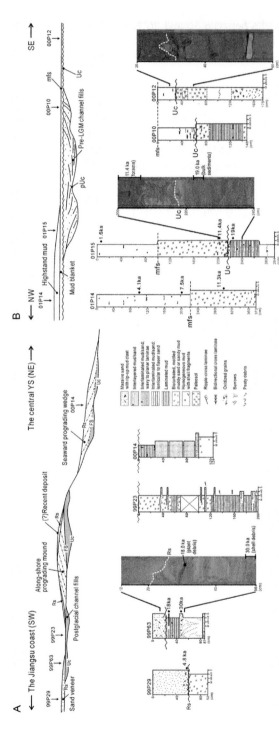

FIGURE 4.36 Sketched cross sections showing distribution of the deposition units and variations of lithology in cores from the western Yellow Sea (A, sand-dominated) and central Yellow Sea (B, mud-dominated). *Modified from Shinn et al. (2007).*

from older deposits or resuspended from the Yellow River mouth and later transported by the complicated regional current system contribute to the mud accumulation.

In contrast to the sand-dominated western Yellow Sea, two mud layers have been identified from the central Yellow Sea (Figure 4.36; Shinn et al., 2007). The older mud layer, or mud blanket, comprises homogeneous or bioturbated mud with sparse silt laminae beginning to accumulate at about 11.4 kyr BP. It is underlain by a subaerial unconformity and channel-fill deposits consisting of partly oxidized, bioturbated muddy sediments or sticky paleosol of 18–37 kyr BP. The mud blanket is overlain by a younger mud wedge largely comprising homogeneous clayey mud with high water contents with an age <8 kyr. These sedimentary facies interdigitate landward with tidal/estuarine sandy mud or sand facies of Holocene transgressive deposits (Figure 4.36).

Although the Yellow River has been playing a key role in supplying muddy sediments to the Yellow Sea, the contribution from other smaller rivers may have been also important, especially for rivers in western Korean Peninsula. To differentiate their specific sediment components is the first step toward understanding their role in deposition of the Yellow Sea. In general, the sediments of Korean rivers (Han, Keum, and Yeongsan) are characterized by higher concentrations of Rb and Th, lower element ratios of Ti/Nb and Cr/Th, and higher REE fractionation parameter $(La/Yb)_N$, compared to those from the Yellow, Yangtze, and Yalu Rivers (Yang et al., 2003). The Yangtze-derived sediment also features higher Ti, whereas Nb, Zr, Ba, and many others are highly enriched in the Yalu sediment and Li, Rb, and Th are enriched in the Yeongsan sediment (Figure 4.37).

Two sediment cores, YS2 (36°22′N, 124°18′E, water depth 78 m) and YS3 (36°18′N, 124°48′E, water depth 82 m), from the central Yellow Sea are used to represent muddier deposition (YS2) and sandier deposition (YS3), respectively (Figure 4.38). Grain size and elemental analyses indicate that

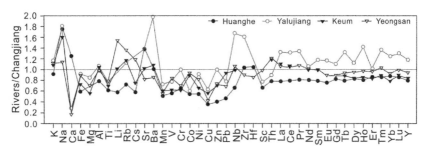

FIGURE 4.37   Averaged elemental concentration in sediments from Chinese and Korean rivers surrounding the greater ECS, as compared with those from the Yangtze (Changjiang). Values >1 indicate relatively higher concentration, and vice versa when values drop below 1. *Modified from Yang et al. (2003).*

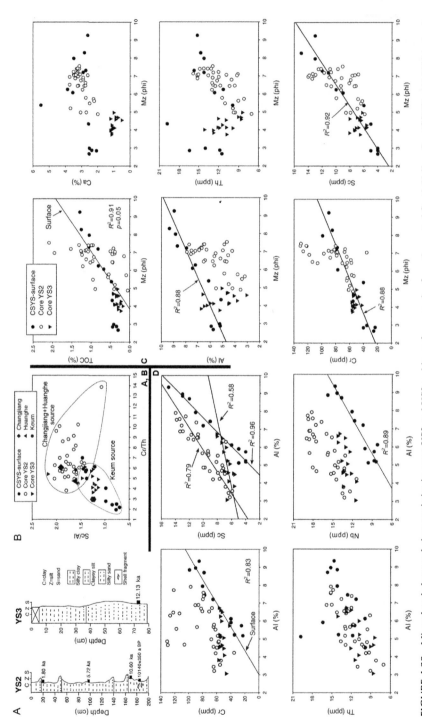

FIGURE 4.38 Mineral and elemental characteristics in cores YS2 and YS3 and in surface samples from the central South Yellow Sea (CSYS). (A) YS2 is dominated by mud and YS3 by fine sand and silt. (B) Sc/Al vs. Cr/Th plots reveal two main groups, indicating different provenances. (C) Element concentrations show generally good correlation with the mean grain size (Mz). (D) Plots between Al and trace elements also show generally good correlations. *Modified from Yang et al.* (2007).

concentrations of Al and most transition elements, as well as TOC, have good correlations with mean sediment grain size. Similarly, positive correlations exist between Al and some trace elements such as Sc, Cr, Nb, and Th, especially for the surface sediment samples. Scatter plots of Cr/Th vs. Sc/Al clearly show two main groups: The muddier core YS2 sediments have geochemical ratios close to the Yellow and Yangtze River sediment averages despite their large variations, and the sandier core YS3 sediments have element ratios similar to those of the Keum River sediment (Figure 4.38; Yang et al., 2003). These results confirm that the central Yellow Sea mud was mainly sourced from the Yellow and Yangtze Rivers, and the associated fine sand and silt in the nearby area or scattered in the mud field was largely from the Keum River. The impact of the Yangtze-derived sediments on the central Yellow Sea mud was probably initiated by a cross shelf turbid water plume developed during winter months and completed by the northward-moving Yellow Sea Warm Current (Lim et al., 2007). This mechanism may have accounted also for the accumulation of the SW Yellow Sea mud and other mud patches in the eastern Yellow Sea-South Korean region (Lim et al., 2007).

4.4.1.6   Pearl River Delta (Zhujiang Delta)

The Pearl River Delta covers an area of >40,000 km^2, or 2/3 of the size of the Yangtze River Delta. The river's sediment discharge was estimated to be ~70 Mt/yr (Table 2.2), but a compilation of the historical data from 1954 to 2000 shows the total suspended load averaging at 88.72 Mt/yr, with the majority (~71 Mt/yr) being supplied by the Xijiang River (Wei & Wu, 2011). Due to the strong landward coastal currents, most Pearl River sediments are trapped inside the estuary, and only a small portion of sediments is transported to the shelf and along the southwestern coast. Sediment flux shows distinctive seasonal variations, as modeling results indicate that deposition flux in the Pearl River estuary is high with 15.24 Mt/yr in wet seasons, decreasing to 2.33 Mt/yr in dry seasons, whereas oceanic flux is low with 0.24 Mt/yr in wet seasons, increasing to 2.07 Mt/yr in dry seasons (Hu, Li, & Geng, 2011).

As shown in Figure 4.39A, the Holocene depocenter lies inside of the estuary (Liu, Xue, et al., 2009). Silt, clay, and sand are the major sediment types, and the clay mineral assemblage contains abundant kaolinite (46%) and illite +chlorite (51%) and a minimal amount of smectite (~3%) (Liu, Berné, et al., 2007).

Based on the analyses of borehole sediments, five depositional phases (I–V) have been recognized (Wei & Wu, 2011). Phase I (12–8 kyr BP) is basically a sand–gravel valley fill. Phase II (8–6 kyr BP) is a transgressive deposit, with estuary sediments in the lower delta and littoral-swamp sediments in relatively high-lying areas. Phase III (6–4 kyr BP) represents

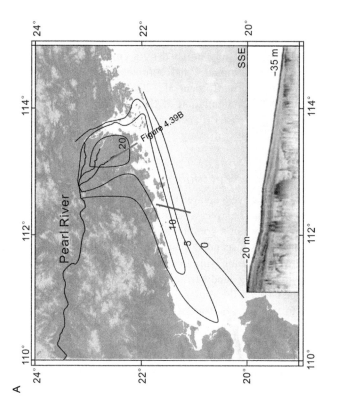

A

Pearl River

Figure 4.39B

20

10

5

0

SSE

−20 m

−35 m

24°

22°

20°

110°   112°   114°

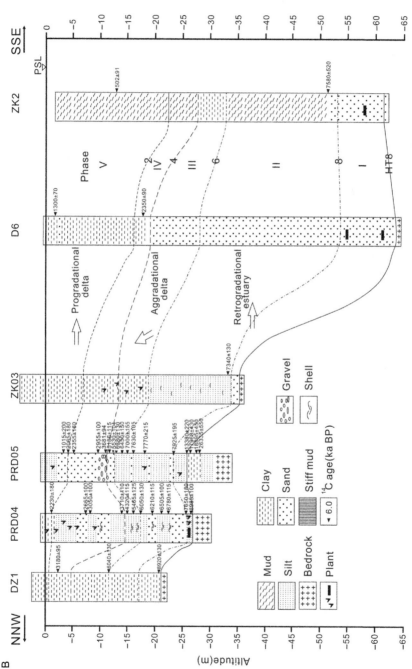

FIGURE 4.39 (A) Isopach map (in m) of the Holocene Pearl River Delta showing a depocenter inside the estuary due to landward tidal currents and a southwest extension driven by the alongshore currents (Liu, Xue, et al., 2009). (B) Sediment characteristics in cores along the Modaomen transect (blue line in A), showing Holocene sediment sequences and the five depositional phases, I–V. *Modified from Wei and Wu (2011).*

maximum transgression and the development of a paleoestuary. Phase IV (4–2 kyr BP) contributes to further provincialism of tributaries with different deposition modes and patterns. Phase V (since 2 kyr BP) endorses the formation of channels and the central subdeltaic plains (Figure 4.39B). Analyses of seismic profiles have also found two Holocene seismic units and three late Pleistocene units for correlation (Tang, Zhou, Endler, Lin, & Harff, 2010).

Zong, Huang, Switzer, Yu, and Yim (2009) proposed three stages for the evolution of the Pearl River Delta (Figure 4.40). Stage 1 (9000–6800 yr BP) was influenced by the early Holocene rapid sea-level rise and strong monsoon runoff, and transgressive deltaic silt and clay replaced the earlier fine sand deposits mainly in the middle and upper parts of the basin, following a change from shallow tidal processes to deep tidal processes. Stage 2 (6800–2000 yr BP) saw the delta starting to grow, although with a reduced progradation rate since ~6000 yr BP when summer monsoon gradually weakened. As the delta plain prograded in the upriver areas, steady vertical aggradation of delta front sedimentation took place in the central and seaward parts of the depocenter, with deposition mainly affected by tidal processes. Stage 3 (since 2000 cal. BP) was marked by increased soil erosion and sediment supply due to an increase in human activities. Sediments were largely trapped in tidal flats

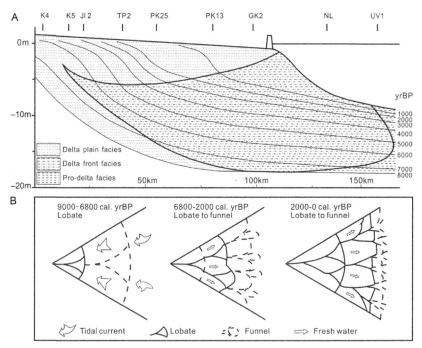

FIGURE 4.40 (A) A schematic cross section of the Pearl River Delta based on core records. (B) A conceptual model showing the development of the Pearl River Delta in three stages (Zong et al., 2009).

and newly reclaimed delta plains and some in paddy fields on hillsides and small floodplains in the catchment area. Because of the particular practice of land reclamation by human, rapid shoreline advance and decreased estuary sediment accretion finally shaped the modern delta (Zong et al., 2009). Respectively, stages 2 and 3 are equivalent to depositional phases III+IV and V of Wei and Wu (2011) shown in Figure 4.39B.

4.4.1.7 Red River Delta

The Red River delta initiated in the vicinity of Hanoi about 9000 yr BP and subsequently prograded and expanded to reach its present area of 10,300 km^2 (Hori et al., 2004; Tanabe, Hori, et al., 2003; see Chapter 2). Sediments in cores change upward from incised-valley fills to fluvial sediments composed of gravelly sand and mottled clay, to tide-influenced estuarine sediments containing shell and wood fragments, and finally to deltaic sediments composed of tide-influenced sand and mud deposits, in which the contents of sand and wood fragments increase upward (Tanabe, Hori, et al., 2003; Tanabe, Saito, et al., 2003; Tanabe et al., 2006). The clay mineral assemblage is dominated by illite+chlorite (70–90%), with moderate kaolinite (10–30%) and rare smectite (mostly <5%) (Figure 4.3; Liu, Berné, et al., 2007). The estuarine sediments developed during the rapid rise in sea level between 11,000 and 8500 yr BP display a retrogradational stacking pattern, forming a thick transgressive system tract. The subsequent delta system is characterized by aggradational as well as progradational stacking with clinoform architecture formed during slow sea-level rise between 8500 and 6500 yr BP. During the last 6500 years, delta progradation has been dominant because the deceleration of sea-level rise resulted in little accommodation being added (Hori et al., 2004). From 5 cal kyr BP, the emerged area evolved into a floodplain and natural levees formed along the abandoned river channels on the western delta plain (Figure 4.41; Funabiki et al., 2007). Based on core data,

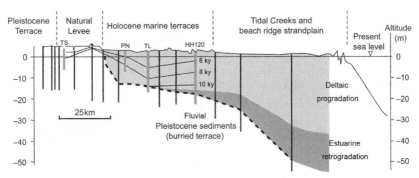

FIGURE 4.41 Sketched stratigraphic cross section of the eastern side of the Red River delta region. *Modified from Funabiki et al. (2007).*

Tanabe, Hori, et al. (2003) provided a series of paleogeographic maps to illustrate the evolution of the Red River over the past 9 kyr (see Chapter 6).

4.4.1.8 Mekong River Delta

The Mekong Delta features a wide, low-lying plain topography with an area of 62,520 to 93,780 km^2 according to different researchers (Nguyen, Ta, & Tateishi, 2000; Ta, Nguyen, Tateishi, Kobayashi, Saito, et al., 2002; Ta, Nguyen, Tateishi, Kobayashi, Tanabe, et al., 2002), which is the third largest in the world and the largest around the China Seas. The majority of the Mekong-derived sediment (~160 Mt/yr) has been transported away from the river mouth toward the southwest and deposited along the shore and around the tip of Cà Mau Peninsula (Figure 4.42; Liu, Xu, et al., 2007; Liu, Zhu, et al., 2007). Modeling indicates that strong wave mixing and downwelling-favorable coastal current associated with the energetic northeast monsoon in the winter season are the main factors controlling the southwestward along-shelf transport (Xue, He, Liu, & Warner, 2012; Xue et al., 2014; Xue, Liu, DeMaster, Nguyen, & Ta, 2010).

The delta sediment is dominated by silt of floodplain and marsh types, becoming sandier in the beach ridges. Ta, Nguyen, Tateishi, Kobayashi, Saito, et al. (2002) and Ta, Nguyen, Tateishi, Kobayashi, Tanabe, et al. (2002) identified four Holocene deltaic sediment facies in cores. In ascending order, they are prodelta mud facies, delta front sandy silt facies, sub- to intertidal flat sandy silt facies, and subaerial delta plain facies (Figure 4.42). (1) The prodelta mud facies is a coarsening-upward succession and consists of dark gray silt and very fine sand, with abundant marine plankton diatom species. (2) The delta front sandy silt facies is a coarsening-upward succession and consists of intercalated greenish gray sandy silt and fine sand, with abundant marine diatoms and common brackish-water species. (3) The sub- to intertidal flat sandy silt facies consists of laminated dark gray sandy silt and fine sand with abundant brackish-water diatoms and less common marine species, characterized also by wave ripples, lenticular bedding, parallel lamination, and discontinuous lamination. (4) The subaerial delta plain facies is represented by two related types: dark silt and sandy silt with rich organic matter, rootlets, and mica flakes at inland localities and well-sorted fine yellowish brown and gray sand at foreshore localities associating with sand dunes or sand ridges (Figure 4.42). As observed by Tjallingii et al. (2010), sudden marine flooding of the Mekong Delta started at ~12 kyr BP at outer shelf sites and completed at ~7 kyr BP (Figure 4.22). For the last 6 kyr, the delta has prograded more than 200 km (Ta, Nguyen, Tateishi, Kobayashi, Saito, et al., 2002; Ta, Nguyen, Tateishi, Kobayashi, Tanabe, et al., 2002). The progradation rate over the last 3 kyr averaged to 16 m/yr in the river mouth region and ~26 m/yr in the southeast (Xue et al., 2010).

Since deltaic-progradational deposition took place only inside the main river valleys, the sedimentary fill of local fluvial channels and wetland

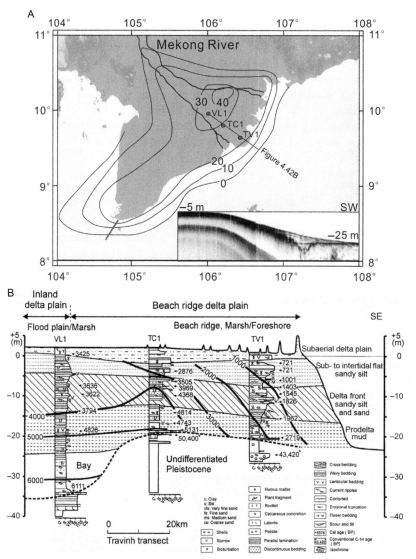

FIGURE 4.42 (A) Isopach map (in m) of the Mekong Delta showing the southwest extension of the delta due to strong tides and waves (Liu, Xue, et al., 2009). (B) Holocene depositional facies inferred from cores with ^{14}C dates along a Travinh transect of the Mekong Delta, with locations shown in A (Ta, Nguyen, Tateishi, Kobayashi, Saito, et al., 2002; Ta, Nguyen, Tateishi, Kobayashi, Tanabe, et al., 2002). Also labeled are isochrons 1000–6000 a BP.

depressions on the abrasion platform have no deltaic character. Therefore, Hanebuth, Proske, Saito, Nguyen, and Ta (2012) introduced a term "estuarine platform" to characterize this stage that has been often referred to "delta initiation" or a "tide-dominated delta" by other workers. The transition from these estuarine platform conditions to true progradational forestepping (clinoforms) took place shortly after 4.8 cal kyr BP when sea level fell below the platform level leaving the platform to exposure and, thus, forcing the system to suddenly migrate seaward (Hanebuth et al., 2012).

The clay mineral assemblage of the Mekong-derived sediment is dominated by illite+chlorite (60–80%), with moderate kaolinite (20–40%) and lesser smectite (<20%) (Figure 4.3; Liu, Berné, et al., 2007; Wan et al., 2010), although smectite gradually increases westward to ~50% in the silty clay-dominated Gulf of Thailand (Xue et al., 2014).

4.4.1.9 Gulf of Thailand

In the Gulf of Thailand, deltas and most part of the gulf are prevailed by mud. At the head of the Gulf of Thailand is the Chao Phraya delta, formed by two large rivers, the Chao Phraya and the Mae Klong. The Holocene evolution of the Chao Phraya delta is similar to other SCS river deltas in experiencing a sudden sea-level rise at 8–7 kyr BP and subsequent delta progradation. After a mud shoal was formed at the paleogulf of Ayutthaya between 7 and 3 kyr BP, the Chao Phraya delta has prograded rapidly particularly during the last 2 kyr (Tanabe, Saito, et al., 2003).

Based on ^{210}Pb data, mass accumulation rates vary between 270 and 490 mg/cm^2 per year in the upper Gulf and between 64 and 190 mg/cm^2 per year in the central Gulf, indicating suspended sediments from rivers are largely trapped in the upper Gulf (Srisuksawad et al., 1997).

In the central Gulf, geoacoustic survey indicates that the late Pleistocene valley incision followed by estuarine to marine transgression was interrupted by deltaic progradation (Puchała, Porębski, Śliwiński, & August, 2011). The valley incision and delta formation are attributed to the Kelantan River draining the Malay Peninsula. The delta growth was halted at 10.4 cal kyr BP by transgressive ravinement. The subsequent emplacement of a thin, condensed mantle of modern marine muds began in the central Sunda Shelf at 11–11.5 cal kyr BP and reached the head of the Thailand Bay at 8–7 cal kyr BP. The presence of deltaic deposits within postglacial transgressive systems can be interpreted to reflect a combination of decelerating sea-level rise (13–10 cal kyr BP) with increased precipitation and sediment yield, coupled with the reactive nature of the Kelantan River catchment (Puchała et al., 2011).

Eight depositional sequences from the central Gulf of Thailand are found to represent marine transgression–regression cycles (corresponding to interglacial–glacial cycles) from 730 to 10 kyr BP. Marine mud features the bottom part of the sequences, while floodplain deposits with soil crust characterize the top, indicating a significant subaerial exposure at the past eight major sea-level lowstands (Feng, Astin, Corbett, & Westerman, 2008).

4.4.1.10   Borneo Deltas

Two river systems in the northwest Borneo–Sarawak are influential on sediment distribution in the southern SCS margin: the Rajang River and the Baram River (see Figure 6.57), both draining from the Central Borneo Massif comprising mainly Cretaceous to Eocene clastic rocks. Rainfall in the region exceeds 370 cm/year, with highest precipitation in the "wet" season associated with the December–March monsoon, resulting in prolific plant growth and widespread peat formation (Staub, Among, & Gastaldo, 2000). Peat swamp forms ombrogenous domed peat accumulations (up to 15–20 m thick) over much of the floodplain, reaching more than 100 km upstream along the main river systems and covering most of the deltaic and coastal pains (Staub et al., 2000; Woodroffe, 2000).

Together with many distributaries, the Rajang River forms a large delta of up to ~6500 km^2, while the Baram River delta only covers an area of ~300 km^2. The wave-influenced, tide-dominated hydrodynamics confines ~80% of the fluvial sediments to building the local deltas. It appears that much of the sediment brought down by the Rajang River is successfully transported through the delta by the active channels and does not lead to extensive overbank sedimentation (Staub & Esterle, 1993). Mud and silt combined can score >80% in abundance, while sand attends only 10–20% (Staub et al., 2000). Among the clay minerals, illite is dominant with ~50%, followed by chlorite (30–40%), kaolinite (10–20%), and negligible smectite (<1%) (Liu et al., 2012).

Like many deltas in the region, the Rajang and Baram deltas did not experience rapid progradation until the end of the Holocene sea-level maximum at ~5400 yr BP (Hutchison, 2005; Lambiase, Abdul Rahim, & Cheong, 2002). Mangrove mud accumulated at ~4300 yr BP, before being replaced by peat swamp facies since ~3850 yr BP.

Outside the river mouths, a sand–silt facies mixed between recent and relict sediments with frequent biogenic components including corals is distributed along the shelf. Most likely, it represents a coastal plain deposition during Pliocene–Pleistocene sea-level lowstands, underlain unconformably by Miocene limestones and open-marine deposits (Hutchison, 2005).

The outer shelf Quaternary sediment succession off the Baram delta is unusually thick, locally reaching >1000 km due to a high sediment supply and growth faulting along the shelf edge (Hiscott, 2001). This succession is largely made up of highstand, forced regressive, and lowstand deposits. The forced regressive sediment tract includes sandy shelf-edge deltas deposited by small rivers after the Baram system began to bypass the shelf through an incised valley at sea-level lowstands. A sequence stratigraphic framework has been established to account for the rapid sediment accumulation controlled by growth faults (Hiscott, 2001).

4.4.2   Relict Sands and Sand Ridges

As a distinctive sedimentary feature, relict sands of early Holocene and late Pleistocene or older ages are widely distributed in the middle and outer shelf of the China Seas. In several regional areas, specifically oriented sand ridges are well developed due to the prevalent tidal currents and waves or cross strait currents (Liu & Xia, 2004; Liu et al., 1998). As shown in Figure 4.43 and listed in Table 4.7, eight major sand ridge fields are located, respectively, (1) off the Yalu River Estuary and West Korean Bay, NE Yellow Sea; (2) off the Han River mouth, NE Yellow Sea; (3) off the Liugu River mouth, N Bohai Sea; (4) at Liaodong bank in the Bohai Strait; (5) at Jiangsu offshore; (6) at the ECS shelf; (7) at Taiwan shoals in the Taiwan Strait; and (8) at the Qiongzhou (Hainan) Strait. Among them, the sand ridge field on the ECS shelf is the largest, while the radial sand ridge field in the SW Yellow Sea is the most distinctive. Sand ridges of smaller scales also occur in other areas, such as the southeast Yellow Sea (Chough et al., 2002) and Korean Strait (Park, Han, & Yoo, 2003). Compared to those from the middle and outer shelf, most ridges immediately from river mouths and on shallow shelf are recently formed and have a modern aspect in containing higher amounts of recent sediments. Muddier ridges may occur off some estuaries or in areas with reduced tidal currents, such as in part of outer Yangtze estuary (Chen, Saito, et al., 2003; Chen, Shiau, et al., 2003; Hori et al., 2001a; Figure 4.32).

During the Last Glacial Maximum, a large part of the greater ECS stretching over 10° of latitude was exposed, and many shelf areas became "deserts" (e.g., Zhao, 1996), mainly with Yellow River-derived loess sediments in the northern region and Yangtze River-derived sandy sediments in the southern region. It was these deposits that constituted the main part of sand ridges formed during different stages of Holocene transgression (e.g., Chen, Saito, et al., 2003; Chen, Shiau, et al., 2003; Liu, Berné, et al., 2007; Liu et al., 1998; Park et al., 2003; Wu et al., 2010). The major sedimentary features of the ECS shelf sand ridges and Jiangsu offshore sand ridges are discussed in the following.

4.4.2.1 East China Sea Sand Ridges

Covering an area of over 50,000 km^2, the ECS shelf sand ridge field is the largest (Figure 4.44). Most ECS sand ridges show internal structures characterized by southwest-dipping beds, indicating that the regional net sediment transport is toward the southwest (Figure 4.45). They were initiated from the last transgression, first as tidal ridges in estuarine or shallow shelf areas. As sea level continued to rise, the sediments on the seafloor were eroded, transported, and redeposited under the tidal current action. The bedforms adjusted gradually to the varying hydrodynamic environment, and the sand ridges probably lost some of their original characteristics. The original or earlier sand ridges might have consisted of relatively muddier sediments because of the huge muddy sediment supply from the Yangtze River. However, reworking during the transgression has produced sand ridges composed of relatively sandier sediments and has transported the fine sediments from the

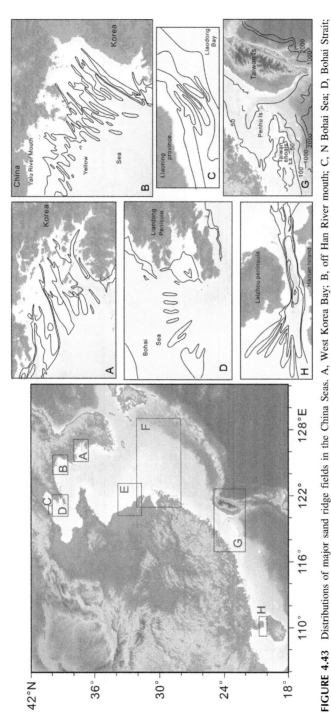

FIGURE 4.43 Distributions of major sand ridge fields in the China Seas. A, West Korea Bay; B, off Han River mouth; C, N Bohai Sea; D, Bohai Strait; E, Jiangsu offshore (shown in Figure 4.48); F, ECS shelf (shown in Figure 4.44); G, Taiwan Strait; H, Qiongzhou (Hainan) Strait. *Modified from Wang, Zhang, Zou, Zhu, and Piper (2012).*

TABLE 4.7 Major Sand Ridge Fields in the China Seas

Sand Ridge Field	Central Position (Lat. N, Long. E)	Stretching Direction	Water Depth (m)	Area (10^3 km^2)	Current Maximum Speed (cm/s)	Selected References
1. W Korea Bay	39°00′, 124°00′	NE	10–50	37	100–150	Liu et al. (1998)
2. Off Han River mouth	37°00′, 125°40′	NE	10–50	40	50–100	Liu et al. (1998)
3. N Bohai Sea	40°20′, 121°30′	N	20	~1		Liu and Xia (2004)
4. Bohai Strait	39°20′, 120°40′	SE	10–36	4	64–115	Marsset et al. (1996), Liu et al. (1998), andShi, Wang, Li, and Pichel (2011)
5. Jiangsu offshore	33°00′, 121°40′	Radial	10–30	>20	100–150	Liu et al. (1998), Li et al. (2001), Wang (2002), and Wang et al. (2012)
6. ECS shelf	29°30′, 123°30′	NW–W	25–130	>50	120	Liu et al. (1998), Liu, Berne, Saito, Lericolais, and Marset (2000), and Liu, Berné, et al. (2007)
7. Taiwan Strait	23°00′, 119°30′	SE–S	20–40	13	100	Liao, Yu, and Su (2008)
8. Qiongzhou Strait	20°20′, 109°30′	E–SE	10–30	5	100–150	Liu et al. (1998)

Modified from Xu et al. (1997) and Liu et al. (1998).

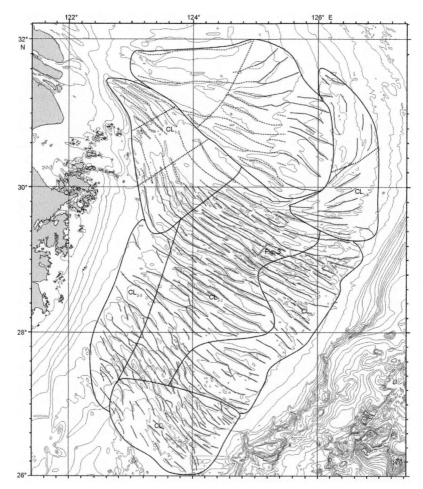

FIGURE 4.44   Bathymetry map of the East China Sea shelf showing the distribution of linear sand ridges superimposed by projected crest lines (red) and (sub)zones (Wu et al., 2010).

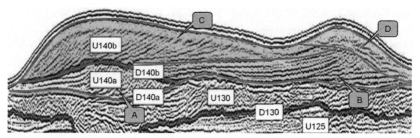

FIGURE 4.45   Multistage buried sand ridges (A–D) shown on a typical seismic profile. Also labeled are sequence units (U125–U140b) and sequence boundaries (D130–D140b) of Berne et al. (2002) shown in Figure 4.16. *From Wu et al. (2010).*

shelf to other areas including the Okinawa Trough in the east and the central Yellow Sea in the north. As such, recycled sand was left for new ridges to be built on shelf.

The modern-day sand ridges are ~5 m at maximum height, but the buried sand ridges in seismic records may reach 26 m (Liu, Berné, et al., 2007), indicating a more dynamic formation process in earlier stages of the Holocene transgression. Migration of sand ridges over the last ca. 2–3 kyr has been observed, and their evolution is marked by continuous changes from muddier sand ridges to sandier sand ridges in response to the shoreline retreat, which resulted in a decrease of riverine muddy sediments and recycling of sandy materials by tidal currents. Although most active sand ridge formation occurred during the last transgression, the sand ridges on the present middle to outer shelf are still being influenced by the modern hydrodynamics (Liu, Berné, et al., 2007).

The sand ridge lithostratigraphy can be summarized as follows and in Figure 4.46, according to Berne et al. (2002). The regressive surface of marine erosion forms during deltaic progradation "forced" by glacioeustatic sea-level falls and marks the limit of erosion by combined wave and tides. It is a sub-horizontal surface, except at the position of deltaic distributaries or estuarine channels where (submarine) tidal scouring creates deep incisions about 50 m below sea level. The deltaic–estuarine facies is subsequently topped by a surface of fluvial incision (sequence boundary) generally not very pronounced because of the low gradient of the shelf. This surface is remodeled by the transgressive surface of marine erosion, which separates early transgressive estuarine deposits within fluvial incision from late transgressive marine deposits (offshore sand ridges). Prodeltaic/offshore fine-grained deposits formed both during highstand (interglacial) condition and more likely during the falling stage of shorter period (20–40 kyr) interstadial glacioeustatic cycles. The sand layer that tops the sand ridge facies corresponds to the veneer of sediment in equilibrium with highstand hydrodynamic regime (Berne et al., 2002).

Based on their distribution patterns, the sand ridges in the ECS can be divided into three zones (CL1, CL2, and CL3), with CL1 and CL2 each further dividable into three subzones (Figure 4.44; Wu et al., 2010). Sand ridges in zones CL1 and CL3 represent river mouth ridges, while those in CL2 characterize normal, open shelf ridges. Although they were formed under similar hydroclimate conditions during transgressive periods, the sand ridges in these zones and subzones differ widely in terms of strikes and other morphologic features because of the influence of submarine topography and distance of the river mouth, as well as the tidal direction and strength. Periods of stable sea level appear to have been more conducive to the growth of sand ridges than times of rapid sea-level rise. From model simulation using all relevant data, especially changes in water depth, sea level, tidal currents, and stratigraphy, the development and evolution of the sand ridges on the ECS shelf can

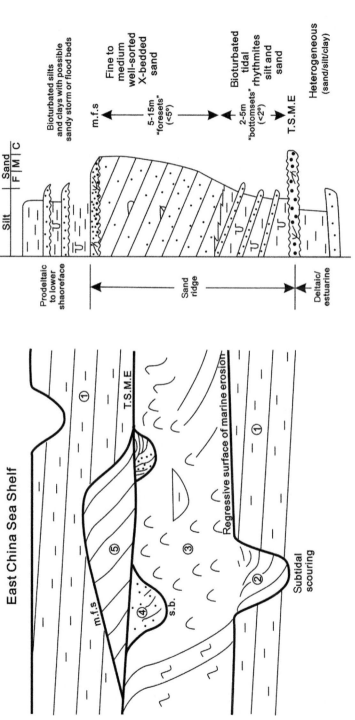

FIGURE 4.46 (A) Sketched depositional sequences on the outer continental shelf of the East China Sea and (B) Late Pleistocene–early Holocene sand ridge lithostratigraphy, based on seismic data, boreholes, and piston cores (Berne et al., 2002). (1) Prodeltaic/offshore fine-grained deposits, (2) deltaic distributaries or estuarine channels, (3) deltaic–estuarine facies, (4) early transgressive estuarine deposits, (5) late transgressive marine deposits (offshore sand ridges); transgressive surface of marine erosion (TSME), sequence boundary (s.b.; fluvial incision surface), maximum flooding surface (m.f.s.).

Geology of the China Seas

be categorized into four main stages (Figure 4.47; Wu et al., 2010). Stage I was for those formed before 14.5 kyr BP in outer shelf depression areas, including sand ridges in CL3 and part of CL2-1. Stage II from 14 to 12 kyr BP saw the development of sand ridges on most parts of the present middle to outer shelf. Stage III from 11.5 to 9.5 kyr BP was marked by further development of sand ridges in most middle shelf areas, after a rapid rise of sea level from −60 to −40 m, although those on the outer shelf gradually reverted into moribund types, while others were buried. Stage IV since 9 kyr BP has seen the growth of sand ridges on the inner shelf and continuous modification of those older ones on the middle and outer shelf.

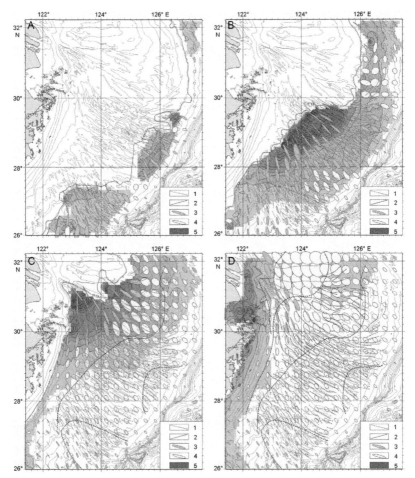

FIGURE 4.47 Simulation results reveal the evolution of sand ridges on the ECS shelf in four stages: (A) stage I, before 14.5 kyr BP; (B) stage II, 14–12 kyr BP; (C) stage III, 11.5–9.5 kyr BP; (D) stage IV since 9 kyr BP. 1, Crest lines of sand ridges; 2, boundaries; 3, ellipticity of M2 less than 0.4; 4, ellipticity of M2 greater than 0.4; 5, active bedform areas. *From Wu et al. (2010).*

4.4.2.2 Jiangsu Offshore Sand Ridges

The Jiangsu offshore tidal sand ridges lie in the coastal area of SW Yellow Sea, at water depths mostly less than 30 m (Figure 4.48). About 70 sand ridges are arranged as a radial pattern with the apex at the port of Qianggang in Jiangsu Province. The longest ridge exceeds 100 km (Liu et al., 1989), probably representing the longest active tidal ridges known in the world.

Sediment composition including heavy minerals and clay minerals in surface ridge samples shows that most of the sand was derived from the (paleo-) Yangtze River, but in the northern part of the field particularly, the clay has a

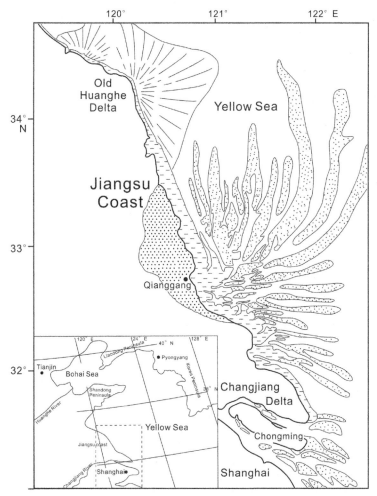

FIGURE 4.48 Distribution of the Jiangsu offshore radial sand ridges in the southwestern Yellow Sea, superimposed by sand and mud facies along the coast. *Simplified from Li, Zhang, Fan, and Deng (2001).*

Yellow River origin (Wang et al., 2012). Seismic profiles reveal late Pleistocene distributary channels of the paleo-Yangtze River underlying much of the southern part of the sand ridge field. The late Pleistocene deposition of fluvial and estuarine sands from paleo-Yangtze River distributaries at times of low sea level appears to have been acting as the favorable ground. The Holocene transgression cut a widespread ravinement surface at subbottom depths of 15–30 m, reworking abundant late Pleistocene sandy sediment that was subsequently recycled again for building large sand ridges by tidal currents. The ravinement surface is often overlain by horizontally bedded strata interpreted as tidal flat sediments. A middle Holocene erosion surface marks the onset of the modern tidal regime and the development of tidal channels and adjacent ridges in the northern and southern parts of the field. The location of the ridges is strongly influenced by relict channels in the southern part of the field and by the radiating tidal currents in the north. In the central part, tidal channels generally follow incised fluvial channels largely formed during the Last Glacial Maximum. Since the middle Holocene, wave action has flattened and broadened the tidal ridges in the shallower water area (Wang et al., 2012). Based on the facies characteristics of the sand ridges and the surrounding region, however, Li et al. (2001) argued for a Holocene regression cause for the radial sand ridges, with a change of the apex from ∼30 km inland to the present position after the Yangtze River Delta progradation over the past 2–1 kyr (Figure 4.49).

Today, significant rectilinear currents are present over the northern sand ridge area, whereas rotary currents prevail over the southern area, the transition between the two being dominated by a locally generated trapped wave, which results in a radial current field and high suspended sediment concentrations (Xing, Wang, & Wang, 2012). Modeling also shows that the deep channels, in general, are undergoing erosion, while the shallow ridges and coastal tidal flats are accreting. The net sediment transport is directed toward the coastal tidal flat and the Yangtze River subaqueous delta.

4.5 BIOGENIC SEDIMENTS

Compared to the high proportion of terrigenous sediments, biogenic sediments only occupy a small portion of the deposition in the China Seas, although they may become richer than their terrigenous counterparts in deposits at specific settings, such as coral reefs and areas with high surface productivity or deep sea with high pelagic biogenic accumulation. Like those from low-latitude marginal sea regions, three major biogenic components are common: carbonate, opal, and organic matter.

4.5.1 Carbonate Buildups and Coral Reefs

4.5.1.1 Modern and Quaternary Carbonates

Distribution of Detrital and Biogenic Carbonate

The suspended load with high detrital $CaCO_3$ from the Yellow River (>8%) and Yangtze River (∼7%) plays a significant role in $CaCO_3$ buildup in shelf

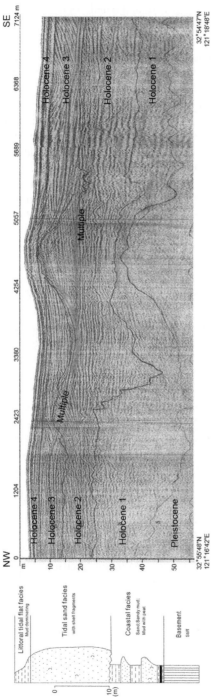

FIGURE 4.49 (A) A sketch showing major depositional facies in cores from the (northern) Jiangsu offshore sand ridges. (B) A seismic profile from the center of the radial sand ridge field showing thick Holocene sandy sequences and the modern sand ridges building on older ones. *Panels (A) from Li et al. (2001) and (B) modified from Wang (2002).*

areas of the Bohai Sea, Yellow Sea, and (shallow) ECS affected by the fluvial sediment load. Thus, it is reasonable to assume their weak influence in areas with relatively lower $CaCO_3$ values, and vice versa when CaCO3 is high. In the Bohai Sea, for example, up to $\sim 10\%$ $CaCO_3$ occurs in the south and west and near the Bohai Strait with large amount of (re-)suspended sediment from the Yellow River (Li, 2012). In the Yellow Sea, $CaCO_3$ reaches 15–50% in the northwest along the north Jiangsu coast, 5–10% in the south and southeast, and <5% in the northeast as well as in the central mud area (Lan, Wang, Li, Lin, & Zhang, 2007). Obviously, the extremely high $CaCO_3$ along the north Jiangsu coast was due to the detrital supply from the Old Yellow River and does not represent the normal marine sedimentation. In the ECS, $CaCO_3$ increases from north to south and from west to east, reaching 20–30% in deeper-water silty facies (Yang, Cui, & Zhang, 2010) and 30–40% or more in the Okinawa Trough (Li & Chang, 2009). High $CaCO_3$ values in sediments from the south and southeast Yellow Sea, the ECS middle to outer shelf, and the Okinawa Trough are mainly contributed by biogenic carbonate, especially skeletons of foraminifera.

Carbonate dissolution intensifies below the lysocline depth. In the southern Okinawa Trough, it appears to have been shoaling over the late Quaternary to the modern depth of 1500–1600 m. In core E017 ($26°34.45'N$, $126°01.38'E$, water depth 1826 m), fragmentation of planktonic foraminiferal tests increases steadily from 15–25% at ~ 17 kyr BP to 60–75% in the late Holocene (Xiang et al., 2003), indicating increasing dissolution through time in the southern Okinawa Trough. However, no obvious carbonate dissolution has been observed at most localities in the central and northern trough due to their <1500 m water depths, which lie above the lysocline, with most samples containing beautifully preserved aragonitic gastropods (Li & Chang, 2009).

Unlike the ECS, the majority of $CaCO_3$ in the SCS is of biogenic origin. As shown in Figure 4.50, coral reef regions have highest $CaCO_3$ (up to 90% or more), followed by 20–30% in the northern and 30–40% in the southern shelf and slope and <10% in many inner shelf areas due to the dominance of terrigenous sediments. The central deep basin may only attend $\sim 2\%$ or lesser $CaCO_3$ below the CCD, which lies at about 3500 m, or ~ 500 m below the lysocline depth (Chapter 2).

On the continental slope of the SCS, seep carbonates formed by the synergistic metabolism of methane-oxidizing archaea and sulfate-reducing bacteria have been found at over thirty sites (Chen, Huang, Yuan, & Cathles, 2005; Chen et al., 2006; Lu et al., 2006; Suess, Huang, Wu, Han, & Su, 2005). Trace and rare earth element data suggest that the seep carbonates were all formed under an anoxic environment (Ge et al., 2010). They appear to have formed in different time intervals, with U/Th ages ranging from 63 to 77 kyr BP at localities in the NE Dongsha Islands and from 152 to 330 kyr BP at localities of Shenhu offshore from the Pearl River Mouth (Tong et al., 2013).

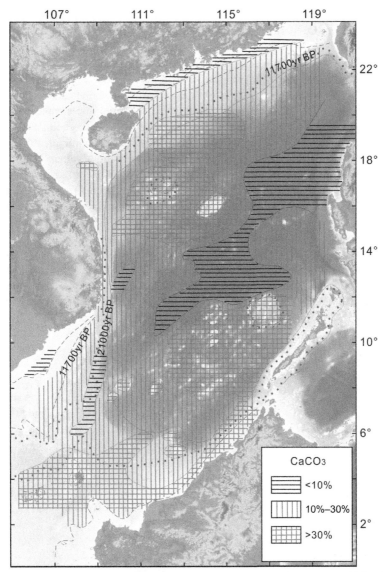

FIGURE 4.50   Distribution of CaCO₃ in the surface sediments of the SCS, showing high values in coral reef areas. *Modified from Xu et al. (1997).*

Quaternary Carbonate Cycles

As a result of response to Quaternary glacial oscillations, deep-sea carbonate production and preservation experienced two different carbonate cycles. The "Atlantic cycles" are similar to the fluctuations of $\delta^{18}O$ curve with low CaCO₃ in glacials and high CaCO₃ in interglacials, while the "Pacific cycles" are marked by minimal CaCO$_3$ not only in MIS 4 but also in MIS 5. In the China Seas, most of the Quaternary carbonate profiles show the "Atlantic cycles," and only those from below the lysocline in the SCS have the imprints of the "Pacific cycles" (Wang, 1995; Wang & Li, 2009).

As shown in Figure 4.51, respectively, in core MD972142 (12°41.133'N, 119°27.9'E, water depth 1557 m) from the eastern SCS and at ODP Site 1143 (9°21.720'N, 113°17.102'E, water depth 2772 m) from the southern SCS, the abundance of CaCO$_3$ achieves 30–40% and 10–37% in sediments from Holocene MIS 1, decreases with strong fluctuations to minimal values of 25% and ~1% in MIS 4, and increases again to new high values in MIS 5, closely following the pattern of the SPECMAP δ^{18}O record. However, in core SO17956 (13°50.9'N, 112°35.3'E, water depth 3387 m) from the central western SCS below the lysocline, CaCO$_3$ fluctuations mirror the δ^{18}O record, with peaks at the two glacial–interglacial transitions (MIS 6/5 and MIS 2/1).

Core data indicate that late Quaternary CaCO$_3$ fluctuations in the Okinawa Trough are all of the "Atlantic cycles," regardless whether the sites are from above or from below the lysocline (Li & Chang, 2009). For example, core 155 (27°02.2'N, 126°10.5'E, water depth 950 m) from the central western slope of the trough contains ~20% CaCO3 during the MIS 3, decreases to 10–15% during MIS 2–4, and increases to 30–40% in MIS 5 (Figure 4.51).

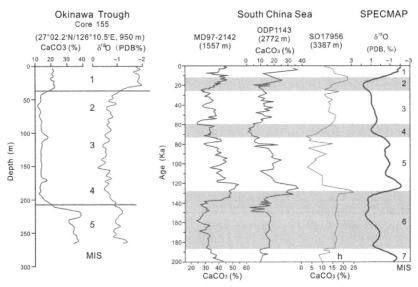

FIGURE 4.51 Late Quaternary carbonate fluctuations in the Okinawa Trough (left; Li & Chang, 2009) and the SCS as compared to the SPECMAP δ^{18}O curve (right; Wang & Li, 2009). Note that core SO17956 from below the lysocline shows the "Pacific cycles" with low CaCO$_3$ in MIS4 to MIS 5, in opposite to the "Atlantic cycles" for all other sites from above the lysocline.

The occurrence of both "Atlantic cycles" and "Pacific cycles" in late Quaternary carbonate preservation in the China Seas can be attributed to a dilution effect by terrigenous sediments in glacials for the former and to an enhanced dissolution effect due to lysocline shoaling at high sea level for the latter. However, the role of biogenic carbonate production remains critical on deep-sea carbonate deposition and preservation over glacial cycles.

A more complete record of Quaternary carbonate cycles from ODP Site 1143 (9°22′N, 113°17′E, water depth 2772 m) shows clear cycles in carbonate deposition and preservation well correlated to glacial cycles defined by the benthic $\delta^{18}O$ (Figure 4.52; Xu, 2004). All the proxies of carbonate preservation, including carbonate MAR, $CaCO_3\%$, absolute abundance of foraminifers, and coarse fraction, show similar patterns with benthic $\delta^{18}O$, and they peak at glacial–interglacial transitions, or terminations, whereas the dissolution index (fragmentation%) varies in an opposite way. Therefore, the late Quaternary carbonate preservation patterns were upheld for the entire Quaternary interval as driven primarily by glacial cycles. Further examination of Figure 4.52 reveals decreased carbonate and enhanced dissolution at MIS 13, MIS 27–29, and MIS 53–57, which may correspond to the three major carbon isotope maximum events ($\delta^{18}Omax$-1 to $\delta^{18}Omax$-3) affected by global carbon reservoir changes (Wang, Tian, Cheng, Liu, & Xu, 2003, 2004).

Long-Term Trends in Carbonate Deposition

Largely based on the results of ODP Leg 184, the long-term patterns of carbonate deposition and preservation since the Miocene in the SCS as relating to global and regional events have been discussed and summarized in Wang and Li (2009). Two periods marked by decreases in carbonate mass accumulation rate at 21–20 Ma and at 5–3 Ma are most significant in a global context (Figure 4.53). The 21–20 Ma carbonate decrease corresponded to the early Miocene CCD rise due to general warming and rise in sea level, resulting in shallow sequestration of $CaCO_3$ as shallow seas expanded and warmed (Lyle, 2003). A period of increased $CaCO_3\%$ from ~10 Ma to ~5 Ma in the late Miocene can be broadly correlated with the "biogenic bloom" widely recorded in the equatorial Pacific. The subsequent $CaCO_3$ decline since the Pliocene has been largely associated with enhanced supply of terrigenous clasts as a result of frequent sea-level variations over glacial cycles. Superimposed on these global patterns, the five major dissolution events over the Neogene observed at ODP Site 1148, respectively, at ~21 Ma, 16–15 Ma, 13–11 Ma, 10–9 Ma, and 5–3 Ma, are likely linked to the stepwise development of deepwater masses in the SCS Basin (Figure 4.53; Li et al., 2006).

4.5.1.2   Coral Reefs

Modern Distribution

Coral reefs are widely distributed across the South China Sea, covering the areas from Zenmuansha (~4° N) in the south, to Leizhou Peninsula, Weizhou

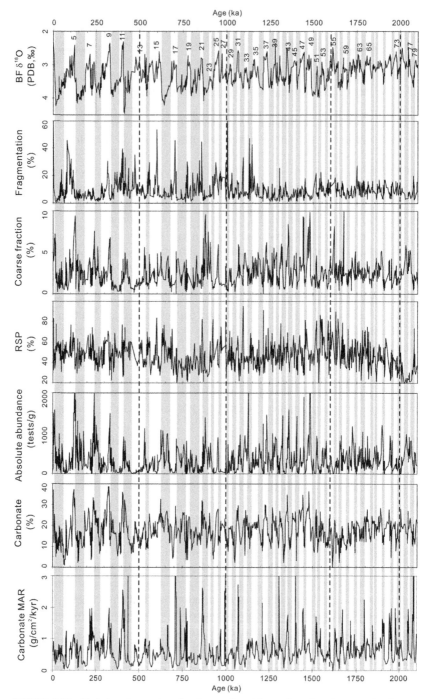

FIGURE 4.52 Variations in proxies of carbonate deposition and preservation over the Quaternary at Site 1143 showing "Atlantic cycles" as relating to glacial–interglacial oscillations represented by the benthic δ18O curve with stages. Dashed lines mark major periods of decreased carbonate and enhanced dissolution. *From Xu (2004).*

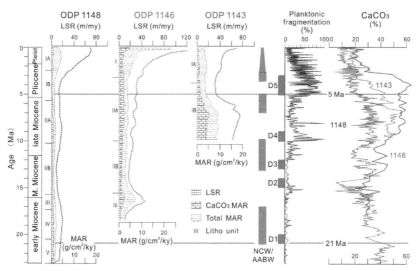

FIGURE 4.53   Variations of linear sedimentation rate (LSR), mass accumulation rate (MAR), planktonic foraminiferal fragmentation%, and CaCO₃%a at ODP Site 1143 (purple; water depth 2772 m), Site 1146 (green; water depth 2091 m), and Site 1148 (blue; water depth 3297 m) from the SCS, as compared with major production periods of northern component water (NCW) and Antarctic bottom water (AABW) and the regional dissolution events (D1–D5). *Modified from Li et al. (2006) and Li et al. (2007).*

Island (~20°–21°N), and islands offshore of northeastern Taiwan (~25°N) in the north, including fringing reefs, platform reefs (with sand cays), and atolls (Figure 4.54; He, 2006; Yu & Zhao, 2009). Coral reefs also occur spottily surrounding several small islands near Diaoyu Island to the northeast of Taiwan, along the southern fringe of the Okinawa Trough. Beyond these areas, such as the southern ECS shelf, solitary coral skeletons may also occur, although some of them could have been relict (Yang et al., 2010). Along the eastern side of the Okinawa Trough, coral reefs also grow surrounding the chain of volcanic islands of Ryukyu up to the Tokara Strait, at ~30º latitude. The Ryukyu reefs have been recently studied for understanding the response of coral ecosystem to Kuroshio variations affected by recent climate change (Hongo, 2012; Hongo & Kayanne, 2010, 2011).

The SCS coral reefs mainly occur in the following geographic regions: South China coast and its offshore islands, Hainan Island and its offshore islands, Taiwan Island and its offshore islands, the Philippine coast and its offshore islands, the Vietnamese coast and its offshore islands, Dongsha (Pratas) Islands, Zhongsha Islands (Macclesfield Bank), Xisha (Paracel) Islands, and Nansha (Spratly) Islands. The coral reefs on Dongsha, Zhongsha, Xisha, and Nansha Islands are dominated by atolls, whereas the other regions consist mainly of fringing reefs (Yu, 2012; Yu & Zhao, 2009). Excluding those along the Vietnamese and Philippine coasts and their offshore islands,

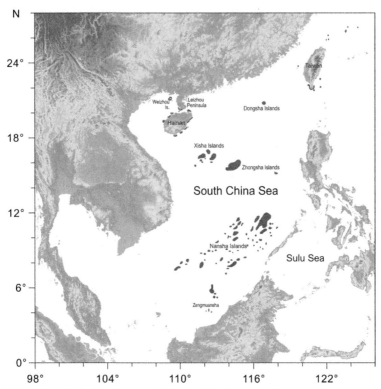

FIGURE 4.54 Distribution of coral reefs in the China Seas, showing two main types: atolls in the open water and fringing reefs along coasts. Note that reefs along the coast of the Philippines and Vietnam, as well as those along the southern margin, are not shown due to insufficient data. *Modified from He (2006) and Yu and Zhao (2009).*

the shallow-water (<50 m) modern coral reefs in the SCS occupy an area of about 8000 km^2. A brief summary of the work by Yu and Zhao (2009) is given in the succeeding text.

Fringing reefs surrounding Leizhou Peninsula, Weizhou Island, Hainan Island, and Taiwan Island account for a total area of ~1150 km^2, with the Taiwan reef area being the largest (940 km^2). About 40 reef-building coral species have been found in Leizhou reefs and ~60 species in Weizhou reefs. The fringing reefs of Hainan consist of ~110 coral species, while ~300 reef-building species live in the Taiwan area due to the flow of the Kuroshio Warm Current (Table 4.8). *Acropora* species appear to be dominant in all of the fringing reefs except those from Leizhou Peninsula where species of *Porites* and *Goniopora* are more abundant.

Atolls on Dongsha Islands, Zhongsha Islands, Xisha Islands, and Nansha Islands occupy a total area of ~6850 km^2, almost half of which is from the Nansha atolls (~3000 km^2). Over 150 reef-building coral species have been

TABLE 4.8 The Estimated Total Area, Number of Reef-Building Coral Genera and Species, and CaCO$_3$ Production for Different Coral Reef Regions in the SCS, with Data Pending for Those from Philippine and Vietnam Waters

Region	Reef Flat (km^2)	Lagoon (km^2)	Fore-Reef (km^2)	Total Area (km^2)	No of Coral Genera (species)	CaCO$_3$ Production ($\times 10^3$ t/yr)
Fringing reefs						
1. South China coast and offshore islands	17	0	0	17.0	25 (~58)	16
2. Hainan and offshore islands	180	0	0	180.0	34 (~110)	330
3. Taiwan and offshore islands	940	0	0	940.0	>55 (~300)	1723
4. Philippine coast and offshore islands				Pending		
5. Vietnamese coast and offshore islands				Pending		
Atolls						
6. Dongsha (Pratas) Islands	125	292	7.5	424.5	34 (~140)	1270
7. Huangyan Dao, Zhongsha Islands	53	77	3.2	133.2		412
8. Zhongsha Islands (Macclesfield Bank)	1495			1495.0	34–50 (>100?)	4036
9. Xisha (Paracel) Islands	221.6	1614.8	13.3	1849.7	38 (~130)	5215
10. Nansha (Spratly) Islands	507.5	2396.8	30.5	2934.8	44 (>150?)	8430
Total	3539.1	4380.6	54.4	7974.1		21,432

Modified from Yu and Zhao (2009).

found from these atolls, although the exact species number for most regions is lacking. The massive corals are represented by *Porites* and *Favia*, and the branching forms mainly include *Pocillopora* and *Acropora*. Although often mixed, corals show preference for specific atoll zones or different water depths. In the internal lagoons, skeletons of the branching corals are completely mixed with those of coralline algae, *Halimeda* plant, mollusks, large benthic foraminifera, and other reef-associating organisms.

CaCO$_3$ Production by Reefs

The accumulation method for estimating CaCO$_3$ production by reefs is based on the vertical deposition rate of reefal sediments and the amount of aragonite precipitated by corals. For fringing reefs, the reef growth rate is 0.07 cm/yr for Leizhou reefs and 0.13 cm/yr for Hainan and Taiwan reefs as calculated from their Holocene reef thickness. Together with the growth rates, a classic carbonate density of 2.9 g cm^{-3} with 50% porosity can be employed for calculating reef carbonate production. For atolls, deposition rates for reef flat, lagoon, and fore-reef areas are treated independently because these areas differ not only in size but also in reef growth or sediment accumulation rate. Based on core data (Figure 4.55), the average deposition rate is 3.85 mm/yr for lagoons and 2.51 mm/yr for reef flat and fore-reef areas. By calculation, carbonate density is set at 0.7 g cm^{-3} for lagoon areas, while the classic density value of 2.9 g cm^{-3} is followed for reef flats and fore reefs with 50% porosity. The mean carbonate production can then be calculated to be 2700 g and 3640 g CaCO$_3$ m^{-2} yr^{-1} for lagoons and reef flats, respectively, and the latter can be also applied to fore-reef areas (Yu & Zhao, 2009). Therefore, the total annual CaCO$_3$ production for all the SCS reef regions listed in Table 4.8 is estimated to be 21.4 Mt/yr, which is about 1.6–3.3% of the global reefal carbonate production estimated by various workers. However, the estimate remains incomplete for reef CaCO$_3$ production in the SCS until data from other coral reef areas, particularly those of Vietnam, the Philippines, and the southern margin, become available.

Carbonate production by coral reefs in the SCS during the Oligocene–Miocene might be even higher, as many pre-Quaternary reef complexes or carbonate platforms have been found from outside the modern coral reef areas (e.g., Qiu & Wang, 2001). For example, the carbonate platform area in the Pearl River Mouth (Zhujiangkou) Basin, northern SCS, is up to 56,750 km^2 (Hu & Wang, 1996), and the reef area is about 2000 km^2 (Hu & Xie, 1987). The reef here started to develop from the early Miocene and reached 560 m in maximum thickness before being buried by late Miocene and younger sediments. With proven results, the carbonate sequence of the Zhujiang Formation has been targeted for oil exploration for over two decades in the basin. Also reported are Oligocene–early Miocene carbonate platform and reef buried in some areas of the southern SCS, such as in the Nansha (Borneo) Trough (Hutchison, 2004, 2010).

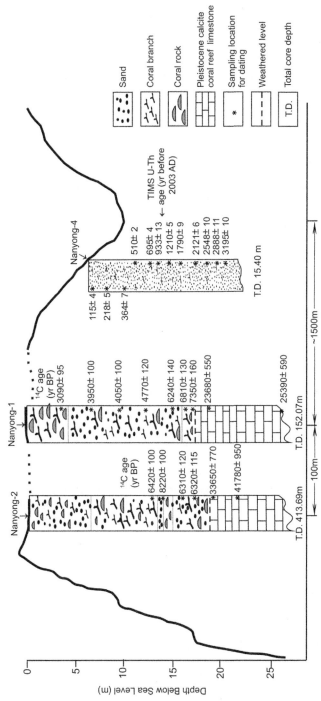

FIGURE 4.55 Holocene reef structure with ^{14}C or U–Th ages in three cores from Yongshu Reef, Nansha (Spratly) Islands (Yu & Zhao, 2009).

SCS Coral Reef History

Studies of many drill holes indicate a Holocene age for the fringing reefs on Hainan Island and on Leizhou Peninsula, mostly starting from 8000 to 7000 yr BP. Although the reef development has been episodic, the period of 7500–6500 yr BP appears to be the most favorable for reef growth in these northern SCS regions.

A longer reef history is found for the buried reefs in northern SCS shelf–slope basins, as well as those underneath the present atolls. The earliest age for these old reef deposits ranges from the late Oligocene to early Miocene. For example, well Sampaguita-1 on the northern Nansha Islands recovered a reef complex of around 2100 m thick initiated from the late Oligocene (Du Bois, 1981; Yao, Liu, et al., 2012). In Palawan, the Nido limestone of early Miocene age unconformably overlies late Oligocene Nido platform carbonates, which are underlain by pre-Nido limestones at some localities of NW Palawan shelf (Steuer et al., 2013), indicating older carbonate buildups in the Eocene and early Oligocene. Well Xiyong-1, on Yongxing Island of Xisha Islands, recovered a reef sequence of 1251 m thick since the early Miocene, unconformably underlain by metamorphic complex of Precambrian to Paleozoic age (Zhu, Sha, Guo, Yu, & Zhao, 1997).

Figure 4.56 summarizes the reef lithofacies in cores from Nansha and Xisha Islands. Clearly, the Miocene and Pliocene reef limestones in both regions, as well as those from the lower part of the early Pleistocene in Nansha Islands, have been strongly dolomitized, leaving only the better-preserved Pleistocene–Holocene sections for any meaningful paleoenvironmental information to be deducted. Nevertheless, the up to 900 m thick Miocene carbonate sequence indicates a period of major reef growth, probably also relating to significant carbonate platform expansion in the region.

In the southern SCS, some carbonate buildups and reefs especially those in the Nansha Trough (Northwest Borneo Trough) had been buried or drowned, mainly because they could not keep up with the postrift thermal subsidence (e.g., Hutchison, 2004, 2010). The nearby reefs on the Dangerous Ground, however, survive and continue active largely as a result of relatively slower subsidence in the NE Nansha Islands region.

4.5.2 Calcareous Components

Apart from corals, the biogenic calcareous components in the sediments of the China Seas include foraminifers, nannoplankton, ostracods, gastropods, bivalves, and calcareous algae. While others are widely distributed, bivalves and calcareous algae mainly occur in shallow waters. In the following, the distribution patterns of the three major components, foraminifers, nannoplankton, and ostracods, are discussed.

4.5.2.1   Foraminifers

The pioneer work on foraminifers in seabed samples of China Seas includes Polski (1959), Waller (1960), and Cheng and Cheng (1960). A systematic account on foraminifer distribution in all the China Seas was given by Zheng and Fu (1994) who, based on about 800 surface sediment samples and 1000 plankton samples, estimated that the total benthic foraminifer species in the China Seas may reach 2000, plus ∼40 planktonic species, although this could have been somewhat overestimated according to current classification standard.

A

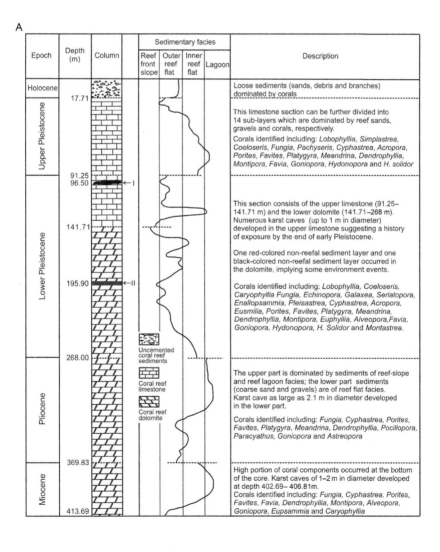

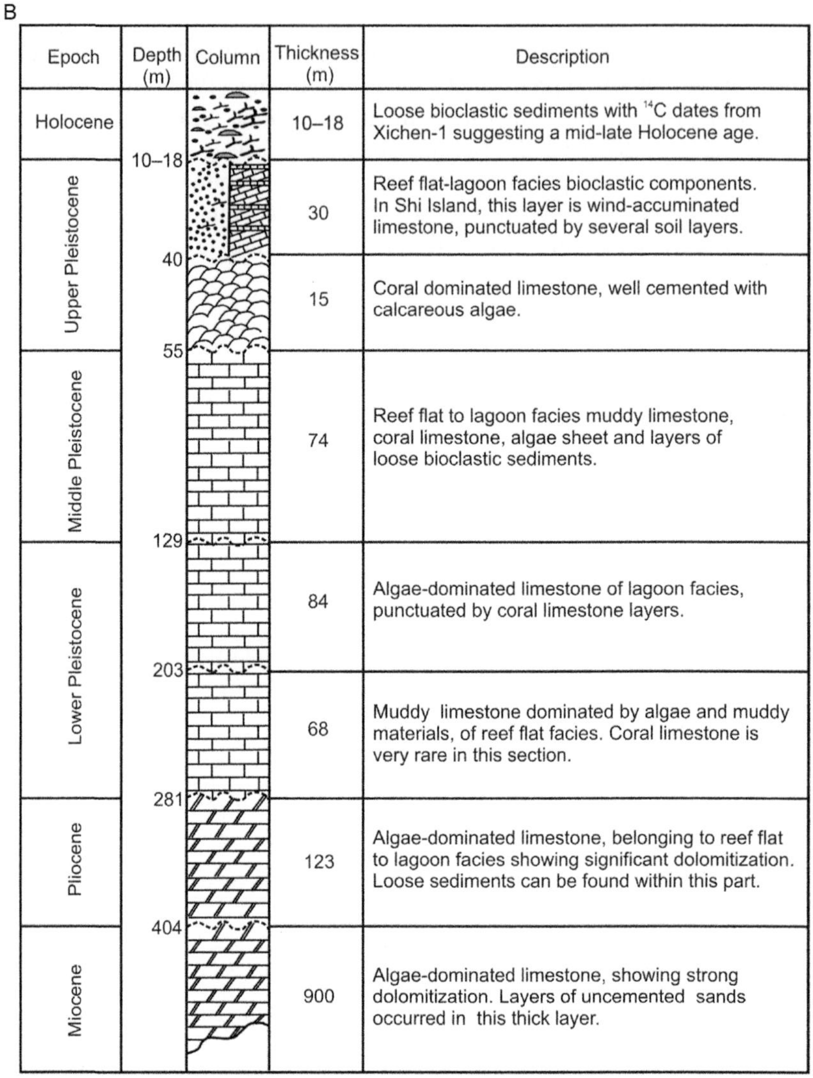

B

Epoch	Depth (m)	Column	Thickness (m)	Description
Holocene	10–18		10–18	Loose bioclastic sediments with ^{14}C dates from Xichen-1 suggesting a mid-late Holocene age.
Upper Pleistocene	40		30	Reef flat-lagoon facies bioclastic components. In Shi Island, this layer is wind-accumulated limestone, punctuated by several soil layers.
Upper Pleistocene	55		15	Coral dominated limestone, well cemented with calcareous algae.
Middle Pleistocene	129		74	Reef flat to lagoon facies muddy limestone, coral limestone, algae sheet and layers of loose bioclastic sediments.
Lower Pleistocene	203		84	Algae-dominated limestone of lagoon facies, punctuated by coral limestone layers.
Lower Pleistocene	281		68	Muddy limestone dominated by algae and muddy materials, of reef flat facies. Coral limestone is very rare in this section.
Pliocene	404		123	Algae-dominated limestone, belonging to reef flat to lagoon facies showing significant dolomitization. Loose sediments can be found within this part.
Miocene			900	Algae-dominated limestone, showing strong dolomitization. Layers of uncemented sands occurred in this thick layer.

FIGURE 4.56 (A) Simplified lithology in core Nanyong-2 of Nansha Islands showing deducted facies changes and vertical truncations by three major discontinuous surfaces (dashed lines), as well as the black color layer (I) and red color layer (II) events (Zhu et al., 1997). (B) A synthesized figure showing major reef lithologies based on cores from Xisha Islands (Zhang et al., 1989).

Modern Distribution

Benthic foraminifers are common in the sediments of the Bohai Sea, Yellow Sea, ECS, and SCS, with increasing diversity from north to south. Planktonic foraminifers are sporadic in the Bohai Sea, frequent in the Yellow Sea, and common to abundant in the ECS and SCS.

The Bohai Sea foraminifer assemblage is dominated by shallow-water to brackish-water species, mainly species of *Ammonia* (*A. beccarii* var.), *Buccella* (*B. frigida*), *Cribrononion* (*C. subincertum*), *Elphidium* (*E. advenum* and *E. magellanicum*), *Protelphidium* (*P. tuberculatum*), and *Quinqueloculina* (*Q. akneriana* and *Q. lamarckiana*), while *Globigerina bulloides* and *Globigerinoides ruber* with very low numbers are so far the only planktonic species found at a number of localities. Among the >200 species of benthic foraminifers identified, only 2–10 species occur in and near river mouths, and the species number increases offshore to a maximum of ~40 in open bay areas in the north, southeast, and southwest (Li, 2012). In the Liaodong Bay, northern Bohai Sea, for example, benthic foraminifers are high in both the species diversity and abundance. The highest abundance, with ~1700 tests per gram of dried sediment, was recorded in samples from the southeast corner of the Bohai Bay (Li, 2012).

Although many species found in the Bohai Sea also occur in the sediments of the Yellow Sea, the foraminifer assemblage in the Yellow Sea shows the following characteristics: (1) Species diversity increases from north to south as well as from nearshore to offshore. (2) Average species abundance is similar to the Bohai Sea with 50–200 individuals per gram of dried sediment. (3) The number and abundance of middle shelf species increase, particularly species of *Bolivina*, *Cribrononion*, *Florilus*, and *Lagena*. (4) Apart from *A. beccarii*, thick-walled, warm-water *Ammonia* and related species become common, including *A. compressiuscula*, *A. ketienziensis*, and *Cavarotalia annectens* (Li, 2012; Liu, Wu, & Wang, 1987). Three *Ammonia* associations have been found to be water depth-related: *A. beccarii* (~20%) in Old Yellow River mouth area, *A. compressiuscula* (~10%) in <50 m, and *A. ketienziensis* (~20%) in >50 m (Liu et al., 1987). In sandier sediments, especially from the ridge area in the northeast, the benthic foraminifers are characterized by agglutinated forms, such as *Ammobaculites* and *Eggerella*.

In the ECS, foraminifer and ostracod assemblages in surficial sediments were studied in details by Wang et al. (1988). For foraminifers, they identified and illustrated 32 planktonic species and more than 264 benthic species, supplemented by distribution maps for common species. Meanwhile, Zheng (1988) systematically described 280 agglutinated and 146 porcelaneous benthic species from the ECS region. According to Wang et al. (1988), the diversity and abundance of both planktonic and benthic species increase from west to east with increasing water depth. Planktonic foraminifers are prevalent in outer shelf and slope, increasing from ~50–70% at ~150 m to over 95% at >700 m in the Okinawa Trough (Figure 4.57). The subtropical planktonic assemblage (Zheng & Fu, 1994) is dominated by *Neogloboquadrina dutertrei*, *Globigerinoides ruber*, *G. sacculifer*, and *Pulleniatina obliquiloculata* in areas influenced by the Kuroshio and by *Globigerinita glutinata* and *Globigerina bulloides* in shallow shelf areas affected by coastal currents. At >150 m localities, especially in the Okinawa Trough, deepwater species *Globorotalia*

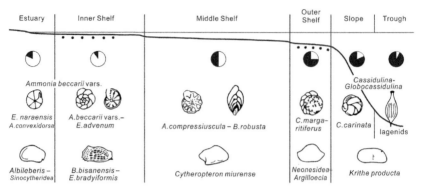

FIGURE 4.57 Foraminifer and ostracod (sub-)assemblages in the shelf–slope sediments of the East China Sea. Relict sediments are marked with black dots. The filled part of circles represents the abundance of planktonic foraminifers relative to benthic species. *Modified from Wang et al. (1988).*

tumida, G. truncatulinoides, and *Sphaeroidinella dehiscens* are frequent. For benthic foraminifers, species of *Ammonia* are characteristic in shallow shelf sediments as in the Yellow Sea. The middle and outer shelf areas are dominated by species of *Bolivina, Bulimina, Astrononion, Textularia, Cibicides, Heterolepa, Hanzawaia,* and *Textularia.* Further offshore, *Cassidulina, Globocassidulina, Epistominella, Cyclammina, Trochammina,* and lagenid species become common in samples from the slope of the Okinawa Trough, but benthic foraminifers, especially those hyaline species, decrease sharply in the bottom sediments of the trough (Figure 4.57; Wang et al., 1988).

In the SCS, numerous studies have been conducted on foraminifer distribution and environmental implications. For example, Wang, Min, and Bian (1985) and Saidova (2007) traced foraminiferal distribution patterns on the northern shelf, Szarek, Kuhnt, Kawamura, and Kitazato (2006) and Szarek, Kuhnt, Kawamura, and Nishi (2009) studied benthic foraminifers from the Sunda Shelf and slope, and Miao and Thunell (1993) and Jian and Zheng (1995) focused on benthic foraminifers in deep-sea sediments. In general, foraminifers are common to abundant in sediments from shelf and slope areas except where the $CaCO_3$ content is low, such as inner shelf, river mouths, relict sandy areas, and below the CCD. The total number of benthic species exceeds 300 in the northern shelf (Wang et al., 1985) and increases to nearly 600 in the south (Szarek et al., 2006; Tu & Zheng, 1991), which includes many relict species with stained tests. In general, benthic foraminifers reach maximum species diversity at upper slope settings, while minimal diversity with <10 species occurs in river mouths or below the CCD >3500 m. Similarly, planktonic foraminifers increase from the middle shelf to deeper water depths and become the dominant component in the pelagic ooze at many slope localities. As a typical tropical assemblage, the SCS planktonic foraminifers feature many warm-water species, including *Globigerinoides sacculifer, G. ruber, G. conglobatus, Globorotalia menardii, G. tumida, G. truncatulinoides, Globoquadrina conglomerosa, Pulleniatina obliquiloculata,* and *Sphaeroidinella dehiscens.* The farther south of the SCS, the stronger the tropical nature of the planktonic foraminifer fauna becomes (Zheng & Fu, 1994). The species *P. obliquiloculata* appears to have a closer link with the Kuroshio Current (e.g., Li, Jian, & Wang, 1997), while *Neogloboquadrina dutertrei* has been considered as relating mostly to monsoon upwelling (Jian et al., 2003). The fragmentation rate of total planktonic foraminifers provides a simple but effective proxy for estimating $CaCO_3$ dissolution in deep-sea sediments from above the CCD.

Reef-associating foraminifers are dominated by large benthic species, as reported by Zheng and Zheng (1978, 1979), Zheng (1980), and Li (1985), among others. The Xisha Islands assemblage contains about 240 benthic species dominated by *Rotalia calcarinoides* and *Calcarina hispida* (Zheng & Zheng, 1978, 1979), the Zhongsha Islands assemblage contains abundant *Baculogypsinoides spinosa* and many others likely endemic to the region (Zheng, 1980), and the Nansha Islands assemblage features species of *Nummulites* and *Amphistegina* (Tu & Zheng, 1991). In fringing reefs of Hainan, however, the benthic foraminifer assemblage is characterized by *Calcarina* (especially *C. hainanensis*), *Amphistegina*, and *Cellanthus* (Li, 1985).

On the northern SCS shelf, Wang et al. (1985) recognized five benthic foraminifer assemblages for different water depth ranges (Table 4.9). As in many parts of the China Seas, species of *Ammonia* are characteristic in inner shelf <20 m, while other hyaline forms, particularly species of *Hanzawaia*, *Elphidium*, *Cavarotalia*, and *Florilus*, are dominant in water depths down to ~50 m. From 50 to 80 m, species of *Bigenerina*, *Heterolepa*, and *Pseudorotalia* are common, while the outer shelf at 80–150 m or more is characterized by species of *Neouvigerina*, *Textularia*, and *Hoeglundina*. The upper slope settings between 200 and 480 m are patronized by such species as *Uvigerina peregrina*, *Karreriella bradyi*, and *Cibicides tenuimago*. In deeper water from 480 to >3500 m, four benthic foraminifer assemblages characterize four depth zones, according to Jian and Zheng (1995, 1997). Generally, agglutinated species increase their relative abundance with increasing depth, while hyaline species decrease from about 2600 to 3000 m until completely absent from below the CCD. Therefore, the >3500 m assemblage contains only a number of agglutinated species characterized by *Eggerella bradyi* (Table 4.9). Together with abundance variations in characteristic species, these assemblages provide a solid ground for studying the deepwater mass evolution in the SCS (e.g., Jian & Wang, 1995, 1997).

The benthic foraminifer assemblages (or biofacies) of the Sunda Shelf and slope are different from those found in the northern SCS region (Table 4.9; Szarek et al., 2006, 2009). Although many metropolitan species remain intact, the Sunda Shelf and slope biofacies contain more frequent agglutinated forms, probably relating to the sandier substrate there.

TABLE 4.9 Distribution of Benthic Foraminifer Assemblages or Biofacies in Different Depth Zones of the SCS

Northern SCS		Southern SCS	
Depth range (m)	Dominant and characteristic species	Dominant and characteristic species	Depth range (m)
<20 (river mouths)	*Ammonia beccarii, A. convexidorsa*	(Pending)	<20
<50	*Hanzawaia nipponica, Elphidium advenum, Cavarotalia annectens, Florilus japonicus*	Atolls: *Amphistegina, Nummulites, Textularia, Pseudorotalia*	<50
50–80	*Bigenerina taiwanica, Heterolepa dutemplei, Pseudorotalia indopacifica*	*Hanzawaia nipponica, Ammomassilina alveoliniformis, Asterorotalia pulchella* High-energy zone: *Heterolepa dutemplei, Textularia lythostrota, Asterorotalia gaimardii,* and miliolids	60–110
80–150 (−200)	*Neouvigerina proboscidea, Textularia pseudocarina, Hoeglundina elegans*	*Facetocochlea pulchra, Bulimina marginata, Cibicides deprimus, Bolivina spathulata*; high energy zone: *Cibicidoides pachyderma, Operculina ammonoides, Poroepistominella decoratiformis, Uvigerina schwageri, Hoeglundina elegans,* plus textulariids and miliolids	110–230
200–480	*Uvigerina peregrina, Karreriella bradyi, Cibicides tenuimago*	*Asterorotalia pulchella, Pararotalia* sp., *Hoeglundina elegans*	200–300
480–1200	*Globocassidulina subglobosa, Cibicidoides bradyi, Pullenia bulloides*	*Uvigerina auberiana, Bolivina robusta, Nuttallides rugosus, Ehrenbergina undulata*	300–1000
1200–2600	*Astrononion novozealandicum, Oridorsalis tener, Sphaeroidina bulloides, Cibicidoides wuellerstorfi*	<1300: *Lagenammina diflugiformis, Uvigerina auberiana, Uvigerina peregrina* >1300: *Paratrochammina challenger, Astrononion novozealandicum*	1000–2000
2600–3500	*Bulimina aculeata, Eggerella bradyi, Epistominella exigua*	*Eggerella bradyi, Gyroidinella profundus, Cibicidoides bradyi*	2600–3500
>3500	*Eggerella bradyi*	*Eggerella bradyi*	>3500

Based on Wang et al. (1985), Tu and Zheng (1991), Jian and Zheng (1995), Chen, Cai, Tu, and Lu (1996), and Szarek (2006, 2009).

As Indicators of Depositional Environments

Based on their modern distribution patterns, foraminifers have been widely used as good indicators of past depositional environments, especially in providing information related to depositional setting, extent of marine influence, productivity, water depth, current and water mass, and transgression–regression cycles. For the Bohai Sea and Yellow Sea, foraminifers are extremely useful for locating marine intervals in the Quaternary alluvial–fluvial–deltaic–estuarine–shelf depositional complex and for defining the nature of all the seven marine layers (M1F–M7F) in the Bohai Sea and the eight marine layers (HI–HVIII) in the Yellow Sea (Figure 4.13; Qin, 1985, 1989; Yang & Lin, 1993; Zheng, 1991). The prevalence of such species as *Ammonia beccarii* var., *A. ketienziensis*, *Elphidium advenum*, *E. magellanicum*, *Buccella frigida*, *Protelphidium tuberculatum*, and *Quinqueloculina akneriana* indicates river mouth to inner shelf environments for these marine layers. With adequate ^{14}C and OSL dating and geomagnetic chronological control, the earliest marine transgressions are now believed to have had taken place at ~2 Ma in the early Pleistocene in the Yellow Sea and at ~0.2 Ma in the later middle Pleistocene in the Bohai Sea. However, foraminifers have been also reported in individual samples from layers presumably of early Pleistocene or older age in several deep cores from suburbs of Beijing and Tianjin, considerably inland from the Bohai Sea (Qin, 1985; Yang & Lin, 1993). Most foraminifers found in these samples are *Ammonia* and *Elphidium* species, and only in one case are planktonic species mentioned. Whether these foraminifers represent older unknown marine influence or other causes remains a mystery (see Chapter 8).

In the deeper SCS and ECS (including the Okinawa Trough), the common occurrence of foraminifers in most Quaternary sections provides not only the biofacies-associated information on depositional environments but also the indispensable material for geochemical analyses (e.g., Wang, 1985, 1995). Among the planktonic species, for example, two bioevents are of regional significance: the extinction of *Globigerinoides ruber* (pink) at 120 kyr BP and the minimal abundance of *Pulleniatina obliquiloculata* at ~4.5–3 kyr. The former event occurs widely in the Indo-Pacific region and has been used as a chronological marker (e.g., Shyu, Chen, Shieh, & Huang, 2001; Wang et al., 2000). The latter event occurs in the NW Pacific but mainly in the Okinawa Trough region and has been related to a period of reduced flow of the Kuroshio presumably due to stronger coastal plume influenced by strengthened winter monsoon (Jian et al., 2000), although this issue is still under debate (Lin et al., 2006).

In the SCS, abundant *Bulimina aculeata*, *Melonis barleeanum*, *Globobulimina* spp., *Chilostomella oolina*, and other benthic foraminifer species are found in several cores from below 1000 m during the Last Glacial Maximum (LGM) (Figure 4.58; Jian & Wang, 1997), likely associated with low

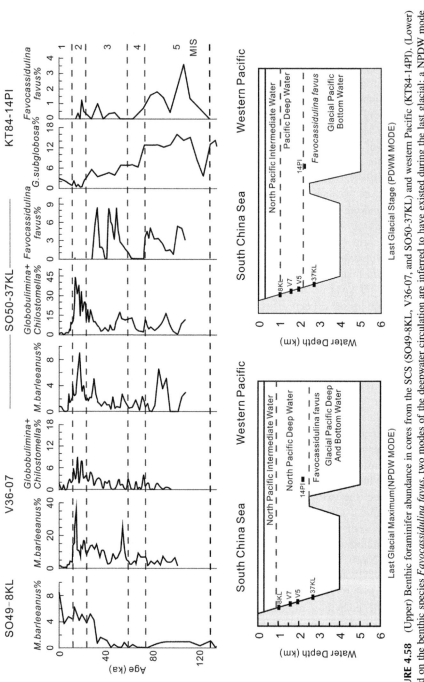

FIGURE 4.58 (Upper) Benthic foraminifer abundance in cores from the SCS (SO49-8KL, V36-07, and SO50-37KL) and western Pacific (KT84-14PI). (Lower) Based on the benthic species *Favocassidulina favus*, two modes of the deepwater circulation are inferred to have existed during the last glacial: a NPDW mode from the LGM (bottom left) and a PDW mode before the LGM (bottom right). *Modified from Jian and Wang (1997).*

dissolved oxygen content conditions as a result of increased surface and export productivity. *Favocassidulina favus* is a species associated with the Pacific Bottom Water (PBW; >2500 m) and has never been found in the surface sediments of the SCS due to the separation by the Bashi Strait (sill depth at ~2400 m) nor in the sediments of sites <2500 m from the western Pacific where the Pacific Deep Water (PDW) is influential. However, many *F. favus* specimens are found in samples from below the LGM interval in cores SO50-37KL (water depth 2695 m) and SO49-41KL (water depth 2120 m) from the northern SCS, as well as in core KT84-14PI (water depth 2200 m) from the western Pacific (Figure 4.58). Therefore, Jian and Wang (1997) inferred that, before the LGM, the PDW/PBW boundary lay at the water depth of about 2000 m, instead of the present depth of 2500 m. However, the absence of *F. favus* in core KT84-14PI and SO50-37KL within the LGM could have resulted from stronger production of the PDW or a new North Pacific Deep Water (NPDW), which subsequently depressed the PDW/PBW boundary to about 2500 m and forced *F. favus* to disappear from the SCS ever since (Figure 4.58). Therefore, two modes of deepwater masses might have existed in the western Pacific and SCS at the last glacial time: (1) a PDWM mode in which there was no formation of deep water from the North Pacific with the PDW/PBW boundary at about 2000 m and (2) a NPDW mode in which the NPDW was formed from a source in the North Pacific beyond the upper limit of the PBWM at about 2500 m (Figure 4.58).

4.5.2.2   Calcareous Nannoplankton

Similar to planktonic foraminifers, both the abundance and diversity of calcareous nannoplankton are low in the semienclosed Bohai and North Yellow Sea but gradually increase southward to the South Yellow Sea and East China Sea. In general, their abundance in the SCS is relatively higher, especially from the slope area (Figure 4.59). The number of coccolith species increases from 2 to 5 in the Bohai Sea, to ~15 species in the South Yellow Sea (Li, 1991; Rui, Liu, Liang, & Zhao, 2011), to 28–35 species in the deepwater ECS, Okinawa Trough, and SCS (Cheng, 1995; Wang & Cheng, 1988; Wang et al., 1985). The river mouth to shallow shelf assemblage comprises only 2–3 species, dominated by *Gephyrocapsa oceanica* (up to 95%) and *Emiliania huxleyi*. The abundance percentages of these two species change toward deeper-water localities, where *E. huxleyi* may reach 60–70%, while *G. oceanica* may drop to minimal values but often become dominant again below the lysocline because of its dissolution-resistant nature (Figure 4.60). Apart from *G. oceanica* and *E. huxleyi*, other species common in sediments from the outer shelf and open water include *Florisphaera profunda*, *Gephyrocapsa muellerae*, *G.* spp. (small types), and *Helicosphaera carteri*. As a species living in the lower mixing zone, *F. profunda* is often considered as a nutricline indicator in the SCS (Liu, Wang, Tian, & Cheng, 2008), where it attends 50% or more

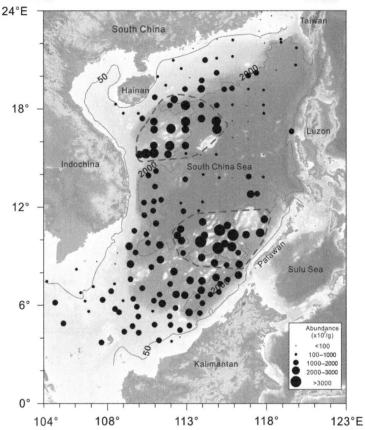

FIGURE 4.59 Distribution of calcareous nannoplankton in the surface sediment of the South China Sea, showing high abundance in Xisha and eastern Nansha areas (Wang et al., 2007).

in sediment samples from >1500 m water depths (Figure 4.60; Wang et al., 2007), especially from the northeastern SCS area (Liu, Wang, et al., 2008). High abundance of calcareous nannoplankton is distributed also along the eastern side of the Okinawa Trough and along the outer shelf–slope of the SCS, with the Xisha Islands area and the Nansha Islands area having the highest individual numbers (Figure 4.59; Cheng, 1995; Wang et al., 2007). Comparatively, however, their distribution in the Okinawa Trough is patchy, probably due to dissolution caused by volcanism-related activities. Below the CCD (>3500 m) in the SCS, individuals of calcareous nannoplankton are rare or completely dissolved except those wrapped up in fecal pellets or in turbiditic clays (Figure 4.59).

Over the years, calcareous nannofossils have been extensively used in stratigraphic and paleoceanographic studies of the China Seas (e.g., Wang, 1985, 1995). For example, based on analyses of the assemblage data and

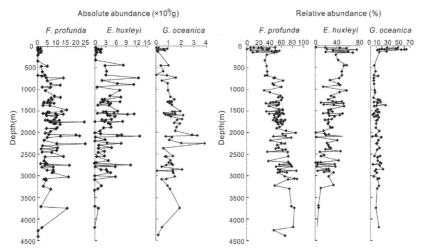

FIGURE 4.60 Distribution of the three most common calcareous nannoplankton species, *Gephyrocapsa oceanica*, *Emiliania huxleyi*, and *Florisphaera profunda*, in the surface sediments of the South China Sea. *Modified from Wang et al. (2007).*

comparison to other paleoclimate records, Cheng and Wang (1997) concluded that the *E. huxleyi/G. oceanica* ratio is a useful proxy of water fertility or surface productivity, but cannot be used as a direct monsoon indicator.

Based on calcareous nannofossils in ODP Site 1202 cores and in surface sediments from the southern Okinawa Trough, Su and Wei (2005) attempted to indicate Kuroshio activities over the last 13 kyr by using the ratio of *F. profunda* against *F. profunda + E. huxleyi + G. oceanica* (F–EG ratio) as a proxy of the Kuroshio Current and the ratio of *G. oceanica* against *F. profunda + E. huxleyi + G. oceanica* (G–FE ratio) as a proxy of near-coast environment. The results from the surface sediments indicate that the F–EG ratio is >15% in the assemblage lying directly under the main route of the Kuroshio Current, dropping to <10% in sediments off the current, whereas >30% G–FE ratio values occur in all samples on the ECS continental shelf and from near-coast cores. In sediment of Site 1202, extremely low F–EG ratio together with very high G–FE ratio from the latest Pleistocene and the earliest Holocene suggest the absence of the Kuroshio Current in the area studied (Figure 4.61). The intrusion of the Kuroshio Current was clearly marked by a dramatical increase of F–EG ratio and a notable reduction in the G–FE ratio around 9 kyr BP. During the Holocene, the variation of these nannofossil ratios depicts three long-term cycles (with a periodicity of ~3000 yr) of the Kuroshio-related activities (Figure 4.61). However, these data do not support the speculation for a very weak Kuroshio at ~4.5–3 kyr BP as implied by the minimum zone of the planktonic foraminifer *P. obliquiloculata*, although the estimated SST at Site 1202 dropped noticeably at the same interval (Figure 4.61).

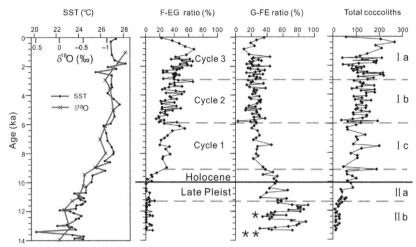

FIGURE 4.61 The $U^{K'}_{37}$-derived SST and *N. dutertrei* $\delta^{18}O$ (Zhao, Huang, & Wei, 2005), as correlated to calcareous nannofossil abundance and ratios at ODP Site 1202 from the southern Okinawa Trough. F-EG refers to *F. profunda* against *F. profunda* + *E. huxleyi* + *G. oceanica* and G-FE to *G. oceanica* against *F. profunda* + *E. huxleyi* + *G. oceanica*. The total coccolith abundance was based on counting the number of coccoliths on 10 view fields. *Modified from Su and Wei (2005).*

4.5.2.3 Ostracods

Like the benthic foraminifers, ostracods are common components in sediments from near shore and shallow water of the China Seas. Freshwater and blackish species are frequent or common in assemblages from river mouths and tidal flats and may also present in shelf sediments due to reworking. Marine species are frequent at shelf and slope settings and reach high diversity and high abundance in 500–1000 m on the upper part of the continental slope although their number is always considerably lower than the benthic foraminifers. The ostracod diversity increases from <10 species in river mouths and in the Bohai Sea to ~180 species in the ECS (Wang et al., 1988) and over 300 species in the SCS (Li, 2012; Zhao & Zhou, 1995, and references).

Five ostracod assemblages were recognized by Zhao and Wang (1990) to characterize ostracod distribution on shelves. (1) The estuarine and coastal water assemblage is mainly made up by euryhaline and eurythermal species. *Sinocytheridea impressa* predominates this assemblage, which also includes such common species as *Neomonoceratina delicata* (in the SCS), *N. cirspata* (in the Bohai Sea), *Albileberis* spp. and *Leptocythere ventriclivosa* (in the Yellow Sea and ECS), and *Aurelia, Callistocythere, Cushmanidea, Keijella,* and *Neosinocythere* species. (2) The shallow sea assemblage is marked by stenohaline and eurythermal species, including many also occurring in the estuarine assemblage. This inner shelf assemblage is dominated by *Cytheropteron miurense, Keijella* spp., *Munseyella* spp., and *Pistocythereis* spp. In addition, abundant warm-water species including *Copytus postero-sulcus*, *Hemikrithe orientalis*, and *Loxoconcha* sp. occur in the SCS sector, while cool water species of *Cytheropteron*, *Cytheromorpha*, *Finmarchinella*, and *Loxoconcha* are common in 20–50 m of the Bohai and Yellow Seas. (3) The >50 m Yellow Sea assemblage is characterized by a low diversity of cold water species, mainly the species of *Howeina*, *Sarsicytheridea*, and *Acanthocythereis* in the northern area and species of *Amphileberis*, *Buntonia*, *Kobayashiina*, and *Krithe* in the southern area. (4) The middle to outer shelf assemblage in the East and South China Seas is dominated by many warm-water species, including *Argilloecia hanaii*, *Cytherelloidea senkakuensis*, *Foveoleberis cypraeoides*, *Keijella apta*, *K., japonica*, *Loxoconcha* sp. A, and *Neocytheretta* spp. The assemblage is dividable into an ECS subassemblage characterized by *Acanthocythereis muneckikai* and a SCS subassemblage by *Bradleya albatrossia*. (5) The coral reef assemblage features shallow-water tropical species, such as *Loxoconcha* spp., *Mutilus* cf *packardi*, *Quadracythere parviloba*, and *Triebelina sertata*. Many warm-water species have affinities with those widely recorded in the tropical Indo-Pacific, while many cold water species are linked to those from regions farther north such as the Sea of Japan and NW Pacific, indicating distinct provincialism in ostracod distribution (Zhao & Wang, 1999).

Many more ostracod assemblages have been recognized for specific areas of the China Seas. For example, three ostracod assemblages were found in the Bohai Sea (*Neomonoceratina*, *Munseyella*, and *Sinocytheridea*) (Li, 2012, and references) and five in the ESC (*Albileberis-Sinocytheridea*, *Bicornucythere–Echinocythereis*, *Cytheropteron*, *Neonesidea–Argilloecia*, and *Krithe product*) (Figure 4.57; Wang et al., 1988). For the deep SCS ostracods, *Neonesidea* spp. are dominant in sediments from the outer shelf and uppermost slope, while *Krithe* spp. predominate >300 water depths (Table 4.10). The diversity and abundance of oceanic deepwater species increase to maximum values at 500–1000 m but decrease in deeper water to a minimum below the CCD at >3500 m. Four ostracod assemblages from the deep SCS were reported as relating to depth-associating water masses at 180–300 m (SCS subsurface water), 300–1000 m (SCS middle water), 1000–2500 m (SCS deep water), and >2500 m (SCS basin water), respectively, as listed in Table 4.10.

These modern distribution data of ostracods provide a good reference for comparison and interpretation of older assemblages recovered from cores, as have been done in many studies (e.g., Zhao, Li, & Jian, 2009 (deep sea); Tanaka, Komatsu, Saito, Nguyen, & Vu, 2011 (Red River delta)).

4.5.3   Siliceous Components

The biogenic siliceous components, or biogenic opal, in the sediments of the China Seas are mainly represented by diatoms and radiolarians. They play a

TABLE 4.10 Characteristics and Mean Abundance (>4%) of Ostracod Species in Four Deepwater Assemblages from the SCS, Which were Recognized as Depth- and Water Mass-Related

	Subsurface Water (180–300 m)	SCS Middle Water (300–1000 m)	SCS Deep Water (1000–2500 m)	SCS Basin Water (>2500 m)
Mean abundance (no. of valves per 10 g dry sediment)	20	36	13	2
Mean diversity (no. of species)	104	395	20	7-Jun
Abrocythereis spp.	1.2	4.6	0	0
Abyssocythere spp.	0	0	3.7	1.4
Abyssocythereis sulcatoperforat	0	0	5	3.6
a Acanthocythereis spp.	13.3	0	0	0
Ambocythere spp.	1.2	5	0.8	0
Argilloecia spp.	1.9	5.7	7.4	10.9
Bradleya spp.	6.7	6.8	3.3	6
Cytherelloidea spp.	12.6	1.2	0	0
Cytheropteron spp.	9.9	8.3	3	0.4
Henryhowella spp.	0	0.4	6.1	4
Krithe spp.	0	20.6	38.7	31.8
Legitomocythere acanthoderma	0	0.8	3.2	4.8
Neonesidea spp.	33.9	2.6	0	0
Parakrithe spp.	0.4	4.5	6.9	1.7
Ragocythereis spp.	0	0	3.4	4.4
Xestoleberis spp.	4.9	9.3	2.1	5

Based on Zhao and Zhou (1995) and Zhao and Zheng (1996).

critical role as direct contributors in surface productivity and in deep-sea opal accumulation. In the surface sediments of the SCS, >6% opal abundance occurs in the southern SCS deep sea, near and to the southwest of the Liyue Bank, 3–6% in most slope area and in the central basin, and <3% in the shelf area and in the northeastern corner close to the Bashi Strait (Figure 4.62). The relatively low abundance in shelf sediments was likely related to dilution by terrigenous particles, while the low abundance near the Bashi Strait was obviously caused by the inflow of the warm and saline Kuroshio water. A comparison between the total opal data and radiolarian data (Figure 4.62) indicates that almost all of the biogenic opal from the shelf and at least half of the opal from the high opal area of the SCS slope and deep sea are made up of diatoms, dinoflagellates, and sponge spicules other than radiolarians. For a longer timescale, Wang (2009) provided an updated account on the distribution and paleoceanographic implications of biogenic opal in the SCS since the Oligocene based on ODP Leg 184 material.

4.5.3.1   Diatoms

Diatoms are common in the sediments of the China Seas, with brackish to tidal flat species dominating assemblages from <20 m and marine planktonic species increasing from 20 to high numbers further offshore. The nearshore assemblage in the Bohai and Yellow Seas is characterized by *Cyclotella*, and the >20 m inner shelf assemblage by *Palaria* in the Bohai Sea and by *Hyalodiscus* and *Coscinodiscus* in the Yellow Sea (Li, 2012, and references). These assemblages can be divided into several subassemblages to represent regional variations. Compared to those in the Bohai and Yellow Seas, diatoms in the ECS are richer and more diverse with >150 taxa of different living modes. Three areas show high concentration of diatoms: coastal area (including river mouths) with 200–300 individuals per slide, the Okinawa Trough with 100–200 individuals per slide, and the middle shelf area with >100 individuals per slide (Li, 2012, and references). In contrast to those from coastal and inner shelf waters, diatoms in the Okinawa Trough are typified by many warm-water species, all showing a decreasing trend from south to north as influenced by the Kuroshio Current.

Among the 300 diatom taxa reported from the SCS (Li, 2012 and references), about 120 species are found in the deep-sea basin (Jiang, Zheng, Ran, & Seidenkrantz, 2004; Zhan, 1987). Their highest concentration with $>10 \times 10^3$ individuals per gram of dry sediment occurs in the central SCS region, decreasing to $5–6 \times 10^3$ in slope sediments. The cosmopolitan species *Thalassionema nitzschioides* and *Azpeitia nodulifera* are most abundant, followed by such warm-water species as *Nitzschia marina*, *Azpeitia* spp., and *Rhizosolenia bergonii*. Although good specimens of the gigantic *Ethmodiscus rex* (up to 1.8 mm) are rarely seen, numerous shell fragments of this species indicate its common presence. Based on statistical analyses of species

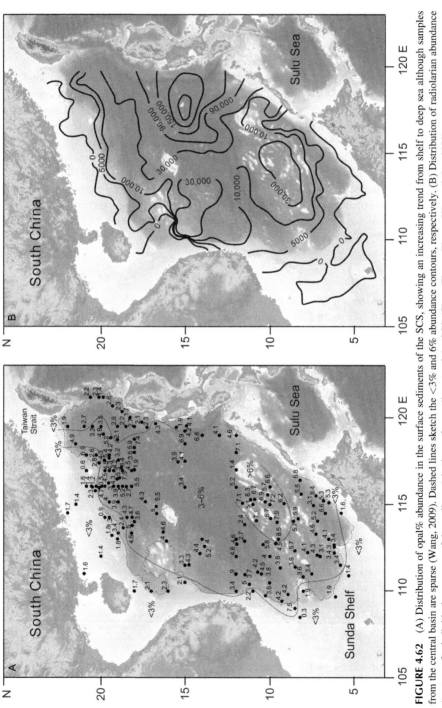

FIGURE 4.62 (A) Distribution of opal% abundance in the surface sediments of the SCS, showing an increasing trend from shelf to deep sea although samples from the central basin are sparse (Wang, 2009). Dashed lines sketch the <3% and 6% abundance contours, respectively. (B) Distribution of radiolarian abundance (contours for individuals per gram of dry sediment) in the surface sediments of the SCS (Chen, Zhang, Zhang, Xiang, & Lu, 2008).

distribution, Jiang et al. (2004) attributed the six assemblages they recognized for the SCS to different hydrologic (current) conditions although the sampling coverage is relatively sparse.

In the Pearl River estuary and shallow water of the SCS, over 70 diatom taxa recorded are dominated by brackish-water species (Figure 4.63). Statistical tests indicate that the diatom distribution in this region is strongly correlated with salinity although other environmental variables such as sand content and water depth are also influential (Zong et al., 2010).

In core MD05-2908 (24°48.04′N, 122°29.35′E, water depth 1275 m) from the southern Okinawa Trough, Li, Jiang, Li, and Zhao (2011) used the following warm-water taxa to represent the flow of the Kuroshio Current: *A. nodulifera*, *A. africana*/*A. neocrenulata*, *N. marina*, and *R. bergonii*. During AD 950–1500, at the time of the Medieval Warm Period (MWP), these species were high in abundance with strong fluctuations and reached maximum values around AD 1450, indicating strengthening of the Kuroshio Current. During AD 1500–1900, at the time of the Little Ice Age, however, warm-water species abundance decreased markedly, indicating a weakened Kuroshio Current at the time. In contrast, high abundance of freshwater taxa from the same interval likely had resulted from enhanced precipitation in northeastern Taiwan due to the southward migration of the western Pacific subtropical high (Li et al., 2011).

4.5.3.2 Radiolarians

Radiolarians in surface sediments and in plankton tows from the most part of the China Seas were systematically studied by Chen and Tan (1996) and Tan and Chen (1999). Most of the ~540 radiolarian taxa reported belong to the Spumellaria and Nassellaria groups. Because of their open oceanic nature, radiolarians generally show a seaward increase trend in both abundance and diversity. In sediment samples, for example, radiolarians may attend <20 species and ~300–400 individuals per gram of dry sediment from the inner to middle ECS shelf, ~20–40 species and 500–4000 individuals from the outer shelf, ~50 species and ~5000 individuals from offshore Northern Taiwan, 40–100 species and $100–500 \times 10^2$ individuals from the slope, and over 100 species and 1000×10^2 or more individuals in the Okinawa Trough. But only 5 species and 39 individuals were found in a sample from the South Yellow Sea (Tan & Chen, 1999). Most of the species entering the Yellow Sea drop sharply at ~30°N and become extremely rare further north although some living species may extend to about 33°N and beyond especially during summer and autumn (Tan & Chen, 1999). Due to their rare occurrence, radiolarian distribution in the Bohai Sea and the northern Yellow Sea remains largely unknown.

On the shelf of the southern Yellow Sea and ECS, *Rhizoplecta trithyris*, *Spongosphaera streptacantha*, *Gazelletta hexanema*, and *Lithomelissa spinosissima* are common in the living fauna, while *Tetrapyle quadriloba*,

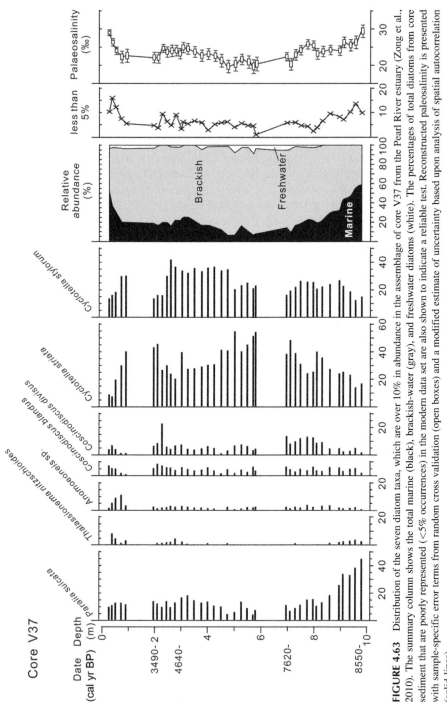

FIGURE 4.63 Distribution of the seven diatom taxa, which are over 10% in abundance in the assemblage of core V37 from the Pearl River estuary (Zong et al., 2010). The summary column shows the total marine (black), brackish-water (gray), and freshwater diatoms (white). The percentages of total diatoms from core sediment that are poorly represented (<5% occurrences) in the modern data set are also shown to indicate a reliable test. Reconstructed paleosalinity is presented with sample-specific error terms from random cross validation (open boxes) and a modified estimate of uncertainty based upon analysis of spatial autocorrelation (solid lines).

Tetrapyle circularis, and *Octopyle actospinosa* are most abundant in sediment samples. These discrepancies may have resulted from the susceptibility of the dominant living forms to dissolution. Other factors, such as circulation and nutrient level, may have also played an important role. Accordingly, high radiolarian abundance often occurs in areas with eddy circulation, high productivity, and fine substrate (Tan & Chen, 1999). Although their diversity and abundance are high in the Okinawa Trough, especially in the northern part with intensive upwelling and volcanism (Chang et al., 2003), the most effective driving force may not have been only the Kuroshio itself but the combined effect of currents, upwelling, and preservation.

As shown in Figures 4.62 and 4.64, radiolarians are very rare in the inner and middle shelf of the SCS, increase rapidly at slope and deepwater localities, and reach very high abundance values surrounding the Huangyan Island, offshore from Luzon, where they may occupy ~30% of the deep-sea deposit (Chen & Tan, 1996; Chen et al., 2008). Among the ~480 radiolarian species identified in the SCS, the dominant taxa include *Tetrapyle quadriloba–T. octacantha* (13.24%), *Ommatartus tetrathalamus* (5.39%), and *Giraffospyris angulata* (2.91%). Areas with high radiolarian abundance also have a high number of species. For example, while most middle and lower slope areas >1800 m are found with ~120 species, the area around the Huangyan Island has over 150 species. They are most enriched in areas affected by upwelling and volcanic ash. In areas affected by turbidites, however, both radiolarian diversity and abundance are low (Figure 4.64).

In core 17957-2 (10°53.9′N, 115°18.3′E, water depth 2195 m) from the southern SCS, biogenic siliceous components show high abundance in glacial intervals, especially after the middle Pleistocene climatic transition at ~0.9 Ma, likely in responding to high productivity due to stronger winter monsoons (Wang & Abelmann, 2002). Radiolarians are mainly composed of tropical and subtropical species (Table 4.11), and species of surface living mode show different abundance fluctuations from their subsurface–intermediate counterparts (Figure 4.65). The most pronounced change occurred at ~600 kyr BP with a big jump in the abundance of subsurface–intermediate dwellers, probably marking thermocline shoaling in the region as reported earlier based on planktonic foraminifers (Figure 4.65).

4.5.4 Organic Carbon

Like many other sea regions, the major portion of total organic carbon (80–95% TOC) in the northern SCS (and the ESC proper) is represented by dissolved organic carbon (DOC) mainly due to the Kuroshio intrusion and continental fluxes (Hung, Wang, & Chen, 2007). The concentration of DOC ranged from 70 to 85 μM in the mixed layer excluding the coastal zone, while the concentration of particulate organic carbon (POC) was relatively low, ranging from 1.6 to 4 μM in the mixed layer. In >1000 m waters, both

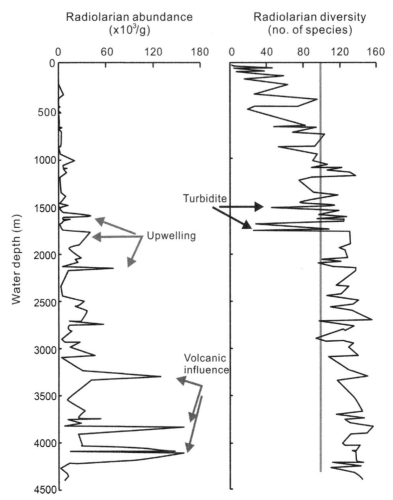

FIGURE 4.64 Depth distribution of radiolarian abundance and diversity in the surface sediments of the SCS, showing the positive effect of upwelling (green arrows) and volcanic ash (red arrows) and the negative effect of turbidites (blue arrows). *Modified from Chen et al. (2008).*

DOC and POC are stable, at 43 ± 3 μM and 1.1 ± 02 μM, respectively. However, DOC jumps up to 132 μM and POC to 13 μM in the Pearl River plume during summer with the highest river discharge. Their relatively high concentration in shelf zones and during summer is obviously related to terrestrial inputs and mixing processes, while in the central basin, biological effect on DOC and POC distributions may be more significant, as shown by their positive correlation with chlorophyll a. In general, the downward flux of DOC is smaller than the sinking flux of C_{org}, but its magnitude (0.27–4.4 mmol C m^{-2} d^{-1}) constitutes a significant flux of TOC transported through the

TABLE 4.11 Typical Radiolarian Taxa from the Southern South China Sea

Tropical taxa	*Tetrapyle octacantha, Octopyle stenozona, Lophospyris* spp. *Tholospira cervicornis, Stylodictya* sp., *Dimelissa monceras, Lithelius* sp.
Subtropical taxa	*Pterocorys zancleus, Cycladophora bicornis, Stylochlamydium asteriscus, Phormospyris stabilis, Spongodiscus resurgens*
Surface dwellers	*Botryocyrtis scutum, Zygocircus productus, Phorticium* spp.
Subsurface and intermediate dwellers	*Botryostrobus auritus–B. australis, Lophophaena hispida, L. variabilis, Larcopyle butschlii*

Based on Wang and Abelmann (2002).

depth of 100–150 m in the northern SCS basin. It is estimated that the downward fluxes of TOC are nearly balanced by new productivity derived from upward fluxes of nutrients, except for fluxes through 100 m in summer. Therefore, DOC appears to play an important role in organic carbon cycling and budgets in the upper layer of the northern SCS (Hung et al., 2007).

Riverine inputs of TOC and suspended particulate matter from the Pearl River to the shallow SCS during 2005–2006 were studied by Ni, Lu, Luo, Tian, and Zeng (2008). Their results indicate that DOC concentration ranges from 1.38 to 2.13 mg/L, or 1.67 mg/L on average. Concentrations of POC ranged from 2.66% to 4.12% of total suspended particulate matter (SPM), and both correspond closely to river runoff, which peaks in June due to summer monsoon (Figure 4.66). The fluxes of TOC and SPM from the Pearl River Delta via the main eight outlets have been estimated to be 9.2×10^5 and 2.5×10^7 t/yr, respectively.

4.5.4.1 TOC in Surface Sediments

Distribution of organic carbon in the surface sediments of the China Seas is closely related to grain size in shallow shelf areas and to productivity in deep waters. For the most parts of the Bohai Sea and Yellow Sea, TOC is low, 0.01% to <0.1%, and increases to 1–2% or more only in muddy sediments sourced mainly from the Yellow River (Li, 2012; Song, 2010, Hu et al., 2013, and references). Similarly, in the inner shelf muddy sediments of the ECS, TOC is relatively high, 0.61% on average, but decreases to an average of 0.35% in sandy sediments from the middle and outer shelf (Figure 4.67; Lin et al., 2002; Qin et al., 1988). As indicated by biomarker analyses, most of the organic carbon from the ECS shelf is sourced from the Yangtze River, while marine productivity basically controls the TOC distribution in upwelling areas (Xing, Zhang, Yuan,

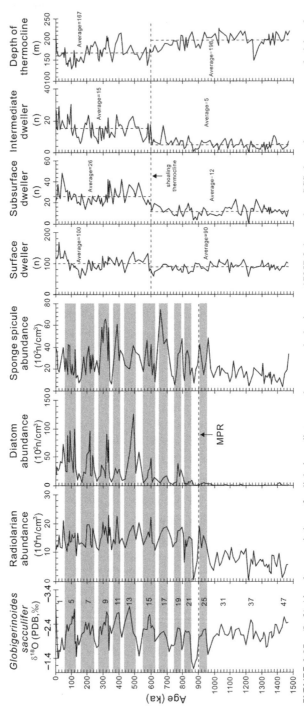

FIGURE 4.65 (Left) Abundance variations of radiolarians, diatoms and sponge spicules in core 17957-2 the southern SCS show close responses to glacial cycles based on *G. sacculifer* $\delta^{18}O$ with odd stages marked. (Right) Abundance variations of surface, subsurface and intermediate living radiolarians show a major change at 600 kyr BP in responding to thermocline shoaling, as compared to planktonic foramimifer data. *Modified from Wang and Abelmann (2002).*

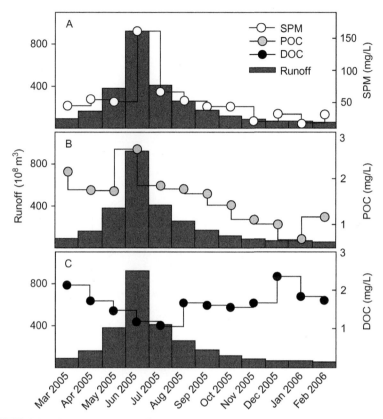

FIGURE 4.66 Correlations of runoff amounts with (A) SPM concentrations (mg/L), (B) POC concentrations (mg/L), and (C) DOC concentrations (mg/L) from the Pearl River Delta (Ni et al., 2008).

Sun, & Zhao, 2011). Terrestrial organic matter contribution decreases seaward, as the influences of the Yangtze River discharge decrease. In the Okinawa Trough, TOC is high, averaging 1.35%, with relatively lower values on the western slope compared to those higher values from the bottom of the trough, largely due to marine inputs (Li & Chang, 2009; Qin et al., 1988).

In the SCS, TOC distribution also shows a clear pattern in having relatively higher values (>0.8%) in coastal regions, lower (0.5–0.6%) on the sandy outer shelf, higher (0.8–0.9%) again on the slope, and lower (<0.7%) in the central basin (Figure 4.68). The area offshore from Borneo–Kalimantan has highest TOC (>1%), and the two upwelling areas (off SE Vietnam and to the west of the Bashi Strait) also register relatively high TOC values (~0.9%).

Kao, Shiah, Wang, and Liu (2006) conducted a study on carbon isotopic compositions of suspended particulate matter (SPM) in different depth localities offshore from southwest Taiwan to determine the influence of terrestrial

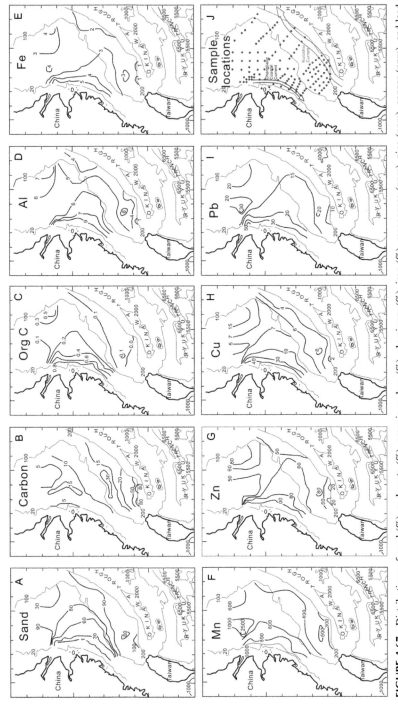

FIGURE 4.67 Distributions of sand (%), carbonate (%), organic carbon (%), aluminum (%), iron (%), manganese (ppm), zinc (ppm), copper (ppm), and lead (ppm) in nearly 130 surface sediment samples from the ECS shelf (Lin, Hsieh, Huang, & Wang, 2002).

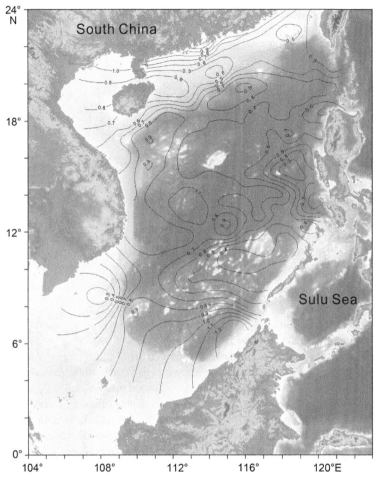

FIGURE 4.68 Distribution of TOC% in the surface sediments of the SCS, showing >1% high values from offshore Borneo–Kalimantan and relatively high values (0.8–1%) in northern inner shelf areas. *Modified from Li (2012).*

and marine end-members of organic matter. Their results show that the concentrations of POC and PN in water column ranged from 105 to 265 µg C l^{-1} and 20 to 41 µg N l^{-1}, respectively, and TOC content in marine sediments offshore from the river mouth ranges from 0.45% to 1.35% with the highest values on the upper slope (Figure 4.69). The TOC/TN ratio of the SPM samples from an offshore station is 6.8 ± 0.6, and the δ^{13} Corg values average at −21.5 ± 0.3%, all indicating their predominantly marine origin, while the riverine SPM samples exhibit typical terrestrial δ^{13}Corg values around −25%. In surficial sediments, however, the δ^{13}Corg values range from −24.8% to −21.2%, showing a distribution pattern influenced by inputs from the

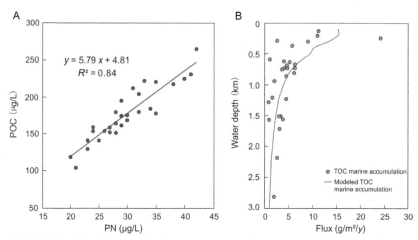

FIGURE 4.69 (A) C/N scatter plot for suspended particulate organic matter collected at the head of the Gaoping Canyon. (B) The observed and modeled accumulation rate of marine organic carbon versus depth in the continental margin off southwestern Taiwan. *Modified from Kao et al. (2006).*

Gaoping River. The relative contributions from marine and terrestrial sources to sedimentary organic carbon can be determined using an isotope mixing model with end-member compositions derived from the riverine and marine SPM. The results also show that the marine organic accumulation rate ranges from 1.6 to 70 g C m^{-2} yr^{-1} with an area-weighted mean of 4.2 g C m^{-2} yr^{-1}, which is on a par with the mean terrestrial contribution and accounts for ~2.3% of mean primary production. Away from the nearshore area, the content of terrigenous organic carbon in surficial sediments decreases with distance from the river mouth, indicating its degradation in marine environments (Figure 4.69; Kao et al., 2006).

4.5.4.2 TOC in Glacial Cycles

Since only a small percentage of the original surface production (mostly <1% for TOC) is preserved in sediments, it is often debatable whether the sedimentary biogenic contents can be used as productivity proxies for the past (Zhao, Li, et al., 2009; Zhao, Wang, et al., 2009). The early work on TOC–productivity relationship includes Thunell et al. (1992) who attributed the higher TOC contents during the LGM in a transect of cores from the southeastern SCS to enhanced productivity. In general, TOC can be viewed as productivity indicator for most part of the SCS, and high TOC content in glacial periods reflects high productivity most likely driven by intensified winter monsoon winds associated with glacial boundary conditions (Chen, Saito, et al., 2003; Chen, Shiau, et al., 2003).

Shiau et al. (2008) presented a multiproxy reconstruction of productivity for core MD97-2142 (12°41.33′N, 119°27.90′E, 1557 m water depth) from the southeastern SCS over the last 870 kyr (Figure 4.70). A composite productivity index (CPI), which was extracted from the principal component analysis of such biogenic components as TOC, carbonate, opal, and alkenone, shows higher total productivity during glacials superimposed on an upward strengthening trend. The alkenone record generally parallels the CPI record, also suggesting higher haptophyte productivity during glacials. However, opal content from this site suggests higher diatom productivity during interglacials and lower productivity during glacials, in agreement with results from ODP Site 1143. Although high diatom productivity during interglacials could have linked with stronger summer monsoon and river inputs, however, a question remains open: Why did diatoms not respond to the winter monsoon fertilization during glacials when total productivity was higher?

Unlike those in core MD97-2142 from the southeastern SCS, TOC and Mn records in core GIK17925-3 (19°51.2′N, 119°2.8′E, 2980 m water depth) from the northeastern slope of the SCS show considerably larger amplitude (Figure 4.71). The development of a stable estuarine circulation characterized

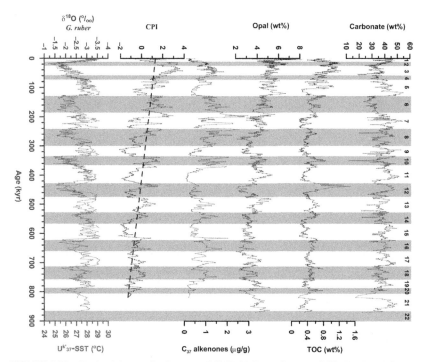

FIGURE 4.70 Productivity proxies from core MD97-2142, southeastern SCS, showing their various responses to glacial–interglacial oscillations (Shiau et al., 2008). CPI is the composite productivity index extracted from the principal component analysis of these biogenic components.

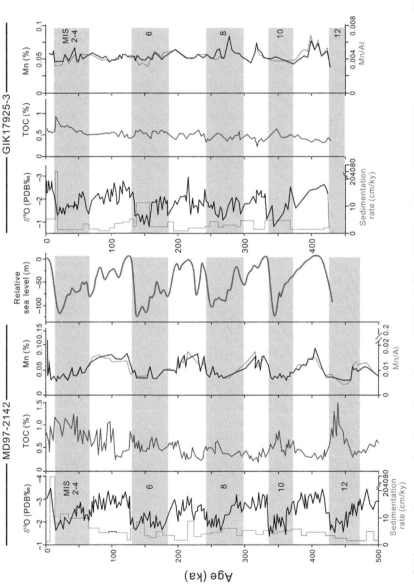

FIGURE 4.71 Stable oxygen isotope stratigraphy, sedimentation rate, TOC%, Mn%, and Mn/Al ratio in core MD97-2142 from the southeastern SCS and core GIK 17925-3 from the northeastern SCS, superimposed by the standard relative sea-level curve (Löwemark et al., 2009). These records show similar glacial–interglacial cyclicity between the two cores but the variation amplitude from MD97-2142 is considerably larger than from GIK 17925-3.

by stagnating water bodies, nutrient recycling, and increased primary productivity in the SCS during glacial intervals was probably the main cause due to the closure of the shallow and narrow straits connecting the SCS in the south and east (Löwemark et al., 2009). Concentrations and Mn/Al ratios of the redox-sensitive element Mn show clear glacial–interglacial cycles with maxima during interglacial periods and minima during glacial periods, indicating ventilation cycles of the bottom water in connecting to sea-level changes. The decreases in Mn correspond to times when sea level dropped 40–60 m below the present level. In contrast, TOC concentrations display an opposite pattern with pronounced maxima during glacial times, especially in core MD972142 from the south. The TOC variations are mostly affected by two factors: (1) variations in primary productivity controlled by variations in the intensity of the winter monsoon and (2) preservation of TOC controlled by variations in ventilation, which in turn is ultimately regulated by sea level. Therefore, variations in TOC represent a superimposition of primarily sea level-influenced preservation control and winter monsoon-driven variations in primary productivity intensity. The larger amplitude of the variations in TOC and Mn at the southern site could have been caused by stronger oxygen depletion and nutrient recycling in a region further away from the only remaining opening to the open Pacific, the Luzon Strait (Figure 4.71; Löwemark et al., 2009).

To the far north, on the western coast of the Bohai Sea, TOC and other geochemical and molecular records in core TP 23 (38°39′54.0″N, 117°24′18.4″E) reveal environmental changes over the last 21 kyr BP. After two peat layers were deposited at ca. 9000–8460 yr BP, the core site has been consistently covered by seawater until recent land reclamation by human. High TOC concentrations at 6000–3800 yr BP can be attributed to enhanced discharge of Corg-rich sediments from the Yellow River, when vegetation density increased and soil was well developed on land under warm and humid middle Holocene climate conditions. Lignin phenol compositions and C31/C29n-alkane ratio suggested the largest expansion of woody plants between ca. 5300 and 4000 yr BP. Since ca. 3800 yr BP, an abrupt increase in the C31/C29n-alkane ratio was clearly related to higher abundance of grasses in relatively dry climate conditions (Sun et al., 2011).

Similarly, the ratio of TOC to TN (total nitrogen concentration) can be used to indicate paleoenvironmental changes. In the Okinawa Trough, for example, the C/N ratio represents the intensity of vascular plant fragment input from continental rivers, as confirmed by its parallel changes with lignin phenol content and ratio of C28 to C16 fatty acid content. Therefore, upward decrease of C/N ratios in the Okinawa Trough cores most likely demonstrates the postglacial retreat of the continental shoreline (Ujiié et al., 2001). However, adjustments of TOC for marine source by long-chain alkenones need to be exercised before any such speculation can be made, especially for the Okinawa Trough region (Li & Chang, 2009).

4.6   VOLCANIC SEDIMENTS

In the China Seas, volcanic sediments occur mainly in two regions: the Okinawa Trough and the SCS due to their active tectonic activities from both on site and near field. Apart from volcanic glasses, mineral components often associating with volcanism include hypersthene, augite, hornblende, magnetite, feldspar with vesicular wall, quartz, and acicular apatite, and most of the volcanic minerals commonly appear as euhedral crystals with complete domes (Qin, 1994). Accordingly, elements such as Ti, V, Si, and Zr found in areas with active volcanism may have been produced by volcanism rather than terrigenous source.

4.6.1   Okinawa Trough Volcanic Deposition

Volcanic ash and glasses are widely distributed in the Okinawa Trough, more frequently in the bottom of the basin and the eastern slope area (Figure 4.6). In general, volcanic ash increases from continental shelf to the trough and is more abundant in the northern and central trough than the southern trough. Together with volcanic glasses, pyroxene, hypersthene, magnetite, and other thermally generated minerals are common (Table 4.1). Similarly, Na_2O and Zr are high, while terrigenous Fe_2O_3 and K_2O and biogenic CaO and Sr are low (Table 4.3; Jiang, Li, & Li, 2010).

Over the last 15 kyr, at least four layers of volcanic ash are recorded in sediment cores from the Okinawa Trough, as shown in Figure 4.72. Core Y127 (30°32.97′N,128°18.26′E, water depth 739 m) in the northern trough contains seven tephra layers, although the youngest three layers since ~12 kyr BP have been laid together pseudoconformably. In core E017 (26°34.45′N, 126°01.38′E, water depth 1826 m) from the southern trough, four tephra layers are quite distinct, bearing ages of 14,355–14,467, 9470–11,021, 7551–9011, and 1458–5050 yr BP, respectively. It is apparent that regional volcanic activities have intensified since ~12 ka, with more ash buildup in the northern trough.

The younger two tephra layers in core E017 were probably related to eruptions of the Kikai-Akahoya (K-Ah) caldera and Aira-Tanzawa (AT) caldera from south Kyushu, which have been largely responsible for widespread Holocene and late Pleistocene tephra covers in the Japan region (e.g., Machida, 1999). The youngest major eruption of K-Ah caldera was centered at 6.3 ^{14}C kyr or 7.3 cal. kyr BP, while the cataclysmic eruption of AT caldera occurred at 25–24 ^{14}C kyr, at the MIS 3/2 transition. In addition, some volcanic ash layers from the northern trough may represent the eruption of the Ulreung-Oki (U-Oki) caldera at 9.3 ^{14}C kyr or 10.7 cal. kyr BP, while the activities of the Pinatubo volcano in Luzon have been recorded in the sediments of the southern Okinawa Trough as well as from the SCS (Li & Chang, 2009). Due to the active hydrothermal field underneath the trough, however, the Pleistocene and older volcanic depositions remain poorly known.

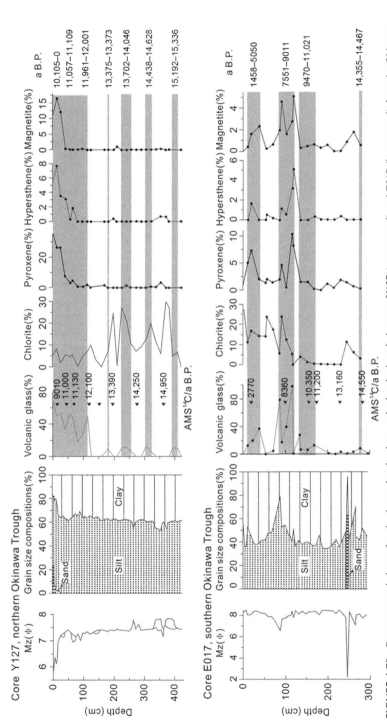

FIGURE 4.72 Downcore variations of grain size, volcanic glass, and related minerals in cores Y127 (upper panel) and E017 (lower panel) from the Okinawa Trough, showing characteristics of tephra layers. *Modified from Jiang et al. (2010).*

During IODP Expedition 331 in 2010, 24 holes at 5 sites were drilled on the Iheya North hydrothermal field in the middle Okinawa Trough to study deep hot biosphere (Expedition 331 Scientists, 2010). Drilling at the top of the active hydrothermal mound failed to recover core, and drilling at the base of the mound yielded only 2.1 m of core with sphalerite-rich black ore from 45 m of penetration. The other four sites yielded interbedded hemipelagic and volcaniclastic sediment and volcanogenic breccias and pumice that are variably hydrothermally altered and mineralized, in the zeolite to greenschist facies. Shipboard analyses of interstitial water and headspace gas reveal complex patterns with depth and laterally at most sites over distances of only a few meters. Apparent hydrothermal processes include formation of brines and vapor-rich fluids by phase separation and segregation, uptake of Mg and Na by alteration minerals in exchange for Ca, leaching of K at high temperature and uptake at low temperature, anhydrite precipitation, microbial oxidation of organic matter and anaerobic oxidation of methane-utilizing sulfate, microbial methanogenesis, abrupt changes in composition with depth that result from sealing by relatively impermeable caprock, and generation of hydrogen at depth. However, the drilling has not confirmed the presence of an active deep hot biosphere, although there is ample evidence for microbial activity supported by sedimentary organic matter in sediments within the upper 10–30 mbsf where temperatures were relatively low (Expedition 331 Scientists, 2010).

4.6.2 South China Sea Volcanic Deposition

Igneous rocks are distributed in many regions surrounding the SCS (Figure 4.73). Volcanism along the Luzon Arc since the middle Miocene is most important, as andesite and basalt younger than 10 Ma are well exposed from southeast Taiwan to Luzon. In central and southern Indochina Peninsula, Neogene eruptions produced a large volcanic rock province. The Indochina basalts mostly comprise shield-building tholeiites capped by small-volume undersaturated types, the latter often bearing mantle xenoliths and "exotic" xenocrysts such as sapphire and zircon. The extrusion was accommodated by left-lateral strike-slip shearing on the Ailao Shan–Red River Fault, coeval with seafloor spreading in the South China Sea. Most of the Indochina basalts first appeared at about 17 Ma, approximating with the cessation of both continental extrusion and seafloor spreading (Hoang et al., 2013). Basalt rocks of smaller scales also occur in South China (mainly Paleogene ages) and in Hainan Island (mainly Pliocene–Quaternary).

In the SCS, volcanic rocks are widely distributed (Figure 4.73A). Although volcanic sediments comprise only ∼1% of the total deposition, high proportion up to 30% has been recorded locally from the eastern slope. In the northern SCS, the Paleocene–Eocene intermediate–acidic extrusives and Oligocene–Miocene basalt and intermediate eruptives to ∼400 m thick with large amounts of tuff and breccia occur in the Pearl River Mouth Basin (Yan et al., 2006). In the central SCS, a chain of high seamounts along the spreading ridge rising up to 4000 m above the abyssal plain are found to consist of basalt with ages ranging from 14 to 3.5 Ma (Yan et al., 2006). In the southern SCS, early Miocene rhyolite and Pleistocene basalt on the northwest Palawan shelf and young volcanoes covered by thin Quaternary deposits on high seamounts from the Nansha region have been reported (Liu, Huang, et al., 2009; Liu, Zhao, et al., 2009; Yan et al., 2006, and references).

As seen in Figure 4.73B, the distribution of volcanic ash shows high concentration in the eastern SCS, indicating a main source from Luzon volcanoes. The characteristic westward ash lobe has resulted obviously from eruption of Mount Pinatubo in June 1991. Of the total eruption of 4.8–6.0 km^3 dense-rock equivalent ash from Pinatubo, more than 80% were deposited in the SCS (Wiesner, Wetzel, Catane, Listanco, & Mirabueno, 2004), subsequently stimulating the bloom of such biota as radiolarians especially in the eastern area (Figure 4.62B).

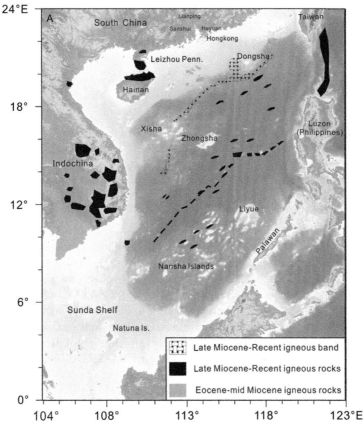

FIGURE 4.73—CONT'D

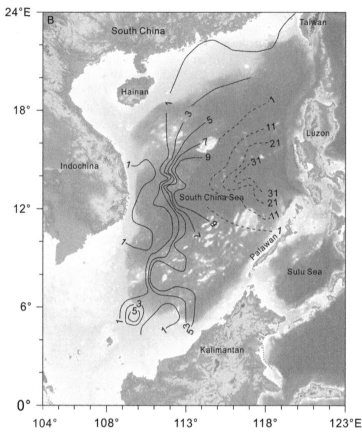

FIGURE 4.73 (A) Distribution of igneous rocks in and surrounding the South China Sea (Yan, Deng, Liu, Zhang, & Jiang, 2006). (B) Distribution of volcanic ash in the SCS, with solid lines indicating the ash weight percentage in the 63–125 mm fraction and dashed lines the volcanic clasts grain percentage in total components (Chen, Huang, et al., 2005; Chen, Xia, et al., 2005).

Tephra layers of Pleistocene age have been found throughout the SCS (Liu et al., 2006; Liu, Huang, et al., 2009; Liu, Zhao, et al., 2009). For example, more than 18 distinct, well-defined ash-rich layers with thickness varying from 0.5 to 10 cm were recognized in core MD97-2142 (12°41.133′N, 119°27.9′E, water depth 1557 m) from offshore Palawan (Figure 4.74). Consisting of vitric, microlite-bearing, and crystalline components, these ash layers were deposited at 10, 26–42, 64–79, 110–290, 390–430, and 788 kyr BP (Wei, Lee, & Shipboard Scientific Party of IMAGES III/MD106-IPHIS Cruise (Leg II), 1998).

All Pleistocene volcanic ash layers found in ODP Leg 184 cores are thin, generally <5 cm, and are light-colored, reflecting a dominant dacitic–rhyolitic composition of the arc's explosive fraction. Most of the ashes were deposited

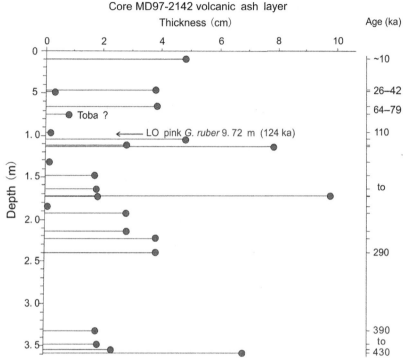

FIGURE 4.74   Positions, thickness, and estimated ages of ash layers in core MD97-2142 from the eastern SCS. LO, last occurrence datum. *Modified from Wei et al. (1998).*

since 1 Ma on the northern margin and since 2 Ma in the south. The uphole increase in volcanic ash may reflect either more volcanic eruptions during the Pleistocene or weaker diagenetic alteration of chemically unstable volcanic glass during burial. This latter explanation may account for much of the pattern on the northern margin, since older ashes in this area tend to be devoid of glass and are simply composed of angular quartz, mica, and other accessory mineral grains. In contrast, at Site 1143 in the southern SCS, fresh glass is found in Miocene-aged ash beds, indicating an older feature for the Philippine Arc (Wang et al., 2000).

Large-scale eruptions of the Toba volcano in northern Sumatra over the Quaternary have been widely recorded and discussed in the region (Lee, Chen, Wei, Iizuka, & Carey, 2004, and references). Early eruptions included the Haranggoal Dacite Tuff at 1.2 Ma, the oldest Toba tuff at 788 kyr BP, and the middle Toba tuff at 501 kyr BP. The youngest Toba eruption that expelled >2800 km^3 dense-rock equivalent (DRE) of rhyolitic magma occurred at ~74 kyr BP, near the MIS 5/4 boundary. Volcanic ash and glass shards from this late Pleistocene eruption cover the southern SCS up to ~12°N, including ODP Site 1143 (Bühring et al., 2000; Liu et al., 2006, and references).

4.7 SEDIMENTOLOGIC SUMMARY

4.7.1 Origin of Sediments

Over their evolutionary history, the China Seas have accumulated voluminous sediments of land and marine origins. Terrigenous sediments, mainly sand, silt, and clay, are widely distributed, showing a decreasing trend from coastal areas to the slope and deep sea, while biogenic sediments, mainly biotic skeletons, decrease from the deep-sea to shallow-water localities except in areas covered by coral reefs.

The contribution of terrigenous sediment to deposition in the China Seas is mainly achieved by river discharge. In the north, the mud-dominated Yellow River plays a controlling role on sedimentation of the Bohai and Yellow Seas at a rate of ~1080 Mt/yr, while the sand–silt-dominated Yangtze River on the ECS proper at a rate of ~420 Mt/yr (Figure 4.6) until the recent human disturbance (Liu, Yang, et al., 2014). Other rivers, especially those discharging into the western and northern Bohai Sea and those into the eastern Yellow Sea, also affect the regional deposition but often on a smaller scale (e.g., Yang & Li, 1999). However, the Okinawa Trough receives sediments not only from the Yangtze River but also from rivers of Taiwan (mainly in the southern trough) and the Yellow River (especially at glacial lowstands in the northern trough).

The SCS shows a more complicated scenario of sediment supply and deposition. The three major rivers control terrigenous sediment supply to the northern and southwestern SCS shelf areas, with the Pearl contributing ~70 Mt/yr to the northern shelf, the Red ~130 Mt/yr to the Tokin Gulf, and the Mekong ~160 Mt/yr to the Sunda Shelf, respectively (Figure 4.9). Rivers of Taiwan are effective in supplying sediments at a total rate of ~300 Mt/yr mainly to the Taiwan Strait, the northeastern SCS, and the southern Okinawa Trough, while rivers from other parts of the SCS periphery make important contributions to local deposition mostly in inner shelf areas.

The sediments of large rivers surrounding the China Seas have different mineral and elemental properties such that differentiation of sediment sources becomes possible. Compared to the Yangtze River sediment, for example, the sediments of the Yellow River origin are muddier, higher in calcite and smectite, and lower in dolomite and kaolinite (Figure 4.75; Table 4.1). A quite extraordinary Eu anomaly also characterizes REE variations in Yellow and Yangtze sediments, particularly when compared to those in Korean rivers with LREEs highly enriched (Figure 4.76; Xu, Lim, et al., 2009; Xu, Milliman, et al., 2009). Apparently, variations in the level of chlorite and monazite determine these REE patterns, as these two minerals are more abundant in Korean river sediments.

Strontium isotopic ratios can also serve as a useful diagnostic tool for tracing the provenance of sediments. For example, the muddy sediments of the Yangtze River's submerged delta have much lower $^{87}Sr/^{86}Sr$ ratios (0.7162–0.7180) relative to those of the Yellow River-derived Shandong

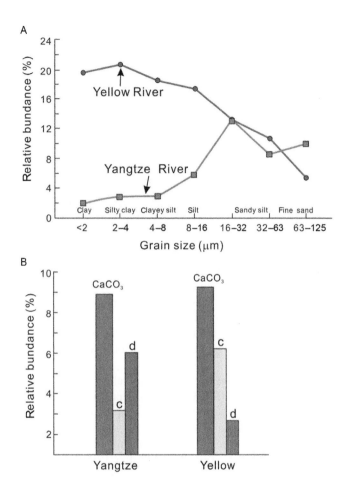

FIGURE 4.75   Abundance variations of detrital carbonate in sediments from the Yellow and Yangtze Rivers. (A) CaCO₃ in different grain sizes. (B) Relative abundance of calcite (c) and dolomite (d). *Modified from Yang et al. (2009).*

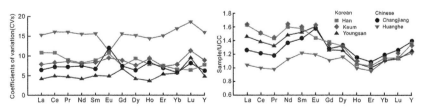

FIGURE 4.76   Coefficients of REE variations (left) and UCC-normalized REE distribution in Chinese and Korean river sediments (right), showing a characteristic positive Eu anomaly especially for the Yellow (Huanghe) River (Xu, Lim, et al., 2009; Xu, Milliman, et al., 2009).

Peninsula mud wedge (0.7216–0.7249). The ^{87}Sr/^{86}Sr ratios of the muddy sediments from the ESC outer shelf, or southwest Chiju Island Mud, are in a wide range between 0.7169 and 0.7216, suggesting mixed sources at least from the Yellow and Yangtze Rivers (Figure 4.77; Youn et al., 2007).

Reconstruction of sediment dispersal and circulation patterns in the SCS and the adjacent part of the ECS has been increasingly relying on the characteristic clay mineral assemblages (Figure 4.2). For sediments in the Taiwan Strait, clay mineral results indicate a main source from the Yangtze River although the contribution of western Taiwan rivers (mainly the Choshui River) at a rate of ∼100 Mt/yr (Liu, Liu, et al., 2008) is more important for the eastern strait (Figure 4.78; Huh et al., 2011; Xu, Lim, et al., 2009; Xu, Milliman, et al., 2009). Rivers in Southern Taiwan, such as the Gaoping River, contribute illite- and chlorite-rich sediments into the NE SCS deep sea at a rate of ∼50 Mt/yr (Liu, Liu, et al., 2008), where they are mixed with smectite-rich sediments from Luzon (Liu, Colin, et al., 2010; Liu et al., 2011). While the Pearl and the Red supply the kaolinite-rich sediments to the northern SCS shelf and the Tonkin Gulf, the Mekong delivers illite- and chlorite-rich sediments mainly to the coastal area outside the river estuary. As indicated by a gradual westward decrease in the illite + chlorite abundance (Figure 4.79), the Mekong's influence on the deposition in the Gulf of Thailand weakens considerably in a region prevailed by local muds with high smectite.

A change in relative abundance of clay minerals at 12.5 kyr was recorded in core KNG5 (19°55.17′N, 115°8.53′E, water depth 1085 m) from the northern SCS slope (Huang et al., 2011). Older samples register 70–90% illite + chlorite, 10–30% kaolinite, and <5% smectite, but younger samples have 85–95%, 5–10%, and <10%, respectively, indicating a provenance shift from Pearl River before 12.5 kyr to Taiwan rivers after 12.5 kyr. The results also imply that the Pearl River plume (see Chapter 2) was stronger before 12.5 kyr BP and the modern circulation in the northern SCS did not establish until that time (Huang et al., 2011).

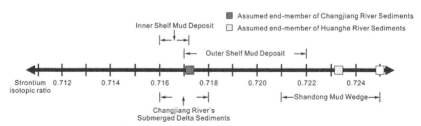

FIGURE 4.77 Ranges of ^{87}Sr/^{86}Sr ratios in carbonate-free fraction finer than 4Φ grain size from the Yangtze (Changjiang) River's submerged delta, Shandong mud wedge, ECS inner shelf mud deposit, and outer shelf mud deposit. *Modified from Youn, Yang, and Park (2007).*

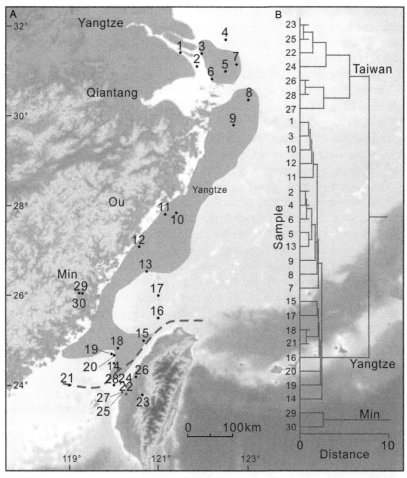

FIGURE 4.78 Clustering clay mineral assemblages in sediment samples along the ECS shelf and from the Taiwan Strait for provenance divisions: Yangtze, Taiwan (Choshui), and Min Rivers. *Modified from Xu, Milliman, et al. (2009) and Xu, Lim, et al. (2009).*

4.7.2 Sediment Transport Dynamics

The sediment distribution and deposition patterns of the China Seas reflect the interplay between sediment input from the surrounding landmass and marine production and the hydrodynamic system in response to sea-level fluctuations. Sediment input from the surrounding land was largely achieved by rivers discharge. The five large rivers, the Yellow, Yangtze, Pearl, Red, and Mekong, together discharge ~1900 Mt of sediment annually, while rivers in Taiwan supply further 300 Mt, and many smaller surrounding rivers also make their contributions. It is estimated that about half of this sediment load from rivers has been deposited in deltas and the other half in the China Seas. Delta evolution has been mainly affected by three driving mechanisms: (1) rising or lowering sea level that influences the available sediment accommodation space, (2) fluvial discharge as influenced by monsoon climate, and (3) human activities that alter landscape and sedimentation within the deltaic system.

The role of the hydrodynamic system (Chapter 2) is more important in sediment transport and deposition in shallow parts of the China Seas. Figure 4.80 provides a summary of the modern sediment transport routes, by emphasizing the near- and far-filed deposition of terrigenous sediment from major rivers, the near-field deposition of small rivers, and the main routes of those mixed terrigenous and marine sediments on shelf and slope.

The suspended load from the Yellow River is mainly transported by the westward alongshore current to the western and northwestern Bohai Sea due to a tidal standing wave node near the modern river mouth. The remaining sediment load was joined by those resuspended and carried eastward by coastal current along the Shandong Peninsula and then southward into the northwestern Yellow Sea, forming a far-field alongshore mud wedge. The resuspended sediment from the Old Yellow River mouth is carried by the southeastward Yellow Sea Cold Current toward the northern Okinawa Trough. Mixing with the Yangtze River sediment may occur along this flowing path. For sediments discharged from the Yangtze River, a large part is transported by the southwestward alongshore current and the remaining to

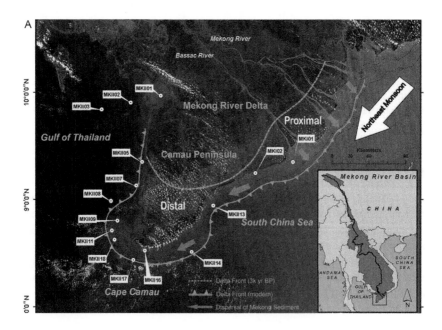

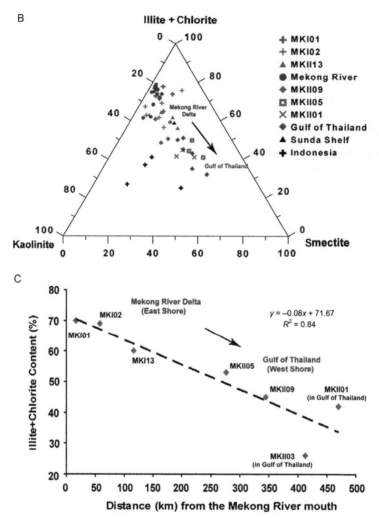

FIGURE 4.79 (A) The Mekong Delta showing site localities for clay mineral analyses. (B) Ternary plot of clay mineral assemblages from site stations in the southwestern SCS. (C) Correlation between illite+chlorite percentage and site stations according to their distance from the Mekong mouth. *From Xue et al. (2014).*

the east and north as the load in summer plumes. The Yangtze plumes often mix with marine and resuspended sediments in their way to the southern Yellow Sea carried by the Yellow Sea Warm Current. In the greater ECS in general, a combined effect of tides and waves with a distinctive seasonality in sedimentary facies is characterized by tide-dominated mud deposition in summer and wave-dominated sand deposition and erosion in winter (e.g., Zhu & Chang, 2000). The complicated tidal wave system has produced

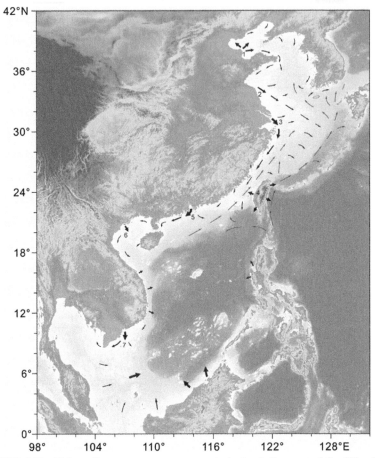

FIGURE 4.80 Major sediment transport routes mostly in the shallow part of the China Seas (based on various sources) for fluvial sediments by coastal currents (blue), for mixed marine and terrigenous sediments by warm summer currents on shelf (red), and for oceanic sediments by the Kuroshio Current (purple). 1, modern Yellow River sediment; 2, Old Yellow River Mouth sediment; 3, Yangtze River sediment; 4, sediments from Taiwan rivers; 5, Pearl River sediment; 6, Red River sediment; 7, Mekong sediment.

spectacular sand ridge fields on the shelf, while mud fields form in areas with relatively weak tidal wave activities particularly in the central Yellow Sea where the southward cold current and the northward warm current meet.

To the south, sediments derived from the Yangtze reach at least the Taiwan Strait and form a distal subaqueous delta before mixing with sediment from western Taiwan. Although most of the Taiwan sediment is confined to the eastern side of the Taiwan Strait, a significant part is carried by the Taiwan Warm Current flowing northward into the ECS and possibly also into the Okinawa Trough by side advection. Terrigenous input to the southern

Okinawa Trough, however, is largely from the northern and eastern Taiwan transported by the vigorous Kuroshio Current (e.g., Diekmann et al., 2008). Rivers in southern and eastern Taiwan empty their load directly into deep sea due to the steep gradient (Figure 4.81). This active sediment dispersal system often builds up large sediment wave fields with channel feeding turbidites, such as those found in the northeastern SCS slope (Figure 4.12).

Sediments discharged from rivers in the northern and western SCS are also largely affected by the southwestward flowing alongshore current, and farfield deposition in middle shelf and beyond occurs only with seasonal plumes affected by the summer monsoon. The Red River, for example, rapidly deposits most of its sediment (~100 Mt/yr) on a prograding delta front (van Maren & Hoekstra, 2005). During the wet season, the surface waters in the coastal zone are strongly stratified with a low density and high sediment concentration caused by low mixing rates of river plumes with ambient water. The river plume is advected to the south by a well-developed coastal current, which originates from the river plumes that enter the Gulf of Tonkin north of the river mouth and are deflected southward by the Coriolis force. Sediment predominantly leaves the surface plume by settling from suspension and less by mixing of fresh and marine water (van Maren & Hoekstra, 2005).

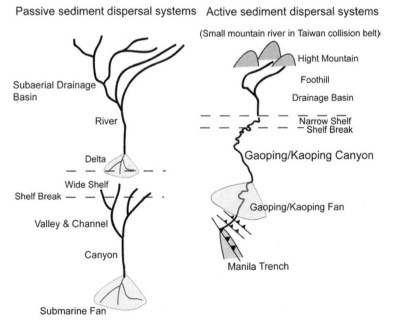

FIGURE 4.81 Comparison between the passive sediment dispersal system for most mainland rivers and the active sediment dispersal system for southern Taiwan rivers (Yu, Chiang, & Shen, 2009).

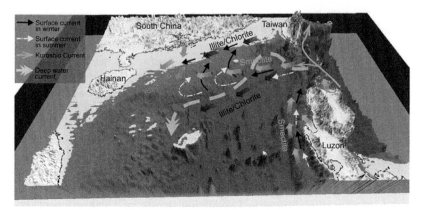

FIGURE 4.82   The sediment transport routes in the northern SCS as implied by the distribution of clay minerals (Liu, Colin, et al., 2010).

Sediment transport in the SCS is also affected by the wave-dominated hydrodynamic system and by a large deep-sea basin where abundant biogenic sediments have accumulated. While the incoming Kuroshio Current supports the proliferation of warm planktonic biota, the westward flow of the incoming deep Pacific water transports Taiwan- and Luzon-sourced sediments further into the northern SCS (Figure 4.82; e.g., Liu, Colin, et al., 2010; Liu et al., 2011). The mesoscale internal waves produced by inflood of the North Pacific Water also cause the development of large-scale sand dunes and sediment waves on the northern SCS upper slope.

REFERENCES

Bao, C. W. (1995). Buried ancient channels and deltas in the Zhujiang River Mouth shelf area. *Marine Geology and Quaternary Geology*, *15*(2), 25–34 (in Chinese).

Berne, S., Vagner, P., Guichard, F., Lericolais, G., Liu, Z. X., Trentesaux, A., et al. (2002). Pleistocene forced regressions and tidal sand ridges in the East China Sea. *Marine Geology*, *188*, 293–315.

Bian, C. W., Jiang, W. S., & Song, D. H. (2010). Terrigenous transportation to the Okinawa Trough and the influence of typhoons on suspended sediment concentration. *Continental Shelf Research*, *30*, 1189–1199.

Boulay, S., Colin, C., Trentesaux, A., Frank, N., & Liu, Z. (2005). Sediment sources and East Asian monsoon intensity over the last 450 ky: Mineralogical and geochemical investigations on South China Sea sediments. *Palaeogeography Palaeoclimatology Palaeoecology*, *228*, 260–277.

Bühring, C., Sarnthein, M., & Leg 184 Shipboard Scientific Party. (2000). Toba ash layers in the South China Sea: Evidence of contrasting wind directions during eruption ca. 74 ka. *Geology*, *28*, 275–278.

Cai, D., Shi, X., Zhou, W., Liu, W., Zhang, S., Cao, Y., et al. (2003). Sources and transportation of suspended matter and sediment in the southern Yellow Sea: Evidence from stable carbon isotopes. *Chinese Science Bulletin*, *48*(1-Suppl.), 21–29.

Chang, F. M., Zhuang, L. H., Li, T. G., Yan, J., Cao, Q. Y., & Cang, S. X. (2003). Radiolarian fauna in surface sediments of the northeastern East China Sea. *Marine Micropaleontology, 48,* 169–204.

Chen, G. D. (1991). *Modern sedimentation and modes in the Huanghe Delta.* Beijing: Geological Publishing House, 110 pp (in Chinese).

Chen, L. R. (2008). *Sedimentary mineralogy of the China Sea.* Beijing: China Ocean Press, 476 pp (in Chinese).

Chen, M. H., Cai, H. M., Tu, X., & Lu, L. H. (Eds.), (1996). *Late quaternary microbiotas and environment of the Nansha Islands and Adjacent Sea. The multidisciplinary Oceanographic expedition team of Academia Sinica to the Nansha Islands.* Beijing: China Ocean Press, 210 pp (in Chinese).

Chen, D., Huang, Y., Yuan, X., & Cathles, L. M., III. (2005). Seep carbonates and preserved methane oxidizing archaea and sulfate reducing bacteria fossils suggest recent gas venting on the seafloor in the Northeastern South China Sea. *Marine and Petroleum Geology, 22*(5), 613–621.

Chen, Z. Y., Saito, Y., Hori, K., Zhao, Y. W., & Kitamura, A. (2003). Early Holocene mud-ridge formation in the Yangtze offshore, China: A tidal-controlled estuarine pattern and sea-level implications. *Marine Geology, 198,* 245–257.

Chen, M.-T., Shiau, L.-J., Yu, P.-S., Chiu, T.-C., Chen, Y.-G., & Wei, K.-Y. (2003). 500,0000-year records of carbonate, organic carbon, and foraminiferal sea-surface temperature from the southeastern South China Sea (near Palawan Island). *Palaeogeography, Palaeoclimatology, Palaeoecology, 197,* 113–131.

Chen, M. H., & Tan, Z. Y. (1996). *Radiolarians in sediments of the Central and Northern South China Sea.* Beijing: China Science Press, 271 pp (in Chinese).

Chen, Z., Xia, B., Yan, W., Chen, M., Yang, H., Gu, S., et al. (2005). Distribution, chemical characteristics and source area of volcanic glass in the South China Sea. *Acta Oceanologia Sinica, 27*(5), 73–81 (in Chinese).

Chen, G. D., & Xue, C. D. (1997). *Sedimentary geology of the Huanghe Delta.* Beijing: Geological Publishing House, 147 pp (in Chinese).

Chen, Z., Yan, W., Cheng, M., Wang, S., Lu, J., Zhang, F., et al. (2006). Discovery of seep carbonate nodules as new evidence for gas venting on the northern continental slope of South China Sea. *Chinese Science Bulletin, 51*(10), 1228–1237.

Chen, M. H., Zhang, L. L., Zhang, L. L., Xiang, R., & Lu, J. (2008). Distributions of radiolarian diversity and abundance in surface sediments of the South China Sea and their environmental implications. *Earth Science—Journal of China University of Geosciences, 33*(4), 431–442 (in Chinese).

Chen, Q. Q., & Zhu, Y. R. (2012). Holocene evolution of bottom sediment distribution on the continental shelves of the Bohai Sea, Yellow Sea and East China Sea. *Sedimentary Geology, 273–274,* 58–72.

Cheng, X. R. (1995). Calcareous nannofossils in surficial sediments of the South China Sea. In P. X. Wang (Ed.), *The South China Sea since 150 ka* (pp. 158–168). Shanghai: Tongji University Press (in Chinese).

Cheng, T. C., & Cheng, S. Y. (1960). The planktonic foraminifera of the Yellow Sea and the East China Sea. *Oceanologia et Limnologia Sinica, 3,* 125–156, (in Chinese).

Cheng, X. R., & Wang, P. X. (1997). Controlling factors of coccolith distribution in surface sediments of the China seas: Marginal sea nannofossil assemblages revisited. *Marine Micropaleontology, 32,* 155–172.

Chough, S. K., Kim, J. W., Lee, S. H., Shinn, Y. J., Jin, J. H., Suh, M. C., et al. (2002). High-resolution acoustic characteristics of epicontinental sea deposits, central–eastern Yellow Sea. *Marine Geology, 188,* 317–331.

Chough, S. K., Lee, H. J., Chun, S. S., & Shinn, Y. J. (2004). Depositional processes of late Quaternary sediments in the Yellow Sea: A review. *Geosciences Journal*, *8*(2), 211–264.

Chough, S. K., Lee, H. J., & Yoon, S. H. (2000). *Marine geology of Korean Seas*. Amsterdam: Elsevier, 313 pp.

Dickinson, J. A., Ware, K., Cosham, S., & Murphy, B. (2012). Slope failure and canyon development along the northern South China Sea margin. In Y. Yamada, K. Kawamura, K. Ikehara, Y. Ogawa, R. Urgeles, D. Mosher, J. Chaytor, & M. Strasser (Eds.), *Submarine mass movements and their consequences, advances in natural and technological hazards research*: Vol. 31. (pp. 223–232). Berlin: Springer Science + Business Media B.V.

Diekmann, B., Hofmann, J., Henrich, R., Fütterer, D. K., Röhl, U., & Wei, K.-Y. (2008). Detrital sediment supply in the southern Okinawa Trough and its relation to sea-level and Kuroshio dynamics during the late Quaternary. *Marine Geology*, *255*, 83–95.

Dou, Y. G., Yang, S. Y., Liu, Z. X., Clift, P. D., Shi, X. F., Yu, H., et al. (2010). Provenance discrimination of siliciclastic sediments in the middle Okinawa Trough since 30 ka: Constraints from rare earth element compositions. *Marine Geology*, *275*, 212–220.

Dou, Y. G., Yang, S. Y., Liu, Z. X., Shi, X. F., Li, J., Yu, H., et al. (2012). Sr–Nd isotopic constraints on terrigenous sediment provenances and Kuroshio Current variability in the Okinawa Trough during the late Quaternary. *Palaeogeography, Palaeoclimatology, Palaeoecology*, *365–366*, 38–47.

Du Bois, E. P. (1981). Review of principal hydrocarbon-bearings of the South China Sea. *Energy*, *6*(11), 1113–1140.

Duc, D. M., Nhuan, M. T., Ngoi, C. V., Tran Nghi, Tien, D. M., van Weering, Tj. C. E., et al. (2007). Sediment distribution and transport at the nearshore zone of the Red River delta, Northern Vietnam. *Journal of Asian Earth Sciences*, *29*, 565–588.

Expedition 331 Scientists. (2010). *Deep hot biosphere: IODP preliminary report, 331*. Texas: Integrated Ocean Drilling Program. http://dx.doi.org/10.2204/iodp.pr.331.2010.

Fan, D. D., Wang, Y. Y., & Wu, Y. J. (2012). Advances in provenance studies of Changjiang riverine sediments. *Advances in Earth Science*, *27*(5), 515–528 (in Chinese).

Feng, Z. Q., Astin, T., Corbett, P. W. M., & Westerman, R. (2008). Climate-driven cyclicity in the Quaternary of the Gulf of Thailand. *Earth Science Frontiers*, *15*(2), 20–25.

Funabiki, A., Haruyama, S., Van Quy, N., Van Hai, P., & Thai, D. H. (2007). Holocene delta plain development in the Song Hong (Red River) delta, Vietnam. *Journal of Asian Earth Sciences*, *30*, 518–529.

Ge, L., Jiang, S., Swennen, R., Yang, T., Yang, J., Wu, N., et al. (2010). Chemical environment of cold seep carbonate formation on the northern continental slope of South China Sea: Evidence from trace and rare earth element geochemistry. *Marine Geology*, *277*, 21–30.

Gong, C. L., Wang, Y. M., Peng, X. C., Li, W. G., Qiu, Y., & Xu, S. (2012). Sediment waves on the South China Sea Slope off southwestern Taiwan: Implications for the intrusion of the Northern Pacific Deep Water into the South China Sea. *Marine and Petroleum Geology*, *32*, 95–109.

Guo, X., Zhang, Y., Zhang, F., & Cao, Q. (2010). Characteristics and flux of settling particulate matter in neritic waters: The southern Yellow Sea and the East China Sea. *Deep-Sea Research II*, *57*, 1058–1063.

Hanebuth, T. J. J., Proske, U., Saito, Y., Nguyen, V. L., & Ta, T. K. O. (2012). Early growth stage of a large delta—Transformation from estuarine-platform to deltaic-progradational conditions (the northeastern Mekong River Delta, Vietnam). *Sedimentary Geology*, *261–262*, 108–119.

Hanebuth, T., Stattegger, K., & Grootes, P. M. (2000). Rapid flooding of the Sunda shelf—A Late-Glacial sea-level record. *Science*, *288*, 1033–1035.

Hanebuth, T. J. J., Voris, H. K., Yokoyama, Y., Okuno, J., & Saito, Y. (2011). Formation, fate, and implications of depocentres along the sedimentary pathway on the Sunda Shelf (Southeast Asia) over the past 140 ka. *Earth-Science Reviews, 104*, 92–110.

He, Q. X. (2006). *Marine sedimentary geology of China.* Beijing: China Ocean Press, 503p (in Chinese).

Hiscott, R. N. (2001). Depositional sequences controlled by high rates of sediment supply, sea-level variations, and growth faulting: The Quaternary Baram Delta of northwestern Borneo. *Marine Geology, 175*, 67–102.

Hoang, N., Flower, M. F. J., Chi, C. T., Xuan, P. T., Van Quya, H., & Son, T. T. (2013). Collision-induced basalt eruptions at Pleiku and Buon Mê Thuôt, south-central Viet Nam. *Journal of Geodynamics, 69*, 65–83.

Hongo, C. (2012). Holocene key coral species in the Northwest Pacific: Indicators of reef formation and reef ecosystem responses to global climate change and anthropogenic stresses in the near future. *Quaternary Science Reviews, 35*, 82–99.

Hongo, C., & Kayanne, H. (2010). Relationship between species diversity and reef growth in the Holocene at Ishigaki Island, Pacific Ocean. *Sedimentary Geology, 223*, 86–99.

Hongo, C., & Kayanne, H. (2011). Key species of hermatypic coral for reef formation in the northwest Pacific during Holocene sea-level change. *Marine Geology, 279*, 162–177.

Hori, K., & Saito, Y. (2007). An early Holocene sea-level jump and delta initiation. *Geophysical Research Letters, 34*, 1–5. http://dx.doi.org/10.1029/2007GL031029, L18401.

Hori, K., Saito, Y., Zhao, Q. H., Cheng, X. R., Wang, P. X., Sato, Y., et al. (2001a). Sedimentary facies of the tide-dominated paleo-Changjiang (Yangtze) estuary during the last transgression. *Marine Geology, 177*, 331–351.

Hori, K., Saito, Y., Zhao, Q. H., Cheng, X. R., Wang, P. X., Sato, Y., et al. (2001b). Sedimentary facies and Holocene progradation rates of the Changjiang (Yangtze) delta, China. *Geomorphology, 41*, 233–248.

Hori, K., Tanabe, S., Saito, Y., Haruyama, S., Nguyen, V., & Kitamura, A. (2004). Delta initiation and Holocene sea-level change: Example from the Song Hong (Red River) delta, Vietnam. *Sedimentary Geology, 164*, 237–249.

Hsu, S.-C., Lin, F.-J., Jeng, W.-L., Chung, Y.-C., Shaw, L.-M., & Hung, K.-W. (2004). Observed sediment fluxes in the southwesternmost Okinawa Trough enhanced by episodic events: Flood runoff from Taiwan rivers and large earthquakes. *Deep-Sea Research Part I, 51*, 979–997.

Hu, J. T., Li, S. Y., & Geng, B. X. (2011). Modeling the mass flux budgets of water and suspended sediments for the river network and estuary in the Pearl River Delta, China. *Journal of Marine Systems, 88*, 252–266.

Hu, L. M., Shi, X. F., Guo, Z. G., Wang, H. J., & Yang, Z. S. (2013). Sources, dispersal and preservation of sedimentary organic matter in the Yellow Sea: The importance of depositional hydrodynamic forcing. *Marine Geology, 335*, 52–63.

Hu, P. Z., & Wang, J. Z. (1996). *Tertiary organic reef in the Pearl River Mouth Basin, organic reef and oil of China.* (pp. 294–323). Beijing: China Ocean Press (in Chinese).

Hu, P. Z., & Xie, Y. X. (1987). Tertiary reef complexes and their relationship with hydrocarbon accumulation in Pearl River Mouth Basin. In G. P. Society (Ed.), *International conference on petroleum geology of the Northern Continental shelf of the South China Sea* (pp. 505–529).

Huang, C.-Y., Chiu, Y.-L., & Zhao, M. (2005). Core description and preliminary sedimentological study of Site 1202D, Leg 195, in southern Okinawa Trough. *Terrestrial, Atmospheric and Oceanic Sciences, 16*(1), 19–44.

Huang, H. Z., Tang, B. G., & Yang, W. D. (1996). *Sedimentary geology of the Changjiang Delta.* Beijing: Geological Publishing House (in Chinese).

Huang, J., Li, A., & Wan, S. (2011). Sensitive grain-size records of Holocene East Asian summer monsoon in sediments of northern South China Sea slope. *Quaternary Research, 75,* 734–744.

Huang, W., & Wang, P. (2006). Sediment mass and distribution in the South China Sea since the Oligocene. *Science China (D), 49*(11), 1147–1155.

Huang, W., & Wang, P. (2007). Accumulation rate characteristics of deep-water sedimentation in the South China Sea during the last glaciation and the Holocene. *Acta Oceanologia Sinica, 29*(5), 69–73 (in Chinese).

Huh, C.-A., Su, C.-C., Liang, W.-T., & Ling, C.-Y. (2004). Linkages between turbidites in the southern Okinawa Trough and submarine earthquakes. *Geophysical Research Letters, 31,* L12304. http://dx.doi.org/10.1029/2004GL019731.

Huh, C.-A., Su, C.-C., Wang, C.-H., Lee, S.-Y., & Lin, I.-T. (2006). Sedimentation in the southern Okinawa Trough—Rates, turbidites and a sediment budget. *Marine Geology, 231,* 129–139.

Huh, C.-A., Chen, W., Hsu, F.-H., Chiu, J.-K., Lin, S., Liu, C.-S., et al. (2011). Modern (<100 years) sedimentation in the Taiwan Strait: Rates and source-to-sink pathways elucidated from radionuclides and particle size distribution. *Continental Shelf Research, 31,* 47–63.

Hung, J.-J., Wang, S.-M., & Chen, Y.-L. (2007). Biogeochemical controls on distributions and fluxes of dissolved and particulate organic carbon in the Northern South China Sea. *Deep-Sea Research. Part II, Topical Studies in Oceanography, 54,* 1486–1503.

Hutchison, C. S. (2004). Marginal basin evolution: The southern South China Sea. *Marine and Petroleum Geology, 21,* 1129–1148.

Hutchison, E. S. (2005). *Geology of North West Borneo.* Amsterdam: Elsevier, 421 pp.

Hutchison, C. S. (2010). The North-West Borneo Trough. *Marine Geology, 271,* 32–43.

Jian, Z. M., & Wang, L. J. (1995). Deep water masses of the South China Sea and West Pacific. In P. X. Wang (Ed.), *The South China Sea since 150 ka* (pp. 83–96). Shanghai: Tongji University Press (in Chinese).

Jian, Z. M., & Wang, L. J. (1997). Late quaternary benthic foraminifera and deep-water paleoceanography in the South China Sea. *Marine Micropaleontology, 32,* 127–154.

Jian, Z. M., Wang, P. X., Saito, Y., Wang, J. L., Pflaumann, U., Oba, T., et al. (2000). Holocene variability of the Kuroshio Current in the Okinawa Trough, northwestern Pacific Ocean. *Earth and Planetary Science Letters, 184,* 305–319.

Jian, Z. M., Zhao, Q. H., Cheng, X. R., Wang, J. L., Wang, P. X., & Su, X. (2003). Pliocene-Pleistocene stable isotope and paleoceanographic changes in the northern South China Sea. *Palaeogeography, Palaeoclimatology, Palaeoecology, 193,* 425–442.

Jian, Z. M., & Zheng, L. F. (1995). Deep-sea benthic foraminifera in surficial sediments of the South China Sea. In P. X. Wang (Ed.), *The South China Sea since 150 ka* (pp. 148–158). Shanghai: Tongji University Press (in Chinese).

Jiang, F. Q., Li, A. C., & Li, T. G. (2010). Sedimentary response to volcanic activity in the Okinawa Trough since the last deglaciation. *Chinese Journal of Oceanology and Limnology, 28,* 171–182.

Jiang, H., Zheng, Y. L., Ran, L. H., & Seidenkrantz, M.-S. (2004). Diatoms from the surface sediments of the South China Sea and their relationships to modern hydrography. *Marine Micropaleontology, 53,* 279–292.

Kao, S.-J., Shiah, F.-K., Wang, C.-H., & Liu, K.-K. (2006). Efficient trapping of organic carbon in sediments on the continental margin with high fluvial sediment input off southwestern Taiwan. *Continental Shelf Research, 26,* 2520–2537.

Katayama, H., & Watanabe, Y. (2003). The Huanghe and Changjiang contribution to seasonal variability in terrigenous particulate load to the Okinawa Trough. *Deep-Sea Research. Part II, Topical Studies in Oceanography, 50,* 475–485.

Kuang, Z. G., Zhong, G. F., Wang, L. L., & Guo, Y. Q. (2014). Channel-related sediment waves on the eastern slope offshore Dongsha Islands, northern South China Sea. *Journal of Asian Earth Sciences, 79A,* 540–551.

Lambiase, J. J., Abdul Rahim, bin A.A., & Cheong, Y. P. (2002). Facies distribution and sedimentary processes on the modern Baram Delta, implications for the reservoir sandstones of NW Borneo. *Marine and Petroleum Geology, 19,* 69–78.

Lan, X. H., Wang, H. X., Li, R. H., Lin, Z. H., & Zhang, Z. X. (2007). Major elements composition and provenance analysis in the sediments of the South Yellow Sea. *Earth Science Frontiers, 14*(4), 197–203 (in Chinese).

Lee, M.-Y., Chen, C.-H., Wei, K.-Y., Iizuka, Y., & Carey, S. (2004). First Toba supereruption revival. *Geology, 32,* 61–64.

Li, W. Q. (1991). The distributive law of calcareous nannofossil in the surficial sediments from the Huanghai Sea and the Bohai Sea and flow way of warm current in the Huanghai Sea. *Journal of Oceanography of Huanghai & Bohai Seas, 9*(1), 7–11 (in Chinese).

Li, J. B. (Ed.), (2008). *Regional geology of East China Sea.* Beijing: China Ocean Press, 631 pp (in Chinese).

Li, J. B. (Ed.), (2012). *Regional oceanography of China Seas—Marine geology.* Beijing: China Ocean Press, 547 pp (in Chinese).

Li, T. G., & Chang, M. F. (2009). *Paleoceanography in the Okinawa Trough.* Beijing: China Ocean Press, 259 pp (in Chinese).

Li, C. X., Chen, Q., Zhang, J., Yang, S., & Fan, D. (2000). Stratigraphy and paleoenvironmental changes in the Yangtze Delta during the late Quaternary. *Journal of Asian Earth Sciences, 18,* 453–469.

Li, C. X., Guo, X., Xu, S., Wang, J., & Li, P. (1979). The characteristics and distribution of Holocene sand bodies in Changjiang, River Delta area. *Acta Oceanologica Sinica, 1,* 252–268 (in Chinese).

Li, B. H., Jian, Z. M., & Wang, P. X. (1997). *Pulleniatina obliquiloculata* as a paleoceanographic indicator in the southern Okinawa Trough during the last 20,000 years. *Marine Micropaleontology, 32,* 59–69.

Li, D. L., Jiang, H., Li, T. G., & Zhao, M. X. (2011). Late Holocene paleoenvironmental changes in the southern Okinawa Trough inferred from a diatom record. *Chinese Science Bulletin, 56,* 1131–1138.

Li, C. X., & Li, P. (1983). The characteristics and distribution of Holocene sand bodies in the Changjiang delta area. *Acta Oceanologica Sinica, 2,* 54–66.

Li, C. X., Li, P., & Cheng, X. (1983). The influence of marine factors on sedimentary characteristics of Yangtze River channel below Zhenjiang. *Acta Geographica Sinica, 38,* 128–140 (in Chinese).

Li, C. X., & Wang, P. X. (1998). *Studies on late quaternary stratigraphy of Changjiang Estuary.* Beijing: China Science Press, 222 pp (in Chinese).

Li, X. J., Wang, P. X., Liao, Z. L., Chen, F., & Huo, Z. H. (2008). Distribution of clastic minerals of surface sediments in the western China Sea and their provenance. *Geology in China, 35*(1), 123–130 (in Chinese).

Li, C. X., Wang, P., Sun, H. P., Zhang, J. Q., Fan, D. D., & Deng, B. (2002). Late Quaternary incised-valley fill of the Yangtze delta (China): Its stratigraphic framework and evolution. *Sedimentary Geology, 152,* 133–158.

Li, Q. Y. (1985). Foraminiferal assemblages and distribution in coral reef regions of Hainan Island. In *Collected Papers of Micropaleontology* (pp. 27–35). Beijing: Science Press.

Li, Q. Y., Wang, P. X., Zhao, Q. H., Shao, L., Zhong, G. F., Tian, J., et al. (2006). A 33 Ma lithostratigraphic record of tectonic and paleoceanographic evolution of the South China Sea. *Marine Geology, 230*, 217–235.

Li, H., Wang, Y. M., Zhu, W. L., Xu, Q., He, Y. B., Tang, W., et al. (2013). Seismic characteristics and processes of the Plio-Quaternary unidirectionally migrating channels and contourites in the northern slope of the South China Sea. *Marine and Petroleum Geology, 43*, 370–380.

Li, C. X., Zhang, J. Q., Fan, D. D., & Deng, B. (2001). Holocene regression and the tidal radial sand ridge system formation in the Jiangsu coastal zone, east China. *Marine Geology, 173*, 97–120.

Li, Q. Y., Zhao, Q. H., Zhong, G. F., Jian, Z. M., Tian, J., Cheng, X. R., et al. (2007). Deep water ventilation and stratification in the Neogene South China Sea. *Journal of China University of Geosciences, 18*(2), 95–108.

Li, Q. Y., Zhong, G. F., & Tian, J. (2009). Stratigraphy and sea level changes. In P. X. Wang & Q. Y. Li (Eds.), *The South China Sea–Paleoceanography and Sedimentology* (pp. 75–170). Dordrecht: Springer.

Liao, H.-R., Yu, H.-S., & Su, C.-C. (2008). Morphology and sedimentation of sand bodies in the tidal shelf sea of eastern Taiwan Strait. *Marine Geology, 248*, 161–178.

Lim, D. I., Choi, J. Y., Jung, H. S., Rho, K. C., & Ahn, K. S. (2007). Recent sediment accumulation and origin of shelf mud deposits in the Yellow and East China Seas. *Progress in Oceanography, 73*, 145–159.

Lim, D. I., Jung, H. S., Choi, J. Y., Yang, S., & Ahn, K. S. (2006). Geochemical compositions of river and shelf sediments in the Yellow Sea: Grain-size normalization and sediment provenance. *Continental Shelf Research, 26*, 15–24.

Lin, S., Hsieh, I.-J., Huang, K.-M., & Wang, C.-H. (2002). Influence of the Yangtze River and grain size on the spatial variations of heavy metals and organic carbon in the East China Sea continental shelf sediments. *Chemical Geology, 182*, 377–394.

Lin, Y.-S., Wei, K.-Y., Lin, I.-T., Yu, P.-S., Chiang, H.-W., Chen, C.-Y., et al. (2006). The Holocene *Pulleniatina* Minimum Event revisited: Geochemical and faunal evidence from the Okinawa Trough and upper reaches of the Kuroshio Current. *Marine Micropaleontology, 59*, 153–170.

Liu, F., Yang, Q. S., Chen, S. L., Luo, Z. F., Yuan, F., & Wang, R. T. (2014). Temporal and spatial variability of sediment flux into the sea from the three largest rivers in China. *Journal of Asian Earth Sciences, 87*, 102–115.

Liu, X. Q. (1996). Sedimentary division in marginal seas of China. *Marine Geology and Quaternary Geology, 16*(3), 1–11 (in Chinese).

Liu, Z. X. (1997). Yangtze Shoal—A modern tidal sand sheet in the northwestern part of the East China Sea. *Marine Geology, 137*, 321–330.

Liu, Z. X., Berne, S., Saito, Y., Lericolais, G., & Marsset, T. (2000). Quaternary seismic stratigraphy and paleoenvironments on the continental shelf of the East China Sea. *Journal of Asian Earth Sciences, 18*, 441–452.

Liu, Z. X., Berné, S., Saito, Y., Yu, H., Trentesaux, A., Uehara, K., et al. (2007). Internal architecture and mobility of tidal sand ridges in the East China Sea. *Continental Shelf Research, 27*, 1820–1834.

Liu, J. G., Chen, M. H., Chen, Z., & Yan, W. (2010). Clay mineral distribution in surface sediments of the South China Sea and its significance for sediment sources and transport. *Chinese Journal of Oceanology and Limnology, 28*, 407–415.

Liu, Z. F., Colin, C., Huang, W., Le, K. P., Tong, S., Chen, Z., et al. (2007). Climatic and tectonic controls on weathering in South China and the Indochina Peninsula: Clay mineralogical and geochemical investigations from the Pearl, Red, and Mekong drainage basins. *Geochemistry, Geophysics, Geosystems, 8.* http://dx.doi.org/10.1029/2006GC001490, Q05005.

Liu, Z. F., Colin, C., Li, X. J., Zhao, Y. L., Tuo, S. T., Chen, Z., et al. (2010). Clay mineral distribution in surface sediments of the northeastern South China Sea and surrounding fluvial drainage basins: Source and transport. *Marine Geology, 277,* 48–60.

Liu, Z. F., Colin, C., & Trentesaux, A. (2006). Major element geochemistry of glass shards and minerals of the Youngest Toba Tephra in the southwestern South China Sea. *Journal of Asian Earth Sciences, 27,* 99–107.

Liu, Z. F., Colin, C., Trentesaux, A., Blamart, D., Bassinot, F., Siani, G., et al. (2004). Erosional history of the eastern Tibetan Plateau over the past 190 kyr: Clay mineralogical and geochemical investigations from the southwestern South China Sea. *Marine Geology, 209,* 1–18.

Liu, Z. F., Huang, W., Li, J. R., Wang, P. X., Wang, R. J., Yu, K. F., et al. (2009). Sedimentology. In P. X. Wang & Q. Y. Li (Eds.), *The South China Sea—Paleoceanography and sedimentology* (pp. 171–295). Dordrecht: Springer.

Liu, J., Kong, X., Saito, Y., Liu, J. P., Yang, Z., & Wen, C. (2013). Subaqueous deltaic formation of the Old Yellow River (AD 1128–1855) on the western South Yellow Sea. *Marine Geology, 344,* 19–33.

Liu, J. P., Liu, C. S., Xu, K. H., Milliman, J. D., Chiu, J. K., Kao, S. J., et al. (2008). Flux and fate of small mountainous rivers derived sediments into the Taiwan Strait. *Marine Geology, 256,* 65–76.

Liu, J. P., Milliman, J. D., & Gao, S. (2002). The Shandong mud wedge and post-glacial sediment accumulation in the Yellow Sea. *Geo-Marine Letters, 21,* 212–218.

Liu, J. P., Milliman, J. D., Gao, S., & Cheng, P. (2004). Holocene development of the Yellow River's subaqueous delta, North Yellow Sea. *Marine Geology, 209,* 45–67.

Liu, J., Saito, Y., Kong, X. H., Wang, H., Wen, C., Yang, Z. G., et al. (2010). Delta development and channel incision during marine isotope stages 3 and 2 in the western South Yellow Sea. *Marine Geology, 278,* 54–76.

Liu, Z. F., Tuo, S. T., Colin, C., Liu, J. T., Huang, C.-Y., Selvaraj, K., et al. (2008). Detrital fine-grained sediment contribution from Taiwan to the northern South China Sea and its relation to regional ocean circulation. *Marine Geology, 255,* 149–155.

Liu, Z. F., Wang, H., Hantoro, W. S., Sathiamurthy, E., Colin, C., Zhao, Y. L., et al. (2012). Climatic and tectonic controls on chemical weathering in tropical Southeast Asia (Malay Peninsula, Borneo, and Sumatra). *Chemical Geology, 291,* 1–12.

Liu, C. L., Wang, P. X., Tian, J., & Cheng, X. R. (2008). Coccolith evidence for Quaternary nutricline variations in the southern South China Sea. *Marine Micropaleontology, 69,* 42–51.

Liu, M. H., Wu, S. Y., & Wang, Y. J. (1987). *Late quaternary sedimentology of the Yellow Sea.* Beijing: China Ocean Press, 431 pp (in Chinese).

Liu, Z. X., & Xia, D. X. (2004). *Tidal Sands in the China Seas.* Beijing: China Ocean Press, 222 p (in Chinese).

Liu, Z. X., Xia, D. X., Berne, S., Wang, K. Y., Marsset, T., Tang, Y. X., et al. (1998). Tidal deposition systems of China's continental shelf, with special reference to the eastern Bohai Sea. *Marine Geology, 145,* 225–253.

Liu, Z. X., Huang, Y. C., & Zhang, Q. N. (1989). Tidal current ridges in the southwestern Yellow Sea. *Journal of Sedimentary Petrology, 59,* 432–437.

Liu, J. G., Xiang, R., Chen, M. H., Chen, Z., Yan, W., & Liu, F. (2011). Influence of the Kuroshio Current intrusion on depositional environment in the northern South China Sea: Evidence from surface sediment records. *Marine Geology, 285,* 59–68.

Liu, J. P., Xu, K. H., Li, A. C., Milliman, J. D., Velozzi, D. M., Xiao, S. B., et al. (2007). Flux and
    fate of Yangtze River sediment delivered to the East China Sea. *Geomorphology*, *85*(3),
    208–224.

Liu, J. P., Xue, Z., Ross, K., Wang, H. J., Yang, Z. S., Li, A. C., et al. (2009). Fate of sediments
    delivered to the sea by Asian large rivers: Long-distance transport and formation of remote
    alongshore clinothems. *The Sedimentary Record*, *7*(4), 4–9.

Liu, Z. F., Zhao, Y. L., Colin, C., Siringan, F. P., & Wu, Q. (2009). Chemical weathering in
    Luzon, Philippines from clay mineralogy and major-element geochemistry of river sediments.
    Applied Geochemistry, *24*, 2195–2205.

Liu, Z. S., Zhao, H. T., Fan, S. Q., & Chen, S. Q. (2002). *Geology of the South China Sea*.
    Beijing: China Science Press, 502 pp (in Chinese).

Liu, Z. F., Zhao, Y. L., Li, J., & Colin, C. (2007). Late Quaternary clay minerals off Middle Viet-
    nam in the western South China Sea: Implications for source analysis and East Asian mon-
    soon evolution. *Science in China Series D-Earth Sciences*, *50*, 1674–1684.

Liu, J., Zhu, R. X., Li, T. G., Li, A. C., & Li, J. (2007). Sediment-magnetic signature of the mid-
    Holocene paleoenvironmental change in the central Okinawa Trough. *Marine Geology*, *239*,
    19–31.

Long, Y. Z. (1997). *Sedimentary geology of the Zhujiang Delta*. Beijing: Geological Publishing
    House, 165 pp (in Chinese).

Löwemark, L., Steinke, S., Wang, C.-H., Chen, M.-T., Müller, A., Shiau, L.-J., et al. (2009). New
    evidence for a glacioeustatic influence on deep water circulation, bottom water ventilation
    and primary productivity in the South China Sea. *Dynamics of Atmospheres and Oceans*,
    47, 138–153.

Lu, H., Chen, F., Liu, J., Liao, Z., Sun, X., & Su, X. (2006). Characteristics of authigenic carbon-
    ate chimneys in Shenhu area, northern South China Sea: Recorders of hydrocarbon-enriched
    fluid activity. *Geological Review*, *52*(3), 352–357 (in Chinese).

Lüdmann, T., Wong, H. K., & Wang, P. (2001). Plio-Quaternary sedimentation processes and
    neotectonics of the northern continental margin of the South China Sea. *Marine Geology*,
    172, 331–358.

Lyle, M. (2003). Neogene carbonate burial in the Pacific Ocean. *Paleoceanography*, *18*, 1059.
    http://dx.doi.org/10.1029/2002PA000777.

Machida, H. (1999). The stratigraphy, chronology and distribution of distal marker-tephras in and
    around Japan. *Global and Planetary Change*, *21*, 71–94.

Marsset, T., Xia, D., Berne, S., Liu, Z., Bourillet, J. F., & Wang, K. (1996). Stratigraphy and sed-
    imentary environments during the Late Quaternary, in the Eastern Bohai Sea (North China
    Platform). *Marine Geology*, *135*, 97–114.

Meng, X. W., Liu, Y. G., Du, D. W., & Shi, X. F. (2009). Terrestrial flux in sediments from the
    Okinawa Trough estimated using geochemical compositional data and its response to climate
    changes over the past 35 000 a. *Acta Oceanologica Sinica*, *28*, 47–54.

Miao, Q., & Thunell, R. C. (1993). Recent deep-sea benthic foraminiferal distribution in the South
    China Sea. *Marine Micropaleontology*, *22*, 1–32.

Nguyen, V. L., Ta, T. K. O., & Tateishi, M. (2000). Late Holocene depositional environments and
    coastal evolution of the Mekong River Delta, Southern Vietnam. *Journal of Asian Earth
    Sciences*, *18*(4), 427–439.

Ni, H. G., Lu, F. H., Luo, X. L., Tian, H. Y., & Zeng, E. Y. (2008). Riverine inputs of total
    organic carbon and suspended particulate matter from the Pearl River Delta to the coastal
    ocean off South China. *Marine Pollution Bulletin*, *56*, 1150–1157.

Oiwane, H., Tonai, S., Kiyokawa, S., Nakamura, Y., Suganuma, Y., & Tokuyama, H. (2011). Geomorphological development of the Goto Submarine Canyon, northeastern East China Sea. *Marine Geology, 288*, 49–60.

Park, S.-C., Han, H.-S., & Yoo, D.-G. (2003). Transgressive sand ridges on the mid-shelf of the southern sea of Korea (Korea Strait): Formation and development in high-energy environments. *Marine Geology, 193*, 1–18.

Polski, W. (1959). Foraminiferal biofacies off the North Asiatic coast. *Journal of Paleontology, 33*(4), 569–587.

Puchała, R. J., Porębski, S. J., Śliwiński, W. R., & August, C. J. (2011). Pleistocene to Holocene transition in the central basin of the Gulf of Thailand, based on geoacoustic survey and radiocarbon ages. *Marine Geology, 288*, 103–111.

Qiao, S. Q., Shi, X. F., Saito, Y., Li, X. Y., Yu, Y. G., Bai, Y. Z., et al. (2011). Sedimentary records of natural and artificial Huanghe (Yellow River) channel shifts during the Holocene in the southern Bohai Sea. *Continental Shelf Research, 31*(13), 1336–1342.

Qin, Y. S. (Ed.), (1985). *Geology of the Bohai Sea.* Beijing: China Science Press, 231 pp (in Chinese).

Qin, Y. S. (Ed.), (1989). *Geology of the Yellow Sea.* Beijing: China Ocean Press, 289 pp (in Chinese).

Qin, Y. S. (1994). Sedimentation in the northern China seas. In D. Zhou, Y. B. Liang, & C. K. Zheng (Eds.), *Oceanology of China Seas* (pp. 395–406). Dordrecht: Kluwer Academic Publishers.

Qin, Y. S., Zhao, Y. Y., & Chen, L. R. (1988). *Geology of the East China Sea.* Beijing: China Science Press, 263 pp (in Chinese).

Qiu, Y., Huang, Y. Y., & Zhong, H. X. (1999). Sedimentation and chronostratigraphy discussion of the western area since Late Pleistocene. In B. C. Yao & Y. Qiu (Eds.), *Geological tectonic characteristics and Cenozoic sedimentation of the Western South China Sea* (pp. 71–84). Beijing: Geological Publishing House, (in Chinese).

Qiu, Y., & Wang, Y.-M. (2001). Reefs and paleostructure and paleoenvironment in the South China Sea. *Marine Geology and Quaternary Geology, 21*, 65–73 (in Chinese).

Reeder, D. B., Ma, B. B., & Yang, Y. J. (2011). Very large subaqueous sand dunes on the upper continental slope in the South China Sea generated by episodic, shoaling deep-water internal solitary waves. *Marine Geology, 279*, 12–18.

Ren, M. E. (1994). *The three larger deltas of China.* Beijing: High Education Press, 332 p (in Chinese).

Rui, X. Q., Liu, C. L., Liang, D., & Zhao, M. X. (2011). Distribution of calcareous nannofossils in the surface sediments of the southern Yellow Sea. *Marine Geology and Quaternary Geology, 31*(5), 89–93, (in Chinese).

Saidova, Kh. M. (2007). Benthic foraminiferal assemblages of the South China Sea. *Oceanology, 47*(5), 653–659.

Saito, Y., Wei, H., Zhou, Y., Nishimura, A., Sato, Y., & Yokota, S. (2000). Delta progradation and chenier formation in the Huanghe (Yellow River) delta, China. *Journal of Asian Earth Sciences, 18*, 489–497.

Saito, Y., Yang, Z. S., & Hori, K. (2001). The Huanghe (Yellow River) and Changjiang (Yangtze River) deltas: A review on their characteristics, evolution and sediment discharge during the Holocene. *Geomorphology, 41*, 219–231.

Shi, W., Wang, M. H., Li, X. F., & Pichel, W. G. (2011). Ocean sand ridge signatures in the Bohai Sea observed by satellite ocean color and synthetic aperture radar measurements. *Remote Sensing of Environment, 115*, 1926–1934.

Shiau, L. J., Yu, P. S., Wei, K. Y., Yamamoto, M., Lee, T. Q., Yu, E. F., et al. (2008). Sea surface temperature, productivity, and terrestrial flux variations of the southeastern South China Sea over the past 800,000 years (IMAGES MD972142). *Terrestrial, Atmospheric and Oceanic Sciences, 19*(4), 363–376.

Shinn, Y. J., Chough, S. K., Kim, J. W., & Woo, J. (2007). Development of depositional systems in the southeastern Yellow Sea during the postglacial transgression. *Marine Geology, 239*, 59–82.

Shyu, J.-P., Chen, M.-P., Shieh, Y.-T., & Huang, C.-K. (2001). A Pleistocene paleoceanographic record from the north slope of the Spratly Islands, southern South China Sea. *Marine Micropaleontology, 42*, 61–93.

Song, J. M. (2010). *Biogeochemical processes of biogenic elements in China marginal seas.* Berlin: Zhejiang University Press, Hangzhou and Springer-Verlag, 662 pp.

Srisuksawad, K., Porntepkasemsan, B., Mouchpramool, S., Yamkate, P., Carpenter, R., Petersom, M. L., et al. (1997). Radionuclide activities, geochemistry, and accumulation rates of sediments in the Gulf of Thailand. *Continental Shelf Research, 17*, 925–965.

Staub, J. R., & Esterle, J. S. (1993). Provenance and sediment dispersal in the Rajang River delta/ coastal plain system, Sarawak, East Malaysia. *Sedimentology Geology, 85*, 191–201.

Staub, J. R., Among, H. L., & Gastaldo, R. A. (2000). Seasonal sediment transport and deposition in the Rajang River delta, Sarawak, East Malaysia. *Sedimentary Geology, 133*, 249–264.

Steuer, S., Franke, D., Meresse, F., Savva, D., Pubellier, M., Auxietre, J.-L., et al. (2013). Time constraints on the evolution of southern Palawan Island, Philippines from onshore and offshore correlation of Miocene limestones. *Journal of Asian Earth Sciences, 76*, 412–427.

Su, G. Q., Fan, S. Q., & Chen, S. M. (1989). *The sedimentary atlas of Northern and Central South China Sea.* Guangzhou: Guangdong Science and Technology Press, 68 pp (in Chinese).

Su, G. Q., & Wang, T. X. (1994). Basic characteristics of modern sedimentation in the South China Sea. In D. Zhou, Y. B. Liang, & C. K. Zheng (Eds.), *Oceanology of China Seas* (pp. 407–418). Dordrecht: Kluwer Academic Publishers.

Su, X., & Wei, K.-Y. (2005). Calcareous nannofossils and variation of the Kuroshio Current in the Okinawa Trough during the last 13,000 years. *Terrestrial, Atmospheric and Oceanic Sciences, 16*(1), 95–111.

Suess, E., Huang, Y., Wu, N., Han, X., & Su, X. (2005). *South China Sea continental margin: Geological methane budget and environmental effects of methane emissions and gas hydrates: RV SONNE cruise report SO 177, Sino-German cooperative project.* Kiel: IFM-GEOMAR.

Sun, D. Y., Tan, W. B., Pei, Y. D., Zhou, L. P., Wang, H., Yang, H., et al. (2011). Late Quaternary environmental change of Yellow River Basin: An organic geochemical record in Bohai Sea (North China). *Organic Geochemistry, 42*, 575–585.

Szarek, R., Kuhnt, W., Kawamura, H., & Kitazato, H. (2006). Distribution of recent benthic foraminifera on the Sunda Shelf (South China Sea). *Marine Micropaleontology, 61*, 171–195.

Szarek, R., Kuhnt, W., Kawamura, H., & Nishi, H. (2009). Distribution of recent benthic foraminifera along continental slope of the Sunda Shelf (South China Sea). *Marine Micropaleontology, 71*, 41–59.

Ta, T. K. O., Nguyen, V. L., Tateishi, M., Kobayashi, I., Saito, Y., & Nakamura, T. (2002). Sediment facies and Late Holocene progradation of the Mekong river delta in Bentre province, southern Vietnam: An example of evolution from a tide-dominated to a tide- and wave-dominated delta. *Sedimentary Geology, 152*(3–4), 313–325.

Ta, T. K. O., Nguyen, V. L., Tateishi, M., Kobayashi, I., Tanabe, S., & Saito, Y. (2002). Holocene delta evolution and sediment discharge of the Mekong River, southern Vietnam. *Quaternary Science Reviews, 21*(16–17), 1807–1819.

Tan, Z. Y., & Chen, M. H. (1999). *Offshore radiolarians in China.* Beijing: China Science Press, 404 pp (in Chinese).

Tanabe, S., Hori, K., Saito, Y., Haruyama, S., Vu, V. P., & Kitamura, A. (2003). Song Hong (Red River) delta evolution related to millennium-scale Holocene sea-level changes. *Quaternary Science Reviews, 22,* 2345–2361.

Tanabe, S., Saito, Y., Sato, Y., Suzuki, Y., Sinsakul, S., Tiyapairach, S., et al. (2003). Stratigraphy and Holocene evolution of the mud-dominated Chao Phraya delta, Thailand. *Quaternary Science Reviews, 22,* 789–807.

Tanabe, S., Saito, Y., Vu, Q. L., Hanebuth, T. J. J., Ngo, Q. L., & Kitamura, A. (2006). Holocene evolution of the Song Hong (Red River) delta system, northern Vietnam. *Sedimentary Geology, 187,* 29–61.

Tanaka, G., Komatsu, T., Saito, Y., Nguyen, D. P., & Vu, Q. L. (2011). Temporal changes in ostracod assemblages during the past 10,000 years associated with the evolution of the Red River delta system, northeastern Vietnam. *Marine Micropaleontology, 81,* 77–87.

Tang, B. G. (1996). Quaternary stratigraphy on the shelf of the East China Sea. In Z. G. Yang & H. M. Liu (Eds.), *Quaternary stratigraphy in China and its International Correlation* (pp. 56–75). Beijing: Geology Press (in Chinese).

Tang, C., Zhou, D., Endler, R., Lin, J. Q., & Harff, J. (2010). Sedimentary development of the Pearl River Estuary based on seismic stratigraphy. *Journal of Marine Systems, 82,* S3–S16.

Thunell, R., Miao, Q., Calvert, S., Calvert, S., & Pedersen, T. (1992). Glacial-Holocene biogenic sedimentation patterns in the South China Sea: productivity variations and surface water pCO$_2$. *Paleoceanography, 7,* 143–162.

Tjallingii, R., Stattegger, K., Wetzel, A., & Phung Van, P. (2010). Infilling and flooding of the Mekong River incised valley during deglacial sea-level rise. *Quaternary Science Reviews, 29,* 1432–1444.

Tong, H. P., Feng, D., Cheng, H., Yang, S. X., Wang, H. B., Min, A. G., et al. (2013). Authigenic carbonates from seeps on the northern continental slope of the South China Sea: New insights into fluid sources and geochronology. *Marine and Petroleum Geology, 43,* 260–271.

Tu, X., & Zheng, F. (1991). Foraminifera in surface sediments of the Nansha sea area. In The multidisciplinary oceanographic expedition team of academia Sinica to the Nansha Islands, Quaternary biological groups of the Nansha Islands and the neighboring waters (pp. 129–198). Guangzhou: Zhongshan University Publishing House.

Ujiié, H., Hatakeyama, Y., Gu, X. X., Yamamoto, S., Ishiwatari, R., & Maeda, L. (2001). Upward decrease of organic C/N ratios in the Okinawa Trough cores: Proxy for tracing the postglacial retreat of the continental shore line. *Palaeogeography, Palaeoclimatology, Palaeoecology, 165,* 129–140.

van Maren, D. S., & Hoekstra, P. (2005). Dispersal of suspended sediments in the turbid and highly stratified Red River plume. *Continental Shelf Research, 25,* 503–519.

Waller, H. O. (1960). Foraminiferal biofacies of the South China coast. *Journal of Paleontology, 34*(6), 1164–1182.

Wan, S. M., Li, A. C., Clift, P. D., Wu, S. G., Xu, K. H., & Li, T. G. (2010). Increased contribution of terrigenous supply from Taiwan to the northern South China Sea since 3 Ma. *Marine Geology, 278,* 115–121.

Wan, X. N., Li, J. F., & Shen, H. T. (2009). Distribution and fluxes of suspended sediments in the offshore waters of the Changjiang (Yangtze) Estuary. *Acta Oceanologica Sinica, 28*(4), 86–95.

Wan, S. M., Li, A. C., Xu, K. H., & Yin, X. M. (2008). Characteristics of clay minerals in the northern South China Sea and its implications for evolution of East Asian Monsoon since Miocene. *Journal of the China University of Geosciences, 19*(1), 95–108.

Wang, P. X. (Ed.), (1985). *Marine micropaleontology of China.* Beijing: China Ocean Press, 370 pp.

Wang, P. X. (Ed.), (1995). *The South China Sea since 150 ka.* Shanghai: Tongji University Press, 184 p (in Chinese).

Wang, P. X. (1999). Response of Western Pacific marginal seas to glacial cycles: Paleoceanographic and sedimentological features. *Marine Geology, 156,* 5–39.

Wang, Y. (Ed.), (2002). *Radial sand ridges on the Huanghai Shelf.* Beijing: China Environmental Science Press, 433 p (in Chinese).

Wang, R. J. (2009). Opal. In P. X. Wang & Q. Y. Li (Eds.), *The South China Sea—Paleoceanography and sedimentology* (pp. 217–229). Dordrecht: Springer.

Wang, R. J., & Abelmann, A. (2002). Radiolarian responses to paleoceanographic events of the southern South China Sea during the Pleistocene. *Marine Micropaleontology, 46,* 25–44.

Wang, Y. J., Chen, M. H., Lu, J., Xiang, R., Zhang, L. L., & Zhang, L. L. (2007). Distribution of calcareous nannofossils in surface sediments of South China Sea. *Journal of Tropical Oceanography, 26*(5), 26–34 (in Chinese).

Wang, P. X., & Cheng, X. R. (1988). Distribution of calcareous nannoplankton in bottom sediments of the East China Sea. *Acta Oceanologica Sinica, 10*(1), 76–85 (in Chinese).

Wang, P. X., & Li, Q. Y. (Eds.), (2009). *The South China sea paleoceanography and sedimentology.* Dordrecht: Springer, 506 pp.

Wang, P. X., Min, Q. B., & Bian, Y. H. (1985). Foraminiferal biofacies in the northern continental shelf of the South China Sea. In P. X. Wang (Ed.), *Marine micropaleontology of China* (pp. 151–175). Beijing: China Ocean Press.

Wang, P. X., Prell, W. L., & Blum, P. (Eds.), (2000). *Proceedings of ocean drilling program, initial reports, 184.* College Station, TX: Ocean Drilling Program.

Wang, H. J., Saito, Y., Zhang, Y., Bi, N. S., Sun, X. X., & Yang, Z. S. (2011). Recent changes of sediment flux to the western Pacific Ocean from major rivers in East and Southeast Asia. *Earth-Science Reviews, 108,* 80–100.

Wang, P. X., Tian, J., Cheng, X. R., Liu, C. L., & Xu, J. (2003). Carbon reservoir change preceded major ice-sheets expansion at Mid-Brunhes Event. *Geology, 31,* 239–242.

Wang, P. X., Tian, J., Cheng, X. R., Liu, C. L., & Xu, J. (2004). Major Pleistocene stages in a carbon perspective: The South China Sea record and its global comparison. *Paleoceanography, 19,* 1–16. http://dx.doi.org/10.1029/2003PA000991.

Wang, P. X., Zhang, J. J., Zhao, Q. H., Min, Q. B., Bian, Y. H., Zheng, L. F., et al. (1988). *Foraminifera and Ostracoda in bottom sediments of the East China Sea.* Beijing: China Ocean Press, 438 pp (in Chinese).

Wang, Y., Zhang, Y. Z., Zou, X. Q., Zhu, D. K., & Piper, D. (2012). The sand ridge field of the South Yellow Sea: Origin by river–sea interaction. *Marine Geology, 291–294,* 132–146.

Wei, K. Y., Lee, T. Q., & Shipboard Scientific Party of IMAGES III/MD106-IPHIS Cruise (Leg II). (1998). Late Pleistocene volcanic ash layers in core MD972142, offshore from northwestern Palawan, South China Sea: A preliminary report. *Terrestrial, Atmospheric and Oceanic Sciences, 9*(1), 143–152.

Wei, K.-Y., Mii, H., & Huang, C.-Y. (2005). Age model and oxygen isotope stratigraphy of Site ODP1202 in the southern Okinawa Trough, northwestern Pacific. *Terrestrial, Atmospheric and Oceanic Sciences, 16*(1), 1–17.

Wei, X., & Wu, C. Y. (2011). Holocene delta evolution and sequence stratigraphy of the Pearl River Delta in South China. *Science China D—Earth Science, 54,* 1523–1541.

Wetzel, A., & Unverricht (2013). A muddy megaturbidite in the deep central South China Sea deposited ~350 yrs BP. *Marine Geology, 346*, 91–100.

Wiesner, M. G., Wetzel, A., Catane, S. G., Listanco, E. L., & Mirabueno, H. T. (2004). Grain size, areal thickness distribution and controls on sedimentation of the 1991 Mount Pinatubo tephra layer in the South China Sea. *Bulletin of Volcanology, 66*, 226–242.

Woodroffe, C. D. (2000). Deltaic and estuarine environments and their Late Quaternary dynamics on the Sunda and Sahul shelves. *Journal of Asian Earth Sciences, 18*, 393–413.

Wu, Z. Y., Jin, X. L., Cao, Z. Y., Li, J. B., Zheng, Y. L., & Shang, J. H. (2010). Distribution, formation and evolution of sand ridges on the East China Sea shelf. *Science China D—Earth Science, 53*(1), 101–112.

Xiang, R., Li, T. G., Yang, Z. S., Li, A. C., Jiang, F. Q., Yan, J., et al. (2003). Geological records of marine environmental changes in the southern Okinawa Trough. *Chinese Science Bulletin, 48*(2), 194–199.

Xing, F., Wang, Y. P., & Wang, H. V. (2012). Tidal hydrodynamics and fine-grained sediment transport on the radial sand ridge system in the southern Yellow Sea. *Marine Geology, 291–294*, 192–210.

Xing, L., Zhang, H. L., Yuan, Z. N., Sun, Y., & Zhao, M. X. (2011). Terrestrial and marine biomarker estimates of organic matter sources and distributions in surface sediments from the East China Sea shelf. *Continental Shelf Research, 31*, 1106–1115.

Xu, J. (2004). *Quaternary planktonic foraminiferal assemblages in the southern South China Sea and paleoclimatic variations.* PhD thesis, Shanghai: Tongji Univ. (in Chinese).

Xu, K. H., Li, A. C., Liu, J. P., Milliman, J. D., Yang, Z. S., Liu, C. S., et al. (2012). Provenance, structure, and formation of the mud wedge along inner continental shelf of the East China Sea: A synthesis of the Yangtze dispersal system. *Marine Geology, 291–294*, 176–191.

Xu, Z. K., Lim, D., Choi, J. Y., Yang, S. Y., & Jung, H. (2009). Rare earth elements in bottom sediments of major rivers around the Yellow Sea: Implications for sediment provenance. *Geo-Marine Letters, 29*, 291–300.

Xu, D. Y., Liu, X. Q., Zhang, X. H., Li, T. G., & Chen, B. Y. (Eds.), (1997). *China offshore geology.* Beijing: Geological Publishing House, 310 pp (in Chinese)..

Xu, K. H., Milliman, J. D., Li, A. C., Liu, J. P., Kao, S. J., & Wan, S. M. (2009). Yangtze- and Taiwan-derived sediments in the inner shelf of East China Sea. *Continental Shelf Research, 29*, 2240–2256.

Xue, C. T. (1993). Historical changes in the Yellow River delta, China. *Marine Geology, 113*, 321–329.

Xue, Z., He, R. Y., Liu, J. P., & Warner, J. C. (2012). Modeling transport and deposition of the Mekong River sediment. *Continental Shelf Research, 37*, 66–78.

Xue, Z., Liu, J. P., DeMaster, D., Leithold, E. L., Wan, S. M., Ge, Q., et al. (2014). Sedimentary processes on the Mekong subaqueous delta: Clay mineral and geochemical analysis. *Journal of Asian Earth Sciences, 79A*, 520–528.

Xue, Z., Liu, J. P., DeMaster, D., Nguyen, V. L., & Ta, T. K. O. (2010). Late Holocene evolution of the Mekong subaqueous delta, southern Vietnam. *Marine Geology, 269*, 46–60.

Xue, C. T., Zhu, X. H., & Lin, H. M. (1995). Holocene sedimentary sequence, foraminifera and ostracoda in west coastal lowland of Bohai Sea, China. *Quaternary Science Reviews, 14*(5), 521–530.

Yan, P., Deng, H., Liu, H., Zhang, Z., & Jiang, Y. (2006). The temporal and spatial distribution of volcanism in the South China Sea region. *Journal of Asian Earth Sciences, 27*, 647–659.

Yan, Q. S., & Xu, S. Y. (1987). *Studies on modern sedimentology of the Changjiang delta.* Shanghai: East China Normal University Press.

Yang, W. D., Cui, Z. K., & Zhang, Y. B. (Eds.), (2010). *Geology and mineral resources of the East China Sea*. Beijing: China Ocean Press, 780 pp (in Chinese).

Yang, S. Y., Jung, H. S., Lim, D. I., & Li, C. X. (2003). A review on the provenance discrimination of sediments in the Yellow Sea. *Earth-Science Reviews, 63*(1), 93–120.

Yang, S. M., & Kou, X. Q. (1996). Quaternary stratigraphy of the northern South China Sea shelf. *HaiYangDiZhi (Marine Geology), 1996*(3), 14–27 (in Chinese).

Yang, S. Y., & Li, C. X. (1999). Characteristic element compositions of the Yangtze and the Yellow river sediments and their geological background. *Marine Geology and Quaternary Geology, 19*(2), 19–26 (in Chinese).

Yang, S. Y., Li, C. X., & Cai, J. G. (2006). Geochemical compositions of core sediments in eastern China: Implication for late Cenozoic palaeoenvironmental changes. *Palaeogeography, Palaeoclimatology, Palaeoecology, 229*, 287–302.

Yang, S. Y., Li, C. X., & Yokoyama, K. (2006). Elemental compositions and monazite age patterns of core sediments in the Changjiang Delta: Implications for sediment provenance and development history of the Changjiang River. *Earth and Planetary Science Letters, 245*, 762–776.

Yang, Z. G., & Lin, H. M. (1993). *Quaternary processes in eastern China and their international correlation*. Beijing: China Ocean Press, 125 pp (in Chinese).

Yang, Q. H., Lin, Z. H., Zhang, F. Y., Lin, X. T., & Ji, F. W. (2002). Mineral assemblage provinces for surficial sediments of the central-eastern South China Sea and their geological significance. *Oceanologia et Limnologia Sinica, 33*(6), 591–599 (in Chinese).

Yang, Z. S., & Liu, J. P. (2007). A unique Yellow River-derived distal subaqueous delta in the Yellow Sea. *Marine Geology, 240*, 169–176.

Yang, Z. S., Wang, H. C., & Qiao, S. Q. (2009). Carbonate mineral in estuary sediments of the Changjiang (Yangtze River) and Huanghe (Yellow River): The content, morphology and influential factors. *Oceanologia et Limnologia Sinica, 40*(6), 674–681 (in Chinese).

Yang, S. Y., & Youn, J.-S. (2007). Geochemical compositions and provenance discrimination of the central south Yellow Sea sediments. *Marine Geology, 243*, 229–241.

Yao, Y. J., Liu, H. L., Yang, C. P., Han, B., Tian, J. J., Yin, Z. X., et al. (2012). Characteristics and evolution of Cenozoic sediments in the Liyue Basin, SE South China Sea. *Journal of Asian Earth Sciences, 60*, 114–129.

Yi, L., Lai, Z., Yu, H. J., Xu, X. Y., Su, Q., Yao, J., et al. (2012). Chronologies of sedimentary changes in the south Bohai Sea, China: constraints from luminescence and radiocarbon dating. *Boreas*. http://dx.doi.org/10.1111/j.1502-3885.2012.00271.x, ISSN 0300-9483.

Yi, L., Yu, H. J., Ortiz, J. D., Xu, X. Y., Chen, S. L., Ge, J. Y., et al. (2012). Late Quaternary linkage of sedimentary records to three astronomical rhythms and the Asian monsoon, inferred from a coastal borehole in the south Bohai Sea, China. *Palaeogeography, Palaeoclimatology, Palaeoecology, 29-330*, 101–117.

Youn, J. S., Yang, S. Y., & Park, Y. A. (2007). Clay minerals and geochemistry of the bottom sediments in the northwestern East China Sea. *Chinese Journal of Oceanology and Limnology, 25*, 235–246.

Yu, K. F. (2012). Coral reefs in the South China Sea: Their response to and records on past environmental changes. *Science China Earth Sciences, 55*(8), 1217–1229.

Yu, H.-S., Chiang, C.-S., & Shen, S.-M. (2009). Tectonically active sediment dispersal system in SW Taiwan margin with emphasis on the Gaoping (Kaoping) Submarine Canyon. *Journal of Marine Systems, 76*, 369–382.

Yu, K. F., & Zhao, J. X. (2009). Coral reefs. In P. X. Wang & Q. Y. Li (Eds.), *The South China Sea—Paleoceanography and sedimentology* (pp. 229–254). Dordrecht: Springer.

Yun, C. X. (2004). *The basic law of the Changjiang estuary evolution.* Beijing: China Ocean Press, 290 pp (in Chinese).

Zhan, Y. F. (1987). A preliminary study on the distribution of diatom in surface sediments of the middle South China Sea. *Donghia Marine Science, 5*(1), 48–59 (in Chinese).

Zhang, M., He, Q., Ye, Z., Han, C., Li, H., Wu, J., et al. (1989). *Sedimentary geology of Xisha reef carbonates.* Beijing: China Science Press, 117 pp (in Chinese).

Zhao, H. T. (1990). *Evolution of the Zhujiang Estuary.* Beijing: China Ocean Press, 357 p (in Chinese).

Zhao, S. L. (1996). *Shelf desertization.* Beijing: China Ocean Press, 194 pp (in Chinese).

Zhao, M. X., Huang, C.-Y., & Wei, K.-Y. (2005). A 28,000 year U37K' sea surface temperature record of ODP Site 1202B, Southern Okinawa Trough. *Terrestrial, Atmospheric and Oceanic Sciences, 16*(1), 45–56.

Zhao, Q. H., Li, Q. Y., & Jian, Z. M. (2009). Deep waters and oceanic connection. In P. X. Wang & Q. Y. Li (Eds.), *The South China Sea—Paleoceanography and sedimentology* (pp. 395–437). Dordrecht: Springer.

Zhao, Q. H., & Wang, P. X. (1999). Progress in Quaternary paleoceanography of the South China Sea: a review. *Quaternary Science, 6,* 481–501 (in Chinese).

Zhao, B. C., Wang, Z. H., Chen, J., & Chen, J. Y. (2008). Marine sediment records and relative sea level change during late Pleistocene in the Changjiang delta area and adjacent continental shelf. *Quaternary International, 186,* 164–172.

Zhao, M. X., Wang, P. X., Tian, J., & Li, J. (2009). Biogeochemistry and the carbon reservoir. In P. X. Wang & Q. Y. Li (Eds.), *The South China Sea—Paleoceanography and sedimentology* (pp. 439–483). Dordrecht: Springer.

Zhao, Y. Y., & Yan, M. C. (1994). *Geochemistry of shallow sea sediments of China.* Beijing: Science Press, 200 pp (in Chinese).

Zhao, Q. H., & Wang, P. X. (1990). Modern ostracoda in shelf seas off China: Zoogeographical zonation. *Oceanologica Limnologica Sinica, 21,* 458–464 (in Chinese).

Zhao, Q. H., & Zheng, L. F. (1996). Distribution of deep sea ostracods in surface sediments of the South China Sea. *Acta Oceanologica Sinica, 18*(1), 61–72.

Zhao, Q. H., & Zhou, B. C. (1995). Deep sea ostracods in surface sediments of the South China Sea. In P. X. Wang (Ed.), *The South China Sea since 150 ka* (pp. 140–148). Shanghai: Tongji University Press (in Chinese).

Zheng, S. Y. (1980). The recent foraminifera of the Zhongsha Islands I. *Studia Marina Sinica, 16,* 143–152 (in Chinese).

Zheng, S. Y. (1988). *The agglutinated and porcellanous foraminifera from the East China Sea.* Beijing: China Science Press, 337 pp (in Chinese).

Zheng, G. Y. (Ed.), (1991). *Quaternary geology in the Yellow Sea.* Beijing: China Science Press, 164 pp (in Chinese).

Zheng, X. M. (1999). *Changjiang delta and dust deposition and environment in Sea.* Shanghai: East China Normal University Press, 174 pp (in Chinese).

Zheng, S. Y., & Fu, Z. X. (1994). Foraminiferal faunal trends in China Seas. In D. Zhou (Ed.), *Oceanology of China Seas* (pp. 255–274). Dordrecht: Kluwer Academic Publishers.

Zheng, Z. Z., & Zheng, S. Y. (1978). The recent foraminifera of the Xisha Islands I. *Studia Marina Sinica, 12,* 149–266 (in Chinese).

Zheng, S. Y., & Zheng, Z. Z. (1979). The recent foraminifera of the Xisha Islands II. *Studia Marina Sinica, 15,* 101–232 (in Chinese).

Zhu, Y., & Chang, R. (2000). Preliminary study of the dynamic origin of the distribution pattern of bottom sediments on the continental shelves of the Bohai Sea, Yellow Sea and East China Sea. *Estuarine, Coastal and Shelf Science, 51*(5), 663–680.

Zhu, Y. Z., Sha, Q. A., Guo, L. F., Yu, K. F., & Zhao, H. T. (1997). *Cenozoic coral reef geology of Yongshu reef, Nansha Islands.* Beijing: China Science Press, 134 pp (in Chinese).

Zhuang, Z., Xu, W., Liu, D., Zhuang, L., Liu, B., Cao, Y., et al. (1999). Division and environmental evolution of Late Quaternary marine beds of S3 hole in the Bohai Sea. *Marine Geology and Quaternary Geology, 19*(2), 27–35 (in Chinese).

Zong, Y., Huang, G. Q., Switzer, A. D., Yu, F. L., & Yim, W. W.-S. (2009). An evolutionary model for the Holocene formation of the Pearl River delta, China. *The Holocene, 19*(1), 129–142.

Zong, Y. Q., Kemp, A. C., Yu, F. L., Lloyd, J. M., Huang, G. Q., & Yim, W. W.-S. (2010). Diatoms from the Pearl River estuary, China and their suitability as water salinity indicators for coastal environments. *Marine Micropaleontology, 75,* 38–49.

Chapter 5

Basins and Stratigraphy

5.1 INTRODUCTION

The monograph "Chinese Sedimentary Basins" compiled by Xia Zhu (1989) represents one of the earlier systematic studies on China basins. Its Chinese edition with a slightly different title "Mesozoic and Cenozoic Sedimentary Basins of China" appeared about 1 year later (Zhu, 1990). Almost all the earlier studies published by Chinese workers, including Zhu (1989, 1990), focused on lithostratigraphy and resource accumulation and distribution in land and nearshore basins. Representing probably the first textbook on Petroleum Geology of China, for example, Wang (1983) described 16 China petroliferous basins, including the tectonic and petroleum history of the Bohai Basin and Subei–South Yellow Sea Basin, and some brief accounts on the East China Sea and South China Sea due to a lack of data at the time. Comparatively, the monographs by Zhu (1989, 1990) provided some detailed descriptions of these basins with data primarily from industrial exploration up to that time. Meanwhile, Wang (1990) gave an in-depth review on the progress of Neogene stratigraphy and paleoenvironmental studies in the region. More detailed accounts on lithobiostratigraphy of various China Cenozoic basins were given in a series of monographs entitled "Tertiary in Petroliferous Regions of China." The first volume of that series was by Ye et al. (1993) who overviewed the general patterns of Tertiary sediment sequences, fossil assemblages, and depositional environments, with an English edition published a few years later (Ye et al., 1997).

The 1990s also saw the publication of 15 volumes of "Stratigraphical Lexicon of China," which categorized all lithostratigraphic units (formations) used by Chinese geologists for various time intervals, including the Tertiary (Zheng, He, Liu, & Li, 1999) and Quaternary (Zhou, Bian, & Wang, 2000). As a new update, "Geological Formation Names of China (1866–2000)" was published by Zhang (2009), briefing the historical and stratigraphic characteristics of various Chinese formations. All these compilations provide critical information on lithostratigraphy of land and offshore basins of China, especially in clarifying some confusing terminologies and names.

This chapter deals with lithostratigraphy in basins of the China Seas, that is, those currently being bathed in seawater. We start from an overview of the basins and their basement types, Cenozoic stratigraphy, and the generalized biostratigraphic framework. Following the overview is the main part of this chapter, in which the major characteristics of lithostratigraphy and sequence stratigraphy in 13 large basins (or basin groups) are to be deliberated. The isotopic and astronomical stratigraphy for deep-sea deposits will then be briefly described. Finally, a summary on the major lithostratigraphy patterns and their implications for the evolution of the China Seas will be given.

5.2   AN OVERVIEW OF LITHOSTRATIGRAPHY

5.2.1   Basins and Pre-Cenozoic Basements

5.2.1.1   Basins of the China Seas

A series of early Cenozoic extensional basins lie along the eastern margin of China–Indochina with initial rifting probably triggered by the subduction roll-back of the oceanic Pacific Plate in the east (Allen, Macdonald, Zhao, Vincent, & Brouet-Menzies, 1997) and the collision of Australia with the Philippine Sea Plate in the south (Hall & Morley, 2004). Largely, in responding to the Indo–Asia collision, which started about 50 Ma, most basins of the China Seas developed likely as part of a gigantic linked, dextral pull-apart basin system due to the NNE- to ENE-ward motion of east Eurasia (Xu, Ben-Avraham, Kelty, & Yu, 2014). On a smaller scale, activities of major fault systems were decisive in controlling the formation of the basins in both the East China Sea (ECS) and South China Sea (SCS; Ren, Tamaki, Li, & Zhang, 2002). As described in Chapter 3, for example, the NNE-striking Tan-Lu fault in the Northeast China was a major transcurrent fault system activated in the Mesozoic due to the collision between the North and South China blocks directly related to the oblique northward subduction of the Kula Plate under the Asian continent. Two major episodes of strike-slips of the Tan-Lu fault are significant for basin formation in Northeast China region: a sinistral slip up to 740 km during the Triassic–early Cretaceous (Xu & Zhu, 1994) and a dextral slip of about 40 km in the early Cenozoic (Hsiao, Graham, & Tilander, 2004). Under the influence of the Tan-Lu fault, Songliao, Bohai, Hefei, and Jianghan Basins developed on its west, while Subei–Yellow Sea Basins formed on its east (Figures 3.2 and 5.1). More specifically, the Tan-Lu fault was directly responsible for the formation of the Bohai Basin, the Jiaxiang fault for the Yellow Sea Basin, and the Donghai fault for the East China Sea Shelf Basin. Similarly, in the SCS, the sinistral strike-slip Red River Fault directly controlled the formation of the Yinggehai (Song Hong) and Qiongdongnan Basins (Figures 3.2 and 5.1).

As summarized in Figure 5.1 and Table 5.1, about 20 large Cenozoic basins developed in the China Seas. Together with many other smaller ones scattering around particularly in the SCS sector, they cover ~50% of the modern shelf-slope area of the China Seas. From northeast to southwest, they are Bohai Basin, North and South Yellow Sea Basins, ECS Shelf Basin,

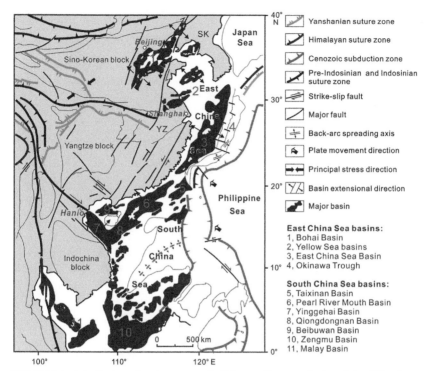

FIGURE 5.1 Distribution of major near- and offshore sedimentary basins of the China Seas, superimposed by regional tectonic framework. Note that only the mainland blocks are shaded with light gray. *Modified from Zhu (2009).*

Okinawa Trough Basin, Taixinan Basin, Zhujiangkou (Pearl River Mouth) Basin, Beibuwan Basin, Yinggehai (Song Hong) Basin, Qiongdongnan Basin, Wan'an (Nam Con Son) Basin, Zengmu (East Natuna and Sarawak) Basin, Liyue (Reed Bank) Basin, Malay Basin, and Thai Basin Groups (Table 5.1). Unlike others with a continental rifting origin, the Okinawa Trough is an active back-arc basin since the late Miocene, while the ECS Shelf Basin may represent a group of relict back-arc basins (Sibuet & Hsu, 1997), as probably also some small basins along the eastern margin of the SCS.

In general, the formation age was older for basins in the far north and younger toward the south, with rifting in the middle and late Mesozoic and since the Paleogene or early Neogene, respectively. However, the Bohai and Yellow Sea Basins in the Paleozoic–Mesozoic appeared to have been functioning similarly to those outcropping landmass along their margin until the late Mesozoic and early Cenozoic when they started to develop into the modern shape (e.g., Li, 2012b and references). During the Paleogene, extensional block faulting led to depressions in rifted half grabens that

TABLE 5.1 Characteristics of Major Cenozoic Sedimentary Basins of the China Seas

Basin	Area (10³·km²)	Formation Character	Basement	Synrift Sediment	Postrift Sediment	Other Phases	Cenozoic strata Thickness (km)
East China Sea basins							
Bohai (Bohaiwan)	80	Fault depression	Pz–Mz clastic rocks	Paleogene	Since late Oligocene		12
North Yellow Sea	15	Fault depression	Pz–Mz clastic and marine rocks	Jurassic–Pg	Late Pg to Neogene	Inversion (Oligocene–Miocene)	6
South Yellow Sea	75	Fault depression	Pz–Mz clastic and marine rocks	Jurassic–Pg	Late Pg to Neogene	Inversion (Oligocene–Miocene)	8
ECS Shelf	260	Marginal rifting	(Pre-)Pz meta rocks and Mz igneous rocks	1 Late Cretaceous–Paleogene–Eocene; 2 early Oligocene	Oligocene–Miocene	Inversion (Miocene)	12
Okinawa Trough	150	Back-arc	Paleozoic	Since late Miocene	Pleistocene		10
Northern South China Sea basins							
Taixinan (SW Taiwan)	66	Extensional to foreland	Mz clastics	Paleogene	Oligocene–Miocene	Foreland (latest Miocene)	12
Zhujiangkou (Pearl River Mouth)	147	Extensional faulting	Mesozoic granite	Late Cretaceous–early Oligocene	Late Oligocene to the present		10
Beibuwan (Beibu Gulf)	38	Extensional faulting	Pz–Mz granite and clastics	Paleogene	Neogene		9
Yinggehai (Song Hong)	113	Transverse to sheared	Mz granite and clastics	Paleogene	Neogene	Shear (late Miocene)	>12
Qiongdongnan (SE Hainan)	45	Extensional faulting	Mz granite and clastics	Paleogene	Neogene	Shear (late Miocene)	10
Southern South China Sea basins							
Cuu Long	25	Extensional faulting	Late Mz granite	Eocene–Oligocene	Late Oligocene to the present		10

				Eocene–Miocene	Pliocene to the present		
Wan'an (Nam Con Son)	90	Extensional to sheared	Mz granite, volcanics and clastics		Pliocene to the present		>12
Zengmu (East Natuna and Sarawak)	170	Foreland to compressed	Cretaceous–Eocene turbidite			Foreland (Oligocene to the present)	10
Liyue (Reed Bank)	20	Foreland to compressed	Cretaceous–Eocene turbidite			Foreland (Oligocene to the present)	6
West Natuna	70	Extensional to inversional	Pre-Cenozoic meta volcanics	Late Eocene–early Oligocene	Late Oligocene to the present	Inversion (late Oligocene–Miocene)	6
Penyu	10	Extensional faulting	Mesozoic volcanics	Oligocene	Miocene to present	Inversion (middle–late Miocene)	>5
Malay	80	Extensional to transverse	Pz–Mz igneous, meta clastics	Late Eocene–early Oligocene	Late Oligocene–early Miocene	Inversion (middle–late Miocene)	8
Thai Basin group (including Pattani)	75	Extensional to pull-apart	Pz–Mz igneous, clastics	Late Eocene–early Miocene	Middle Miocene to the present		>5
Sabah–Borneo	50	Compressed to foreland	Mz–Pg marine sediment	Oligocene		Foreland (middle Miocene to the present)	10
Palawan	40	Compressed to foreland	Mz–Pg marine sediment	Oligocene		Foreland (middle Miocene to the present)	7
Nansha Trough (NW Borneo Trough)	25	Compressed to foreland	Mz–Pg marine sediment	Oligocene		Foreland (M Miocene to the present)	6

Pz, Paleozoic; Mz, Mesozoic; Pg, Paleogene; meta, metamorphic.
Compiled mainly from ASCOPE (Asean Council on Petroleum) (1981), Du Bois (1985), Jin (1989), Zhu (1989), Fraser et al. (1997), Gong and Li (1997), Madon (1999a, 1999b, 1999c, 1999d), Madon, Abolins, et al. (1999), Madon, Karim, et al. (1999), Madon, Leong, et al. (1999), Chen (2002), Binh, Tokunaga, Son, and Binh (2007), Zhong and Li (2009), Zhu (2009), and Li (2012b).

subsequently grew to basins including the Bohai, Subei–South Yellow Sea, Beibuwan, and Pearl River Mouth Basins (Li, 2012a). In the ECS proper, stepwise rifting occurred from west to east to form the ECS Shelf Basin during the Paleogene, before entering the newest phase since the late Miocene for the opening of the Okinawa Trough (Lin, Sibuet, & Hsu, 2005). In the south, the formation of basins has been closely related to the opening and development of the SCS since the Oligocene, although their relationship with the ECS-Okinawa back-arc basins is under debating. Therefore, the (northern) ECS basins are characterized by older ages, multiple growth stages, and lesser marine influence, compared with those in the SCS with younger ages and more intensive marine influence (Li, 2012b; Zhu, 1989; and references).

5.2.1.2 Pre-Cenozoic Basements

Various basement types, ranging from late Proterozoic, Paleozoic, to Meso-zoic sedimentary, igneous, or metamorphic rocks, have been found under-neath the Cenozoic sequence in many basins of the China Seas (see Section 3.1.2). Mainly, from outcrop data of basin margins, the Paleozoic and Mesozoic sequences in the Bohai and Yellow Sea Basins appear to be more complete compared with basins in the far south. For example, shallow-marine carbonate facies are dominant in early Paleozoic sequences of the Bohai Basin and surrounding areas and in Paleozoic to early Mesozoic sequences of the Yellow Sea region, while nonmarine sequences prevail in other periods in these regions (Chough, Lee, & Yoon, 2000; Li, 2012b; and references). The basement of the ECS Shelf Basin is dominated by Mesozoic and older igneous and metamorphic rocks, largely as the seaward extension of the bedrocks along the eastern coast of China. In addition to terrestrial, igne-ous, and metamorphic rocks, Mesozoic marine facies has been reported as likely existing in some area of the ECS (e.g., Li, Gong, et al., 2012; Liu, Li, & Fang, 2005), but solid evidence needs to support this. During the Ceno-zoic, the basement topography of the newer structural units along the eastern ECS is much more developed than the older belts in the west, possibly in rela-tion to the late Eocene/early Oligocene change in steepness of the Ryukyu slab (Lin et al., 2005).

In northern SCS basins, Cretaceous or older igneous granites, Jurassic to Cretaceous marine shales and sandstones, and metamorphic rocks represent the main basement rock types, comprising 55%, 25%, and <20%, respectively (Gong & Li, 1997). The oldest sedimentary rock so far recovered in SCS basins is limestone of Devonian–Carboniferous age (Gong & Li, 1997), which appears to have a Paleo-Tethys affinity (Liu et al., 2011). Early Mesozoic marine sandstone and mudstone mainly occur in areas near Taiwan and to the east of the Dangerous Grounds in the far south. Wider occurrence of Jurassic and Cretaceous marine shale and sandstone may indicate the exis-tence of a broad shallow "Proto-South China Sea" during the late Mesozoic and early Cenozoic (Wang & Li, 2009). In southern SCS basins, the dominant basement rocks include granite and diorite, and volcanic and Mesozoic to Paleogene shallow-marine sediments are also common, indicating a more complex prerift lithotopography (ASCOPE (Asean Council on Petroleum), 1981; Du Bois, 1985) relative to those in the northern SCS.

Not only the basement type but also the basement depth differs greatly between basins and between different structural units in a basin. The NE–SW stretching Pearl River Mouth Basin, for example, has a basement averaging at 3000–4000 m depths or up to 10,000 m in deep sags, mainly with granite in the east and metamorphic quartzose sandstones and siltstones in the west (Gong & Li, 1997).

5.2.2 Cenozoic Stratigraphy

5.2.2.1 Shelf-Slope Basins

Due to their continental margin nature, most basins of the China Seas received nonmarine sedimentation in the rifting phase of basin evolution. Although alluvial and fluvial facies were widespread, paleolakes occurred mostly in the ECS basins during the late Cretaceous–Paleocene and later extended to cover most shelf-slope basins of the northern SCS during the Eocene (Figure 5.2), providing ideal sources and space for hydrocarbon buildup (Zhu, 2009).

For the northern SCS, earlier work on Tertiary stratigraphy dated back to the 1970s, about a decade later than that in the Bohai and Yellow Seas. As a result, a monograph by Zeng et al. (1981) summarized various aspects of lithostratigraphy of strata recovered in wells, which laid a solid foundation for stratigraphic correlation and subsequent systematic research in the region.

Work over decades has confirmed that many basins contain at least two packages of sediments: synrift and postrift sequences. As marking the transition between synrift and postrift sediment successions, a regional unconformity occurred in most basins sometimes associating with missing of strata up to millions of years. For example, the entire Oligocene is missing from many localities in Yellow Sea Basins and from some localities in the ECS Shelf Basin (Hsiao, Graham, & Tilander, 2010; Yi, Yi, Batten, Yun, & Park, 2003; Zhu, 2009). Unconformities also truncated other synrift sediment successions, indicating major shifts in sedimentary environments affected by intensive structural activities, although their exact duration may be hard to measure due to the limitation of methods, which mainly involve palynology and invertebrate fossils such as bivalves and ostracods. Nevertheless, most basins have likely undergone various stages of evolution due to changes of tectono-environmental regimes with time, and their multihistory nature needs to be identified for better understanding of the processes of basin evolution and resource accumulation (Yang et al., 2012).

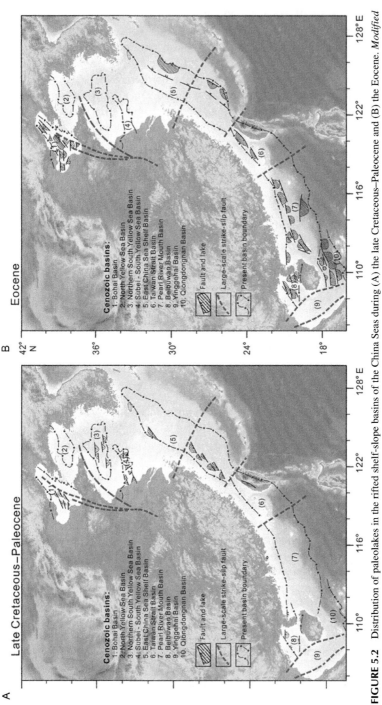

FIGURE 5.2 Distribution of paleolakes in the rifted shelf-slope basins of the China Seas during (A) the late Cretaceous–Paleocene and (B) the Eocene. *Modified from Zhu (2009).*

East China Sea Sector

The ECS sector includes three large basins or basin groups: the Bohai Basin, the Yellow Sea Basins, and the ECS Shelf Basin (Figure 5.3; Table 5.1). Table 5.2 lists major stratigraphic units in these ECS basins, emphasizing the synrift and postrift sediment sequences. Clearly, the synrift and postrift sediments in the Bohai and Yellow Sea Basins are characterized by nonmarine facies, ranging from alluvial, fluvial, flood plain, and lacustrine to deltaic lithologies. The synrift sequences of the Paleogene age include the Kongdian, Shahejie, and Dongying Formations in the Bohai Basin (Figure 5.4); the unnamed Paleogene sequences in the North Yellow Sea Basin (Figure 5.5);

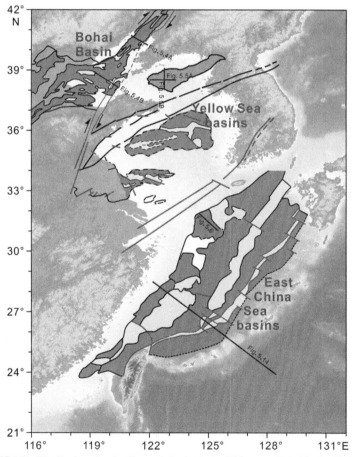

FIGURE 5.3   Distribution of major Cenozoic basins in the ECS sector: Bohai Basin, Yellow Sea Basins (including North Yellow Sea Basin (NYSB), Northern South Yellow Sea Basin (NSYSB), and Southern South Yellow Sea Basin (SSYSB)), and East China Sea Shelf Basin (ECSSB). Depression areas within the basins are shaded with gray.

TABLE 5.2 Cenozoic Stratigraphic Units (Formations) of the Major ECS Basins, with Synrift Sediments Marked with Gray Color

Epoch	Bohai Basin	North Yellow Sea	South Yellow Sea	ECS Shelf	Okinawa Trough
Pleistocene	Pingyuan (deltaic–littoral)	Quaternary (deltaic–littoral)	Donghai (neritic)	Donghai (~500 m, neritic)	Pleistocene (neritic, volcanic)
Pliocene	Minghuazhen (alluvial)	Pliocene (fluvial)	Upper Yancheng (alluvial–fluvial)	Santan (~800 m, detaic–neritic)	Pliocene (neritic, volcanic)
Miocene	Guantao (alluvial to flood plain)	Miocene (fluvial)	Lower Yancheng (fluvial–lacustrine)	Liulang (0–830 m, lacustrine–neritic)	Middle–upper Miocene (fluvial-marine, volcanic)
				Yuquan (0–1300 m, lacustrine–fluvial) Longjing (0–1330 m, lacustrine–fluvial)	
Oligocene	Dongying 1–3 (alluvial to lacustrine) Shahejie 1 (?2) (lacustrine to littoral)	Oligocene (fluvial, flood plain, lacustrine)	(Absent)	Huanggang (0–1300 m, fluvial–deltaic) (absent)	
Eocene	Shahejie 2–4 (lacustrine to littoral) Upper Kongdian (lacustrine)	Eocene (fluvial, flood plain, lacustrine)	(Upper Eocene absent) Sanduo (alluvial–fluvial) Dainan (fluvial–lacustrine)	Pinghu (~1700 m, semienclosed bay) Oujiang (~1000 m, neritic)	
Paleocene	Lower Kongdian (lacustrine) (lower Pg absent)	(Absent)	Funing (lacustrine)	Mingyuefeng (~700 m, coastal) Lingfeng (~870 m, neritic) Yueguifeng (~750 m, fluvial–coastal)	
basement	Archaean to Mz igneous, metamorphic, and sedimentary rocks	Cretaceous mudstone, sandstone, and locally volcanics	Upper Cretaceous Taizhou (fluvial)	Mz Shimentan (fluvial)	Mz metamorphic

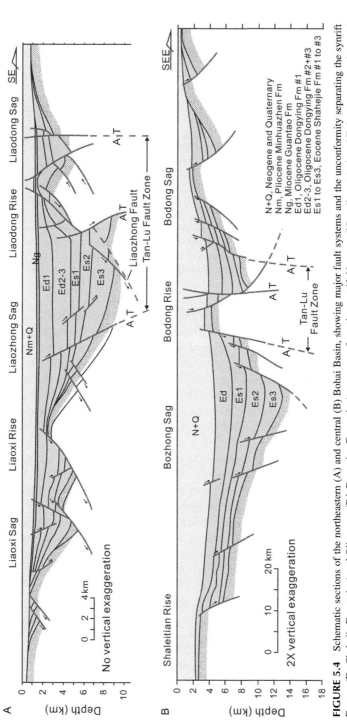

FIGURE 5.4 Schematic sections of the northeastern (A) and central (B) Bohai Basin, showing major fault systems and the unconformity separating the synrift Eocene (Es, Shahejie Formation) and Oligocene (Ed, Dongying Formation) sequences from the postrift Neogene (N) sediments). A and T denote away and toward the viewer, respectively. See Figure 5.3 for line locations. *Modified from Qi, Zhou, Deng, and Zhang (2008).*

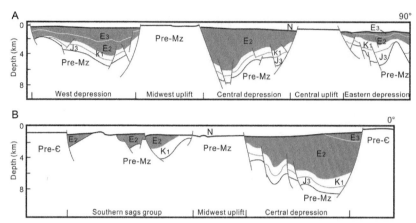

FIGURE 5.5 Schematic sections of the North Yellow Sea Basin, showing the unconformity (black line) separating the synrift late Jurassic (J3), Cretaceous (K), Eocene (E2), and Oligocene (E3) sequences from the postrift Neogene (N) sediments. See Figure 5.3 for line locations. *Modified from Li, Lu, et al. (2012).*

the Funing, Dainan, and Sanduo Formations in the South Yellow Sea Basin; and the Yueguifeng, Mingyuefeng, Oujiang, Pinghu, and Huanggang Formations in the ECS Shelf Basin (Figure 5.6). The Neogene postrift sequences include the Guantao and Minghuazhen Formations in the Bohai Basin and the Yancheng and Dongtai Formations in the South Yellow Sea Basin. These nonmarine sediment successions indicate a wide range of lacustrine and fluvial–alluvial environments in Northeast China during most Cenozoic periods and the absence of sea influence from these regions. In fact, marine sediments did not develop in Bohai and Yellow Seas until the late Pliocene or early Pleistocene, indicating a region-wide subsidence and marine transgression only during the latest episode of the basin history. In contrast, in the ECS Shelf Basin, marine sediments occurred in all periods since the Paleocene, including the synrift Lingfeng, Mingyuefeng, Oujiang, Pinghu, and Huanggang Formations and the postrift Yuquan, Liulang, and Santan Formations. The marine facies developed better in deep depressions in the southern and eastern ECS than in the north and west where lacustrine, deltaic, and littoral deposits often prevail (Yang, Cui, & Zhang, 2010).

South China Sea Sector

Over 30 Cenozoic basins developed in the SCS sector, including the Taixinan Basin, Pearl River Mouth Basin, Beibuwan Basin, Yinggehai Basin, and Qiongdongnan Basin in the northern part and Wan'an (Nam Con Son) Basin, Zengmu Basin, and Malay Basin in the southern part (Figure 5.7; Table 5.1). Rifting started earlier in the Paleocene in the northeastern and northern SCS and in the Eocene or later in southern areas (Table 5.3). Many half grabens

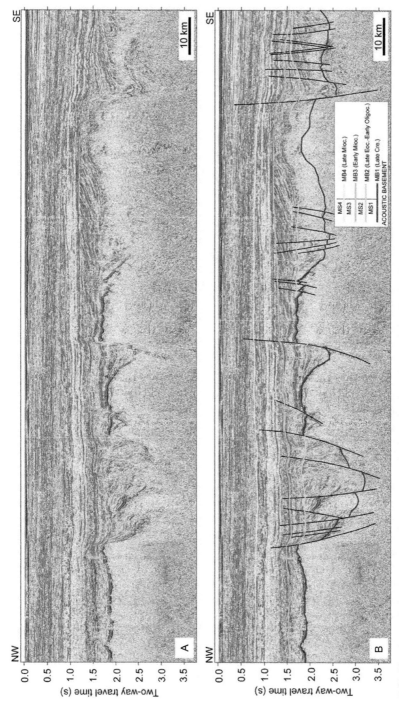

FIGURE 5.6 (A) Uninterpreted and (B) interpreted seismic section from the East China Sea Shelf Basin, showing the unconformity MB3 separating the synrift Oligocene and older sequences from the postrift Miocene sediments (Cukur, Horozal, Kim, & Han, 2011). See Figure 5.3 for line location.

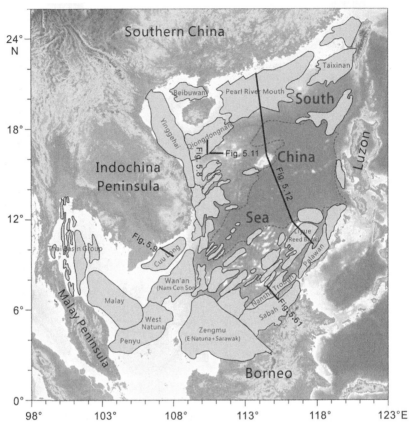

FIGURE 5.7   Distribution of Cenozoic basins in the South China Sea. *Modified from Zhong and Li (2009).*

developed in areas now occupied by the present shelf and slope, and the syn-rift sediments are characterized by alluvial clastic, lacustrine to littoral marine facies, occasionally interrupted by volcaniclastic rocks. Unconformably over-lain by the postrift Neogene sequences of mainly marine origin, these synrift sequences show similar features across basins in both lithologic features and sediment packaging (Figures 5.8 and 5.9). The synrift sequences include the Changliu, Liushagang, and Weizhou Formations in the Beibuwan Basin; the Yacheng and Lingshui Formations in the Qiongdongnan and Yinggehai Basins; and the Shenhu, Wenchang, and Enping Formations in the Pearl River Mouth Basin (Table 5.3). These fluvial to lacustrine deposits show various thicknesses in different parts of the basins. Often unconformably overlying the pre-Cenozoic granites or dolomites, for example, the upper Oligocene Yacheng and Lingshui Formations in the Yinggehai Basin may attain a thick-ness of over 3000 m.

TABLE 5.3 Cenozoic Stratigraphic Units (Formations) of the Major SCS Basins, with Synrift Sediments Marked with Gray Color

Epoch	Taixinan Basin	Zhujiangkou	Beibuwan	Yinggehai–Qiongdongnan	Malay	Zengmu	Wan'an (Nan Con Son)
Pleistocene	Pleistocene (neritic)	Pleistocene (neritic)	Pleistocene (deltaic to neritic)	Pleistocene (neritic)	Pleistocene (neritic)	Pleistocene (neritic to semiabyssal)	Upper Bien Dong (neritic)
Pliocene	Lower Gutinkeng (foreland neritic)	Wanshan (neritic)	Wanglougang (neritic)	Upper Yinggehai–Huangliu (neritic)	Pliocene (neritic, volcanic)	Pliocene (littoral, neritic, semiabyssal)	Lower Bien Dong (neritic)
Miocene	Chihwang (neritic)	Yuehai (neritic)	Dingloujiao (neritic)	Lower Yinggehai–Huangliu (neritic)	Upper Miocene	Upper Miocene	Nan Con Son (neritic)
	Chingyuan (neritic)	Hanjiang (neritic)	Jiaowei (neritic)	Meishan (neritic)	Middle Miocene (littoral)	Middle Miocene (littoral)	Mang/Thong (neritic)
	Cherngan (neritic)	Zhujiang (neritic)	Xiayang (littoral to neritic)	Shanya (neritic)	Late Miocene (alluvial, fluvial, lacustrine)	Late Miocene (fluvial, littoral)	Dua (lacustrine)
Oligocene	Chihchang (littoral, neritic)	Zhuhai (neritic)	(Absent)	Lingshui (littoral–neritic)	Oligocene (alluvial, fluvial)	Oligocene (littoral, neritic)	Cau (alluvial, fluvial, lacustrine)
		Upper Enping (fluvial to littoral)	Weizhou (fluvial–lacustrine)	Yacheng (alluvial, fluvial, lacustrine)			
		Lower Enping Wenchang (fluvial to lacustrine)	Liushagang (alluvial–fluvial)				
Eocene	Jianfeng (fluvial)			Eocene (alluvial, fluvial, lacustrine)			
Paleocene	?	Shenhu (alluvial, fluvial)	Changliu (alluvial–fluvial)	?			
Basement (Mesozoic or older)	Mz (fluvial)	Mz (igneous, metamorphic)	Mz (metamorphic, neritic)	Mz (metamorphic, igneous, neritic)	Mz (igneous, metamorphic)	Mz	Mz

         Geology of the China Seas

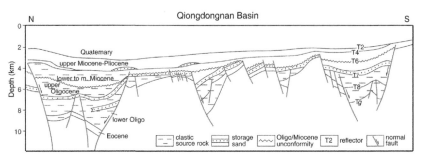

FIGURE 5.8 Schematic section showing the regional unconformity T6 separating the synrift Paleogene sequences from the overlying postrift Neogene sequences in the southern Qiongdongnan Basin, northern SCS. See Figure 5.7 for line location. *Modified from Gong and Li (1997).*

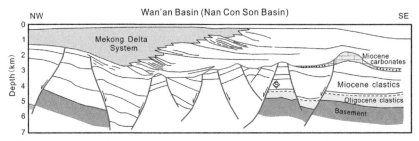

FIGURE 5.9 Schematic section showing the regional unconformity separating the synrift Oligocene to early Miocene sequences from the overlying postrift middle to late Miocene sequences in the nearshore part of the Wan'an Basin, southern SCS. See Figure 5.7 for line location. *Modified from Matthews, Fraser, Lowe, Todd, and Peel (1997).*

The postrift sediments in northern SCS basins represent a mixture of terrigenous and shallow-marine origin, with seaward increasing in the proportion of marine biotic particles, especially foraminifera. They include the Zhujiang, Hanjiang, Yuehai, and Wanshan Formations in the Pearl River Mouth Basin and the Shanya, Meishan, Yinggehai, and Huangliu Formations in the Yinggehai and Qiongdongnan Basins. Comparatively, the marine influence on the postrift sedimentation in the Beibuwan Basin is restricted due to its proximity to the coast. Therefore, the postrift lithofacies in individual basins remain highly variable in space and time.

In the southern SCS, the synrift sequence accumulated from the late Eocene to early Miocene or earlier mainly in basins in and to the south of the Dangerous Grounds (Mohd Idrus, Abdul, Abdul Manaf, Sahalan, & Mahendran, 1995; Xia & Zhou, 1993). The "mid-Miocene unconformity (MMU)" or "deep regional unconformity (DRU)" (Figure 5.10; ASCOPE (Asean Council on Petroleum), 1981; Yumul, Dimalanta, Tamayo, & Maury, 2003; Hutchison, 2004) separates the synrift sequence from the overlying postrift sequence, forming a distinct regional feature in many southern SCS basins and in areas of the Nansha (Spratly) Islands (Hutchison & Vijayan, 2010).

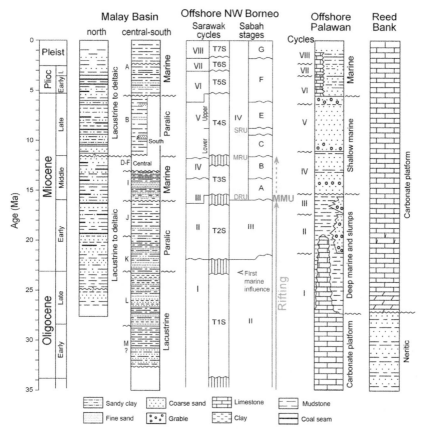

FIGURE 5.10 Oligocene to Pleistocene lithostratigraphy of the southern SCS basins, showing major sediment types, sediment cycles, and unconformities, including the shallow, middle, and deep regional unconformities (SRU, MRU, and DRU). The middle Miocene unconformity, or MMU (DRU), separates the synrift Oligocene–early Miocene sequences from the postrift sequences. *Modified from ASCOPE (Asean Council on Petroleum) (1981), Madon, Abolins, et al. (1999), Madon, Karim, et al. (1999), and Mat-Zin and Tucker (1999).*

From a compilation of ASCOPE (Asean Council on Petroleum) (1981), Du Bois (1985), Fraser et al. (1997), Madon, Abolins, Hoesni, and Ahmad (1999), Madon, Karim, and Fatt (1999), Hutchison (2004), and Hutchison and Vijayan (2010), it is clear that the synrift Oligocene–early Miocene sequences in southern SCS basins show a rifting–spreading process distinct from northern basins by the following characteristics: (1) Synrift sequences are younger in age, Eocene to early Miocene, (2) lacustrine and deltaic sediments are dominant, (3) more clastic deposition occurred in basins from the southwest than the southeast, and (4) carbonate platforms started to form offshore Palawan and Reed Bank during the rifting period.

Except in areas with carbonate platforms, the synrift and postrift lithostratigraphy in most southern SCS basins is prevailed by cycles of paralic sediments presumably resulting from a combination of intensive tectonic
activities and frequent sea level fluctuations (Figure 5.10). The (I to VIII)
cycle concept, first proposed by the Shell Oil Company for Oligocene to
Pleistocene sequences from offshore of Sarawak in the 1970s, has been
widely applied to the region. Due to structural differences and complex depositional environments between basins strongly affected by syndepositional
tectonic deformation, however, these "cycles" rarely show a clear cyclic relationship between the contemporary sediment packages within or between
basins, as illustrated in several sections from offshore Palawan by Schlüter
et al. (1996). Particularly, the postrift Miocene–Pliocene sequences in several
southeastern and southwestern basins have been strongly deformed due to
inversion and other neotectonic activities associating with collision and subduction along the eastern SCS margin, which altered the sediment properties
(Ingram et al., 2004) and made cycle distinction difficult. Therefore, supersequences or stages bounded by finite unconformities (seismic reflectors) have
been introduced to replace the earlier cycle concept for a better correlation
between the clastic-dominated lithofacies in southern SCS basins (e.g.,
Lovell, 1987; Mat-Zin & Tucker, 1999). However, to unify the complex stratigraphic nomenclatures currently used in the southern region for a more easily
identified, well-architected correlation scheme still requires further efforts
(Madon, Karim, et al., 1999).

Carbonate Platforms

Eocene to earliest Oligocene carbonate platforms first developed in offshore
areas from Palawan and neighboring islands in the southern SCS
(Figure 5.10; ASCOPE (Asean Council on Petroleum), 1981; Jiang, Zeng,
Li, & Zhong, 1994; Qiu & Wang, 2001). By late Oligocene time, the center
of carbonate growth shifted to the Reed Bank (Liyue) area and further north
(Taylor & Hayes, 1980). These carbonate platforms continued to grow and
expand until the early Miocene before being partly drowned due to rifting
and sea floor spreading since the middle Miocene (Figure 5.10). The most
extensive Miocene platforms in the southern SCS include the Laconia–NW
Sabah platform and Kudat platform from offshore Sabah (Madon & Hassan,
1999b; Madon, Leong, & Anuar, 1999). The drowned platforms may form a
prominent feature under the Dangerous Grounds, Nansha (NW Borneo)
Trough, and Palawan slope (Hutchison, 2004, 2010; Hutchison & Vijayan,
2010). In the Zengmu (East Natuna) Basin, a distinct middle to late Miocene
carbonate complex called the Terumbu Formation L-structure is mainly composed of boundstones, grainstones, and packstones that have been homoge-
nized by organic activity (May & Eyles, 1985). During the middle and late
Miocene, carbonate platforms were widely developed also in the southeastern

Wan'an (Nam Con Son) Basin not affected by the Mekong Delta (Figure 5.9; Matthews et al., 1997), as well as offshore Sabah (Madon & Hassan, 1999b; Madon, Leong, & Anuar, 1999), Xisha, and Nansha Islands. Apart from the Reed Bank, reef platforms now scattering in the southern region all appear to have been building on a Miocene or younger base (Figure 5.11; Ma et al., 2011). However, rapid subsidence in the region in the late Miocene and early Pliocene had caused large-scale drowning of carbonate platforms.

In the northern SCS, carbonate platforms widely developed along the shelf edge since the late Oligocene and achieved a maximum distribution during the early Miocene (Gong & Li, 1997; Li, Zhong, & Tian, 2009). Like their southern counterparts, however, most of these northern platforms were subsequently drowned during the early middle Miocene, in response to further local subsidence along the northern margin and a general rise in sea level.

5.2.2.2 Deep-Water Stratigraphy
South China Sea

As revealed by seismic data, the SCS deep basin is covered by hemipelagic sediments up to ~1000 m in thickness except for areas close to seamounts (Figure 5.12). During ODP Leg 184 in 1999, a sediment sequence of 853 m was recovered at ODP Site 1148 (with ~74% core recovery) from the lower slope of the northern SCS (18°50′N, 116°34′E), at water depth of 3292 m. Covering the early Oligocene to Quaternary, the sediment succession represents the deepest section so far drilled for scientific research in the region (Wang, Prell, & Blum, 2000). Based on sediment composition (especially clay vs. nannofossils), depositional facies, and color variations, seven lithostratigraphic units were identified: unit I (0–194.02 m) for the Pliocene–Holocene, units II–V (194.02–457.22 m) for the Miocene, and units VI–VII (457.22–859.45 m) for the Oligocene (Figure 5.13; Wang, Prell, et al. 2000). Due to deep water depths, post-Oligocene biogenic components of Site 1148 had been affected by dissolution, as indicated by the high abundance of the fragmented foraminiferal tests (Figure 5.13).

The hemipelagic sediments recovered at ODP Leg 184 sites contain 40–85% terrigenous particles and up to ~60% biogenic components, particularly the calcareous nannofossils and foraminifera. Trace fossils including *Zoophycos* and *Chondrites* are also common. Biogenic components attain highest proportions in Miocene–early Pliocene intervals, matching well with wider marine deposition and carbonate platform development on shelf in a period of intensified regional subsidence (Figure 5.13).

At Site 1148, the Oligocene section exceeds 390 m, dominated by monotonous grayish to olive green, quartz-rich clay that accumulated during the early Oligocene (unit VII, 364.53 m) with sedimentation rate of over 60 m/Ma. However, the upper Oligocene (unit VI, 37.70 m) is a thin slumped unit masked by unconformities probably erasing a sediment record up to

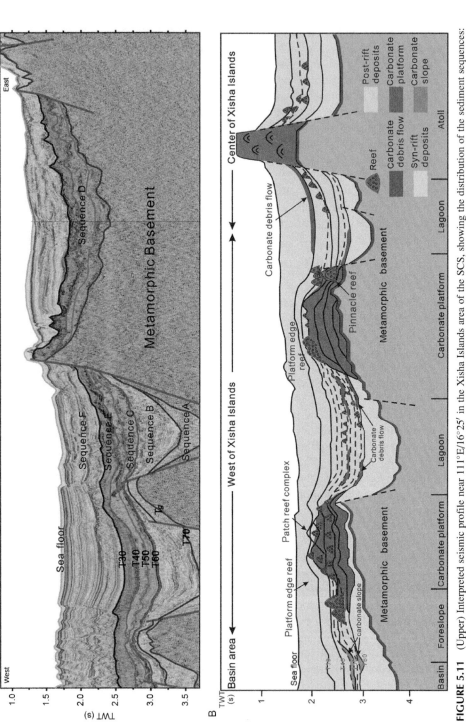

FIGURE 5.11 (Upper) Interpreted seismic profile near 111°E/16°25′ in the Xisha Islands area of the SCS, showing the distribution of the sediment sequences: A and B, Oligocene sequences; C–E, Miocene (carbonate) sequences; and F, Pliocene–Quaternary sequence. See Figure 5.7 for line location. (Lower) Schematic profile of the Xisha Island region showing the development of carbonate platforms. (*Lower panel*) *Modified from Ma et al.* (2011).

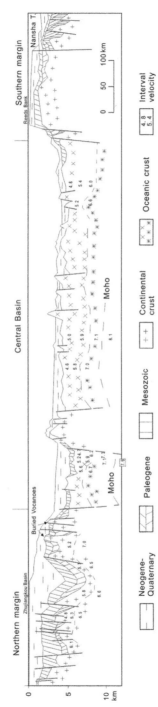

FIGURE 5.12 Simplified cross N–S section across the central SCS showing variations in sediment cover. See Figure 5.7 for line location. *Modified from Yan, Zhou, and Liu (2001).*

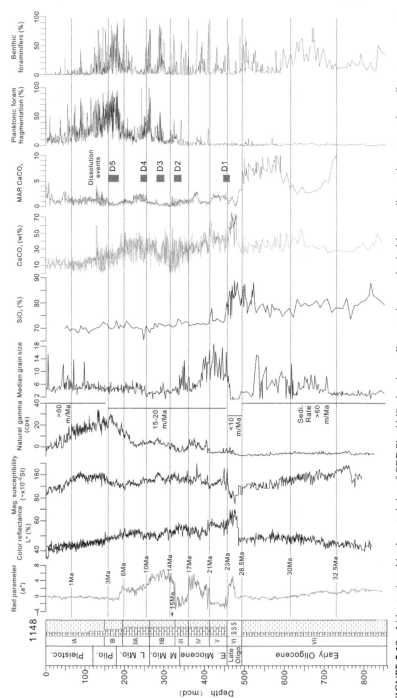

FIGURE 5.13 Lithostratigraphic characteristics of ODP Site 1148, showing sediment units, geophysical logs, sedimentation rate, major sediment components, and foraminiferal abundance, dissolution events D1–D5. *Modified from Wang, Prell, et al. (2000) and Li et al. (2006).*

about 3 myr. This event signals the large-scale tectonic transition from rifting to spreading at ∼25 to 23 Ma (Clift, Lin, & ODPLeg 184 Scientific Party, 2001; Li, Jian, & Su, 2005). The Miocene units II–V consist mainly of greenish to grayish brown nannofossil clay mixed sediments, with common iron sulfide particles.

Better-preserved Miocene to Pleistocene hemipelagic successions were recovered at northern Site 1146 (19°27′N, 116°16′E, w.d. 2092 m, 18.6–0 Ma) and southern Site 1143 (9°22′N, 113°17′E, w.d. 2772 m, ∼11–0 Ma; Wang, Prell, et al., 2000). These better-preserved sediments have been used intensively in many high-resolution paleoceanographic studies (e.g., Holbourn, Kuhnt, Schulz, & Erlenkeuser, 2005; Holbourn, Kuhnt, Schulz, Flores, & Andersen, 2007; Tian, Wang, & Cheng, 2004a,2004b,2004c). Unlike those from the northern slope, the late Miocene section at Site 1143 contains many thin turbidite layers, similar to those reported from the Nansha (NW Borneo) Trough offshore from Sabah and Palawan (Hutchison, 2004; Mohd Idrus et al., 1995). Much thicker turbidite deposits, up to 2600 m, have been found overlying a thin pelagic–hemipelagic layer and oceanic basement in the northern part of the Manila Trench, likely carried from the uplifted collision zone of Taiwan into the trench by gravity flows (Lewis & Hayes, 1984). However, thinner turbidite deposits in the southern end of the trench appeared to have been transported northward from the collision zone involving the north Palawan block and the central Philippines (Lewis & Hayes, 1984).

Other turbidite deposits include those in the Taixinan Basin and in Taiwan island (Lee, Tang, Ting, & Hsu, 1993). The two distinct turbidite systems in Taiwan, namely, the late Miocene system and the Pliocene–Quaternary system, appear to lack any obvious kinetic relationship, as the late Miocene turbidites found in the Hengchun Peninsula were deposited in the northeastern SCS passive margin and later accreted into the precollision accretionary prism when subduction was active in the late Miocene, while the huge Pliocene–Pleistocene turbidite layers were deposited in the former and modern North Luzon Trough forearc basin before being uplifted as part of the Coastal Range (Huang et al., 1997c). Apparently, the concentration of turbidite deposition along the eastern margin was responding to intensive collision and subduction in the eastern region since the late Miocene time (Lin, Watts, & Hesselbo, 2003; Sibuet, Hsu, Le Pichon, Le Formal, & Reed, 2002; Sibuet, Hsu, & Debayle, 2004). Turbidites also characterize the late Miocene sediment sequence in the Qiongdongnan Basin (Wang, Wu, Qin, Spence, & Lü, 2013). It is not clear, however, whether the thin turbidite layers at ODP Site 1143 were generated by periodic intrusions of turbidity fronts further afield from the eastern margin or by slope failures from the Sunda Shelf due to rapid discharge of the Mekong and other rivers especially at sea level lowstands.

Thin volcanic ash layers and thin green clay layers also occur in the hemipelagic sediments recovered at ODP sites, mostly confined to the Pliocene–Pleistocene intervals (Wang, Prell, et al., 2000). While the volcanic ashes represent the frequency and intensity of regional volcanism, the green layers are probably linked to the former presence of organic matter or some reducing conditions (Tamburini, Adatte, & Föllmi, 2006; Wang, Prell, et al., 2000).

Okinawa Trough

Covering an area of about 150,000 km^2, the Okinawa Trough (Figures 3.2 and 5.3) is an active back-arc rifting basin between the ECS Shelf Basin/Taiwan–Sinzi belt and the Ryukyu arc since at least the late Miocene (Gungor et al., 2012; Park, Tokuyama, Shinohara, Suyehiro, & Taira, 1998), as the youngest back-arc rifting belt in the East China Sea (Figures 5.14 and 5.15; Lin et al., 2005). Based on seismic data, Park et al. (1998) identified three stages of tectonic evolution of the Okinawa Trough (Figure 5.15). Stage 1 from the late Miocene to earliest Pleistocene is marked by widespread prerift deposition in the northern forearc region with sediments from the ECS continental shelf. Stage 2 is defined by a series of tectonic processes involving crustal doming, erosion, subsidence, and sedimentation, in association with initial rifting of the southern Okinawa Trough during most of early Pleistocene time. Stage 3 since the late Pleistocene is a period of back-arc rifting still in progress with synrift sedimentation. A similar 3-stage history was recognized by Li et al. (2004) who, however, considered stage 2 as ranging from the late Pliocene (2 Ma) to the late Pleistocene (~130 kyr BP). The northern trough has an earlier developing history from the (middle? to) late Miocene with relative slow deposition, while the southern trough developed later in the late Pliocene with relative rapid deposition (Chapter 3). The Miocene–Quaternary sediment succession in the Okinawa Trough varies between 2000 and 6000 m but may reach a maximum thickness of ~10,000 m in the shelf front depression (Yang et al., 2010) or the Ho Basin (Figure 5.14; Lin et al., 2005; Yang et al., 2010).

Although considered as mainly containing igneous/volcanic rocks and Miocene sedimentary rocks, the basement of the trough, especially its southern part, may comprise Paleogene and Mesozoic sedimentary sequences similar to those lying beneath the ECS Shelf Basin and outcropped in Taiwan (Liu, Gao, Fang, & Wu, 2008). Due to its high-thermal regime and lack of drilling, however, the basement and older sediment sequence of the trough can only be projected by those available from the nearby arc islands (Ujiie & Nishimura, 1992) or by seismic data. Based on seismic correlation, for example, Park et al. (1998) recognized the equivalence of the Paleogene to early Miocene Yaeyama group and the Miocene to Pliocene Shimajiri group in the eastern trough underlain by Pleistocene and Holocene hemipelagic sediments (Figure 5.15). The early Miocene Yaeyama group is dominated by sandstones and siltstones deposited in brackish-water to shallow-marine environments (Saitoh & Masuda, 2004 and references). According to Tanaka and Nomura (2009), the Shimajiri group in the region is mainly

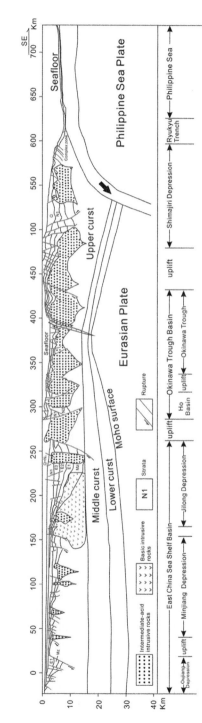

FIGURE 5.14 Schematic profile from the East China Sea to the Ryukyu Trench, showing the thickness variations of sediments and crust and the dominance of basic igneous rock basement in the East China Sea Shelf Basin and intermediate–felsic igneous rock basement in the Okinawa Trough and Ryukyu arc. See Figure 5.3 for line location. *Modified from Yang et al. (2010)*.

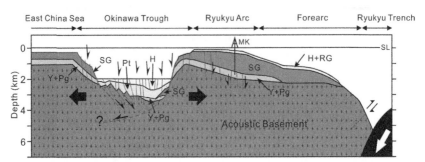

FIGURE 5.15 Schematic profile from the East China Sea outer shelf to the Ryukyu Trench, showing the dominance of Pleistocene and Holocene sediments in the Okinawa Trough. H, Holocene; Pt, early Pleistocene; Pg, Paleogene; RG, Ryukyu group (late Pleistocene); SG, Shimajiri group (late Miocene to Pliocene); Y, Yaeyama group (early Miocene); SL, sea level. *Modified from Park et al. (1998).*

composed of the late Miocene Maja formation (7.8–7.2 Ma, littoral clastic facies) and the middle Pliocene Aka formation (about 3.2 Ma, shallow-marine facies). In Ryukyu group limestones, Ujiie (1994) identified 11 planktonic foraminiferal zones, many of which represent regional zonations with age ranging from the late Pliocene to earliest Pleistocene as marking the birth of the Okinawa Trough.

Lying at intermediate depths (>1000 m or maximum of ~2700 m), the Okinawa Trough today receives both hemipelagic (basin-wide) and volcanic sediments (mainly in the northern and central parts). The volcanic rocks are dominated by intermediate–felsic types, while basic types occur only in some local areas. For the hemipelagic deposits, the proportion of terrigenous particles in a belt along the western slope and in the southern trough is higher than elsewhere, indicating the major sources of sediment input from ECS shelf and Taiwan. Accordingly, the Pleistocene and Holocene sediments in the trough mainly consist of terrigenous sand, silt, and clay from the ECS shelf mixed or interbedded with biogenic and volcanic sediments. Numerous short cores have been obtained from the trough for paleoceanographic studies (e.g., Li & Chang, 2009). In July 2002, drilling at ODP Site 1202 (24°48.24′N, 122°30.01′E, water depth 1274 m) in the southern Okinawa Trough recovered a 410 m hemipelagic sequence representing the last ~68 kyr deposition (Wei, 2006). With the highest sedimentation rate of up to 4.2 m kyr^{-1} for the region, the deposition at Site 1202 can be attributed to the Lanyang River load from Taiwan. In addition, submarine volcanic eruptions supplied pumice and volcanic grains to the site mainly during 65–52 kyr BP (Figure 5.16). Distal turbidites in the form of fine-sandy layers became more frequent during 52–44 kyr BP, with turbidite layers reaching up to 15 cm in thickness enriched with mica flakes and quartz grains of metamorphic origin. Microfossils, if present, are almost all displaced biota from the continental shelf (Huang et al., 2005). Turbidite deposition declined substantially from 44 to 36 kyr BP and finally decreased to zero by ~32 kyr BP (Figure 5.16).

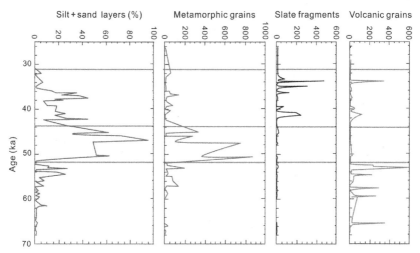

FIGURE 5.16   Sediment characteristics in samples from ODP Site 1202 in the southern Okinawa Trough, showing frequent turbidite deposition especially during 52–44 kyr BP and frequent volcanic deposition before 52 kyr BP. *Modified from Huang, Chiu, and Zhao (2005).*

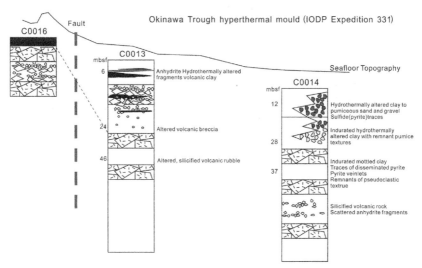

FIGURE 5.17   Characteristics of cores recovered at three sites of IODP Expedition 331 in the middle Okinawa Trough, showing various volcanic clastics mostly altered by the hyperthermal regime (Expedition 331 Scientists, 2010).

In a search for deep hot biosphere, IODP Expedition 331 uncovered various volcanic sediments at five sites (C0013–C0017) on a hyperthermal mound in the middle Okinawa Trough (Expedition 331 Scientists, 2010). The oldest and deepest stratigraphic unit is a volcaniclastic rock formed during explosive submarine volcanism (Figure 5.17). These volcanic sedimentary rocks were subsequently altered by hydrothermal activity, with much of their original composition altered to or replaced by silica (quartz) and phyllosilicates, including kaolinite, illite/muscovite, and Mg chlorite. Overlying the silicified volcaniclastic rocks are scree, mostly unconsolidated deposits comprising large pumice clasts interbedded with hemipelagic mud. While Site C0017 deposition is dominated by volcaniclastic sediment derived from a variety of pumice types, Site C0015 contains clastic material derived from the continental shelf. However, hydrothermal activity not only silicified the deeper stratigraphic units but also formed new rock types at and near the surface, producing thick anhydrite-bearing horizons as veins or nodules. At Sites C0013 and C0016, massive sulfide-rich and sulfide-rich sediment occurred, and hydrothermally altered clay is present at the surface at Site C0013 and in the shallow subsurface at Site C0014 (Figure 5.17).

5.2.3 Biostratigraphic Framework

Starting from the 1960s and 1970s, with a strong push from the oil and gas industries, Chinese paleontologists embarked on studies of various fossil groups in samples from wells drilled in onshore and shallow offshore basins. This has resulted in the publication of many monographs especially from the late 1970s to the 1990s, including six volumes on various fossil groups for the Bohai Basin (He & Hu, 1978; Hou, Huang, et al., 1978; Song, Cao, Zhou, Guan, & Wang, 1978; Song, He, Qian, Pan, Zheng, & Zheng, 1978; Wang, Huang, Yang, & Li, 1978; Yu, Mao, Chen, & Huang, 1978) and two volumes for the ECS Shelf Basin (Yang, Sun, Li, & Liu, 1989a,1989b). A monograph entitled "Tertiary Paleontology of North Continental Shelf of South China Sea" (Hou, Li, Jin, & Wang, 1981) was probably the first systematic work on different microfossils in northern SCS basins, as a companion volume to "Tertiary of North Continental Shelf of South China Sea" (Zeng et al., 1981) dealing with lithostratigraphy. The most common and significant paleontological groups include spores and pollen (Guan, Fan, & Song, 1989; Song et al., 1985; Song, He, et al., 1978; Yang et al., 1989a), dinoflagellates (Song, He, et al., 1978), foraminifera (Hao & Ding, 1988; Hao, Xu, & Xu, 1996; He & Hu, 1978; Hou et al., 1981; Yang et al., 1989b; Zhong, Jiang, Li, & Wang, 2006), nannofossils (Hao & Ding, 1988; Hao et al., 1996; Hou et al., 1981; Yang et al., 1989a; Zhong et al., 2006), and ostracods (Hou et al., 1978; Yang et al., 1989b; Zhong et al., 2006). With moderate abundance especially in the Bohai–Yellow Sea–ECS region, charophytes and gastropods also attracted attention (Wang et al., 1978; Yang et al., 1989b; Yu et al., 1978). However, most of the published works are in Chinese except "Marine Micropaleontology of China" edited by Wang (1985), "Chinese Sedimentary Basins" by Zhu (1989), and "Tertiary in Petroliferous Regions of China" by Ye et al. (1997), among a few others.

Decades of stratigraphic research indicate that microfossils remain as one of the most powerful tools in correlating the regional stratigraphy. Specifically, palynology and ostracods are more important for stratigraphy and environmental interpretations of terrestrial and shallow neritic deposits, especially in the Bohai, Yellow Sea, and ECS Shelf Basins, while foraminifera and nannofossils are more useful in offshore basins of the SCS and eastern ECS. In southern SCS basins with frequent shallow platform and paralic facies, larger benthic foraminifera and palynomorphs are widely employed in stratigraphic division and correlation.

5.2.3.1   Nannofossil and Planktonic Foraminiferal Biostratigraphy

Normal marine facies are found in most SCS basins and in some sections of the ECS Shelf–Okinawa Trough Basins. At most localities, well-preserved nannofossils and planktonic foraminifera so far recovered are mainly of Oligocene and younger ages, providing a standard base for biostratigraphic correlation of these marine sequences. Some easily identified and widely accepted biotic events, mostly the first or last occurrences of the age-diagnostic species as listed in Figure 5.18, become the key datums for biostratigraphy. However, to apply these global standards to local stratigraphy may be difficult when only the ditch cutting or dragged samples are available. Poor preservation, lacking index species, mixing by drilling, or inadequate sampling may also contribute to errors in dating sediment age. Therefore, work over the last three decades has generated a regional scheme (Figure 5.19) for marine sequences of the SCS by using mainly the last occurrence events of species that can be readily recognized. By compiling industrial well data, for example, Jiang et al. (1994) published a monograph on the Cenozoic litho-biostratigraphy of the northern SCS basins. For the Pearl River Mouth Basin, biostratigraphy of the Oligocene–Miocene units was examined in a group of papers edited by Hao et al. (1996).

Eocene and older marine sediments with well-preserved planktonic foraminifera and nannofossils have been found in wells from the southern ECS and the northeastern SCS, in a conjunction between the present ECS and SCS sectors (see Chapter 8). For example, an early Eocene planktonic foraminiferal assemblage with species of *Acarinina* and *Morozavella* occurred in Well DP21-1-1 from the Taixinan Basin (unpublished results). In Well HJ15-1-1 of the NE Pearl River Mouth Basin, Huang (1997a, 1997b) reported a middle Eocene nannofossil assemblage characterized by *Sphenolithus furcatolithoides* and *Helicosphaera seminulum*. Further north, in the Oujiang and Xihu Depressions of the ECS Shelf Basin, Paleocene and Eocene planktonic foraminifera and nannofossils have been reported (Wu & Zhou, 2000; Yang et al., 1989a,1989b, 2010; detailed in Chapter 8). It is most likely, therefore, that these microfossils represent faunal/floral associations in some deep troughs connecting the two areas before the opening of the South China Sea.

5.2.3.2   Spores and Pollen Assemblages

In coastal marine and nonmarine sediments, the stratigraphic significance of spores and pollen assemblages increase. In the northern SCS, for example, 13 palynological zones have been identified, including eight for the Paleocene

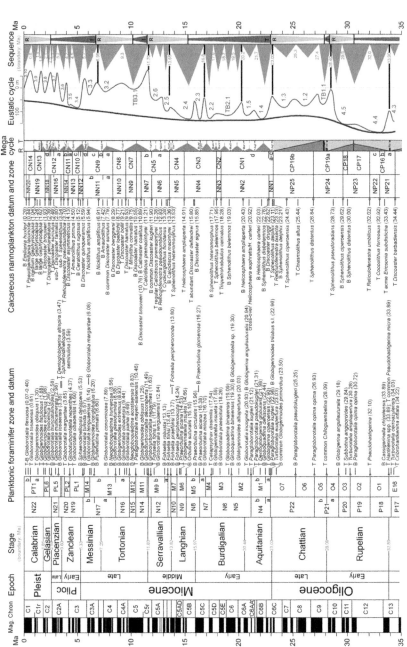

FIGURE 5.18 Standard Oligocene to Pleistocene planktonic foraminiferal and nannoplankton biostratigraphy on the timescale of Gradstein, Ogg, Schmitz, and Ogg (2012), superimposed by the sea level curves of Haq, Hardenbol, and Vail (1987) and sequences of Hardenbol et al. (1998). B, base; T, top. Note that the age for sequence boundaries was adjusted to the timescale of Gradstein, Ogg, and Smith (2004) for comparison. *From Li, Zhong, et al. (2009).*

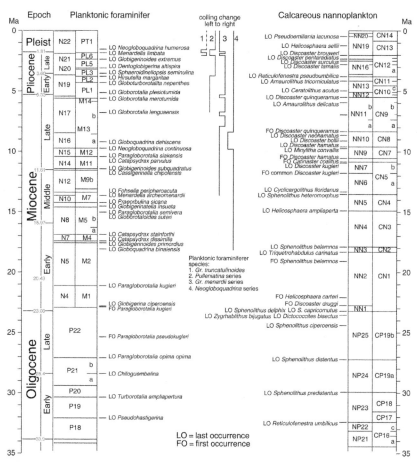

FIGURE 5.19 Regional datums of planktonic foraminifera and nannofossils in the northern SCS based on the timescale of Gradstein et al. (2004), superimposed by coiling direction changes of some important planktonic species. Refer to Figure 5.18 for the new timescale of Gradstein et al. (2012). *From Li, Zhong, et al. (2009); based on Jiang et al. (1994).*

and five for the Neogene of the Pearl River Mouth Basin and neighboring region (Figure 5.20; Jiang et al., 1994). Recently, further work in the Beibuwan Basin has refined the Eocene Liushagang Formation into four palynomorph assemblages and the Oligocene Weizhou Formation into three palynomorph assemblages. In an ascending order, they are *Monocolpopollenites tranquillus–Crassoretitriletes*, *Quercoidites–Polypodiaceaesporites–Osmundacidites–Momipites*, *Quercoidites-Ulmipollenites–Pentapollenites*, and *Quercoidites–Leiotriletes–Granodiscus* assemblages of the Eocene and *Trilobapollis–Gothanipollis–Polypodiaceresporites*, *Quercoidites–Magnastriatites–Leiotriletes–Lycopodiumsporites*, and *Quercoidites–Pinuspollenites–Pediastrum* assemblages of the Oligocene (Xie et al., 2012).

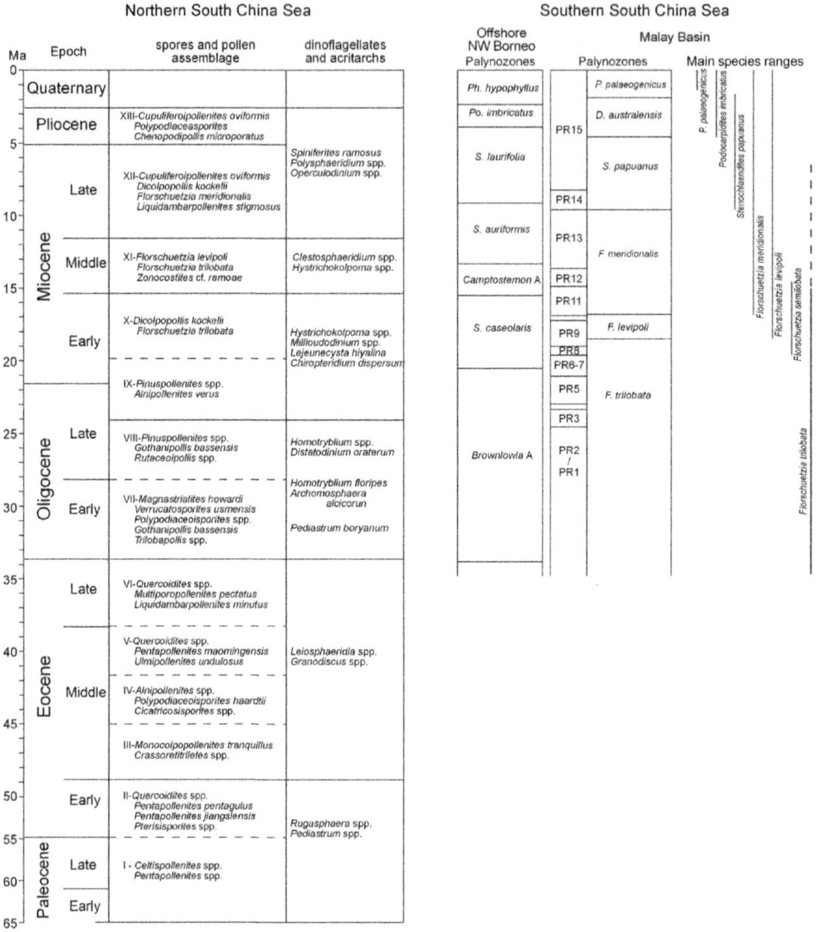

Northern South China Sea

Southern South China Sea

FIGURE 5.20   Palynological assemblages or zones found in sediments of northern and southern SCS basins. *Based on Jiang et al. (1994), Madon, Abolins, et al. (1999), Madon, Karim, and Fatt (1999b), and Mat-Zin and Tucker (1999).*

In the southern SCS, palynological assemblages show distinctive tropical characteristics, with 7 palynozones identified for the Oligocene–Pleistocene strata of Sabah and up to 15 palynozones for the contemporary strata in the Malay Basin (Figure 5.20; Madon, Abolins, et al., 1999; Madon, Karim, et al., 1999; Mat-Zin & Tucker, 1999). These palynological zones provide a fairly reliable biostratigraphic framework for the paralic sequences in the southern SCS that are otherwise lithologically too similar to be differentiated (Morley, 1991; Woodruff, 2003). From the Malay Basin, West Natuna Basin, Nan Con Son Basin, to basins offshore of Borneo and Sabah, warm assemblages characterized by *Retitriporites curvimurati* occurred in the late Eocene, by *Florschuetzia trilobata* in the Oligocene, and by *Florschuetzia levipoli* in the early and early middle Miocene (Figure 5.20; ASCOPE (Asean Council on Petroleum), 1981; Jiang et al., 1994; Madon, Karim, et al., 1999). During the early and early middle Miocene, the *F. trilobata–F. levipoli* assemblage achieved a wider distribution not only in the southern but also in the northern SCS (Hao et al., 1996), indicating the effects of the middle Miocene climate optimum.

In basins of the ECS, palynomorphs found in all sediment sequences show subtropical to temperate affinities. As listed in Table 5.4, Song, Cao, et al. (1978) and Guan et al. (1989) described palynomorph assemblages in the Bohai Basin, and Wu and Zhou (2000) and Yang et al. (2010) in the ECS Shelf Basin. In a more detailed study, Yi et al. (2003) provided palynological evidence of climate and paleoenvironmental changes in the Northern South Yellow Sea Basin since the Jurassic–Cretaceous (Figure 5.21). These authors found prevalent subtropical climate conditions in the Yellow Sea region during most Cretaceous periods before changing to temperate conditions since the Paleocene.

In several basins of the ECS, the clastic sequences contain common to abundant algae, including charophytes and acritarchs. Nonmarine dinoflagellates with abundance up to 70% characterized some Paleogene intervals, while green algae and sphere-shaped acritarchs also occupied a considerable proportion. The high abundance of algae demonstrates the existence of good source rocks in these basins (Zhu, 2009).

To the south, in the shallow part of the shelf-slope basins of the northern SCS, the freshwater algae remained frequent in some synrift sequences of fluvial to lacustrine origin. With increased marine influence, dinoflagellates and benthic foraminifera became important components of the sediment accumulated at various littoral–neritic settings. For example, Mao et al. (in Hao et al., 1996) described three dinoflagellate assemblages in Well BY7-1-1 from the Pearl River Mouth Basin: *Homotryblium tenuispinosum–Hystrichosphaeridium tubiferum* assemblage for the late Eocene–early Oligocene Enping Formation (3441–3117 m), *Homotryblium plectilum–Corfodosphaeridium gracile* assemblage for the late Oligocene Zhuhai Formation (3099–2817 m), and *Polysphaeridium zoharryi–Lingulodinium machaerophorum* assemblage for the early Miocene Zhujiang Formation (2685–2142 m). Together with spores and pollen, the dinoflagellate and acritarch assemblages found in the Pearl River Mouth and Beibuwan Basins of the northern SCS are summarized in Figure 5.20.

5.2.3.3   Ostracod Assemblages

Like algae and palynomorphs, ostracods are frequent in synrift and postrift sediments of many basins (Table 5.5), providing an important means for interpreting sediment facies and depositional water depths. Clearly, most ostracod assemblages with many endemic species were confined to local basin settings, although some species may become widely distributed when water depth increased to 50 m or more. In Oligocene–Pleistocene sediments from ODP

TABLE 5.4 Spores and Pollen Assemblages in the ECS Shelf, Northern South Yellow Sea, and Bohai Basins

Epoch	ECS Shelf Basin		Northern South Yellow Sea Basin		Bohai Basin	
Pleistocene					*Chenopodipollis* *Artemisiaepollenites* *Polypodiaceaesporites*	Littoral
Pliocene	*Polypodiaceaesporites* Gramineae *Polygonum*	Deltaic–neritic	*Graminidites* *Persicarioipollis*	Fluvial	*Persicarioipollis* *Chenopodipollis* *Magnastriatites*	
Late Miocene	*Liquidambarpollenites* *Magnastriatites*	Lacustrine–neritic	Hiatus		*Ulmipollenites* Herb *Sporotrapoidites*	Alluvial–flood plain
Middle Miocene	*Rutaceoipollis* *Miliaceoidites*	Lacustrine–neritic	*Liquidambarpollenites* *Fupingopollenites*	Deltaic–lacustrine	*Magnastriatites* *Fupingopollenites* *Liquidambarpollenites*	
Early Miocene	*Sporotrapoidites minor* *Pinuspollenites* *Alietineaepollenites*	Lacustrine–neritic	As in the preceding text	Deltaic–lacustrine	Betulaceae *Sporotrapoidites* Pinaceae	
Late Oligocene	*Magnastriatites* *Pinuspollinites* *Quercoidites* *Graminidites*	Fluvial–deltaic	Hiatus		*Ulmipollenites* *Quercoidites*	Fluvial–lacustrine
Early Oligocene	*Pinuspollinites* Taxodiaceae *Quercoidites* *Trilobapollis*	Fluvial (or hiatus)	Hiatus		*Quercoidites* *Ulmipollenite*	Fluvial–lacustrine

Eocene	*Alnipollenites* *Magnastriatites* *Abietineaepollenites* *Taxodiaceaepollentes* *Gothanipollis*	Fluvial–neritic	*Quercoidites* *Pinuspollenites*	Fluvial	*Quercoidites* *Ulmipollenites* *Alnipollenites*	Lacustrine–littoral
	Myricaceae *Cupuliferoipollenties*	Neritic	*Caryapollenites* *Inaperturopollenites*	Fluvial	As in the preceding text	Lacustrine–littoral
			As in the preceding text	Fluvial	*Ephedripites* *Schizaeoisporites*	Lacustrine
Paleocene	*Myricaceae* *Casuarinaepollenites* *Nyssapollenties*	Coastal marine	*Momipites* *Coryluspollenites*	Fluvial–alluvial	*Taxodiaceaepollenites* *Inaperturopollenites*	Fluvial–lacustrine
Authors	Yang et al. (1989a), Wu and Zhou (2000), and Yang et al. (2010)		Yi et al. (2003)		Song, Cao, et al. (1978) and Guan et al. (1989)	

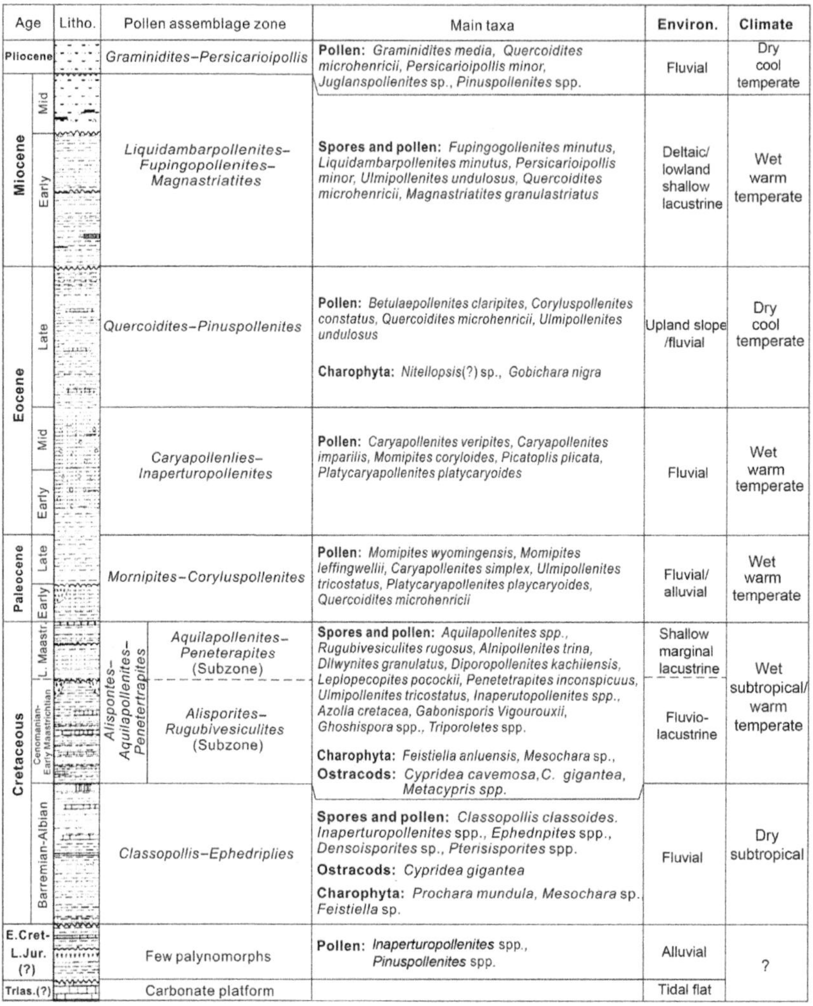

Age	Litho.	Pollen assemblage zone	Main taxa	Environ.	Climate
Pliocene		Graminidites–Persicarioipollis	**Pollen:** *Graminidites media, Quercoidites microhenricii, Persicarioipollis minor, Juglanspollenites* sp., *Pinuspollenites* spp.	Fluvial	Dry cool temperate
Miocene (Mid, Early)		Liquidambarpollenites–Fupingopollenites–Magnastriatites	**Spores and pollen:** *Fupingogollenites minutus, Liquidambarpollenites minutus, Persicarioipollis minor, Ulmipollenites undulosus, Quercoidites microhenricii, Magnastriatites granulastriatus*	Deltaic/lowland shallow lacustrine	Wet warm temperate
Eocene (Late)		Quercoidites–Pinuspollenites	**Pollen:** *Betulaepollenites claripites, Coryluspollenites constatus, Quercoidites microhenricii, Ulmipollenites undulosus* **Charophyta:** *Nitellopsis*(?) sp., *Gobichara nigra*	Upland slope /fluvial	Dry cool temperate
Eocene (Mid, Early)		Caryapollenlies–Inaperturopollenites	**Pollen:** *Caryapollenites veripites, Caryapollenites imparilis, Momipites coryloides, Picatoplis plicata, Platycaryapollenites platycaryoides*	Fluvial	Wet warm temperate
Paleocene (Late, Early)		Mornipites–Coryluspollenites	**Pollen:** *Momipites wyomingensis, Momipites leffingwellii, Caryapollenites simplex, Ulmipollenites tricostatus, Platycaryapollenites playcaryoides, Quercoidites microhenricii*	Fluvial/ alluvial	Wet warm temperate
Cretaceous (Maastr./Early Maastrichtian; Cenomanian–Early Maastrichtian)		Alisportes–Aquilapollenites–Penetetrapites → Aquilapollenites–Peneterapites (Subzone) / Alisporites–Rugubivesiculites (Subzone)	**Spores and pollen:** *Aquilapollenites* spp., *Rugubivesiculites rugosus, Alnipollenites trina, Dilwynites granulatus, Diporopollenites kachiiensis, Leplopecopites pocockii, Penetetrapites inconspicuus, Ulmipollenites tricostatus, Inaperutopollenites* spp., *Azolla cretacea, Gabonisporis Vigourouxii, Ghoshispora* spp., *Triporoletes* spp. **Charophyta:** *Feistiella anluensis, Mesochara* sp., **Ostracods:** *Cypridea cavemosa, C. gigantea, Metacypris* spp.	Shallow marginal lacustrine; Fluvio-lacustrine	Wet subtropical/ warm temperate
Cretaceous (Barremian–Albian)		Classopollis–Ephedriplies	**Spores and pollen:** *Classopollis classoides, Inaperturopollenites* spp., *Ephednpites* spp., *Densoisporites* sp., *Pterisisporites* spp. **Ostracods:** *Cypridea gigantea* **Charophyta:** *Prochara mundula, Mesochara* sp. *Feistiella* sp.	Fluvial	Dry subtropical
E.Cret- L.Jur. (?)		Few palynomorphs	**Pollen:** *Inaperturopollenites* spp., *Pinuspollenites* spp.	Alluvial	?
Trias.(?)		Carbonate platform		Tidal flat	

FIGURE 5.21 Lithostratigraphy, palynological and microfossil assemblages, and inferred environments of the northern South Yellow Sea Basin (Yi et al., 2003).

Leg 184, for example, the deep-sea ostracod assemblages are characterized by *Krithe* species, as reported by Zhao (2005a, 2005b) and Zhao, Wang, Yuan, Wang, and Zhu (2009).

5.2.3.4   Benthic Foraminiferal Assemblages

The synrift and postrift sediments in basins of the China Seas contain numerous benthic foraminifera, which are good indicators of paleoenvironmental changes. In many cases, both benthic and planktonic species of foraminifera

TABLE 5.5 Typical Ostracod Assemblages Found in Sediments of Different China Sea Basins

Epoch	Bohai Basin	ECS Shelf (West) Basin	ECS Shelf (East) Basin	Beibuwan Basin	Ying–Qiong Basins
Quaternary		*Echinocythereis* *Alocopocythere* *Neocytheretta*	*Echinocythereis* *Leguminocythereis* *Neocytheretta*		
Pliocene		*Echinocythereis* *Leguminocythereis*	*Alocopocythere* *Ambocythere* *Abrocythereis*	*Hemikrithe foveata*– *Neomonoceratina delicata*	*Neomonoceratina delicate*– *Ambocythere elliptica*
Late Miocene		Rare	Rare	*Spinileberis inflexicostata*– *Cytheropteron striatituberculata*	*Argilloecia–Krithe–* *Xestoleberis*
Middle Miocene		*Amplocypris*	*Nanocoquimba* *Munscyells*	*Spinileberis longicaudata*	
Early Miocene				*Puriana? Nanhaiensis*– *Psammocythere? Luminosa*	
Late Oligocene	*Dongyingia*	*Chinocythere* *Candoniella*	*Chinocythere* *Candoniella*		
Early Oligocene	*Phacocypris* *Guangbeinia* *Camarocypris*			*Chinocythere*	
Middle to late Eocene	*Phacocypris* *Huabeinia* *Austrocypris*	*Sinocypris* *Candona*	*Loxoconcha* *Paijenborchella*	*Chinocythere* and *Eucypris stagnalis*	
Early Eocene and older	*Cyprinotus* *Eucypris* *Cypris* *Cyclocypris*		*Neomonoceratina* *Cytherella* *Neomonoceratina*	*Sinocypris*	
Authors	Hou et al. (1978)	Yang et al. (1989a)	Yang et al. (1989a) and Yang et al. (2010)	Jiang et al. (1994)	Jiang et al. (1994)

are used to infer the existence of typical marine environment and the extent of marine transgression or sea level fluctuations (He & Hu, 1978; Hou et al., 1981; Thompson & Abbott, 2003; Yang et al., 1989b). For example, the occurrence of benthic foraminifera *Triloculina, Discorbis, Nonion,* and *Florilus* in the limestone and mudstone of the fourth member of the Shahejie Formation may indicate a near-bay or lagoonal setting during the early Eocene in the northern part of Bohai Basin (Shandong) (He & Hu, 1978). In the sandy mudstone of Minghuazhen Formation from Shandong, a benthic assemblage typified by *Elphidium, Ammonia, Quinqueloculina,* and *Cribroelphidium* implies coastal marine environment in the Pliocene. However, wider marine influence in the Bohai region started only since the late Pleistocene (see Chapter 4).

Compared with small benthic species, larger benthic foraminifera live only in warm shallow-water environments. Based on the range of some key larger species, the carbonate platform sequences in many SCS basins can be subdivided into "Letter Stages" for stratigraphic correlation (Adams, 1970; BouDagher-Fadel, 2008). Respectively, the Paleogene is defined by Ta-1 (Paleocene), Ta-2 to Tb (Eocene), and Tc to Te-4 (Oligocene) and the Neogene by Te-5 to lower Tf-1 (early Miocene), upper Tf-1 to Tf-3 (middle Miocene), and Tg-Th (late Miocene; Figure 5.22). Clearly, these "Letter Stages" only provide a relatively coarse biostratigraphic resolution, compared to planktonic foraminiferal zonation (Figures 5.18 and 5.19).

5.2.3.5   Other Fossil Groups

Radiolarians, mollusks, gastropods, and charophytes may also become useful for stratigraphy and environmental interpretation of specific strata when their abundance is high (e.g., Hou et al., 1981; Tan & Chen, 1999). Based on the census data of 237 gastropod species, for example, 9 gastropod assemblages have been recognized for the Paleogene sequences in the Bohai Basin (Yu et al., 1978). The Paleocene–Eocene Kongdian Formation is characterized by a freshwater *Physa* assemblage, the Eocene–Oligocene Shahejie Formation by six freshwater brackish-water assemblages (*Lymnaea, Sinoplanorbis, Liratina, Tulotomoides, Bohaispira supracarinata,* and *Bohaispira spiralifera*), and the middle to upper Oligocene Dongying Formation by the mixed fresh/brackish-water *Viviparus* and *Tianjinospira* assemblages. From the same strata, Wang et al. (1978) identified three charophyte assemblages: *Peckichara–Neochara* for the Kongdian Formation, *Gyrogona–Obtusochara* for the fourth member of the Shahejie Formation, and *Grovesichara–Maedlerisphaera* for the first–third members of the Shahejie and Dongying Formations.

Relatively high abundance of siliceous microfossils, mainly radiolarians, diatoms, and dinoflagellates, occurred in the late Miocene and late Quaternary sections of ODP Leg 184 sites in the SCS (Wang, Li, & Li, 2004), probably as

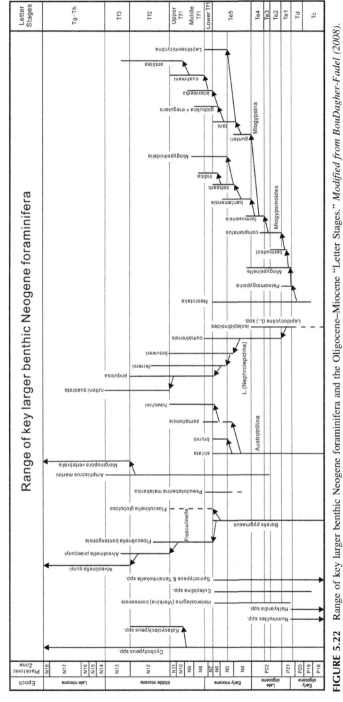

FIGURE 5.22 Range of key larger benthic Neogene foraminifera and the Oligocene–Miocene "Letter Stages." *Modified from BouDagher-Fadel (2008).*

the local responses to the global biogenic bloom during 12–5.8 Ma and the intensified monsoon since the 0.6 Ma, respectively. However, the stratigraphic significance of radiolarians and diatoms in the China Seas on a long-term scale has never been properly explored, although efforts have been made on some Quaternary radiolarian events (Wang & Abelmann, 1999, 2002).

5.2.3.6 Quaternary Stratigraphic Events

For stratigraphic division of the Quaternary sequence, especially in the SCS region, the litho-, bio-, and geomagnetostratigraphic events listed in Table 5.6 have been frequently used. Younger Quaternary strata are often dated by ^{14}C dating and/or by matching to marine isotope stages. While geomagnetic chrons and subchrons give reliable ages for both the terrestrial and marine sediments, microtektite and volcanic ash events have been also proven reliable in dating the base of Brunhess at ~0.8 Ma and the youngest Toba eruption at ~74 kyr BP (Zhao & Wang, 1999).

TABLE 5.6 Important Quaternary Litho-, Bio-, and Geomagnetostratigraphic Events

Type	Datum/Event	Age (Ma)
Volcanic ash	Youngest Toba ash (MIS4/5 transition)	0.074
Coccolith	FO *Emiliania huxleyi* (acme)	0.090
Planktonic foraminifera	LO *Globigerinoides ruber* (pink)	0.12
Radiolaria	FO *Buccinosphaera invaginata*	0.21
Radiolaria	FO *Collosphaera tuberosa*	0.42
Planktonic foraminifera	FO *Globigerinoides ruber* (pink)	0.42
Coccolith	LO *Pseudoemiliania lacunosa*	0.44
Benthic foraminifera	LO *Stilostomella*	0.75
Geomagnetism	Brunhes/Matuyama transition	0.78
Volcanic ash	Oldest Toba ash (MIS19/20 transition)	0.788
Microtektite	Microtektite (MIS20)	0.80
Coccolith	LO/FO *Recticulofenestra asanoi*	0.91/1.14
Geomagnetism	Jaramillo event (top/base)	0.99/1.07
Geomagnetism	Cobb Mountain event (base)	1.24
Coccolith	FO *Gephyrocapsa* (>0.004 mm)	1.69
Geomagnetism	Olduvai Event (top/base)	1.77/1.95
Coccolith	LO *Discoaster brouweri*	1.93
Planktonic foraminifera	FO *Globorotalia truncatulinoides*	1.93
Planktonic foraminifera	LO *Globigerinoides extremus*	1.98
Geomagnetism	Reunion event (top/base)	2.14/2.23
Coccolith	LO *Discoaster pentaradiatus*	2.39

FO, first occurrence; LO, last occurrence.
From Zhao and Wang (1999), Wang and Abelmann (1999), Wang and Abelmann (2002), Li, Zhong, et al. (2009), Li, Zeng, et al. (2009), and Anthonissen and Ogg (2012).

For deep-sea Quaternary stratigraphy, bioevents of planktonic foraminifera, nannofossils, and radiolarians are more frequently used. As many bioevents have been astronomically tuned (Anthonissen & Ogg, 2012), they provide a good base for age dating and correlation. However, care should be exercised when applying the events listed in Table 5.6, as errors may arise from poor preservation, hiatus, and disturbed range in some species due to changes in climate conditions and habitat preference.

5.3   REGIONAL STRATIGRAPHY AND SEQUENCE STRATIGRAPHY

Due to the combined effect of the (pre-)Indosinian orogeny (W–E direction) and the Himalayan orogeny (NE direction), almost all basins in the China Seas region are characterized by a NE-stretching elongate shape with belts of depressions and uplifts and by structural units of nearly W–E direction as swells or sags within each basins (Figure 5.1). Over several decades, numerous studies have been made on depositional sequences and their implications for structural evolution of many basins. In a synthetic study of many Chinese marine and outcrop sediment sections of Mesoproteozoic to Cenozoic age, for example, Wang, Shi, et al. (2000) proposed a sequence chronostratigraphic scale, including a chapter devoting to sequence stratigraphy and sea level change during the Mesozoic and Cenozoic. In this section, we summarize the major sediment packages in large basins in three China Sea sectors starting from the Bohai Basin in the northeast. For basins of the SCS, we largely follow the work of Zhong and Li (2009) with some updates.

5.3.1   East China Sea Sector

5.3.1.1   Bohai Basin

The Bohai Basin, also known as Northeast China Basin, or Bohai Gulf Basin, or Bohaiwan Basin, stretches northeasterly across the Bohai Sea to a maximum range of 1080 km and contains six major depressions, or (sub)basins, and three major uplifts, with a combined area of $\sim$80,000 km^2 (Figure 5.23). The central and southern parts of the modern Bohai Sea are occupied by Bozhong Depression while the northern Bohai (or Liaodong Bay) by the offshore portion of Liaohe Depression or Xialiao Basin. Secondary normal faults controlled 55 sags and 44 minor swells within the six depressions (Li, 2012a). The Bohai Basin is a Mesozoic–Cenozoic rift developed on the North China Platform with the Archean Taishan crystalline complex constituting the basement.

After the deposition of the huge middle and late Proterozoic carbonate megasequences (up to $\sim$9.4 km), the Cambrian to middle Ordovician marine sediments accumulated on the North China Platform before being uplifted during the late Ordovician–early Carboniferous. The deposition of the

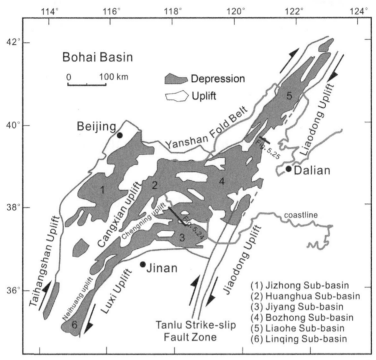

FIGURE 5.23 Major depressions (subbasins) and uplifts of the Bohai Basin. *Modified from Jiang, Liu, Zhang, Su, and Jiang (2011).*

alternating transgressive and regressive coal-bearing sequences took place in the middle and late Carboniferous and the intracratonic sequences in the Permian. Regional uplift in the late Permian marking the Hercynian tectonic phase of the Indosinian orogeny caused folding of all Proterozoic to Permian strata throughout northern China. Deposition resumed in the early Jurassic with sandstones, mudstones, and conglomerates intercalated with coal seams and basalt and in the middle and late Jurassic with volcanic agglomerates, sandstones, mudstones, conglomerates, and volcanic tuffs intercalated with oil shales and coal seams, to a total maximum thickness of ~10 km. The Cretaceous strata are characterized by tuffs, andesitic agglomerates, andesites, basalts, tuffaceous conglomeratic sandstones, and siltstones in the lower part (Qingshan Formation) and sandstones, conglomerates, siltstones, and mudstones in the upper part (Wangsi Formation).

Rifting in the late Cretaceous–Paleogene formed sags of various sizes with asymmetry half graben structures and multiple rifting characteristics (Hsiao et al., 2010; Hu, O'Sullivan, Raza, & Kohn, 2001; Huang, Liu, Zhou, & Wang, 2012; Huang, Yen, Zhao, & Lin, 2012; Qi et al., 2008), which shaped the basic features of the Bohai Basin. Rifting was accompanied by the deposition of the Kongdian, Shahejie, and Dongying Formations, mainly comprising sandstones, mudstones, oil shales, and conglomerates of alluvial, fluvial, and lacustrine origins (Table 5.2; Figure 5.24; Zhao et al., 2009). The Paleocene to Eocene Kongdian Formation includes three members, the Eocene to Oligocene Shahejie Formation four members, and the Oligocene Dongying Formation three members reaching over 2000 m thick. The total thickness of the synrift sequences may attain a maximum of 5000 m with considerable variations between depressions and between sags. Because of lacking accurate dating, however, some part of these sequences remains stratigraphically uncertain. With rich petroleum resource, these synrift sequences have been under industrial exploration over several decades (Figure 5.24; Tian et al., 2000; Zhao et al., 2009).

Detailed work on the sedimentary sequences in the northeastern part of the Bohai Basin, or Xialiao Basin, was carried out by Hsiao et al. (2010) and Hsiao and Graham (2012), among others. The Xialiao Basin extends for nearly 400 km and is 80–120 km wide, including the Liaodong Bay as the offshore (southern) part and the Liaohe subbasin as the onshore (northern) part. Based on 3-D seismic data, Qi, Li, Yu, & Yu (2013) provided an updated account on the structural evolution of the Liaohe subbasin. Like other parts of the Bohai Basin, the synrift Kongdian, Shahejie, and Dongying sequences in the Xialiao Basin correspond to different phases of rifting. Data indicate that sedimentation kept pace with tectonic subsidence during the deposition of the Kongdian and Dongying sequences but was outpaced by tectonic subsidence during Shahejie deposition (Hsiao et al., 2010). The initial rift Kongdian sequence consists of alluvial conglomerate near the faulted basin margins, whereas deposits of lacustrine, dry-pan, or fluvial environments dominated the basin center. The rift climax Shahejie sequence is characterized by deep lacustrine mudstone and marginal lacustrine mudstone and siltstone toward the basin center and margins, respectively, and the latter may include sublacustrine fan or surfaces of subaerial exposure. The late rift Dongying sequence consists of upward-coarsening deltaic deposits, indicated by large southward-prograding clinoform complexes, overlain by fluvial strata (Figure 5.25; Hsiao et al., 2010). In the geodynamic history of the Bohai Basin, the Dongying Formation may represent deposition in a period marked by the Tan-Lu fault's slip reversal from left lateral to right lateral at ~38 Ma and an enhancement of right-lateral slip at ~24 Ma (Huang, Liu, et al., 2012; Huang, Yen, et al., 2012).

For the synrift sequences of the Xialiao Basin, a series of sediment facies can be recognized (Hsiao et al., 2010). The *alluvial facies* is characterized by conglomerate and mudstone with textures/structures that suggest deposition by debris flows and traction currents at alluvial fan to fluvial settings. Gravel clasts are subrounded to subangular quartzite, volcanic, and metamorphic rocks. The *fluvial channel facies* is composed of well-sorted, medium-grained sandstone. Sedimentary structures include trough and low-angle planar cross-bedding,

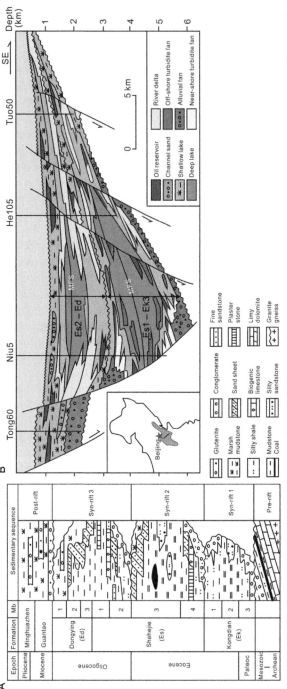

FIGURE 5.24 (A) Generalized lithostratigraphic characteristics of the Bohai Basin, showing the lithology and stratigraphy of the Paleogene and Neogene formations and their members. (B) Variations of lithofacies with oil reservoirs in a fault-controlled basin based on seismic and cored data. See Figure 5.23 for line location. *Modified from Zhao et al. (2009).*

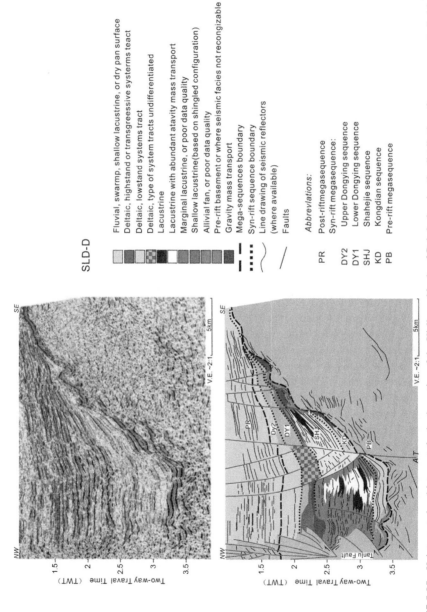

FIGURE 5.25 Uninterpreted and interpreted seismic section of the Xialiao subbasin of Bohai, showing the distribution of sequences controlled by faults. See Figure 5.23 for line location. *Simplified from Hsiao et al. (2010).*

horizontal bedding, and faint to structureless bedding. Thinly bedded sandstone interlayered with mud lamination is occasionally present, as is fully bioturbated sandstone. The *flood plain facies* is composed of mudstone with interlayering medium-grained sandstone. Mudstone is deposited from suspension in quiet, relatively low-energy environment, such as swamp or oxbow lakes. The *wetland facies* includes three lithofacies: laminated mudstone, massive mudstone, and interlayered sandstone. The *sublacustrine fan facies* includes structureless, parallel-laminated, or Bouma-sequenced fine-grained sandstone, massive mudstone, and pebbly mudstone. Dominated by black massive mudstone with occasional fine-sand laminations, this fairly homogeneous, mud-rich facies probably represents background sedimentation in a relatively deep, quiet, and poorly oxygenated lacustrine setting. The *profundal lacustrine facies* is dominated by black massive mudstone with occasional fine-sand laminations and mainly occurs at the Shahejie/Dongying sequence boundary or in association with sublacustrine fan facies (Hsiao et al., 2010).

The temporospatial variations of lithofacies between different synrift formations and members are better revealed in well records from various depressions of the Bohai Basin, as summarized by Tian et al. (2000). In particular, shallow-water carbonate and algal reef facies characterize the upper fourth member of the Shahejie Formation in the Jiyang Depression. Reefal limestone and dolomite contain abundant algae and ostracods with porosity up to 35.49%. Each reef body may cover an area of several 10 km^2 and reach nearly 100 m in thickness.

Apart from these characteristic facies, the three synrift sequences also exhibit other distinct large-scale seismic features. For the Kongdian sequence (Paleocene–early Eocene), high-amplitude reflections near the basin center suggest a relatively stable depositional environment, such as lacustrine and saline lacustrine and/or dry-pan environments. Although not yet encountered in the Xialiao Basin, evaporites deposited in saline lacustrine environments have been documented in other Bohai subbasins. The Shahejie sequence (middle Eocene–early Oligocene) shows both continuous and less continuous reflections, with the former representing relatively stable lacustrine environments and the latter implying frequent soft-sediment deformation or gravity mass transport. Parallel reflections correspond to mud-prone sections characterized also by irregular logging records. Most marginal lacustrine seismic facies occur toward the lake margin, including subaqueous talus deposits or fan deltas near the faulted basin margin and slump, slide, debris flow, and turbidity current deposits near the flexural margin. The Dongying sequence (late Oligocene) in the lower part is dominated by clinoforms associated with deltaic environments, as revealed by cores (Hsiao et al., 2010). Each clinoform started from a lake bottomset and completed with a delta plain topset. Nevertheless, the members of the Kongdian, Shahejie, and Dongying Formations may represent the second-order sequences, with sediment packages within each members corresponding to third-order sequences. For example, the third-order sequences Sq1 to Sq4 recognized by Feng, Li, and Lu (2013) to characterize the deposition at high lake level of the late Eocene in the Dongying Depression appear to correspond fully to the lower second to third members of the Shahejie Formation (Figure 5.26).

After the end of rifting and extension in the latest Oligocene, the Bohai Basin underwent a period of slow thermal subsidence through the early and middle Miocene, followed by rapid subsidence since 12 Ma in the Bozhong Depression and since 2 Ma in the Liaodong Bay (Hu et al., 2001; Keen et al., 2001). During this period, basin-wide subsidence led to the deposition of the fluviodeltaic Guantao Formation of Miocene age and the Minhuazhen Formation of the Pliocene. These postrift sequences are mainly composed of conglomerates, conglomeratic sandstones, sandstones, and mudstones with a thickness of 800–3000 m (Feng et al., 2013; Li, 2012a). Overlying the Neogene strata is the Quaternary Pingyuan Formation, mainly comprising conglomerates, sand, and muddy aeolian loess-type deposits, interbedded by silty layers with marine microfossils.

5.3.1.2   Yellow Sea Basins

Three large, NE-extending basins developed in the Yellow Sea: from north to south, the North Yellow Sea Basin, the Northern South Yellow Sea Basin, and the Southern South Yellow Sea Basin, separated by structural highs (Figure 5.27; Li, 1995; Li, Lu, Liu, & Xu, 2012). Like the Bohai Basin, these Yellow Sea basins developed on the North China Platform constituted by Proterozoic to Triassic mainly paralic to full-marine carbonate sequences (Cai, 2005), except in the Devonian that was dominated by nonmarine strata. In several wells, Carboniferous, Permian, and Triassic carbonates and shelf sequences are confirmed to exist in the Yellow Sea (Chough et al., 2000).

All three basins of the Yellow Sea shared a similar formation history under the combined influence of the Tan-Lu fault system and subduction of the Philippine Sea Plate during the late Mesozoic–Paleogene (Cai, 2005; Yao et al., 2010). They show a tendency of sequential development from north to south, with rifting starting in the Jurassic–Cretaceous in the Northern Yellow Sea Basin and in the Cretaceous–Paleocene in the South Yellow Sea Basins (Hou et al., 2008; Li, 1995). Deposition in these basins is characterized by fluviolacustrine sediments of Cretaceous to Eocene age. After a period of region-wide erosion during the Oligocene, postrift Neogene sediments of mainly nonmarine clastic facies accumulated and overlain unconformably on older Cenozoic deposits. Although brackish-water microfossils have been reported from some intervals of late Cretaceous to Neogene strata, typical marine environment did not establish in the region until the Quaternary time.

North Yellow Sea Basin

The North Yellow Sea Basin extends from the central North Yellow Sea to the North Korean coast. It covers an area of 15,000 km^2 and consists of six

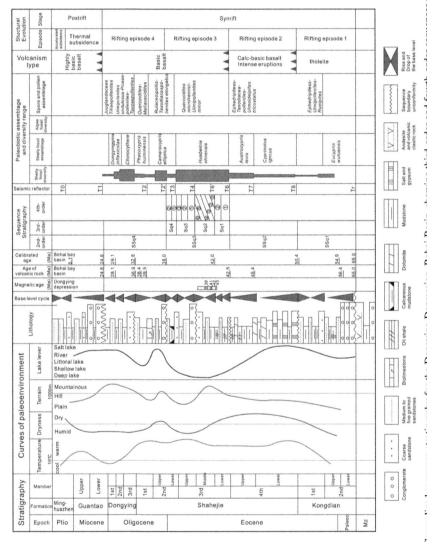

FIGURE 5.26 Generalized sequence stratigraphy for the Dongying Depression, Bohai Basin, showing third-order and fourth-order sequences formed during the late Eocene with a maximum lake level. *Modified from Feng et al. (2013).*

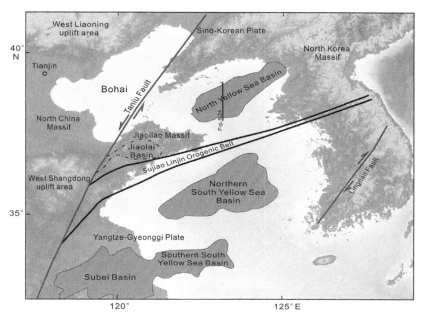

FIGURE 5.27 Distribution of the three large basins in the Yellow Sea. *Modified from Li, Lu, et al. (2012) and Dong, Cao, Li, Zhao, and Li (2013).*

secondary structural units, including three depressions, one sag group, and two uplifts (Figure 5.28). The eastern depression of the North Yellow Sea Basin is also called the West Bay Basin (Li, Lu, et al., 2012) or the Sun Basin (Chen, Jin, Liu, Wang, & Wang, 2008; Dong et al., 2013). According to Chen, Liu, Jin, Wang, and Yuan (2008), the structural evolution in the region can be divided into five stages: sag depression in the late Jurassic–early Cretaceous, thermal upwelling in the late Cretaceous–Paleocene, extensional fault depression in the Paleogene, compression in the late Oligocene–early Neogene, and thermal subsidence in the Neogene. Although it has long been considered as a residue basin (e.g., Yao et al., 2010), the North Yellow Sea Basin is attributed by Li, Zeng, and Huang (2009) to a superimposed basin because of its two rifting stages: the first in the late Jurassic–early Cretaceous and the second in the Eocene–Oligocene.

The North Yellow Sea Basin is dominated by lacustrine mudstone and alluvial fan deposits (Cai, 2005; Cai, Dai, Chen, Li, & Sun, 2005; Li, Lu, et al., 2012; Wang, Wang, & Hu, 2010). In the western depression, the maximum thickness of the Mesozoic and Cenozoic sequences combined is about 10,200 m (Wang et al., 2010). In the eastern depression, drilling confirmed the Jurassic sequence reaching 2500 m, the lower Cretaceous 3000 m, the Eocene–Oligocene 3000 m, and the Neogene–Quaternary ~600 m (Cai et al., 2005). For most part of the basin, however, the upper Cretaceous, Paleocene, and part of the Oligocene

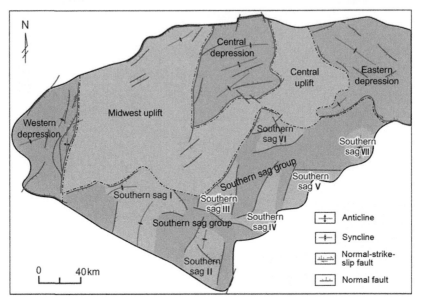

FIGURE 5.28 Structural units of the North Yellow Sea Basin (Li, Lu, et al., 2012), including four main depressions: eastern depression, western depression, central depression, and southern depression (or sag group).

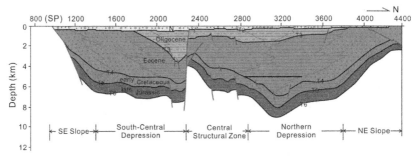

FIGURE 5.29 A S–N cross section of the east depression of the North Yellow Sea Basin, showing the dominance of two synrift sequence sets (late Jurassic–early Cretaceous and Eocene–Oligocene) dominated by fluvial fan sandstones and lacustrine mudstones (Wang et al., 2012). See Figure 5.27 for line location.

did not develop presumably due to erosion and nondeposition (Figures 5.4 and 5.29; Li, Lu, et al., 2012; Wang, Guo, & Nie, 2012).

Seismic–stratigraphic mapping in the central depression of the North Yellow Sea Basin has produced facies maps as shown in Figure 5.30. Clearly, the late Jurassic, early Cretaceous, and Eocene were the three periods with maximum extensions and variations of sedimentary facies. During all the periods, the deposition of alluvial fan sandstones and lacustrine mudstones prevailed.

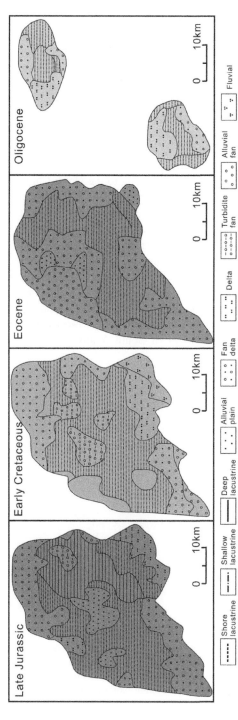

FIGURE 5.30 Distribution of sedimentary facies in the central depression of the North Yellow Sea Basin during the late Jurassic, early Cretaceous, Eocene, and Oligocene (Li, Lu, et al., 2012).

Northern and Southern South Yellow Sea Basins

Located in the central Yellow Sea is the Northern South Yellow Sea Basin (Figure 5.27), also known as the Central Yellow Sea Basin and (for the eastern part) the Gunsan Basin (Lee, 2010; Shinn, Chough, & Hwang, 2010). It is bounded on the north by the Qianliyan fault, which bifurcates northeastward away from the Tan-Lu fault, and on the south by the Central Massif. Further south of the Central Massif is the Southern South Yellow Sea Basin, an offshore extension of the onshore Subei Basin (or North Jiangsu Basin; Figure 5.28). With a total coverage of 75,000 km^2, these basins comprise a number of depressions or subbasins that are mostly bounded by NE- to E-trending normal faults. They shared a similar early history in tying to the North China Platform during the Paleozoic–early Mesozoic but started to differentiate in the Jurassic–Cretaceous with various rifting styles, resulting in a residue type or superimposed type with single or multiple stages of development (e.g., Li, Zeng, & Huang, 2009; Yang & Chen, 2003; Zheng et al., 2001, Figure 5.31).

Like the North Yellow Sea Basin, the Northern and Southern South Yellow Sea Basins were initially filled with alluvial to fluvial deposits containing volcaniclastic rocks, red beds, and lacustrine deposits of Cretaceous age. Major sedimentation occurred in the Paleogene when rifting intensified. Lacustrine mudstones and fluvial sandstones are most common in all subbasins of the Northern and Southern South Yellow Sea Basins (Cai, 2005;

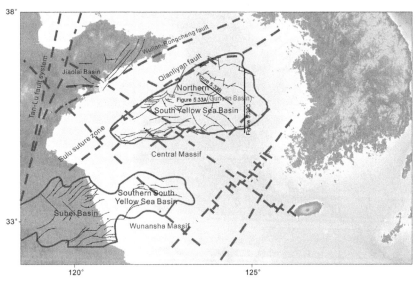

FIGURE 5.31   Major structural characteristics of the Northern South Yellow Sea Basin (including the Gunsan Basin) and the Southern South Yellow Sea Basin, the latter representing an offshore extension of the onshore Subei Basin. Major fault zones (dash line) and normal faults (thin line) that controlled the formation of basins are shown. *Modified from Shinn et al. (2010).*

Shinn et al., 2010). The synrift deposition includes the upper Cretaceous Taizhou Formation, the Paleocene Funing Formation, the early Eocene Dainan
Formation, and the middle Eocene Sanduo Formation. The postrift deposition
includes the Miocene–Pliocene Yancheng Formation and Quaternary Dongtai
Formation, with a combined thickness of ~1000 m (Figure 5.32). These stratal nomenclatures, originally from the Subei Basin, are applicable to both the
Northern and Southern South Yellow Sea Basins. However, no specific

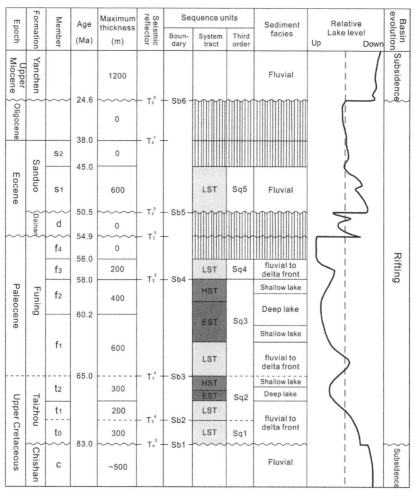

FIGURE 5.32 Sequence stratigraphy of the Baiju Sag, with additional data from neighboring
depressions in the Subei Basin (Wei et al., 2011), showing relative long periods of erosion and
thin Cretaceous–Paleogene sequences compared with those in the offshore Southern and Northern
South Yellow Sea Basins. LST, lowstand system tract; TST, transgressive system tract; HST,
highstand system tract.

formation nomenclature has been in place for strata from either the Gunsan Basin or the North Yellow Sea Basin, where only the standard epoch-period scheme is in use.

The synrift and postrift sequences show some different characteristics between the Northern and Southern South Yellow Sea Basins. For example, the Cretaceous deposition is widespread in the northern basin but only confined to a few depressions or sags in the southern basin. While well developed in both basins, the Paleocene and Eocene strata often achieved a wider coverage and a greater thickness in the north than in the south. For the Neogene and Quaternary deposits, however, their overall thickness in the southern basin (and also the onshore Subei Basin) is always larger than in the north although with considerable variations within and between subbasins (Hou et al., 2008; Yao et al., 2010).

In the Baiju sag of the Subei Basin (33°N, 120°20′E), ∼30 km inland from the South Yellow Sea, Wei et al. (2011) identified five sequences for the upper Cretaceous–Paleogene strata (Figure 5.32): SQ1 and SQ2 for the lacustrine Taizhou Formation representing the initial rifting phase, SQ3 and SQ4 for the lacustrine Funing Formation deposits during the intensified rifting phase, and SQ5 for the fluvial Sanduo Formation strata in the thermal subsidence phase. The entire early Eocene Dainan Formation and the upper parts of the Funing and Sanduo strata, however, are missing from the Baiju sag (Figure 5.32), probably due to a relative high baseline of the basin compared to other depressions further inland or offshore during the Cretaceous–Eocene time. It demonstrates that, although the fluvial–lacustrine environment was prevalent, variations in deposition and subsequent erosion due to tectonic disturbance between (sub)basins were quite large. In association with the fluvial–lacustrine environment occurred many kinds of nonmarine microfossils. Apart from abundant spores and pollen, ostracods, and bivalves, rare benthic foraminifera referable to species of *Discorbis* and *Ammonia* have been found in some intervals of Taizhou and Funing Formations in the Subei Basin (Fu, Li, Zhang, & Liu, 2007), indicating the influence of coastal marine environment at settings such as brackish-water lakes.

In the eastern depression of the Northern South Yellow Sea Basin, or Gunsan Basin, three major seismic units are found corresponding to the Cretaceous–Eocene (SU-1), early Miocene (SU-2), and middle Miocene–Quaternary sequences (SU-3), respectively (Figure 5.33; Shinn et al., 2010). The depositional environmental history of the Gunsan Basin, as reconstructed on palynomorph evidence by Yi et al. (2003), also reveals various stages of basin evolution (Figure 5.34), from rifting in the Cretaceous–Eocene, inversion and erosion in the Oligocene, to postrift regional subsidence in the Miocene–Pliocene. Frequent marine transgressions did not occur until the Pleistocene, as reported in core QC2 from the South Yellow Sea (Chapter 4). Palynomorph data imply that the fluviolacustrine depositional environments were associated with climate variations between semiarid and

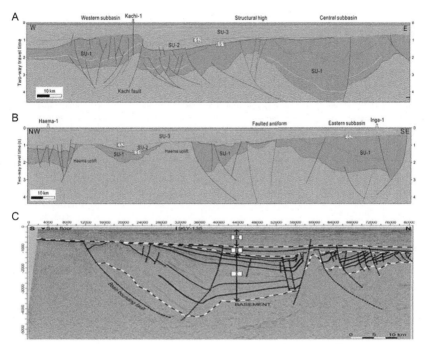

FIGURE 5.33   Regional seismic profiles and interpretations of the Gunsan Basin, or the eastern depression of the Northern South Yellow Sea Basin. (A) W–E direction profile and (B) SW–NE direction profile, showing faults (black line) and seismic units SU-1 (Cretaceous–Eocene), SU-2 (early Miocene), and SU-3 (middle Miocene–Quaternary) separated by two unconformity surfaces ES1 and ES2, respectively (Shinn et al., 2010). Also shown are the approximate base of the Eocene (pink line) and the base of Paleocene (yellow line). (C) S–N direction profile, showing faults (black dash line) and seismic sequences I (Paleocene and older), II (Eocene), and III (middle Miocene to the present) separated by unconformities (dash line) (Lee, 2010). See Figure 5.31 for line locations.

wet and subtropical and warm temperate conditions, except during the late Eocene and Pliocene when a cool-temperate climate prevailed (Figure 5.21; Yi et al., 2003).

Work on the SW subbasin of the Gunsan basin by Lee (2010) has identified two phases of the subsidence history of the basin: a main subsidence phase during the late Cretaceous–Oligocene and a secondary subsidence phase since the middle Miocene. These phases were separated by an uplift and erosion phase, or inversion phase, during early Miocene times. Associated with these structural phases were three seismic sequences: MSQ I (acoustic basement to Paleocene), MSQ II (Eocene), and MSQ III (middle Miocene to the present).

It is not clear, however, whether SU-2 of Shinn et al. (2010) and MSQ II of Lee (2010) both from the Gunsan Basin correspond to each other at all, even though SU-2 was originally supposed to bear the age of early Miocene

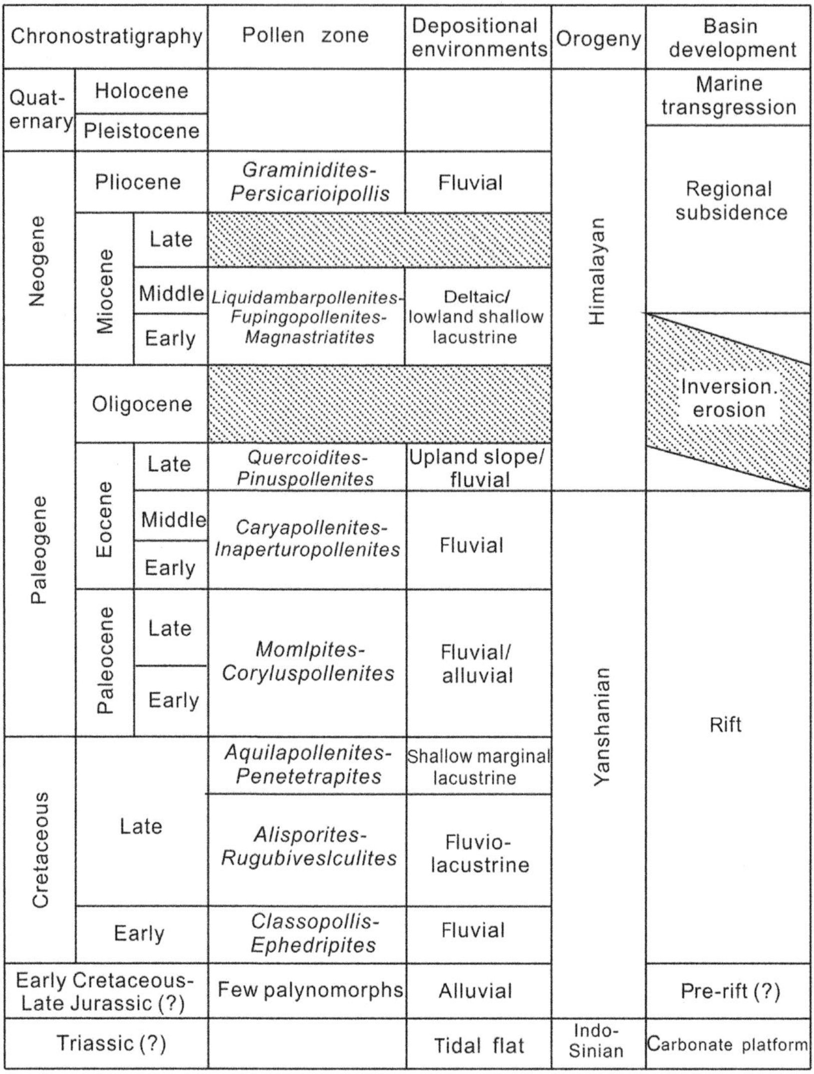

Chronostratigraphy			Pollen zone	Depositional environments	Orogeny	Basin development
Quat-ernary	Holocene				Himalayan	Marine transgression
	Pleistocene					
Neogene	Pliocene		Graminidites-Persicarioipollis	Fluvial		Regional subsidence
	Miocene	Late				
		Middle	Liquidambarpollenites-Fupingopollenites-Magnastriatites	Deltaic/lowland shallow lacustrine		Inversion. erosion
		Early				
Paleogene	Oligocene					
	Eocene	Late	Quercoidites-Pinuspollenites	Upland slope/fluvial		
		Middle	Caryapollenites-Inaperturopollenites	Fluvial	Yanshanian	
		Early				
	Paleocene	Late	Momlpites-Coryluspollenites	Fluvial/alluvial		Rift
		Early				
Cretaceous	Late		Aquilapollenites-Penetetrapites	Shallow marginal lacustrine		
			Alisporites-Rugubiveslculites	Fluvio-lacustrine		
	Early		Classopollis-Ephedripites	Fluvial		
Early Cretaceous-Late Jurassic (?)			Few palynomorphs	Alluvial		Pre-rift (?)
Triassic (?)				Tidal flat	Indo-Sinian	Carbonate platform

FIGURE 5.34 Stratigraphy, palynomorph assemblages, and tectonic history of the eastern depression (Gunsan Basin) of the northern South Yellow Sea Basin (Yi et al., 2003).

and MSQ II the Eocene. In addition, Yi et al. (2003) found the absence of two sections (Oligocene and upper Miocene) from the Gunsan Basin region (Figure 5.34), casting more questions on regional stratigraphic correlation. Likely, a series of inversion-induced erosional event had occurred in the Miocene northern South Yellow Sea, but their accurate age and duration need further analyses.

5.3.1.3 East China Sea Shelf Basin

As the name implies, the East China Sea Shelf Basin (ECSSB) occupies much of the East China Sea shelf and borders the China coast in the west and the Taiwan–Sinzi Belt (or Diaoyudao Ridge) and the Okinawa Trough in the east (Figure 5.35). Covering an area of 260,000 km^2 with an average water depth of 72 m, the NE-extending ECSSB is the largest Cenozoic sedimentary basin in the China Seas today (Table 5.1).

The ECSSB mainly includes a western depression belt, a central low rise belt, and an eastern depression belt (Zhao, 2004, 2005a,2005b) and has a different history from the Okinawa Trough (see Chapter 3). Subordinate tectonic units are controlled by NE longitudinal faults, displaying an alternating distribution of the three basinal and uplift belts. The western depression belt comprises six secondary depressions or basins: from north to south, Changjiang, Qiantang, Oujiang, Min Jiang, Nanridao, and Pengxi, while the eastern depression belt contains four basins: Fujiang, Xihu, Jilong, and Xinzhu (or Fuzhou; Figure 5.35A; Yang et al., 2010). Stratigraphically, basins in the western belt are characterized by having Paleocene deposits but lacking the Oligocene, lower Miocene, and upper Miocene deposits, while basins in the eastern belt by having strata of the middle and upper Eocene, Oligocene, and middle Miocene, indicating tectono-environmental variations between the two belts. However, care should be exercised when dealing with the (sub)basins as many different names have been used by different authors (Figure 5.35B; Cukur et al., 2011; Lin et al., 2005; Xu, Ma, Fu, Liang, & Zheng, 2003; Yang, 1992; Zhou, Zhao, & Yin, 1989). For example, the Chang Jiang Basin from the northern ECS is equivalent to the Sototra Basin by Korean researchers (Cukur et al., 2011), while the Jilong and Xinzhu Basins from the south are often referred to as the Taipei Basin by authors from Taiwan, which may even include a part of the Xihu Basin from the central eastern ECS (Figure 5.35B; Lin et al., 2005).

A comprehensive analysis of available industry seismic data from the East China Sea indicates the existence of six belts of elongated deep back-arc basins and associated arc volcanic ridges (Figure 5.35B; Lin et al., 2005). These belts of basins and ridges are progressively younger from the China shoreline (late Cretaceous–early Paleogene) to the Okinawa Trough (middle Miocene to the present). The basement topography of the three more recent belts of basins and ridges in the east is much more developed than the one of the three older belts in the west, possibly in relation with the late Eocene/early Oligocene change in steepness of the Ryukyu slab (Lin et al., 2005). In general, the evolution of the ECSSB can be divided into five stages: (1) thermal uplift in the late Jurassic to early Cretaceous, (2) rifting during the Paleocene–Eocene, (3) depression in the Oligocene, (4) inversion–depression in the Miocene, and (5) draping since the Pliocene (Lee, Kim, Shin, & Sunwoo, 2006; Yang et al., 2010). The Oligocene and Miocene saw major

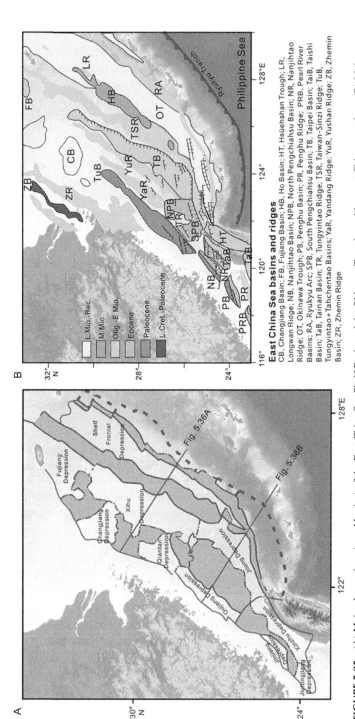

FIGURE 5.35 (A) Major depressions (or subbasins) of the East China Sea Shelf Basin and the Okinawa Trough used by most Chinese workers. (B) Major structural units of the ECSSB according to other researchers. Note that belts of depressions (basins bounded by solid lines) and associated ridges in B are considered as corresponding to six rifting phases from the coastal East China Sea to the Okinawa Trough. (A) Modified from Yang et al. (2010), Li (2012b), and references. (B) Modified from Lin et al. (2005).

East China Sea basins and ridges
CB, Changjiang Basin; FB, Fujiang Basin; HB, Ho Basin; HT, Hsüehshan Trough; LR, Longwan Ridge; NB, Nanjihtao Basin; NPB, North Pengchiahsu Basin; NR, Nanjihtao Ridge; OT, Okinawa Trough; PB, Penghu Basin; PR, Penghu Ridge; PRB, Pearl River Basins; RA, Ryukyu Arc; SPB, South Pengchiahsu Basin; TB, Taipei Basin; TaiB, Taishi Basin; TaB, Tainan Basin; TR, Tungyintao Ridge; TSR, Taiwan-Sinzi Ridge; TuB, Tungyintao+Tahchentao Basins; YaR, Yandang Ridge; YuR, Yushan Ridge; ZB, Zhemin Basin; ZR, Zhemin Ridge

differentiations between the western and eastern belts: uplift and then subsidence in the west but subsidence and then uplift in the east. A major tectonic event occurred at late Miocene (∼5 Ma), causing inversion and erosion in the Xihu Depression and a sharp angular unconformity in the Okinawa Trough, indicating a significant tectono-environmental change in the East China Sea region (Chapter 3).

The basement rock types of the ECSSB include Mesozoic sedimentary rocks and igneous rocks or older metamorphic rocks. In most basins, the upper Cretaceous fluvial Shimentan Formation underlies the synrift Paleogene sequences (Figures 5.14 and 5.34). The Paleocene sequences include the fluvial Yueguifeng Formation (originally as the upper part of the Shimentan, Yang et al., 2010; Wu et al., 2000), neritic Lingfeng Formation, and lacustrine to neritic Mingyuefeng Formation, and the Eocene sequences comprise the Oujiang, Baoshi, and Pinghu Formations with lacustrine to neritic facies. The Oligocene is typified by the fluviolacustrine Huanggang Formation, although its stratigraphic position is still under debate (e.g., Chen et al., 2013). The postrift Neogene sequences include the lacustrine to neritic Longjing, Yuquan, and Luilang Formations of Miocene age and the deltaic to neritic Santan Formation of Pliocene age, overlain by the shallow neritic or clastic Quaternary Donghai Formation.

The Mesozoic–Cenozoic strata together may reach a maximum thickness of 12 km or more in southern and western basins (Figure 5.36; Yang et al., 2010) but only 500–5500 m in some northern basins (Cukur et al., 2011). The strata are characterized by alluvial and fluviolacustrine facies with frequent intervals of marine deposition especially in the south (e.g., Chen, Lu, Zhang, & Zhang, 2013; Xu et al., 2003; Yang et al., 2010). During the late Cretaceous–Paleocene, alternative marine and terrestrial facies first developed in the south of the basin with sea transgression from the southeast (Figure 5.37). During the Eocene, neritic facies developed predominantly in the south and southwest of the basin, while other areas were prevailed by deltaic and semiclosed bay facies. The Oligocene was dominated by fluviolacustrine facies with no deposition or erosion in some areas. Fluviolacustrine or swamp facies continued to dominate the Miocene deposition except in some local areas influenced by shallow-marine environment. Since the Pliocene, the regional deposition changed gradually from fluvial to littoral and, since the Quaternary, to neritic facies (Figure 5.37; Wang & Zhu, 1992; Wu et al., 2000; Yang et al., 2010). Although these pictures have been painted with industrial data, however, the stratigraphy of the upper Cretaceous and Paleogene sequences in the region requires further investigations.

Cukur et al. (2011) identified four megasequences bounded by four regional unconformities including the top of the acoustic basement in the Socotra Basin (Changjiang Depression), Domi Basin (Fujiang Depression), and Jeju Basin (Zhedong Depression) in the northern ECSSB, as a modification of the scheme proposed earlier by Lee et al. (2006). MS1 (late

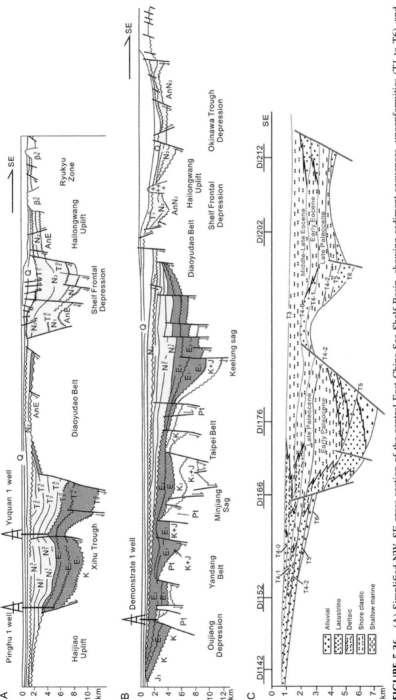

FIGURE 5.36 (A) Simplified NW–SE cross section of the central East China Sea Shelf Basin, showing sediment sequences, unconformities (T1 to T6), and faults. (B) Simplified NW–SE cross section of the southern ECSSB to the Okinawa Trough. (C) An enlarged view of the western portion of B, showing the synrift sequences and sedimentary facies in the lower part of the Oujiang Depression affected by faults. See Figure 5.35 for line locations. (A) *Modified from Yang et al. (2010). (B) modified from Yang et al. (2010). (C) modified from Wu, Lu, and Li (2000).*

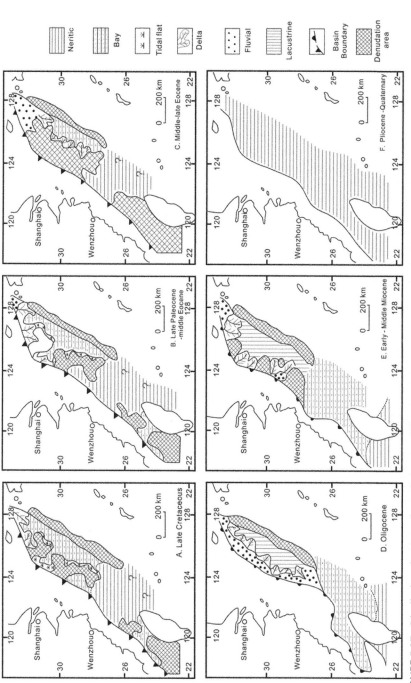

FIGURE 5.37 Distribution of major sediment facies over different periods in the East China Sea Shelf Basin. Note that the extent of marine deposition during the late Cretaceous–Paleogene is highly speculated, and the Pliocene marine records are sparse and limited only to the southern part. *Modified from Zhou et al. (1989) and Wu et al. (2000).*

Cretaceous–late Eocene and early Oligocene) is more than 1100 m thick in the SW Domi Basin and thickens to over 1600 m in the Socotra Basin and up to 2100 m in the Jeju Basin, consisting mainly of fluviolacustrine and some upper coastal deposits interbedded with volcaniclastic rocks and thin coal beds. MS2 (late Eocene and early Oligocene–early Miocene) is less than 300 m thick in the SW Domi Basin and thickens to a maximum thickness of 1050 m in the Jeju Basin, comprising mainly of sandstone and mudstone with interbeds of conglomerate and coal and freshwater limestone deposited in fluvial–alluvial to lacustrine environments. MS3 (early Miocene–late Miocene) reaches 1500 m or more in the southern Jeju Basin, consisting of sandstone and mudstone with coal beds deposited in fluvial environment. The upper MS3 in the eastern Domi and Jeju Basins and in the Socotra Basin was probably deposited during a gradual transition from nonmarine to paralic to shallow-marine environment as the entire northern ECSSB subsided progressively from southeast to northwest. MS4 (late Miocene to the present) is less than 250 m thick in the NW Domi Basin and increases to >750 m in the Socotra Basin and more than 1000 m in the Jeju Basin, dividable into a lower subunit dominated by Pliocene sandstones and mudstones with very minor interbeds of coal and an upper subunit of the Quaternary with fine-grained neritic or shallow-marine sediments containing shell fragments of foraminifera, gastropods, mollusks, nannofossils, and dinoflagellates (Cukur et al., 2011). The dominance of clastic facies in the ECSSB has stimulated attempts to classify them based on their seismic features (e.g., Pigott, Kang, & Han, 2013).

A systematic study on sequence stratigraphy of the Xihu Depression in the central eastern ECSSB was carried out by Wu et al. (2000). A total of 7 supersequences (SS1–SS7) and 19 sequences were identified, corresponding to stratal formations and their members widely used in the region (Figure 5.38). For example, supersequence SS4 and its four third-order sequences IV_A to IV_B represent deltaic to coastal bay deposits of the Pinghu Formation and its four members, respectively, in water depths <80 m during the middle and late Eocene. Based on litho- and biostratigraphic evidence, maximum marine influence with water depths of 80–100 m in the ECSSB occurred only in two periods: late Paleocene and late Pleistocene–Holocene, with the former responding to intensive rifting and the latter to most recent regional subsidence. The work by Zhang, Xu, Zhong, Zhang, and Yu (2012) on the sequence stratigraphy of the Pinghu (with three sequences), Huanggang (2), Longjing (1), Yuquan (2), and Liulang (2) Formations in the Xihu Depression basically confirmed the scheme developed by Wu et al. (2000) (Figure 5.38). The Pinghu Formation with three third-order sequences had also been described by Yu, Chen, Du, and Zhang (2007). Based on a synthesis of palynological and other microfossil data, Chen et al. (2013) attributed the Huanggang Formation in the Xihu Depression to a late Oligocene deposition (rather than the entire Oligocene from earlier opinions), thus implying a hiatus during the early Oligocene in the region.

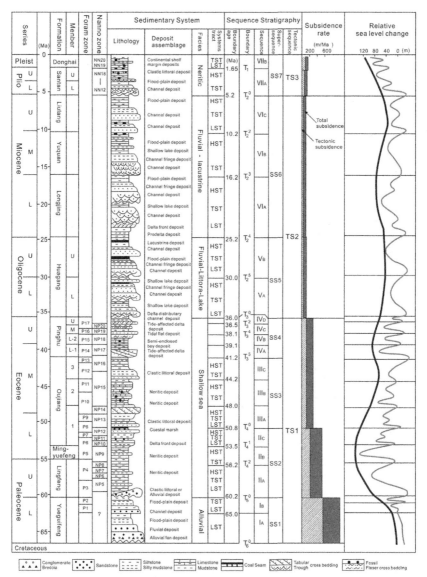

FIGURE 5.38   Generalized sequence stratigraphy and sea level changes of the East China Sea Shelf Basin. Note that the Huanggang Formation is now placed in the late Oligocene, indicating the missing of the early Oligocene in the region (Chen et al., 2013). *Modified from Wu et al. (2000).*

5.3.2   Northern South China Sea Sector

5.3.2.1   Taixinan Basin

In the northeastern corner of the SCS developed four NE–SW-trending basins: from north to south, the Nanjihtao, Taihsi, Penghu, and Taixinan Basins

(SW Taiwan or Tainan; Figure 5.39). The Taixinan Basin is the largest, with area of about 70,000 km², and includes a large offshore part and a small onshore part in southern Taiwan. It is separated to the north from the Penghu and Taihsi Basins by the Penghu-Peikang High (a pre-Cenozoic basement uplift) and to the west from the Pearl River Mouth Basin by the Dongsha Swell. The basement of the Taixinan Basin comprises mainly late Mesozoic continental to paralic clastic rocks as revealed by well data (Jin, 1989; Lee et al., 1993; Lin et al., 2003; Tzeng, Uang, Hsu, & Teng, 1996). Four NE-trending structural belts in the basin can be recognized: the northern depression, central uplift, southern depression, and southern uplift (Figure 5.39; Chang, 1997; Du, 1991, 1994; Tang, Oung, Hsu, & Yang, 1999).

The Taixinan Basin is a Cenozoic rift basin superimposed by a latest Miocene to the present foreland basin, respectively, relating to the opening of the SCS and the later arc-continental collision (Chen et al., 2001; Du, 1991, 1994; Lee et al., 1993; Lin et al., 2003; Yu & Hong, 2006). The total Cenozoic sediments may reach ∼10,000 m thick (Huang, Liang, & Juang, 1998; Lin et al.,

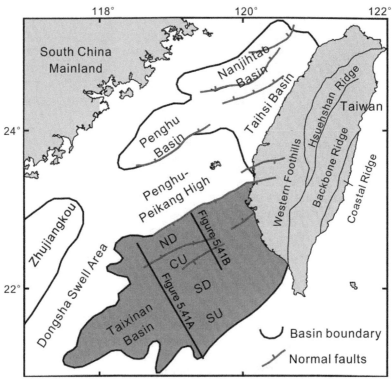

FIGURE 5.39 Sketch map showing major faults and structural units in the Taixinan and neighboring basins. ND, northern depression; CU, central uplift; SD, southern depression; SU, southern uplift. *From Zhong and Li (2009).*

2003). Oligocene and younger sediments have been found in wells drilled on the central uplift (e.g., Lee et al., 1993; Tzeng et al., 1996). Well DP21-1-1 from the SW part of the basin recently uncovered early and middle Eocene marine strata with abundant planktonic foraminifera, representing the only marine early Eocene record found in the SCS (see Chapter 8). This finding has stimulated the imagination that perhaps these marine Eocene strata were once related to the contemporary marine sequences in the southern ECS Shelf Basin.

Oligocene and younger deposits in the Taixinan Basin mainly consist of shallow-marine sandstones, siltstones, shales, and clays (Oung, 2000; Tzeng et al., 1996; Figure 5.40). The Oligocene Chihchang Formation is composed of sandstones and siltstones in the lower part, with minor shales and limestones, and mainly shales in the upper part. The lower Miocene Cherngan Formation consists of shales with minor siltstones, sandstones, and limestones. The middle Miocene Chingyuan Formation includes shales interbedded with siltstones and some thin limestone beds. The upper Miocene Chihwang Formation is composed of shales intercalated with coarse sands, limestone, and thin coal beds. The Pliocene lower Gutinkeng Formation consists of clays with minor sandstone and siltstone interbeds. The Pleistocene includes the upper Gutinkeng, Erchungchi, and Liushuang Formations, mainly comprising clay intercalated with thin-bedded sandstones (Figure 5.40).

The stratigraphic framework for the Taixinan Basin was established on seismic and sequence stratigraphy, supplemented by limited lithostratigraphic and paleontological calibrations (Du, 1991, 1994; Lee et al., 1993; Lin et al., 2003; Tzeng et al., 1996). About eight regional unconformities identified provide a framework for sequence stratigraphy in the basin (Chang, 1997; Du, 1991, 1994; Lee et al., 1993; Oung, 2000; Tzeng et al., 1996; Figure 5.41). Three megasequences characterize the three stages in basin development: the Oligocene (?)–early Miocene synrift, middle to late Miocene postbreakup, and latest Miocene to the present foreland stages (e.g., Ding et al., 2008). Some authors suggest that the breakup unconformity separating the synrift from the postrift megasequences lies either at the Eocene–Oligocene boundary ~34 Ma (Lin et al., 2003) or earlier at 37.8 Ma (Hsu, Yeh, Doo, & Tsai, 2004; Yeh, Sibuet, Hsu, & Liu, 2010). Based on stratigraphic correlation in Taiwan, Huang, Liu, et al. (2012) and Huang, Yen, et al. (2012) considered the breakup of the SCS as early as 39 Ma, in the middle Eocene.

5.3.2.2   Pearl River Mouth Basin

The NE-trending Pearl River Mouth (Zhujiangkou) Basin is located in the shelf and slope area offshore from the Pearl River mouth, reaching 800 km long and 100–300 km wide, with an area of approximately $14.7 \times 10^4 \, km^2$ (Gong, Jin, Qiu, Wang, & Meng, 1989). It is bordered to the north by the

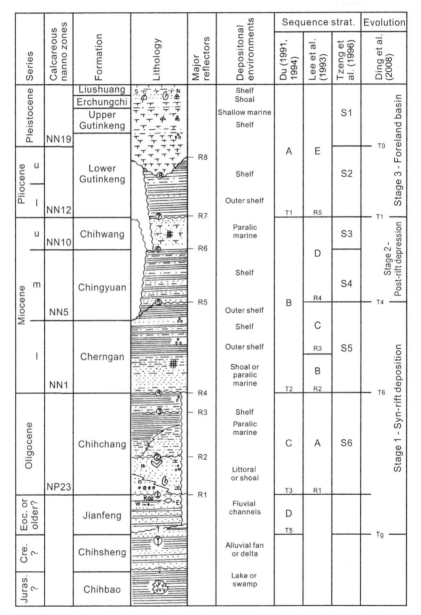

FIGURE 5.40 Taixinan Basin lithostratigraphy, sequence stratigraphy, and major stages of basin development. *From Zhong and Li (2009); based on Oung (2000), Chang (1997), and Ding et al. (2008).*

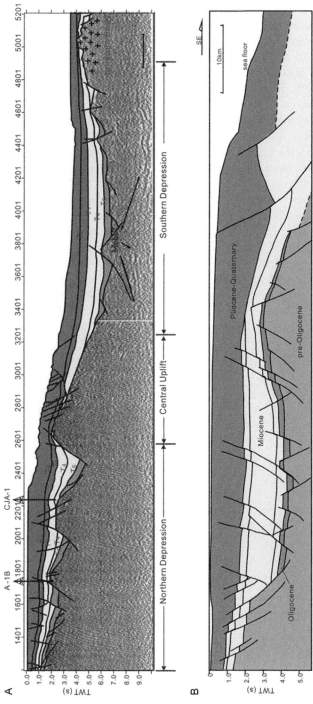

FIGURE 5.41 (A) Taixinan Basin seismic profile and interpretations, showing major faults, stratigraphic units, and major seismic reflectors. (B) Interpreted cross section from the Taixinan Basin. See Figure 5.39 for line locations. (A) *Modified from Ding et al.* (2008). (B) *From Zhong and Li* (2009); *based on Du* (1991, 1994) *and Lee et al.* (1993).

Wanshan Swell area, to the west by the Hainan uplift zone and the Qiong-dongnan Basin, and to the east by the Dongsha Swell Area and the Taixinan Basin. To its south lies the central deep basin of the SCS (Figure 5.7). The basin contains six major structural units: Zhu I, Zhu II, and Zhu III subbasins and Dongsha, Shenhuansha, and Panyu uplifts (Figure 5.42; Chen, 2000; Gong et al., 1989; Jin, 1989).

The Pearl River Mouth Basin is a Cenozoic rift basin developed on a downfaulted basement consisting mainly of Mesozoic and older granites and sedimentary rocks and some volcanic rocks (Chen, 2000; Gong et al., 1989; Jin, 1989). Very thick Jurassic–Cretaceous sedimentary rocks with deep-sea radiolarians have been encountered in a well drilled in the eastern part of the basin (Shao et al., 2007; Wu, Wang, Hao, & Shao, 2007). The overlying Cenozoic sequence may reach a maximum thickness of >10,000 m (Figure 5.43; Chen, 2000; Gong et al., 1989).

Like many other basins in the region, the development of the Pearl River Mouth Basin involved a rifting phase in the Paleogene and a postrift phase in the Neogene (e.g., Chen, Xu, & Sang, 1994), although the breakup unconfor-mity has been considered as close to the early/late Oligocene boundary at 28–30 Ma in earlier studies (Gong et al., 1989; Gong & Li, 1997). Two rifting

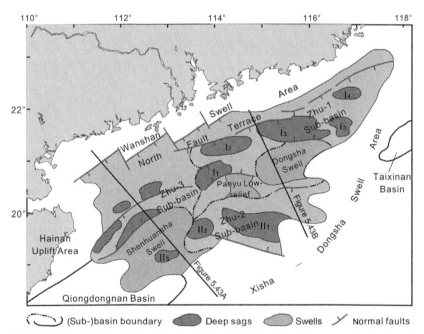

FIGURE 5.42 The Pearl River Mouth (Zhujiangkou) Basin and its major structural units (Zhong & Li, 2009; based on Gong et al., 1989). I1, Enping Sag; I2, Xijiang Sag; I3, Huizhou Sag; I4, Lufeng Sag; II1, Baiyun Sag; II2, Kaiping Sag; II3, Shunde Sag; III1, Wenchang Sag; III2, Qionghai Sag; III3, Yangjiang Sag.

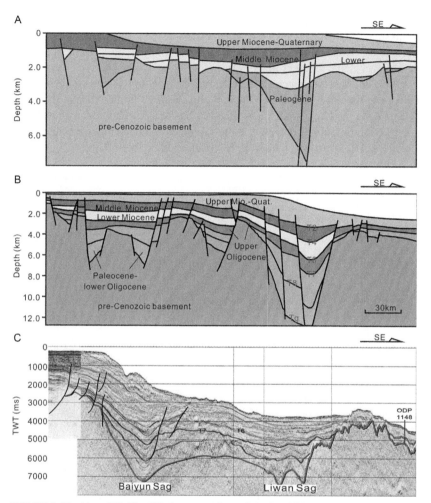

FIGURE 5.43 (A and B) Simplified cross sections of the Pearl River Mouth Basin, showing faults and sediment packages (Zhong & Li, 2009). (C) Seismic profile across the deepwater sags and ODP Site 1148 from the lower slope (modified from Pang et al., 2007), showing the merge of two reflectors T6 and T7 at the ODP 1148 locality due to slump and hiatus in the late Oligocene. See Figure 5.42 for line locations.

stages, from late Cretaceous to Paleocene and from late Eocene to early Oligocene, were most significant for the deposition of mainly alluvial, fluvial, and lacustrine clastic sediments in half grabens and faulted centers (Chen et al., 1994; Li & Rao, 1994; Pigott & Ru, 1994; Ru & Pigott, 1986; Wang & Sun, 1994). After the middle Oligocene unconformity, rapid subsidence caused a rise in local sea level and marine transgressions from south to north. As a result, the previously exposed structural highs (Shenhu, Dongsha, and Panyu) were gradually submerged and became the sites for carbonate platform growth (e.g., Moldovanyi, Waal, & Yan, 1995; Zampetti, Sattler, & Braaksma, 2005; Zhu, 1987). Sediments transported by the paleo-Pearl River were largely trapped in depressions as a huge, mainly destructive delta (paleo-Pearl River delta; Chen et al., 1994).

The sediment sequence of the Pearl River Mouth Basin comprises the Paleogene Shenhu, Wenchang, Enping, and Zhuhai Formations and the Neogene Zhujiang, Hanjiang, Yuehai, and Wanshan Formations, with relative good biostratigraphic controls for the Oligocene and younger marine strata (Figure 5.44; Duan & Huang, 1991; Hou et al., 1981; Jiang et al., 1994;

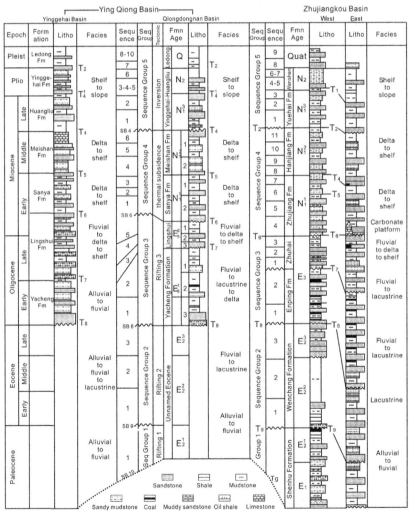

FIGURE 5.44 Lithology and sequence stratigraphy of the Yinggehai, Qiongdongnan, and Pearl River Mouth Basins. *Modified from Wu et al. (2000) and references therein.*

Wang, Min, & Bian, 1985; Wang, Xia, & Chen, 1985; Zeng et al., 1981). Characterized by a lacustrine facies developed during an early rift stage, the Eocene Wenchang Formation shales are good oil source rocks, while the coal-bearing strata in the more widespread Enping Formation (Eocene–early Oligocene) are good gas source. Intensified faulting in this synrift stage created a major unconformity before further subsidence to form a shallow-marine basin and the deposition of the Zhuhai Formation up to 1500 m thick. The Zhuhai Formation features sheetlike sandstones, mainly glauconitic quartzose sandstone and arkosic quartzose sandstone with very little mudstone, indicating a wave- or fluvial-dominated delta–shelf environment (Chen et al., 1994). The overlying early Miocene Zhujiang Formation also accumulated at delta to shelf settings in slightly deeper waters, prograding to carbonate platform reef facies on the Dongsha Swell zone. The middle Miocene Hanjiang Formation is characterized by shallow-marine mudstones and sandstones similar to those of the Zhujiang Formation, probably deposited also in an overall transgressive environment (Figure 5.44; Chen et al., 1994; Jiang et al., 1994).

Intensive exploration over decades in the Pearl River Mouth Basin has stimulated seismic and sequence stratigraphic studies (e.g., Gong & Li, 1997; Wu et al., 2000; Xu, 1999). About 10 seismic reflectors, T0 to T9, and Tg are considered as corresponding to regional unconformities useful for confining megasequences and sequences in the basin (Figures 5.43 and 5.44). The breakup unconformity (T7) separates the three synrift nonmarine megasequences with six sequences from the overlying two postrift marine megasequences with about twenty sequences (Figure 5.44; Wu et al., 2000).

5.3.2.3   Yinggehai and Qiongdongnan Basins

The Yinggehai Basin (Song Hong Basin; Figure 5.45) is a NNW–SSE-trending elongated basin in the southwestern Gulf of Beibu between Hainan Island and Indochina Peninsula in water depths mostly <100 m and extends toward the northwest into the Hanoi Basin. It is about 500 km long and 50–60 km wide and covers an area of approximately 113,000 km^2 (Jiang et al., 1994; Zhang & Zhang, 1991). To its southeast lies the Qiongdongnan Basin (Southeast Hainan Basin), a NE-trending basin situated between Hainan Island in the north and the Xisha Islands in the south and bordered by the Pearl River Mouth Basin in the east. With an area of about 45,000 km^2 in water depths mostly <200 m (Jiang et al., 1994), the Qiongdongnan Basin is separated from the Yinggehai Basin by the so-called No. 1 fault, a southern extension of the NW–SE-trending Red River Fault (Figure 5.45). Seismic data indicate that the No. 1 fault is a normal fault controlling the northeastern boundary of the Yinggehai Basin, and it fades gradually southeastward (Zhang & Zhang, 1991), although it could have been operating as a right-lateral strike-slip fault over various periods.

Both as the Cenozoic rift basins, the Yinggehai and Qiongdongnan Basins consist of a similar synrift to postrift succession. The breakup unconformity,

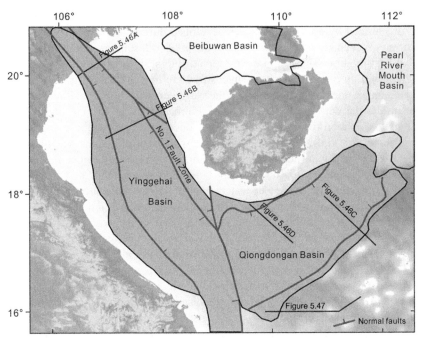

FIGURE 5.45 The major fault systems that controlled the development of the Yinggehai and Qiongdongnan basins. *From Zhong and Li (2009) and references therein.*

however, has been given with different ages: base of the Miocene (T6) by Gong and Li (1997), base of the upper Oligocene by Andersen et al. (2005), and base of the lower Miocene by Xie, Müller, Li, Gong, and Steinberger (2006). Compared with the Qiongdongnan Basin, the Yinggehai Basin has been influenced more strongly by transformed extensional tectonic processes closely relating to activities of the Red River Fault or even representing a strike-slip tensional basin at some developmental stages (e.g., Clift & Sun, 2006; Zhang & Zhang, 1991). Since the Oligocene, the depocenter in the Yinggehai Basin has moved southward due to stepwise greater subsidence in the south and/or sequential availability of basin space further offshore because of rapid accumulation nearshore in the north, close to the Red River mouth (Gong & Li, 1997). Associated with depocenter movement, there was the development of a series of mud diapiric structures in close relation to overpressured shales. About 18 shale diapirs arranged in six N–S-striking vertical en echelon zones in the Yinggehai Basin have been found as related to activity on the offshore continuation of the Red River Fault (Lei et al., 2011). The overpressured system in this basin was caused by rapid sediment load and great burial depths. Now, at burial depths of 3–5 km, for example, even the relatively young Ying–Huang Formations (lower part, 11–5 Ma) may produce high pressure of 60–80 MPa (Hao, Sun, Li, & Zhang, 1995).

Strongly deformed sequences from the northwestern Yinggehai Basin (Figure 5.46A) were most likely affected also by the lateral strike-slip Red River Fault during the late Miocene–Pliocene (Andersen et al., 2005).

Very thick Cenozoic deposits have been found in these two basins: up to 17 km in the Yinggehai Basin and 12 km in the Qiongdongnan Basin (Gong & Li, 1997; Zhang, 1999; Zhang & Zhang, 1991), including a Pleistocene sequence up to 2000 m thick (Wang, Xia, Wang, & Cheng, 1991). Because of similar lithologies, contemporary sediment packages in these two basins are often referred to a single formation name. The synrift strata include the lower Oligocene and older-aged Yacheng Formation and the upper Oligocene Lingshui Formation, while the postrift strata include the lower Miocene Sanya Formation, middle Miocene Meishan Formation, upper Miocene Huangliu Formation, and Pliocene Yinggehai Formation (Figure 5.44; Jiang et al., 1994). The latter two formations are sometimes lumped together as Ying–Huang (or Yinggehai–Huangliu) Formations for their similar neritic facies. Sandstones of the Lingshui, Huangliu, Yinggehai, and even those Neogene Formations often form good oil and gas reservoirs (Jin, 2005).

Lithostratigraphic evidence indicates that the Qiongdongnan Basin has undergone intensive tectono-environmental impacts during the late Miocene. In the southwestern corner of the basin, mass transport deposits (MTD) widely developed and formed part of the late Miocene upper Huangliu Formation (8.2–5.5 Ma; Figure 5.47; Wang et al., 2013). This large-scale buried MTD system covers an area of more than $18,000 \text{ km}^2$ with general flow direction from south to north. The main source areas are believed to have been mainly from the Guangle Uplift in the far south and the Xisha Uplift in the southeast. The dominant period when the mass transport deposits developed corresponds to the phase when the Red River Fault reversed from left-lateral to right-lateral slip, indicating the structural control on the regional deposition (Wang et al., 2013). More interestingly, a large submarine canyon of late Miocene age has been recently found extending through the Qiongdongnan Basin and Yinggehai Basin (e.g., Gong et al., 2011; Li et al., 2013). The central canyon extends NE to NNE subparallel to the slope to over 570 km in length and more than 8 km in width. The sediment fill in the canyon can be divided into four sequences characterized by gravity flow deposits during a period from ~5.5 to ~2 Ma (Gong et al., 2011).

A number of basin-wide unconformities identified in the Yinggehai and Qiongdongnan Basins have been used for local sequence stratigraphy (Figure 5.44; e.g., Hao, Chen, Wan, & Dong, 2000; Lin et al., 2001; Wei et al., 2001; Wu et al., 2000; Xie, Müller, Ren, Jiang, & Zhang, 2008). In the Qiongdongnan Basin, for example, Chen et al. (1993) identified five megasequences bounded by regional unconformities, and Wang, Wu, Zhou, and Li (1998) recognized six sequences for the Paleogene synrift megasequence and eight sequences for the Neogene postrift megasequence. Wei et al. (2001) also identified five supersequences, TA1 to TA5, to characterize

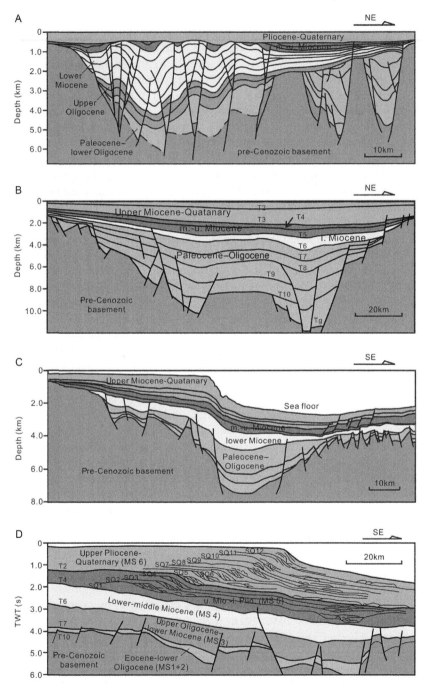

FIGURE 5.46 Interpreted cross sections show major faults and Cenozoic sequences in (A) the northernmost Yinggehai Basin, (B) the northern Yinggehai Basin, (C) the eastern Qiongdongnan Basin, and (D) the central Qiongdongnan Basin. See Figure 5.45 for line locations. *(A) Modified from Andersen et al. (2005), (B) modified from Gong and Li (1997), (C) modified from Xie et al. (2006), and (D) modified from Chen, Chen, and Zhang (1993).*

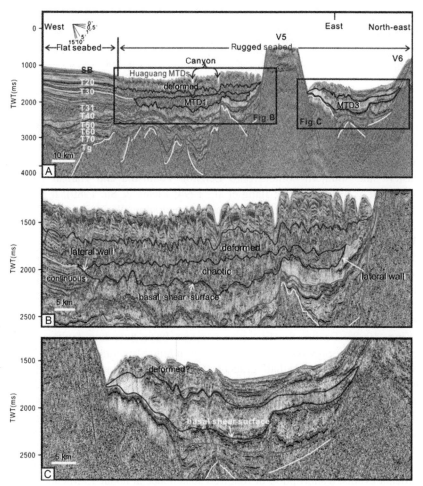

FIGURE 5.47 Interpreted seismic cross sections in the southwestern Qiongdongnan Basin showing the characteristics of mass transport deposits (MTD), which formed part of the Huangliu Formation (late Miocene), between reflectors T20 and T30 (Wang et al., 2013). See Figure 5.45 for line location.

five major stages of basin evolution: (1) TA1, initial rifting in the Eocene; (2) TA2, middle rifting in the early Oligocene; (3) TA3, later rifting during the late Oligocene and earliest Miocene; (4) TB1 indicating an early postrift ramp setting from the late early Miocene to middle Miocene; and (5) TB2 indicating a late postrift passive margin setting during the late Miocene and Pliocene. Similarly, Wu et al. (2000) identified five supersequences: respectively, SS1–SS3 representing the synrift deposition during the Paleocene, Eocene, and Oligocene, while SS4 and SS5 representing the postrift deposition in the early and middle Miocene (SS4) and late Miocene to Quaternary

(SS5; Figure 5.44). Within the postrift Neogene sequences developed a shelf–slope system, from which several lowstand fans have been interpreted from seismic profiles and drilling has revealed that the lowstand fans are mainly composed of turbidites (Jiang, Xie, Ren, Zhang, & Su, 2009).

5.3.2.4 Beibuwan Basin

Lying in the northeastern Gulf of Beibu (Tokin), northwestern SCS, is the Beibuwan Basin (Beibu Gulf Basin) covering an area of about 38,000 km^2. Mainly in modern water depths <50 m, the shallow-water basin extends its eastern and southern corners onto Leizhou Peninsula and Hainan Island, respectively (Figure 5.48A; Jiang et al., 1994; Zhang & Su, 1989). It mainly comprises six W–E stretching subbasins, including Weixinan, Haizhong, Wushi, Haitoubei, Maichen, and Fushan subbasins, separated by uplifted fault blocks (Figure 5.48). The pre-Cenozoic basement consists of slightly meta-morphosed Paleozoic carbonate and clastic rocks, Mesozoic granitic plutonic rocks and andesitic volcanic rocks and Mesozoic sedimentary rocks (Ying, 1998; Zhang & Kou, 1989; Zhang & Su, 1989).

The Beibuwan Basin is a Cenozoic rift basin with two distinct phases of extension during the Paleocene–late Eocene and the Oligocene, respectively (Ying, 1998). During the first phase, major basin-bounding fault systems evolved along previously existing NE-trending structural lineations and resulted in large simple half grabens. In the second phase, the previously existing Eocene graben-bounding faults were reactivated, and the newly gen-erated secondary faults crosscut and rotated the synrift sequence deposited in the first extensional phase, causing structural reconfiguration of the basin and shifting of subsidence centers. Numerous secondary faults developed with increasing extension rates during this later phase of extension. The relative stable postrift structural period since the Miocene was punctuated by two sig-nificant events: a short pulse of renewed subsidence at about 15 Ma and basin inversion since the late Pliocene, especially in the western part of the basin (Rangin, Klein, Roques, Le Pichon, & Trong, 1995; Ying, 1998).

The Cenozoic sequence in the Beibuwan Basin may reach a maximum thickness of 7000 m in depocenters, dominated by Paleogene synrift sediments and relatively thinner Neogene postrift sediments (Figure 5.48; Kang, Zhao, Pan, Zhang, & Chen, 1994; Ying, 1998). The synrift strata are mainly nonmar-ine and include Changliu (upper Paleocene to lower Eocene), Liushagang (mid-dle Eocene to lower Oligocene), and Weizhou (middle and upper Oligocene) Formations. The postrift strata are all shallow marine and include Xiayang (lower Miocene), Jiaowei (middle Miocene), Dengloujiao (upper Miocene), and Wanglougang (Pliocene–Quaternary) Formations (Figure 5.49). The Liushagang, Weizhou, and Jiaowei sandstones have been confirmed to contain good petroleum reservoirs (Jin, 1989, 2005).

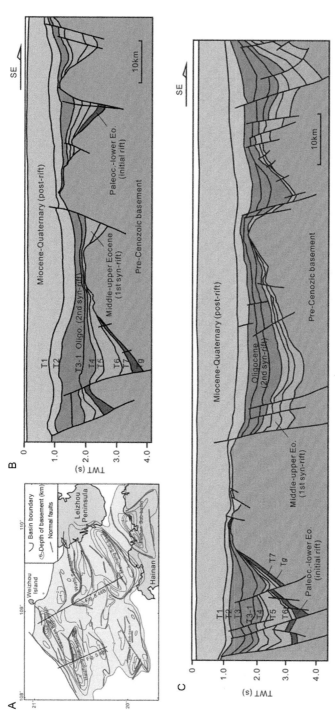

FIGURE 5.48 (A) Major structural units and thickness of Cenozoic sediments (km) in the Beibuwan Basin. (B, C) Interpreted cross sections of the Beibuwan Basin showing synrift sequences (Oligocene (T3) or older) controlled by faults and postrift sequences (blue). *Modified after Ying (1998).*

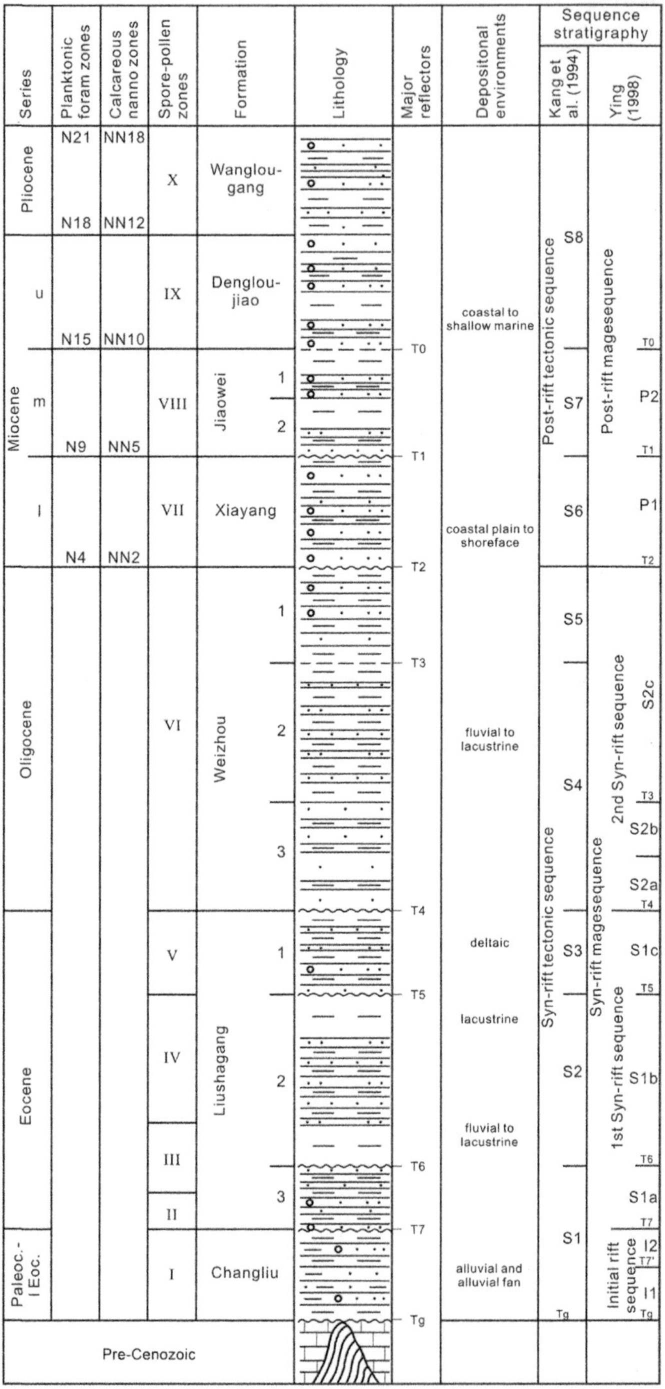

FIGURE 5.49 Composite lithostratigraphy of the Beibuwan Basin, superimposed by sequence stratigraphic schemes of Kang et al. (1994) and Ying (1998). *Modified from Jiang et al. (1994).*

The Paleogene synrift strata are only coarsely dated using spore–pollen assemblages, while the Neogene postrift marine sediments contain age-diagnostic planktonic foraminifera and nannofossils (Hu & Su, 1981; Kou & Ye, 1981; Sun et al., 1981, Wang, 1985; Ying, 1998; Zhang & Kou, 1989). For example, the Changliu Formation consists of reddish sandstones, conglomerates, and shales up to ~900 m thick deposited at alluvial and alluvial fan settings and contains *Celtispollenites triporatus–Pentapollenites* palynomorphs indicating an age of Paleocene to early Eocene. The Liushagang Formation features very thick lacustrine dark gray shales, and its three members can be differentiated by having interbedded sandstones for the upper member Liu-1, having sandstones of the intermediate Liu-2, and having more intercalated sandstones for the lower member Liu-3. The alluvial–fluvial Liushagang Formation may reach a maximum thickness of over 1800 m and contains middle to late Eocene *Quercoidites* palynomorphs, dividable into four assemblages (Xie et al., 2012). The Weizhou Formation is composed of interbedded purple, green, gray variegated mudstone and gray sandstone and conglomerate, with locally distributed coal and marine sandstone interbeds and containing Oligocene *Magnastriatites howardi–Gothanipollis bassensis–Hydrocotaepites pachydermus* palynomorphs. The >700 m thick Xiayang Formation is dominated by gray sandy conglomerate, pebbly sandstone with minor mudstone deposited in coastal to shallow-marine environments with early Miocene planktonic foraminifera. The overlying Jiaowei Formation features up to ~600 m thick greenish gray fine sandstone, muddy sandstone, and (mainly in the upper part) mudstone deposited in neritic environments with middle Miocene foraminifera. The Dengloujiao Formation comprises nearly 600 m thick interbedded gray sandstone, sandy conglomerate, and gray mudstone with late Miocene planktonic foraminifera and nannofossils. The ~300 m thick Wanglougang Formation is dominated by greenish gray to gray mudstone intercalated with sandstones and contains Pliocene marine microfossils. The thin Quaternary sediment layer, ~35 m, is dominated by littoral sand and clay.

Several major seismic reflectors have been identified for marking sequence boundaries of the Cenozoic succession in the Beibuwan Basin (Figure 5.49; Du & Wei, 2001; Kang et al., 1994; Ying, 1998). The breakup unconformity (T2) separates the Paleogene synrift megasequence from the overlying Neogene–Quaternary postrift megasequence (Figure 5.48B and C). The synrift megasequence, consisting of the Changliu, Liushagang, and Weizhou Formations with a combined thickness of several thousand meters, is confined to isolated grabens and half grabens. Ying (1998) further divided the Paleogene synrift megasequence into two initial rift sequences I1 and I2 (Tg–T7) and two synrift sequences S1 (T7–T4) and S2 (T4–T2; Figure 5.49). The Neogene postrift megasequence, including the Xiayang, Jiaowei, Dengloujiao, and Wanglougang Formations, is characterized by coastal to shallow-marine progradational clastic facies and by locally developed lower Miocene carbonate reef buildups.

5.3.3 Southern South China Sea Sector

5.3.3.1 Cuu Long and Wan'an (Nam Con Son) Basins

The Cuu Long and Wan'an (Nam Con Son) Basins are located offshore from SE Vietnam in the shelf-slope area between the Mekong River mouth to the northwest and Nansha Islands (the Dangerous Grounds) to the southeast. Separated by the northeast-trending Con Son Swell zone, the northern Cuu Long Basin is nearshore and covers an area of about 25,000 km^2, while the southern Wan'an Basin is more offshore and covers about 90,000 km^2 (Binh et al., 2007; Figure 5.50A).

The structural evolution of the Cuu Long and Wan'an Basins is characterized by rifting and regional thermal subsidence (Dien, Phung, Nguyen, & Do, 1998; Lee et al., 2001; Matthews et al., 1997). One rifting phase during the Eocene–early Oligocene has been identified in the Cuu Long Basin (Lee et al., 2001), and at least two rifting phases occurred in the Wan'an Basin (Canh, Ha, Carstens, & Berstad, 1994; Fraser, Matthews, Lowe, Todd, & Peel, 1996; Lee et al., 2001; Matthews et al., 1997). Timings for the two rifting phases in the Wan'an Basin have been assigned to Eocene–Oligocene and early to late Miocene (24–8 Ma) by Lee et al. (2001) and to Eocene–early Oligocene and middle to late Miocene by Matthews et al. (1997), respectively. The discrepancy in the age assigned to the two rifting phases is probably due to poor dating, but the existence of a third rifting phase may be also likely if all areas of the Wan'an Basin were fully surveyed, as suggested by Olson and Dorobek (2002).

The Eocene to Quaternary sediment succession varies mostly between 2 and 5 km in thickness (Figure 5.50B and C) but may reach a maximum of about 8 km in the Cuu Long Basin (Binh et al., 2007) and ∼12.5 km in the Wan'an Basin (Chen & Peng, 1995). The basement in the Cuu Long Basin comprises late Jurassic–early Cretaceous granites and granodiorites as revealed by drilling in the Bach Ho field (Areshev, Dong, San, & Shnip, 1992), while the basement in the Wan'an Basin is formed by highly variable pre-Cenozoic granitic, volcanic, and low-grade metasedimentary units, as revealed in wells drilled on structural highs (Matthews et al., 1997).

As shown in Figure 5.51, Wan'an Basin strata include Eocene–lower Oligocene deposits, Oligocene Cau Formation, lower Miocene Dua Formation, middle Miocene Thong and Mang Cau Formations, upper Miocene Nam Con Son Formation, and Pliocene to Quaternary Bien Dong Formation (Binh et al., 2007; Matthews et al., 1997). These formations are mainly composed of clastic rocks, but the middle Miocene to lower Pliocene formations of Thong, Mang Cau, Nam Con Son, and lower Bien Dong from eastern structural highs also feature carbonate platform and shallow-marine facies.

In contrast, all sediment deposits in the Cuu Long Basin are of clastic origin, mainly from Mekong discharge (Figure 5.51). From bottom to top, they are the Tra Cu Formation (late Eocene to early Oligocene), Tra Tan

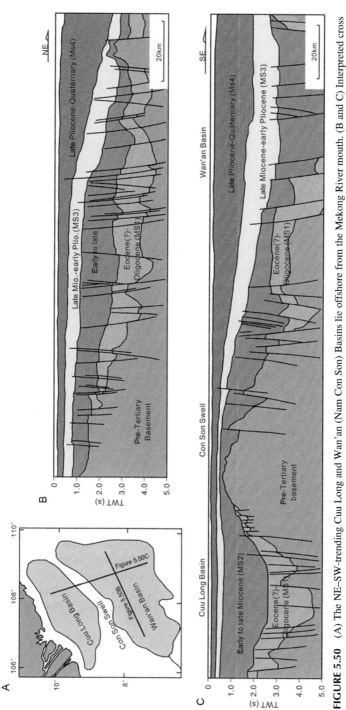

FIGURE 5.50 (A) The NE–SW-trending Cuu Long and Wan'an (Nam Con Son) Basins lie offshore from the Mekong River mouth. (B and C) Interpreted cross sections showing major faults and Cenozoic seismic sequences in these two basins. (A) *Modified from Lee, Lee, and Watkins (2001) and Binh et al. (2007). (B and C) From Zhong and Li (2009), based on Lee et al. (2001).*

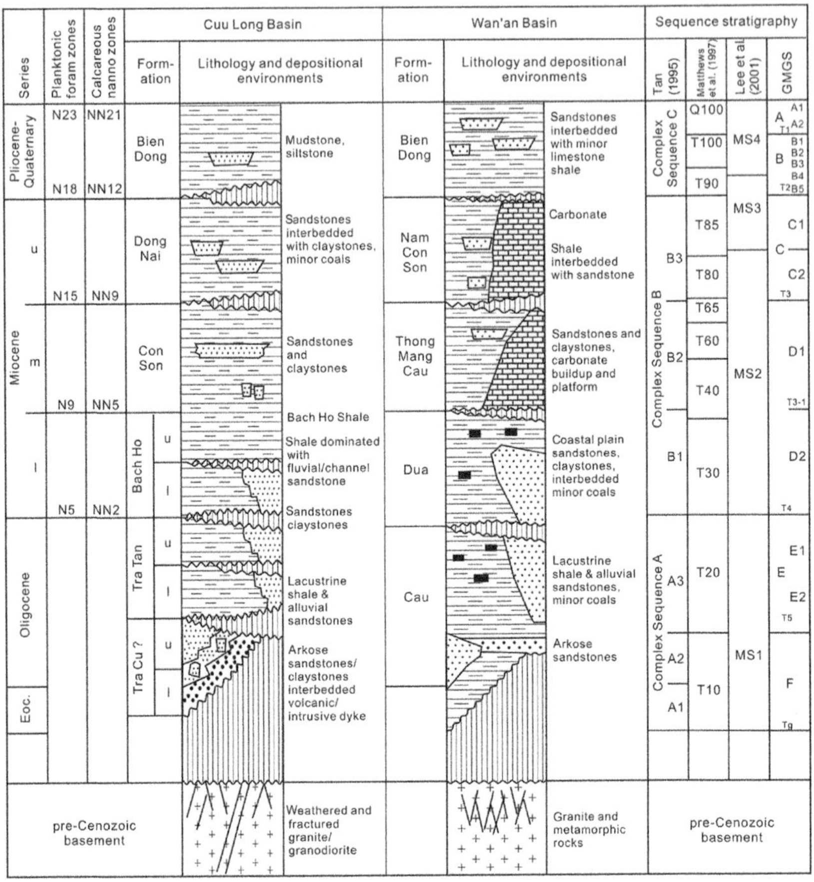

Series	Planktonic foram zones	Calcareous nanno zones	Cuu Long Basin Formation	Cuu Long Basin Lithology and depositional environments	Wan'an Basin Formation	Wan'an Basin Lithology and depositional environments	Tan (1995)	Matthews et al. (1997)	Lee et al. (2001)	GMGS
Pliocene-Quaternary	N23 / N18	NN21 / NN12	Bien Dong	Mudstone, siltstone	Bien Dong	Sandstones interbedded with minor limestone shale	Complex Sequence C	Q100 / T100 / T90	MS4	A1 / A2 / B1 / B2 / B3 / B4 / T2B5
Miocene u	N15	NN9	Dong Nai	Sandstones interbedded with claystones, minor coals	Nam Con Son	Carbonate; Shale interbedded with sandstone	B3	T85 / T80	MS3	C1 / C / C2 / T3
Miocene m	N9	NN5	Con Son	Sandstones and claystones; Bach Ho Shale	Thong Mang Cau	Sandstones and claystones, carbonate buildup and platform	Complex Sequence B / B2	T65 / T60 / T40	MS2	D1 / T3-1
Miocene l	N5	NN2	Bach Ho u / l	Shale dominated with fluvial/channel sandstone; Sandstones claystones	Dua	Coastal plain sandstones, claystones, interbedded minor coals	B1	T30		D2 / T4
Oligocene			Tra Tan u / l	Lacustrine shale & alluvial sandstones	Cau	Lacustrine shale & alluvial sandstones, minor coals	Complex Sequence A / A3	T20		E1 / E / E2 / T5
Eoc.			Tra Cu ? u / l	Arkose sandstones/claystones interbedded volcanic/intrusive dyke		Arkose sandstones	A2 / A1	T10	MS1	F / Tg
			pre-Cenozoic basement	Weathered and fractured granite/granodiorite		Granite and metamorphic rocks		pre-Cenozoic basement		

FIGURE 5.51 Simplified lithobiostratigraphic columns (Binh et al., 2007), interpreted depositional environments, and different sequence stratigraphic schemes for the Cuu Long and Wan'an (Nam Con Son) Basins. The GMGS (Guangzhou Marine Geological Survey) scheme is integrated from various sources as given in Zhong and Li (2009).

Formation (later early and late Oligocene), Bach Ho Formation (early Miocene), Con Son Formation (middle Miocene), Dong Nai Formation (late Miocene), and Bien Dong Formation (Pliocene to Quaternary).

For the sedimentary strata in the Cuu Long and Wan'an Basins, many sequence stratigraphic schemes have been proposed (e.g., Chen & Peng, 1995; Lee et al., 2001; Matthews et al., 1997; Tan, 1995; Wang, Liu, Wu, & Zhong, 1996; Wang, Zhong, & Wu, 1997; Wu, Liu, Zhou, & Liu, 1997; Wu & Yang, 1994; Yang, Wu, Yang, & Duan, 1996; Figure 5.51). The depositional history is characterized by a synrift succession and a postrift succession, although the age separating the two is often considered different between the two basins: at the Oligocene/Miocene contact for the Cuu Long and at the Miocene/Pliocene contact for the Wan'an (Figure 5.51). The synrift succession in the Wan'an Basin can be further divided into an older, Eocene–Oligocene synrift megasequence, which is dominated by alluvial and fluvial sandstones and lacustrine shales intercalated occasionally with coal beds, and a younger, Miocene synrift megasequence, which mainly comprises coastal to shallow shelf sandstones, shales, and platform carbonate rocks (Lee et al., 2001; Matthews et al., 1997). The postrift megasequence in the Wan'an Basin is characterized by shelfal to deep-marine shales intercalated with siltstones and sandstones, but in the Cuu Long Basin, it mainly comprises littoral to shallow-marine clastics and mudstones (Figure 5.51). Sandstones in both basins and caronates in the Wan'an Basin contain good petroleum reservoirs (Lee et al., 2001; Tan, 1995).

5.3.3.2   Zengmu Basin

Zengmu Basin, also known as the Sarawak and East Natuna Basins, is located in the southern margin of the SCS between the Nansha Islands (the Dangerous Grounds) to the north, the Natuna Islands to the west, and Borneo (onshore Sarawak) to the south (Figure 5.52). Covering an area of about 170,000 km^2 (Qiu et al., 2005), the Zengmu Basin is the largest shelf-slope basin in the southern SCS region.

Tectonically, the Zengmu Basin is a late Eocene to the present rift basin on the Zengmu (Locunia) block in response to the Oligocene to early Miocene opening of the SCS (e.g., Huang, 1997a,1997b; Madon, 1999b). Like several other microblocks in the region, such as the Nansha (the Dangerous Grounds), Liyue (Reed Bank), and north Palawan, the Zengmu (Locunia) block is a microcontinental block rifted off and drifted southward from the South China continent and later collided with the northern Borneo margin. Other authors, however, considered the Zengmu block as an eastward extension of the Indochina continental block (e.g., Jin & Li, 2000; Wu, 1991).

The Zengmu Basin is structurally separated from the West Natuna Basin to the west by the Natuna Arch, from the Wan'an Basin to the northwest by Xiya Uplift, and from the Sabah Basin to the east by the West Baram Line, a NW–SE-extending tectonic discontinuity interpreted as a major transform fault by some authors (Figure 5.52; Madon, 1999b). Its southern margin lies in the Rajang Fold–Thrust Belt (Sibu Zone) onshore the northern Borneo, which consists of highly deformed, low-grade metamorphosed late Cretaceous to late Eocene deep-marine shales and turbidites, with lesser amount of radiolarian chert, spilite, and dolerite (Madon, 1999b). Therefore, the early Zengmu Basin probably represents part of the remnants of the late Cretaceous to early Miocene "proto-South China Sea" oceanic basin, although the majority of the proto-SCS may have been since subducted southward beneath the Borneo and along the island arc in the east (Figure 5.53; Madon, 1999d; Madon et al., 2013). Accordingly, it may have initiated as a peripheral

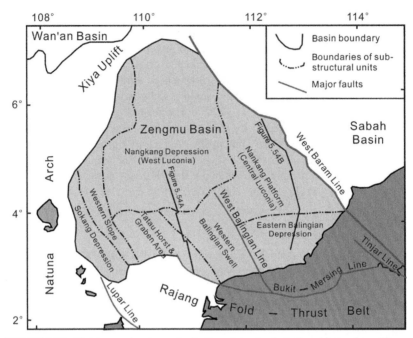

FIGURE 5.52 Sketch map showing major faults and structural–stratigraphic provinces (depressions) in the Zengmu Basin. *From Zhong and Li (2009), based on Qiu et al. (2005) and Madon (1999a, 1999b).*

foreland basin that formed in the north front of Rajang collisional orogen after the collision of the Zengmu block with the West Borneo and the closure of the proto-SCS during the late Eocene (Madon, 1999b). The deformation and uplift of the Rajang Fold–Thrust Belt (Figure 5.53) have provided abundant sediment supply to the Zengmu Basin. The basin's main formation stage since the Oligocene has been mainly extensional, with active extensional and strike-slip tectonics playing an important role in its formation and evolution.

Some authors proposed that the Zengmu Basin was formed by strike-slip movement of major NW–SE faults during the late Oligocene to early Miocene (e.g., Mat-Zin & Swarbrick, 1997), while others suggested at least two phases for the formation of the basin (Chen, 2002; Jin & Li, 2000; Zhong, Xu, & Wang, 1995). The first phase was associated with the formation of a peripheral foreland basin during the late Eocene–early Miocene, and the second phase with a strike-slip or shear extensional pull-apart basin during the middle and late Miocene. Since the middle Miocene or late Miocene, the Zengmu Basin entered a new stage of regional subsidence and outbuilding of a passive continental margin (Chen, 2002; Jin & Li, 2000; Zhong et al., 1995). Based on well and seismic data, Madon et al. (2013) put forward a conceptual model of basin evolution, as summarized in Figure 5.53.

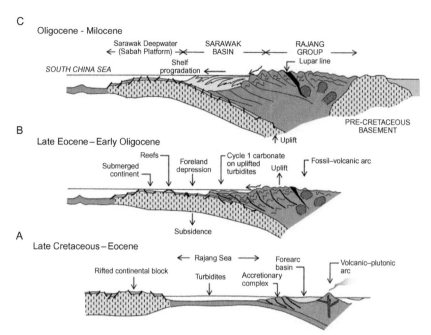

C

Oligocene - Milocene

Sarawak Deepwater   SARAWAK   RAJANG
← (Sabah Platform) →⊗—   BASIN   —⊗—   GROUP   →
                     Shelf   ┌ Lupar line
SOUTH CHINA SEA   progradation ←

                                            PRE-CRETACEOUS
                                            BASEMENT
B
Late Eocene – Early Oligocene   ↑ Uplift

            Reefs ┐   ┌ Foreland ┌ Cycle 1 carbonate ┌ Fossil–volcanic arc
   Submerged   │   │ depression │ on uplifted
   continent ┐   │   │   │ turbidites Uplift
                                                   ↑

                        ↓
                     Subsidence
A
Late Cretaceous – Eocene

            ← Rajang Sea →   Forearc   ┌ Volcanic–plutonic
   Rifted continental block   basin ┐   │ arc
                        Turbidites Accretionary
                                    complex ┐

FIGURE 5.53   Schematic tectonic model for the evolution of the Zengmu (Sarawak) Basin. (A) Late Cretaceous–Eocene: rifting of southern margin of South China coupled with subduction of the proto-South China Sea (Rajang Sea) produced an accretionary complex, which was later uplifted onshore. (B) Late Eocene–early Oligocene ("flysch"): collision of Luconia Block with Borneo at end Eocene and cessation of subduction and deep-marine sedimentation. Foreland basin developed on the rifted lithosphere while undergoing subduction. Spreading of South China Sea began. (C) Early Oligocene–early Miocene ("molasse"): spreading of the South China Sea caused an accelerated strike-slip movement and a major uplift and erosion in the Rajang fold belt, which supplied sediment to the foreland basin. The marine foreland basin became shallower where deltaic and lower coastal plain sediments were deposited. Deposition in NW Sabah continued until about 15 Ma when the Dangerous Grounds block collided with the NW Sabah Block. *From Madon, Kim, and Wong (2013).*

The Zengmu Basin has been divided into several major structural–stratigraphic provinces, or depressions, formed mainly by basement-involved extensional, strike-slip, or gravity-driven detached tectonics with different phases of structural and sedimentologic evolution (ASCOPE (Asean Council on Petroleum), 1981; Chen, 2002; Jiang et al., 1994; Madon, 1999a,1999b; Qiu et al., 2005). For example, PETRONAS (Madon, 1999b) identifies seven provinces for the Sarawak Basin: the SW Sarawak, West Luconia, Tatau, Balingian, Tinjar, Central Luconia, and North Luconia Provinces. Other schemes recognize Sokang Depression, Tatau Horst and Graben Area, Western Balingian Swell, Eastern Balingian Depression (Balingian and Tinjar Provinces), Nankang Platform (Central Luconia Province), Nankang Depression, and Western Slope (Figure 5.52; ASCOPE (Asean Council on Petroleum),

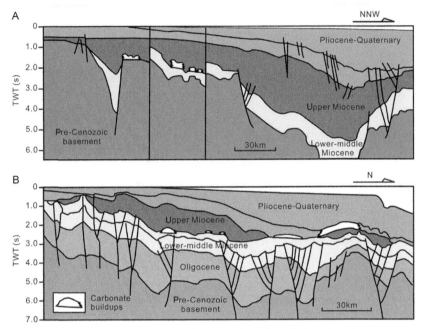

FIGURE 5.54 Interpreted cross sections showing major faults and Cenozoic sediment packages in the Zengmu (Sarawak) Basin. See Figure 5.52 for line locations. *Modified after Mat-Zin and Swarbrick (1997).*

1981; Qiu et al., 2005). Most provinces are dominated by clastic rocks, but the Central Luconia Province features a huge thickness of carbonate rocks (Figure 5.54; Ali & Abolins, 1999; Ho, 1978; Ismail & Hassan, 1999; Madon, 1999c; Madon & Hassan, 1999a,1999b). The maximum thickness of the late Eocene to Quaternary sediments in the Zengmu Basin is estimated by different authors as varying from 12 to more than 15 km (Chen, 2002; Jin & Li, 2000; Madon, 1999b).

The stratigraphic subdivision in the Malaysian side of the Zengmu Basin was established by Shell Oil Company in the 1970s. The Eocene and younger sedimentary successions in offshore Sarawak were divided into eight regional regressive cycles separated by brief transgressive phases based on a sedimentary cycle concept (Ho, 1978; Figure 5.55). The cycles are determined by marine transgressive units using biostratigraphic assemblages or zones (foraminiferal zonation for marine sediments or palynological zonation for coastal and nonmarine deposits; see Section 5.2.3 in the preceding text). Each cycle starts with a transgressive unit dominated by marine shale, followed by a regressive unit in which sand percentage increases upward before terminated by deposits indicative of coastal conditions.

According to ASCOPE (Asean Council on Petroleum) (1981), Doust (1981), and Madon, Karim, et al. (1999b), the sediment cycles bear the

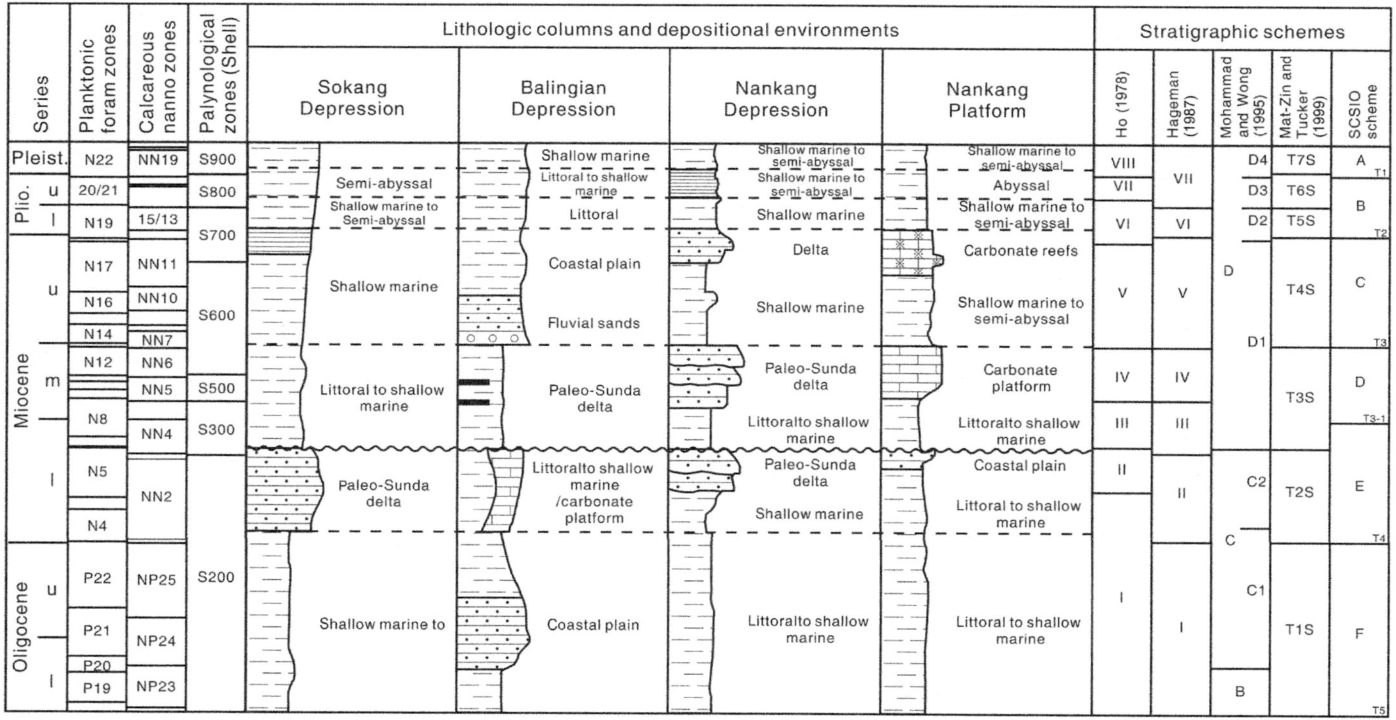

FIGURE 5.55 Correlation between various litho-, bio-, and sequence stratigraphic schemes and interpreted depositional environments for different sediment packages or cycles (I–VIII) in the Zengmu (Sarawak) Basin. SCSIO scheme is integrated from various sources (Zhong & Li, 2009). Palynomorph zones of Shell were adopted from Mat-Zin and Tucker (1999). Modified from Qiu et al. (2005).

following characteristics (Figure 5.55): Cycles I and II, upper Eocene to lower
Miocene, are mainly fluvial and estuarine channel sands, overbank clays and
coals, and minor carbonate sediments. Cycle III, lower to middle Miocene,
consists of shale with thin limestone and sandstone beds, with patchy but
thicker limestones in the upper part. Cycle IV, middle Miocene, represents
open shallow-marine conditions throughout most parts of the Sarawak shelf,
coeval to a period of extensive carbonate buildups in the Central Luconia
Province. Cycle V, middle to upper Miocene, was developed with extensive
reefal carbonates in Central Luconia and with clastics in the east and west
of Central Luconia, as a result of progradation of the Baram and Rajang
paleodeltas. Cycles VI to VIII, upper Miocene to the present, comprise open
marine to coastal clays and sands during delta progradation, sometimes
completely covering carbonate buildups in the Central Luconia Province
(Figure 5.55). Probably representing the earliest attempt of sequence stratigraphy in the southern SCS, Shell's concept of cycles is equivalent to the "T-R
sequence" of Emery and Meyers (1996).

However, the limitations of Shell's cycle concept have been noted
(Hageman, 1987; Mat-Zin & Tucker, 1999) because of the difficulty in defining cycles in coastal and nonmarine sediments where age-diagnostic microfossils especially foraminifera are often absent and in open marine
successions where cyclic rhythms are less developed. In addition, since a
cycle is bounded by marine flooding surfaces, the possible presence of a
sequence boundary within it can be easily overlooked. Practically, cycle
boundaries associating with major erosional events along basin margins are
more easily identified on seismic sections and well logs than those relating
to marine flooding surfaces. With an attempt to resolve these issues,
Hageman (1987) modified Shell's scheme by defining new cycle boundaries
at the known unconformities for the bases of cycles II, III, and V (Figure 5.55).
However, this modified scheme incidentally falls into a hybrid of the T-R
cycle concept and the Exxon sequence concept (Mat-Zin & Tucker, 1999).

The sediment succession in the Zengmu Basin can be divided and correlated using seismic data tied to well control, although the number of
sequences may vary between authors. For example, Mohammad and Wong
(1995) identified eight seismic horizons in the offshore Sarawak deepwater
area for grouping the local sequences into four supersequences
(Figure 5.55): A for the upper Cretaceous–middle Eocene, B for the upper late
Eocene–upper Oligocene, C for the upper Oligocene–lower Miocene, and
D for the lower Miocene to the present. These supersequences are definable
also by biofacies and lithofacies data and GR logs. Mat-Zin and Swarbrick
(1997) and Mat-Zin and Tucker (1999) identified seven seismic sequences
for the Sarawak Basin: T1S (mainly upper Oligocene), T2S (mainly lower
Miocene), T3S (mainly middle Miocene), T4S (mainly upper Miocene),
T5S (mainly Pliocene), T6S (Pliocene–Pleistocene), and T7S (Pleistocene to
the present; Figure 5.55). Mat-Zin and coworkers suggested that all the

Oligocene to Miocene sequence boundaries were probably tectonically induced but Pliocene–Pleistocene sequence boundaries were likely derived from eustatic sea level changes or a combined eustasy and tectonics forcing. Based on a large number of seismic profiles, researchers in the South China Sea Institute of Oceanology (SCSIO) and Guangzhou Marine Geological Survey (GMGS) recognized six regional unconformities in the Zengmu Basin and neighboring areas (Figure 5.55): T1 (base of the Pleistocene), T2 (base of the Pliocene), T3 (base of the upper Miocene), T3-1 (base of the middle Miocene), T4 (base of lower Miocene), and T5 or Tg (top of pre-Cenozoic basement). Therefore, six seismic sequences A to F bounded by these horizons were defined to represent various sedimentary stages in the Zengmu Basin (Zhong & Li, 2009, and references therein). Over several decades, industrial exploration has led to the finding of oil and gas reservoirs in cycle II and III sandstones in the Balingian Province and in Miocene carbonate rocks of cycles IV and V from the Nankang Platform (Mahmud & Saleh, 1999).

5.3.3.3   Malay and West Natuna Basins

Malay Basin

The Malay Basin is situated in the southern part of the Gulf of Thailand between Indochina Peninsula to the north and Malay Peninsula to the south. The NW–SE-trending elongate basin extends to approximately 500 km long and 200 km wide and covers an area of approximately 80,000 km^2 (Bishop, 2002; Madon, Abolins, et al., 1999). The basin runs almost parallel to the Malay Peninsula but perpendicular to the NE–SW-trending West Natuna and Penyu Basins to its southeast. To the northwest, the Malay Basin is separated from the N–S-trending Pattani Basin by the Narathiwat Ridge in the Gulf of Thailand (Figure 5.56; Bishop, 2002; Madon, Abolins, et al., 1999). Its pre-Cenozoic basement is made up by Carboniferous to Mesozoic igneous, metamorphic, and sedimentary rocks (Leo, 1997; Liew, 1994; Madon, Abolins, et al., 1999).

The Malay Basin is an extensional basin, but its development may have also involved a significant strike-slip component of extension (Madon & Watts, 1998). Several models have been proposed for the basin formation, including back-arc basin, extrusive pull-apart basin, regional thinning of the continental crust, crustal extension above a hot spot, and failed rift basin (Morley & Westawayw, 2006; Zhong & Li, 2009). The extrusion model emphasizes a direct link between the formation of the Malay Basin and the Three Pagodas Fault, a major strike-slip fault zone running through the head of the Gulf of Thailand (Figure 5.56; Madon, Abolins, et al., 1999). Faulting formed numerous grabens and half grabens, most of which have been rarely drilled because of their great depths (Madon, Abolins, et al., 1999). Magnetic, gravity, and seismic data indicate that the basement normal faults largely occur in the southern and central parts of the basin. These faults extend east–westly, oblique to the

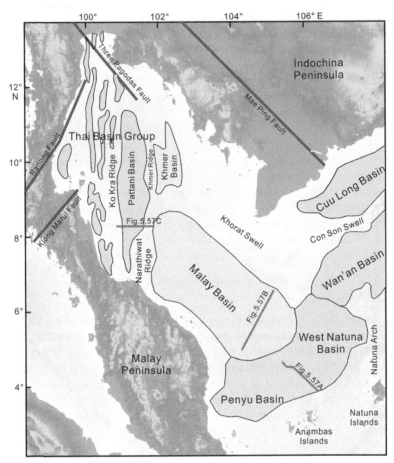

FIGURE 5.56 Distribution of the West Natuna, Malay, and Thai Basins in the Gulf of Thailand, western SCS. *From Zhong and Li (2009), based on ASCOPE (Asean Council on Petroleum) (1981), Phillips, Little, Michael, and Odell (1997), Madon, Abolins, et al. (1999a), Bishop (2002), and Jardine (1997).*

overall basin trend, and align en echelon into two large basement-involved fault zones, that is, the Axial Malay Fault Zone (AMFZ) in the central basin and the Western Hinge Fault (WHF) in the southwestern margin (Madon, 1997; Madon, Abolins, et al., 1999).

The late Eocene–early Oligocene was the time rifting started in the Malay Basin, followed by a period of thermal subsidence since the Miocene (Madon, 1999a). Similar to the West Natuna and Penyu Basins, the Malay Basin experienced basin inversion during the middle and late Miocene, producing major anticlinal structures that trap large oil and gas reservoirs (Madon, 1997; Tjia, 1994). Basin inversion was likely caused by a dextral shear regime following a change in the regional stress field (Madon, 1997; Tjia & Liew, 1996).

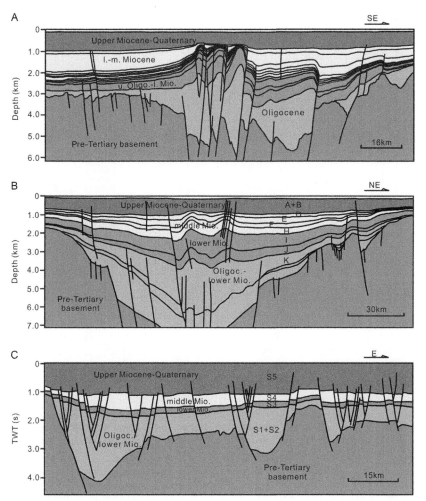

FIGURE 5.57 Interpreted cross sections showing major seismic sequences controlled by faults in (A) West Natuna Basin, (B) Malay Basin, and (C) Pattani Basin. See Figure 5.56 for line locations. *(A) Redrawn after Phillips et al. (1997), (B) Redrawn after Madon (1999a), and (C) redrawn after Watcharanantakul and Morley (2000).*

The maximum thickness of Cenozoic sediments in the Malay Basin is estimated as varying between 12 and 15 km (Madon, 1997; Madon, Abolins, et al., 1999;Madon & Watts, 1998; Morley & Westaway, 2006). In general, the lacustrine Oligocene deposits are overlain by an extensive fluvial to deltaic Miocene succession (Figure 5.57B). After the middle to late Miocene inversion, the basin has been prevailed by fully marine environments.

Stratigraphy of the Malay Basin was initially established by Esso Production Malaysia Inc. in the late 1960s (Figures 5.10 and 5.58A; Madon,

A   B

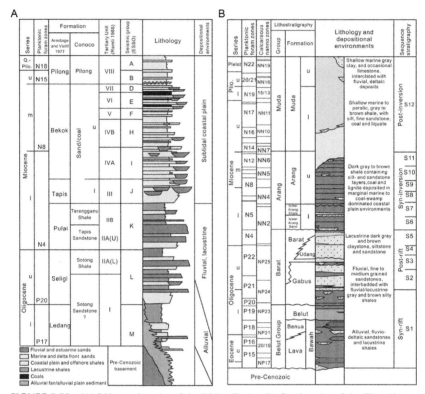

FIGURE 5.58 (A) Lithostratigraphy of the Malay Basin. (B) Stratigraphy of the West Natuna Basin, including lithostratigraphy (Darmadi, 2005), lithology and depositional environments (ASCOPE, Asean Council on Petroleum, 1981, Meirita, 2003; Darmadi, Willis, & Dorobek, 2007), and sequence stratigraphy (Phillips et al., 1997). *(A) From Zhong and Li (2009), based on Madon, Abolins, et al. (1999a) and Madon, Karim, et al. (1999).*

Karim, et al., 1999b). Sediment succession was subdivided into units called "groups" (later as formations by Armitage & Viotti, 1977), A to M from top to bottom, bounded by unconformities, which are seismically defined packages of strata similar to "depositional sequences" defined by Mitchum, Vail, and Thompson (1977) in modern sequence stratigraphy. Units M and older represent the synrift succession, consisting of alternating sand- and shale-dominated fluvial and lacustrine sequences. Units L and younger are interpreted as postrift deposits of nonmarine and blackish water origin (Madon & Watts, 1998). Sandstones of these units contain good hydrocarbon reservoirs (Madon, Abolins, et al., 1999).

Some petroleum companies developed their own stratigraphic nomenclatures for their respective exploration acreages. While drilling in the southern part of the Malay Basin during the late 1960s, Conoco established a different stratigraphic scheme using formations, roughly corresponding to Esso's seismic groups, but the Conoco scheme was later superseded by a new scheme introduced by PETRONAS Carigali when they assumed the operatorship of Conoco's exploration blocks in 1979 (Ramli, 1988). In the Carigali scheme, the sediment sequence is subdivided into eight "Tertiary units," I to VIII upsections (Figure 5.58A) described by Ramli (1988), defined either by the top of regional shale markers (as in units I and II) or by micropaleontologically defined marine transgressive pulses (as in units III and VIII) or a combination of both, in the same way as the "genetic stratigraphic sequence" of Galloway (1989). However, this scheme has been used only in PETRONAS Carigali's exploration blocks and in some neighboring areas (Madon, Karim, et al., 1999).

Biostratigraphy for all sediment sections of the Malay Basin relies heavily on palynomorphs (Figure 5.20). Since paleowater depths rarely reached 100 m, reliable index species of planktonic foraminifera and nannofossils are rare. Thus, their occurrence in high abundance likely indicates water deepening or even a marine flooding event.

West Natuna Basin

To the west of Natuna Island and north of the Anambas Islands lie two intensely faulted basins (Figure 5.56): the NE-trending West Natuna Basin and the W–E-trending Penyu Basin, mostly in present-day water depths of 50–60 m. Structurally, the West Natuna Basin belongs to the Sundaland, the cratonic core of SE Asia, and is enclosed by a series of fault-bounded basement highs, including the Khorat Swell to the north, the Natuna Arch to the east, shallow parts of the Sunda "Craton" to the south, and a gradual transition into the Malay Basin to the west (ASCOPE (Asean Council on Petroleum), 1981; Pollock, Hayes, Williams, & Young, 1984).

The West Natuna Basin is generally considered as originating from a rift or a pull-apart basin. It is characterized by a series of small, almost W–E-oriented depocenters with intervening basement ridges that formed during Paleogene rifting (Daines, 1985; Ginger, Ardjakusumah, Hedley, & Pothecary, 1993; Wongsosantiko & Wirojudo, 1984). Many half graben depocenters in the basin experienced significant contraction during the late Oligocene and Miocene, leading to the formation of "Sunda Folds," some classical structural examples of inverted basins (Darmadi et al., 2007; Ginger et al., 1993). The Miocene inversion reactivated preexisting faults and caused large-scale folding/faulting and uplift in the West Natuna and neighboring basins (Figure 5.57A).

The West Natuna Basin is filled by a Cenozoic sediment succession of >5 km thick, unconformably overlying the pre-Cenozoic basement of mainly metamorphosed lavas, amphibolites, and "gray effusive rocks" (ASCOPE, 1981). The Cenozoic strata include several formations, including the lower Oligocene Belut Formation, uppermost lower Oligocene to upper Oligocene

Gabus Formation, upper Oligocene to lower Miocene Barat Formation, lower to middle Miocene Arang Formation, and upper Miocene to the present Muda Formation (Figure 5.58B; ASCOPE (Asean Council on Petroleum), 1981; Darmadi et al., 2007; Phillips et al., 1997; Pollock et al., 1984). As in other southern SCS basins, to correlate the highly variable strata in the West Natuna Basin is difficult. Micropaleontology is only of occasional use, palynological markers are long ranging and imprecise, and lithology is subject to rapid vertical and horizontal variations. Therefore, stratigraphic correlation depends heavily on seismic interpretation, supplemented by limited paleontology, palynology, well log, and lithology data (ASCOPE (Asean Council on Petroleum), 1981).

The widely distributed early Oligocene Gabus Formation consists of fluvial, fine- to medium-grained sandstones interbedded with gray and brown silty shales, forming a primary exploration target in the basin (Figure 5.58B). The overlying Barat Formation mainly includes dark gray and brown claystone, as well as clean sandstones (or Udang Formation). The upper limit of the Barat is truncated locally by an unconformity (reflector "e"), but more often, it is transgressively overlain by the Arang Formation with several locally distinct members. The unconformably overlying Muda Formation is characterized by a late Miocene to lower Pliocene gray to brown shale, with silt, fine sandstone, coal, and liquate, and a late Pliocene and younger-aged soft gray clay with intercalations of silt and occasional coal and limestone. Recent studies indicate that the upper Muda Formation is dominantly a fluvial succession, which changes upward from offshore shelf facies, to a thin interval of deltaic facies and a series of fluvial facies, and finally to a thin transgressive facies lying on the modern sea floor (Darmadi et al., 2007).

In the West Natuna Basin, four megasequences have been recognized as marking its four tectonic stages: synrift, postrift, syninversion, and postinversion (Figure 5.58B). Respectively, the Oligocene synrift megasequence is equivalent to the Belut Formation; the late Oligocene to early Miocene postrift megasequence with four sequences to the Gabus, Udang, and Barat Formations; the early to middle Miocene syninversion megasequence with six sequences to the Arang Formation; and the late Miocene to the present postinversion megasequence to the Muda Formation (Ginger et al., 1993; Phillips et al., 1997; Figure 5.58B).

5.3.3.4 Thai Basin Group

About 12 basins of various sizes developed in the northwestern Gulf of Thailand (maximum water depth 86 m), mostly narrowly elongated and neatly in parallel N–S trending (Figure 5.56). Structurally, these basins were formed by a series of longitudinally arranged fault-bounded troughs (grabens) and ridges (horsts), in a sharp contrast to the adjacent NW–SE-extending Malay Basin. The central Ko Kra Ridge divides the group into a western area and an eastern area (Pigott & Sattayarak, 1993; Polachan & Sattayarak, 1991; Watcharanantakul & Morley, 2000). The western graben area consists of about ten grabens, namely, Sakhon, Paknam, Hua Hin, Prachuap, N. Western, Western, Kra, Chumphon, Nakhon, and Songkhla Basins, and the east graben area includes two larger basins: Pattani Basin and Khmer Basin. As the largest and the most hydrocarbon-rich in the group, the Pattani Basin is separated to the east from the Khmer Basin by the Khmer Ridge and to the southeast from the Malay Basin by the Narathiwat Ridge (Figure 5.56).

According to Jardine (1997) (Figure 5.59), the evolution of the Pattani Basin involved the following stages: (1) prerift folding and uplift of pre-Cenozoic accreted basement terranes, (2) initial rifting and creation of loca-lized subbasins (half grabens) from late Eocene to Oligocene, (3) structural inversion and erosion at the end of the Oligocene, (4) rifting and basin forma-tion in the early Miocene, (5) postrift collapse and basin subsidence in the middle Miocene, (6) widespread erosion from middle to early late Miocene, and (7) continued basin subsidence from the late Miocene to the present.

The Cenozoic sequence varies greatly in thickness between basins, from >8 km in the Pattani Basin, 4 km in basins from the western graben area, to <300 m over the shallow subcropping pre-Cenozoic highs (Lian & Bradley, 1986; Pigott & Sattayarak, 1993; Watcharanantakul & Morley, 2000). The sediment consists almost exclusively of nonmarine, fluvial to deltaic facies and becomes shallow marine only in the youngest intervals (Figure 5.59; Pigott & Sattayarak, 1993). The dominance of nonmarine deposits and the lack of diagnostic microfossil species have made stratigraphic correlation in the basin group rely heavily on palynology and sediment cycles, as do so in basins lying immediately to the southeast.

5.3.3.5   Other South China Sea Basins

To the northeast of the Zengmu (Sarawak) Basin, in and surrounding the Dan-gerous Grounds along the southern margin of the SCS, lie several smaller basins such as the Sabah Basin, Palawan Shelf Basin, Nansha (NW Palawan) Trough, and Reed Bank Basin(Figure 5.7), some of which are known for their rich hydrocarbon resource (Table 5.1).

The Sabah Basin forms part of the NW Sabah Shelf, and together with the neighboring Zengmu (Sarawak) Basin, it represents a major segment of the greater Sarawak–Brunei–Sabah continental margin that evolved since at least the late Cretaceous time. The Sabah Basin area was part of the proto-South China Sea or "Rajang Sea" until the Eocene (Leong, 1999; Tongkul, 1991). The subduction of the proto-SCS under the NW margin of Borneo caused deformation and uplift in the late Eocene, subsequently giving birth to a series of new depocenters in the so-called Oligocene basins filled by Oligocene–early Miocene turbidites and deepwater sediments often with a chaotic char-acter. The younger Sabah Basin started to develop since the middle Miocene

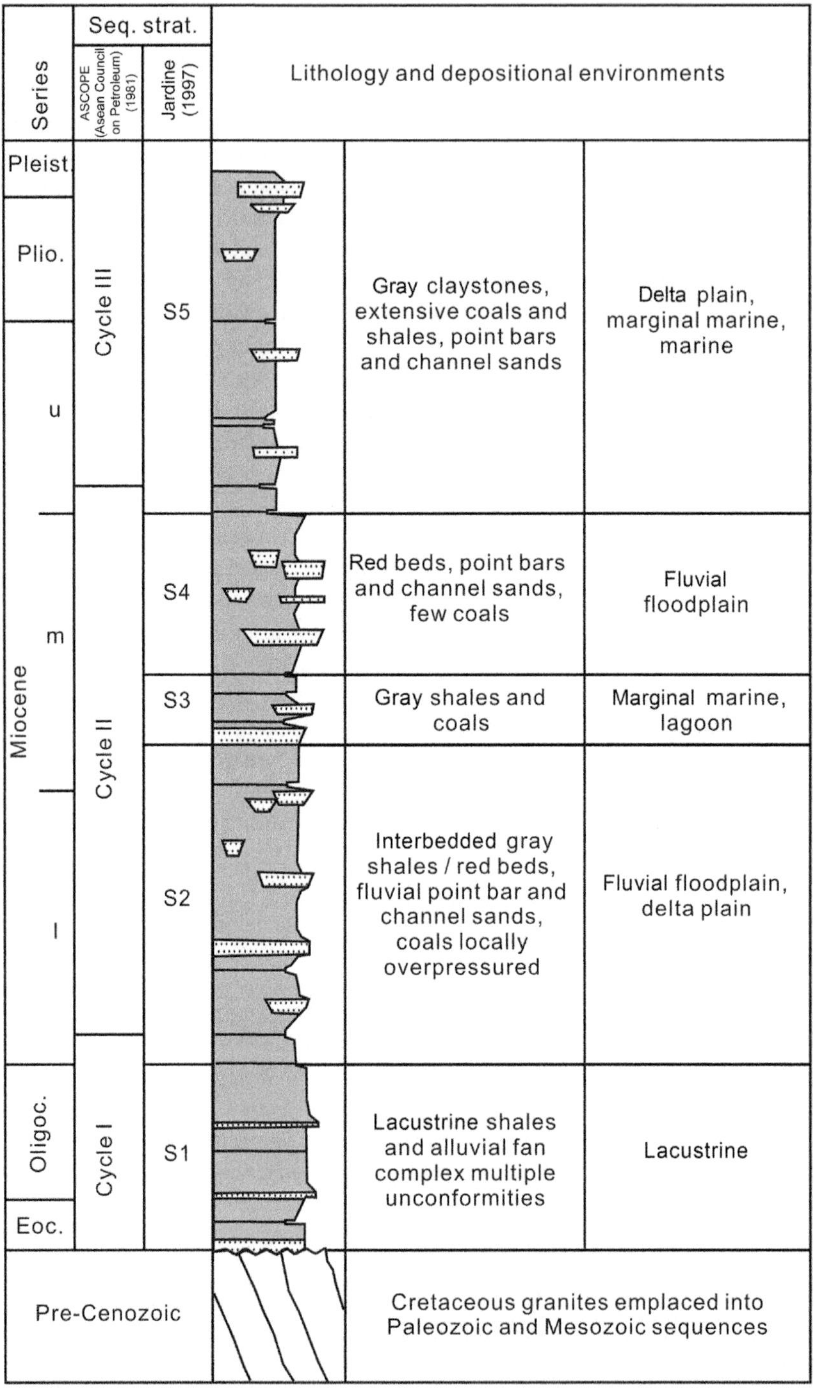

FIGURE 5.59 Generalized lithostratigraphy in the Thai Basin group, showing sediment sequences and cycles, major lithology, and depositional environments (Zhong & Li, 2009, and references).

probably as a foreland basin following the collision of an SCS microcontinental fragment (now forming the NW Sabah Platform) with western Sabah (Madon et al., 2013; Madon, Leong, et al., 1999). Subsidence was rapid (>500 m/Ma) between 15 and 9 Ma, followed by a generally slower rate accompanied by pronounced uplift along the eastern margin of the basin (Madon et al., 2013; Madon, Leong, et al., 1999). The Neogene Sabah Basin is mainly composed of middle Miocene and younger-aged fluvial to shallow-marine sandstone, siltstone, and mudstone/shale, truncated by several major unconformities, that is, the shallow, middle, and deep regional unconformities (or SRU, MRU, and DRU) at about 9, 11–12, and 15–16 Ma, respectively (Figure 5.10). Limestone developed mainly on the Kudat platform in the north since the late Miocene. The lithology in the southern Sabah Basin progradationally integrates into the Baram Delta, offshore Brunei and Sarawak (Du Bois, 1985). The post-middle Miocene succession has been subdivided into seven substages, A to G upsections, bounded by unconformities (Figure 5.10). Similar to the neighboring basins along the southern margin, the Sabah Basin and basins of older-age onshore Sabah are rich in hydrocarbon resource. Madon, Leong, et al. (1999) provided a detailed summary on tectonostratigraphy and hydrocarbon resource in the region, and Madon et al. (2013) provided an updated tectonic model for the southern SCS based on data from deepwater Sarawak (Figure 5.53).

The western Palawan shelf parallels the NE–SW-trending Palawan Island and borders to the northwest with the Palawan Trough (part of the Nansha Trough) (Figure 5.7). Across the trough to further northwest is the Reed Bank, an area of active carbonate platform growth like many atolls in the Nansha Islands since the late Oligocene time (Figure 5.10). Several NE-trending fault systems control the development of the Palawan Shelf Basin as part of the NW Borneo Geosyncline. The pre-Cenozoic basement in the area mainly consists of Mesozoic marine facies of shallow to bathyal environments. Shallow-marine deposition continued into the Paleogene, leading to carbonate platform buildups first on NW Palawan shelf and the Dangerous Grounds from late Eocene to early Miocene (Figure 5.60; Ding et al., 2013 and references) and from the Oligocene in the Reed Bank area (Figures 5.10; ASCOPE (Asean Council on Petroleum), 1981; Du Bois, 1985). The shallow Paleogene sea represented by the carbonate facies likely formed part of the proto-SCS before the area was rifted and subsided and subsequently drowning the platform on the NW Palawan shelf in the early Miocene. The drowning of the carbonate buildups was more evident in the trough due to its rapid subsidence (Figure 5.61; Hutchison, 2010), but those in the Nansha Islands continued to the present day. The breakup unconformity lies close to the early/middle Miocene boundary in Palawan–Dangerous Grounds area (Figure 5.60), while the Reed Bank area remained relatively stable as a microcontinent not affected by collision and compressional extension. The post-middle Miocene sediment sequences in the NW Palawan shelf in general bear a close affinity with those

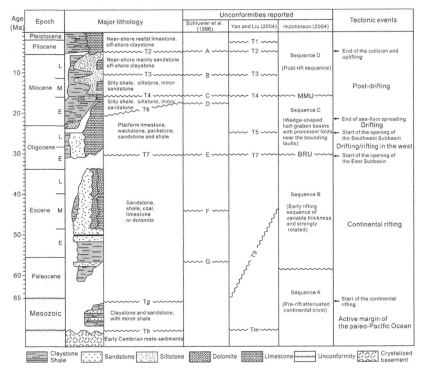

FIGURE 5.60 Seismic stratigraphy, lithology, and major tectonic events of the Dangerous Grounds area (Ding, Franke, Li, & Steuer, 2013, and references).

in the southwest, in Sabah and Sarawak Basins (Figure 5.10), and those in the Dangerous Grounds region (Figure 5.60).

5.4 ISOTOPIC AND ASTRONOMICAL STRATIGRAPHY

Continuous $\delta^{18}O$ and $\delta^{13}C$ sequences have been obtained from ODP Site 1148 for the entire Neogene with 5–10 kyr resolution (Tian, Zhao, Wang, Li, & Cheng, 2008; Zhao, Jian, et al., 2001; Zhao, Wang, et al., 2001), from Site 1143 for the last 5 Ma with 2 kyr resolution (Tian et al., 2004a,2004b; Tian, Wang, Cheng, & Li, 2002), and from Site 1146 for the middle Miocene interval between 17 and 12 Ma (Holbourn, Kuhnt, & Schulz, 2004; Holbourn et al., 2005, 2007). In addition to these long sequences, there are a 200-year resolution record for the last 0.7 Ma (Bühring et al., 2004) and a 370-year resolution record for the 0.8–1.06 Ma from Site 1144 (20°03′N, 117°25′E, w.d. 2037 m; Jin & Jian, 2013), as well as many high-resolution records for various late Quaternary intervals (mostly 0.3 Ma or younger; Wang & Li, 2009). These continuous records are more advantageous over the composite curves of Zachos, Pagani, Sloan, Thomas, and Billups (2001), Zachos, Shackleton, Revenaugh, Pälike, and Flower (2001), and Zachos, Dickens, and Zeebe

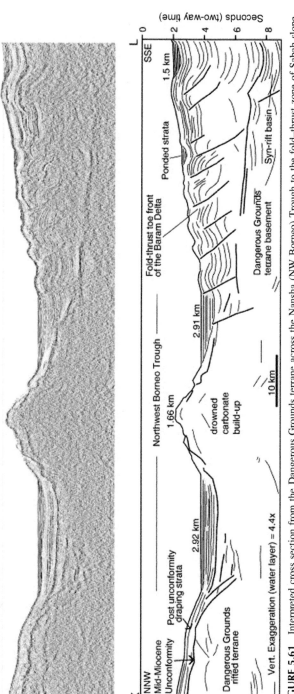

FIGURE 5.61 Interpreted cross section from the Dangerous Grounds terrane across the Nansha (NW Borneo) Trough to the fold-thrust zone of Sabah slope, showing the dominant middle Miocene unconformity, the drowned carbonate buildup, and other features (Hutchison, 2010). See Figure 5.7 for line location.

(2008) from some 40 DSDP/ODP isotopic profiles as they provide more direct paleoenvironmental information for interpolating long- and short-term paleo-oceanographic trends of global or regional significance such as the opening and closing of the South China Sea, East Asian monsoon variability, and global cooling at 13.9 Ma (Holbourn et al., 2005, 2007; Tian et al., 2004b,2004c). Largely as an update from Tian and Li (2009), this section provides a brief account on isotopic stratigraphy of ODP Sites 1148 and 1143, which are becoming the standard references for deep-sea research in the region.

5.4.1   Pliocene–Pleistocene Isotopic Records

ODP Site 1143 lies within the northern part of the Dangerous Grounds or Nansha Islands area, in the southern SCS. The Pliocene–Pleistocene sediments recovered at the site consist mostly of olive, greenish, and light gray-green and greenish gray clayey nannofossil mixed sediment, clay with nannofossils, and clay, with average linear sedimentation rates of 30–70 m/Ma (Wang et al., 2000). Planktonic and benthic foraminifers in 1992 samples at 2 cm spacing from 0 to 190.77 mcd (meter composite depth) were measured for stable isotopes (Tian et al., 2002, 2004a,2004b).

An initial age model was based on biostratigraphy of 14 planktonic foramin-ifer datums, as well as the Brunhes/Matuyama paleomagnetic polarity reversal at 43.2 mcd of Hole 1143C that represents an age of 780 kyr. By comparing the Site 1143 $\delta^{18}O$ curve to the composite curve of Shackleton, Hall, and Pate (1995) for the last 5 myr, the corresponding marine isotopic stages (MIS 1 to MIS T1) were identified (Figure 5.62). Apart from these distinct glacial cycles, an increasing trend in isotopic values from MIS MG2 to MIS 96 marks the growth of Northern Hemisphere ice sheet between 3.3 and 2.5 Ma.

Orbital tuning of the Site 1143 benthic $\delta^{18}O$ record was performed using the astronomical solution of Laskar (1990) for the obliquity and precession as the tuning target (Tian et al., 2002). Because previous studies have found $\delta^{18}O$ lagging the Northern Hemisphere 65° summer insolation maxima by ~69° at the obliquity band and ~78° at the precession band, equivalent to 7.8 and 5 kyr, respectively, an 8 kyr lag for the obliquity curve and 5 kyr lag for the pre-cession curve from Laskar (1990) were applied for tuning using the dynamic opti-mization technique (Yu & Ding, 1998). A final age model (Figure 5.62) was obtained by repeatedly tuning the initial age model to obliquity until coherencies between the filtered $\delta^{18}O$ signals and orbital signals reached a maximum fit.

Cross spectral analyses of the tuned $\delta^{18}O$ record with the sum of normal-ized eccentricity (E), normalized obliquity (T), and negative normalized pre-cession (P) (or ETP) for intervals of 0–1, 1–2, 2–3, 3–4, and 4–5 Ma all show good coherency (most >95%) at the 41 kyr obliquity band and 23 kyr and 19 kyr precession bands (Figure 5.63). The spectral densities of 100, 23, and 19 kyr periodicities in $\delta^{18}O$ increase significantly through time, as also their corresponding coherencies. A strong 41 kyr cycle highly coherent with

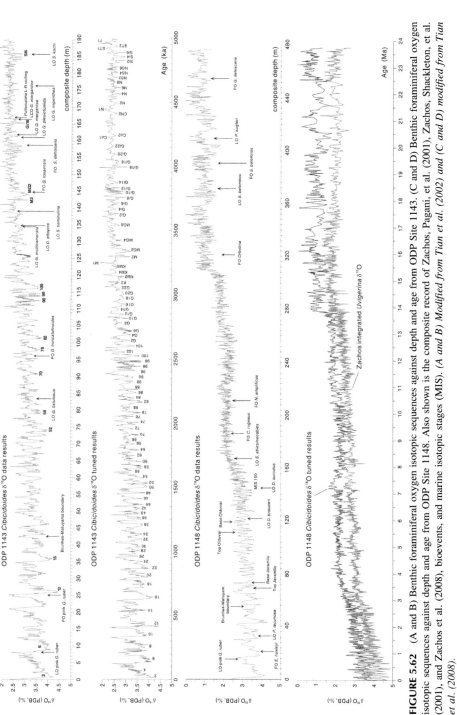

FIGURE 5.62 (A and B) Benthic foraminiferal oxygen isotopic sequences against depth and age from ODP Site 1143. (C and D) Benthic foraminiferal oxygen isotopic sequences against depth and age from ODP Site 1148. Also shown is the composite record of Zachos, Pagani, et al. (2001), Zachos, Shackleton, et al. (2001), and Zachos et al. (2008), bioevents, and marine isotopic stages (MIS). *(A and B) Modified from Tian et al. (2002) and (C and D) modified from Tian et al. (2008).*

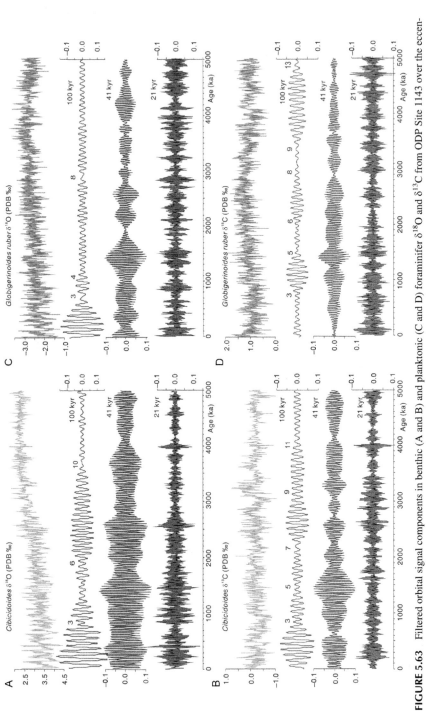

FIGURE 5.63 Filtered orbital signal components in benthic (A and B) and planktonic (C and D) foraminifer δ^18O and δ^13C from ODP Site 1143 over the eccentricity (100 kyr), obliquity (41 kyr), and precession (21 kyr) bands. Note the discontinuous distribution of the 400 kyr long-eccentricity cycles numbered 3–13. *Modified from Tian et al. (2004b).*

ETP in all the intervals clearly demonstrates its importance over the Pliocene–Pleistocene. The dominance of the 41 kyr cycle was replaced by the 100 kyr eccentricity cycle at about 0.9 Ma, marking the middle Pleistocene climatic transition (Li et al., 2008, and references; Jin & Jian, 2013). The ~100 kyr short- and ~400 kyr long-eccentricity cycles, however, appear to be discontinuous on these $\delta^{18}O$ records (Figure 5.63), although the ~400 kyr cycle often forms a distinct feature on many $\delta^{13}C$ records including the one from Site 1143 in association with oceanic carbon reservoir changes (Wang, Li, et al., 2004; Wang, Li, & Tian, 2014; Wang, Tian, Cheng, Liu, & Xu, 2003, 2004).

The tuned age model can be assessed by comparing the filtered 41 kyr component of $\delta^{18}O$ with the lagged obliquity and comparing the filtered 23 kyr component of $\delta^{18}O$ with the lagged precession. As shown in Figure 5.63, good matches between the filtered signals and the orbital parameters confirm the accuracy of the tuned timescale. Some minor discrepancies exist between the filtered 23 kyr component and the lagged precession especially at ages of ~0.4, 2.8–2.9, 3.95–4.3, and 4.65–5.0 Ma. Although further studies are needed to clarify the cause of these discrepancies, the tuning strategy by fixing on to a good 41 kyr obliquity cycle may have forced some mismatch on the 23 kyr precession cycle (Tian et al., 2004b).

5.4.2 Neogene Isotopic Records

The stable oxygen and carbon isotopic series from ODP Site 1148 in the northern SCS involved 1755 benthic foraminiferal measurements from the upper 476.68 mcd and 975 planktonic foraminiferal measurements from the upper 409.58 mcd, with average sampling resolution of ~16 kyr for the Miocene and ~8 kyr for the Pliocene–Holocene (Jian et al., 2003; Tian et al., 2008; Zhao, Jian, et al., 2001; Zhao, Wang, et al., 2001). In most cases, benthic species *Cibicidoides wuellerstorfi* and *Uvigerina* sp. and planktonic species *Globigerinoides ruber* (above 165 mcd) and *Globigerinoides sacculifer* (below 165 mcd) were used, and their results were adjusted to represent that of *C. wuellerstorfi* and *G. ruber*, respectively.

The chronobiostratigraphic timescale of Gradstein et al. (2004) was used as an initial age model, which was then slightly modified by comparing between the benthic $\delta^{18}O$ curves of Site 1148 and that of Zachos, Pagani, et al. (2001), Zachos, Shackleton, et al. (2001), and Zachos et al. (2008). Increase by >1 per mil of the benthic $\delta^{18}O$ from the interval of 305–316 mcd at Site 1148 marks the relative abrupt global cooling at ~14 Ma, when East Antarctic ice sheet expanded (Figure 5.62). Similarly, gradual $\delta^{18}O$ increase from the interval of 143.36–161.98 mcd denotes the middle Pliocene global cooling and the enlargement of the Northern Hemisphere ice sheet. Other prominent features of the Site 1148 benthic $\delta^{18}O$ curve include strong fluctuating cycles from the interval of 0–44.51 mcd and much stronger fluctuations below 320 mcd

(Figure 5.62), respectively, indicating a shift to 100 kyr climate cycle dominance since the middle Pleistocene and more variable climate and bottom conditions in the early Miocene. Because of many well-studied records now available for comparison (Shackleton et al., 1995; Tian et al., 2002; Tiedemann, Sarnthein, & Shackleton, 1994), post-8 Ma benthic $\delta^{18}O$ stratigraphy at Site 1148 is more reliable than for older time intervals. To choose some well-dated, consistent litho-, bio-, and chemostratigraphic datums as the control points is a prerequisite for establishing an initial age model to be tuned for a Neogene orbital timescale.

Apart from nannofossil and planktonic datums (Li, Jian, & Li, 2004; Su, Xu, & Tu, 2004), several $\delta^{18}O$ and $\delta^{13}C$ events correlate well with their counterparts identified originally from other localities, including Mi (Miocene oxygen isotopic maxima) events and CM (carbon maxima) events (Miller, Fairbanks, & Mountain, 1987; Miller, Wright, & Fairbanks, 1991; Woodruff & Savin, 1991; Zhao, Jian, et al., 2001; Zhao, Wang, et al., 2001). Compared to Zachos, Pagani, et al. (2001) and Zachos, Shackleton, et al. (2001), the Site 1148 benthic $\delta^{18}O$ results show a shift at 16–15 Ma, characterized by overall lighter $\delta^{18}O$ values before 15 Ma and lighter $\delta^{13}C$ values after 15 Ma, probably as a response to the termination of seafloor spreading in the South China Sea. Lighter $\delta^{18}O$ values from ~23 to 15 Ma appear to support the paleobotanical evidence for an early Miocene development of the East Asian monsoon (Sun & Wang, 2005) and relative humid conditions surrounding ODP Site 1148 (Clift, 2006). In contrast, lighter $\delta^{13}C$ values after ~15 Ma could have been caused by bottom-water ventilation or by a newly established local bottom water since the end of spreading.

Following the tuning procedure of Shackleton et al. (1999), a tuned orbital age model was generated for Site 1148 natural gamma ray (NGR) data to the Earth's orbital solution of Laskar et al. (2004), an update from the solution of Laskar et al. (1999). The NGR record was supplemented by color spectral reflectance (CSR, or lightness, L*) record since it is inversely correlated to the NGR record. The tuning was done by aligning the cycles of the tuning material data (NGR and CSR) with that of the tuning target data (obliquity or precession), and further constrained by matching between the 100 or 400 kyr eccentricity components of the tuning material records and the eccentricity. Shackleton et al. (1999) generated a Northern Hemisphere insolation curve as a tuning target, which contains a dominant obliquity signal and a weaker but discernible precession signal. For evaluation, these authors used the 1.2 Ma amplitude modulation of obliquity and 400 kyr amplitude modulation of climatic precession to evaluate the fidelity of the generated age model.

Temporal variability of the basic cycles (400 and 100 kyr) of the NGR record was detected using wavelet analyses. The results show that the 41 kyr cycle stands out as having higher spectral density than the precession band-related cycle but not as strong as the eccentricity-related cycles.

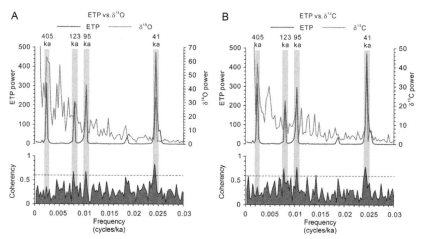

FIGURE 5.64 Blackman–Tukey spectral cross correlation of ETP with (A) benthic foraminiferal $\delta^{18}O$ and (B) benthic foraminiferal $\delta^{13}C$ from ODP Site 1148 (0–23 Ma). Horizontal dashed lines denote 80% confidence level of coherency. Bandwidth is 0.000216. *From Tian et al. (2008)*.

The marked 400, 100, and 41 kyr cycles identified from the Blackman–Tukey and wavelet spectra in the depth domain reveal the astronomical imprint on the NGR record, which forms the base of this astronomical tuning exercise (Tian et al., 2008). The visually identified NGR and CSR cycles to the obliquity cycles can be directly matched. Following this tuning scheme, the maxima in obliquity are correlated to minima in NGR or maxima in CSR.

The evaluation of the tuned age model was done using cross spectral analyses to reveal its coherencies relative to ETP. The results show high coherencies exceeding 80% statistical level at 100, 41, 23, and 19 kyr bands between NGR and ETP records. However, the coherency between the NGR record and the ETP record is below 80% statistical level for the 400 kyr band. Similarly, Blackman–Tukey cross spectral analyses were carried out to reveal the coherency between the ETP from La2004 (normalized eccentricity, tilt, and precession mix) and the isotopic records over the entire Neogene period for Site 1148. As seen in Figure 5.64, the 400 and 41 kyr cycles are highlighted from the "red noise" spectra of both the $\delta^{18}O$ and the $\delta^{13}C$. The 100 kyr cycle is also highly significant in both the $\delta^{18}O$ and the $\delta^{13}C$ spectra, exhibiting two separated peaks of 95 and 123 kyr. Both the $\delta^{18}O$ and the $\delta^{13}C$ are coherent with the ETP at the eccentricity (95 and 123 kyr) and obliquity (41 kyr) bands, with coherency exceeding 80% confidence level. However, the extremely low coherencies at the 400 kyr for both the $\delta^{18}O$ and the $\delta^{13}C$ at ODP Site 1148 somehow contradict with some early studies (e.g., Pälike, Frazier, & Zachos, 2006; Pälike, Norris, et al., 2006).

Evolutive spectral analyses on the benthic $\delta^{18}O$ and $\delta^{13}C$ records of Site 1148 (Figure 5.65) clearly reveal that the 100 kyr climate cycle is prominent in the late Pleistocene and distinctive for the whole Neogene. For the 400 kyr

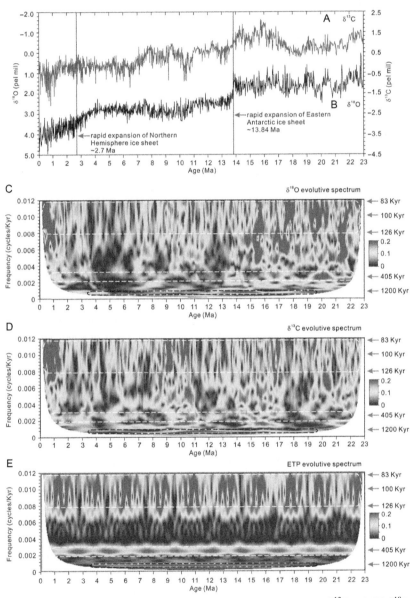

FIGURE 5.65 (A) ODP Site 1148 benthic foraminiferal isotopic records, $\delta^{13}C$ and (B) $\delta^{18}O$, 0–23 Ma. (C) Evolutive spectrum of $\delta^{18}O$. (D) Evolutive spectrum of $\delta^{13}C$. (E) Evolutive spectrum of ETP. *From Tian et al. (2008).*

cycle, the physical property and benthic $\delta^{18}O$ and $\delta^{13}C$ records also show its strong presence, with special strength during the early and middle Miocene. But its low coherency with the ETP and weaker presence in some time intervals likely originated from a nonlinear response to solar radiation forcing. Therefore, the signals of strong 100, 41, and 19–23 kyr cycles demonstrate the accuracy and reliability of this orbital tuned isotopic record for the SCS Neogene (Figure 5.65; Tian et al., 2008).

In summary, the orbitally tuned timescales from ODP Leg 184 sites provide some unprecedented means for high-resolution paleoceanographic studies in the SCS region. These timescales lay a solid foundation not only for more accurate stratigraphic correlation of deep-sea sediment sequences but also for exploring the regional paleoceanographic response to global climate change over various Neogene time intervals.

5.5   STRATIGRAPHIC SUMMARY

5.5.1   Lithostratigraphy Patterns

Intensive industrial drilling and geologic and geophysical survey over decades in various parts of the China Seas have accumulated a large amount of data from which lithostratigraphic changes in different basins can be extrapolated. As discussed in the preceding text in this chapter, lithostratigraphic variations are large between and within these basins, largely affected by changes in depositional environments at different stages of basin development. Major characteristics of regional lithostratigraphy can be summarized as follows:

(1) Basement types of basins are diverse. Although Mesozoic sedimentary rocks are the most common type of basement in many basins, Mesozoic and older igneous and metamorphic rocks are also common. Igneous or metamorphic basement rocks are often unconformably underlain by the synrift early Cenozoic sequence in some SCS basins, such as the Yinggehai Basin and many parts of the Pearl River Mouth Basin and Wan'an–Cuu Long Basins (Table 5.3; Figure 5.51). Sedimentary rocks are mainly of nonmarine origin, except in parts of the (southern) ECS Shelf Basin and basins along the SE margin of the SCS where shallow-marine deposits may be present. The contact between these Mesozoic sedimentary rocks and the overlying early Cenozoic sequence is either continuous (as in Bohai Basin and Yellow Sea Basins) or discontinuous and unconformable (as in ECS Shelf Basin and many SCS basins). Apparently, the prerift topography and the onset time of rifting together determined the nature of the basement contact in most basins.

(2) Nonmarine deposits, mainly of alluvial, fluvial, and lacustrine facies, are prevalent in basins now lying in shallow water depths, particularly those immediately offshore from China–Indochina coast. Sandstones, siltstones, and mudstones characterize the synrift sequence and many intervals of the

postrift sequence in these basins. Stratigraphy of these nonmarine deposits relies heavily on palynomorphs and shelly fossils that can only provide a coarsely estimated age for stratigraphic correlation. Therefore, many non-marine and paralic sequences with different assigned ages still await fur-ther efforts to clarify their stratigraphic position. The problematic strata include the Kongdian Formation (Paleocene–Eocene? Bohai Basin), Sanduo Formation (middle Eocene? South Yellow Sea Basin), Huanggang Formation (late Oligocene? ECS Shelf Basin), Enping Formation (Eocene–Oligocene? Pearl River Mouth Basin), and Cau Formation (Eocene–Oligocene? Wan'an–Nan Con Son Basin), among others.

(3) Thick layers of mudstones are found in many basins of the China Seas, mostly in association with synrift deposition. These mudstones of lacus-trine origin often form part of Kongdian, Shahejie, and Dongying Formations (Bohai Basin), Funing and Dainan Formations (South Yellow Sea Basin), Wenchang and Enping Formations (Pear River Mouth Basin), Weizhou Formation (Beibuwan Basin), Yacheng Formation (Yinggehai–Qiongdongnan Basins), and Oligocene to Miocene sequences in some southern SCS basins. The wide occurrence of lacustrine facies indicates the extensive development of paleolakes during various time intervals of rifting (Figure 5.2). Many of the synrift lacustrine mudstones are good source rocks (see Chapter 7).

(4) Marine sequences are restricted to deeper parts of basins in the ECS proper and in the SCS, mainly as the postrift deposition in the Neogene. Synrift Paleocene–Eocene strata with marine fossils are known from two regions: (i) from southern ECS Shelf Basin, to Taiwan, to the Taxinan Basin in the northeastern SCS and (ii) along the southeastern margin, in Palawan–Sabah–Reed Bank Basins. These regions appear to have been related, respectively, to back-arc subduction and to the influence of the proto-SCS in the Paleogene, although whether they had once connected to each other in representing the proto-SCS is not clear. Shallow-marine beds of mainly Oligocene age also occur in other northern SCS basins, such as Lingshui Formation (Yinggehai–Qiongdongnan basins), Zhuhai Formation (Pearl River Mouth Basin), and Cau Formation (Wan'an–Nan Con Son Basin). These Oligocene marine beds may represent the deposition either during the transition from rifting to postrifting or even the earliest phase of postrifting.

(5) Limestones and carbonate platforms characterize the synrift and postrift deposition in many SCS basins. Shallow-marine carbonate buildups first developed in Dangerous Grounds–offshore Palawan–Reed Bank areas of the southern SCS during the Paleocene–Eocene and expanded to the northern SCS (Yinggehai–Qiongdongnan–Pearl River Mouth–Taixinan Basins) during the Oligocene–Miocene. Most of the carbonate platforms, however, were drowned in the later part of middle Miocene and late Mio-cene due to rapid sinking of basins caused by regional thermal subsidence

and subduction along the southeastern margin, particularly in the Nansha–NW Palawan Troughs. In contrast, carbonate buildups on such stable microblocks as Nansha (the Dangerous Grounds)–Reed Bank and Xisha Islands areas continued into the present day.

(6) Although marine deposition became widespread in most parts of the SCS and ECS since the early Miocene, marine influence in the Yellow Sea and Bohai Sea occurred much later, since the Pleistocene. Marine beds in Yellow Sea and Bohai Basins accumulated largely during interglacial intervals, similar to those from the nearshore part of many SCS and ECS basins. There are reports of marine microfossils in (mostly single) samples from older (probably Pliocene) strata in Yellow Sea and Bohai Basins, but the sporadic nature of these brackish-water assemblages casts doubts on their *in situ* occurrence or their indication of full-marine deposition (see Chapter 8).

(7) Numerous sequence stratigraphy schemes have been proposed for synrift and postrift depositions in various China Sea basins. Bounded by unconformities or major seismic reflectors, the assigned sequences often fall in corresponding formations or members recognized earlier based on lithofacies and biofacies. The sequences provide critical information on the relationship between stratal packaging and environmental variations especially sea level change, but sophisticated schemes such as those for the Taixinan and Pearl River Mouth Basins (Figures 5.40 and 5.44) sometimes appear to be unpractical and confusing.

(8) Depositional successions in most China Sea basins are truncated by unconformities of various durations, marked also by major and secondary seismic reflectors. The most significant basin-wide unconformities occurred during the Oligocene and Miocene, collectively as the breakup unconformity marking the rifting/postrifting transition. Representing these are the Oligocene/Miocene contact unconformity, such as T1 in the Bohai Basin, T2 in Yellow Sea Basins, T30–T24 in the ECS Shelf Basin, and T60–T70 in the Pearl River Mouth Basin, or the middle Miocene unconformity in many southern SCS basins. The breakup unconformity may erase a sediment record up to several million years, particularly in the South Yellow Sea Basins and western ECS Shelf Basin (missing Oligocene and part of upper Eocene), Beibuwan Basin (missing upper Oligocene), and offshore Sarawak (missing lower Miocene and part of the Oligocene and upper Eocene). At ODP Site 1148 from the northern SCS lower slope, this event is recorded by a slumped unit with the missing of sediment record of >3 myr.

5.5.2   Stratigraphy and Evolution of the China Seas

Stratigraphic evidence indicates that the Cenozoic evolution of the China Seas has involved different stages in basin development: rifting and postrifting deformation in ECS basins and rifting, spreading, and postspreading deformation in SCS basins. Rifting occurred earlier during the late Cretaceous–Paleogene in the Bohai and Yellow Seas, during the Paleogene in most ECS and northern SCS basins, during the Paleogene–early Neogene in southern SCS basins, and since the late Miocene in the Okinawa Trough. Synrift sequences in most basins are similarly composed of alluvial, fluvial, and lacustrine to deltaic sediments, except those along the southeastern SCS margin where shallow-marine facies were common likely in relation to the proto-SCS. The contact between synrift and postrift sequences often coincides with a basin-wide unconformity, marking significant changes in depositional environments during the rifting/spreading or rifting/postrifting transition.

Nonmarine deposition continuous from the synrift (late Cretaceous to) Paleogene to the postrift Neogene in Bohai and Yellow Sea Basins, while marine deposition established in the later synrift stage during the Paleocene–Eocene in (southern) ECS and northern SCS basins, indicating major differences in synrift and postrift environments between these basin regions. It can be argued that the Bohai and Yellow Sea Basins prevailed by nonmarine sequences represent the land-based basins of the similar type in the Northeast China with a sea character developed only since the recent past. Similar situation existed in the far south and far west when the Sunda Shelf and the Gulf of Thailand were a land with basins receiving mainly terrestrial sediments during most of the Paleogene and Neogene time.

During the synrift Paleocene–Eocene, there were two regions under marine influence: (1) the area now in the south and along the SE margin of the SCS linking to the proto-SCS and (2) the southern ECS likely as an early belt of back-arc subduction. Full-marine Eocene sediments recently recovered in the Taixinan Basin probably represent the deposition in the southwest extension of an ECS back-arc trough. The link between the ECS and SCS since then was interrupted by tectonic activities related to the accretion of Taiwan. Since the Eocene–Oligocene, while the SCS opened and spread, the ECS also continued to expand by adding more back-arc belts of troughs, as indicated by increases in hemipelagic sediments accumulated in corresponding northern SCS and ECS basins. Therefore, the sea area of the China Seas was very small in the synrift Paleogene. It increased substantially during the Neogene largely due to spreading of the SCS, increased more significantly in the later Neogene when the Okinawa Trough opened up, and reached the present size in the Quaternary.

Since the Oligocene, deformation and basin conversion leading to the accumulation of lacustrine to deltaic facies characterize the ECS history, which was shared at least partly by basins of the Yellow Sea. In contrast, in the SCS, carbonate components increased and hemipelagic sediment became widespread due to accelerated spreading and rise in regional sea level since the early Miocene. Carbonate reefs, which were mainly confined to the Dangerous Grounds and offshore Palawan, now expanded to include the northern

SCS continental margin. Further subsidence and accumulation of full-marine deposits continued in many northern SCS basins even after spreading ceased near the early/middle Miocene boundary. However, southern SCS basins entered a period of postrift deposition since that time, mainly with clastic and paralic sediments except at more offshore settings along the SE margin (near and in the Nanshan–NW Borneo Trough) due to subduction and rapid subsidence. Further subduction and collision during the late Miocene and Pliocene caused basin inversion, deformation, and frequent turbidite deposition not only in the southeastern SCS but also in basins of the greater ECS, while Okinawa Trough started to take shape.

REFERENCES

Adams, C. G. (1970). A reconsideration of the East Indian Letter Classification of the Tertiary. *Bulletin of British Museum (Natural History)*, *19*(3), 87–137.

Ali, M. Y. B., & Abolins, P. (1999). Central luconia province. In K. M. Leong (Ed.), *The petroleum geology and resources of Malaysia* (pp. 371–392). Kuala Lumpur: PETRONAS.

Allen, M. B., Macdonald, D. I. M., Zhao, X., Vincent, S. J., & Brouet-Menzies, C. (1997). Early Cenozoic two-phase extension and late Cenozoic thermal subsidence and inversion of the Bohai Basin, North China. *Marine and Petroleum Geology*, *14*, 951–972.

Andersen, C., Mathiesen, A., Nielsen, L. H., Tiem, P. V., Petersen, H. I., & Dien, P. T. (2005). Distribution of source rocks and maturity modelling in the northern Cenozoic Song Hong Basin (Gulf of Tonkin), Vietnam. *Journal of Petroleum Geology*, *28*, 167–184.

Anthonissen, E., & Ogg, J. G. (2012). Biochronology of Paleogene and Neogene microfossils. In F. M. Gradstein, J. G. Ogg, M. Schmitz, & G. Ogg (Eds.), *The geologic time scale 2012* (pp. 1083–1127). The Netherlands: Elsevier B.V.

Areshev, E. G., Dong, T. L., San, N. T., & Shnip, O. A. (1992). Reservoirs in fractured basement on the continental shelf of southern Vietnam. *Journal of Petroleum Geology*, *15*, 451–464.

Armitage, J. H., & Viotti, C. (1977). Stratigraphic nomenclature–southern end Malay Basin. In *Proceedings of the Indonesian petroleum association 6th annual convention, May 1977* (pp. 69–94).

ASCOPE (Asean Council on Petroleum). (1981). *Tertiary sedimentary basins of the Gulf of Thailand and South China Sea: Stratigraphy, structure and hydrocarbon occurrences.* Jakarta, Indonesia: ASCOPE, 72 pp.

Binh, N. T. T., Tokunaga, T., Son, H. P., & Binh, M. V. (2007). Present-day stress and pore pressure fields in the Cuu Long and Nam Con Son Basins, offshore Vietnam. *Marine and Petroleum Geology*, *24*, 607–615.

Bishop, M. G. (2002). *Petroleum systems of the Malay Basin Province, Malaysia: Open-File Report 99-50T. U.S. Geological Survey.*

BouDagher-Fadel, M. K. (2008). Evolution and Geological Significance of Larger Benthic Foraminifera. *Developments in Palaeontology and Stratigraphy*, *21*, 1–548.

Bühring, C., Sarnthein, M., & Erlenkeuser, H. (2004). Toward a high resolution stable isotope stratigraphy of the last 1.1 m.y.: Site 1144, South China Sea. In W. L. Prell, P. Wang, P. Blum, D. K. Rea, & S. C. Clemens (Eds.), *Proceedings of ODP scientific results: Vol. 184.* (pp. 1–29), Online.

Cai, Q. Z. (2005). *Oil and gas geology in China Seas.* Beijing: China Ocean Press, 406 pp (in Chinese).

Cai, F., Dai, C., Chen, J., Li, G., & Sun, P. (2005). Hydrocarbon potential of pre-Cenozoic strata in the North Yellow Sea Basin. *Marine Science Bulletin, 7*, 21–36.

Canh, T., Ha, D. V., Carstens, H., & Berstad, S. (1994). Vietnam-attractive plays in a new geological province. *Oil and Gas Journal, 92*(11), 78–83.

Chang, M. (1997). Studies on the unconformities of the Tainan Basin, Offshore Southwestern Taiwan. *Petroleum Geology of Taiwan, 31*, 155–168.

Chen, C. (2000). Petroleum geology and conditions for hydrocarbon accumulation in the eastern Pearl River Mouth Basin. *China Offshore Oil and Gas (Geology), 14*(2), 73–83 (in Chinese).

Chen, L. (2002). Geologic structural feature in west of Zengmu Basin, Nansha Sea Area. *Geophysical Prospective Petroleum, 37*, 354–362 (in Chinese).

Chen, P. P. H., Chen, Z. Y., & Zhang, Q. M. (1993). Sequence stratigraphy and continental margin development of the northwestern shelf of the South China Sea. *AAPG Bulletin, 77*, 842–862.

Chen, L., Jin, Q., Liu, Z., Wang, H., & Wang, L. (2008). Features and genetic mechanism of faults in the Sun Basin in Mesozoic and Cenozoic times. *Marine Geology and Quaternary Geology, 28*(2), 53–60 (in Chinese).

Chen, L., Liu, Z., Jin, Q., Wang, Y., & Yuan, S. (2008). Meso-Cenozoic tectonic evolution of the east depression of North Yellow Sea. *Geotectonica et Metallogenia, 32*, 308–316 (in Chinese).

Chen, Z. Y., Lu, F. W., Zhang, J. P., & Zhang, T. (2013). Age of Cenozoic Sedimentary Formations of the Xihu Sag, East China Sea Continental Shelf. *Shanghai Land and Resources, 34*, 42–45 (in Chinese).

Chen, L., & Peng, X. (1995). Preliminary study of seismic stratigraphy of Wan'an Basin in Nansha waters. *Geophysical Prospective Petroleum, 34*(2), 57–70 (in Chinese).

Chen, W. S., Ridgway, K. D., Hong, C. S., Chen, Y. K., Shea, K. S., & Yeh, M. G. (2001). Stratigraphic architecture, magnetostratigraphy, and incised-valley systems of the Pliocene-Pleistocene collisional marine foreland basin of Taiwan. *GSA Bulletin, 113*, 1249–1271.

Chen, J., Xu, S., & Sang, J. (1994). The depositional characteristics and oil potential of paleo Pearl River delta systems in the Pearl River Mouth Basin, South China Sea. *Tectonophysics, 235*, 1–11.

Chough, S. K., Lee, H. J., & Yoon, S. H. (2000). *Marine Geology of Korean Seas.* Amsterdam: Elsevier, 313 p.

Clift, P. (2006). Controls on the erosion of Cenozoic Asia and the flux of clastic sediment to the ocean. *Earth and Planetary Science Letters, 241*, 571–580.

Clift, P. D., Lin, J., & ODPLeg 184 Scientific Party. (2001). Patterns of extension and magmatism along the continent-ocean boundary, South China Margin. In R. C. L. Wilson, R. B. Whitmarsh, B. Taylor, & N. Froitzheim (Eds.), *Non-volcanic rifting of continental margins: A comparison of evidence from land and sea: Vol. 187* (pp. 489–510). London: Geological Society Special Publication.

Clift, P. D., & Sun, Z. (2006). The sedimentary and tectonic evolution of the Yinggehai-Song Hong basin and the southern Hainan margin, South China Sea: Implications for Tibetan uplift and monsoon intensification. *Journal of Geophysical Research, 111*. http://dx.doi.org/10.1029/2005JB004048, B06405.

Cukur, D., Horozal, S., Kim, D. C., & Han, H. C. (2011). Seismic stratigraphy and structural analysis of the northern East China Sea Shelf Basin interpreted from multi-channel seismic reflection data and cross-section Restoration. *Marine and Petroleum Geology, 28*, 1003–1022.

Daines, S. R. (1985). Structural history of the West Natuna Basin and the tectonic evolution of the Sunda region. In *Proceedings of 14th annual convention of Indonesian petroleum association, Jakarta* (pp. 39–61).

Darmadi, Y. (2005). *Three-dimensional fluvial-deltaic sequence stratigraphy Pliocene-Recent Muda Formation, Belida field, West Natuna Basin, Indonesia*. Texas A&M University MSc Thesis.

Darmadi, Y., Willis, B. J., & Dorobek, S. L. (2007). Three-dimensional seismic architecture of fluvial sequences on the low-gradient Sunda Shelf, offshore Indonesia. *Journal of Sedimentary Research, 77*, 225–238.

Dien, P. T., Phung, S. T., Nguyen, V. D., & Do, V. N. (1998). Late Mesozoic to Tertiary basin evolution along the southeast continental shelf of Vietnam. *AAPG Annual Convention Abstracts*, CD-ROM.

Ding, W., Franke, D., Li, J., & Steuer, S. (2013). Seismic stratigraphy and tectonic structure from a composite multi-channel seismic profile across the entire Dangerous Grounds, South China Sea. *Tectonophysics, 582*, 162–176.

Ding, W., Li, J., Li, M., Qiu, Q., Fang, Y., & Tang, Y. (2008). A Cenozoic tectono-sedimentary model of the Tainan Basin, the South China Sea: Evidence from a multi-channel seismic profile. *Journal of Zhejiang University (Science), A9*, 702–713.

Dong, J., Cao, J., Li, B., Zhao, L., & Li, K. (2013). Tectonic characteristics and evolution of Meso-Cenozoic of Sun Basin in North Huanghai Sea. *Journal of Southwest Petroleum University (Science and Technology), 35*(3), 59–66.

Doust, H. (1981). Geology and exploration history of offshore central Sarawak. In M. T. Halbouty (Ed.), *AAPG Studies in Geology: Vol. 12. Energy resources of the Pacific Region* (pp. 117–132).

Du, D. (1991). Characteristics of geologic structure and hydrocarbon potential of the Southwest Taiwan Basin. *Marine Geology and Quaternary Geology, 11*(3), 21–33 (in Chinese).

Du, D. (1994). Tectonic evolution and analysis of oil-gas accumulation in Southwest Taiwan Basin. *Marine Geology and Quaternary Geology, 14*(3), 5–18 (in Chinese).

Du Bois, E. P. (1985). Review of principal hydrocarbon-bearing basins around the South China Sea. *Bulletin of Geological Society of Malaysia, 18*, 167–209.

Du, Z., & Wei, K. (2001). Sequence stratigraphic framework and its characteristics of the Weizhou Formation in North Sag of Beibuwan Basin. *Acta Sedimentologica Sinica, 19*, 563–568 (in Chinese).

Duan, W., & Huang, Y. (1991). Tertiary calcareous nannofossil biostratigraphy in the north part of the South China Sea. *Acta Geologica Sinica, 65*, 86–101 (in Chinese).

Emery, D., & Meyers, K. J. (1996). *Sequence stratigraphy*. Oxford, UK: Blackwell Science, 297 pp.

Expedition 331 Scientists. (2010). *Deep hot biosphere: IODP preliminary report, 331*. http://dx. doi.org/10.2204/iodp.pr.331.2010.

Feng, Y., Li, S., & Lu, Y. (2013). Sequence stratigraphy and architectural variability in late Eocene lacustrine strata of the Dongying Depression, Bohai Bay Basin, eastern China. *Sedimentary Geology, 295*, 1–26.

Fraser, A. J., Matthews, S. J., Lowe, S., Todd, S. P., & Peel, F. J. (1996). Structure, stratigraphy and petroleum geology of the South East Nam Con Son Basin, offshore Vietnam. *AAPG Annual Convention Program and Abstracts*, A49.

Fraser, A. J., Matthews, S. J., & Murphy, R. W. (Eds.), (1997). *Petroleum geology of Southeast Asia*. In *Geological Society London Special Publication: Vol. 126*. 442 pp.

Fu, Q., Li, Y., Zhang, G., & Liu, Y. (2007). Evidence of transgression lake of Subei Basin during late Cretaceous and Paleocene and its geological significance. *Acta Sedimentolgica Sinica, 25*(3), 380–385 (in Chinese).

Galloway, W. E. (1989). Genetic stratigraphic sequence in basin analysis I: Architecture and genesis of flooding-surface bounded depositional units. *AAPG Bulletin, 73*, 125–142.

Ginger, D. C., Ardjakusumah, W. O., Hedley, R. J., & Pothecary, J. (1993). Inversion history of the West Natuna Basin: Examples from the Cumu-Cumi PSC. In *Proceedings of 22nd annual convention of Indonesian petroleum association, Jakarta* (pp. 635–658).

Gong, Z., Jin, Q., Qiu, Z., Wang, S., & Meng, J. (1989). Geology tectonics and evolution of the Pearl River Mouth Basin. In X. Zhu (Ed.), *Chinese sedimentary basins* (pp. 181–196). Amsterdam: Elsevier.

Gong, Z., & Li, S. (Eds.), (1997). *Continental margin basin analysis and hydrocarbon accumulation of the Northern South China Sea*. Beijing: China Science Press, 510 pp (in Chinese).

Gong, C., Wang, Y., Zhu, W., Li, W., Xu, Q., & Zhang, J. (2011). The Central Submarine Canyon in the Qiongdongnan Basin, northwestern South China Sea: Architecture, sequence stratigraphy, and depositional processes. *Marine and Petroleum Geology, 28*, 1690–1702.

Gradstein, F. M., Ogg, J. G., Schmitz, M., & Ogg, G. (Eds.), (2012). *The geologic time scale 2012*. The Netherlands: Elsevier, 1176 pp.

Gradstein, F., Ogg, J., & Smith, A. (Eds.), (2004). *A geologic time scale 2004*. Cambridge: Cambridge University Press, 589 pp.

Guan, X., Fan, H., & Song, Z. (1989). *Research on late Cenozoic Palynology of the Bohai Sea*. Nanjing: Nanjing University Press, 152 pp + 40 pls (in Chinese).

Gungor, A., Lee, G. H., Kim, H. J., Han, H. C., Kang, M. H., Kim, J., et al. (2012). Structural characteristics of the northern Okinawa Trough and adjacent areas from regional seismic reflection data: Geologic and tectonic implications. *Tectonophysics, 522–523*, 198–207.

Hageman, H. (1987). Palaeobathymetrical changes in NW Sarawak during Oligocene to Pliocene. *Bulletin of Geological Society of Malaysia, 21*, 91–102.

Hall, R., & Morley, C. K. (2004). Sundaland basins. In P. Clift, P. Wang, W. Kuhnt, & D. Hayes (Eds.), *Continent-Ocean interactions within East Asian Marginal Seas. AGU geophysical monograph: Vol. 149* (pp. 55–85).

Hao, Y., Chen, P., Wan, X., & Dong, J. (2000). Late Tertiary sequence stratigraphy and sea level changes in Yinggehai-Qiongdongnan Basins. *Earth Science—Journal of China University of Geosciences, 25*, 237–245 (in Chinese).

Hao, Y., & Ding, P. (Eds.), (1988). *Quaternary microbiotas of the Okinawa Trough and their geological significance*. Beijing: Geological Publishing House, 510 pp + 86 pls (in Chinese).

Hao, F., Sun, Y. C., Li, S. T., & Zhang, Q. M. (1995). Overpressure retardation of organic-matter maturation and hydrocarbon generation: A case study from the Yinggehai and Qiongdongnan basins, offshore South China Sea. *AAPG Bulletin, 79*, 551–562.

Hao, Y., Xu, Y., & Xu, S. (1996). *Research on micropalaeontology and paleoceanography in Pearl River Mouth Basin, South China Sea*. Beijing: China University of Geoscience Press, 136 pp (in Chinese).

Haq, B. U., Hardenbol, J., & Vail, P. R. (1987). Chronology of fluctuating sea levels since the Triassic (250 million years ago to present). *Science, 235*, 1156–1167.

Hardenbol, J., Thierry, J., Farley, M. B., Jacquin, T., De Graciansky, P. C., & Vail, P. R. (1998). Cenozoic sequence chronostratigraphy. In P. C. De Graciansky, J. Hardenbol, T. Jacquin, & P. R. Vail (Eds.), *SEPM Special Publication: Vol. 60. Mesozoic and Cenozoic Sequence chronostratigraphic framework of European Basins* (pp. 3–13), Charts 1–8.

He, Y., & Hu, L. (1978). *Cenozoic foraminifera from the coastal Region of Bohai*. Beijing: China Science Press, Research Institute of Ministry of Petroleum and Chemical Industries and Nanjing Institute of Geology and Paleontology CAS. 48 pp + 9 pls (in Chinese).

Ho, K. F. (1978). Stratigraphic framework for oil exploration in Sarawak. *Bulletin of Geological Society of Malaysia, 10*, 1–14.

Holbourn, A., Kuhnt, W., & Schulz, M. (2004). Orbitally paced climate variability during the middle Miocene: High resolution benthic stable-isotope records from the tropical western Pacific. In P. D. Clift, P. Wang, D. Hayes, & W. Kuhnt (Eds.), *AGU Geophysical Monograph*: *Vol. 149. Continent-ocean interactions in the East Asian Marginal Seas* (pp. 321–337).

Holbourn, A., Kuhnt, W., Schulz, M., & Erlenkeuser, H. (2005). Impacts of orbital forcing and atmospheric carbon dioxide on Miocene ice-sheet expansion. *Nature, 438*, 483–487.

Holbourn, A., Kuhnt, W., Schulz, M., Flores, J. A., & Andersen, N. (2007). Orbitally-paced climate evolution during the middle Miocene "Monterey" carbon-isotope excursion. *Earth and Planetary Science Letters, 261*, 534–550.

Hou, Y., Huang, B., Geng, L., Li, Y., Dan, H., & Cai, Z. (Eds.), (1978). *Early tertiary Ostracod fauna from the Coastal Region of Bohai*. Beijing: China Science Press, Research Institute of Ministry of Petroleum and Chemical Industries and Nanjing Institute of Geology and Paleontology CAS. 205 pp + 79 pls (in Chinese).

Hou, Y., Li, Y., Jin, Q., & Wang, P. (Eds.), (1981). *Tertiary paleontology of the Northern Continental Shelf of South China Sea*. Guangzhou: Guangdong Science and Technology Press, 274 pp + 108 pls (in Chinese).

Hou, F., Zhang, Z., Zhang, X., Li, S., Li, G., Guo, X., et al. (2008). Geological evolution and tectonic styles in the South Yellow Sea Basin. *Marine Geology and Quaternary Geology, 28*(5), 61–68 (in Chinese).

Hsiao, L. Y., & Graham, S. A. (2012). Xialiao, North China Basin. In D. G. Roberts & A. W. Bally (Eds.), *Regional geology and tectonics: Phanerozoic Rift systems and Sedimentary Basins* (pp. 237–257). Elsevier B.V.

Hsiao, L. Y., Graham, S. A., & Tilander, N. (2004). Seismic reflection imaging of a major strike-slip fault zone in a rift system: Paleogene structure and evolution of the Tan-Lu fault system, Liaodong Bay, Bohai, offshore China. *AAPG Bulletin, 88*, 71–97.

Hsiao, L. Y., Graham, S. A., & Tilander, N. (2010). Stratigraphy and sedimentation in a rift basin modified by synchronous strike-slip deformation: Southern Xialiao basin, Bohai, offshore China. *Basin Research, 22*, 61–78.

Hsu, S.-K., Yeh, Y.-C., Doo, W.-B., & Tsai, C.-H. (2004). New bathymetry and magnetic lineations in the northernmost South China Sea and their tectonic implications. *Marine Geophysical Research, 25*29–44. http://dx.doi.org/10.1007/s11001-005.0731-7.

Hu, S., O'Sullivan, P. B., Raza, A., & Kohn, B. P. (2001). Thermal history and tectonic subsidence of the Bohai Basin, northern China: A Cenozoic rifted and local pull-apart basin. *Physics of the Earth and Planetary Interiors, 126*, 221–235.

Hu, Z., & Su, H. (1981). Tertiary of the Beibu Gulf and Yinggehai depressions. In D. Zeng, B. Guo, C. Huo, S. Zhong, X. Huang, P. Hu, & H. Su (Eds.), *Tertiary system of the Northern Continental Shelf of South China Sea* (pp. 35–43). Guangzhou: Guangdong Science and Technology Press (in Chinese).

Huang, C. (1997a). Tectonic elements and analysis of Cenozoic sedimentary basins in the South China Sea and its surrounding areas. In Z. Gong & S. Li (Eds.), *Continental margin basin analysis and hydrocarbon accumulation of the Northern South China Sea* (pp. 44–52). Beijing: China Science Press (in Chinese).

Huang, L. (1997b). Calcareous nannofossil biostratigraphy in the Pearl River Mouth Basin, South China Sea, and Neogene reticulofenestrid coccoliths size distribution pattern. *Marine Micropaleontology, 32*, 31–57.

Huang, C.-Y., Chiu, Y.-L., & Zhao, M. (2005). Core description and preliminary sedimentological study of Site 1202D, Leg 195, in southern Okinawa Trough. *Terrestrial, Atmospheric and Oceanic Sciences, 16*(1), 19–44.

Huang, C.-Y., Wu, W. Y., Chang, C. P., Tsao, S., Yuan, P. B., Lin, C. W., et al. (1997c). Tectonic evolution of accretionary prism in the arc-continent collision terrane of Taiwan. *Tectonophysics*, *281*, 31–51.

Huang, F. F. W., Liang, S. C., & Juang, H. J. (1998). Depositional model for the Oligocene sandstone in the southwest Taiwan offshore. *Petroleum Geology of Taiwan*, *32*, 69–86.

Huang, L., Liu, C. Y., Zhou, X. H., & Wang, Y. B. (2012). The important turning points during evolution of Cenozoic basin offshore the Bohai Sea: Evidence and regional dynamics analysis. *Science China Earth Sciences*, *55*, 476–487.

Huang, C.-Y., Yen, Y., Zhao, Q. H., & Lin, C.-T. (2012). Cenozoic stratigraphy of Taiwan: Window into rifting, stratigraphy and paleoceanography of South China Sea. *Chinese Science Bulletin*, *57*, 3130–3149.

Hutchison, C. S. (2004). Marginal basin evolution: The southern South China Sea. *Marine and Petroleum Geology*, *21*, 1129–1148.

Hutchison, C. S. (2010). The North-West Borneo Trough. *Marine Geology*, *271*, 32–43.

Hutchison, C. S., & Vijayan, V. R. (2010). What are the Spratly Islands? *Journal of Asian Earth Sciences*, *39*, 371–385.

Ingram, G. M., Chisholm, T. J., Grant, C. J., Hedlund, C. A., Stuart-Smith, P., & Teasdale, J. (2004). Deepwater North West Borneo: Hydrocarbon accumulation in an active fold and thrust belt. *Marine and Petroleum Geology*, *21*, 879–887.

Ismail, M. I. B., & Hassan, R. B. Abu (1999). Tinjar Province. In K. M. Leong (Ed.), *The petroleum geology and resources of Malaysia* (pp. 395–410). Kuala Lumpur: PETRONAS.

Jardine, E. (1997). Dual petroleum systems governing the prolific Pattani Basin, offshore Thailand. In J. V. C. Howes & R. A. Noble (Eds.), *Proceedings of international conference on petroleum systems of SE Asia and Australia* (pp. 351–363). Jakarta: Indonesian Petroleum Association.

Jian, Z., Zhao, Q., Cheng, X., Wang, J., Wang, P., & Su, X. (2003). Pliocene-Pleistocene stable isotope and paleoceanographic changes in the northern South China Sea. *Palaeogeography, Palaeoclimatology, Palaeoecology*, *193*, 425–442.

Jiang, Z. X., Liu, H., Zhang, S. W., Su, X., & Jiang, Z. L. (2011). Sedimentary characteristics of large-scale lacustrine beach-bars and their formation in the Eocene Boxing Sag of Bohai Bay Basin, East China. *Sedimentology*, *58*, 1087–1112.

Jiang, T., Xie, X., Ren, J., Zhang, C., & Su, M. (2009). Deepwater turbidites and their implications for hydrocarbon exploration in Qiongdongnan Basin, Northern South China Sea. *Journal of Geochemical Exploration*, *101*, 52.

Jiang, Z., Zeng, L., Li, M., & Zhong, Q. (Eds.), (1994). *The North Continental shelf region of South China Sea. Tertiary in petroliferous regions of China: Vol. 8*. Beijing: Petroleum Industry Publishing House, 145 pp (in Chinese).

Jin, Q. (Ed.), (1989). *The Geology and petroleum resources in the South China Sea*. Beijing: Geological Publishing House, 417 pp (in Chinese).

Jin, Q. (2005). Overview on offshore peteoleum exploration of the northern South China Sea. *Fisheries Science and Technology*, *3*, 1–5 (in Chinese).

Jin, H., & Jian, Z. (2013). Millennial-scale climate variability during the mid-Pleistocene transition period in the northern South China Sea. *Quaternary Science Reviews*, *70*, 15–27.

Jin, Q., & Li, T. (2000). Regional geologic tectonics of the Nansha Sea Area. *Marine Geology and Quaternary Geology*, *20*, 1–8 (in Chinese).

Kang, X., Zhao, W., Pan, Z., Zhang, Q., & Chen, Z. (1994). Study on architecture of sequence stratigraphic framework of Beibuwan Basin. *Earth Science*, *19*, 493–502 (in Chinese).

Keen, C. E., Potter, P., Hu, S., O'Sullivan, P. B., Raza, A., & Kohn, B. P. (2001). Thermal history and tectonic subsidence of the Bohai Basin, northern China: A Cenozoic rifted and local pull-apart basin. *Physics of the Earth and Planetary Interiors, 126*, 221–235.

Kou, C., & Ye, G. (1981). Tertiary of the Leizhou Peninsula. In D. Zeng, B. Guo, C. Huo, S. Zhong, X. Huang, P. Hu, & H. Su (Eds.), *Tertiary system of the Northern Continental Shelf of South China Sea* (pp. 149–177). Guangzhou: Guangdong Science and Technology Press.

Laskar, J. (1990). The chaotic motion of the solar system: A numerical estimate of the size of the chaotic zones. *Icarus, 88*, 266–291.

Laskar, J., Robutel, P., Joutel, F., Gastineau, M., Correia, A. C. M., & Levrard, B. (2004). A long term numerical solution for the insolation quantities of the Earth. *Astronomy Astrophysics, 428261–285.* http://dx.doi.org/10.1051/0004-6361:20041335.

Lee, E. Y. (2010). Subsidence history of the Gunsan Basin (Cretaceous-Cenozoic) in the Yellow Sea, offshore Korea. *Australian Journal of Earth Sciences, 103*, 111–120.

Lee, G. H., Kim, B., Shin, K. S., & Sunwoo, D. (2006). Geologic evolution and aspects of the petroleum geology of the northern East China Sea Shelf Basin. *AAPG Bulletin, 90*, 237–260.

Lee, G. H., Lee, K., & Watkins, J. S. (2001). Geologic evolution of the Cuu Long and Nam Con Son basins, offshore southern Vietnam, South China Sea. *AAPG Bulletin, 85*, 1055–1082.

Lee, T.-Y., Tang, C.-H., Ting, J.-S., & Hsu, Y.-Y. (1993). Sequence stratigraphy of the Tainan Basin, offshore southwestern Taiwan. *Petroleum Geology of Taiwan, 28*, 119–158.

Lei, C., Ren, J., Clift, P. D., Wang, Z., Li, X., & Tong, C. (2011). The structure and formation of diapirs in the Yinggehai–Song Hong Basin, South China Sea. *Marine and Petroleum Geology, 28*, 980–991.

Leo, C. T. A. (1997). Exploration in the Gulf of Thailand in deltaic reservoirs, related to the Bongkot Field. In A. J. Fraser, S. J. Matthews, & R. W. Murphy (Eds.), *Geological Society of London Special Publication: Vol. 126. Petroleum geology of Southeast Asia* (pp. 77–87).

Leong, K. M. (1999). Geological setting of Sabah. In K. M. Leong (Ed.), *The petroleum geology and resources of Malaysia* (pp. 475–497). Kuala Lumpur: PETRONAS.

Lewis, S. D., & Hayes, D. E. (1984). A geophysical study of the Manila Trench, Luzon, Philippines: 2. Fore arc basin structural and stratigraphic evolution. *Journal of Geophysical Research: Solid Earth, 89*(B11), 9196–9214.

Li, N. (1995). Tectonic evolution of three structural basins in the Yellow Sea. *Oceanologia et Limnologia Sinica, 26*(4), 355–362 (in Chinese).

Li, D. S. (2012a). Cenozoic rifts of eastern China. In D. G. Roberts & A. W. Bally (Eds.), *Regional geology and tectonics: Phanerozoic Rift Systems and Sedimentary Basins* (pp. 196–234). Elsevier B.V.

Li, J. B. (Ed.), (2012b). *Regional oceanography of China Seas—Marine geology.* Beijing: China Ocean Press, 547pp (in Chinese).

Li, T. G., & Chang, M. F. (2009). *Paleoceanography in the Okinawa Trough.* Beijing: China Ocean Press, 259 pp (in Chinese).

Li, X. Q., Fairweather, L., Wu, S. G., Ren, J. Y., Zhang, H. J., Quan, X. Y., et al. (2013). Morphology, sedimentary features and evolution of a large palaeo submarine canyon in Qiongdongnan Basin, northern South China Sea. *Journal of Asian Earth Sciences, 62*, 685–696.

Li, G., Gong, J., Yang, C., Yang, C., Wang, W., Wang, H., et al. (2012). Stratigraphic features of the Mesozoic "Great East China Sea"—A new exploration field. *Marine Geology and Quaternary Geology, 32*(3), 97–104 (in Chinese).

Li, Q., Jian, Z., & Li, B. (2004). Oligocene-Miocene planktonic foraminifer biostratigraphy, Site 1148, northern South China Sea. In W. L. Prell, P. Wang, P. Blum, D. K. Rea, & S. C. Clemens (Eds.), *Proceedings of ODP scientific results: Vol. 184.* (pp. 1–26), [Online].

Li, Q., Jian, Z., & Su, X. (2005). Late Oligocene rapid transformations in the South China Sea. *Marine Micropaleontology, 54*, 5–25.

Li, W., Lu, W., Liu, Y., & Xu, J. (2012). Superimposed versus residual basin: The North Yellow Sea Basin. *Geoscience Frontiers, 3*, 33–39.

Li, P., & Rao, C. (1994). Tectonic characteristics and evolution history of the Pearl River Mouth Basin. *Tectonophysics, 235*, 13–25.

Li, Q., Wang, P., Zhao, Q., Shao, L., Zhong, G., Tian, J., et al. (2006). A 33 Ma lithostratigraphic record of tectonic and paleoceanographic evolution of the South China Sea. *Marine Geology, 230*, 217–235.

Li, Q., Wang, P., Zhao, Q., Tian, J., Cheng, X., Jian, Z., et al. (2008). Paleoceanography of the mid-Pleistocene South China Sea. *Quaternary Science Reviews, 27*, 1217–1233.

Li, W., Zeng, X., & Huang, J. (2009). Meso-Cenozoic North Yellow Sea: Residual basin or superimposed basin? *Acta Geologica Sinica, 83*, 1269–1275.

Li, X. S., Zhao, Y. X., Liu, B. H., Liu, C. G., Zheng, Y. P., & Wang, K. Y. (2004). Spatiotemporal characteristics of depositional evolution of the Okinawa Trough. *Advances in Marine Science, 22*, 472–479 (in Chinese).

Li, Q., Zhong, G., & Tian, J. (2009). Stratigraphy and sea level changes. In P. Wang & Q. Li (Eds.), *The South China Sea Paleoceanography and Sedimentology* (pp. 75–170). Springer.

Lian, H. M., & Bradley, K. (1986). Exploration and development of natural gas, Pattani Basin, Gulf of Thailand. In *Transactions of 4th circum-pacific energy and mineral resource conference, Singapore* (pp. 171–181).

Liew, K. K. (1994). Structural development at the west-central margin of the Malay Basin. *Geological Society of Malaysia Bulletin, 36*, 67–80.

Lin, C., Liu, J., Cai, S., Zhang, Y., Lu, M., & Li, J. (2001). Depositional architecture and developing settings of large-scale incised valley and sub-marine gravity flow systems in the Yinggehai and Qiongdongnan basins, South China Sea. *Chinese Science Bulletin, 46*, 690–693.

Lin, J.-Y., Sibuet, J. C., & Hsu, S. K. (2005). Distribution of the East China Sea continental shelf basins and depths of magnetic sources. *Earth, Planets and Space, 57*, 1063–1072.

Lin, A. T., Watts, A. B., & Hesselbo, S. P. (2003). Cenozoic stratigraphy and subsidence history of the South China Sea margin in the Chinese Taipei region. *Basin Research, 15*, 453–479.

Liu, H., Zheng, H., Wang, Y., Lin, Q., Wu, C., Zhao, M., et al. (2011). Basement of the South China Sea area: Tracing the Tethyan realm. *Acta Geologica Sinica, 85*, 637–655.

Liu, J. H., Gao, J. Y., Fang, Y. X., & Wu, S. G. (2008). Analysis on the basement constitution of the southern Okinawa Trough. *Marine Science Bulletin, 10*(1), 75–86.

Liu, J. H., Li, M. B., & Fang, Y. X. (2005). Mesozoic strata in East China Sea Shelf Basin and their relationship with adjacent paleo-seas. *Journal of Tropical Oceanography, 32*(3), 1–7 (in Chinese).

Lovell, B. K. (1987). The nature and significance of regional unconformity in the hydrocarbon-bearing Neogene sequence offshore West Sabah. *Bulletin of Geological Society of Malaysia, 21*, 55–90.

Ma, Y., Wu, S., Lv, F., Dong, D., Sun, Q., Lu, Y., et al. (2011). Seismic characteristics and development of the Xisha carbonate platforms, northern margin of the South China Sea. *Journal of Asian Earth Sciences, 40*, 770–783.

Madon, M. B. Hj. (1997). The kinematics of extension and inversion in the Malay Basin, offshore Peninsular Malaysia. *Bulletin of Geological Society of Malaysia, 41*, 127–138.

Madon, M. B. Hj. (1999a). Basin types, tectono-stratigraphic provinces, and structural styles. In K. M. Leong (Ed.), *The petroleum geology and resources of Malaysia* (pp. 77–111). Kuala Lumpur: PETRONAS.

Madon, M. B. Hj. (1999b). Geological setting of Sarawak. In K. M. Leong (Ed.), *The petroleum geology and resources of Malaysia* (pp. 275–290). Kuala Lumpur: PETRONAS.

Madon, M. B. Hj. (1999c). North Luconia Province. In K. M. Leong (Ed.), *The petroleum geology and resources of Malaysia* (pp. 443–454). Kuala Lumpur: PETRONAS.

Madon, M. B. Hj. (1999d). Plate tectonic elements and the evolution of Southeast Asia. In K. M. Leong (Ed.), *The petroleum geology and resources of Malaysia* (pp. 59–76). Kuala Lumpur: PETRONAS.

Madon, M. B. Hj, Abolins, P., Hoesni, M. J. B., & Ahmad, M. B. (1999). Malay Basin. In K. M. Leong (Ed.), *The petroleum geology and resources of Malaysia* (pp. 173–217). Kuala Lumpur: PETRONAS.

Madon, M. B. Hj, & Hassan, R. B. Abu. (1999a). Tatau Province. In K. M. Leong (Ed.), *The petroleum geology and resources of Malaysia* (pp. 413–426). Kuala Lumpur: PETRONAS.

Madon, M. B. Hj, & Hassan, R. B. Abu. (1999b). West Luconia Province. In K. M. Leong (Ed.), *The petroleum geology and resources of Malaysia* (pp. 429–439). Kuala Lumpur: PETRONAS.

Madon, M. B. Hj, Karim, R. Bt. Abd, & Fatt, R. W. H. (1999). Tertiary stratigraphy and correlation schemes. In K. M. Leong (Ed.), *The petroleum geology and resources of Malaysia* (pp. 113–137). Kuala Lumpur: PETRONAS.

Madon, M., Kim, C. L., & Wong, R. (2013). The structure and stratigraphy of deepwater Sarawak, Malaysia: Implications for tectonic evolution. *Journal of Asian Earth Sciences, 76*, 312–333.

Madon, M. B. Hj, Leong, K. M., & Anuar, A. (1999). Sabah Basin. In K. M. Leong (Ed.), *The petroleum geology and resources of Malaysia* (pp. 501–542). Kuala Lumpur: PETRONAS.

Madon, M. B., & Watts, A. B. (1998). Gravity anomalies, subsidence history and the tectonic evolution of the Malay and Penyu Basins (offshore Peninsular Malaysia). *Basin Research, 10*, 375–392.

Mahmud, O. A. B., & Saleh, S. B. (1999). Petroleum resources, Sarawak. In K. M. Leong (Ed.), *The petroleum geology and resources of Malaysia* (pp. 457–472). Kuala Lumpur: PETRONAS.

Matthews, S. J., Fraser, A. J., Lowe, S., Todd, S. P., & Peel, F. J. (1997). Structure, stratigraphy, and petroleum geology of the SE Nam Con Son Basin, offshore Vietnam. In A. J. Fraser, S. J. Matthews, & R. W. Murphy (Eds.), *Geological Society London Special Publication: Vol. 126. Petroleum geology of Southeast Asia* (pp. 89–106).

Mat-Zin, I. C., & Swarbrick, R. E. (1997). The tectonic evolution and associated sedimentation history of Sarawak Basin, eastern Malaysia: A guide for future hydrocarbon exploration. In A. J. Fraser, S. J. Matthews, & R. W. Murphy (Eds.), *Geological Society London Special Publication: Vol. 126. Petroleum geology of Southeast Asia* (pp. 237–245).

Mat-Zin, I. C., & Tucker, M. E. (1999). An alternative stratigraphic scheme for the Sarawak Basin. *Journal of Asian Earth Sciences, 17*, 215–232.

May, J. A., & Eyles, D. R. (1985). Well log and seismic character of tertiary terumbu carbonate, South China Sea, Indonesia. *AAPG Bulletin, 69*, 1339–1358.

Meirita, M. F. (2003). *Structural and depositional evolution, KH Field, West Natuna Basin, offshore Indonesia.* Texas A&M University MSc thesis, 56 p.

Miller, K. G., Fairbanks, R. G., & Mountain, G. S. (1987). Tertiary oxygen isotope synthesis, sea level history, and continental margin erosion. *Paleoceanography, 2*, 1–19.

Miller, K. G., Wright, J. D., & Fairbanks, R. G. (1991). Unlocking the ice house: Oligocene-Miocene oxygen isotopes, eustasy, and margin erosion. *Journal of Geophysical Research, 96*(B4), 6829–6848.

Mitchum, R. M., Jr., Vail, P. R., & Thompson, S.III, (1977). Seismic stratigraphy and global changes of sea level, Part 2: Depositional sequence as a basic unit for stratigraphic analysis. In C. E. Payton (Ed.), *AAPG Memoir: Vol. 26. Seismic stratigraphy—Applications to hydrocarbon exploration* (pp. 53–62).

Mohammad, A. M., & Wong, R. H. F. (1995). Seismic sequence stratigraphy of the Tertiary sediments offshore Sarawak deepwater area, Malaysia. *Bulletin of Geological Society of Malaysia, 37*, 345–361.

Mohd Idrus, B. I., Abdul, R. E., Abdul Manaf, M., Sahalan, A. A., & Mahendran, B. (1995). The geology of Sarawak deepwater and surrounding areas. *Bulletin of Geological Society of Malaysia, 37*, 165–178.

Moldovanyi, E. P., Waal, F. M., & Yan, Z. J. (1995). Regional exposure events and platform evolution of Zhuijang Formation carbonates, Pearl River Mouth Basin: Evidence for primary and diagenetic seismic facies. In D. A. Budd, A. H. Saller, & P. M. Harris (Eds.), *AAPG Memoir: Vol. 63. Unconformities and porosity in carbonate strata* (pp. 133–145).

Morley, R. J. (1991). Tertiary stratigraphic palynology in Southeast Asia: Current status and new directions. *Bulletin of Geological Society of Malaysia, 28*, 1–36.

Morley, C. K., & Westawayw, R. (2006). Subsidence in the super-deep Pattani and Malay basins of Southeast Asia: A coupled model incorporating lower-crustal flow in response to post-rift sediment loading. *Basin Research, 18*, 51–84.

Olson, C. C., & Dorobek, S. L. (2002). Comparison of structural styles and regional subsidence patterns across the Nam Con Son and Cuu Long Basins, offshore Southeast Vietnam. In *Chapman conference on continent-ocean interactions within the East Asian Marginal Seas (Abstracts), San Diego, CA.*

Oung, J.-N. (2000). Two-dimensional basin modeling—A regional study on hydrocarbon generation, offshore Taiwan. *Petroleum Geology Taiwan, 34*, 33–54.

Pälike, H., Frazier, J., & Zachos, J. C. (2006). Extended orbitally forced palaeoclimatic records from the equatorial Atlantic Ceara Rise. *Quaternary Science Reviews, 25*, 3138–3149.

Pälike, H., Norris, R. D., Herrle, J. O., Wilson, P. A., Coxall, H. K., Lear, C. H., et al. (2006). The heartbeat of the Oligocene climate system. *Science, 314*, 1894–1898.

Pang, X., Chen, C., Peng, D., Zhu, M., Shu, Y., He, M., et al. (2007). Sequence stratigraphy of Pearl River deep-water fan system in the South China Sea. *Earth Science Frontiers, 14*, 220–229 (in Chinese).

Park, J.-O., Tokuyama, H., Shinohara, M., Suyehiro, K., & Taira, A. (1998). Seismic record of tectonic evolution and backarc rifting in the southern Ryukyu island arc system. *Tectonophysics, 294*, 21–42.

Phillips, S., Little, L., Michael, E., & Odell, V. (1997). Sequence stratigraphy of Tertiary petroleum systems in the West Natuna Basin, Indonesia. In J. V. C. Howes & R. A. Noble (Eds.), *Proceedings of international conference on petroleum systems in SE Asia and Australia, Indonesian Petroleum Association Jakarta* (pp. 381–401).

Pigott, J. D., Kang, M. H., & Han, H. C. (2013). First order seismic attributes for clastic seismic facies interpretation: Examples from the East China Sea. *Journal of Asian Earth Sciences, 66*, 34–54.

Pigott, J. D., & Ru, K. (1994). Basin superposition on the northern margin of the South China Sea. *Tectonophysics, 235*, 27–50.

Pigott, J. D., & Sattayarak, N. (1993). Aspects of sedimentary basin evolution assessed through tectonic subsidence analysis. Example: Northern Gulf of Thailand. *Journal of Southeast Asian Earth Sciences, 8*, 407–420.

Polachan, S., & Sattayarak, N. (1991). Development of Cenozoic Basins in Thailand. *Marine and Petroleum Geology, 8*, 84–87.

Pollock, R. E., Hayes, J. B., Williams, K. P., & Young, R. A. (1984). The petroleum geology of the KH Field, Kakap, Indonesia. In *Proceedings of 13th annual convention of Indonesian petroleum association, Jakarta*: Vol. 1. (pp. 407–423).

Qi, J., Li, X., Yu, F., & Yu, T. (2013). Cenozoic structural deformation and expression of the "Tan-Lu Fault Zone" in the West Sag of Liaohe Depression, Bohaiwan basin province, China. *Science China Earth Sciences, 56*, 1707–1721.

Qi, J., Zhou, X., Deng, R., & Zhang, K. (2008). Structural characteristics of the Tan-Lu fault zone in Cenozoic basins offshore the Bohai Sea. *Science China Earth Sciences, 51*(Supp. II), 20–31.

Qiu, Y., Chen, G., Xie, X., Wu, L., Liu, X., & Jiang, T. (2005). Sedimentary filling evolution of Cenozoic strata in Zengmu Basin, Southwestern South China Sea. *Tropical Oceanology, 24* (5), 43–52 (in Chinese).

Qiu, Y., & Wang, Y.-M. (2001). Reefs and paleostructure and paleoenvironment in the South China Sea. *Marine Geology & Quaternary Geology, 21*, 65–73 (in Chinese).

Ramli, Md. N. (1988). Stratigraphy and palaeofacies development of Carigali's operating areas in the Malay Basin, South China Sea. *Bulletin of Geological Society of Malaysia, 22*, 153–188.

Rangin, C., Klein, M., Roques, D., Le Pichon, X., & Trong, L. V. (1995). The Red River fault system in the Tonkin Gulf, Vietnam. *Tectonophysics, 243*, 209–222.

Ren, J. Y., Tamaki, K., Li, S. T., & Zhang, J. X. (2002). Late Mesozoic and Cenozoic rifting and its dynamic setting in Eastern China and adjacent areas. *Tectonophysics, 344*, 175–205.

Ru, K., & Pigott, J. D. (1986). Episodic rifting and subsidence in the South China Sea. *AAPG Bulletin, 70*, 1136–1155.

Saitoh, Y., & Masuda, F. (2004). Miocene sandstone of 'continental' origin on Iriomote Island, southwest Ryukyu Arc, Eastern Asia. *Journal of Asian Earth Sciences, 24*, 137–144.

Schlüter, H. U., Hinz, K., & Block, M. (1996). Tectoni-stratigraphic terranes and detachment faulting of the South China Sea and Sulu Sea. *Marine Geology, 130*, 39–78.

Shackleton, N. J., Hall, M. A., & Pate, D. (1995). Pliocene stable isotope stratigraphy of Site 846. *Proceedings of ODP Scientific Results, 138*, 337–355.

Shao, L., You, H., Hao, H., Wu, G., Qiao, P., & Lei, Y. (2007). Petrology and depositional environments of Mesozoic strata in the northeastern South China Sea. *Geological Review, 53*, 164–169 (in Chinese).

Shinn, Y. J., Chough, S. K., & Hwang, I. G. (2010). Structural development and tectonic evolution of Gunsan Basin (Cretaceous–Tertiary) in the central Yellow Sea. *Marine and Petroleum Geology, 27*, 500–514.

Sibuet, J.-C., & Hsu, S.-K. (1997). Geodynamics of the Taiwan arc-arc collision. *Tectonophysics, 274*, 221–251.

Sibuet, J.-C., Hsu, S.-K., & Debayle, E. (2004). Geodynamic context of the Taiwan orogen. In P. Clift, P. Wang, W. Kuhnt, & D. Hayes (Eds.), *Continent-Ocean interactions within East Asian Marginal Seas. AGU geophysical monograph*: Vol. 149 (pp. 127–158).

Sibuet, J.-C., Hsu, S.-K., Le Pichon, X., Le Formal, J. P., & Reed, D. (2002). East Asia plate tectonics since 15 Ma: Constraints from the Taiwan region. *Tectonophysics, 344*, 103–134.

Song, Z., Cao, L., Zhou, H., Guan, X., & Wang, H. (1978). *Early tertiary spores and pollen from the Coastal Region of Bohai*. Beijing: China Science Press, Research Institute of Ministry of Petroleum and Chemical Industries and Nanjing Institute of Geology and Paleontology CAS. 177 pp + 62 pls (in Chinese).

Song, Z., Guan, X., Li, Z., Zheng, Y., Wang, W., & Hu, Z. (1985). *A research on Cenozoic Palynology of the Longjing structural area in the Shelf Basin of the East China Sea (Donghai) region*. Hepei: Anhui Science and Technology Publishing House, 209 pp + 55 pls (in Chinese).

Song, Z., He, S., Qian, Z., Pan, Z., Zheng, G., & Zheng, Y. (1978). *Paleogene dinoflagellates and acritarchs from the coastal region of Bohai*. Beijing: China Science Press, Research Institute of Ministry of Petroleum and Chemical Industries and Nanjing Institute of Geology and Paleontology CAS. 190 pp + 49 pls (in Chinese).

Su, X., Xu, Y., & Tu, Q. (2004). Early Oligocene–Pleistocene calcareous nannofossil biostratigraphy of the northern South China Sea (Leg 184, Sites 1146–1148). In W. L. Prell, P. Wang, P. Blum, D. K. Rea, & S. C. Clemens (Eds.), *Proceedings of ODP scientific results: Vol. 184*. (Online).

Sun, X., Li, M., Zhang, Y., Lei, Z., Kong, Z., Li, P., et al. (1981). Palynology and plant fossils. In Y. Hou, Y. Li, Q. Jin, & P. Wang (Eds.), *Tertiary paleontology of the Northern Continental Shelf of South China Sea* (pp. 1–59). Guangzhou: Guangdong Science and Technology Press (in Chinese).

Sun, X., & Wang, P. (2005). How old is the Asian monsoon system? - Palaeobotanical records from China. *Palaeogeography, Palaeoclimatology, Palaeoecology, 222*, 181–222.

Tamburini, F., Adatte, T., & Föllmi, K. B. (2006). Origin and nature of green clay layers, ODP Leg 184, South China Sea. In W. L. Prell, P. Wang, P. Blum, D. K. Rea, & S. C. Clemens (Eds.), *Proceedings of ODP, Scientific results: Vol. 184* (pp. 1–23) [Online].

Tan, M. (1995). Seismic-stratigraphic studies of the continental of Southern Vietnam. *Journal of Petroleum Geology, 18*, 345–354 (in Chinese).

Tan, Z., & Chen, M. (1999). *Offshore radiolarians in China*. Beijing: China Science Press, 404 pp (in Chinese).

Tanaka, G., & Nomura, S. (2009). Late Miocene and Pliocene Ostracoda from the Shimajiri Group, Kume-jima Island, Japan: Biogeographical significance of the timing of the formation of back-arc basin (Okinawa Trough). *Palaeogeography, Palaeoclimatology, Palaeoecology, 276*, 56–68.

Tang, F. S. L., Oung, J. N., Hsu, J. Y. Y., & Yang, C. N. (1999). Elementary study of structural evolution in Tainan Basin in southwest Taiwan Strait. *Petroleum Geology Taiwan, 33*, 125–149.

Taylor, B., & Hayes, D. E. (1980). The tectonic evolution of the South China Sea Basin. In D. E. Hayes (Ed.), *The Tectonic and geologic evolution of southeast Asian Seas and Islands. AGU geophysical monograph* (pp. 89–104).

Thompson, P. R., & Abbott, W. H. (2003). Chronostratigraphy and microfossil-derived sea-level history of the Qiongdongnan and Yinggehai basins, South China Sea. In H. C. Olson & R. M. Leckie (Eds.), *SEPM Special Publication, Tulsa: Vol. 75. Microfossils as Proxies of sea-level change and Stratigraphic discontinuities* (pp. 97–117).

Tian, J., & Li, Q. (2009). Isotopic and astronomical stratigraphy. In P. Wang & Q. Li (Eds.), *The South China Sea paleoceanography and sedimentology* (pp. 100–109). Berlin: Springer.

Tian, J., Wang, P., & Cheng, X. (2004a). Time-frequency variations of the Plio–Pleistocene foraminiferal isotopes: A case study from the southern South China Sea. *Earth Science, 15*(3), 283–289.

Tian, J., Wang, P., & Cheng, X. (2004b). Responses of foraminiferal isotopic variations at ODP Site 1143 in the southern South China Sea to orbital forcing. *Science China (D)*, *47*(10), 943–953.

Tian, J., Wang, P., & Cheng, X. (2004c). Development of the East Asian monsoon and Northern Hemisphere glaciation: Oxygen isotope records from the South China Sea. *Quaternary Science Reviews*, *23*, 2007–2016.

Tian, J., Wang, P., Cheng, X., & Li, Q. (2002). Astronomically tuned Plio-Pleistocene benthic δ18O records from South China Sea and Atlantic-Pacific comparison. *Earth and Planetary Science Letters*, *203*, 1015–1029.

Tian, K., Yu, Z., Peng, M., Yang, C., Liao, Q., Zhou, J., et al. (2000). *Oil and gas geology and exploration of the deep buried lower tertiary in the Bohaiwan Basin*. Beijing: Petroleum Industry Publishing House, 292 pp (in Chinese).

Tian, J., Zhao, Q., Wang, P., Li, Q., & Cheng, X. (2008). Astronomically modulated Neogene sediment records from the South China Sea. *Paleoceanography*, *23*, PA3210, 1–20. http://dx.doi.org/10.1029/2007PA001552.

Tiedemann, R., Sarnthein, M., & Shackleton, N. J. (1994). Astronomic timescale for the Pliocene Atlantic δ 18O and dust flux records from Ocean Drilling Program Site 659. *Paleoceanography*, *9*, 619–638.

Tjia, H. D. (1994). Inversion tectonics in the Malay Basin: Evidence and timing of events. *Bulletin of Geological Society of Malaysia*, *36*, 119–126.

Tjia, H. D., & Liew, K. K. (1996). Changes in tectonic stress field in northern Sunda Shelf basins. In R. Hall & D. Blundell (Eds.), *Geological Society of London Special Publication: Vol. 106. Tectonic evolution of Southeast Asia* (pp. 291–306).

Tongkul, F. (1991). Tectonic evolution of Sabah, Malaysia. *Journal of Southeast Asian Earth Sciences*, *6*, 395–405.

Tzeng, J., Uang, Y. C., Hsu, Y. Y., & Teng, L. S. (1996). Seismic stratigraphy of the Tainan Basin. *Petroleum Geology Taiwan*, *30*, 281–307.

Ujiie, H. (1994). Early Pleistocene birth of the Okinawa Trough and Ryukyu Island Arc at the northwestern margin of the Pacific: Evidence from Late Cenozoic planktonic foraminiferal zonation. *Palaeogeography Palaeoclimatology Palaeoecology*, *108*, 457–474.

Ujiie, H., & Nishimura, Y. (1992). Transect of central to southern Ryukyu Island Arcs. In *29th IGC field trip guide book, Geological survey of Japan: Vol. 5* (pp. 337–361).

Wang, X. W. (Ed.). (1983). *Petroleum geology of China*. Beijing: Petroleum Industry Publishing House, 348 p. (in Chinese).

Wang, P. (Ed.), (1985). *Marine micropaleontology of China*. Beijing: China Ocean Press, 370 pp.

Wang, P. (1990). Neogen stratigraphy and paleoenvironments of China. *Palaeogeography Palaeoclimatology Palaeoecology*, *77*, 315–334.

Wang, R., & Abelmann, A. (1999). Pleistocene radiolarian biostratigraphy in the South China Sea. *Science China (D)*, *42*, 536–543.

Wang, R., & Abelmann, A. (2002). Radiolarian responses to paleoceanographic events of the southern South China Sea during the Pleistocene. *Marine Micropaleontology*, *46*, 25–44.

Wang, L., Guo, L., & Nie, X. (2012). Structural features and evolutionary stages of the western depression of the North Yellow Sea Basin. *Marine Geology & Quaternary Geology*, *32*(3), 55–62 (in Chinese).

Wang, S., Huang, R., Yang, C., & Li, H. (1978). *Early tertiary charophytes from the coastal regions of Bohai*. Beijing: China Science Press, Research Institute of Ministry of Petroleum and Chemical Industries and Nanjing Institute of Geology and Paleontology CAS. 49 pp + 21 pls (in Chinese).

Wang, P., & Li, Q. (Eds.), (2009). *The South China Sea paleoceanography and sedimentology.* Berlin: Springer, 506 pp.

Wang, R., Li, J., & Li, B. (2004). Data report: Late Miocene–Quaternary biogenic opal accumulation at ODP Site 1143, southern South China Sea. In W. L. Prell, P. Wang, P. Blum, D. K. Rea, & S. C. Clemens (Eds.), *Proceedings of ODP, scientific results: Vol. 184.* (Online).

Wang, P., Li, Q., & Tian, J. (2014). Pleistocene paleoceanography of the South China Sea: Progress over the past 20 years. *Marine Geology,* (in press). http://dx.doi.org/10.1016/j.margeo.2014.03.003.

Wang, L., Liu, Z., Wu, J., & Zhong, G. (1996). Depositional history and its relationship to hydrocarbon in Wan'an Basin. *China Offshore Oil and Gas (Geology), 10*(3), 144–152.

Wang, P., Min, Q., & Bian, Y. (1985). Foraminiferal biofacies in the northern continental shelf of the South China Sea. In P. Wang (Ed.), *Marine micropaleontology of China* (pp. 151–175). Beijing: China Ocean Press.

Wang, P., Prell, W. L., & Blum, P. (Eds.), (2000). *vol. 184. Proceedings of ODP, initial reports.* College Station, TX: Ocean Drilling Program, Texas A&M University, (CD-ROM).

Wang, H., Shi, X., Wang, X., Yin, H., Qiao, X., Liu, B., et al. (2000). *Research on the sequence stratigraphy of China.* Guangzhou: Guangdong Science and Technology Press, 457 pp (in Chinese).

Wang, C., & Sun, Y. (1994). Development of Paleogene depressions and deposition of lacustrine source rocks in the Pearl River Mouth Basin, northern margin of the South China Sea. *AAPG Bulletin, 78,* 1711–1728.

Wang, P., Tian, J., Cheng, X., Liu, C., & Xu, J. (2003). Carbon reservoir change preceded major ice-sheet expansion at the mid-Brunhes event. *Geology, 31,* 239–242.

Wang, P., Tian, J., Cheng, X., Liu, C., & Xu, J. (2004). Major Pleistocene stages in a carbon perspective: The South China Sea record and its global comparison. *Paleoceanography, 19.* http://dx.doi.org/10.1029/2003PA000991.

Wang, L., Wang, Y., & Hu, X. (2010). Stratigraphy and sedimentary characteristics of the western depression, North Yellow Sea Basin. *Marine Geology and Quaternary Geology, 30*(3), 97–104 (in Chinese).

Wang, D., Wu, S., Qin, Z., Spence, G., & Lü, F. (2013). Seismic characteristics of the Huaguang mass transport deposits in the Qiongdongnan Basin, South China Sea: Implications for regional tectonic activity. *Marine Geology, 346,* 165–182.

Wang, G., Wu, C., Zhou, J., & Li, S. (1998). Sequence stratigraphic analysis of the Tertiary in the Qiongdongnan Basin. *Experimental Petroleum Geology, 20,* 124–128 (in Chinese).

Wang, P., Xia, L., & Chen, X. (1985). Neogene biostratigraphy in the northern shelf of the South China Sea. In P. Wang (Ed.), *Marine micropaleontology of China* (pp. 291–303). Beijing: China Ocean Press.

Wang, P., Xia, L., Wang, L., & Cheng, X. (1991). Lower boundary of the marine Pleistocene in northern shelf of the South China Sea. *Acta Geologica Sinica, 1992*(2), 176–187 (in Chinese).

Wang, L., Zhong, G., & Wu, J. (1997). Sequence stratigraphic analysis of Wan'an Basin. *Geological Research of South China Sea, 9,* 67–77 (in Chinese).

Wang, G., & Zhu, W. (1992). Cenozoic sedimentary environment in East China Sea Basin. *Acta Sedimentologica Sinica, 10,* 100–108 (in Chinese).

Watcharanantakul, R., & Morley, C. K. (2000). Syn-rift and post-rift modelling of the Pattani Basin, Thailand: Evidence for a ramp-flat detachment. *Marine and Petroleum Geology, 17,* 937–958.

Wei, K.-Y. (2006). Leg 195 synthesis: Site 1202—Late Quaternary sedimentation and paleoceanography in the southern Okinawa Trough. In M. Shinohara, M. H. Salisbury, & C. Richter (Eds.), *Proceedings of ODP, scientific results: Vol. 195.* (pp. 1–31) (Online).

Wei, K., Cui, H., Ye, S., Li, D., Liu, T., Liang, J., et al. (2001). High-precision sequence stratigraphy in Qiongdongnan Basin. *Earth Science*, *26*, 59–66 (in Chinese).

Wei, X., Zhang, T., Huang, J., Liang, X., Yao, Q., & Tang, X. (2011). Sequence stratigraphy characteristics and filling evolution models of Paleogene in Baiju Sag, Subei Basin. *Acta Geoscientica Sinica*, *32*, 427–437 (in Chinese).

Wongsosantiko, A., & Wirojudo, G. K. (1984). Tertiary tectonic evolution and related hydrocarbon potential in the Natuna Area. In *Proceedings of 13th annual convention of Indonesian petroleum association, Jakarta: Vol. 1.* (pp. 161–183).

Woodruff, D. S. (2003). Neogene marine transgressions, palaeogeography and biogeographic transitions on the Thai-Malay Peninsula. *Journal of Biogeography*, *30*, 551–567.

Woodruff, F., & Savin, S. M. (1991). Mid-Miocene isotope stratigraphy in the deep sea: High-resolution correlations, paleoclimatic cycles and sediment preservation. *Paleoceanography*, *6*, 755–806.

Wu, J. (1991). Structural characteristics and perspective on petroleum resources in the Nansha Islands. *Geological Research of South China Sea, Memoir*, *3*, 24–38 (in Chinese).

Wu, J., Liu, B., Zhou, W., & Liu, B. (1997). Sequence stratigraphic characteristics and prediction of non-structural traps in the Wan'an Basin. In H. Xu (Ed.), *From seismic stratigraphy to sequence stratigraphy* (pp. 65–72). Beijing: Petroleum Industrial Press (in Chinese).

Wu, F., Lu, Y., & Li, S. (2000). Cenozoic sequence stratigraphy and sea-level changes in the eastern and southern sea regions of China. In H. Wang, X. Shi, H. Yin, X. Qiao, B. Liu, S. Li, & J. Chen (Eds.), *Research on the sequence stratigraphy of China* (pp. 330–351). Guangzhou: Guangdong Science and Technology Press (in Chinese).

Wu, G. X., Wang, R. J., Hao, H. J., & Shao, L. (2007). Microfossil evidence for development of marine Mesozoic in the north of South China Sea. *Marine Geology and Quaternary Geology*, *27*, 79–85 (in Chinese).

Wu, J., & Yang, M. (1994). Seismic sequence analysis in the southern South China Sea. *Geological Research of South China Sea*, *6*, 16–29 (in Chinese).

Wu, F., & Zhou, P. (Eds.), (2000). *Analyses of tertiary sequence stratigraphy and depositional systems in the Xihu Depression of East China Sea Shelf Basin*. Beijing: Geological Publishing House, 179 pp (in Chinese).

Xia, K. Y., & Zhou, D. (1993). The geophysical characteristics and evolution of northern and southern margins of the South China Sea. *Geological Society of Malaysia Bulletin*, *33*, 223–240.

Xie, J., Li, J., Mai, W., Zhang, H., Cai, K., & Liu, X. (2012). Palynofloras and age of the Liushagang and Weizhou formations in the Beibuwan Basin, South China Sea. *Acta Palaeontologica Sinica*, *51*(3), 385–394 (in Chinese).

Xie, X., Müller, R. D., Li, S., Gong, Z., & Steinberger, B. (2006). Origin of anomalous subsidence along the Northern South China Sea margin and its relationship to dynamic topography. *Marine and Petroleum Geology*, *23*, 745–765.

Xie, X., Müller, R. D., Ren, J., Jiang, T., & Zhang, C. (2008). Stratigraphic architecture and evolution of the continental slope system in offshore Hainan, northern South China Sea. *Marine Geology*, *247*, 129–144.

Xu, S. (1999). Sequence stratigraphic theory and practice in exploration prospect prediction: Example from the Pearl River Mouth Basin. *China Offshore Oil and Gas (Geology)*, *13*(3), 152–158 (in Chinese).

Xu, J., Ben-Avraham, Z., Kelty, T., & Yu, H.-S. (2014). Origin of marginal basins of the NW Pacific and their plate tectonic reconstructions. *Earth Science Reviews*, *130*, 154–196. http://dx.doi.org/10.1016/j.earscirev.2013.10.002.

Xu, H., Ma, H. F., Fu, Q. N., Liang, R. B., & Zheng, J. (2003). The Cenozoic stratigraphic characteristics and reservoir potential of the East China Sea Shelf Basin. *Marine Geology Letters*, *19*(4), 22–25 (in Chinese).

Xu, J., & Zhu, G. (1994). Tectonic models of the Tan-Lu Fault Zone, Eastern China. *International Geology Review*, *36*, 771–784.

Yan, P., Zhou, D., & Liu, Z. (2001). A crustal structure profile across the northern continental margin of the South China Sea. *Tectonophysics*, *338*, 1–21.

Yang, Q. (1992). Geotectonic framework of the East China Sea. In J. S. Watkins, Z. Feng, & K. J. McMillen (Eds.), *AAPG Memoir: Vol. 53. Geology and geophysics of continental margins* (pp. 17–25).

Yang, Q., & Chen, H. (2003). Tectonic evolution of the North Jiangsu-South Yellow Sea Basin. *Petroleum Geology and Experiment*, *25*(suppl), 562–565.

Yang, W., Cui, Z., & Zhang, Y. (Eds.), (2010). *Geology and minerals of East China Sea*. Beijing: China Ocean Press, 780 pp (in Chinese).

Yang, F., Gao, D., Sun, Z., Zhou, Z., Wu, Z., & Li, Q. (2012). The evolution of the South China Sea Basin in the Mesozoic–Cenozoic and its significance for oil and gas exploration: A review and overview. *AAPG Memoir*, *100*, 397–418.

Yang, J., Sun, M., Li, Z., Liu, Z., & Research Party of Marine Geology, Ministry of Geology and Mineral Resources, and Geological Institute CAS, (Eds.), (1989a). *Cenozoic paleobiota of the continental shelf of the East China Sea (Donghai) Micropaleobotanical Volume*. Beijing: Geological Publishing House, 324 pp + 107 pls (in Chinese).

Yang, J., Sun, M., Li, Z., Liu, Z., & Research Party of Marine Geology, Ministry of Geology and Mineral Resources, and Geological Institute CAS, (Eds.), (1989b). *Cenozoic paleobiota of the continental shelf of the East China Sea (Donghai) Paleozoological Volume*. Beijing: Geological Publishing House, 280 pp + 75 pls (in Chinese).

Yang, M., Wu, J., Yang, R., & Duan, W. (1996). Stratigraphic division and nomenclature of the southwestern Nansha sea area. *Geological Research of South China Sea*, *8*, 37–46 (in Chinese).

Yao, Y., Chen, C., Feng, Z., Zhang, S., Hao, T., & Wan, R. (2010). Tectonic evolution and hydrocarbon potential in northern area of the South Yellow Sea. *Journal of Earth Science*, *21*, 71–82.

Ye, D., Zhong, X., Yao, Y., Yang, F., Zhang, S., Jiang, Z., et al. (1993). *Tertiary in petroliferous regions of China I—An overview*. Beijing: Petroleum Industry Press, 407 pp (in Chinese).

Ye, D., Zhong, X., Yao, Y., Yang, F., Zhang, S., Jiang, Z., et al. (1997). *Tertiary in petroliferous regions of China*. Beijing: Petroleum Industry Press, 374 pp.

Yeh, Y.-C., Sibuet, J.-C., Hsu, S.-K., & Liu, C.-S. (2010). Tectonic evolution of the Northeastern South China Sea from seismic interpretation. *Journal of Geophysical Research*, *115*. http://dx.doi.org/10.1029/2009JB006354, B06103.

Yi, S. H., Yi, S. S., Batten, D. J., Yun, H. S., & Park, S.-J. (2003). Cretaceous and Cenozoic non-marine deposits of the Northern South Yellow Sea Basin, offshore western Korea: Palynostratigraphy and palaeoenvironments. *Palaeogeography, Palaeoclimatology, Palaeoecology*, *191*, 15–44.

Ying, D. (1998). *Syn-rift structural style, basin fill and sequence stratigraphy and their control on development of organic facies in the Beibu Gulf Basin, South China Sea*. Stanford University, PhD dissertation, 336 pp.

Yu, C. F., Chen, J. W., Du, Y. S., & Zhang, Y. G. (2007). Division of sequence stratigraphy of Pinghu Formation in Xihu Sag in East China Sea. *Marine Geology and Quaternary Geology*, *27*(5), 85–90 (in Chinese).

Yu, Z., & Ding, Z. (1998). An automatic orbital tuning method for paleoclimate records. *Geophysical Research Letters*, *25*, 4525–4528.

Yu, H. S., & Hong, E. (2006). Shifting submarine canyons and development of a foreland basin in SW Taiwan: Controls of foreland sedimentation and longitudinal sediment transport. *Journal of Asian Earth Sciences*, *27*, 922–932.

Yu, W., Mao, X., Chen, Z., & Huang, L. (1978). *Early tertiary gastropod fossils from the coastal region of Bohai*. Beijing: China Science Press, Research Institute of Ministry of Petroleum and Chemical Industries and Nanjing Institute of Geology and Paleontology CAS. 157 pp + 33 pls (in Chinese).

Yumul, G. P., Jr., Dimalanta, C. B., Tamayo, R. A., Jr., & Maury, R. C. (2003). Collision, subduction and accretion events in the Philippines: A synthesis. *Island Arc*, *12*, 77–91.

Zachos, J. C., Dickens, G. R., & Zeebe, R. E. (2008). An early Cenozoic perspective on greenhouse warming and carbon-cycle dynamics. *Nature*, *451*, 281–283.

Zachos, J. C., Pagani, M., Sloan, L., Thomas, E., & Billups, K. (2001). Trends, rhythms, and aberrations in global climate 65 Ma to present. *Science*, *292*, 686–693.

Zachos, J. C., Shackleton, N. J., Revenaugh, J. S., Pälike, H., & Flower, B. P. (2001). Climate response to orbital forcing across the Oligocene-Miocene boundary. *Science*, *292*, 274–278.

Zampetti, V., Sattler, U., & Braaksma, H. (2005). Well log and seismic character of Liuhua 11-1 Field, South China Sea: Relationship between diagenesis and seismic reflections. *Sedimentary Geology*, *175*, 217–236.

Zeng, D., Guo, B., Huo, C., Zhong, S., Huang, X., & Hu, P., et al. (Eds.). (1981). *Tertiary system of the Northern Continental Shelf of South China Sea*. Guangzhou: Guangdong Science and Technology Press, 263 pp (in Chinese).

Zhang, Q. (1999). Evolution of the Ying-Qiong basin and its tectonic-thermal system. *Natural Gas Industry*, *19*, 12–18 (in Chinese).

Zhang, S. (Ed.), (2009). *Geological formation names of China (1866-2000)*. Beijing: Higher Education Press of China, 1537 pp.

Zhang, Q., & Kou, C. (1989). Petroleum geology of Cenozoic basins in the northwestern continental shelf, South China Sea. In X. Zhu (Ed.), *Chinese sedimentary basins: Sedimentary basins of the world: Vol. 1.* (pp. 197–206): Elsevier.

Zhang, Q., & Su, H. (1989). Petroleum geology of the Beibu Gulf Basin. *Marine Geology and Quaternary Geology*, *9*(3), 73–81 (in Chinese).

Zhang, J. P., Xu, F., Zhong, T., Zhang, T., & Yu, Y. F. (2012). Sequence stratigraphic models and sedimentary evolution of Pinghu and Huagang formations in Xihu Trough. *Marine Geology and Quaternary Geology*, *32*(1), 35–41 (in Chinese).

Zhang, Q., & Zhang, Q. (1991). A distinctive hydrocarbon basin—Yinggehai Basin, South China Sea. *Journal of Southeast Asian Earth Sciences*, *6*, 69–74.

Zhao, J. (2004). The forming factors and evolvement of the Mesozoic and Cenozoic basins in the East China Sea Part A. *Offshore Oil*, *2004*(4), 6–14 (in Chinese).

Zhao, J. (2005a). The forming factors and evolvement of the Mesozoic and Cenozoic basins in the East China Sea Part B. *Offshore Oil*, *2005*(1), 1–10 (in Chinese).

Zhao, Q. (2005b). Late Cainozoic ostracod faunas and paleoenvironmental changes at ODP Site 1148, South China Sea. *Marine Micropaleontology*, *54*, 27–47.

Zhao, Q., Jian, Z., Wang, J., Cheng, X., Huang, B., Xu, J., et al. (2001). Neogene oxygen isotopic stratigraphy, ODP Site 1148, northern South China Sea. *Science China (D)*, *44*(10), 934–942.

Zhao, Q., & Wang, P. (1999). Progress in Quaternary paleoceanography of the South China Sea: A review. *Quaternary Science Reviews*, *6*, 481–501 (in Chinese).

Zhao, Q., Wang, P., Cheng, X., Wang, J., Huang, B., Xu, J., et al. (2001). A record of Miocene carbon excursions in the South China Sea. *Science China (D)*, *44*, 943–951.

Zhao, W., Wang, H., Yuan, X., Wang, Z., & Zhu, G. (2009). Petroleum systems of Chinese non-marine basins. *Basin Research*, *22*, 4–16.

Zheng, J., He, X., Liu, S., & Li, Z. (1999). *Stratigraphical Lexicon of China: Tertiary*. Beijing: Geological Publishing House, 163 pp (in Chinese).

Zheng, Y. P., Liu, B. H., Wu, J. L., Wang, H. Y., Wang, S. G., Han, G. Z., et al. (2001). Seismic stratigraphic characteristics and structural facies analyses of the Cenozoic basin of the South Yellow Sea. *Chinese Science Bulletin*, *46*(Suppl), 52–58 (in Chinese).

Zhong, S., Jiang, L., Li, B., & Wang, J. (2006). *High-resolution Paleogene Biostratigraphy and Sequence Stratigraphy of the Taipei Depression, East China Sea Shelf Basin*. Beijing: Petroleum Industry Publishing House, 162 pp (in Chinese).

Zhong, G., & Li, Q. (2009). Stratigraphy of major shelf and slope basins. In P. Wang & Q. Li (Eds.), *The South China Sea paleoceanography and sedimentology* (pp. 109–147). Berlin: Springer.

Zhong, G., Xu, H., & Wang, L. (1995). Structure and evolution of Cenozoic basins in the south-west area of South China Sea. *Marine Geology and Quaternary Geology*, *15*(Suppl), 87–94 (in Chinese).

Zhou, M., Bian, L., & Wang, S. (2000). *Stratigraphical Lexicon of China: Quaternary*. Beijing: Geological Publishing House, 122 pp (in Chinese).

Zhou, Z., Zhao, J., & Yin, P. (1989). Characteristics and tectonic evolution of the East China Sea. In X. Zhu (Ed.), *Chinese sedimentary basins. Sedimentary basins of the world: Vol. 1.* (pp. 165–179). Amsterdam: Elsevier.

Zhu, W. (1987). Study of Miocene carbonate and reef facies in the Pearl River Mouth Basin. *Marine Geology and Quaternary Geology*, *7*, 11–19 (in Chinese).

Zhu, X. (Ed.), (1989). *Chinese sedimentary basins. Sedimentary basins of the world: Vol. 1.* Amsterdam: Elsevier.

Zhu, X. (Ed.). (1990). *Mesozoic and Cenozoic sedimentary basins of China*. Beijing: Petroleum Industry Publishing House, 399 p. (in Chinese).

Zhu, W. L. (2009). *Paleolimnology and source rock studies of Cenozoic Hydrocarbon-bearing Offshore Basins in China*. Beijing: Geological Publishing House, 239 pp (in Chinese).

Chapter 6

Paleoceanography and Sea-Level Changes

6.1 INTRODUCTION

This chapter introduces research progress in two major fields of environmental changes in the China Seas: paleoceanography in deeper waters and sea-level changes in coastal waters. Paleoenvironmental change was the starting point for geology of the China Seas, with the earliest papers published in the 1930s dealing with coral reefs. When China developed its exploration programs in coastal plains in the 1960s and then offshore drilling in the 1980s, sea-level changes were among the primary research topics. Slightly different was paleoceanography. It was the international collaborative cruises in the 1990s that initiated paleoceanographic studies in the deepwater parts of the China Seas.

Sea–land interactions represent the core of environmental changes in marginal seas. The China Seas are featured by huge drainage basins and extensive continental shelves. The Sunda Shelf in the southern South China Sea (SCS) and the East China Sea (ECS) shelf are among the largest low-latitude shelves in the world. Consequently, the China Seas are particularly sensitive to sea-level changes. In oceanographic and climatic aspects, the Asian monsoon and the Kuroshio are the two major features that determine environmental changes of the China Seas. Since the ECS is predominated by shallow-water shelves, its environmental changes have been largely controlled by sea-level fluctuations, and the history is largely related to the influence of the Kuroshio and runoff from large rivers such as the Yangtze. Meanwhile, our discussion on the paleoceanographic part will focus on the SCS, especially the deepwater history related to its connection with the Pacific, and the upper water changes largely driven by monsoon variations.

In this chapter, we start with paleoceanographic responses to basin evolution in the SCS and in the Okinawa Trough, before moving on to paleoenvironmental changes over Quaternary glacial cycles in both the continental shelves and deepwater basins.

Developments in Marine Geology, Vol. 6. http://dx.doi.org/10.1016/B978-0-444-59388-7.00006-8

6.2 PALEOCEANOGRAPHY: LONG-TERM TRENDS

6.2.1 East China Sea: Okinawa Trough

The Okinawa Trough is a back-arc rifting basin behind the Ryukyu Islands and segmented into three parts with different timing of formation (Figure 3.39). The middle and northern Okinawa Trough may be related to the opening of the Japan Sea with initial opening in the Late Miocene, whereas the southern trough opened later, in the earliest Pleistocene (see Section 3.3.2). Since the sediment records so far recovered from the Okinawa Trough (Section 4.3.3.2) are restricted to the late Quaternary, the early history of the trough has to be reconstructed on the basis of the stratigraphic archives from the uplifted Ryukyu Islands.

The Ryukyu Islands extend from Kyushu in the north to Taiwan in the south and exceed 1200 km in length. The islands' simplified Neogene stratigraphy is shown in Table 6.1. The Miocene sandstones of the Yaeyama Group are of nonmarine origin, mostly from the Eurasian continent as evinced by heavy mineral components, thus excluding the existence of the Okinawa Trough at that time (Saitoh & Masuda, 2004). The Shimajiri Group comprises siltstones and sandstones accompanied by tuff and contains a rich marine fauna indicative of littoral to bathyal depositional environments (e.g., MacNeil, 1960; Tanaka & Nomura, 2009; Ujiié, 1994) during the late Miocene to earliest Pleistocene, as dated by planktonic foraminifera (Iryu et al., 2006). Up section, the limestones of the Ryukyu Group overlie unconformably the Shimajiri or Yaeyama Groups, showing significant variations between islands in thickness, elevation, and dating range from ~1.65 or 1.45 Ma to 0.3 Ma (Humblet, Iryu, & Nakamori, 2009; Iryu et al., 2006; Sagawa, Nakamori, & Iryu, 2001).

The Ryukyu group is composed of plural reef complex deposits, each comprising proximal water coral limestone and distal rhodolith and detrital limestones (Sagawa et al., 2001). The earliest Pleistocene transition from

---

TABLE 6.1 Neogene Stratigraphy of the Ryukyu Islands
(Based on Iryu et al., 2006)

Stratigraphic Unit	Age	Lithology	Environment
Ryukyu Group	Pleistocene (~1.6–0.3 Ma)	Limestone	Coral reef
Shimajiri Group	Late Miocene–earliest Pleistocene	Siltstone, sandstone, with tuff	Marine
Yaeyama Group	Miocene	Sandstone with shale	Nonmarine

Shimajiri Group to Ryukyu Group in the Ryukyu Islands is of great paleocea-nographic significance. The replacement of clastic sediments by coral reef facies was considered to imply (i) a reduction, if not cessation, in the supply of terrigenous material from the Asian continent, to be trapped in the subsided Okinawa Trough, and (ii) a warm, tropical/subtropical climate. The beginning of coral reef development on the Ryukyu Islands at the earliest Pleistocene, however, was counterintuitive to the global climate indicated by large Arctic ice-sheet development and intensified glacial cycles. Thus, the only possible explanation of the "warm feeling" is intrusion of the Kuroshio warm current during this time, ~1.5–1.6 Ma (Iryu et al., 2006; Osozawa et al., 2012; Ujiié, 1994).

As the most powerful western boundary current of the North Pacific, the Kuroshio is crucial to the regional climate, as its present course of running through the Okinawa Trough controls the oceanographic features of the ECS and the growth of coral reefs along the Okinawa Trough. Marking the entry of the Kuroshio into the trough was the transition from the Shimajiri Group to Ryukyu Group at ~1.6 Ma, hence the onset of its present course flowing through the ECS (Figure 6.1; Iryu et al., 2006; Osozawa et al., 2012). Today's northern limit of reef growth is at 30°44′N, on the coast of island Tanegashima, coinciding with the northernmost latitude of the Kuroshio inside the Okinawa Trough (Ikeda, Iryu, Sugihara, Ohba, & Yamada, 2006). In the Miocene, the Ryukyu arc was the eastern margin of the Asian continent, and the Kuroshio was flowing in the Pacific, to the east of the present Ryukyu Islands (Figure 6.1A). With the opening of its southern segment around 1.6 Ma, the

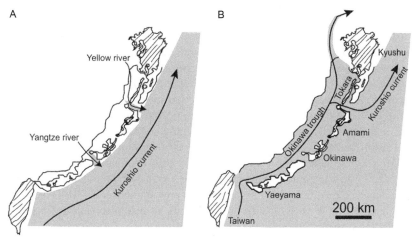

FIGURE 6.1 Opening of the Okinawa Trough and course change of the Kuroshio. (A) Miocene, the Kuroshio flew outside the trough. (B) Pleistocene, the Kuroshio flows through the trough. *Modified from Osozawa et al. (2012).*

Okinawa Trough rifted and subsided along the full length, allowing the entry and flow of the Kuroshio through it (Figure 6.1 B) (Iryu et al., 2006; Osozawa et al., 2012).

6.2.2 South China Sea: Ocean Connection and Basin Evolution

6.2.2.1 Environmental Response to Sea-Floor Spreading

The 5500 m high-quality sediment cores recovered during ODP Leg 184 provide an unprecedented record of paleoceanographic evolution in the SCS. Of particular significance is the record from ODP Site 1148 (18°50′N, 116°34′E, water depth 3292 m). Its 853 m long sediment sequence covers the 33 myr history of the SCS basin since its opening and reveals how sedimentologic and paleoceanographic processes were responding to major tectonic events. As shown in Figure 6.2, the deepwater SCS basin increased in size from its opening around 32 Ma to the end of seafloor spreading at ∼16 Ma (see Section 3.3.1). Nearly about that time, the SCS basin began its eastward subduction along the Manila Trench and reduced its size with the closure of its eastern margin (Figures 3.35 and 3.37).

This tectonic history is recorded in deep-sea sediments. At ODP Site 1148, the deposition rate reached its maximal values (>60 m/Ma) in the early Oligocene at the beginning of the basin opening, followed by the Quaternary, while the lowest rates occurred in the middle Miocene. Four tectonics-induced hiatuses occurred between 28.5 and 23 Ma and erased sediment records of 3 myr at the least, with the main unconformity at the base of slump deposits dated at ∼25 Ma (Li et al., 2005). These records render some solid evidence that the northern SCS experienced tremendous environmental changes during the late Oligocene. The succession of benthic faunas from ODP Site 1148 indicates drastic deepening from upper bathyal zone (<1500 m) in the early Oligocene to lower bathyal zone (>2500 m) in the early Miocene (Zhao, 2005). A similar trend was observed in the sediment sequence from the Baiyun Depression of the Pearl River Mouth (Zhujiangkou) Basin on the northern upper slope, indicating a transition from shallow- to deepwater conditions (Pang et al., 2007). Other concurring changes include abrupt reductions of opal and carbonate accumulation rates (Figure 6.3) and a sediment source shift from the southern to the northern provenance after the event (Li et al., 2003). The late Oligocene event recorded in SCS sediments not only represents the largest tectonic change but also corresponds to an extensive reorganization of geologic and climatological environments in SE Asia, as exemplified by the transition from a planetary to a monsoonal atmospheric circulation system (Sun & Wang, 2005).

Another aspect is carbonates. Thanks to its low-latitude position and intensive ventilation of deep waters, the SCS has the best deepwater carbonate preservation in the western Pacific region. Aside from the deep part below

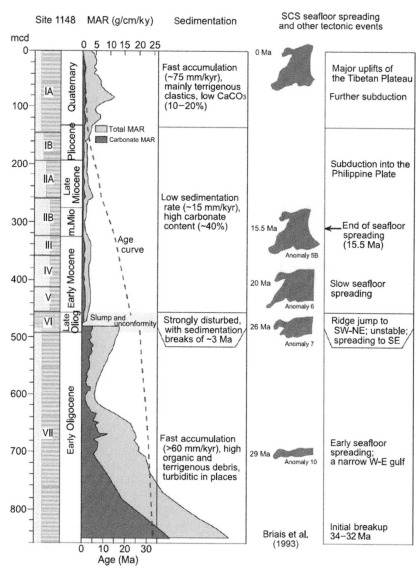

FIGURE 6.2   Sedimentologic response to basin evolution of the SCS recorded at ODP Site 1148. From left to right: Major lithologic units, mass accumulation rates (MAR, in g/cm²/kyr), and sea-floor spreading events with the shapes of oceanic-crust basin. *Li, Jian, and Su (2005).*

3500 m, the rest of the SCS contains no less than 10% carbonate in bottom sediments. The ocean drilling has revealed remarkable variations in carbonate accumulation over the last 33 Ma, including changes common to the low-latitude Pacific: the carbonate dissolution events (D1–D5 in Figure 6.3), the

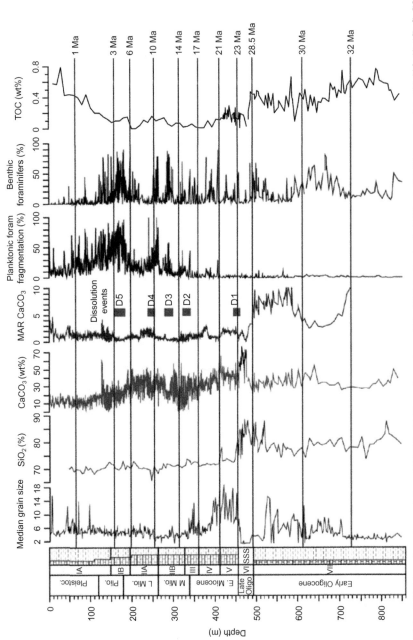

FIGURE 6.3 A 33 Ma sediment record from ODP Site 1148, northern slope of the SCS (water depth 3297 m). (A) Main grain size. (B) SiO$_2$%. (C) CaCO$_3$%. (D) Mass accumulation rate of carbonate in g/cm/kyr; D1–D5 denote deep-sea dissolution events. (E) Planktonic foraminifera fragmentation (%). (F) Benthic foraminifera (%). (G) Total organic carbon. *Wang (2012) and Fang (2002).*

"carbonate crash" after 11 Ma, and the "biogenic bloom" after 8 Ma (Figure 6.3) (Wang & Li, 2009). As described in Section 4.5.1.2, the modern SCS basin is studded with numerous coral reefs, with an estimated total modern reef area of ca. 8000 km^2 and an estimated total annual regional reef carbonate production in the order of 2.1×10^7 t, which is about 1.6–3.3% of global reefal carbonate production (Yu & Zhao, 2009). According to the geologic data, coral reefs in the SCS began to develop in the SCS since the middle Miocene when the supply of terrigenous material was minimal (Figure 6.2).

6.2.2.2   The Deepwater Oligocene

At ODP Site 1148, the 400 m thick section between 860 and 460 m represents the 9 myr long Oligocene between 32.8 and 23.9 Ma (Figure 6.4), mainly comprising nannofossil-rich clay with deepwater trace fossils such as *Zoophycos*. The CaCO$_3$% averages from 30% to 60%, with much higher percentages in the upper than the lower Oligocene (Figures 6.2 and 6.3). Foraminiferal assemblage is predominated by planktonic species, with rare to common deepwater benthic species such as the calcareous *Cibicidoides* and various agglutinated species, which comprise nearly 50% of the total foraminiferal biofacies in the early Oligocene (Kuhnt, Holbourn, & Zhao, 2002). Ostracods are characterized by deepwater species including *Krithe* with 30–40% abundance. These foraminiferal and ostracod faunas indicate deepwater environments up to 1500 m, corresponding to the modern upper to middle continental slope, which already existed in the Oligocene before seafloor spreading (Wang, Tian, Cheng, Liu, & Xu, 2003).

The Oligocene sediment at Site 1148 is also enriched with pollen and nearshore phytoplankton, and its organic carbon content (0.4–0.6%) is significantly higher than in the Neogene (<0.2%) (Figure 6.3), inferring a nearshore environment. Therefore, the concurrence of deep- and shallowwater microfossils may signify that the very early Oligocene SCS must have been a narrow gulf with fairly steep slopes that facilitated rapid accumulation of the thick lower Oligocene sequence (P. Wang et al., 2003).

Noteworthy is the occurrence of thin-layered sandstones close to the base of the drilling holes at Site 1148, at a level with turbidite-associated agglutinated foraminifers (Kuhnt et al., 2002) and with accumulation rates as high as 60 g/cm^2/kyr in a stage of rapid sedimentation. Of particular interest is the clear reflector at about 710 m in the seismic profile corresponding to a negative excursion of bulk sample δ^{13}C (Figure 6.4B). At about the same depth, the proportion of freshwater phytoplankton increases (Figure 6.4G), and pollen assemblages show the replacement of tropical and subtropical low mountain and lowland evergreen vegetation (such as *Taxodiaceaepollenites*) by a temperate forest featuring high-mountain conifer and deciduous broad-leaved trees (such as *Cedripites*, *Tsugaepollenites*, *Abiespollenites*, and *Piceaepollenites*), marking a drop in air temperature (Figure 6.5; Wu et al., 2003).

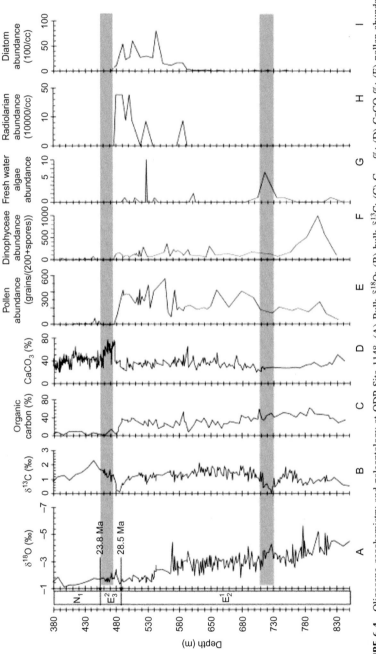

FIGURE 6.4 Oligocene geochemistry and paleontology at ODP Site 1148. (A) Bulk δ¹⁸O; (B) Bulk δ¹³C; (C) C$_{org}$ %; (D) CaCO$_3$%; (E) pollen abundance (grains/200 exotic *Lycopodium* spores); (F) dinoflagellate abundance; (G) freshwater algae: *Pediastrum*, *Botryoccus*, and *Concentricystis*; (H) radiolarian abundance (10⁴/cc); (I) diatom abundance (10²/cc). *Wang, Tian, et al. (2003)*.

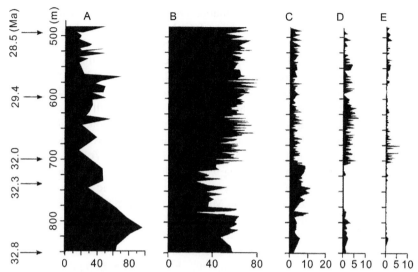

FIGURE 6.5   Pollen diagrams of the Oligocene deposits, ODP Site 1148. (A) Phytoplankton/ pollen ratio (compiled by Fang, 2002). (B) *Pinuspollenites %*. (C) *Taxodiaceaepollenites %*. (D) *Cedripites %*. (E) *Tsugaepollenites, Abiespollenites,* and *Piceaepolenites %*. Based on Wu, Qin, and Mao (2003).

All these may have responded to a major global marine regression in the early Oligocene (Haq, Hardenbol, & Vail, 1987). In the later part of the early Oligocene, from 600 to 475 m at Site 1148, the abundance of diatoms and radiolarian increased remarkably (Figure 6.4h and i), similar to the record from the equatorial Pacific (Wang et al., 2001). This increase of siliceous microfossils can be attributed to the middle Oligocene fractionation in water chemistry between the Pacific and the Atlantic.

6.2.2.3   Connection with Open Ocean

As discussed in Section 2.4.4.2, the Pacific is today the only deepwater source of the SCS. The incoming Pacific water enters the SCS through the Bashi Strait over the 2600 m deep sill, and the deepwater overflow gradually gets mixed with ambient and then returns back to the Pacific in intermediate layers (Tian, Yang, et al., 2006). As a result, the modern SCS is characterized by extraordinary deepwater ventilation, with a fast flushing time of 40–50 years (Chen, Wang, Wang, & Pai, 2001) or even 30 years (Su, 2004). Here, the key element in geography is the Bashi Strait (Figures 2.42 and 3.35). Numerous internal waves are produced in the Bashi Strait and propagate into the SCS, enhancing deepwater mixing in the SCS by two orders of magnitude higher than in the Pacific. Diapycnal mixing drives the cyclonic circulation in the deep SCS and modulates the deepwater renewal there (Tian & Qu, 2012).

In the geologic past, deepwater ventilation could be much weaker in the SCS before the Bashi Strait was formed or when the sill depth was deep.

In the Oligocene and early Miocene, the newly formed SCS was completely open to the East with free water exchanges with the Pacific. Neither any "deepwater overflow" was then needed, nor was the deep circulation necessarily cyclonic. Together with the northerly migration of the Luzon Volcanic Arc and closure of the SCS basin, the submarine sill restricted interbasinal water exchanges and enhanced deepwater ventilation, giving rise to extremely young and well-oxygenized deep waters in the SCS. Specifically, the accretionary prism of the Manila Trench and the Luzon Volcanic Arc represent the two main sills obstructing water exchanges. As shown in geologic records from Taiwan, the accretionary prism and the volcanic arc were formed some 15–16 Ma in the early middle Miocene and subaerially emerged by the end of Miocene about 6.5 Ma (Huang et al., 1997, 2006). Interestingly, these are exactly the two time intervals when deepwater oxygen concentration and negative excursion of carbon isotope remarkably increased in deep-sea drilling records. As seen from Figure 6.6, the benthic foraminiferal succession of ODP Site 1148 displays a two-step increase in the high-oxygen group, and the carbon isotopes show significant lighter values at the corresponding stratigraphic levels in comparison to the global average (Wang & Li, 2009).

Stepwise increases in deepwater ventilation are also recorded at other localities from the northern and southern SCS (Figures 6.7 and 6.8) (Hess & Kuhnt, 2005; Huang et al., 2007; Kawagata, Hayward, & Kuhnt, 2007). At ~2.1 Ma, sudden decline in the abundance of *Bulimina alazanensis* and other infaunal foraminiferal species was recorded at both Site 1143 (9°22′N, 113°17′E, w.d. 2772 m) and Site 1146 (19°27′N, 116°16′E, w.d. 2091 m) (Figure 6.7). This 2.1 Ma infaunal decline event was followed by significant increase in the abundance of epifaunal species such as *Cibicidoides wuellerstorfi* commonly living in oxic to suboxic bottom conditions, suggesting increased oxygenation in the deep water (Huang et al., 2007). Similarly, during the mid-Pleistocene climate transition between 1.2 and 0.6 Ma, most species of the elongate, cylindrical benthic species (mainly nodosariids, pleurostomellids, and stilostomellids) disappeared from Sites 1143, 1146, and 1148 (Figure 6.8). The extinction of these infaunal species ~0.9 Ma ago was again responding to increased glacial cooling and consequent increased ventilation of the deep-sea water masses (Kawagata et al., 2007), probably also associating with stronger seasonal carbon flux after 0.9 Ma (Hess & Kuhnt, 2005).

6.2.3   South China Sea: Upper Waters and Monsoon Climate

Climate and hydrography in the SCS are largely controlled by the monsoonal circulation system. Today, strong southwesterly winds during summer and northeasterly winds during winter drive a semiannual reversal in surface-water

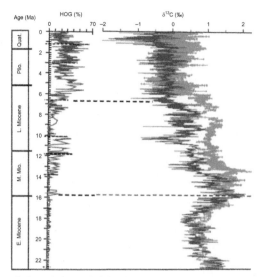

FIGURE 6.6   Downhole variations of (left) high-oxygen group (HOG) of benthic foraminifera and (right) benthic δ^{13}C (black) at ODP Site 1148, as compared to the global ocean average (grey). Dotted lines show development stages of the submarine sill in the Bashi Strait. *Redrawn from Wang and Li (2009).*

circulation from approximately clockwise to anticlockwise (Figures 2.29 and 2.31). The monsoon system and the oceanographic features of the upper ocean in the SCS were highly sensitive to changes in climate boundary conditions, such as the growth of the polar ice sheets. Since the 1990s, paleomonsoon has become a major topic in the SCS research, and numerous expeditions, such as "Monitor Monsoon" expedition in 1994 (Sarnthein, Pflaumann, Wang, & Wong, 1994) and ODP Leg 184 (Wang, Prell, & Blum, 2000), have been organized. Now, the SCS and the Arabian Sea have, respectively, become the foci of the East and South Asian monsoon studies in both modern and paleoclimatology (Wang, Clemens, & Liu, 2003), and Jian, Tian, and Sun (2009) recently reviewed the progress in paleomonsoon research in the SCS.

Here, the evolution of upper waters and monsoon in the SCS is summarized in three aspects: (i) SST and thermocline, (ii) monsoon climate weathering rate, and (iii) productivity and carbon cycling. These will be followed by a discussion, in Section 6.3.1, on monsoon variations over the late Pleistocene glacial cycles.

6.2.3.1   SST and Thermocline

SST was the starting point of the paleomonsoon study in the SCS since the 1990s (e.g., Wang & Wang, 1990). Three methods have been commonly

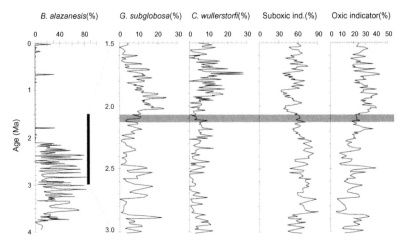

FIGURE 6.7 Dramatic reduction in infaunal species was accompanied by increases in oxic species at ~2.1 Ma from ODP Site 1146, likely responding to increased oxygenation of the deep water. *Huang et al. (2007).*

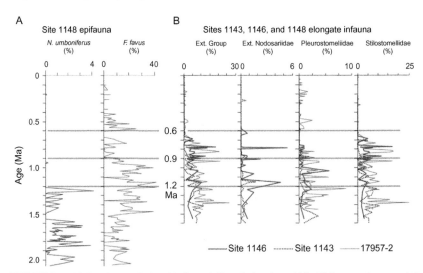

FIGURE 6.8 Major changes in benthic foraminifera during the middle Pleistocene between 1.2 and 0.6 Ma are recorded in (A) epifaunal and (B) infaunal taxa at ODP sites. *Wang and Li (2009).*

applied to reconstructing past SST variations in the SCS: transfer function method based on paleoecology of planktonic foraminifera, alkenone unsaturation index ($U_{37}^{K'}$) of sediments, and Mg/Ca ratios of planktonic foraminiferal tests. The recent reconstruction of $U_{37}^{K'}$-based SST at ODP Site 1143 in the southern SCS reveals three stages over the past 5 myr: a high and stable temperature state in the middle Pliocene (4.0–2.7 Ma), a long-term increasing temperature fluctuation trend across the Pliocene and Pleistocene (2.7–0.9 Ma), and a stage with large glacial–interglacial temperature oscillations up to $\sim$3 °C since the middle Pleistocene (0.9–0 Ma) (Figures 6.9G and 6.10C; Li et al., 2011).

The SST evolution in the SCS is well in line with the history of the Arctic ice-sheet development since the late Pliocene. The onset of major Northern Hemisphere glaciation from $\sim$3.3 to 2.5 Ma was accompanied by remarkable changes in the stable isotopes and SST of the SCS. During this time interval, the deepwater $\delta^{18}O$ in both the northern (ODP Site 1146; Figure 6.9L) and southern (ODP Site 1143, Figure 6.9A) sites increased by approximately 1.25‰, while Mg/Ca-based SST at ODP Site 1143 declined by 2–3 °C (Figure 6.9K; Tian, Pak, et al., 2006; Tian, Yang, et al., 2006), which is close to the result of winter SST estimation based on foraminiferal transfer function (TF) at ODP Site 1146 (Figure 6.9O; Huang, 2002). However, SST reconstructions based on $U_{37}^{K'}$ in the southern Site 1143 (Figure 6.9J; Li et al., 2011) and northern Site 1148 (see fig. 2 in Jia, Chen, & Peng, 2008) showed only $\sim$1 °C decrease in SST over the same period. Although different methodologies and materials used may have contributed to these SST discrepancies, the real source of the difference between proxies remains unclear (Wang et al., 2014).

When the SST records between ODP 1143 in the far western Pacific and ODP 846 in the eastern Pacific (3°5'S, 90°49'W, w.d. 3296 m) are compared, the mean temperature gradient ($\Delta SST_{1143-846}$) between the two sites increased from $\sim$3 °C in the middle Pliocene to 5–6 °C in the late Pleistocene (Figure 6.10A). This lower zonal temperature gradient in the tropical Pacific during the early and middle Pliocene may have contributed to long-lasting El Niño-like conditions and a super warm pool (e.g., Li et al., 2011).

The Asian monsoon has enhanced the SST contrast between the northern and southern SCS. The modern SST varies between 28.5 and 29.5 °C for the whole SCS in summer and between 26 °C for the north and 28 °C for the south in winter. The annual mean ΔSST between the northern and southern SCS is $\sim$2 °C today, but ΔSST increases with intensified winter monsoon. Therefore, the meridional SST contrast between the northern (ODP Sites 1146 and 1148) and southern SCS (Site 1143) is indicative of winter monsoon intensity. As shown in Figure 6.9B, the mean value and the fluctuation amplitudes of $\Delta SST_{1143-1146}$ and $\Delta SST_{1143-1148}$ remarkably increased since the late Pliocene, implying winter monsoon strengthening in the SCS region, obviously in response to the increase of global ice volume (Li et al., 2011).

The same trend is also observed in the changing contrast between the benthic and planktonic geochemistry records. The period from 3.3 to 2.5 Ma in the late Pliocene is characterized by a continuous increase of global ice volume as recorded by a trend of positive shift in benthic $\delta^{18}O$. The planktonic $\delta^{18}O$ records, however, remain relatively constant, resulting in an increasing

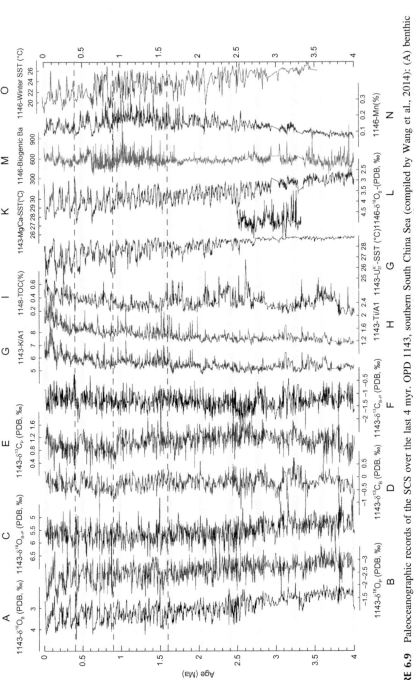

FIGURE 6.9 Paleoceanographic records of the SCS over the last 4 myr. OPD 1143, southern South China Sea (compiled by Wang et al., 2014): (A) benthic $\delta^{18}O$, (B) planktonic $\delta^{18}O$, (C) difference between benthic and planktonic $\delta^{18}O$, (D) benthic $\delta^{13}C$, (E) planktonic $\delta^{13}C$, (F) difference between benthic and planktonic $\delta^{13}C$ (Tian, Wang, Cheng, & Li, 2002), (G) K/Al ratio of bulk sediment, (H) Ti/Al ratio (Tian, Xie, Ma, Jin, & Wang, 2011), (I) total organic carbon in sediment (Li et al., 2013), (J) U^K_{37}-derived SST (Li et al., 2011), (K) Mg/Ca-derived SST of 3.3–2.5 Ma (Tian, Pak, et al., 2006; Tian, Yang, et al., 2006). ODP 1146 (Northern SCS): (L) benthic $\delta^{18}O$. M. biogenic Ba in sediments (Clemens, Prell, Sun, Liu, & Chen, 2008), (N) Mn in sediments (Clemens, Prell, Sun, Liu, & Chen, 2008), (O) winter SST based on transfer function FP-12E (Huang, 2002). Bars indicate oceanic carbon maximum ($\delta^{13}Cmax$, Wang, Tian, & Lourens, 2010).

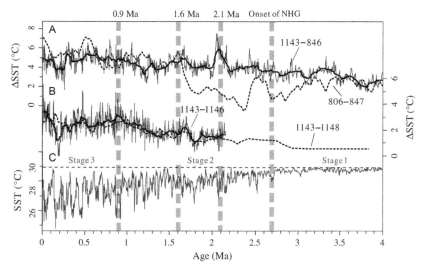

FIGURE 6.10 Alkenone-based SST in the South China Sea over the last 4 myr. (A) Zonal SST variations between tropical eastern and western Pacific: $\Delta SST_{1143-846}$; (B) Meridional SST variations between northern and south SCS sites: $\Delta SST_{1143-1146}$ (solid fine line) and $\Delta SST_{1143-1148}$ (dashed smoothing line); (C) SST at Site 1143. *Li et al. (2011)*.

difference between benthic and planktonic isotopes, as exemplified by the Site 1143 records (Figure 6.11A and B) (Jian et al., 2003; Tian, Pak, et al., 2006; Tian, Yang, et al., 2006). To test whether the differences originated from monsoon-related salinity changes, the ice volume and water temperature and salinity effects in the isotope signals need to be discriminated. The Mg/Ca-based SST decreased stepwise from 3.3 to 2.5 Ma (Figure 6.11C), and the surface-water $\delta^{18}O_{SW}$ can be calculated from planktonic $\delta^{18}O$ and SST. Since the $\delta^{18}O_{SW}$ depends on both the global ice volume and water salinity, the regional sea surface salinity (SSS) variations can be estimated by removing the ice volume effects from the $\delta^{18}O_{SW}$, that is, the residual $\Delta\delta^{18}O_{SW\text{-}B}$. By subtracting the benthic foraminiferal $\delta^{18}O$ from the calculated $\delta^{18}O_{SW}$, the resulted $\Delta\delta^{18}O_{SW\text{-}B}$ for Site 1143 shows a stepwise decrease during the period of significant ice-sheet growth 3.3–2.5 Ma (Figure 6.11D), suggesting a decline of surface salinity owing to intensified monsoon precipitation. Because the primary productivity of the SCS also depends on monsoon intensity, the increase of TOC as a productivity proxy supports the aforementioned conclusion (Figure 6.11E) (Tian, Pak, et al., 2006; Tian, Yang, et al., 2006).

Another monsoon proxy is the thermocline depth. In oceans, the monsoon-driven upwelling can lead to increased productivity and shoaled thermocline. Important faunal signals of strengthened monsoons include increases in abundance of productivity-indicative planktonic foraminifera due to upwelling and decreases in the percentage of planktonic foraminifera living in the mixed layer because of a shoaled thermocline. In the SCS, species of *Globigerinoides*,

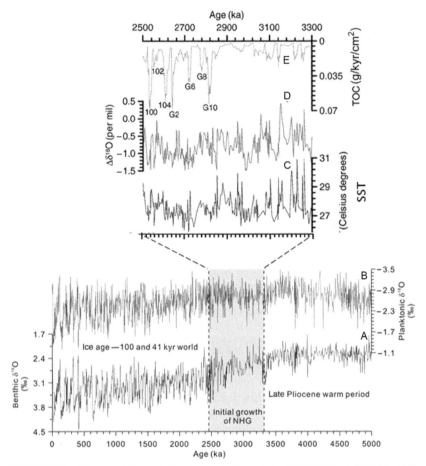

FIGURE 6.11 Monsoon-driven salinity changes in the period of 2.5–3.3 Ma at ODP Site 1143 when Northern Hemisphere glaciation (NHG) started. (A) Benthic $\delta^{18}O$ over the past 5 myr. (B) Planktonic $\delta^{18}O$ over the past 5 myr. (C) Mg/Ca-derived SST. (D) $\Delta\delta^{18}O_{SW-B}$, the difference between seawater $\delta^{18}O$ and benthic $\delta^{18}O$, indicating sea surface salinity changes. (E) total organic carbon mass accumulation rate, with marine isotope stages indicated. *From Tian, Pak, et al. (2006) and Tian, Yang, et al. (2006).*

Globigerinita, and *Globigerina* comprise the major part of shallow-water dwellers in the mixed layer, and *Neogloboquadrina dutertrei* represents the deep-dwelling assemblage, with its higher percentage implying higher productivity. Figure 6.12 plots the variations of *N. dutertrei*% and shallow-water dwellers% at Site 1146 in the northern SCS over the past 12 myr, showing its abrupt increases at 7.6 Ma and further increases from 3.2 to 2.0 Ma. Opposite to this trend are abrupt decrease in the shallow-water dweller % after 8 Ma and their further decrease from 3.2 to 2.0 Ma, although their abundance has also been affected by carbonate dissolution around 11 Ma (Wang, Jian, et al., 2003).

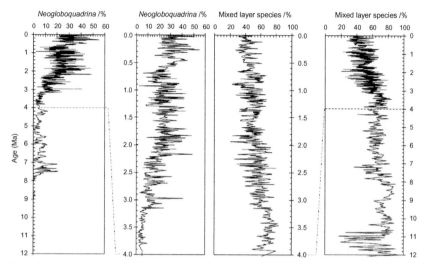

FIGURE 6.12   Downcore variations of planktonic foraminifers % at Site 1146 since 12 Ma. (A, B) *Neogloboquadrina dutertrei* %. (C, D) Percentage of the mixed-layer shallow-water species. *P. Wang et al. (2003).*

Together, these planktonic foraminiferal results indicate paleomonsoon enhancements at 8–7 and 3.2–2.0 Ma (Wang, Tian, et al., 2003).

Because monsoon-driven upwelling can lead to shoaled thermocline, the N–S thermocline gradient is also monsoon-indicative. The 28 °C isotherm, which constrains the northern boundary of the western Pacific warm pool (WPWP), separates the SCS into two parts from the northeast to the southwest. The thermocline in the modern SCS shoals from ∼175 m in the southern part belonging to the WPWP to ∼125 m in the monsoon-prevailed northern part. Variations in the relative abundances of the deep-dwelling planktonic foraminifera at ODP Site 1146 and at ODP Site 1143 reflect the evolution of the thermocline depth gradient across the northern and southern SCS since the late Miocene (Jian et al., 2006) (Figure 6.13). The N–S thermocline gradient appeared to enhance for the first time during 11.5–10.6 Ma, as indicated by an opposite change in the relative abundance of deep-dwelling planktonic foraminifera between Sites 1146 and 1143. Between 10.6 and 4.0 Ma was a period with weakened thermocline gradient between the two sites. More significant increases in the thermocline gradient occurred at about 4.0–3.2 Ma, as marked by a jump in the abundance of deep-dwelling species at Site 1146 but a major decrease at Site 1143 (Jian et al., 2006). The faunal evidence of thermocline deepening from ODP Site 1143 at about 4.0–3.2 Ma matches well with the record of thermocline shoaling in the east equatorial Pacific around ∼3.5 Ma (Wara, Ravelo, & Delaney, 2005). This finding implies enhanced asymmetry of the equatorial Pacific and development of the WPWP on one hand and intensification of the monsoon circulation on the other.

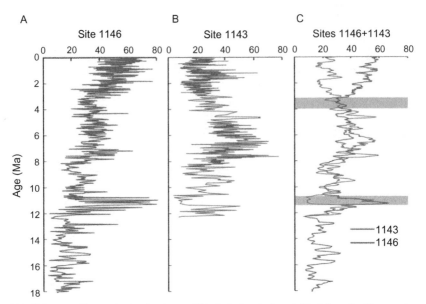

FIGURE 6.13 Changes in the abundances (%) of planktonic foraminiferal deep-dwelling species at ODP Sites 1146 (A) and 1143 (B) show increased south–north thermocline gradient (C, five-point moving average) at ~11 Ma and since ~3 Ma (horizontal bars) in the SCS. *Jian et al. (2006).*

6.2.3.2 Monsoon Climate and Weathering Rate

The SCS sediments recorded environmental changes not only in the marine basin but also in river catchments from the surrounding region. Three groups of proxies were used for analyzing the monsoon variations over the Neogene and Quaternary at ODP Site 1148: the black carbon and its isotope (Jia, Peng, Zhao, & Jian, 2003); the element ratios sensitive to chemical weathering such as Al/K, Ca/Ti, Rb/Sr, and CIA (chemical index of alteration) (Wei, Li, Liu, Shao, & Liang, 2006); and the mineralogical ratio C_{RAT} (chlorite/chlorite +hematite+goethite) (Clift et al., 2008). Marine-deposited black carbon originates from regional terrestrial biomass burning, and its isotope $\delta^{13}C_{BC}$ provides valuable information about humidity and seasonality, which are related to monsoon intensity. Because $\delta^{13}C$ of C4 plants is heavier than that of C3 plant and C3 plants increase their $\delta^{13}C$ values when moisture decreases, the increasing $\delta^{13}C$ in black carbon ($\delta^{13}C_{BC}$) may suggest decreasing humidity or reducing monsoon precipitation. As geochemical proxies of weathering rate, the higher CIA and Al/K values are indicative of increased humidity.

At ODP Site 1148 in the northern SCS, all the curves of $\delta^{13}C_{BC}$, CIA, and Al/K ratio are generally parallel to the benthic $\delta^{18}O$, and the lightest $\delta^{13}C_{BC}$, the highest CIA, and Al/K ratio occur at the earliest Miocene (~23 Ma), suggesting a very humid climate (Figure 6.14). Since then, all the three curves show

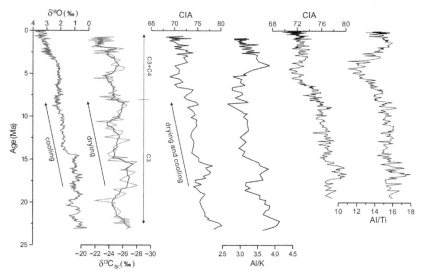

FIGURE 6.14   Weathering rates over the last 23 myr in northern South China Sea records. (A–D) ODP Site 1148. (A) benthic $\delta^{18}O$; (B) black carbon $\delta^{13}C$ (Jia et al., 2003); (C) CIA; (D) Al/K (Wei et al., 2006). (E–F) ODP Site 1146. (E) CIA; (F) Al/K (Wan, Clift, Li, Li, & Yin, 2010).

a similar trend toward drier conditions in the Pleistocene, with several fluctuations centering at approximately 15.7, 8.4, and 2.5 Ma, corresponding to global cooling since the middle Miocene (Figure 6.14A–D) (Wei et al., 2006).

A similar trend toward cooling and drying climates from the Miocene to the present is seen at ODP Site 1146 (Figure 6.14E and F; Wan, Kürschner, Clift, Li, & Li, 2009). Of particular interest is the middle Miocene climatic optimum of 17–15 Ma when the SCS experienced a warm and humid climate, and the terrigenous flux at ODP Sites 1146 and 1148 was about three to four times higher than those before and after this period in the Miocene (Wan et al., 2009). The middle Miocene warm and humid conditions, however, were not restricted to this region but global in scale presumably related to higher CO_2 concentration (Foster, Lear, & Rae, 2012). The next climate fluctuations around 8.4 and 2.5 Ma are well correlated to micropaleontological evidence (Figure 6.12), indicating paleomonsoon enhancements at 8–7 and 3.2–2.0 Ma (Wang, Jian, et al., 2003). A recent work on pollen records of 3.15–0.67 Ma from ODP Site 1145, northern SCS, found doubled pollen influx around 2.8 Ma and during 2.0–1.8 Ma, suggesting enhancement of winter monsoon in response to ice-sheet growth in the Northern Hemisphere (Luo & Sun, 2013).

Along with the secular changes over millions of years, the chemical weathering records also exhibit cyclic changes on 10^5-year timescale. For example, the K/Al ratios at Sites 1148 and 1143 show 400 kyr cycles corresponding to the $\delta^{13}C$ record (Figure 6.15). This is a signal of global monsoon variations in response to long-eccentricity rhythms in the orbital forcing (Wang et al., 2014), as discussed in more detail in the succeeding text.

6.2.3.3 Carbon Cycling

One of the merits of the ODP Leg 184 drilling in the SCS is the finding of the long-term changes in oceanic carbon reservoir at the 10^5-year timescale. The $\delta^{13}C$ curve is divided by a series of $\delta^{13}C$ maxima, denoting heavy values over 400–500 kyr cycles (Wang, Jian, et al., 2003; Wang, Tian, Cheng, Liu, & Xu, 2004). These $\delta^{13}C$ maximum events can be correlated globally, marking long-term changes in the carbon reservoir of the global ocean. The finding in the SCS of the long-term changes in oceanic carbon reservoir and their connection with environmental variations has stimulated further discussions and researches (Köhler & Bintanja, 2008; Ma, Tian, Li, & Wang, 2011; Russon, Paillard, & Elliot, 2010). The long-eccentricity cycles is inherent to the Earth system and has been functioning in the oceanic carbon system through the global monsoon since the geologic deep time but modified in the Quaternary, probably in response to the development of polar ice sheets (Wang et al., 2004; Wang, Tian, & Lourens, 2010).

In the SCS, the 400 kyr cyclicity in $\delta^{13}C$ is characteristic of the entire Neogene archive, with the $\delta^{13}C$ records of the early to middle Miocene representing the best example. Figure 6.16 shows a 4 myr record of the benthic foraminiferal $\delta^{13}C$ in the middle Miocene (16.5–12.5 Ma) from ODP Site 1146 with long-term cycles all corresponding to the long-eccentricity periodicity, but the long-eccentricity forcing of $\delta^{13}C$ stopped after 13.9 Ma (Figure 6.16;

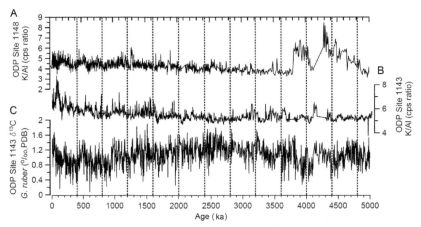

FIGURE 6.15 400 kyr long-eccentricity cycles in 5 myr K/Al records of the South China Sea. (A) K/Al from ODP 1148, northern SCS; (B) K/Al from ODP 1143, southern SCS; (C) plankton $\delta^{13}C$ from ODP 1143. Note that the cycles in $\delta^{13}C$ sequence become longer after 1.6 Ma. *Wang et al. (2014), based on Tian et al. (2011).*

Holbourn et al., 2005, 2007), presumably in response to the amplification of the Antarctic ice sheet.

The long-eccentricity cycles are not restricted to the carbon isotope records. Actually, the long-eccentricity cycle in the SCS was first presented in the K/Al ratio record as a proxy of chemical weathering in the late Pliocene at ODP Site 1145 (19°35′N, 117°38′E, w.d. 3175 m) (Wehausen & Brumsack, 2002), reconfirmed by the new data from ODP Sites 1143 and 1148 showing the K/Al ratio responses to the long eccentricity throughout the last 5 myr (Figure 6.16) (Tian et al., 2011). This is a convincing evidence for the monsoon origin of the 400 kyr cycles in the carbon reservoir.

A global compilation shows that all $\delta^{13}C$ time series in the various oceans display clear 400 kyr cycles in the Pliocene part and a total of 13 long-term $\delta^{13}C$ maximum events corresponding to long-eccentricity minima could be recognized in the last 5 Ma. But the rhythmic occurrence of $\delta^{13}C$ max at long-eccentricity minimum stopped at 1.6 Ma. The later $\delta^{13}Cmax$ events, however, are out of phase with the eccentricity signal, with the interval between $\delta^{13}C$ max extending to 500 kyr in the last million years (Figure 6.17) (Wang et al., 2014, 2010).

As demonstrated by our recent review (Wang et al., 2014), the long-term changes in $\delta^{13}C$ are related to monsoon-driven biogeochemical processes. The 400 kyr cycles in the oceanic $\delta^{13}C$ sequence before the Quaternary can be hypothetically explained by the changes in the ratio between particulate and dissolved organic carbon (POC/DOC) in the ocean depending on the monsoon-controlled nutrient supply. However, the ocean restructuring at 1.6 Ma marked by the isolation of a sluggish abyss under the Southern Ocean has obscured the long-eccentricity 400 kyr signal in the oceanic $\delta^{13}C$.

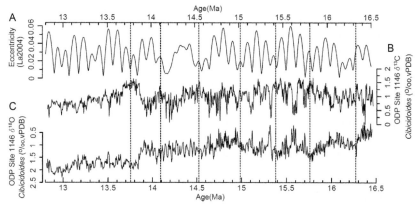

FIGURE 6.16   400 kyr long-eccentricity cycles in the middle Miocene records from ODP Site 1146, northern SCS. (A) Eccentricity cycles; (B) benthic $\delta^{13}C$, dotted vertical lines denote $\delta^{13}C$ maxima occurring at the long-eccentricity minima; (C) benthic $\delta^{18}O$. *Based on Holbourn, Kuhnt, Schulz, and Erlenkeuser (2005), Holbourn, Kuhnt, Schulz, Flores, and Andersen (2007), and P. Wang et al. (2014).*

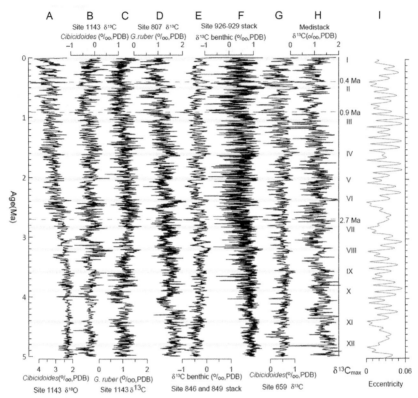

FIGURE 6.17 Carbon isotopic records from the South China Sea and global ocean over the past 5 myr. (A–C) ODP Site 1143, South China Sea: (A) benthic $\delta^{18}O$; (B) benthic $\delta^{13}C$; (C) $\delta^{13}C$. (D–E) Pacific: (D) planktonic $\delta^{13}C$ of ODP Site 807; (E) stacked benthic $\delta^{13}C$ of ODP Sites 846 and 849. (F–G) Atlantic: (F) stacked benthic $\delta^{13}C$ of ODP Leg 154 (Sites 926–929); (G) benthic $\delta^{13}C$ of ODP Site 659. (H) Mediterranean: stacked planktonic $\delta^{13}C$. Yellow bars indicate $\delta^{13}C$ maximum; red dotted lines denote major events of ice-sheet development at 2.7, 0.9, and 0.4 Ma. *Wang et al. (2014).*

6.2.3.4 Pleistocene Paleoceanography

The long-term trend in the late Cenozoic paleoceanography is associated with the growth polar ice sheets. This has continued in the Pleistocene and manifested as three major events in environmental changes: the 1.6 Ma event, the mid-Pleistocene transition, and the mid-Brunhes Events (MBEs).

1.6 Ma Event

The 1.6 Ma event is clearly represented in the SCS. At ODP Site 1143 from the southern SCS, for example, both the benthic and planktonic $\delta^{18}O$ show a turning point at MIS 52 (1.55 Ma) that divides between the increasing stage of glacial $\delta^{18}O$ since $\sim$1.9 Ma and its depleting stage toward $\sim$1.3 Ma

(Figure 6.9A and B). Related to this turning point is a heavy peak of the benthic $\delta^{13}C$ at 1.65–1.55 Ma, labeled as δ^{13}Cmax-IV (Figure 6.9D) (Wang et al., 2010).

At 1.6 Ma, biological productivity and many climate attributes also experienced critical changes. As indicators of biological productivity, for example, the contents of Mn and biogenic Ba at ODP Site 1146, northern SCS, started to fluctuate strongly after MIS 52, ~1.55 Ma, in responding to glacial cycles on the background of a general increasing trend (Figure 6.9M and N; Clemens et al., 2008). In the southern SCS, the K/Al and Ti/Al ratios of sediments at ODP Site 1143 also increased drastically from MIS 58 at about 1.65 Ma, suggesting enhancement of monsoon precipitation and chemical weathering till MIS 52 (Tian et al., 2011). The long-term trend of organic carbon (TOC) at the same site displays a high-value stage from 2.5 to 1.6 Ma, followed by a remarkable decline after 1.6 Ma (Figure 6.9I; Li et al, MS), in agreement with a period of low productivity implied by the growing proportion of deep-dwelling nannoplankton *Florisphaera profunda* (%) (Figure 6.18C; Liu et al., 2002).

The 1.6 Ma event not only marks a turn in the long-term cycling of the oceanic carbon reservoir, as rhythmic occurrence of δ^{13}Cmax at long-eccentricity minimum (400 kyr) stopped at 1.6 Ma (Figure 6.17) (Wang et al., 2014, 2010), but also was a time of low carbonate percentage and high foraminiferal fragmentation (Figure 6.18E–F; Xu et al., 2005), as well as a shift to enhanced fluctuations in SST and carbonate preservation (Figure 6.18D–F).

Mid-Pleistocene Transition

The mid-Pleistocene transition (MPT) has attracted increasing attention from the paleoclimate community. It was the turning point when the dominant climate periodicity extended from 41 to 100 kyr and lasted from 1.25 to 0.7 Ma (centered at MIS 23, ~0.9 Ma) according to Elderfield et al. (2012). The transition is well recorded in the SCS. At ODP Site 1143, the glacial $U^{K'}_{37}$-based SST remarkably decreased by 2 °C from ~1.5 to ~1.1 Ma. After a relatively stable stage around δ^{13}Cmax-III, the SST warmed up toward 0.5 Ma (Figure 6.9J; Li et al., 2011). The relatively warm stable stage lasted from MIS 30 to MIS 23, or shortly after the superinterglacial MIS 31, when the western Antarctic ice sheet collapsed (Scherer et al., 2008), until the eve of the major expansion of the Arctic ice sheet at MIS 23/22 (Figure 6.8D; Li et al., 2011).

The transitional stage from MIS 31 to MIS 23 is of particular interest also in the plankton aspect. The proportion of the deep-dwelling nannofossil *F. profunda* (%) experienced rapid changes over the period, with a 30% decrease from MIS 29 to MIS 26 and then a >20% increase from MIS 26 to MIS 22 (Figure 6.18C; Liu, Wang, et al., 2008). Another conspicuous

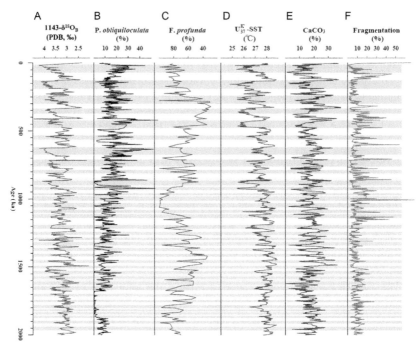

FIGURE 6.18 Paleoceanographic records from ODP Site 1143, southern South China Sea, over the last 2 myr. (A) Benthic $\delta^{18}O$, (B) planktonic foraminifera *Pulleniatina obliquiloculata* % (Xu, Wang, Huang, Li, & Jian, 2005), (C) nannoplankton *Florisphaera profunda* % (Liu et al., 2002; Liu, Wang, Tian, & Cheng), (D) U^K_{37}-derived SST (Li et al., 2011), (E) $CaCO_3$%, (F) fragmentation % (Xu et al., 2005).

change in plankton community is the abundance variations of *Pulleniatina obliquiloculata*. Before MIS 22, this warm-water thermocline foraminifer is more abundant during interglacials basically following the oxygen isotope curve, but it reversed its cyclicity after ∼850 kyr BP by having high abundances during glacials (Figure 6.18B; Xu et al., 2005). In addition, siliceous microfossils diatom and radiolarian were very rare before the MPT but increased significantly after ∼0.9 Ma, though mostly at interglacials (Li et al., 2008). The major biotic changes were not restricted to plankton, but also occurred in the benthic community. Investigations of benthic foraminifers in sediments of ODP Sites 1143 and 1146 have found remarkable extinction of elongated cylindrical forms within the MPT from 1.2 to 0.6 Ma, centered at 0.9 Ma (Figure 6.8) (Kawagata et al., 2007; Li et al., 2008), apparently related to enhanced ventilation of deep water in the SCS.

Probably, the SCS was more sensitive to the MPT than the open ocean. For example, the planktonic $\delta^{18}O$ shift from MIS 23 to MIS 22 showed more significant values in the SCS (+1.6‰ at ODP Site 1146 and +1.0‰ at ODP Site 1143; Figure 6.9B) than in the Pacific (∼+0.5‰ at ODP Site 806)

(Li et al., 2008). The MPT precursor δ^{13}Cmax-IV (MIS 25–27) is also well represented in the SCS (Figure 6.9D and E) and by the carbonate preservation record at Site 1143 (Figure 6.18E and F; Wang et al., 2004).

Mid-Brunhes Event

Another pivotal time interval in the Quaternary history is the MBE centered at the MIS 12/11 turn (~0.43 Ma), marked by enhanced warmth during the interglacials and intensified carbonate dissolution in the deep ocean (Droxler & Farrell, 2000; Jansen, Kuijpers, & Troelstra, 1986). In the southern SCS, the MIS 12/11 shift in benthic δ^{18}O at ODP Site 1143 exceeds 2‰ (Figure 6.19A), which was preceded by a ~1‰ negative excursion in benthic δ^{13}C over MIS 12 (Figure 6.19B), immediately after the oceanic carbon maximum δ^{13}Cmax-II in MIS 13 (Figure 6.9 D and E; Wang, Jian, et al., 2003; Wang et al., 2004). A remarkable feature in the $U_{37}^{K'}$ paleotemperature record at Site 1143 is the warming trend of glacial SST from MIS 21 to MIS 15, following the great cooling at MIS 23/22 turn (Figure 6.18D; Li et al., 2011). The last stage of the warming trend, at MIS 15–13, displays abnormally reduced glacial–interglacial contrast, which has been hypothetically attributed to the collapse of the West Antarctic Ice Sheet (Hillenbrand, Kuhn, & Frederichs, 2009). After the MBE, the glacial–interglacial SST amplitude increased more than double of that before MIS 12 (Figure 6.18D;

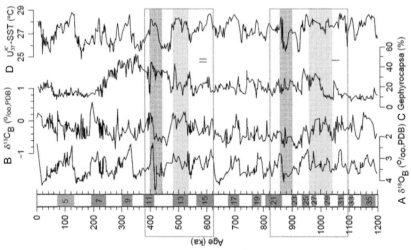

FIGURE 6.19   Late Pleistocene major events recorded at ODP Site 1143, southern South China Sea. (A) Benthic δ^{18}O. (B) Benthic δ^{13}C. (C) Small coccoliths of *Gephyrocapsa*. (D) $U_{37}^{K'}$-based SST. Red boxes represent two groups of events: (I) δ^{13}Cmax-III/MPT and (II) δ^{13}Cmax-II/MBE. Yellow bars denote δ^{13}Cmax-II events; blue bars denote MPT and MBE, respectively. *Wang et al. (2014).*

Li et al., 2011), as observed also in the transfer function-based SST record from the same site (Xu, 2004).

In terms of biological productivity, the MBE occurred near the center of a high-productivity stage lasting from MIS 12 to MIS 8. Nannofossil analysis of ODP Site 1143 has revealed low percentages of the deep-dwelling species *F. profunda* from MIS 12 to MIS 8 (Figure 6.18C; Liu, Wang, et al., 2008), marking a stage of high productivity of small shallow-dwelling coccoliths, as also supported geochemically by high accumulation rate of alkenone (Li, Wang, et al, 2009). All these findings are well in line with the recent suggestion that the global ocean experienced high productivity from ~0.6 to ~0.2 Ma, or from MIS 15 to MIS 8, resulting in enhanced carbonate dissolution (Barker et al., 2006; Rickaby et al., 2007). However, the timing of the MBE in the SCS records is slightly different than in the open ocean, as the high productivity of coccolithophores started later (since MIS 12) and the maximal carbonate dissolution occurred at MIS 13 and MIS 11 (Figure 6.18E and F; Wang, Jian, et al., 2003; Wang et al., 2004).

To sum up, the last million-year period has experienced two major changes in the climate regime: the mid-Pleistocene transition (MPT) centered at 0.9 Ma and the MBE around 0.4 Ma. The MPT and MBE were preluded by δ^{13}Cmax-III about 1.0 Ma and δ^{13}Cmax-II about 0.5 Ma, respectively. Together with other similar hydroclimatic phenomena over corresponding glacial cycles, the two groups of hydrologic and biogeochemical events appear to have been largely driven by oceanographic changes in the Southern Ocean (Figure 6.19; Wang et al., 2014).

6.3 LATE QUATERNARY PALEOCEANOGRAPHY AND SEA-LEVEL CHANGES

6.3.1 East China Sea

6.3.1.1 Introduction

Paleoenvironments were the starting point of extensive marine geology researches in the ECS, resulting in the publication of three synthesizing monographs in the 1980s: "Geology of the Bohai Sea" (Qin, 1985), "Geology of the East China Sea" (Qin, Zhao, Chen, & Zhao, 1988), and "Geology of the Yellow Sea" (Qin, 1990). With the major part of all the three devoted to the late Quaternary sedimentology and environmental changes, they symbolized the first stage of marine geology research in China. In addition to offshore investigations, substantial contributions to the reconstruction of late Quaternary coastline migrations were made by extensive drilling along China Coast. During the LGM, the ECS shelf with a total area of 850,000 km^2 was exposed subaerially. In the last deglaciation, within some 8000 years, the coastline migrated about 1200 km landward, from the western border of the Okinawa Trough to the western coast of the modern Bohai Gulf, corresponding to coastline retreat of >0.4 m per day in average (Wang, 1999). The drastic changes in sea area must have had remarkable influence on the environment not only for the ECS but also for the entire East Asia region.

The hydrologic system in a marginal sea is highly sensitive to changes in both the continent and the ocean, because it is the place where river runoff and oceanic currents come to mix. As one of the largest rivers (Yangtze) meets one of the strongest boundary currents (Kuroshio) in the ECS, large-scale sea–land interactions have exerted profound impacts on the Quaternary environments of the East Asia–west Pacific. In the following sections, we first introduce the Kuroshio variations in the Okinawa Trough and then discuss the environmental changes induced by sea-level fluctuations over late Quaternary glacial–interglacial cycles.

6.3.1.2   Late Quaternary Paleoceanography of the Okinawa Trough

Sea Surface Temperature Changes

A variety of proxies have been applied to estimate SST fluctuations in the Okinawa Trough during glacial cycles, including paleoecological, organic, and inorganic geochemical methods. Transfer functions based on census data of plankton foraminifers were first used to reconstruct the winter and summer SST (Jian, Wang, Saito, et al., 2000; Li, Jian, & Wang, 1997; Li, Liu, et al., 2001; Liu et al., 2001; Ujiié, Ujiié, Taira, Nakamura, & Oguri, 2003; Xiang et al., 2003; Xu & Oda, 1999; among others). New techniques subsequently employed mainly include alkenone-based $U_{37}^{K'}$-based SST (Ijiri et al., 2005; Meng, Du, Liu, & Liu, 2002; Yu et al., 2009; Zhao, Huang, & Wei, 2005; Zhou et al., 2007), TEX_{86}-based SST (Wu, Tan, Zhou, Yang, & Xu, 2012; Yamamoto, Kishizaki, Oba, & Kawahata, 2013), and Mg/Ca-based SST estimations (e.g., Lin et al., 2006; Sun, Oppo, Xiang, Liu, & Gao, 2005). Despite the differences in details, all these methods reveal remarkable SST fluctuations in the glacial cycles. A comparison of transfer function-based SST and $U_{37}^{K'}$-based SST records over the last 40 kyr is displayed in Figure 6.20. The downcore variations in $U_{37}^{K'}$-based SST show a range of about 3–4 °C over the last glacial–interglacial cycle, with the minimum value of 22 °C at the LGM (Figure 6.20B and C; Meng et al., 2002; Yu et al., 2009), well correlated with the winter SST estimations based on microfaunal data (Figure 6.20D; Li, Liu, et al., 2001; Liu et al., 2001). These SST reconstructions are also supported by the Mg/Ca-based SST estimations (Sun et al., 2005).

All the high-resolution SST sequences from the Okinawa Trough also exhibit variabilities at the millennial scale (Figure 6.20). The essential synchrony between some millennial-scale climate events and YD and Heinrich events in the ice core from the northern high latitudes and between the

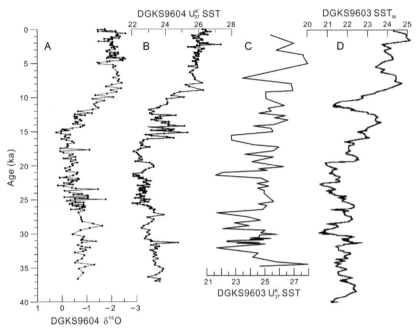

FIGURE 6.20 SST variations over the last 40 kyr in the middle Okinawa Trough. Core DGKS 9604 (28°17′N, 127°01′E): (A) planktonic $\delta^{18}O$; (B) $U_{37}^{K'}$-based SST (Yu et al., 2009). Core DGKS 9603 (28°09′N, 127°16′E): (C) $U_{37}^{K'}$-based SST (Meng et al., 2002); (D) planktonic foraminiferal TF-12E winter SST (Liu et al., 2001) (see Figure 6.22 for core site locations).

regional deglacial warming and the Bølling–Allerød warming suggests a strong climate teleconnection, most likely through the Asian monsoon atmospheric circulation. The Holocene millennial-scale SST oscillations have been observed in more detail. Xiang et al. (2007), for example, recognized 7 cold events (at about 1.7, 2.3–4.6, 6.2, 7.3, 8.2, 9.6, and 10.6 cal kyr BP, respectively) superimposed on a Holocene warming trend. At a higher resolution, TEX_{86}-based SST variations for the last 2700 years have been compared with the historical records (Wu et al., 2012).

Kuroshio Migration

Aside from signaling global climate variability, SST changes in the Okinawa Trough over late Quaternary glacial cycles were largely controlled by the Kuroshio warm current. As a major western boundary current of the Pacific, the Kuroshio carries abundant heat from equator to mid-latitudes and strongly influences climate in the northwestern Pacific region. Today, it enters the southern Okinawa Trough and flows along the outer edge of the ECS continental shelf, bringing heat and salinity to the ECS and beyond (see Section 2.4.2.1). In paleoceanography, planktonic foraminifer *Pulleniatina obliquiloculata* has extensively been used as a Kuroshio indicator, and its abundance variations have been used to indicate migrations and intensity changes of the Kuroshio in space and time (Jian, Saito, Wang, Li, & Chen, 1998; Li et al., 1997; Ujiié & Ujiié, 1999).

The tropical species *P. obliquiloculata* is very abundant (exceeds 10% of the planktonic foraminiferal assemblage) in surface sediments beneath the main axis of the Kuroshio Current in the modern ECS (Wang, Zhang, & Min, 1985). In the Okinawa Trough, relative abundance of *P. obliquiloculata* was extremely low during the glacial time but increased since ~15 kyr BP and more drastically between 7 and 5 kyr BP and after ~3 kyr BP (Figure 6.21; Li, Liu, et al., 2001; Wang, Satito, Oba, Jian, & Wang, 2001; Xiang et al., 2007). This can be attributed to the reentry of the Kuroshio into the Okinawa Trough, although the time of its reentry varies slightly between authors (Jian, Wang, Saito, et al., 2000; Liu, Berné, et al., 2007).

Of special interest is the "*Pulleniatina* Minimum Event" (Figure 6.21). This event about 4.5 and 3 kyr BP is characterized by very low abundance of *P. obliquiloculata* and recorded widely in the Okinawa Trough and the northern SCS, ascribable to weakening of the Kuroshio Current (Li et al., 1997; Jian, Li, Huang, et al., 2000; Jian, Wang, Chen, et al., 2000; Jian, Wang, Saito, et al., 2000; Li, Liu, et al., 2001; Ujiié et al., 2003; Xiang et al., 2007).

Substantial geologic evidence points to a shift of the Kuroshio main axis outside the Okinawa Trough during the glacial time (Figure 6.22). It remains unclear, however, what was the mechanism for the shift. At least two hypotheses have been proposed to explain the reduction if not the absence of the

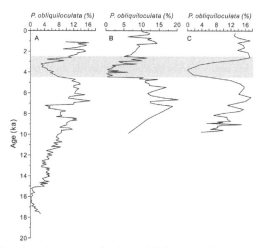

FIGURE 6.21 Variations of percentage abundance of *Pulleniatina obliquiloculata* in planktonic foraminiferal assemblage from the Okinawa Trough. (A) Core A7 (27°49′N, 126°59′E, water depth 1264 m; Xiang et al., 2007). (B) Core 255 (123°07′E, 25°12′N, water depth 1575 m; Jian, Li, Huang, et al., 2000; Jian, Wang, Chen, et al., 2000; Jian, Wang, Saito, et al., 2000). (C) Core MD01-2403 (123.2°E, 25.3°N, water depth 1420 m; Lin et al., 2006) (see Figure 6.22 for core site locations).

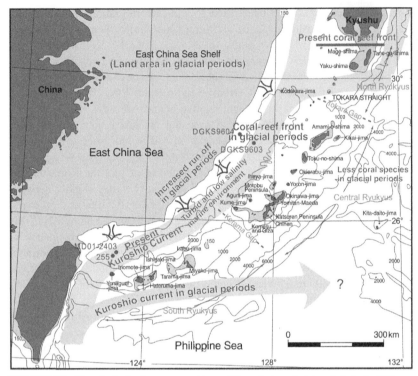

FIGURE 6.22 Hypothetical paleoceanographic conditions during glacial periods in the Okinawa Trough, with core site locations (red dots) for Figs. 6.20 and 6.21 (modified from Iryu et al., 2006). Extensive areas of the present East China Sea shelf emerged (light gray), which caused runoff-induced turbid and low-salinity waters (dark blue open arrows) at the modern shelf break. The Kuroshio Current, which currently flows to the west of the Ryukyu Islands (pink pattern), may have changed its streamline to the east (large light blue arrow). These environmental changes should have resulted in a southward migration of the "coral reef front" (the northern limit of coral reef formation).

Kuroshio in the Okinawa Trough: the "land bridge" hypothesis and the "sea-level" hypothesis. The first was put forward by Ujiié who argued that the "Ryukyu–Taiwan land bridge" blocked the entry gateway into the Okinawa Trough at the southern end of the Ryukyu arc and caused eastward deflection of the Kuroshio Current during the glacial time (Ujiié et al., 1991; Ujiié & Ujiié, 1999; Ujiié et al., 2003). This hypothesis required neotectonic deformations around the Pleistocene/Holocene boundary to open the southern end of the Okinawa Trough but has so far gained little support from geologic observations. Besides, it is even more difficult by using this hypothesis to explain the *Pulleniatina* Minimum Event in the late Holocene.

For the second hypothesis, marked reduction in the Kuroshio inflow into the Okinawa Trough was more likely caused by eustatic sea-level changes corresponding to glacial cycles. As shown by numerical simulation using a 3D ocean model with sea-level drop by 135 m at the LGM, topographic high at the southernmost Okinawa Trough partially blocks the Kuroshio through-flow, resulting in a 43% reduction of the Kuroshio water inflow (Kao, Wu, Hsin, & Dai, 2006). The modeling results also indicate that, at low sea level, the main axis of the Kuroshio flow shifted eastward to the open Pacific, and its main axis would not return back to the Okinawa Trough before the sea level rose to −40 m around the beginning of the Holocene (Figure 6.23; Kao et al., 2006).

Obviously, the idea of sea-level-drop-induced reduction in volume transport of the Kuroshio reconciles better with geologic observations, but further studies are needed to build up systematic understanding of the Kuroshio behavior in the glacial cycles. For example, the *Pulleniatina* Minimum Event was not related to sea-level changes and cannot be explained by any of the two hypotheses, because this event was not restricted to marginal seas, but widespread in the open western North Pacific (Lin et al., 2006). Moreover, contemporary reduction of boundary currents in the glacial time occurred not only in the western but also in the eastern Pacific. Thus, it is enigmatic how different from today was the glacial-time Pacific subtropical gyre as a whole (Wang, 2004). This exemplifies one fundamental problem in paleo-ceanography: the limitations of proxies. All the proxies used to indicate the Kuroshio, including microfossils *P. obliquiloculata* and *F. profunda* (Ahagon, Tanaka, & Ujiié, 1993; Ujiié et al., 1991), are based on modern observations, but there is no guarantee on their suitability for the geologic past when boundary conditions were very different.

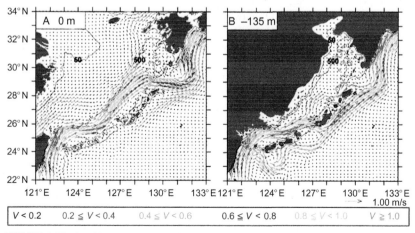

FIGURE 6.23 Modeled annual mean flow patterns of the Kuroshio in the upper 100 m at (A) modern sea level and (B) sea level at −135 m. Arrows point to the flow direction and colors represent velocity categories. *From Kao et al. (2006).*

River Runoff

There are two major hydrologic controls of the Okinawa Trough: the Kuroshio from the equatorial Pacific and the river runoff from Asia and the island arc. Today, deposition of hemipelagic sediments is focused in the southern Okinawa Trough, where a small area below 1000 m isobaths equivalent to less than 10% of the total trough area receives more sediment than the rest of the Okinawa Trough (Huh, Su, Wang, Lee, & Lin, 2006). It was very different in the glacial time, however, when large rivers from East Asia were directly emptying into the trough. An example is the Goto Submarine Canyon in the northern Okinawa Trough, the largest submarine canyon in the area. The canyon head is located 20 km south of the Goto Islands at a depth of approximately 300 m. The canyon extends almost straightly in a SE orientation for ~50 km, reaching the northern Okinawa Trough at a depth of ~700 m (Figure 6.24). The 10 km wide canyon was formed during the glacial sea-level lowstand and incised by hyperpycnal current associated with large river runoff from the Asian continent, functioning as the main conduit from the ECS to the Okinawa Trough (Oiwane et al., 2011).

The role of continental river runoff in the Okinawa Trough declined with the postglacial transgression, as observed in deepwater sediment cores from the trough. For example, the ratio of total organic carbon (TOC) to total nitrogen concentration (C/N ratio) gradually reduced from the LGM toward today, suggesting the decrease of vascular plant fragment input to the trough from continental rivers (Ujiié et al., 2001). However, there are reports showing an opposite trend in C/N. Thus, a progressive shift in TOC $\delta^{13}C$ toward more negative values with rising sea level in the last deglaciation can be interpreted as an increasing fractional contribution of terrestrial organics to the buried organic pool (Kao, Dai, Wei, Blair, & Lyons, 2008). As a basin with active sea–land interactions, the Okinawa Trough is of particular significance for comparison between terrestrial and oceanic processes and for their integration. A recent study discovered asynchronous climate change in the deglacial warming between land vegetation and SST in the Okinawa Trough (Xu et al., 2013). All these controversial findings reflect our insufficient knowledge and underscore the necessity of further investigations.

6.3.1.3 Late Quaternary Environment Changes in the Shelf

The ECS is featured by a broad continental shelf. Together with the Yellow Sea and Bohai Gulf, the ECS shelf covers ~970,000 km^2 in area, with its bottom morphology shaped by the coastal migrations during the glacial cycles. Mud patches and sand ridges are two characteristic features of the ECS shelf (Figure 6.23), and the rich information archived in the muddy and sandy sediments has become the core of research for sea-level and environmental changes over many years, especially in China.

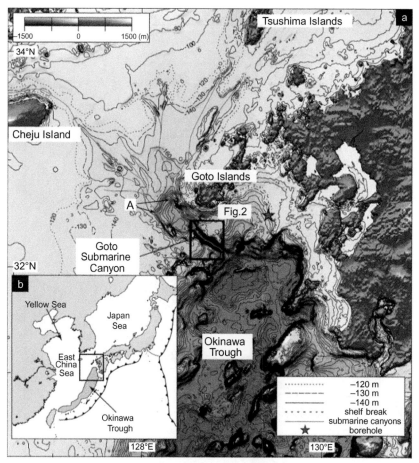

FIGURE 6.24   Bathymetric map on the northernmost Okinawa Trough showing the Goto Submarine Canyon. *Oiwane et al. (2011)*

Coastline Migration and Sea-Level Changes

Coastline migration in the ECS shelf during the last glacial cycle has been systematically investigated since the early 1980s (e.g., Wang, Min, Bian, & Cheng, 1981; Wang & Wang, 1980). Resulting from the extensive hydrogeologic survey on the coastal plains and the beginning of offshore exploration in China since the late 1970s, nearly a hundred of papers were published by Chinese scientists in 5 years from 1979 to 1983. Major findings include that the coastline of the ECS migrated from the "LGM littoral zone" (Figure 6.25) along the shelf break of the western Okinawa Trough to the North China Plain during the Holocene maximal transgression (Figure 6.26; Wang, Zhang, & Min, 1985). As mentioned in the preceding text, the last

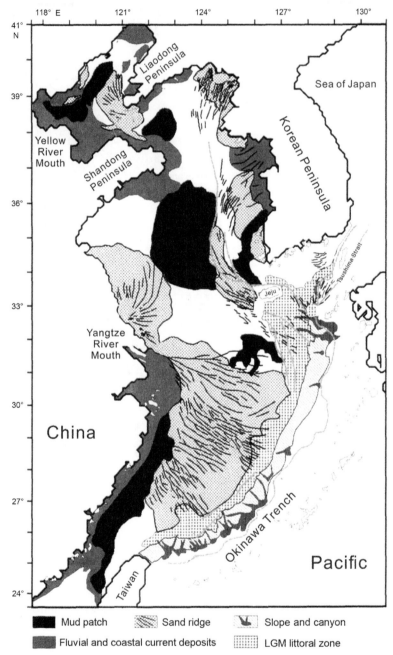

FIGURE 6.25 A sketch map of bottom sediment patterns in the East China Sea. *(Li, Liu, and Yang (2010).)*

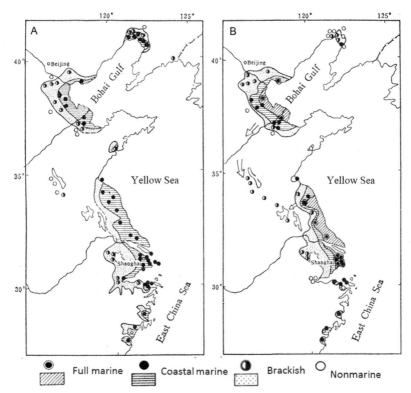

FIGURE 6.26   Distribution of marine deposits in East China during the late Quaternary transgression. (A) Holocene transgression; (B) late Pleistocene transgression. *Wang, Zhang, and Min (1985).*

deglacial coastline migration indicates a coastline retreat of >0.4 m per day in average, which must have had its profound environmental impact.

Numerous shallow seismic profiles and sediment cores have been analyzed for reconstructing coastline migration in response to the late Quaternary sea-level changes (Berné et al., 2002; Liu, Berne, Saito, Lericolais, & Marsset, 2000; Saito et al., 1998). Along a NE–SW across-shelf transect from the Yangtze Estuary to the Okinawa Trough, three sedimentary facies were recognized: regressive, estuarine, and transgressive facies. Transgressive and regressive surfaces of marine erosion are displayed as major discontinuities, traceable over the entire outer continental shelf to distances of hundreds of kilometers (Berné et al., 2002). Interpretations of shallow seismic profiles also revealed that, since MIS 8, the ECS has experienced the development of three stages of tidal sand ridges and four stages of deltas (Liu, Trentesaux, et al., 2003).

Consequently, the surface sediments of the modern ECS are diverse in depositional age and sedimentation rate. Based on measurements of excess ^{210}Pb activities, sediment accumulation rates decrease from 1.8 g/cm^2/year in the inner shelf to 0.049 g/cm^2/year in the Okinawa Trough (Oguri et al., 2003). The seafloor of the modern ECS shelf is covered by sediment systems formed mainly during three intervals: the glacial sea-level lowstand, deglacial transgression, and the Holocene highstand (Li, Sun, Liu, Bickert, & Ma, 2009).

Sand Ridges

Tidal sand ridges, as discussed in Section 4.2.2, form a major topographic and sedimentologic feature of the ECS and the southern Yellow Sea (see Figures 2.6 and 6.25). Sand ridge fields consist of large, elongate sand bodies and channels formed by tidal dynamics in shallow waters with abundant sandy sediment. The ridges are linear-shaped deposits, aligned parallel to the direction of the tidal currents, and reach widths of several kilometers, lengths of 10–200 km, and up to 20 m in thickness (Wu et al., 2010; Yang, 1989). Eight major sand ridge fields are distributed in the broad continental shelf from the Bohai Gulf to the ECS shelf break, as shown in Figure 4.43 and Table 4.7. As the sedimentologic aspect of the sand ridges have been discussed in Chapter 4, here, we focus on the sea-level implications of the sand ridge fields mainly on the ECS shelf and the SW Yellow Sea.

In China, offshore sand ridges began to attract scientific attention in the early 1980s (e.g., Li & Li, 1981; Liu & Xia, 1983; Zhu, Zheng, & Feng, 1984). Later on, multibeam sounding, shallow seismic profiling, and sediment coring were used to explore their distribution, structure, and dynamics of formation. Research activities then were focused on radiant sand ridges in the southern Yellow Sea and the linear sand ridges in the ECS. For example, the linear sand ridge field on the outer shelf of the ECS has been investigated by several Chinese and French-Chinese cruises (Berné et al., 2002; Liu et al., 2000; Wu et al., 2010). The sand ridges are found arranging parallel to each other, mostly trending NW–SE, and distributing from the 60 m isobath in the middle shelf to the 120 m (NE part) or 150 m isobath in the outer shelf (Figure 6.27A and B; Wu et al., 2010, 2005). Over the years, thousands of kilometers of high-resolution seismic data correlated to sediment cores have revealed the late Quaternary depositional history and the relationship between sand ridges and sea-level fluctuations. The results show that sand ridge sequences comprise three elementary sedimentary facies: regressive marine prodeltaic prograding wedges, estuarine and deltaic facies, and transgressive sand ridges (Figure 6.27C; Berné et al., 2002).

The sand ridge layer with high content of shell debris was formed in the last deglacial transgression. Four stages of ridge formation are recognized: 14.5, 12–14, 1.5–9.5, and after 9 kyr BP (Wu et al., 2010). The river-borne sand forming the main ridge body mainly accumulated during the last transgression and reworked afterward by tidal currents. Today, the sand ridges on the ECS are still being influenced by the modern hydrodynamics and gradually migrate toward the southwest. Therefore, these sand ridges on the ECS

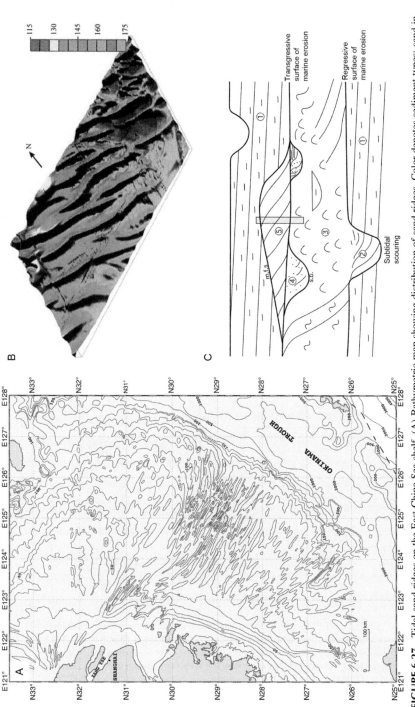

FIGURE 6.27 Tidal sand ridges on the East China Sea shelf. (A) Bathymetric map showing distribution of sand ridges. Color denotes sediment types: sand in yellow, silt and clay in green. (B) Linear sand ridges in the outer continental shelf of the East China Sea. Colors show water depth in meter (Wu, Jin, Li, Zheng, & Wang, 2005). C. Scenarios of depositional sequences on the outer continental shelf in the East China Sea. ① Prodeltaic/offshore fine-grained deposits; ② deltaic distributaries or estuarine channels; ③ deltaic estuarine facies; ④ early transgressive estuarine deposits; ⑤ offshore sand ridges (late transgressive marine deposits) (Berné et al., 2002).

shelf should be referred to as "quasiactive sand ridges" rather than as moribund or relict sand ridges (Liu, Colin, et al., 2007). Nevertheless, there are buried sand ridges in the middle shelf of the ECS, which were formed during 320–200 kyr BP, mostly within the deglaciation following MIS 8 (Berné et al., 2002; Wu et al., 2009).

To correlate sand ridge formation with eustatic sea-level changes, a schematic composite stratigraphic column for the middle to outer shelf of the ECS is displayed in Figure 6.28A. As seen from the tentative correlation of the sand ridge occurrences with sea-level changes over the last 400 kyr, the formation of sand ridges (SR in Figure 6.28B) occurred during sea transgressions at deglaciation (Figure 6.27B; Berné et al., 2002). Restructuring by successive glacial cycles were responsible for the overlapping stratigraphic sequences in the outer shelf where all complicated seismic units are present even at a single location (Figure 6.28A).

Located off the north Jiangsu coast in the southern Yellow Sea is the largest ridge complex along the Chinese coast ridge complex, covering an area of >22,000 km². Unlike the linear sand ridge field in the ECS, the South Yellow Sea sand ridge field radiates from the coast to water depth up to 25 m, with some tidal channels reaching 50 m in depth (Figure 6.29; Li, Zhang, Fan, &

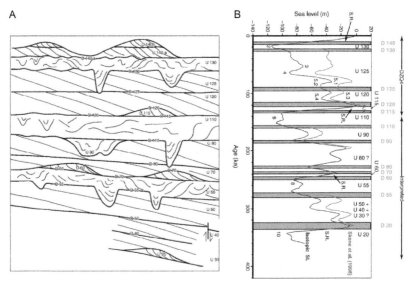

FIGURE 6.28 Tentative correlation between the sand ridge occurrences and sea-level changes in the outer continental shelf of the East China Sea over the past 400 kyr (Berné et al., 2002). (A) Schematic composite stratigraphic column for the middle and outer continental shelf. (B) Correlation between sea-level changes and the depositional sequences on the outer continental shelf in the East China Sea. U: Major seismic unit; D: discontinuities in sediment sequences; S.R.: sand ridge. DZ04: Sediment core (29°25′N, 125°22′E, w.d. 89 m). Gray bars denote discontinuities; sea-level curves.

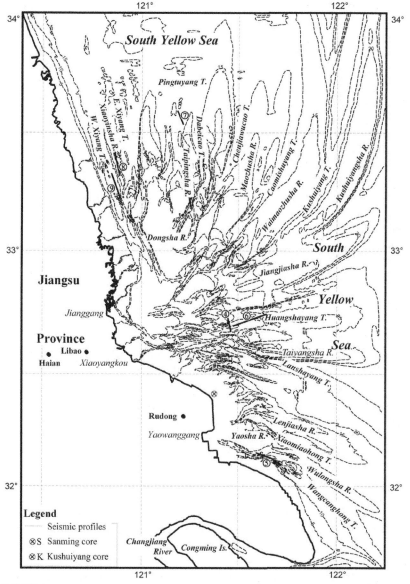

FIGURE 6.29   The radiating sand ridge field in the South Yellow Sea. T: Trough; R: ridge. *Wang, Zhang, et al. (2012).*

Deng, 2001; Wang, Zhang, Zou, Zhu, & Piper, 2012; Wang, Zhu, et al., 1999). Although some hydrologic and geomorphological survey was done in the early 1980s, its systematic investigations with seismic profiling and drilling did not commence until the 1990s. As shown from the geologic records, the coastal area of East China was inundated by transgression in late MIS 3 with fairly warm and humid climate (Shi, Yu, Liu, Li, & Yao, 2001; Xia et al., 2013). This was followed by a river-dominated stage leading to the accumulation of large amounts of fluvial sediments in the area. The sediments were sorted by tidal currents during the Holocene transgression, obviously since ∼9 kyr BP, giving rise to the present sand ridge field (Wang, Zhang, et al., 2012; Xia et al., 2013). According to Li, Zhang, et al. (2001), the radiate sand ridge system of the southern Yellow Sea consists of both onshore and offshore parts, with the onshore part being a buried system formulated in the north Jiangsu coastal plain 5–6 kyr ago and the offshore system developed in the late Holocene, only 1–2 kyr ago (Figure 6.29).

Aside from its paleoenvironmental interest, the extensive sand ridge field off the Jiangsu coast offers huge potential for land reclamation, which has stimulated active researches in the recent years on sediment dynamics of the sand bodies and associated mud flats. Many arrays of evidence indicate the primary transport mechanism as forced by tidal currents and monsoon winds (Gong, Wang, Stive, Zhang, & Chu, 2012; Wang, Gao, et al., 2012; Wang, Zhang, et al., 2012; Zhang et al., 2013).

Mud Areas

In eastern and southern Asia, about one-third to one-half of the river-derived sediments are trapped in the river's low reaches and thus contribute to extensive floodplain and delta plain development, while the remaining sediments are delivered to the sea (Liu et al., 2009). This is the material source of mud patches in the inner shelves of the ECS. While the sand ridge fields discussed in the preceding text on the shelves are formed by river-borne sediments deposited during the Pleistocene glacial, the mud areas are mainly resulted from the Holocene output from large rivers particularly the Yellow and Yangtze Rivers.

The total Yangtze-derived sediments accumulated in its deltaic system and ECS inner shelf have amounted to about 1.7×10^{12} t. A majority of the mass has been deposited in the Yangtze Delta (area 1 in Figure 6.30A), while ∼5.4×10^{11} t of sediment has formed an elongated (∼800 km) distal subaqueous mud wedge extending from the Yangtze River mouth southward alongshore off the Zhejiang and Fujian coasts into the Taiwan Strait (area 2 in Figure 6.30A; Liu et al., 2006, Liu, Xu, et al., 2007), as also described in Chapter 4. Recent investigations show that inner continental shelf sediments in the ECS represent a mixing of not only clays derived from the Yangtze River but also silts/sands from small western Taiwanese rivers. Yangtze clays are widely dispersed southward about 1000 km to the western Taiwan Strait, but silts and sands from Taiwan are mainly deposited in the Taiwan Strait (Xu et al., 2009).

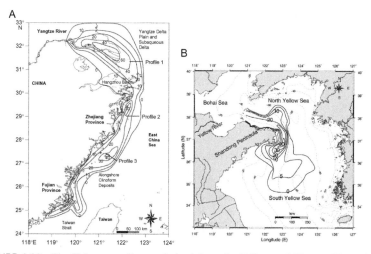

FIGURE 6.30   The Holocene sediments derived from the Yellow and Yangtze Rivers in the inner shelves of the East China Sea and Yellow Sea. (A) Isopach map of the Yangtze-derived sediment deposited over the last 7000 years (Liu et al., 2006). (B) Isopach map of the Holocene mud in the Yellow Sea (Yang & Liu, 2007).

A similar mud wedge wraps around the end of the Shandong Peninsula in the Yellow Sea. The subaqueous clinoform has a maximum thickness of 40 m and generally thins seaward to less than 1 m (Figure 6.30B; Yang & Liu, 2007). It was deposited in a subtidal nearshore environment in the last deglaciation, when the Yellow River discharged into the Bohai Sea near the Bohai Strait (Liu, Saito, et al., 2007). Together with other mud areas (Figure 6.25), these fine-grained sediments yield valuable records of sea-level and oceanographic changes in the region.

Over the recent years, a number of sediment cores and drill holes taken from the mud areas were analyzed for understanding postglacial environmental changes in the continental shelf of the ECS. Ranging from the Bohai Gulf in the north to the offshore of Zhejiang in the south (Figure 6.31; Table 6.2), these cores provide valuable information on the history of the sea–land interactions of the region. In general, the research focus has been laid on the southern Yellow Sea where muddy deposits are better developed.

An outstanding event marking the first establishment of modern oceanographic pattern in the Yellow Sea occurred at ~6.5 kyr BP. Microfossil sequences in sediment cores indicate that the postglacial deposits began with coastal facies characterized by the brackish-water benthic foraminifer *Ammonia beccarii* var., which, however, was quickly replaced by the *Ammonia ketienziensis* assemblage around ~6.5 kyr BP. In modern Yellow Sea bottom sediments, the *A. ketienziensis* assemblage is distributed in the area prevailed by the Yellow Sea Warm Current, a branch of the Kuroshio (Figure 6.32C; Wang, Min, & Bian, 1985). Therefore, this 6.5 kyr BP event is attributed to the inflow of the Yellow River Warm Current and to the establishment of the modern marine conditions in the Yellow Sea (Kong et al., 2006; Liu, Saito, et al., 2007). Contemporary faunal and lithologic changes are clearly recorded in the central part of the Yellow Sea, such as in cores C02 (Figure 6.32A; Fang et al., 2013), CC02 (Figure 6.32B; Kim & Kennett, 1998; Kim & Kucera, 2000), and YE-2 (Xiang et al., 2008). Today, the Yellow Sea Warm Current has an extension entering the Bohai Gulf, and sediment cores from

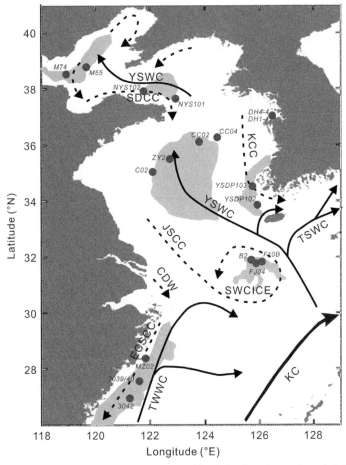

FIGURE 6.31 Location map of sediment cores and boreholes from the continental shelf of the East China Sea studied for paleoenvironments (see Table 6.2). Superimposed are winter circulations: YSWC, Yellow Sea Warm Current; TSWC, Tsushima Current; TWWC, Taiwan Warm Current; KC, Kuroshio Current; SDCC, Shandong Coastal Current; JSCC, Jiangsu Coastal Current; KCC, Korea Coastal Current; ECSCC, East China Sea Coastal Current; CDW, Changjiang Diluted Water; SWCICE, Southwestern Cheju Island Cold Eddy. Light gray color marks mud deposit areas. *Circulation after Wang et al. (2011).*

TABLE 6.2 Selected Paleoenvironmentally Studied Sediment Cores and Boreholes from the Continental Shelf of the East China Sea (see Figure 6.31)

Area	Core or drill Hole	Longitude (E)	Latitude (N)	Water Depth (m)	Length (m)	References[a]
East China Sea						
South	MZ 02	28°10′	121°53′	32	35.30	Li, Jian, Shi, and Liu (2012)
	MD06-3039	27°43′	121°47′	47	8.11	Wang, Zheng, Prins, and Zheng (2008)
	MD06-3040	27°43′	121°47′	46	19.22	Zheng, Zheng, and Wang (2010)
	MD06-3042	27°05′	121°24′	62	34.10	Zheng et al. (2010)
	YQ-1	~29°12′	~125°30′	~90	25.75	Yang (1989)
	DZQ4	29°25′	125°22′	89	51.65	Liu et al. (2000)
North	F10B	31°44′	126°07′	79	1.67	Xing et al. (2013)
	FJ04	31°41′	125°49′	67	1.41	Li, Sun, et al. (2009), Li, Wang, et al. (2009)
	B2	31°45′	125°45′	64	4.03	Xiang et al. (2006)
Yellow Sea						
S South	YSDP 102	33°49′	125°45′	62	60.65	Li, Li, Cang, Liu, and Jeong (2000)
	YSDP 103	34°29′	125°29′	55	34.0	Kong, Park, Han, Chang, and Mackensen (2006)
	YE-2	35°30′	123°20′	75	6.1	Xiang et al. (2008)
	CC02	36°08′	123°49′	77	2.78	Kim and Kennett (1998)

Continued

TABLE 6.2 Selected Paleoenvironmentally Studied Sediment Cores and Boreholes from the Continental Shelf of the East China Sea (see Figure 6.31)—Cont'd

Area	Core or drill Hole	Longitude (E)	Latitude (N)	Water Depth (m)	Length (m)	References
	CC04	36°18'	124°30'	85	2.25	Kim and Kucera (2000)
	DH1-4	37°03'	126°29'	2	4.70	Kim and Kucera (2000)
	DH4-1	37°01'	126°28'	2	3.50	Kim and Kennett (1998)
	ZY2	35°31'	122°39'	69	3.42	Wang et al. (2011)
	C02	35°00'	122°00'	53	3.00	Fang et al. (2013)
North	NYS-101	37°42'	122°48'	49	70.2	Liu, Berné, et al. (2007), Liu, Chen, et al. (2007), Liu, Colin, et al. (2007), Liu, Saito, et al. (2007), Liu, Xu, et al. (2007)
	NYS-102	37°56'	121°44'	34	70.1	Liu, Berné, et al. (2007), Liu, Chen, et al. (2007), Liu, Colin, et al. (2007), Liu, Saito, et al. (2007), Liu, Xu, et al. (2007)
Bohai Gulf						
	M5-5	38°41'	119°46'	28	5.50	Liu, Li, Chen, Xiao, and Wan (2008), Liu, Wang, et al. (2008)
	M7-4	38°38'	119°08'	22	4.30	Liu, Li, et al. (2008), Liu, Wang, et al. (2008)

aOnly the early reference is given here.

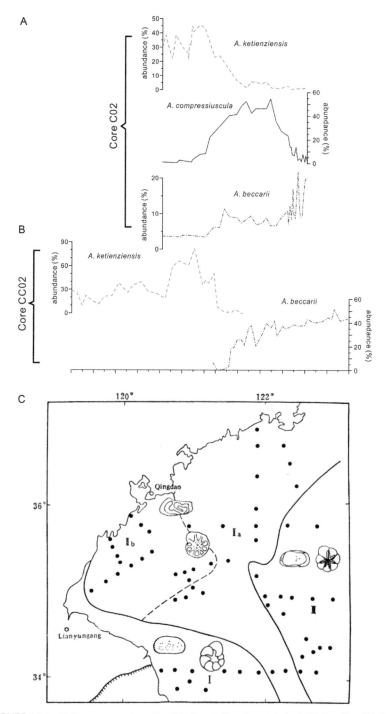

FIGURE 6.32 Holocene succession of benthic foraminifera *Ammonia* in the Yellow Sea. (A) Core C02 (35°N, 122°E) (Fang et al., 2013). (B) Core CC02 (36°08′, 123°49′) (Kim & Kennett, 1998). (C) Distribution of foraminiferal and ostracod assemblages in the surface sediments of the modern southern Yellow Sea: I. *Ammonia beccarii–Sinocytheridea*; II. *Ammonia compressiuscula–Munseyella*; III. *Ammonia ketienziensis–Krithe* assemblages (Wang, Min, & Bian, 1985).

the gulf also provide evidence showing that its inflow into the Bohai began at 6.4 kyr BP, well correlated to the Yellow Sea records (Liu, Li, et al., 2008).

Because of the virtual absence of planktonic foraminifers, paleo-SST attempts for the Yellow Sea have been scarce. A recent work on the $U_{37}^{K'}$-based SST for the late Holocene in the central Yellow Sea showed a significant SST decline at 4–2.3 kyr BP (see fig. 5a in Wang et al., 2011). This cooling event corresponds not only to the "*Pulleniatina* minimum" in the ECS (Li et al., 1997; see Section 6.3.1.2) but also to a depositional hiatus or erosional surface in some areas of the Yellow Sea such as in Core C02 (Fang et al., 2013).

A comprehensive paleotemperature reconstruction has been made in the northeastern ECS, to the south of Cheju Island. Here, at Core F10B, two organic geochemical methods were applied for Holocene paleotemperature reconstruction: the $U_{37}^{K'}$ method for surface and TEX_{86} method for subsurface temperature. As expected, the TEX_{86} temperature is lower than the $U_{37}^{K'}$-based SST, with a general trend correlated well between the two sequences. The results show a higher temperature stage from ~9 to ~6 kyr BP indicative of the global Holocene climate optimum and a cooling stage from 4.5 to ~2 kyr BP corresponding to the "*Pulleniatina* Minimum Event" (Figure 6.33; Xing et al., 2013). The cooling stage has been confirmed by a high-resolution record recovered from the nearby Core FJ04 (Figure 6.33; Li, Sun, et al., 2009).

The mud wedge off the Zhejiang coast has been cored and studied in the recent years (Figure 6.30; Table 6.2). Similar to the Yellow Sea, the early Holocene brackish-water microfauna also gave way to a more saline, normal marine assemblage, but starting relatively earlier, at ~8 kyr BP (Li, Li, et al., 2012). This observation is well in line with the reentry of the Kuroshio axis into the ECS around 8 kyr BP (e.g., Xiang et al., 2007; see Section 6.3.1.2), and the development of the Taiwan Warm Current around the same time (Zhao, Jian, et al., 2009). The depositional hiatus from 13 to 4 kyr BP in Core MD06-3042 might be related to the cooling event at ~4 kyear BP (Zheng et al., 2010), but more data are required for better understanding the history of the mud wedge.

In sum, the Quaternary of the ECS shelf was actively investigated in the 1980s, resulting in a number of boreholes drilled up to a hundred of meters and analyzed for reconstructing paleoenvironments in the region (e.g., Liu, Wu, & Wang, 1987; Qin, Zhao, Chen, & Zhao, 1989; Yang, 1985; Zheng, 1991). However, it was not until the 1990s that Chinese investigations started to match the international standards and make relevant research data publicly available. Recent progresses show that the late Quaternary history of the ECS shelf is dominated by sea–land interactions, mainly between the Kuroshio and the large rivers, and is well correlated with the global records over glacial cycles. New development is expected from the current extensive survey in the area.

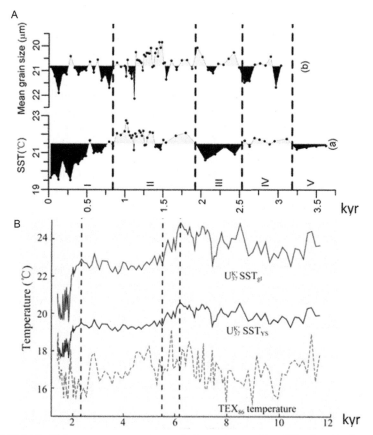

FIGURE 6.33   Holocene temperature changes in the northeastern East China Sea. (A) Core FJ04 (31°41′N, 125°49′E), $U^{K'}_{37}$-based SST over the last 3600 years (Li, Sun, et al., 2009); (B) Core F10B (31°44′N, 126°07′); $U^{K'}_{37}$-based SST and TEX$_{86}$-based temperature of the Holocene. $U^{K'}_{37}$-based SST$_{gl}$ denotes values calculated using the global core-top equation; $U^{K'}_{37}$-based SST$_{YS}$ denotes those using the Yellow Sea core-top equation. *Xing et al. (2013).*

6.3.2   South China Sea

6.3.2.1   Introduction

Late Pleistocene paleoceanography heavily relies on carbonate microfossils for isotopic geochemical and micropaleontological analyses. The hemipelagic sediments of the SCS are favorable for high-resolution paleoceanographic studies because of high sedimentation rates and good carbonate preservation. Unlike the Atlantic, the western Pacific bottom is bathed with more corrosive waters that deteriorate carbonate preservation even at relative shallow depths of 2000–3000 m. With a large area above CCD (~3500 m²), however, the SCS is unique in the region to yield well-preserved sediment sequences suitable for paleoenvironmental reconstructions. It may not be merely a coincidence that two cores from the southern SCS were among the first several cores in the world ocean used by AMS ^{14}C dating for high-resolution stratigraphy (Andree et al., 1986; Broecker et al., 1988).

Late Pleistocene deepwater sedimentologic and paleoceanographic studies in the SCS started in the 1980s. Aside from piston coring and shallow profiling by the US R/V Vema and Conrad (e.g., Damuth, 1979, 1980), extensive investigations then were mainly performed by Chinese expeditions. Numerous volumes and papers describing sediment patterns, coral reefs, and late Quaternary history, among other topics, were published in Chinese. Noticeable contributions are serial reports from the Nansha expeditions of the Chinese Academy of Sciences to the coral reef area in the southern SCS (Nansha Scientific Expedition 1989a, 1989b, 1991, 1992, 1993a, 1993b, 1994); from expeditions of the Ministry of Geology to the Xisha, that is, Paracel Islands area (He & Zhang, 1986; Zhang et al., 1989); from the South China Sea Institute of Oceanology (1982, 1985, 1987; Su et al., 1989); and from the State Oceanic Administration (1988; Zheng and Chen, 1993). At the same time, the isotope-based paleoceanographic studies were first published by Chinese scientists from both sides of the Taiwan Strait (Wang, Chen, & Lo, 1986; Wang, Min, Bian, & Feng, 1986), although substantial international publications from SCS studies did not begin until the early 1990s (e.g., Miao, Thunell, & Andersen, 1994; Schönfeld & Kudrass, 1993; Thunell, Miao, Calvert, Calvert, & Pedersen, 1992; Wang, 1990; Wang & Wang, 1990).

Since the 1990s, the international scientific interest in the SCS has drastically increased. Numerous international expeditions were sent to the region to study topics ranging from climate and sea-level changes in Quaternary glacial cycles, to monsoon evolution and variations, to volcanic ash distribution (Table 1.2). Of particular significance is the first paleoceanographic expedition, Sonne-95 cruise under the logo "Monitor Monsoon," which collected 48 piston and gravity cores at 46 sites from the SCS (Sarnthein et al., 1994) and revealed the regional late Quaternary paleoceanographic history for the first time (Sarnthein & Wang, 1999). Since then, the SCS has become a hot spot in paleomonsoon and paleoceanographic researches, culminated with ODP Leg 184 drilling in 1999 targeting on the East Asian monsoon history (Wang et al., 2000). Up to now, at least 16 international cruises have taken place in the SCS with more than 200 sites drilled or cored for scientific purposes (Table 1.2; Figure 6.34).

In terms of sea-level changes, the best study area is the Sunda Shelf. Surrounded by the Malay Peninsula, Sumatra, Borneo, Java, and other islands, the Sunda Shelf is the largest tropical continental shelf emerged during the LGM. Today, water depths over the shelf rarely exceed 50 m, and the area less than 20 m is quite extensive, resulting in an extreme sensitivity to sea-level changes during glacial cycles. With its high sedimentation rates caused by high precipitation in the region and by steep gradient in the surrounding

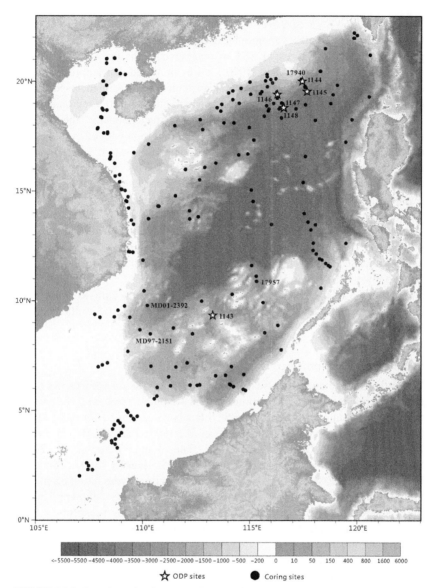

FIGURE 6.34   Locations of coring and drilling sites for paleoceanographic studies up to 2012 on topographic map of the South China Sea. Black dots—coring sites; white stars—ocean drilling sites (Wang et al., 2014). For information of individual sites, see also Table A1-2 in Wang and Li (2009).

hinterland, the Sunda Shelf has archived valuable sediment records for paleoreconstructions. In the following sections, we will discuss SCS deep-sea paleoceanography first and then sea-level change based on the Sunda Shelf.

6.3.2.2 Paleoceanography in Glacial Cycles

One of the earliest paleoceanographic findings in the SCS was its enhanced glacial–interglacial contrast of SST as compared to the open Pacific at the same latitudes (Wang & Wang, 1990). It was proposed that amplification of environmental signals in the SCS can be attributed to the semienclosed nature of the marginal sea and the prevailing monsoon climate of the region (Wang, 1999; Wang, Wang, Bian, & Jian, 1995). Characteristically, the SCS is located in a region prevailed by the East Asian monsoon with alternating predominance of northeastern winds in winter and southwestern winds in summer. The rhomboid shape of the SCS basin with a NE–SW topographic axis and surrounded by coastal ranges is favorable to the development of seasonal reversal in the surface circulation pattern, with a basically cyclonic gyre in winter and an anticyclonic gyre in summer (Figure 2.31). The East Asian monsoon is characterized by a very intensive winter component that substantially drives the biological productivity and the N–S surface temperature gradient in the SCS. Since the winter monsoon often enhanced in glacials, the N–S contrast in SST was also increased there (Wang & Li, 2009). Therefore, a combination of the semiclosed basin and the monsoon climate is accountable for the amplification of environmental signals in the SCS in its response to glacial cycles.

Sea Surface Temperature

While the hemipelagic deep-sea sediments of the SCS provide ideal conditions for high-resolution paleoclimate reconstructions, the quality of various SST proxies commonly used in the SCS has been examined for many times (e.g., Chen et al., 2005; Huang, Liew, et al., 1997; Jian, Li, Huang, et al., 2000; Jian, Wang, Chen, et al., 2000; Jian, Wang, Saito, et al., 2000; Steinke, Kienast, Pflaumann, Weinelt, & Stattegger, 2001; Steinke, Kienast, et al., 2008). Early paleo-SST reconstructions were based on planktonic foraminiferal transfer function, but in the recent years, the application of geochemical proxies has generated a number of high-resolution SST sequences in the late Quaternary, comparable with the ice-core and speleothem records.

The first attempt to reconstruct paleo-SST in the SCS was in 1990, when transfer function FP-12 was applied to foraminiferal census data over the last glacial cycle in three cores from the northern SCS (Wang & Wang, 1990). It was found that the glacial–interglacial SST contrast was much more significant in the SCS than the adjacent open Pacific, indicating an amplifying effect of glacial signals in the marginal sea, which was confirmed by later studies (i.e., Chen et al., 2005). However, the transfer function FP-12E was a regional version of Imbrie–Kipp transfer function, based on census data of planktonic foraminifers in 165 core tops from the open western Pacific Ocean, with the standard errors of 2.48 °C for winter SST and 1.46 °C for summer SST (Thompson, 1981). The SST reconstruction using FP-12E has inevitable biases in the marginal seas where upper ocean hydrology is different from that of the open Pacific.

To avoid these drawbacks, Pflaumann and Jian (1999) developed a calibrated SIMMAX-28 transfer function to estimate past SST in the SCS. SIMMAX method used modern analog technique with a similarity index (Pflauamnn et al., 1996). Results from this regional SIMMAX method offer a slightly better understanding of the marginal sea conditions of the SCS than FP-12E, but this method is also biased toward the tropical temperature regime, like FP-12E, because of the very limited data from temperate and sub-polar regions. The application of SIMMAX-28 transfer function on sediment Core 17940 (20°07′N, 117°23′E, w.d. 1727 m), northern SCS, revealed nearly unchanged summer temperatures but 8.5 °C of glacial–interglacial difference for winter temperatures (Pflaumann & Jian, 1999). When the two different methods are applied to Core 17957-2 (10°54′N, 115°18′E, w.d. 2195 m), southern SCS, glacial–interglacial cycles over the past 1.5 myr are well presented in the FP-12E-derived SST records but are somewhat obscured in the SIMMAX-28-derived SST records (Jian, Li, Huang, et al., 2000; Jian, Wang, Chen, et al., 2000; Jian, Wang, Saito, et al., 2000).

Despite the uncertainty of the transfer function technique, its application has provided interesting information about the SCS's response to the glacial cycles, such as winter monsoon intensity. The south–north temperature gradient in the winter SCS exists throughout the entire Quaternary and can be used to indicate changes in the strength of the East Asian winter monsoon. Reconstructions of SST at ODP Sites 1146 and 1143 reveal increased SST gradient between the northern and southern SCS during late Pleistocene glacials due to strengthened winter monsoon (Figure 6.35; Jian et al., 2009). This is consistent with other records from comparing cores between the northern and southern SCS that displays increased S–N gradients during the last glacial stage and decreased values during the Holocene, indicating strengthened winter monsoon in glacials (Pflaumann & Jian, 1999).

Another significant finding of the transfer function-based SST reconstruction in the SCS is the amplifying effect of glacial signals in the marginal seas (Wang & Wang, 1990). Wang (1999) summarized all FP-12E-based paleo-SST data available then from the SCS and ECS, Sulu Sea, and the adjacent western Pacific during the LGM. The results show much cooler winter SST at the LGM in the western Pacific marginal seas than in the open ocean, whereas summer SST was similar between the marginal seas and ocean, thus causing much more intensive seasonality in the LGM marginal seas. In general, the winter SST at the LGM was at least 3–4 °C lower in the SCS and Sulu Sea than in the open Pacific, and the LGM seasonality in the seas is about 4 °C higher than in the open ocean. Enhanced SST amplitude in glacial cycles is supported by subsequent works in the SCS (e.g., Chen et al., 2005).

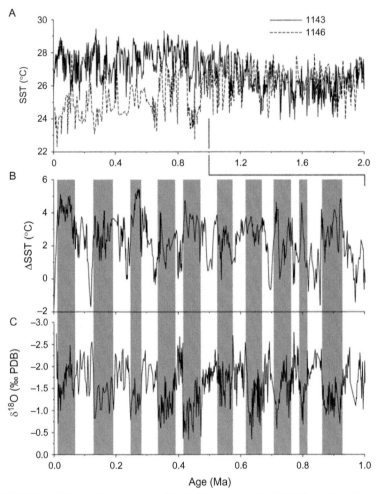

FIGURE 6.35　SST variations are revealed in (A) the northern Site 1143 (dashed line, Xu et al., 2005) and southern Site 1146 (solid line, Huang, Cheng, Jian, & Wang, 2003) derived from foraminiferal transfer function, (B) SST difference (ΔSST) between these two sites, and (C) *G. ruber* $\delta^{18}O$ at Site 1146. Vertical gray bars indicate glacial stages (Jian et al., 2009).

In the aspect of geochemical SST proxy, Pelejero and Grimalt (1997) established a calibrated linear relationship between $U^{K'}_{37}$ in surface sediment samples and SST for the SCS. The equation of $U^{K'}_{37}$ is believed to be valid for temperatures of the annually averaged water mass between 0 and 30 m in the SCS and was widely applied for the late Pleistocene. In two sediment cores from the southwestern and southern SCS, $U^{K'}_{37}$-derived SST increased abruptly by 1 °C or more at the Termination I (Kienast, Hanebuth,

Pelejero, & Steinke, 2003; Kienast, Steinke, Stattegger, & Calvert, 2001), concurrent with the Bølling warming about 14.6 kyr BP ago initially observed in the Greenland ice core. In IMAGES Core MD97-2151 (8°44′N, 109°52′E, w.d. 1589 m) from the southern SCS, Zhao, Huang, Wang, and Wei (2006) used U_{37}^K to reconstruct the SST history for the past 150 kyr on a millennial scale, suggesting glacial–interglacial SST changes by 4 and 5 °C for Termination I and Termination II, respectively.

Figure 6.36 shows a high-resolution $U_{37}^{K'}$-based SST record for the past 35 kyr from core 17940, northern SCS, by calibrating to annual mean SST in 0–30 m water depth, together with the FP-12E-based summer and winter SST estimates (Wang, Sarenthein, Grootes, & Erlenkeuser, 1999). While the summer SST remained almost constant over the entire last glacial cycle, the winter SST decreased with a number of short-term negative oscillations by ~2 °C during the MIS 3, paralleling some "warm" $\delta^{18}O$ minima. Other features include a good match between the MIS 2 SST minima and heavy $\delta^{18}O$ values toward the end of the LGM and prominent declines in winter SST in the late Holocene.

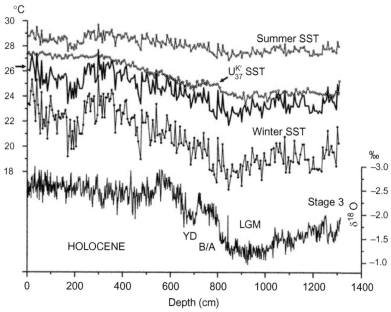

FIGURE 6.36   SST estimates for Core 17940, northern SCS, based on FP-12E transfer function (circles for summer, dots for winter, and coarse line for the annual mean) and $U_{37}^{K'}$ (diamonds for annual mean SST), as compared with *G. ruber* $\delta^{18}O$ curves. YD = Younger Dryas; B/A = Bølling–Allerød. Horizontal arrow marks modern annual mean SST value at 0–30 m. *Modified from Wang, Sarenthein, Grootes, and Erlenkeuser (1999).*

Beginning from the new century, Mg/Ca ratio of planktonic foraminiferal shells has also been frequently used for SST reconstruction in the SCS. Hastings, Kienast, Steinke, and Whitko (2001) developed an empirical equation based on core-top sample calibrations and then compared three independent paleotemperature estimates in Core 18287-3 (5°40′N, 110°40′E, w.d. 598 m) from the southern SCS: Mg/Ca ratio from planktonic foraminifera (*Globigerinoides ruber* and *Globigerinoides sacculifer*), alkenone thermometry ($U_{37}^{K'}$), and different foraminiferal transfer functions (SIMMAX28, RAM, and FP-12E). Results from these three different methods show similar average glacial–interglacial temperature differences of about 2.5 °C. The most important shift present in all the three records is an abrupt SST increase by about 1.3–1.8 °C at the end of the last glacial period, synchronous with the 14.6 kyr BP warming observed in the GISP2 ice core. The Mg/Ca-derived SST estimations were also performed for the last 260 kyr at ODP Site 1144 (20°03′N, 117°25′E, w.d. 2036 m) (Wei, Deng, Liu, & Li, 2007) and for the last 145 kyr for Site 1145 (Oppo & Sun, 2005).

However, the SST estimations based on Mg/Ca ratio of foraminiferal shells are found to be species-dependent. The planktonic foraminifer *G. ruber* (white) has two morphotypes, *G. ruber* sensu stricto (s.s.) living in the upper 30 m of the water column and *G. ruber* sensu lato (s.l.) living at depths also below 30 m (L. Wang, 2000). Steinke et al. (2005) showed that *G. ruber* s.s. often register significantly higher Mg/Ca ratios than *G. ruber* s.l., as the latter precipitated its shells in slightly colder waters, and this can explain some bias in the SST reconstructions for the past 22 kyr in the southern SCS (Steinke et al., 2006; Steinke, Kienast, et al., 2008; Steinke, Vogt, et al., 2008).

In the recent years, remarkable progress has been made on SST reconstructions on millennial timescales in the SCS. Representing these are the alkenone-derived SST record in IMAGES Core MD97-2151 from the southern SCS (Zhao et al., 2006) and the Mg/Ca-derived SST record at ODP Site 1145 from the northern SCS (Oppo & Sun, 2005). Figure 6.37 displays these two SST sequences over the last 140 kyr, both with a time resolution of some hundred years (150–200 years for MD97-2151 and 500 years for ODP 1145). Similar to over a dozen of sequences elsewhere in the SCS, the SST difference at the two sites between the LGM and Holocene was measured to about 4 °C, and millennial-scale SST oscillations of ~1.5 °C over a few hundred years occurred frequently during the entire glacial period, fairly well correlating with the Greenland ice-core and Chinese speleothem records (Figure 6.37A and B). Noteworthy is the phase relationship between the ice-sheet melting and SST warming during the deglaciation. An abrupt increase of at least 1 °C was found at 14.6 kyr BP in the southern SCS, synchronous with the Bølling warming in the ice-core record (Kienast et al., 2001), and the synchronism was confirmed by the newer works shown in

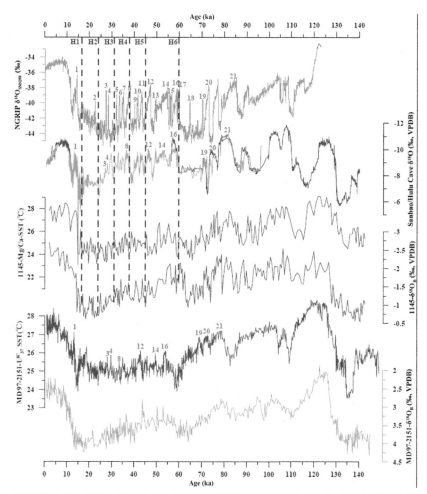

FIGURE 6.37 Millennial-scale variations of SST over the last 150 kyr in the SCS. (A) Greenland ice-core δ^{18}O (NGRIP members, 2004). (B) Speleothem δ^{18}O from Sanbo/Hulu caves, South China (Wang et al., 2008). (C) Mg/Ca-based SST from ODP 1145, northern SCS. (D) Planktonic δ^{18}O from ODP 1145 (Oppo & Sun, 2005). (E) $U_{37}^{K'}$-derived SST from MD97-2151, southern SCS. (F) Benthic δ^{18}O from MD97-2151 (Zhao et al., 2006). Numbers denote D/O events; H numbers denote Heinrich events. *From Wang et al. (2014).*

Figure 6.37. The correlations described in the preceding text at orbital and millennial timescales suggest a possible teleconnection between high-latitude atmospheric forcing and low-latitude climate oscillations in the SCS region probably via the Asian monsoon system (Oppo & Sun, 2005; Zhao et al., 2006).

Now, geochemically based SST sequences for the late Quaternary at many sites in various parts of the SCS have been generated. Table 6.3 is a compilation of SST ranges for the Holocene, last glacial, and MIS 5e from international and Chinese publications on the basis of high-resolution $U_{37}^{K'}$-derived and Mg/Ca-derived SST (Li, Li, et al., 2012).

Productivity

The SCS is an oligotrophic sea and its primary production is limited primarily by the availability of nutrients. High-production regions in the SCS are associated with monsoon-driven upwelling, so the SCS productivity features strong temporal and spatial variations (Gong, Liu, & Liu, 1992; Ning et al., 2004; Wong et al., 2007), with annual productivity ranging from 100 to 1000 mgC/m^2/d by estimates. As revealed by surface chlorophyll distribution, the basin-wide SCS primary production in winter is the highest, followed by a secondary peak in summer. Three upwelling regions, northwest Luzon, east of Vietnam, and north of the Sunda Shelf, can all be recognized by relative lower sea surface temperature and higher chlorophyll concentration in satellite and field measurements. According to modeled shipboard measurement and satellite (CZCS–SeaWiFS) data, east Vietnam upwelling is the most intense in summer and the weakest during spring and autumn, while upwelling in both northwest Luzon and north of the Sunda Shelf is strong in winter (Figure 6.38; Liu, Chen, et al., 2007). This geographic heterogeneity of productivity distribution explains the large spatial variability of paleoproductivity records in the SCS.

Many paleoproductivity proxies have been applied to the SCS reconstruction, resulting in some general consensus on glacial–interglacial productivity changes for the northern SCS. Earlier works used TOC content and mass accumulation rate (MAR) as productivity indicators, revealing higher productivity during glacials (Huang, Liew, et al., 1997; Huang, Wu, et al., 1997). Later studies confirm this pattern for the northern SCS, including using TOC and chlorins (Higginson, Maxwell, & Altabet, 2003), diatoms (Lan, Cheng, & Liu, 1995), benthic foraminifera (Kuhnt, Hess, & Jian, 1999; Jian et al., 1999, 2001), opal (Jia, Jian, Peng, Wang, & Fu, 2000; Wang & Li, 2003; Wang et al., 2007), alkenones (Huang, Liew, et al., 1997; Huang, Wu, et al., 1997; Kienast et al., 2001; Lin et al., 1999), and other proxies.

For the southern SCS, earlier studies gave different patterns on glacial–interglacial productivity changes. MAR of TOC in a transect of cores from the southeastern SCS revealed high TOC contents during the LGM, indicating higher productivity by about two times than today (Thunell et al., 1992). This conclusion was supported by more records based on TOC flux (Fang, Jian, & Wang, 2000; Jian et al., 1999), opal content (Jia et al., 2000), and benthic foraminiferal assemblage (Kuhnt et al., 1999), all showing higher productivity during the LGM for the southern SCS. However, a different pattern was

TABLE 6.3 A List of High-Resolution $U^{K'}_{37}$-Derived and Mg/Ca-Derived SST Records from the SCS (Li, Li, et al., 2012)

Site	Location N	E	w.d. m	Proxy	Average Resolution kyr	SST Range Holocene	Last Glacial	MIS 5e	References
Northern									
SCS90-36	18.00	111.49	2 050	$U^{K'}_{37}$	0.8	25.4	23.3		Huang, Wu, et al. (1997)
SO50-31 KL	18.76	115.87	3 360	$U^{K'}_{37}$	0.2–0.4	26.2–26.5	22.5		Huang, Liew, et al. (1997)
69	20.12	119.22	2 971	$U^{K'}_{37}$	–	23.3–26.9	22.2–23.8		Zhang, Zhou, Lu, et al. (2005)
MD97-2146	20.07	117.23	1 720	$U^{K'}_{37}$	0.3	27.5	25		Shintani, Yamamoto, and Chen (2008)
MD05-2904	19.27	116.15	2 066	$U^{K'}_{37}$	0.5	26	21.5–23.7	28.3	He, Zhao, Li, Wang, and Wang (2008)
ODP1145	19.35	117.38	3 175	Mg/Ca	0.5	28	24	29	Oppo and Sun (2005)
ODP1144	20.03	117.25	2 037	Mg/Ca	0.2–0.5	26.7	23.1	27.8	Wei et al. (2007)
17940	20.12	117.38	1 727	$U^{K'}_{37}$	0.22	27	23–24		Pelejero, Grimalt, Heilig, Kienast, and Wang (1999)
17954	14.8	111.53	1 520	$U^{K'}_{37}$	5	27.1	23.4	28.4	Pelejero, Grimalt, Heilig, et al. (1999)
Southern									
17964	6.16	112.21	1 556	$U^{K'}_{37}$	0.46	27.8	25.0		Pelejero, Grimalt, Heilig, et al. (1999)

Continued

TABLE 6.3 A List of High-Resolution $U_{37}^{K'}$-Derived and Mg/Ca-Derived SST Records from the SCS (Li, Li, et al., 2012)—Cont'd

Site	Location N	E	w.d. m	Proxy	Average Resolution kyr	SST Range Holocene	Last Glacial	MIS 5e	References
17961	8.3	112.33	1 968	$U_{37}^{K'}$	0.64	28	25.2	28.6–29	Pelejero, Grimalt, Sarnthein, et al. (1999)
				Mg/Ca	0.8	29.4	24.7–27.8	26.7–30.6	Jian, Wang, and Qiao (2008)
18252-3	9.14	109.23	1 273	$U_{37}^{K'}$	0.26	26.5–28	25.2		Kienast et al., 2001
18287-3	5.39	110.39	598	$U_{37}^{K'}$	0.19	27.2–28.3	25.9		Kienast et al. (2001)
MD97-2151	8.73	109.87	1 589	$U_{37}^{K'}$	0.15–0.2	27.9	24.4–26	28.9	Zhao et al. (2006)
MD97-2142	12.41	119.27	1 557	Mg/Ca	2	28–29	24.2–25.3	27.5–28	Wei et al. (2004)
MD01-2390	6.63	113.4	1 545	Mg/Ca	0.22	28	25		Steinke et al. (2006)
MD05-2896	8.49	111.26	1 657	Mg/Ca	0.5	28.5	25.2		Tian, Huang, and Pak (2010)
294	13.13	118.07	3 657	$U_{37}^{K'}$	–	24.5–27.4	23.8–25.3		Yao et al. (2007)
SW									
MD01-2392	9.51	110.12	1 966	$U_{37}^{K'}$	0.72	27.8	24.5	28.6	Xie, Jia, Peng, & Shao (2007)
MD05-2901	14.22	110.44	1 454	$U_{37}^{K'}$	1	25–28	23–25.5	28.4	Li, Wang, et al. (2009)

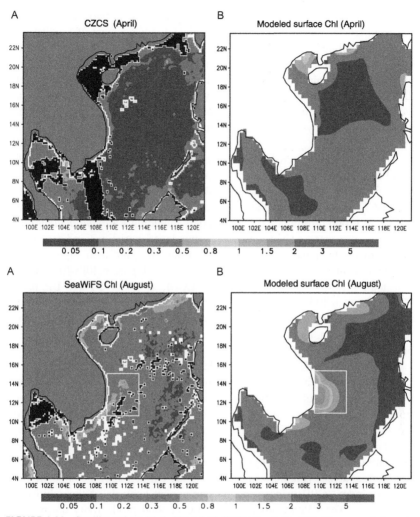

FIGURE 6.38   Satellite and modeled chlorophyll distribution in the SCS. Upper panel, August; lower panel, December. *From Liu, Cheng, Zhu, Tian, and Xia (2002) and Liu, Chen, et al. (2007).*

revealed in a core taken from the southwestern SCS, where TOC MAR during interglacials was higher than glacials, implying higher productivity in inter-glacials (Jian, Huang, Lin, & Kuhnt, 2001). The MARs of authigenic ele-ments, such as Sr, Ba, P, and Ca, also indicated higher productivity during interglacials for the southwestern SCS (Wei, Liu, Li, Chen, & Wei, 2003). The SW–SE difference in productivity might be accounted for by monsoon effect, as the southwestern SCS is dominated by upwelling associated with summer monsoon, which intensified during the interglacials.

The N–S difference in the SCS can be illustrated with two examples. The first is alkenone-based productivity. The glacial enhancement of productivity is clearly exhibited in the northern SCS (Figure 6.39A; Huang, Liew, et al., 1997), but glacial productivity records from the southern SCS show ambiguous cyclicity signals (Figure 6.39B; Zhao et al., 2006, Zhao, Wang, et al., 2009). The observed certain variability in productivity response to glacial cycles in the southern SCS may be related to its tropical position, as precessional forcing plays more significant role in the southern than the northern SCS.

Another example used benthic foraminifera as a productivity proxy. The use of benthic fauna to indicate productivity is based on the observations that infaunal detritus feeders increase their abundance with enhanced productivity. In the SCS, high organic carbon flux intervals in the late Quaternary are dominated by a group of infaunal detritus feeders including *Bulimina aculeata* and *Uvigerina peregrina*. Their downcore abundance variations indicate close links with productivity-induced organic carbon flux, but their high picks in the south occurred before the LGM (Core 17964; Figure 6.40D–F), whereas their high picks in the north after the LGM, at about 10 kyr BP (Core 17940; Figure 6.40A–C) (Jian et al., 1999). Other studies on benthic foraminifera also indicated that in the northern SCS, the productivity peak occurred in the deglaciation, although overall productivity was higher during the LGM than in the Holocene (Jian et al., 2001; Kuhnt et al., 1999).

The observed differences in the paleoproductivity records, particularly in the southern SCS, emphasize the hydrogeochemical complexity in this marginal sea, but some of the differences may be attributed to the use of different proxies, which may have biased by differential preservation or different production quantities. In this sense, paleoreconstruction based on multiproxy records surely has its advantages. For example, TOC, carbonate, opal, and alkenone over the last 870 kyr in Core MD97-2142 from the southeastern SCS were analyzed (Figure 6.41), allowing the extraction of a composite productivity index (CPI) using principal component analysis of these biogenic components. The CPI record indicates higher total productivity during glacials, caused probably by strengthened winter monsoon. The alkenone record generally parallels the CPI record, also indicating higher haptophyte productivity during glacials. This parallelism, to a first-order approximation, may suggest that calcareous planktons were the major producers for this site. However, opal content from this site suggests higher diatom productivity during interglacials and lower productivity during glacials, in agreement with the results from ODP Site 1143 (Wang & Li, 2003) that increase in diatom productivity during interglacials was probably linked with stronger summer monsoon and river inputs. It is not clear, however, why diatoms did not respond to the winter monsoon fertilization during glacials when total productivity was higher. The face values of these records suggest major phytoplankton community structure changes at the site of MD97-2142 (12°41′N, 119°28′E, w.d. 1557 m) during glacial–interglacial cycles. Interestingly, the relative diatom

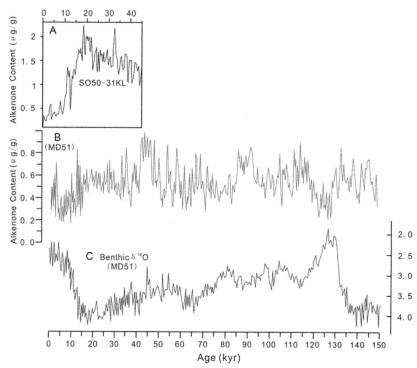

FIGURE 6.39 (A) Alkenone content in core SO50-31KL from the northern SCS (Huang, Liew, et al., 1997). (B) Alkenone content and (C) benthic $\delta^{18}O$ records for Core MD97-2151 from the southwestern SCS (Zhao et al., 2006).

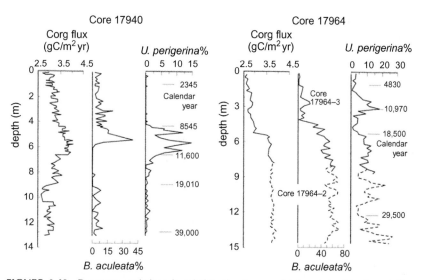

FIGURE 6.40 Downcore variations in relative abundances of infaunal benthic foraminifers in comparison with organic carbon flux. Core 17940 (20°07′N, 117°23′E), northern SCS: (A) organic carbon flux (g/m²/year); (B) *Bulimina aculeata* (%); (C) *Uvigerina peregrina* (%). Core 17964 (6°10′N, 112°13′E), southern SCS: (D) organic carbon flux (g/m²/year); (E) *B. aculeata* (%); (F) *U. peregrina* (%). *Modified from Jian et al. (2009).*

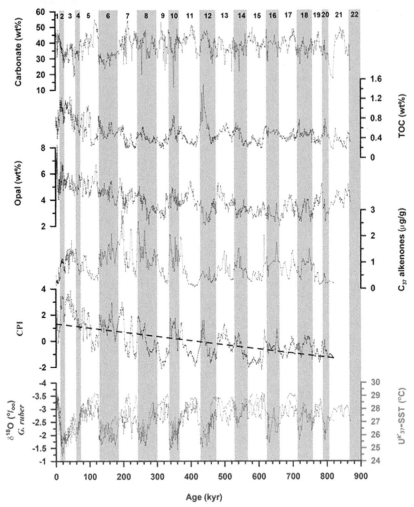

FIGURE 6.41 Productivity proxy records are compared with the planktonic $\delta^{18}O$ curve for Core MD97-2142 from the southeastern SCS. *Shiau et al. (2008).*

contribution to the total productivity was lower during glacials when the total productivity was higher, and its relative contribution was higher during interglacials when total productivity was lower (Zhao, Jian, et al., 2009; Zhao, Wang, et al., 2009).

The phytoplankton community issue was recently addressed in a study of biomarkers in Core MD05-2904 (19°27′N, 116°15′E, w.d. 2066 m), northern SCS. Various biomarkers as proxies of different phytoplankton groups were measured, including C_{37} alkenones for haptophytes, C_{30} diols for eustigmatophytes, dinosterol for dinoflagellates, and brassicasterol for diatoms (He et al.,

2013). The abundances and MARs of these biomarkers over the last 260 kyr indicate comparatively higher phytoplankton productivity during glacial periods (MIS 2, 4, and 6), with relatively weak glacial–interglacial variations in the phytoplankton community structure (Figure 6.42). Although no clear glacial–interglacial contrast has been detected in the relative abundance of diatoms over coccolithophores, the relative abundance of diatoms over coccolithophores certainly raised from MIS 5 to the Holocene (Figure 6.42) (He et al., 2013).

Upper Water Structure

As discussed in Chapter 2, monsoon-related seasonal changes are the most prominent oceanographic feature of the SCS. Not only the surface circulation in the SCS has a seasonally reversal characteristic driven by the East Asian monsoon, but also the upper water structure and productivity are controlled by the monsoon winds. While the long-term changes of upper water features are discussed in Section 6.2.3 and the late Quaternary variations of monsoon-driven latitudinal SST gradient and upwelling-related productivity are presented immediately above (Figures 6.35 and 6.41), thermocline changes driven by monsoon-driven upwelling are being examined here as evidence of variations in upper water structure. For estimations of thermocline depth variations in the glacial cycles, ecological and isotopic proxies have been used, such as the proportion of the very deep-dwelling *Globorotalia truncatulinoides* in the plankton foraminiferal assemblage (Jian, Li, Huang, et al., 2000; Jian, Wang, Chen, et al., 2000; Jian, Wang, Saito, et al., 2000) and the oxygen isotope gradient between the shallow- and deep-dwelling planktonic foraminifers (Tian, Wang, & Cheng, 2004). The $\delta^{18}O$ gradient between the thermocline dweller *P. obliquiloculata* and the mixed-layer species *G. ruber*, or $\delta^{18}O_{(P-G)}$, at ODP Site 1143 has been calculated. The results show reductions of $\delta^{18}O_{(P-G)}$ values by 0.5–1.0‰ during glacials or stadials compared to the adjacent interglacials or interstadials, suggesting that the thermocline depth was deeper during glacial or stadial periods over the last 1.56 myr, in contrast to increased $\delta^{18}O_{(P-G)}$ values during interglacial or interstadial periods, implying a shallower thermocline (Figure 6.43) (Tian et al., 2005).

Monsoon variations on the millennial timescale have been extensively investigated in the northern and southern SCS. Wang, Sarenthein, Erlenkeuser, et al. (1999) were the first to report glacial millennial-scale Dansgaard–Oeschger (D–O) events from the SCS, based on the grain size of terrigenous sediment as proxy of winter monsoon and planktonic $\delta^{18}O$ as proxy of summer monsoon. As shown in Figure 6.44, their results reveal intensified winter monsoon in the glacial period and reduced summer monsoon precipitation in the cold stadials in D–O cycles, well correlating with the Greenland ice-core records. Since then, millennial-scale variations of monsoon intensity have been reported in numerous publications

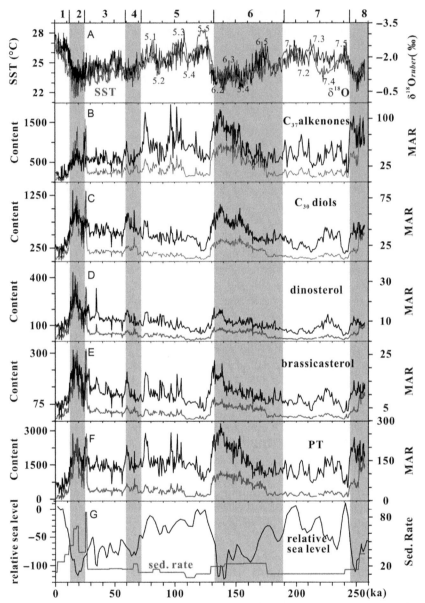

FIGURE 6.42 Biomarkers records over the last 260 kyr in Core MD05-2904 (19°27′N, 116°15′E), northern SCS. (A) Planktonic foraminifer *G. ruber* $\delta^{18}O$ (gray) and $U^{K'}_{37}$ SST (black). (B–F) Biomarker concentrations (ng/g, black) and MARs (10^{-3} mg/cm^2/kyr, gray) of (B) C$_{37}$ alkenones (representing haptophytes); (C) C$_{30}$ diols (eustigmatophytes); (D) dinosterol (dinoflagellates); (E) brassicasterol (diatoms); (F) sum of all other biomarkers (phytoplankton total, PT); and (G) normalized brassicasterol/C$_{37}$ alkenones + brassicasterol ratio as an indicator of the relative abundance of diatoms over coccolithophores. *He, Zhao, Li, and Wang (2013).*

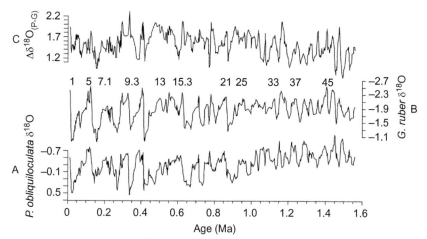

FIGURE 6.43   Last 1.56 myr $\delta^{18}O$ records at ODP Site 1143 from the southern SCS after three-point Gaussian smoothing show variations in the upper water structure: (A) $\delta^{18}O$ of *P. obliquiloculata* (PDB, ‰), (B) *G. ruber* $\delta^{18}O$ (PDB, ‰), and (C) $\Delta\delta^{18}O_{(P-G)}$ (*P. obliquiloculata* $\delta^{18}O$ minus *G. ruber* $\delta^{18}O$) (PDB, ‰). *From Tian, Wang, Chen, and Cheng (2005).*

(e.g., Bühring, Sarnthein, & Erlenkeuser, 2004; He et al., 2013; Higginson et al., 2003; Jian & Huang, 2003; Zhao et al., 2006).

Given the primary forcing of monsoon, other factors also influence the upper water structure of the SCS. As already pointed out in Section 2.2.3, the semienclosed SCS was almost isolated during the last glaciation, with the Bashi Strait as its only connection to the open ocean. Because of the closed southern passageways and the strong supply of river runoff, the upper water in the southern SCS could have been highly stratified during glacial periods with a more brackish surface layer and a highly saline subsurface layer, thus affecting the glacial planktonic fauna, especially the tropical subsurface planktonic foraminifer *P. obliquiloculata* with some peculiar distribution patterns. During glacial time, this warm-water species decreased its proportion in the foraminiferal fauna in the northern SCS (Figure 6.45E), as elsewhere in the ocean, but significantly increased in the southern SCS (Figure 6.45B–D) (Li, Zheng, et al., 2010).

This specific feature of glacial occurrence of tropical species is probably related to the stratification of the upper water column in the southern SCS, a result of freshwater input from the paleo-Sunda River and the cutoff of water exchanges along the southern boundary of the SCS. On a longer timescale, the percentages of *P. obliquiloculata* displayed a "normal" pattern with higher values during interglacials before 0.9 Ma but become more proliferous in glacials afterward (Figure 6.46; Xu et al., 2005). Therefore, upper water stratification during glaciations appeared to have established after 0.9 Ma, possibly associated with the formation of the Sunda River systems.

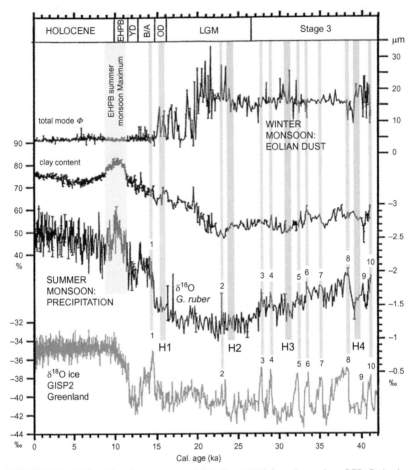

FIGURE 6.44 Millennial-scale events recorded in Core 17940 from the northern SCS. Grain size, clay content, and planktonic δ¹⁸O sequences show Dansgaard–Oeschger (D–O in numbers) and Heinrich events (H1–H4). OD = Oldest Dryas; B/A = Bølling–Allerød; YD = Younger Dryas; EHPB = early Holocene/Preboreal. *Modified from Wang, Sarenthein, Grootes, and Erlenkeuser (1999)*.

Land Climate

Along with the oceanographic response to the monsoon climate, the marine sediments of the SCS also contain monsoon signals of terrestrial vegetation and chemical weathering. Pollen grains in marine sediments originated from vegetation on land and then were buried in the sea after being transported by winds or currents. Deep-sea palynology yields terrestrial climate signals among the ocean records and bridges paleoenvironmental studies across the land and the sea. As land vegetation from around the SCS has sensitively responded to monsoon climate changes, pollen sequences in cored marine sediments can be used to indicate vegetation successions affected by climate change (Sun, Li, & Beug, 1999).

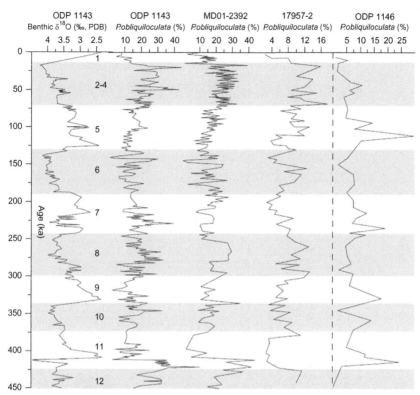

FIGURE 6.45   Downcore variations of planktonic foraminifer *Pulleniatina obliquiloculata* over the last 450 kyr. Southern SCS: (A) benthic $\delta^{18}O$ at ODP 1143, (B) *P. obliquiloculata* % at ODP 1143, (C) MD01-2392, (D) 17957-2; northern SCS: (E) *P. obliquiloculata* % at ODP 1146. *Modified from Li et al. (2010).*

The best Pleistocene palynological record in the SCS is from ODP Site 1144, showing relatively arid climate in the glacial times yet with evergreen broad-leaved forests still surviving in southern coastal China even in the LGM. Generally, pollen influx values are very low ($\sim$4 grains/cm^2/year) before MIS 16, ca. 670 kyr BP. Since then, tree pollen influx varies with glacial cycles, showing higher values in glacials when the intensified winter monsoon transported more pollen grains from the north (Figure 6.47B). The pollen ratio between herbs and *Pinus* (H/P) may indicate the relative distance of the studied site from the coast, with higher values implying shorter distance to the coast. At ODP 1144, the H/P variations before MIS 6 were small but reached maximum amplitudes at the LGM (Figure 6.47C), suggesting that the northern continental shelf of the SCS was more narrow and steep before MIS 6 but greatly broadened since then (Sun et al., 2003).

Of great interest is the vegetation type of the exposed continental shelf during the glacials. The pollen diagram from Core 17940 in the northern slope

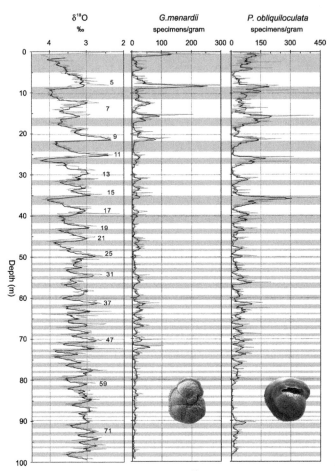

FIGURE 6.46 The last 2 Ma records of the benthic δ^{18}O, % fragmentation, % *Pulleniatina obliquiloculata*, and % *Globorotalia menardii*. Coarse curves indicate the six-point running average. Horizontal dashed line displays the age of 850 kyr BP. Gray bars mark glacial intervals, and numbers are marine oxygen isotope stages (MIS). *Xu et al. (2005)*.

shows alternating dominance of montane conifers and upper montane rain forest taxa with herb taxa during MIS 2–3 (Figure 6.48 upper panel; Sun & Li, 1999), denoting cooler climate than the present day. The occurrence of large amount of herb pollens dominated by *Artemisia* indicates a remarkable decline in temperature and humidity, together with much higher amount of charcoal than during the Holocene, suggesting frequent fire due to the dry climate. Presumably, the exposed continental shelf during MIS 2–3 was partially occupied by grassland, with montane conifers growing on nearby mountains (Sun, Li, Luo, & Chen, 2000). However, the pollen diagrams in the southern slope show different characteristics. As illustrated by Core 17964 near the

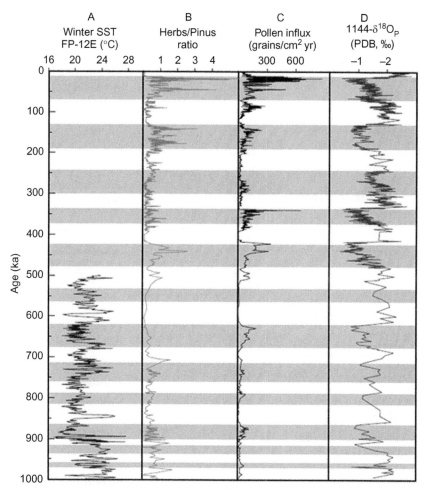

FIGURE 6.47   Paleoclimatological records from ODP Site 1144, northern SCS, over the past 1 myr. (A) Planktonic $\delta^{18}O$ (Bühring et al., 2004). (B) Pollen influx. (C) Herb/*Pinus* ratio in pollen assemblages (Sun, Luo, Huang, Tian, & Wang, 2003). (D) Winter SST based on transfer function FP-12E for 1.0–0.5 Ma (Zheng et al., 2005). *From Jian et al. (2009).*

Sunda Shelf, the glacial–interglacial contrast, which is clearly observed in the north, is virtually absent from the southern record (Figure 6.46, lower panel; Li & Sun, 1999).

Thus, the pollen data outlined a distinct picture of north–south contrast of vegetation in the late Quaternary SCS, especially for the glacial time. During the LGM, grassland vegetation, mainly composed of *Artemisia*, partially covered the exposed northern shelf, indicating colder and drier climate relative to the present day. But the subaerially exposed southern continental shelf, the Sundaland, was covered by lowland rain forest and lower mountain rain

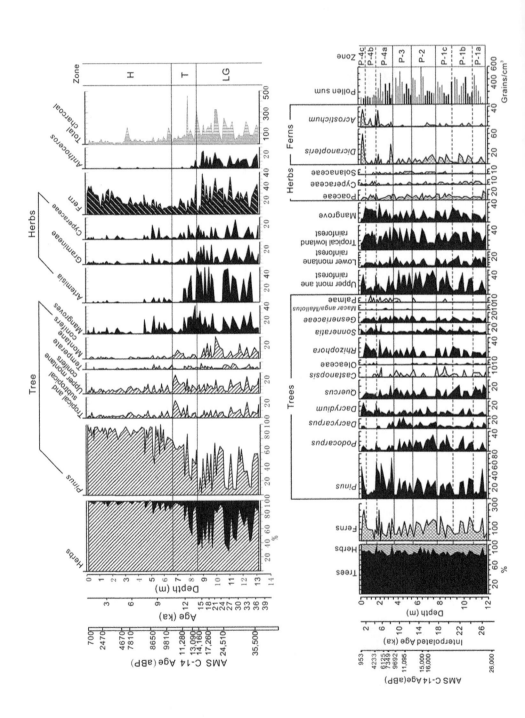

forest, indicating cool but humid climate during the LGM. The difference in humidity between the northern and southern SCS was probably caused by changes in the East Asian monsoon system (Sun, Li, Luo, et al., 2000). The intensification of Siberian High over the East Asian continent during the glacial period strengthened the East Asian winter monsoon, which in turn led to the lower temperature and reduced humidity in the region including the northern SCS. By contrast, the strengthened boreal winter monsoon absorbed moisture when crossing the SCS and provided more precipitation to the Sundaland, leading to the continued growth of lowland rain forest and mangroves there.

Weathering rate and the related mineral and elemental ratios provide another group of monsoon proxies (see Section 6.2.3.2). For the late Quaternary, for example, the ratio of hematite to goethite (Hm/Gt) has been used as a precipitation proxy over the last 600 kyr in the southern SCS (Zhang, Ji, Balsam, Liu, & Chen, 2007). The concentrations of these two climatically significant Fe oxide minerals exerted a good correlation between variations in the Hm/Gt ratio and in stalagmite oxygen isotopes from South China, indicating that the Hm/Gt ratio provides a useful record of monsoon precipitation in Southeast Asia.

In terms of clay minerals, the ratio of smectite to illite + chlorite is used to estimate the relative strength of the summer versus winter monsoons based on the observation that modern illite and chlorite in the SCS mainly come from the mainland of China and Taiwan island, while smectite largely originates from Luzon and other islands of the southern SCS. The record of this ratio at ODP Site 1146 demonstrates distinctive glacial–interglacial cycles for the past 2 myr, with high values implying strong summer monsoons during interglacials and low values implying strong winter monsoons during glacials (Liu, Trentesaux, et al., 2003), consistent with the results of palynological analysis.

In the modern southern SCS, the largest terrigenous source is the Mekong that drains the eastern Tibetan Plateau, and the clay mineralogy of sediment cores from the southwest SCS has been used to explore the weathering history of the Mekong Basin (Liu et al., 2004, 2005; Liu, Colin, et al., 2007). Here, increases of smectites/illite + chlorite and smectites/kaolinite ratios may suggest intensified chemical weathering in the Mekong plain soils, relative to physical erosion of the highland Mekong Basin. Variations in the chemical weathering/erosion history are mainly controlled by changes in the summer monsoon rainfall: wet periods of summer monsoon reinforcement are characterized by enhanced chemical weathering of the Mekong plain soils. The relatively higher ratios developed during interglacials throughout the past 190

FIGURE 6.48   Pollen percentage diagram representing the last glacial to Holocene from the continental slopes of the SCS. Upper panel: Core 17940, northern slope, with three pollen zones recognized: LG (15–37 kyr BP), T (10–15 kyr BP), and H (0–10 kyr BP) (from Sun & Li, 1999). Lower panel: Core 17964 (6° 10′N, 112° 13′E), southern slope, with four pollen zones recognized: P1 (18.3–26.5 kyr BP), P2 (13.8–18.3 kyr BP), P3 (9.9–13.8 kyr BP), and P4 (0–9.9 kyr BP). *Modified from Li and Sun (1999).*

kyr suggest strengthened chemical weathering and weak physical erosion; by contrast, the lower ratios during glacial periods indicate strongly intensified physical erosion and weakened chemical weathering (Figure 6.49; Liu et al., 2004).

In general, clay mineral distributions indicate strong glacial–interglacial cyclicity in the SCS, suggesting significant influence of the winter monsoon affected by Northern Hemisphere ice sheet. However, monsoon variability should be considered not only in the context of glacial–interglacial variability but also in the broader context of tropical and subtropical variability involving ocean–atmosphere interaction (Wang et al., 2005). For example, a clear peak of summer monsoon from 11.5 to 7.7 kyr BP with maximum intensity at 10 kyr BP was found in the northern SCS (Hu et al., 2012). This finding may underscore the role of precession forcing of the East Asian monsoon in the SCS region. The 20 kyr precession cycles are remarkable in records of various monsoon proxies in the late Quaternary records, including monsoon-related productivity (Figure 6.50, upper panel; Zhao et al., 2006) and chemical weathering (Figure 6.50, lower panel; Boulay et al., 2005).

Arguments for a direct response of the monsoon signal to the orbital forcing also come from palynology, as vegetation changes in the low-latitude SCS may precede ice-sheet-related climate changes in high latitudes. Figure 6.51 shows smoothed curves of high-resolution pollen records in Core MD05-2904 from the northern SCS. The vegetation changes reflecting deglacial warming occurred earlier than marine $\delta^{18}O$ that indicates the retreat of boreal ice sheet (Chang et al., 2013).

6.3.2.3   Sea-Level Changes in Sunda Shelf

The Sunda Shelf in the southwestern SCS is one of the largest shelf areas in the world. Much of the shelf surface is flat, with an average depth of 50 m. At the sea-level lowstand during the glaciation, its extensively exposed surface connected Sumatra, Java, and Borneo to form a large landmass about 1.85 million km^2 in area. Because of its enormous lateral extension of up to 800 km and the extremely low gradient of its bottom morphology, the Sunda Shelf offers well-suited conditions for high-resolution studies on sea-level fluctuations in the late Quaternary (Hanebuth, Stattegger, & Bojanowski, 2009; Tjia, 2001).

River System

The drowned valleys on the modern shelf, incised up to 20 m deep, testify the development of the "paleo-Sunda River" system in the glacial. The exposed shelf, or Sundaland, was drained by three gigantic river systems: the East Sunda, the North Sunda, and the South Sunda Rivers. Of those, the North Sunda or "Molengraaff" River drained from the large mountainous islands,

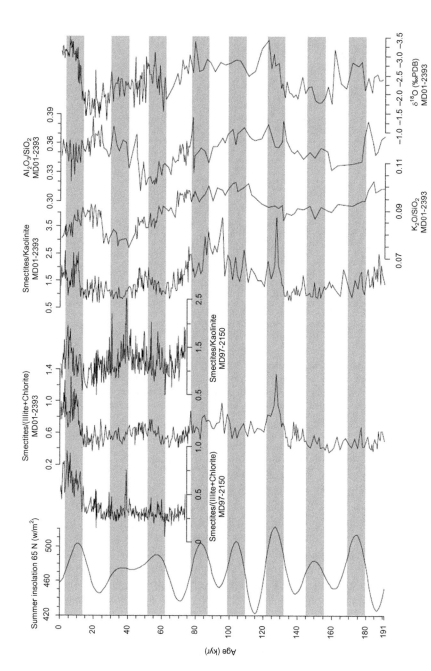

FIGURE 6.49 Variations of smectite/illite+chlorite and smectite/kaolinite ratios in Cores MD01-2393 and MD97-2150, southern SCS, together with K_2O/SiO_2 and Al_2O_3 ratios of Core MD01-2393, as compared to the summer insolation of the Northern Hemisphere (high-amplitude periods marked by horizontal gray bars) and planktonic $\delta^{18}O$ record. *Liu et al. (2004).*

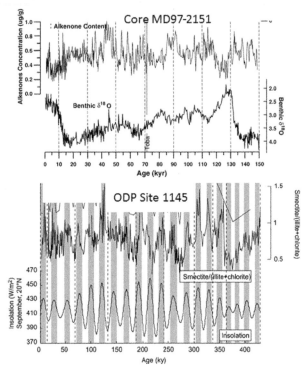

FIGURE 6.50　20 kyr prerecession cycles in late Quaternary records of the South China Sea. Upper panel: Alkenone content and benthic $\delta^{18}O$ in Core MD97-2151; dotted lines mark 20 kyr cycles. Lower panel: Smectite/illite+chlorite ratio compared to the September insolation curve; gray bars show precession cycles. *Upper panel: Zhao et al. (2006); lower panel: Boulay, Colin, Trentesaux, Frank, and Liu (2005).*

the so-called Greater Sunda Islands, into the SCS (Figure 6.52; Hanebuth, Voris, Yokoyama, Saito, & Okuno, 2011; Tjia, 2001).

Located in the region with high precipitation rates and surrounded by a steep-gradient hinterland, the Sunda Shelf is characterized by one of the largest sediment discharge rates worldwide (Hanebuth et al., 2011). With the glacial emergence of the Sunda Shelf, the tremendous amount of terrigenous sediments delivered by the paleo-Sunda River system was transported directly to the southern deep basin. As a result, the highest sedimentation rate in the deepwater SCS occurred in its SW corner directly off the Sunda Shelf during MIS 2 but shifted to the NE corner in the Holocene due to sediment contributions by the Pearl River and mountainous rivers from Taiwan (Figure 6.53; Huang & Wang, 2007).

The high deepwater sedimentation rate near the Sunda Shelf is related to high river discharge driven by humid climate, as indicated by pollen (Sun, Li, & Chen, 2000; Sun, Li, Luo, et al., 2000) and mineralogical and

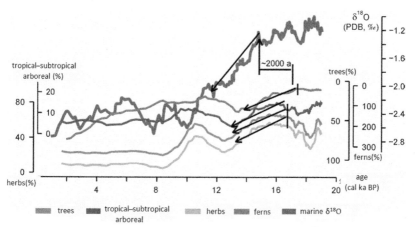

FIGURE 6.51 A comparison of smoothed pollen records with marine $\delta^{18}O$ from Core MD05-2904, northern South China Sea. *Chang, Luo, and Sun (2013).*

geochemical data (Tamburini, Adatte, Follmi, Bernasconi, & Steinmann, 2003). Recent palynological analyses of the cores along the North Sunda River valley confirmed that rain forest remained dominant on the shelf during the glacial, once again underscoring the enhanced N–S climate contrast in the SCS during glaciation (Wang, Sun, Wang, & Stattegger, 2009). The LGM pollen assemblages there are characterized by high percentages of pollen from the lowland rain forests and lower montane rain forests, suggesting that the exposed shelf was covered with humid vegetation. A marshy vegetation (with plants of sedges, reeds, bamboo, etc.) developed in the river valley, and around the marshes were distributed palms and a variety of ferns including tree ferns (*Cyathea*). The climate during the LGM inferred from the vegetation was cooler than that at the present day, but no significant decrease in humidity has been observed (Wang et al., 2009), in contrast to a dry LGM in the region of Sahul Shelf near Australia (e.g., Kershaw, van der Kaars, Moss, & Wang, 2002; van der Kaars, Wang, Kershaw, Guichard, & Setiabudi, 2000).

Active neotectonic deformation in the catchment regions is another reason of the high sedimentation rates of the Sunda Shelf. The hinterland of high elevation surrounding the shelf from south and west like Borneo, and their mountainous rivers, though small, contributes large amounts of sediments into the SCS. The Baram River in NW Borneo, for example, is responsible for the post-Eocene deposits of 8–9 km thickness under the shelf off the Baram Delta and for the >1 km thick Quaternary succession under the outer shelf (Hiscott, 2001), likely as the direct contributor to the thick MIS 2 sequence from the southern deep water shown in Figure 6.53. Active deformation in this region is exemplified by 4–6 mm of plate-scale shortening a year today in NW Borneo, where the Baram Delta system on the outer shelf is the only geologic structure to accommodate ongoing shortening (King, Backé, Morley, Hillis, & Tingay, 2010).

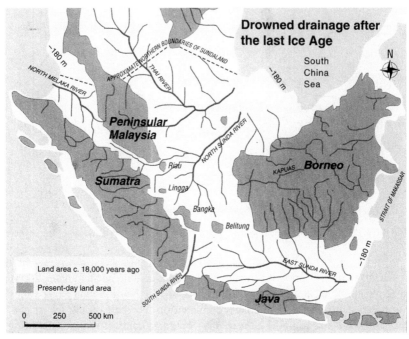

FIGURE 6.52   Sunda Shelf and paleo-Sunda River system in the last glaciation. *Tjia (2001).*

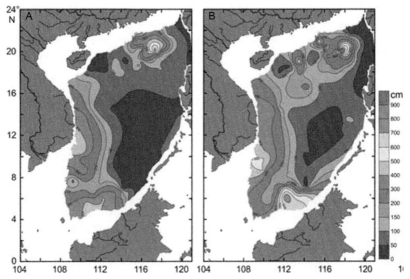

FIGURE 6.53   Distribution of sediment thickness on the slope and deepwater SCS at (A) MIS 1 and (B) MIS 2. *Huang and Wang (2007).*

Sea Level

The Sunda Shelf provides an ideal place for reconstructing sea-level history because of two reasons. First, the shelf is underlain by a huge package of terrigenous deposits accumulated in response to sea-level changes. Second, the thick pre-Holocene terrigenous sediments are covered by a thin Holocene sediment drape, hence offering unique conditions for high-resolution reconstructing of sea-level fluctuations in the late Pleistocene. Due to a combination of high sedimentation supply with an extremely broad (800 km) and very gently inclined (~1:9000) shelf, the Sunda Shelf has preserved the almost complete early history of sea-level changes in the last glaciation, as depicted by Hanebuth et al. (2009). Moreover, a long-term eustatic sea-level curve since the Pliocene has been deduced from high-resolution seismic data in the northern Sunda Shelf, with 36 fourth-order sea-level cycles recognized (Zhong, Geng, Wong, Ma, & Wu, 2004).

In the recent years, a large number of publications have come out on the postglacial sea-level rise (see Hanebuth et al., 2011 for a review) and on the formation of the major river deltas around the SCS. The most critical issue in the matter of debate has been the Holocene sea-level highstand. Because of the complicate topography and neotectonic deformation of the region and unequal quality of dating results, opinions diverge significantly on the sea-level highstand in the southern SCS above the modern level. Some early works have proposed two to three highstands between 6 and 2.7 kyr BP (Tjia, 1996), but not confirmed by further studies. Now generally accepted is a sea-level highstand by +2–+3 m from 6 to 4 kyr BP both on river deltas (Figure 6.54B; Hori & Saito, 2007; Tanabe et al., 2003, 2006) and on the Sunda Shelf (Figure 6.54A; Hanebuth et al., 2011).

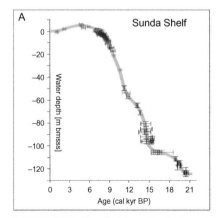
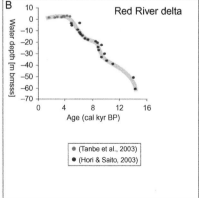

FIGURE 6.54 Postglacial sea-level changes in the SCS. (A) Sunda Shelf (Hanebuth et al., 2011), (B) Red River Delta (Hori & Saito, 2007; Tanabe et al., 2003).

Remarkable is the rapid sea-level rise in the Sunda Shelf during meltwater pulse (MWP) 1a, with 16 meter rise within 300 years (14.6–14.3 kyr BP; Hanebuth, Stattegger, & Grootes, 2000). In the northern SCS, a drastic increase of SST was found at 14.7 kyr BP, corresponding to the contemporary Bølling warming. The abrupt warming was accompanied by a sudden drop of terrigenous organic matter, implying a short-term response of local rivers to rapid sea-level rise before the MWP 1a (Kienast et al., 2001, 2003). If the sea-level rise and warming in the SCS preceded the ice melting in the polar North Atlantic region, this discovery challenges the common wisdom that the deglacial climate changes in the low latitude are just a response to the high-latitude processes in the Northern Hemisphere.

Because of its very gentle slope, remarkable coastline migrations on the Sunda Shelf over Quaternary glacial cycles have attracted many recent studies. As learned from shallow-acoustic profiling and sediment cores, the paleo-topography of the Sundaland in the last deglaciation is only sparsely covered by younger sediments, and the approximate shoreline configurations of the Sunda Shelf can be generated on the basis of modern bathymetry (Hanebuth et al., 2011). The reconstructions for the LGM at 21 kyr BP and for 9.5 kyr BP after the MWP 1B are shown in Figure 6.55. Since the Sunda Shelf region is structurally well developed and extremely wide in size, a locally differential response to the hydroisostatic effect on the shelf should have occurred. In addition, because the tropical Sunda region has been far away from glacial ice sheets, the dominant isostatic component was only the changing water loading over past sea-level cycles. At sea-level lowstand when the area was widely exposed, progressive and laterally differential loading of seawater could be expected during any episode of rising in sea level. Therefore, the differences in magnitude of the LGM sea-level lowstand between the middle and the outer Sunda Shelf could reach as much as 10–15 m due to glaciohydroisostatic adjustment (Hanebuth et al., 2011).

In the 1990s, the German R/V SONNE cruises 115 (Stattegger, Kuhnt, Wong, & Scientific Party, 1997) and 95 (Sarnthein et al., 1994) retreated a large number of sediment cores from the shelf and slope areas of the Sunda Shelf. Sedimentologic and geochemical analyses of the cores have generated a wealth of data about coastline migrations and environmental changes in the LGM and deglaciation (e.g., Hanebuth et al., 2003; Steinke, Kienast, & Hanebuth, 2003; Steinke, Kienast, et al., 2008; Steinke, Vogt, et al., 2008). As discussed in the preceding text, pollen analyses also contribute greatly to paleoenvironmental reconstructions in this region. Among various plants, mangroves thrive in brackish-water environments along tropical coasts and reach the highest diversity in the Sunda and Sahul Shelves, so mangrove pollen can be used as indicator of coastal zone to trace coastline migrations (Grindrod, Moss, & van der Kaars, 2002). Sudden and significant decrease or even disappearance of mangrove pollen at the LGM in the pollen profiles should indicate the process of sea-level lowering, coastline downward

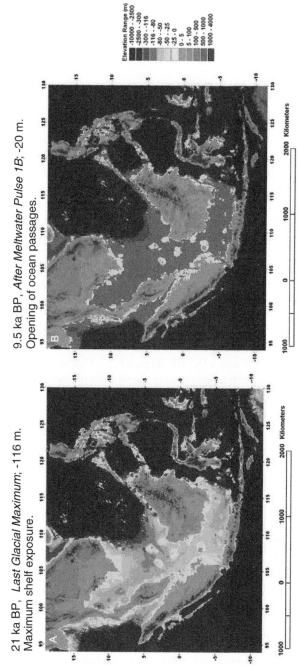

FIGURE 6.55 Reconstructions of paleocoastlines based on present-day bathymetric depth contours. (A) LGM, 21 kyr BP. (B) After MWP 1B, 9.5 kyr BP. *Hanebuth et al. (2011).*

migration, and emergence of the shelf above seawater. Figure 6.56 displays the coastline migration on the basis of the abrupt decreasing or disappearance of mangrove pollen in three cores during the seawater regression in the last glacial (Wang, Sun, Wang, & Stattegger, 2008; Wang et al., 2009).

Delta Growth

In the recent years, numerous drill holes and shallow seismic profiling on the river deltas of the SCS have been used to reconstruct the detailed history of sea-level rising. Two types of deltas are found to have developed in the Sunda Shelf: deltas of small mountainous rivers from islands like the Baram Delta and deltas of large rivers from the continent such as the Mekong Delta. Off the north Borneo, the Baram River forms a lobate, wave-influenced, tide-dominated delta. Its subaerial portion comprises a broad area of $\sim$300 km^2, and the active delta lobe with 120 m in estimated thickness forms a prominent coastal protrusion at the point where the Baram River enters the SCS. Modern sedimentation occurs mainly on the large mouth bar. Because of the high precipitation and steep topography of the catchment area, the Baram River discharges about 1590 m^3/s and large amounts of sediments to the SCS. As a result, the delta grows about 10 m per year in the late Holocene (Hiscott, 2001; Lambiase, Bin Abdul Rahim, & Peng, 2002). Specifically, the Baram Delta system was forced backward during the last Holocene sea-level rise until approximately 5.4 kyr BP. The following relative sea-level falls initiated its rapid progradation, and the present-day delta was largely deposited between approximately 5 and 2 kyr BP (Figure 6.57; Caline & Huong, 1992).

The Mekong is the ninth largest among the world's rivers in its annual sediment discharge, giving rise to the formation of an extensive delta over the past 8 kyr (see Section 4.4.1.8 for its sedimentologic features). The Holocene

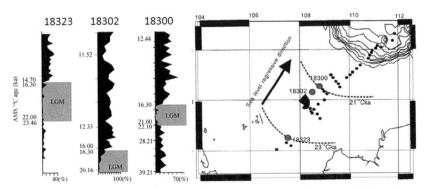

FIGURE 6.56 LGM coastline migration indicated by mangrove pollen along the North Sunda River valley (Wang, Cheng, et al., 2008; Wang et al., 2009). Gray color in the left panel denotes the LGM level with minimum abundance of mangrove pollen in the three cores. Dotted lines in the right panel denote paleocoastlines, with arrow showing direction of regression.

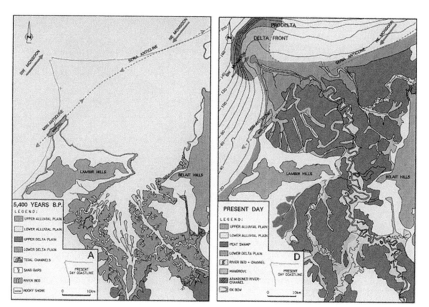

FIGURE 6.57   The old and new Baram Delta. Left: 5.4 kyr BP. Right: Modern. *From Caline and Huong (1992).*

Mekong Delta sediments are typically between 20 and 30 m thick and overlay older marine and estuarine deposits (Nguyen, Ta, & Tateishi, 2000; Ta, Nguyen, Tateishi, Kobayashi, & Saito, 2002; Tamura et al., 2009). Before the modern delta formation, nearly continuous sediment fill of the Mekong incised valley recorded the deglacial flooding history of the southeast Vietnamese shelf between 14 kyr BP and the middle Holocene sea-level highstand (Tjallingii, Stattegger, Wetzel, & Phach, 2010).

The geometry of the modern Mekong Delta is characterized by several shore-perpendicular elongate delta plains, where sequences of beach ridges occur (Figure 6.58B). Recent investigations revealed the architecture and temporal evolution of the subaqueous delta sediment (Figure 6.58C) (Tamura et al., 2012). Between 13.0 and 9.5 kyr BP, the Mekong valley was nearly completely filled with fluvial sediments when sea level rose by approximately 10 mm/year. A rapid flooding occurred between 9.5 and 8.5 kyr BP, and the coastline migrated upstream and reached the Cambodian lowland (Tjallingii et al., 2010).

As evidenced by deposits from the Cambodian lowland, the delta progradation initiated at around 8–7.5 kyr BP as a result of the sea-level stillstand.

Since 6.3 kyr BP, a fluvial system has characterized the delta plain and contributed to rapid delta progradation (Tamura et al., 2009). Around 3.5 kyr BP, the beach shoreline initiated in the lower delta plain by aggradation of basement shoals. The delta plain propagated laterally during the late

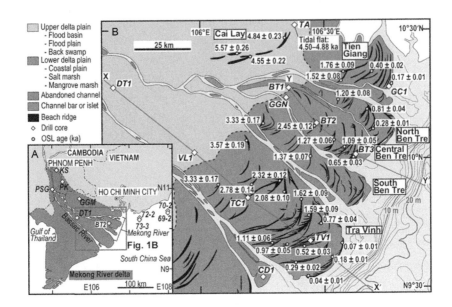

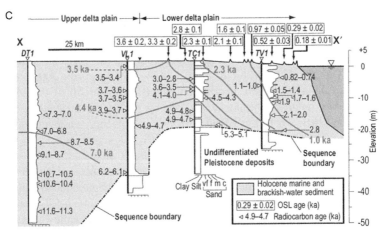

FIGURE 6.58 The Mekong Delta. (A) Location map. (B) Geomorphology and bathymetry, beach ridges (black), and core locations. (C) Cross sections along shore-perpendicular transects X–X′ (with progradation times marked red). Figures denote optically stimulated luminescence (OSL) and radiocarbon ages in kyr. *Tamura et al. (2012).*

Holocene, evolving from bars that resulted in asymmetrical bifurcation of the river mouth, due to the southwestward alongshore sediment drift enhanced by the winter monsoon. Bars were successively formed on the wider side of the bifurcated river mouths and subsequently accreted seaward (Figure 6.58B) (Tamura et al., 2012).

Approximately 80% of Mekong delivered sediment has been trapped within the delta area, which, together with a falling sea level, resulted in a rapidly prograding delta over the past 3000 years. The evolution of the Mekong Delta is characterized by a morphological asymmetry, with the delta apex at Cape Cà Mau that lies >200 km southwest from its present river mouth (Figure 6.58A) (Xue, Liu, DeMaster, Nguyen, & Ta, 2010). As revealed by comparison between topographic maps and satellite images, the shoreline of the Mekong Delta coast has changed asymmetrically over the last 70 years, largely associated to the monsoon climate. The sediments discharged in summer are accumulated in the estuary area and then transported in winter to SW by the dry winter monsoon from NE (Tamura et al., 2010).

Similar evolution has been reported from large river deltas in the northern SCS, such as the Pearl and Red Rivers (see Sections 4.4.1.6 and 4.4.1.7 in Chapter 4 for sedimentology). For the Pearl River Delta, the formation of deltaic sequences was initiated as a consequence of rapid sea-level rise between 9 and 7 kyr BP, after which a switch from transgressive to regressive sedimentation occurred in responding to declining sea-level rise. Accordingly, delta progradation rate gradually slowed down between 6.8 and 2 kyr BP as monsoon-driven runoff weakened. Rapid shoreline advances during the last 2000 years and the evolution of the delta are no longer a simple natural process but rather a common remodeling process accomplished jointly by nature and humans (Wei and Wu, 2011; Zong et al., 2009).

The Red River Delta also initiated about 8.5 kyr BP in response to deceleration of sea-level rise, marked by a transition from estuarine to deltaic deposition. The delta prograded into the drowned valley as a result of early Holocene inundation until 6 kyr BP. The sea-level highstand at +2–+3 m from 6 to 4 kyr BP allowed widespread mangrove to develop on the delta plain. During sea-level lowering after 4 kyr BP, the former delta plain emerged as a marine terrace, and the delta changed into the present tide- and wave-influenced delta with accompanying beach ridges (Hori et al., 2004; Tanabe et al., 2003).

In summary, the morphology, depositional patterns, and sedimentary facies of deltas are all closely related to Holocene sea-level changes. A sea-level jump between ~9 and 8.5 kyr BP forced the estuarine system and its depocenter to move rapidly landward, as seen most obviously in the Red River, and sedimentary facies drastically changed from estuarine sand and mud to shelf or prodelta mud with a remarkable decrease in sediment accumulation rates (Hori & Saito, 2007). Consequently, the modern deltas of the major rivers in the SCS have all been growing since ~ 6 kyr BP (Figure 6.59), despite their diverse development histories (Wang et al., 2014).

6.4   SUMMARY

The review in the preceding text demonstrated the substantial progress in multifaceted paleoenvironmental studies in the China Seas. High depositional rate

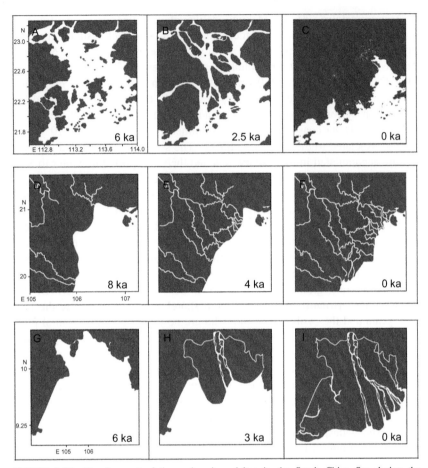

FIGURE 6.59 Development of the major river deltas in the South China Sea during the Holocene. Pearl River Delta: (A) 6.0 kyr BP, (B) 2.5 kyr BP, (C) 0 kyr BP; Red River Delta: (D) 8.0 kyr BP, (E) 4.0 kyr BP, (F) 0 kyr BP (based on Tanabe et al., 2006); Mekong Delta: (G) 6.0 kyr BP, (H) 3.0 kyr BP, (I) 0 kyr BP. *Panels (A–C): Based on Wu et al. (2006). Panels (D–I): Based on Nguyen et al. (2000), Ta et al. (2002), and Wang et al. (2014).*

and favorable carbonate preservation have rewarded the deepwater China Seas with some best paleoceanographic archives in the low-latitude Pacific region, while the extensive continental shelves have recorded the dynamic processes of the sea-level changes. To sum up, the China Seas are distinguished by three specific features in their environmental and paleoceanographic evolution:

1. Sea–land interactions. Located between the largest continent and the largest ocean, the China Seas are constrained by sea–land interactions in their environmental changes. Oceanic water and river runoff comprise the two

factors determining their oceanographic history. In particular, the Kuroshio invasion at ~1.6 Ma initiated coral reef growth along the Okinawa Trough, while the extensive catchment area of the paleo-Red River gave rise to the unusually thick deposits in its offshore deltaic basin. The role of sea–land interactions has been significantly enhanced in the glacial cycles when the repeated long-distance migrations of the coastline on the extensive shelves have not only caused drastic environmental changes but also redistributed the shelf sediments, generating sand ridges and mud patches.

2. East Asian monsoon. Monsoon-related seasonal changes are the most prominent feature of the China Seas' oceanography. Not only the surface circulation in the SCS has a seasonally reversal characteristic driven by the East Asian monsoon, but also the ECS circulation has seasonal patterns constrained by the East Asian monsoon through its effects on the Kuroshio Current and on the Yangtze outflow. As a result, paleoenvironmental variations of the China Seas have been heavily monsoon-dependent. Not only the physical features such as SST, SSS, and thermocline respond to monsoon changes, but also the biogeochemistry and productivity of the China Seas are largely controlled by the monsoon evolution especially its cyclic variability.

3. Amplification of environmental signals. The semienclosed nature of the basin and the prevailing monsoon climate of the region have granted the SCS high sensitivity to external forces in environmental changes, and the presence of broad continental shelves has made the China Seas extremely sensitive to sea-level fluctuations. The climate signals in the glacial–interglacial cycles are amplified in this marginal basin, with increased amplitudes in SST and $\delta^{18}O$ variations as compared to the open ocean at the same latitudes. The closure of the southern margin during the glacial times led to stratification of the upper water and specific faunal changes in the southern SCS. The reversal of monsoon winds drives not only the surface circulation but also the changes in biological production, the N–S gradient in oceanography, and the sediment transport such as the asymmetrical growth of the Mekong Delta.

REFERENCES

Ahagon, N., Tanaka, Y., & Ujiié, H. (1993). *Florisphaera profunda*, a possible nannoplankton indicator of late Quaternary changes in sea-water turbidity at the northwestern margin of the Pacific. *Marine Micropaleontology, 22*, 255–273.

Andree, M., Oeschger, H., Broecker, W. S., Beavan, N., Mix, A., Bonani, G., et al. (1986). AMS radiocarbon dates on foraminifera from deep sea sediments. *Radiocarbon, 28*(2A), 424–428.

Barker, S., Archer, D., Booth, L., Elderfield, H., Henderiks, J., & Rickaby, R. E. M. (2006). Globally increased pelagic carbonate production during the mid-brunhes dissolution interval and the CO2 paradox of MIS 11. *Quaternary Science Reviews, 25*, 3278–3293.

Berné, S., Vagner, P., Guichard, F., Lericolais, G., Liu, Z., Trentesaux, A., et al. (2002). Pleistocene forced regressions and tidal sand ridges in the East China Sea. *Marine Geology, 188*, 293–315.

Boulay, S., Colin, C., Trentesaux, A., Frank, N., & Liu, Z. (2005). Sediment sources and East Asian monsoon intensity over the last 450 ky: Mineralogical and geochemical investigations on South China Sea sediments. *Palaeogeography, Palaeoclimatology, Palaeoecology, 228*, 260–277.

Broecker, W. S., Andree, M., Klas, M., Bonani, G., Wolfli, W., & Oeschger, H. (1988). New evidence from the South China Sea for an abrupt termination of the last glacial period. *Nature, 333*, 156–158.

Bühring, C., Sarnthein, M., & Erlenkeuser, H. (2004). Toward a high resolution stable isotope stratigraphy of the last 1.1 M.y.: Site 1144, South China Sea. In W. L. Prell, P. Wang, P. Blum, D. K. Rea, & S. C. Clemens (Eds.), *Proceedings of the Ocean Drilling Program—Scientific Results: 184*. (pp. 1–29).

Caline, B., & Huong, J. (1992). New insight into the recent evolution of the Baram Delta from satellite imagery. *Geological Society of Malaysia Bulletin, 32*, 1–13.

Chang, L., Luo, Y. L., & Sun, X. J. (2013). Paleoenvironmental change based on a pollen record from deep sea core MD05-2904 from the northern South China Sea during the past 20000 years. *Chinese Science Bulletin, 58*, 3079–3087 (in Chinese).

Chen, M., Li, Q., Zheng, F., Tan, X., Xiang, R., & Jian, Z. (2005). Variations of the Last Glacial Warm Pool: Sea surface temperature contrasts between the open western Pacific and South China Sea. *Paleoceanography, 20*. http://dx.doi.org/10.1029/2004PA001057, PA2005.

Chen, C. T. A., Wang, S. L., Wang, B. J., & Pai, S.-C. (2001). Nutrient budgets for the South China Sea basin. *Marine Chemistry, 75*, 281–300.

Clemens, S. C., Prell, W. L., Sun, Y., Liu, Z., & Chen, G. (2008). Southern Hemisphere forcing of Pliocene $\delta^{18}O$ and the evolution of Indo-Asian monsoons. *Paleoceanography, 23*. http://dx.doi.org/10.1029/2008PA001638, PA4210.

Clift, P. D., Hodges, K., Heslop, D., Hannigan, R., Hoang, L. V., & Calves, G. (2008). Greater Himalayan exhumation triggered by early Miocene monsoon intensification. *Nature Geoscience, 1*, 875–880. http://dx.doi.org/10.1038/ngeo351.

Damuth, J. E. (1979). Migrating sediment waves created by turbidity currents in the northern South China Basin. *Geology, 7*, 520–523.

Damuth, J. E. (1980). Quaternary sedimentation process in the South China Sea basin as revealed by echo-character mapping and piston-core studies. In D. E. Hayes (Ed.), *Geophysical monograph series—American Geophysical Union: 23. The tectonics and geophysical evolution of Southeast Asian Seas and Islands* (pp. 105–125).

Droxler, A., & Farrell, J. W. (2000). Marine isotope stage 11 (MIS 11): New insights for a warm future. *Global and Planetary Change, 24*, 1–5.

Elderfield, H., Ferretti, P., Greaves, M., Crowhurst, S., McCave, I. N., Hodell, D., et al. (2012). Evolution of ocean temperature and ice volume through the mid-Pleistocene climate transition. *Science, 337*, 704–709.

Fang, D. (2002). *Deep-water Oligocene in the South China Sea: Sedimentological characteristics and hydrocarbon potential*. PhD Thesis, Tongji University, 59 pp (in Chinese with English abstract).

Fang, D., Jian, Z., & Wang, P. (2000). The paleoproductivity recorded in the southern Nansha sea area for about 30 ka. *Chinese Science Bulletin, 45*, 227–230.

Fang, L., Xiang, R., Zhao, M., Zhou, L., Liu, J., & Zhang, L. (2013). Phase evolution of Holocene paleoenvironmental changes in the Southern Yellow Sea: Benthic foraminiferal evidence from Core C02. *Journal of Ocean University of China (Oceanic and Coastal Sea Research), 12*(4), 629–638.

Foster, G. L., Lear, C. H., & Rae, J. W. B. (2012). The evolution of pCO2, ice volume and climate during the middle Miocene. *Earth and Planetary Science Letters, 341–344*, 243–254.

Gong, G. C., Liu, K. K., & Liu, C. T. (1992). Chemical hydrography of the South China Sea and a comparison with the West Philippine seas. *Terrestrial, Atmospheric and Oceanic Sciences, 3*, 587–602.

Gong, Z., Wang, Z., Stive, M. J. F., Zhang, C., & Chu, A. (2012). Process-based morphodynamic modeling of a schematized mudflat dominated by a long-shore tidal current at the Central Jiangsu Coast, China. *Journal of Coastal Research, 28*(6), 1381–1392.

Grindrod, J., Moss, P., & van der Kaars, S. (2002). Late Quaternary mangrove pollen records from continental shelf and ocean cores in the North Australian-Indonesian region. In P. Kershaw, B. David, N. Tapper, D. Penny, & J. Brown (Eds.), *Advance in Geoecology, Catena: 34. Bridging Wallace's Line: The environmental and cultural history and dynamics of the SE-Asian-Australian region* (pp. 119–146).

Hanebuth, T. J. J., Stattegger, K., & Bojanowski, A. (2009). Termination of the Last Glacial Maximum sea-level lowstand: The Sunda-Shelf data revisited. *Global and Planetary Change, 66*, 76–84.

Hanebuth, T., Stattegger, K., & Grootes, P. M. (2000). Rapid flooding of the Sunda shelf—A Late-Glacial sea-level record. *Science, 288*, 1033–1035.

Hanebuth, T. J. J., Stattegger, K., Schimanski, A., Lüdmann, T., & Wong, H. J. (2003). Late Pleistocene forced-regressive deposits on the Sunda Shelf (Southeast Asia). *Marine Geology, 199*, 139–157.

Hanebuth, T. J. J., Voris, H. K., Yokoyama, Y., Saito, Y., & Okuno, J. (2011). Formation and fate of sedimentary depocentres on Southeast Asia's Sunda Shelf over the past sea-level cycle and biogeographic implications. *Earth Science Reviews, 104*, 92–110.

Haq, B. U., Hardenbol, J., & Vail, P. R. (1987). Chronology of fluctuating sea levels since the Triassic (250 million years ago to present). *Science, 235*, 1156–1167.

Hastings, D., Kienast, M., Steinke, S., & Whitko, A. A. (2001). A comparison of three independent paleotemperature estimates from a high resolution record of deglacial SST records in the tropical South China Sea. *EOS, 82*, 12B–B.

He, Q., & Zhang, M. (1986). *Geology of Xisha Reef Facies, China*. Beijing: China Science Press (in Chinese).

He, J., Zhao, M. X., Li, L., & Wang, P. X. (2013). Changes in phytoplankton productivity and community structure in the northern South China Sea during the past 260 ka. *Palaeogeography, Palaeoclimatology, Palaeoecology, 392*, 312–323.

He, J., Zhao, M., Li, L., Wang, H., & Wang, P. (2008). Biomarker evidence of relatively stable community structure in the Northern South China Sea during the last glacial and Holocene. *Terrestrial, Atmospheric and Oceanic Sciences, 19*, 377–387.

Hess, S., & Kuhnt, W. (2005). Neogene and Quaternary paleoceanographic changes in the southern South China Sea (Site 1143): The benthic foraminiferal record. *Marine Micropaleontology, 54*, 63–87.

Higginson, M. J., Maxwell, J. R., & Altabet, M. A. (2003). Nitrogen isotope and chlorin paleoproductivituy records from the Northern South China Sea: Remote vs. local forcing of millennial- and orbital-scale variability. *Marine Geology, 201*, 223–250.

Hillenbrand, C.-D., Kuhn, G., & Frederichs, T. (2009). Record of a Mid-Pleistocene depositional anomaly in West Antarctic sediments: An indicator for ice-sheet collapse? *Quaternary Science Reviews, 28*, 1147–1159.

Hiscott, R. N. (2001). Depositional sequences controlled by high rates of sediment supply, sea-level variations, and growth faulting: The Quaternary Baram Delta of northwest Borneo. *Marine Geology, 175*, 67–102.

Holbourn, A., Kuhnt, W., Schulz, M., & Erlenkeuser, H. (2005). Impacts of orbital forcing and atmospheric carbon dioxide on miocene ice-sheet expansion. *Nature*, *438*, 483–487.

Holbourn, A., Kuhnt, W., Schulz, M., Flores, J. A., & Andersen, N. (2007). Orbitally-paced climate evolution during the middle Miocene "monterey" carbon-isotope excursion. *Earth and Planetary Science Letters*, *261*, 534–550.

Hori, K., & Saito, Y. (2007). An early Holocene sea-level jump and delta initiation. *Geophysical Research Letters*, *34*. http://dx.doi.org/10.1029/2007GL031029, L18401.

Hori, K., Tanabe, S., Saito, Y., Haruyama, S., Nguyen, V., & Kitamura, A. (2004). Delta initiation and Holocene sea-level change: Example from the Song Hong (Red River) delta, Vietnam. *Sedimentary Geology*, *164*, 237–249.

Hu, D., Böning, P., Köhler, C. M., Hillier, S., Pressling, N., Wan, S., et al. (2012). Deep sea records of the continental weathering and erosion response to East Asian monsoon intensification since 14 ka in the South China Sea. *Chemical Geology*, *326–327*, 1–18.

Huang, B. (2002). *Late Plio-Pleistocene evolution of the East Asian monsoon recorded by foraminiferal fauna in the northern South China Sea*. PhD Thesis, Tongji University, 87 pp (in Chinese with English abstract).

Huang, B., Cheng, X., Jian, Z., & Wang, P. (2003). Response of upper ocean structure to the initiation of the North Hemisphere glaciation in the South China Sea. *Palaeogeography, Palaeoclimatology, Palaeoecology*, *196*, 305–318.

Huang, B., Jian, Z., & Wang, P. (2007). Benthic foraminiferal fauna turnover at 2.1 Ma in the northern South China Sea. *Chinese Science Bulletin*, *52*(6), 839–843.

Huang, C.-Y., Liew, P. M., Zhao, M., Chang, T. C., Kuo, C. M., Chen, M. T., et al. (1997). Deep sea and lake records of the Southeast Asian paleomonsoons for the last 25 thousand years. *Earth and Planetary Science Letters*, *146*, 59–72.

Huang, W., & Wang, P. (2007). Accumulation rate characteristics of deep-water sedimentation in the South China Sea during the last glaciation and the Holocene. *Acta Oceanologica Sinica*, *29*(5), 69–73 (in Chinese).

Huang, C. Y., Wu, W. Y., Chang, C. P., et al. (1997). Tectonic evolution of accretionary prism in the arc-continent collision terrane of Taiwan. *Tectonophy*, *281*, 31–51.

Huang, C.-Y., Wu, S. F., Zhao, M., Chen, M. T., Wang, C. H., Tu, X., et al. (1997). Surface ocean and monsoon climate variability in the South China Sea since the last glaciation. *Marine Micropaleontology*, *32*, 71–94.

Huang, C. Y., Yuan, P. B., & Tsao, S. J. (2006). Temporal and spatial records of arc-continent collision in Taiwan: A synthesis. *Geological Society of America Bulletin*, *118*, 274–288.

Huh, C.-A., Su, C.-C., Wang, C.-H., Lee, S.-Y., & Lin, I.-T. (2006). Sedimentation in the Southern Okinawa Trough—Rates, turbidites and a sediment budget. *Marine Geology*, *231*, 129–139.

Humblet, M., Iryu, Y., & Nakamori, T. (2009). Variations in Pleistocene coral assemblages in space and time in southern and northern Central Ryukyu Islands, Japan. *Marine Geology*, *259*, 1–20.

Ijiri, A., Wang, L., Oba, T., Kawahata, H., Huang, C.-Y., & Huang, C.-Y. (2005). Paleoenvironmental changes in the northern area of the East China Sea during the past 42,000 years. *Palaeogeography, Palaeoclimatology, Palaeoecology*, *219*, 239–261.

Ikeda, E., Iryu, Y., Sugihara, K., Ohba, H., & Yamada, T. (2006). Bathymetry, biota and sediments on the Hirota Reef, Tane-ga-shima—The northernmost coral reef in the Ryukyu Islands. *Island Arc*, *15*, 407–419.

Iryu, Y., Matsuda, H., Machiyama, H., Piller, W. E., Quinn, T. M., & Mutti, M. (2006). An introductory perspective on the COREF Project. *Island Arc*, *15*, 393–406.

Jansen, J. H. F., Kuijpers, A., & Troelstra, S. R. (1986). A mid-Brunhes climatic event: Long-term changes in global atmosphere and ocean circulation. *Science*, *232*, 619–622.

Jia, G. D., Chen, F. J., & Peng, P. A. (2008). Sea surface temperature differences between the western equatorial Pacific and northern South China Sea since the Pliocene and their paleo-climatic implications. *Geophysical Research Letters, 35.* http://dx.doi.org/10.1029/2008GL034792, L18609.

Jia, G., Jian, Z., Peng, P., Wang, P., & Fu, J. (2000). Biogenic silica records in Core 17962 from southern South China Sea and their relation to paleoceanographical events. *Geochimica, 29,* 293–296 (in Chinese).

Jia, G. D., Peng, P., Zhao, Q., & Jian, Z. (2003). Changes in terrestrial ecosystem since 30 Ma in East Asia: Stable isotope evidence from black carbon in the South China Sea. *Geology, 31,* 1093–1096.

Jian, Z., & Huang, W. (2003). Rapid climate change and high resolution deep-sea sedimentary records. *Advances in Earth Science, 18*(5), 673–680 (in Chinese).

Jian, Z., Huang, B., Lin, H., & Kuhnt, W. (2001). Late Quaternary upwelling intensity and East Asian monsoon forcing in the South China Sea. *Quaternary Research, 55,* 363–370.

Jian, Z., Li, B., Huang, B., & Wang, J. (2000). *Globorotalia truncatulinoides* as indicator of upper-ocean thermal structure during the Quaternary: Evidences from the South China Sea and Okinawa Trough. *Palaeogeography, Palaeoclimatology, Palaeoecology, 162,* 287–298.

Jian, Z., Saito, Y., Wang, P., Li, B., & Chen, R. (1998). Shifts of the Kuroshio axis over the last 20 000 years. *Chinese Science Bulletin, 43*(5), 532–536.

Jian, Z., Tian, J., & Sun, X. (2009). Upper water structure and paleo-monsoon. In P. Wang & Q. Li (Eds.), *The South China Sea: Paleoceanography and sedimentology* (pp. 297–394): Springer.

Jian, Z., Wang, P., Chen, M. P., Li, B., Zhao, Q., Bühring, C., et al. (2000). Foraminiferal response to major Pleistocene paleographic changes in the southern South China Sea. *Paleoceanography, 15,* 229–243.

Jian, Z., Wang, L., Kienast, M., Sarnthein, M., Kuhnt, W., Lin, H., et al. (1999). Benthic forami-niferal paleoceanography of the South China Sea over the last 40,000 years. *Marine Geology, 156*(1–4), 159–186.

Jian, Z., Wang, B., & Qiao, P. (2008). Late Quaternary changes of sea surface temperature in the southern South China Sea and their comparison with the paleoclimatic records of polar ice cores. *Quaternary Sciences, 28*(3), 391–398 (in Chinese with English abstract).

Jian, Z., Wang, P., Saito, Y., Wang, L., Pflaumann, U., Oba, T., et al. (2000). Holocene variability of the Kuroshio Current in the Okinawa Trough, northwestern Pacific Ocean. *Earth and Planetary Science Letters, 184,* 305–319.

Jian, Z., Yu, Y., Li, B., Wang, J., Zhang, X., & Zhou, Z. (2006). Phased evolution of the south-north hydrographic gradient in the South China Sea since the middle Miocene. *Palaeogeography, Palaeoclimatology, Palaeoecology, 230,* 251–263.

Jian, Z., Zhao, Q., Cheng, X., Wang, J., Wang, P., & Su, X. (2003). Pliocene-Pleistocene stable isotope and paleoceanographic changes in the northern South China Sea. *Palaeogeography, Palaeoclimatology, Palaeoecology, 193,* 425–442.

Kao, S.-J., Dai, M. H., Wei, K.-Y., Blair, N. E., & Lyons, W. B. (2008). Enhanced supply of fossil organic carbon to the Okinawa Trough since the last deglaciation. *Paleoceanography, 23.* http://dx.doi.org/10.1029/2007PA001440, PA2207.

Kao, S.-J., Wu, C. R., Hsin, Y. C., & Dai, M. H. (2006). Effects of sea level change on the upstream Kuroshio Current through the Okinawa Trough. *Geophysical Research Letters, 33.* http://dx.doi.org/10.1029/2006GL026822, L16604.

Kawagata, S., Hayward, B. W., & Kuhnt, W. (2007). Extinction of deep-sea foraminifera as a result of Pliocene-Pleistocene deep-sea circulation changes in the South China Sea (ODP Sites 1143 and 1146). *Quaternary Science Reviews, 26,* 808–827.

Kershaw, P., van der Kaars,, S., Moss, P., & Wang, S. (2002). Quaternary record of vegetation, biomass burning, climate and possible human impact in the Indonesian-Australian region. In P. Kershaw, B. David, N. Tapper, D. Penny, & J. Brown (Eds.), *Advance in Geoecology, Catena: 34. Bridging Wallace's Line: The environmental and cultural history and dynamics of the SE-Asian-Australian region* (pp. 97–118).

Kienast, M., Hanebuth, T. J. J., Pelejero, C., & Steinke, S. (2003). Synchroneity of meltwater pulse 1a and the Bølling warming: New evidence from the South China Sea. *Geology, 31,* 67–70.

Kienast, M., Steinke, S., Stattegger, K., & Calvert, S. E. (2001). Synchronous tropical South China Sea SST change and Greenland warming during deglaciation. *Science, 291,* 2132–2134.

Kim, J. M., & Kennett, J. P. (1998). Paleoenvironmental changes associated with the Holocene marine transgression. Yellow Sea (Hwanghae). *Marine Micropaleontology, 34,* 71–89.

Kim, J. M., & Kucera, M. (2000). Benthic foraminifer record of environmental changes in the Yellow Sea (Hwanghae) during the last 15,000 years. *Quaternary Science Reviews, 19,* 1067–1085.

King, R. C., Backé, G., Morley, C. K., Hillis, R. R., & Tingay, M. R. P. (2010). Balancing deformation in NW Borneo: Quantifying plate-scale vs. gravitational tectonics in a delta and deepwater fold-thrust belt system. *Marine and Petroleum Geology, 27,* 238–246.

Köhler, P., & Bintanja, R. (2008). The carbon cycle during the mid Pleistocene transition: The southern ocean decoupling hypothesis. *Climate of the Past, 4,* 311–332.

Kong, G. S., Park, S.-C., Han, H.-C., Chang, J. H., & Mackensen, A. (2006). Late Quaternary paleoenvironmental changes in the southeastern Yellow Sea, Korea. *Quaternary International, 144,* 38–52.

Kuhnt, W., Hess, S., & Jian, Z. M. (1999). Quantitative composition of benthic foraminiferal assemblages as a proxy indicator for organic carbon flux rates in the South China Sea. *Marine Geology, 156,* 123–157.

Kuhnt, W., Holbourn, A., & Zhao, Q. (2002). The early history of the South China Sea: Evolution of Oligocene-Miocene deep water environments. *Revue de Micropaleontologie, 45,* 99–159.

Lambiase, J. J., Bin Abdul Rahim, A. A., & Peng, C. Y. (2002). Facies distribution and sedimentary processes on the modern Baram Delta: Implications for the reservoir sandstones of NW Borneo. *Marine and Petroleum Geology, 19,* 69–78.

Lan, D. Z., Cheng, Z. D., & Liu, S. C. (1995). *Diatom in late Quaternary sediments from the South China Sea.* Beijing, China: Ocean Press, 138 pp (in Chinese).

Li, X., Jian, Z., Shi, X., & Liu, S. (2012). Holocene foraminifera from the mud area of the inner shelf, East China Sea, and their paleoenvironmental significance. *Marine Geology & Quaternary Geology, 32*(4), 61–71 (in Chinese with English abstract).

Li, Q., Jian, Z., & Su, X. (2005). Late Oligocene rapid transformations in the South China Sea. *Marine Micropaleontology, 54,* 5–25.

Li, B., Jian, Z., & Wang, P. (1997). *Pulleniatina obliqueloculata* as a paleoceanographic indicator in the southern Okinawa Trough during the last 20,000 years. *Marine Micropaleontology, 32,* 59–69.

Li, C. Z., & Li, B. C. (1981). Studies on the formation of Subei sand cays (in Chinese with English abstract). *Oceanologica et Limnologica Sinica, 12,* 321–331.

Li, T. G., Li, S. Q., Cang, S. X., Liu, J., & Jeong, H. C. (2000). Paleo-hydrological reconstruction of the southern Yellow Sea inferred from foraminiferal fauna in core YSDP102. *Oceanologia et Limnologia Sinica, 31*(6), 588–595 (in Chinese with English abstract).

Li, L., Li, Q., Tian, J., Wang, P., Wang, H., & Liu, Z. (2011). A 4-Ma record of thermal evolution in the tropical western pacific and its implications on climate change. *Earth and Planetary Science Letters, 309*, 10–20.

Li, Q., Li, Q. Y., & Wang, R. J. (2012). Progress in the paleoceanography of the South China Sea over the last 200 ka: A review. *Advances in Earth Science, 27*(2), 224–239 (in Chinese with English abstract).

Li, L., Li, Q., Tian, J., Wang, H., & Wang, H. (2013). Low latitude hydro-climatic changes during the Plio-Pleistocene: Evidence from high resolution alkane records in the southern South China Sea. *Quaternary Science Reviews, 78*, 209–224.

Li, T., Liu, Z., Hall, M., Berne, S., Saito, Y., Cang, S., et al. (2001). Heinrich event imprints in the Okinawa Trough: Evidence from oxygen isotope and planktonic foraminifera. *Paleogeography, Paleoclimatology, Paleoecology, 176*, 133–146.

Li, G., Liu, Y., & Yang, Z. (2010). Sea-level rise and sedimentary environment response in the East China Sea continental shelf since the last glacial maximum. *Marine Geology & Quaternary Geology, 29*(4), 13–19 (in Chinese with English abstract).

Li, X., & Sun, X. (1999). Palynological records since Last Glacial Maximum from a deep-sea core in the southern South China Sea. *Quaternary Sciences, 4*, 526–535 (in Chinese).

Li, G. X., Sun, X. Y., Liu, Y., Bickert, T., & Ma, Y. Y. (2009). Sea surface temperature record from the north of the East China Sea since late Holocene. *Chinese Science Bulletin, 54*, 4507–4513.

Li, L., Wang, H., Li, J., Zhao, M., & Wang, P. (2009). Changes in sea surface temperature in western South China Sea over the past 450 ka. *Chinese Science Bulletin, 54*, 3335–3343.

Li, Q. Y., Wang, P. X., Zhao, Q. H., Tian, J., Cheng, X. R., Jian, Z. M., et al. (2008). Paleoceanography of the mid-Pleistocene South China Sea. *Quaternary Science Reviews, 27*, 1217–1233.

Li, X., Wei, G., Shao, L., Liu, Y., Liang, X., Jian, Z., et al. (2003). Geochemical and Nd isotopic variations in sediments of the South China Sea: A response to Cenozoic tectonism in SE Asia. *Earth and Planetary Science Letters, 211*, 207–220.

Li, C., Zhang, J., Fan, D., & Deng, B. (2001). Holocene regression and the tidal radial sand ridge system formation in the Jiangsu coastal zone, east China. *Marine Geology, 173*, 97–120.

Li, Q., Zheng, F., Chen, M., Xiang, R., Qiao, P., Shao, L., et al. (2010). Glacial paleoceanography off the mouth of the Mekong River, southern South China Sea, during the last 500 ka. *Quaternary Research, 73*, 563–572.

Lin, H. L., Lai, C. T., Ting, H. C., Wang, L. J., Sarnthein, M., & Hung, J. J. (1999). Late Pleistocene nutrients and sea surface productivity in the South China Sea: A record of teleconnections with northern hemisphere events. *Marine Geology, 156*, 197–210.

Lin, Y. S., Wei, K. Y., Lin, I. T., Yu, P. S., Chiang, H. W., Chen, C. Y., et al. (2006). The Holocene *Pulleniatina* Minimum Event revisited: Geochemical and faunal evidence from the Okinawa Trough and upper reaches of the Kuroshio current. *Marine Micropaleontology, 59*, 153–170.

Liu, Z. X., Berne, S., Saito, Y., Lericolais, G., & Marsset, T. (2000). Quaternary seismic stratigraphy and paleoenvironments on the continental shelf of the East China Sea. *Journal of Asian Earth Sciences, 18*, 441–452.

Liu, Z. X., Berné, S., Saito, Y., Yu, H., Trentesaux, A., Uehara, K., et al. (2007). Internal architecture and mobility of tidal sand ridges in the East China Sea. *Continental Shelf Research, 27*, 1820–1834.

Liu, K.-K., Chen, Y.-J., Tseng, C.-M., Lin, I.-I., Liu, H.-B., & Snidvongs, A. (2007). The significance of phytoplankton photo-adaptation and benthic-pelagic coupling to primary production in the South China Sea: Observations and numerical investigations. *Deep-Sea Research II, 54*, 1546–1574.

Liu, C., Cheng, X., Zhu, Y., Tian, J., & Xia, P. (2002). Oxygen and carbon isotope records of calcareous nannofossils for the past 1 Ma in the southern South China Sea. *Chinese Science Bulletin, 47*, 798–803.

Liu, Z., Colin, C., Huang, W., Le, K. P., Tong, S., Chen, Z., et al. (2007). Climatic and tectonic controls on weathering in South China and the Indochina Peninsula: Clay mineralogical and geochemical investigations from the Pearl, Red, and Mekong drainage basins. *Geochemistry, Geophysics, Geosystems, 8*. http://dx.doi.org/10.1029/2006GC001490, Q05005.

Liu, Z., Colin, C., Trentesaux, A., Blamart, D., Bassinot, F., Siani, G., et al. (2004). Erosional history of the eastern Tibetan Plateau since 190 kyr ago: Clay mineralogical and geochemical investigations from the southwestern South China Sea. *Marine Geology, 209*, 1–18.

Liu, Z., Colin, C., Trentesaux, A., Siani, G., Frank, N., Blamart, D., et al. (2005). Late Quaternary climatic control on erosion and weathering in the eastern Tibetan Plateau and the Mekong Basin. *Quaternary Research, 63*, 316–328.

Liu, J. G., Li, A., Chen, M., Xiao, S., & Wan, S. (2008). Sedimentary changes during the Holocene in the Bohai Sea and its paleoenvironmental implication. *Continental Shelf Research, 28*, 1333–1339.

Liu, Z., Li, T., Li, P., Huang, Q., Cheng, Z., Wei, G., et al. (2001). The paleoclimatic events and cause in the Okinawa Trough during 50 ka BP. *Chinese Science Bulletin, 46*, 153–157.

Liu, J. P., Li, A. C., Xu, K. H., Yang, Z. S., Velozzi, D. M., Milliman, J. D., et al. (2006). Sedimentary features of the Yangtze River-derived along-shelf clinoform deposit in the East China Sea. *Continental Shelf Research, 26*(17–18), 2141–2156.

Liu, J., Saito, Y., Wang, H., Yang, Z. G., & Nakashima, R. (2007). Sedimentary evolution of the Holocene subaqueous clinoform off the Shandong Peninsula in the Yellow Sea. *Marine Geology, 236*(3–4), 165–187.

Liu, Z., Trentesaux, A., Clemens, S. C., Colin, C., Wang, P., Huang, B., et al. (2003). Clay mineral assemblages in the northern South China Sea: Implications for East Asian monsoon evolution over the past 2 million years. *Marine Geology, 201*, 133–146.

Liu, C., Wang, P., Tian, J., & Cheng, X. (2008). Coccolith evidence for Quaternary nutricline variations in the southern South China Sea. *Marine Micropaleontology, 69*, 42–51.

Liu, M. H., Wu, S. Y., & Wang, Y. J. (1987). *Late Quaternary sediments in the Yellow Sea*. Beijing: China Ocean Press, 433 pp (in Chinese).

Liu, Z. X., & Xia, D. X. (1983). A preliminary study of tidal current ridges. *Oceanologica et Limnologica Sinica, 14*, 286–295 (in Chinese with English abstract).

Liu, J. P., Xu, K. H., Li, A. C., Milliman, J. D., Velozzi, D. M., Xiao, S. B., et al. (2007). Flux and fate of Yangtze River sediment delivered to the East China Sea. *Geomorphology, 85*(3–4), 208–224.

Liu, J. P., Xue, Z., Ross, K., Wang, H. J., Yang, Z. S., Li, A. C., et al. (2009). Fate of sediments delivered to the sea by Asian large rivers: Long-distance transport and formation of remote alongshore clinothems. *The Sedimentary Record, 7*(4), 4–9.

Luo, Y. L., & Sun, X. J. (2013). Vegetation evolution and its response to climatic change during 3.15–0.67 Ma in deep-sea pollen record from northern South China Sea. *Chinese Science Bulletin, 58*, 364–372.

Ma, W., Tian, J., Li, Q., & Wang, P. (2011). Simulation of long eccentricity (400-kyr) cycle in ocean carbon reservoir during miocene climate optimum: Weathering and nutrient response to orbital change. *Geophysical Research Letters, 38*, L10701. http://dx.doi.org/10.1029/2011GL047680.

MacNeil, F. S. (1960). Tertiary and Quaternary Gastropoda of Okinawa. *USGS Professional Paper, 339*, 131.

Meng, X. W., Du, D. W., Liu, Y. G., & Liu, Z. X. (2002). Molecular biomarker record of paleoo-ceanographic environments in the East China Sea during the last 35000 years. *Science in China Series D Earth Science, 45*(2), 184–192.

Miao, Q., Thunell, R. C., & Andersen, D. M. (1994). Glacial-Holocene carbonate dissolution and sea surface temperatures in the South China and Sulu seas. *Paleoceanography, 9,* 269–290.

Nansha Scientific Expedition, Chinese Academy of Science. (1989a). *Atlas of sediments in Nansha and its Neighboring Waters.* Wuhan: Hubei Science and Technology Press (in Chinese).

Nansha Scientific Expedition, Chinese Academy of Science. (1989b). *Reports of Multidisciplinary Investigations in Nansha and its Neighboring Waters (I).* Beijing: China Science Press, 820 pp (in Chinese).

Nansha Scientific Expedition, Chinese Academy of Science. (1991). *Contributions to geology and geophysics and Reef Islands Studies in Nansha and its Neighboring Waters (I).* Beijing: China Ocean Press, 265 pp (in Chinese).

Nansha Scientific Expedition, Chinese Academy of Science. (1992). *Quaternary Coral Reef Geology of Yongshu Reef, Nansha Islands.* Beijing: China Ocean Press, 264 pp (in Chinese).

Nansha Scientific Expedition, Chinese Academy of Science. (1993a). *Atlas of sediments in Nansha and its Neighboring Waters.* Wuhan: Hubei Science and Technology Press, 94 pp (in Chinese).

Nansha Scientific Expedition, Chinese Academy of Science. (1993b). *Quaternary Sedimentary Geology of Nansha and its Neighboring Waters.* Wuhan: Hubei Science and Technology Press, 383 pp (in Chinese).

Nansha Scientific Expedition, Chinese Academy of Science. (1994). *Contributions to geology and geophysics and Reef Islands studies in Nansha and its Neighboring Waters (II).* Beijing: China Ocean Press, 239 pp (in Chinese).

Nguyen, V. L., Ta, T. K. O., & Tateishi, M. (2000). Late Holocene depositional environments and coastal evolution of the Mekong River Delta, southern Vietnam. *Journal of Asian Earth Sciences, 18,* 427–439.

Ning, X., Chai, F., Xue, H., Chai, Y., Liu, C., & Shi, J. (2004). Physical-biological oceanographic coupling influencing phytoplankton and primary production in the South China Sea. *Journal of Geophysical Research, 109.* http://dx.doi.org/10.1029/2004JC002365, C10005.

North Greenland Ice Core Project (NGRIP) Members. (2004). High-resolution record of Northern Hemisphere climate extending into the last interglacial period. *Nature, 431,* 147–151.

Oguri, K., Matsumoto, E., Yanada, M., Saito, Y., & Iseki, K. (2003). Sediment accumulation rates and budgets of depositing particles of the East China Sea. *Deep Sea Research Part II: Topical Studies in Oceanography, 50,* 513–528.

Oiwane, H., Tonai, S., Kiyokawa, S., Nakamura, Y., Suganuma, Y., & Tokuyama, H. (2011). Geomorphological development of the Goto Submarine Canyon, northeastern East China Sea. *Marine Geology, 288,* 49–60.

Oppo, D. W., & Sun, Y. (2005). Amplitude and timing of sea-surface temperature change in the northern South China Sea: Dynamic link to the East Asian monsoon. *Geology, 33,* 785–788.

Osozawa, S., Shinjo, R., Armid, A., Watanabe, Y., Horiguchi, T., & Wakabayashi, J. (2012). Palaeo-geographic reconstruction of the 1.55 Ma synchronous isolation of the Ryukyu Islands, Japan, and Taiwan and inflow of the Kuroshio warm current. *International Geology Review, 54*(12), 1369–1388.

Pang, X., Chen, C., Shao, L., Wang, C., Zhu, M., He, M., et al. (2007). Baiyun movement, a great tectonic event on the Oligocene-Miocene boundary in the northern SCS and its implications. *Geological Review, 53*(2), 145–151 (in Chinese).

Pelejero, C., & Grimalt, J. O. (1997). The correlation between the $U_{37}^{K'}$ index and sea surface temperatures in the warm boundary: The South China Sea. *Geochemistry, Geophysics, Geosystems*, *61*, 4789–4797.

Pelejero, C., Grimalt, J. O., Heilig, S., Kienast, M., & Wang, L. (1999). High-resolution $U_{37}^{K'}$ temperature reconstructions in the South China Sea over the past 220 ka. *Paleoceanography*, *14*(2), 224–231.

Pelejero, C., Grimalt, J. O., Sarnthein, M., Wang, L., & Flores, J.-A. (1999). Molecular biomarker record of sea surface temperature and climatic change in the South China Sea during the last 140,000 years. *Marine Geology*, *156*, 109–121.

Pflauamnn, U., Duprat, J., Pujol, C., & Labeyrie, L. D. (1996). SIMMAX: A modern analog technique to deduce Atlantic sea surface temperatures from planktonic foraminifera in deep-sea sediments. *Paleoceanography*, *11*, 15–35.

Pflaumann, U., & Jian, Z. (1999). Modern distribution patterns of planktonic foraminifera in the South China Sea and western Pacific: A new transfer technique to estimate regional sea-surface temperatures. *Marine Geology*, *156*, 41–83.

Qin, Y. (1985). *Geology of the Bohai Sea*. Beijing: Science Press, 231 pp (in Chinese) 1995, China Ocean Press, 354 pp (in English).

Qin, Y. (1990). *Geology of the Yellow Sea*. China Ocean Press, 289 pp (in Chinese).

Qin, Y., Zhao, Y., Chen, L., & Zhao, S. (1988). *Geology of the East China Sea*. Beijing: Science Press, 263 pp (in Chinese) 1996, Science Press, Beijing, 356 pp (in English).

Qin, Y., Zhao, Y., Chen, L., & Zhao, S. (1989). *Geology of the Yellow Sea*. Beijing: China Ocean Press, 289 pp (in Chinese).

Rickaby, R., Bard, E., Sonzogni, C., Rostek, F., Beaufort, L., Barker, S., et al. (2007). Coccolith chemistry reveals secular variations in the global ocean carbon cycle? *Earth and Planetary Science Letters*, *253*, 83–95.

Russon, T., Paillard, D., & Elliot, M. (2010). Potential origins of 400–500 kyr periodicities in the ocean carbon cycle: A box model approach. *Global Biogeochemical Cycles*, *24*, 16. GB2013. http://dx.doi.org/10.1029/2009GB003586.

Sagawa, N., Nakamori, T., & Iryu, Y. (2001). Pleistocene reef development in the southwest Ryukyu Islands, Japan. *Palaeogeography, Palaeoclimatology, Palaeoecology*, *175*, 303–323.

Saito, Y., Katayama, H., Ikehara, K., Kato, Y., Matsumoto, E., Oguri, K., et al. (1998). Transgressive and highstand systems tracts and post-glacial transgression, the East China Sea. *Marine Geology*, *122*, 217–232.

Saitoh, Y., & Masuda, F. (2004). Miocene sandstone of 'continental' origin on Iriomote Island, southwest Ryukyu Arc, Eastern Asia. *Journal of Asian Earth Science*, *24*, 137–144.

Sarnthein, M., Pflaumann, U., Wang, P., & Wong, H. K. (Eds.), (1994). *Geol. Palaont. Inst. Univ. Kiel:* Vol. 48. Preliminary report on SONNE-95 Cruise "Monitor Monsoon" to the South China Sea. Berichte-Reports (pp. 1–225).

Sarnthein, M., & Wang, P. (Eds.), (1999). Response of west Pacific marginal seas to global climate change. *Marine Geology*, *156*, 1–308.

Scherer, R. P., Bohaty, S. M., Dunbar, R. B., Esper, O., Flores, J.-A., Gersonde, R., et al. (2008). Antarctic records of precession-paced insolation-driven warming during early Pleistocene Marine Isotope Stage 31. *Geophysical Research Letters*, *35*. http://dx.doi.org/10.1029/2007GL032254, L03505.

Schönfeld, J., & Kudrass, H. R. (1993). Hemipelagic sediment accumulation rates in the South China Sea related to late Quaternary sea-level changes. *Quaternary Research*, *40*, 368–379.

Shi, Y., Yu, G., Liu, X., Li, B., & Yao, T. (2001). Reconstruction of the 30–40 ka BP enhanced Indian monsoon climate based on geological records from the Tibetan Plateau. *Palaeogeography, Palaeoclimatology, Palaeoecology, 169,* 69–83.

Shiau, L.-J., Yu, P.-S., Wei, K.-Y., Yamamoto, M., Lee, T.-Q., Yu, E.-F., et al. (2008). Sea surface temperature, productivity, and terrestrial flux variations of the southeastern South China Sea over the past 800000 years (IMAGESMD972142). *Terrestrial, Atmospheric and Oceanic Sciences, 19,* 363–376.

Shintani, T., Yamamoto, M., & Chen, M. T. (2008). Slow warming of the northern South China Sea during the last deglaciation. *Terrestrial, Atmospheric and Oceanic Sciences, 19,* 341–346.

South China Sea Institute of Oceanology (1982). *Reports of Multidisciplinary Investigations in the South China Sea (I).* China Sci. Press: Beijing (in Chinese).

South China Sea Institute of Oceanology (1985). *Reports of Multidisciplinary Investigations in the South China Sea (II).* China Sci. Press: Beijing, 432 pp (in Chinese).

South China Sea Institute of Oceanology (1987). *Zeng-Mu Reef-Report of Multidisciplinary Investigations.* China Sci. Press: Beijing, 245 pp (in Chinese).

State Oceanic Administration of China (1988). *Reports of multidisciplinary investigations in central part of South China Sea for resources and environment* (419 p.). Beijing: China Ocean Press. (in Chinese).

Stattegger, K., Kuhnt, W., Wong, H. K., & Scientific Party. (1997). *Cruise report SONNE 115 SUNDAFLUT. Berichte-Report 86.* Institüt für Geowissenschaften, Univ. Kiel, 211 pp.

Steinke, S., Chiu, H. Y., Yu, P. S., Shen, C. C., Erlenkeuser, H., Löwemark, L., et al. (2006). On the influence of sea level and monsoon climate on the southern South China Sea freshwater budget over the last 22,000 years. *Quaternary Science Reviews, 25,* 1475–1488.

Steinke, S., Chiu, H. Y., Yu, P. S., Shen, C. C., Löwemark, L., Mii, H. S., et al. (2005). Mg/Ca ratios of two *Globigerinoides ruber* (white) morphotypes: Implications for reconstructing past tropical/subtropical surface water conditions. *Geochemistry, Geophysics, Geosystems, 6,* 12. Q11005. http://dx.doi.org/10.1029/2005GC000926.

Steinke, S., Kienast, M., Groeneveld, J., Lin, L. C., Chen, M. T., & Rendle-Bühring, R. (2008). Proxy dependence of the temporal pattern of deglacial warming in the tropical South China Sea: Toward resolving seasonality. *Quaternary Science Reviews, 27,* 688–700.

Steinke, S., Kienast, M., & Hanebuth, T. J. J. (2003). The importance of sea-level variations and shelf palaeo-morphology in governing shelf margin and slope sedimentation: Examples from the southern South China Sea during the last deglaciation. *Marine Geology, 201,* 179–206.

Steinke, S., Kienast, M., Pflaumann, U., Weinelt, M., & Stattegger, K. (2001). A high resolution sea-surface temperature record from the tropical South China Sea (16,500-3000 B.P.). *Quaternary Research, 5,* 353–362.

Steinke, S., Vogt, C., Hanebuth, T. J. J., & Stattegger, K. (2008). Sea level induced variations in clay mineral composition in the southwestern South China Sea over the past 17,000 years. *Marine Geology, 250*(3–4), 199–210.

Su, J. (2004). Overview of the South China Sea circulation and its influence on the coastal physical oceanography outside the Pearl River Estuary. *Continental Shelf Researches, 24,* 1745–1760.

Su, G., Fan, Q., & Chen, S. (1989). *Sediment atlas of the central and northern South China Sea* (68 p.). Guangzhou: Guangzhou Sci. Tech. Press. (in Chinese)

Sun, X., & Li, X. (1999). A pollen record of the last 37 ka in deep sea core 17940 from the northern South China Sea. *Marine Geology, 156,* 227–244.

Sun, X., Li, X., & Beug, H. J. (1999). Pollen distribution in hemipelagic surface sediments of the South China Sea and its relation to modern vegetation distribution. *Marine Geology, 156,* 211–226.

Sun, X., Li, X., & Chen, H. (2000). Evidence for natural fire and climate history since 37 ka BP in the northern part of the South China Sea. *Science in China (Series D)*, *43*(5), 487–493.

Sun, X., Li, X., Luo, Y., & Chen, X. (2000). The vegetation and climate at the last glaciation on the emerged continental shelf of the South China Sea. *Palaeogeography, Palaeoclimatology, Palaeoecology*, *160*, 301–316.

Sun, X., Luo, Y., Huang, F., Tian, J., & Wang, P. (2003). Deep-sea pollen from the South China Sea: Pleistocene indicators of East Asian monsoon. *Marine Geology*, *201*, 97–118.

Sun, Y. B., Oppo, D. W., Xiang, R., Liu, W., & Gao, S. (2005). Last deglaciation in the Okinawa Trough: Subtropical northwest Pacific link to Northern Hemisphere and tropical climate. *Paleoceanography*, *20*, A4005. http://dx.doi.org/10.1029/2004PA001061.

Sun, X., & Wang, P. (2005). How old is the Asian monsoon system? Palaeobotanical records from China. *Palaeogeography, Palaeoclimatology, Palaeoecology*, *222*, 181–222.

Ta, T. K. O., Nguyen, V. L., Tateishi, M., Kobayashi, I., & Saito, Y. (2002). Holocene delta evolution and sediment discharge of the Mekong River, southern Vietnam. *Quaternary Science Reviews*, *21*, 1807–1819.

Tamburini, F., Adatte, T., Follmi, K., Bernasconi, S. M., & Steinmann, P. (2003). Investigating the history of East Asian monsoon and climate during the last glacial-interglacial period (0-140 000 years): Mineralogy and geochemistry of ODP Sites 1143 and 1144, South China Sea. *Marine Geology*, *201*, 147–168.

Tamura, T., Horaguchi, K., Saito, Y., Nguyen, V. L., Tateishi, M., Ta, T. K. O., et al. (2010). Monsoon-influenced variations in morphology and sediment of a mesotidal beach on the Mekong River delta coast. *Geomorphology*, *116*, 11–23.

Tamura, T., Saito, Y., Nguyen, V. L., Ta, T. K. O., Bateman, M. D., Matsumoto, D., et al. (2012). Origin and evolution of interdistributary delta plains; insights from Mekong River delta. *Geology*, *40*(4), 303–306.

Tamura, T., Saito, Y., Sieng, S., Ben, B., Kong, M., Sim, I., et al. (2009). Initiation of the Mekong River delta at 8 ka: Evidence from the sedimentary succession in the Cambodian lowland. *Quaternary Science Reviews*, *28*, 327–344.

Tanabe, S., Hori, K., Saito, Y., Haruyama, S., Vu, V. P., & Kitamura, A. (2003). Song Hong (Red River) delta evolution related to millennium-scale Holocene sea-level changes. *Quaternary Science Reviews*, *22*, 2345–2361.

Tanabe, S., Saito, Y., Vu, Q. L., Hanebuth, T. J. J., Ngo, Q. L., & Kitamura, A. (2006). Holocene evolution of the Song Hong (Red River) delta system, northern Vietnam. *Sedimentary Geology*, *187*, 29–61.

Tanaka, G., & Nomura, S. (2009). Late Miocene and Pliocene Ostracoda from the Shimajiri Group, Kume-jima Island, Japan: Biogeographical significance of the timing of the formation of back-arc basin (Okinawa Trough). *Palaeogeography, Palaeoclimatology, Palaeoecology*, *276*, 56–68.

Thompson, P. R. (1981). Planktonic foraminifer in the western North Pacific during the past 150,000 years: Comparison of modern and fossil assemblages. *Palaeogeography, Palaeoclimatology, Palaeoecology*, *35*, 441–479.

Thunell, R., Miao, Q., Calvert, S., Calvert, S., & Pedersen, T. (1992). Glacial-Holocene biogenic sedimentation patterns in the South China Sea: Productivity variations and surface water pCO_2. *Paleoceanography*, *7*, 143–162.

Tian, J., Huang, E., & Pak, D. K. (2010). East Asian winter monsoon variability over the last glacial cycle: Insights from a latitudinal sea surface temperature gradient across the South China Sea. *Palaeogeography, Palaeoclimatology, Palaeoecology*, *292*, 319–324.

Tian, J., Pak, D. K., Wang, P., Lea, D., Cheng, X., & Zhao, Q. (2006). Late Pliocene monsoon linkage in the tropical South China Sea. *Earth and Planetary Science Letters, 252,* 72–81.

Tian, J. W., & Qu, T. D. (2012). Advances in research on the deep South China Sea circulation. *Chinese Science Bulletin, 57,* 3115–3120.

Tian, J., Wang, P., Chen, R., & Cheng, X. (2005). Quaternary upper ocean thermal gradient variations in the South China Sea: Implications for east Asian monsoon climate. *Paleoceanography, 20.* http://dx.doi.org/10.1029/2004PA001115, PA4007.

Tian, J., Wang, P., & Cheng, X. (2004). Development of the East Asian monsoon and Northern Hemisphere glaciation: Oxygen isotope records from the South China Sea. *Quaternary Science Reviews, 23,* 2007–2016.

Tian, J., Wang, P., Cheng, X., & Li, Q. (2002). Astronomically tuned Plio-Pleistocene benthic $\delta^{18}O$ records from South China Sea and Atlantic-Pacific comparison. *Earth and Planetary Science Letters, 203,* 1015–1029.

Tian, J., Xie, X., Ma, W., Jin, H., & Wang, P. (2011). X-ray fluorescence core scanning records of chemical weathering and monsoon evolution over the past 5 Myr in the southern South China Sea. *Paleoceanography, 26.* http://dx.doi.org/10.1029/ 2010PA002045, PA4202.

Tian, J. W., Yang, Q., Liang, X., Xie, L., Hu, D., Wang, F., et al. (2006). Observation of Luzon Strait transport. *Geophysical Research Letters, 33,* L19607. http://dx.doi.org/10.1029/ 2006GL026272.

Tjallingii, R., Stattegger, K., Wetzel, A., & Phach, P. V. (2010). Infilling and flooding of the Mekong River incised valley during deglacial sea-level rise. *Quaternary Science Reviews, 29,* 1432–1444.

Tjia, H. D. (1996). Sea-level changes in the tectonically stable Malay-Thai Peninsula. *Quaternary International, 31,* 95–101.

Tjia, H. D. (2001). Sundaland. In O. J. Eong & G. W. Khoon (Eds.), *The Encyclopedia of Malaysia. The Sea.* (pp. 32–33). Singapore: Archipelago Press.

Ujiié, H. (1994). Early Pleistocene birth of the Okinawa Trough and Ryukyu Island Arc at the northwestern margin of the Pacific: Evidence from Late Cenozoic planktonic foraminiferal zonation. *Palaeogeography, Palaeoclimatology, Palaeoecology, 108,* 457–474.

Ujiié, H., Hatakeyama, Y., Gu, X. X., Yamamoto, S., Ishiwatari, R., & Maeda, L. (2001). Upward decrease of organic C/N tracing the post-glacial retreat of the continental shore line. *Palaeogeography, Palaeoclimatology, Palaeoecology, 165,* 129–140.

Ujiié, H., Tanaka, Y., & Ono, T. (1991). Late Quaternary paleoceanographic record from the middle Ryukyu Trench slope, northwest Pacific. *Marine Micropaleontology, 18,* 115–128.

Ujiié, H., & Ujiié, Y. (1999). Late Quaternary course changes on the Kuroshio Current in the Ryukyu Arc region, northwestern Pacific Ocean. *Marine Micropaleontology, 37,* 23–40.

Ujiié, Y., Ujiié, H., Taira, A., Nakamura, T., & Oguri, K. (2003). Spatial and temporal variability of surface water in the Kuroshio source region, Pacific Ocean, over the past 21,000 years: Evidence from planktonic foraminifera. *Marine Micropaleontology, 49,* 335–364.

van der Kaars, S., Wang, X., Kershaw, P., Guichard, F., & Setiabudi, D. A. (2000). A Late Quaternary palaeoecological record from the Banda Sea, Indonesia: Patterns of vegetation, climate and biomass burning in Indonesia and northern Australia. *Palaeogeography, Palaeoclimatology, Palaeoecology, 155,* 135–153.

Wan, S., Clift, P. D., Li, A., Li, T., & Yin, X. (2010). Geochemical records in the South China Sea: Implications for East Asian summer monsoon evolution over the last 20 Ma. In P. D. Clift, R. Tada, & H. Zheng (Eds.), *Geological Society London Special Publications 342. Monsoon evolution and tectonics–climate linkage in Asia* (pp. 245–263).

Wan, S., Kürschner, W. M., Clift, P. D., Li, A., & Li, T. (2009). Extreme weathering/erosion during the Miocene Climatic Optimum: Evidence from sediment record in the South China Sea. *Geophysical Research Letters*, *36*, L19706. http://dx.doi.org/10.1029/2009GL040279.

Wang, P. (1990). Neogene stratigraphy and paleoenvironments of China. *Palaeogeography, Palaeoclimatology, Palaeoecology*, *77*, 315–334.

Wang, P. (1999). Response of Western Pacific marginal seas to glacial cycles: Paleoceanographic and sedimentological features. *Marine Geology*, *156*, 5–39.

Wang, L. (2000). Isotopic signals in two morphotypes of *Globigerinoides ruber* (white) from the South China Sea: Implications for monsoon climate change during the last glacial cycle. *Palaeogeography, Palaeoclimatology, Palaeoecology*, *161*, 381–394.

Wang, P. (2004). Cenozoic deformation and the history of sea-land interactions in Asia. In P. Clift, P. Wang, W. Kuhnt, & D. Hayes (Eds.), *Continent-ocean interactions in the East Asian Marginal Seas* (pp. 1–22). Washington, D.C: AGU, Geophysical Monograph 149.

Wang, P. (2012). Tracing the life history of a marginal sea—On "The South China Sea Deep" Research Program. *Chinese Science Bulletin*, *57*(24), 3093–3114.

Wang, C. H., Chen, M.-P., & Lo, S.-C. (1986). Stable isotope records of late Pleistocene sediments from the South China Sea. *Bulletin of the Institute of Earth Science, Academica Sinica*, *6*, 185–195.

Wang, Y. J., Cheng, H., Edwards, R. L., Kong, X. G., Shao, X. H., Chen, S. T., et al. (2008). Millennial-and orbital-scale changes in the East Asian monsoon over the past 224,000 years. *Nature*, *451*, 1090–1093.

Wang, P., Clemens, S., Beaufort, L., Braconnot, P., Ganssen, G., Jian, Z., et al. (2005). Evolution and variability of the Asian monsoon system: State of the art and outstanding issues. *Quaternary Science Reviews*, *24*, 595–629.

Wang, B., Clemens, S. C., & Liu, P. (2003). Contrasting the Indian and East Asian monsoons: Implications on geological timescales. *Marine Geology*, *201*, 5–21.

Wang, R., Fang, D., Shao, L., Chen, M., Xia, P., & Qi, J. (2001). Oligocene biogenetic siliceous deposits on the slope of the northern South China Sea. *Science in China (Series D)*, *44*(10), 912–918.

Wang, Y. P., Gao, S., Jia, J., Thompson, C. F. L., Gao, J., & Yang, Y. (2012). Sediment transport over an accretional intertidal flat with influences of reclamation, Jiangsu coast, China. *Marine Geology*, *291–294*, 147–161.

Wang, R., Jian, Z., Xiao, W., Tian, J., Li, J., Chen, R., et al. (2007). Quaternary biogenic opal records in the South China Sea: Linkages to East Asian monsoon, global ice volume and orbital forcing. *Science in China (Series D)*, *50*, 710–724.

Wang, P., Jian, Z., Zhao, Q., Li, Q., Wang, R., Liu, Z., et al. (2003). Evolution of the South China Sea and monsoon history revealed in deep-sea records. *Chinese Science Bulletin*, *48*(23), 2549–2561.

Wang, R., & Li, J. (2003). Quaternary high-resolution opal record and its paleoproductivity implications at ODP Site 1143, southern South China Sea. *Chinese Science Bulletin*, *48*, 363–367.

Wang, P., & Li, Q. (Eds.), (2009). *The South China Sea–Paleoceanography and sedimentology*: Springer, 506 pp.

Wang, P. X., Li, Q. Y., Tian, J., Jian, Z. M., Liu, C. L., Li, L., et al. (2014). Long-term cycles in carbon reservoir of the Quaternary Ocean: A Perspective from the South China Sea. *National Science Review*, *1*, 119–143. China Science Press and Oxford University, Press, Oxford.

Wang, P., Min, Q., & Bian, Y. (1985). Distribution of foraminifera and ostracoda in bottom sediments of the northwestern part of the South Huanghai (Yellow) Sea and its geological significance. In P. Wang (Ed.), *Marine micropaleontology of China* (pp. 93–114). Beijing: China Ocean Press and Springer.

Wang, P., Min, Q., Bian, Y., & Cheng, X. (1981). Strata of the Quaternary transgressions in East China: A preliminary study. *Acta Geologica Sinica*, *55*(1), 1–13 (in Chinese with English abstract).

Wang, P., Min, Q., Bian, Y., & Feng, W. (1986). Planktonic foraminifera in the continental slope of the northern South China Sea during the last 130,000 years and their paleoceanographic implications. *Acta Geologica Sinica*, *60*, 1–11.

Wang, P., Prell, W. L., & Blum, P. (Eds.), (2000). *Vol. 184 [CD-ROM]. Proc. ODP, Init. Repts.* College Station, TX: Ocean Drilling Program, Texas A&M University.

Wang, L., Sarenthein, M., Erlenkeuser, H., Grimalt, J., Grootes, P., Heilig, S., et al. (1999). East Asian monsoon Climate during the late Pleistocene: High-resolution sediment records from the South China Sea. *Marine Geology*, *156*, 245–284.

Wang, L., Sarenthein, M., Grootes, P., & Erlenkeuser, H. (1999). Millennial reoccurrence of century-scale abrupt events of East Asian monsoon: A possible heat conveyor for the global deglaciation. *Paleoceanography*, *14*, 725–731.

Wang, J., Satito, Y., Oba, T., Jian, Z., & Wang, P. (2001). High-resolution records of thermocline in the Okinawa Trough since about 10000 a BP. *Science in China (Series D)*, *44*(3), 193–200.

Wang, X., Sun, X., Wang, P., & Stattegger, K. (2008). The records of coastline changes reflected by mangroves on the Sunda Shelf since the last 40 ka. *Chinese Science Bulletin*, *53*, 2069–2076.

Wang, X. M., Sun, X. J., Wang, P. X., & Stattegger, K. (2009). Vegetation on the Sunda Shelf, South China Sea, during the Last Glacial Maximum. *Palaeogeography, Palaeoclimatology, Palaeoecology*, *278*, 88–97.

Wang, P., Tian, J., Cheng, X., Liu, C., & Xu, J. (2003). Carbon reservoir change preceded major ice-sheets expansion at Mid-Brunhes Event. *Geology*, *31*, 239–242.

Wang, P. X., Tian, J., Cheng, X., Liu, C., & Xu, J. (2004). Major Pleistocene stages in a carbon perspective: The South China Sea record and its global comparison. *Paleoceanography*, *19*. http://dx.doi.org/10.1029/ 2003PA000991, PA4005.

Wang, P., Tian, J., & Lourens, L. J. (2010). Obscuring of long eccentricity cyclicity in Pleistocene oceanic carbon isotope records. *Earth and Planetary Science Letters*, *290*, 319–330.

Wang, J., & Wang, P. (1980). Relationship between sea-level changes and climatic fluctuations in East China since late Pleistocene. *Acta Geographica Sinica*, *35*(4), 299–312 (in Chinese with English abstract).

Wang, L., & Wang, P. (1990). Late Quaternary paleoceanography of the South China Sea: Glacial-interglacial contrasts in an enclosed basin. *Paleoceanography*, *5*, 77–90.

Wang, P., Wang, L., Bian, Y., & Jian, Z. (1995). Late Quaternary paleoceanography of the South China Sea: Surface circulation and carbonate cycles. *Marine Geology*, *127*, 145–165.

Wang, L. B., Yang, Z. S., Zhang, R. P., Fan, D. J., Zhao, M. X., & Hu, B. Q. (2011). Sea surface temperature records of core ZY2 from the central mud area in the South Yellow Sea during last 6200 years and related effect of the Yellow Sea Warm Current. *Chinese Science Bulletin*, *56*(15), 1588–1595.

Wang, P., Zhang, J., & Min, Q. (1985). Distribution of foraminifera in surface sediments of the East China Sea. In P. Wang (Ed.), *Marine micropaleontology of China* (pp. 34–69). Beijing: China Ocean Press and Springer-Verlag.

Wang, Y., Zhang, Y., Zou, X., Zhu, D., & Piper, D. (2012). The sand ridge field of the South Yellow Sea: Origin by river–sea interaction. *Marine Geology*, *291–294*, 132–146.

Wang, K., Zheng, H., Prins, M., & Zheng, Y. (2008). High-resolution paleoenvironmental record of the mud sediments of the East China Sea inner shelf. *Marine Geology & Quaternary Geology*, *28*(4), 1–10 (in Chinese with English abstract).

Wang, Y., Zhu, D., You, K. Y., Pan, S. M., Zhu, X. D., Zou, X. Q., et al. (1999). Evolution of radiating sand ridge field of the south Yellow Sea and its sedimentary characteristic. *Science in China (Series D)*, *42*(1), 97–112.

Wara, M. W., Ravelo, A. C., & Delaney, M. L. (2005). Permanent El Niño-like conditions during the Pliocene warm period. *Science, 309,* 758–761.

Wehausen, R., & Brumsack, H.-J. (2002). Astronomical forcing of the east Asian monsoon mirrored by the composition of Pliocene south china sea sediments. *Earth and Planetary Science Letters, 201,* 621–636.

Wei, G. J., Deng, W. F., Liu, Y., & Li, X. H. (2007). High-resolution sea surface temperature records derived from foraminiferal Mg/Ca ratios during the last 260 ka in the northern South China Sea. *Palaeogeography, Palaeoclimatology, Palaeoecology, 250,* 126–138.

Wei, G. J., Li, X.-H., Liu, Y., Shao, L., & Liang, X. R. (2006). Geochemical record of chemical weathering and monsoon climate change since the early Miocene in the South China Sea. *Paleoceanography, 21.* http://dx.doi.org/10.1029/2006PA001300, PA4214.

Wei, G. J., Liu, Y., Li, X. H., Chen, M. H., & Wei, W. C. (2003). High-resolution elemental records from the South China Sea and their paleoproductivity implications. *Paleoceanography, 18,* 1054. http://dx.doi.org/10.1029/2002PA000826.

Wei, K.-Y., The, E. E., Shiao, L. J., et al. (2004). Sea surface hydrographic variations during the past 165,000 years in the Southeastern South China Sea (near Palawan Island). Scientific Research Report of National Science Council, Project number NSC92-2116-M-002-008.

Wei, X., & Wu, C. Y. (2011). Holocene delta evolution and sequence stratigraphy of the Pearl River Delta in South China. *Science China Earth Sciences, 54,* 1523–1541.

Wong, G. T. F., Ku, T.-L., Tseng, C.-M., Mulholland, M., Tseng, C.-M., & Wang, D.-P. (2007). The Southeast Asian Time-series Study (SEATS) and the biogeochemistry of the South China Sea—An overview. *Deep-Sea Research II, 54,* 1434–1447.

Wu, Z. Y., Jin, X. L., Cao, Z. Y., Li, J. B., Zheng, Y. L., & Shan, J. H. (2009). Space-time contrast of two stages sand ridges on the East China Sea shelf. *Acta Oceanologica Sinica, 31*(5), 69–79 (in Chinese with English abstract).

Wu, Z. Y., Jin, X. L., Cao, Z. Y., Li, J. B., Zheng, Y. L., & Shan, J. H. (2010). Distribution, formation and evolution of sand ridges on the East China Sea shelf. *Science China-Earth Sciences, 53*(1), 101–112.

Wu, Z. Y., Jin, X. L., Li, J. B., Zheng, Y. L., & Wang, X. B. (2005). Linear sand ridges on the outer shelf environment in the East China Sea during the last 35,000 years. *Chinese Science Bulletin, 50*(21), 2517–2528.

Wu, G., Qin, J., & Mao, S. (2003). Deep-water Oligocene pollen record from South China Sea. *Chinese Science Bulletin, 48*(22), 2511–2515.

Wu, W., Tan, W., Zhou, L., Yang, H., & Xu, Y. (2012). Sea surface temperature variability in southern Okinawa Trough during last 2700 years. *Geophysical Research Letters, 39.* http://dx.doi.org/10.1029/2012GL052749, L14705.

Xia, F., Zhang, Y., Wang, Q., Yin, Y., Wegmann, K. W., & Liu, J. P. (2013). Evolution of sedimentary environments of the middle Jiangsu coast, South Yellow Sea since late MIS 3. *Journal of Geographical Sciences, 23*(5), 883–914.

Xiang, R., Li, T. G., Yang, Z. S., Li, A., Jiang, F., & Cao, Q. (2003). Geological records of marine environmental changes in the southern Okinawa Trough. *Chinese Science Bulletin, 48*(2), 194–199.

Xiang, R., Sun, Y., Li, T., Oppo, D. W., Chen, M., & Zheng, F. (2007). Paleoenvironmental change in the middle Okinawa Trough since the last deglaciation: Evidence from the sedimentation rate and planktonic foraminiferal record. *Palaeogeography, Palaeoclimatology, Palaeoecology, 243,* 378–393.

Xiang, R., Yang, Z., Saito, Y., Fan, D., Chen, M., Guo, Z., et al. (2008). Paleoenvironmental changes during the last 8400 years in the southern Yellow Sea: Benthic foraminiferal and stable isotopic evidence. *Marine Micropaleontology*, *67*, 104–119.

Xiang, R., Yang, Z., Saito, Y., Guo, Z., Fan, D., Li, Y., et al. (2006). East Asia Winter Monsoon changes inferred from environmentally sensitive grain size component records during the last 2500 years in mud area southwest off Cheju Island, ECS. *Science in China (Series D)*, *49*(6), 604–614.

Xie, H., Jia, G., Peng, P., & Shao, L. (2007). Sea surface temperature variations in the southwestern South China Sea over the past 160 ka. *Acta Oceanologica Sinica*, *26*(2), 49–55.

Xing, L., Jiang, Y., Yuan, Z., Zhang, H., Li, L., Zhou, L., et al. (2013). Holocene temperature records from the East China sea mud area southwest of the Cheju Island reconstructed by the $U^{K'}_{37}$ and TEX$_{86}$ paleothermometers. *Journal of Ocean University of China*, *12*(4), 599–604.

Xu, J. (2004). *Quaternary planktonic foraminiferal assemblages in the southern South China Sea and Paleoclimatic variations*. Tongji University, Dissertation, 98 pp.

Xu, D., Lu, H., Wu, N., Liu, Z., Li, T., Shen, C., et al. (2013). Asynchronous marine-terrestrial signals of the last deglacial warming in East Asia associated with low- and high-latitude climate changes. *Proceedings of the National Academy of Sciences of the United States of America*, *110*(24), 9657–9662.

Xu, K., Milliman, J. D., Li, A., Liu, J. P., Kao, S. J., & Wan, S. M. (2009). Yangtze- and Taiwan-derived sediments on the inner shelf of East China Sea. *Continental Shelf Research*, *29*, 2240–2256.

Xu, X., & Oda, M. (1999). Surface-water evolution of the eastern East China Sea during the last 36,000 years. *Marine Geology*, *156*, 285–304.

Xu, J., Wang, P., Huang, B., Li, Q., & Jian, Z. (2005). Response of planktonic foraminifera to glacial cycles: Mid-Pleistocene change in the southern South China Sea. *Marine Micropaleontology*, *54*, 89–105.

Xue, Z., Liu, J. P., DeMaster, D., Nguyen, V. L., & Ta, T. K. O. (2010). Late Holocene evolution of the Mekong subaqueous delta, southern Vietnam. *Marine Geology*, *269*, 46–60.

Yamamoto, M., Kishizaki, M., Oba, T., & Kawahata, H. (2013). Intense winter cooling of the surface water in the northern Okinawa Trough during the last glacial period. *Journal of Asian Earth Sciences*, *69*, 86–92.

Yang, Z. G. (1985). Sedimentology and environment in south Huanghai Sea shelf since Late Pleistocene. *Marine Geology and Quaternary Geology*, *5*(4), 1–19 (in Chinese with English abstract).

Yang, C. S. (1989). Active moribund and buried tidal sand ridges in the East China Sea and the southern Yellow Sea. *Marine Geology*, *88*, 97–116.

Yang, Z. S., & Liu, J. P. (2007). A unique Yellow River-derived distal subaqueous delta in the Yellow Sea. *Marine Geology*, *240*, 169–176.

Yao, L., Xue, B., Yao, M., Chen, R., Zhao, Q., & Yang, D. (2007). By using $U^{K'}_{37}$ indes to study sea surface paleotemperature in South China Sea since the Last Glacial Maximum (in Chinese). *Journal of Marine Science*, *25*(4), 24–31.

Yu, H., Liu, Z. X., Berné, S., Jia, G., Xiong, Y., Dickens, G. R., et al. (2009). Variations in temperature and salinity of the surface water above the middle Okinawa Trough during the past 37 kyr. *Palaeogeography, Palaeoclimatology, Palaeoecology*, *281*, 154–164.

Yu, K. F., & Zhao, J. X. (2009). Coral reefs. In P. Wang & Q. Li (Eds.), *The South China Sea: Paleoceanography and sedimentology* (pp. 229–255): Springer.

Zhang, M. S., He, Q. X., Han, C. R., Ye, Z. Z., Ju, L. J., Li, H., et al. (1989). *Sedimentary geology of reef carbonate, Xisha Islands.* Beijing: China Science Press, 113 pp.

Zhang, Y., Ji, J., Balsam, W. L., Liu, L., & Chen, J. (2007). High resolution hematite and goethite records from ODP 1143, South China Sea: Co-evolution of monsoonal precipitation and El Niño over the past 600,000 years. *Earth and Planetary Science Letters*, *264*, 136–150.

Zhang, C., Yang, Y., Tao, J., Chen, Y., Yao, P., & Su, M. (2013). Suspended sediment fluxes in the radial sand ridge field of South Yellow Sea. *Journal of Coastal Research Special Issue*, *65*, 624–629.

Zhang, H., Zhou, H., Lu, B., Chen, R., & Zhang, F. (2005). Environmental change of paleo-oceanography recorded in the sediment stratum in the northern part of South China Sea since the post glacial period. *Acta Oceanologica Sinica*, *25*(3), 52–58 (in Chinese).

Zhao, Q. (2005). Late Cainozoic ostracod faunas and paleoenvironmental changes at ODP Site 1148, South China Sea. *Marine Micropaleontology*, *54*, 27–47.

Zhao, M., Huang, C.-Y., Wang, C. C., & Wei, G. (2006). A millennial-scale $U^{K'}_{37}$ sea-surface temperature record from the South China Sea (8°N) over the last 150 kyr: Monsoon and sea-level influence. *Palaeogeography, Palaeoclimatology, Palaeoecology*, *236*, 39–55.

Zhao, M. X., Huang, C.-Y., & Wei, K.-Y. (2005). A 28,000 year $U^{K'}_{37}$ sea-surface temperature record of ODP Site 1202B, the southern Okinawa Trough. *Terrestrial, Atmospheric and Oceanic Sciences*, *16*, 45–56.

Zhao, Q., Jian, Z., Zhang, Z., Cheng, X., Wang, K., & Zheng, H. (2009). Holocene paleoenvironmental changes of the inner-shelf mud area of the East China Sea: Evidence from foraminiferal faunas. *Marine Geology and Quaternary Geology*, *29*(2), 75–82 (in Chinese with English abstract).

Zhao, M., Wang, P., Tian, J., & Li, J. (2009). Biogeochemistry and the carbon reservoir. In P. Wang & Q. Li (Eds.), *The South China Sea: Paleoceanography and Sedimentology* (pp. 439–483): Springer.

Zheng, G. (Ed.), (1991). *Quaternary geology in the Yellow Sea*. Beijing: Science Press, 164 pp (in Chinese).

Zheng, L., & Chen, W. (Eds.). (1993). *Marine sedimentation process and geochemical studies in the South China Sea* (201 p.). Beijing: China Ocean Press. (in Chinese).

Zheng, F., Li, Q., Li, B., Chen, M., Tu, X., Tian, J., et al. (2005). A millennial scale planktonic foraminiferal record of mid-Pleistocene climate transition in the northern South China Sea. *Palaeogeography, Palaeoclimatology, Palaeoecology*, *223*, 349–363.

Zheng, Y., Zheng, H., & Wang, K. (2010). Sea-level changes since the last glacial as recorded in inner shelf sediments of the East China Sea. *Journal of Tongji University (Natural Science)*, *38*(9), 1381–1386 (in Chinese).

Zhong, G., Geng, J., Wong, H. K., Ma, Z., & Wu, N. (2004). A semi-quantitative method for the reconstruction of eustatic sea level history from seismic profiles and its application to the southern South China Sea. *Earth and Planetary Science Letters*, *223*, 443–459.

Zhou, H., Li, T., Jia, G., Zhu, Z., Chi, B., Cao, Q., et al. (2007). Sea surface temperature reconstruction for the middle Okinawa Trough during the last glacial–interglacial cycle using C37 unsaturated alkenones. *Palaeogeography, Palaeoclimatology, Palaeoecology*, *246*, 440–453.

Zhu, Y. Q., Zheng, C. K., & Feng, Y. (1984). The shelf geomorphic features of the East China Sea. *Donghai Marine Science*, *2*, 1–13 (in Chinese with English abstract).

Zong, Y., Huang, G., Switzer, A. D., Yu, F., & Yim, W. W.-S. (2009). An evolutionary model for the Holocene formation of the Pearl River delta, China. *The Holocene*, *19*, 129–142.

Chapter 7

Hydrocarbon and Mineral Resources

7.1 INTRODUCTION

A large amount of petroleum geology data have been accumulated over the last decades with the rapid development of the oil industry in the China Seas (Figure 7.1), and this chapter will overview offshore oil and gas fields in the Bohai Sea, Yellow Sea, East China Sea (ECS), and South China Sea (SCS), including the newly discovered deepwater hydrocarbon fields and the gas hydrate occurrences. The gas hydrate potentials and mineral resources such as coastal placers and hydrothermal mineral deposits will also be outlined.

Offshore hydrocarbon exploration in China started in the late 1950s and early 1960s. Early efforts focused on discovering and defining regional sedimentary basins (Jin, 1989). The first commercial discovery was made in 1967 in the Bohai Basin. But it was not until the 1980s that large offshore oil and gas fields were discovered in offshore basins, with the advances in new technologies and joint ventures. Currently, the Bohai Basin, ECS basin, and the basins in the continental margins of the SCS are the most productive basins (Figure 7.1). The present new exploration fronts are deepwater and deep pre-Cenozoic plays.

Today, over 1000 wells have been drilled (Figure 7.2), and more than 120 commercial oil and gas fields and over 120 petroleum structures have been found in basins offshore China (Zhu, 2010; Zhu & Mi, 2010). Since 2005, the exploration pace has been much accelerated, with rapid increase in annual reserve discovered (Figure 7.3). The total annual production is over 5000×10^8 m^3 now.

Jin (1989) summarized thoroughly the petroleum geology and hydrocarbon potentials of basins in the SCS region. Gong (1997) and Gong, Li, and Xie (1997) were among the earliest in synthesizing petroleum geology of the China Seas. Reflecting much deepened understanding based on extensive research activities, the first decade of the twenty-first century saw the publication of many books on hydrocarbon potentials of sedimentary basins in the China Seas (Cai, 2005; Chen, Shi, & Xu, 2003; Gong & Li, 2004; Jia & Gu, 2002; Liu & Li, 2001; Tian & Yu, 2000; Zhang, Sui, & Lin, 2010; Zhu, 2009).

Developments in Marine Geology, Vol. 6. http://dx.doi.org/10.1016/B978-0-444-59388-7.00007-X

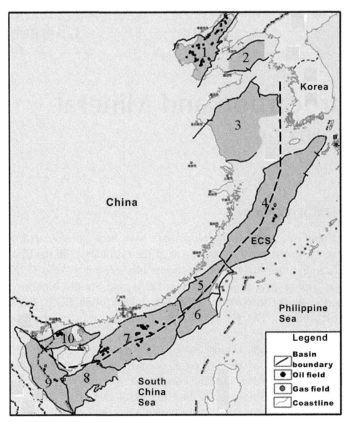

FIGURE 7.1 Oil and gas fields in Cenozoic sedimentary basins off the China's coasts. 1. Bohai Basin; 2. North Yellow Sea Basin; 3. South Yellow Sea Basin; 4. East China Sea basin; 5. Taixi Basin; 6. Taixinan Basin; 7. Pearl River Mouth Basin; 8. Qiongdongnan basin; 9. Yinggehai basin; 10. Beibuwan Basin. *From Zhu (2010).*

7.2 HYDROCARBON RESOURCES

From Figure 7.1, we observe a belt of offshore sedimentary basins along the continental margin of China. Although floored by basement of different tectonic affiliations, these basins all formed by Cenozoic continental margin rifting and faulting. As a result, the geometry and subsidence styles of these basins are controlled essentially by large-scale regional faults.

7.2.1 Bohai Basin

The Bohai Sea is the first explored offshore area (Gong & Wang, 2001) and is the most proliferous sea in hydrocarbon production in China. The readers are to be reminded that the Bohai Basin is much larger than the Bohai Sea and the onshore areas of the basin are larger than the offshore part.

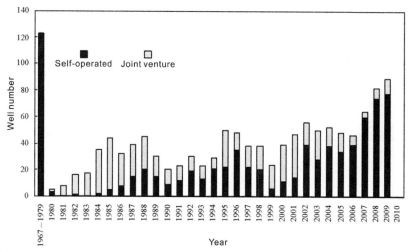

FIGURE 7.2 Statistics of self-operated and joint-ventured wells (Zhu, 2010).

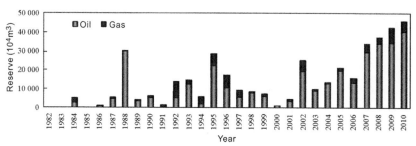

FIGURE 7.3 Annual reserve discovered in China offshore basins (Zhu, 2010).

The Bohai Basin, including both its onshore and offshore sectors, can be subdivided into eleven extensional systems and three strike-slip systems (Figure 7.4). The extensional systems consist of normal faults and transfer faults. The normal faults predominantly trend NNE and NE, and their attitudes vary in different tectonic settings. Three groups of half-graben are identified according to the orientation and dip of boundary faults between graben and horst (Zhang, Zhu, & Shao, 2001). The first group occurred in the southwestern part, and most of boundary faults dip eastward; the second happened in the middle and south, with most of boundary faults dipping southward; and most of boundary faults dip westward in the northeastern area in the third group.

The high exploration potential is thanks to the thick Eocene and Oligocene source rocks deposited in half-grabens formed during the early rifting phase (Zhu & Ge, 2001). Eocene mudstones deposited in a deep lacustrine

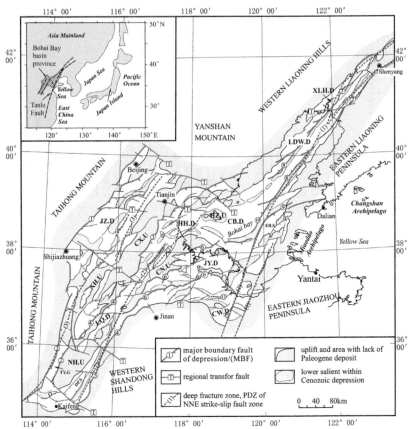

FIGURE 7.4 Cenozoic structures of the Bohai Basin (Qi & Yang, 2010). JZ.D, Jizhong depression; CX.U, Cangxian uplift; HH.D, Huanghua depression; LQ.D, Linqin Depression; XH.U, Xunheng Uplift; DP.S, Dongpu sag; NH.U, Neihuang uplift; TY.G, Tangyin graben; XLH.D, Xialiaohe depression; LDW.D, Liaodongwan depression; HZ.U, Haizhong uplift; CB.DvCentral Bohai depression; EB.S, E. Bohai sag; CN.U, Chenning uplift; JY.D, Jiyang depression; CW.D, Changwei depression. ①, ②, ③, ④, ⑤, ⑥, ⑦, ⑧, ⑨, and ⑩ are master boundary faults (MBFs). The names of master right-lateral strike-slip faults: (1) the Shenyang–Weifang strike-slip fault zone (northern part of the Tanlu fault zone), (2) the Huanghua–Dongming strike-slip fault zone (Lanliao fault zone), and (3) the Baxian–Shulu–Handan strike-slip fault zone (Shulu–Handan fault zone).

environment (Allen, MacDonald, Zhao, Vincent, & Brouet-Menzies, 1997; Chang, 1991; Figure 7.5). Since the Oligocene, rapid subsidence in the offshore Bohai Basin not only provided favorable environments for accommodating source rocks but also enhanced hydrocarbon generation (Chi, 2001; Gong & Wang, 2001). Recently, deeper Paleogene source rocks with good oil-generation potential are also suggested to produce deeper high waxy oil (Li, Pang, Liu, & Jin, 2006). Crude oils produced from the Jiyang depression

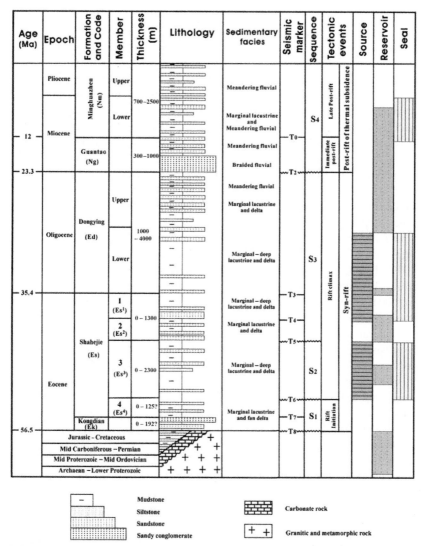

FIGURE 7.5 Generalized stratigraphy and source–reservoir–seal assemblage in the offshore Bohai Basin (Yang & Xu, 2004).

are high in sulfur and differ from most lacustrine oils in this basin and elsewhere (Wang, Wang, Simoneit, Zhang, & Zhang, 2010). Physical properties, biomarker ratios, and aromatic hydrocarbon assemblages reveal that these sulfur-rich heavy oils originate from spatially limited Eocene lacustrine carbonates and marls at the early oil-generation window. Sulfur, associated with evaporite and carbonate environments, is also a significant factor in petroleum formation at abnormally low thermal stress levels (Wang, Wang, Simoneit, et al., 2010).

Genetically different reservoir rocks have been formed in the offshore part of the basin in different periods, including sandstones deposited in fan deltaic, deltaic, turbiditic, meandering river, braided river, and marginal lacustrine environments, and shallow-water carbonate platforms (Yang & Xu, 2004).

Gas reservoirs include Neogene and Paleogene units and the underlying Mesozoic and Paleozoic basement. Pre-Cenozoic buried hill reservoirs consist of Archean migmatites, Proterozoic and Paleozoic carbonates, and Mesozoic pyroclastic rocks. Reservoir porosities are from various dissolution and fractures formed by weathering, erosion, karstification, and tectonic activities (Yang & Xu, 2004). The buried hills are long-lived positive features that experienced karstification at the top and were sealed by organic-rich source rocks. Sandstones deposited during the synrift stage in turbiditic and fan deltaic environments form Eocene and Oligocene reservoirs, and channel sandstones deposited in meandering and braided river environments in the transition from rifting to thermal subsidence form Oligocene–Neogene reservoirs.

The Paleogene tectonic depocenter in the Bohai Basin migrated from the basin fringe to the center and lastly concentrated in the offshore area (Yang & Xu, 2004). This has made rather regular areal distribution of oil and gas reservoirs (Figure 7.6). Since the Oligocene, especially during the Neogene, the right-lateral strike-slip movement of the Tanlu fault was relatively intense in the offshore part of the basin, causing differences between the offshore and onshore sectors in sedimentation, structure, and petroleum geology.

In onshore depressions, oil and gas fields are mainly discovered on the slopes and central anticlinal belts within the Paleogene half-grabens, and only a small number of small fields have been discovered on the rises between sags (Yang & Xu, 2004). But in the offshore sector, most major discoveries in the Neogene reservoirs are in the drape anticlines on the rises, while the Paleogene reservoirs within the half-grabens are secondary. Four types of offshore hydrocarbon accumulation belts have been identified, namely, the drape anticlinal belt on the rise, structural belt on the gentle slope, structural belt on the steep slope, and central faulted anticlinal belt in the central part of the depression (Figure 7.5).

Two rapid tectonic subsidence events during the Paleogene rifting formed two regional mudstone seals (Figure 7.5). Thermal subsidence during the middle–late Miocene formed the third regional seal for the Neogene reservoirs, being fine-grained shallow lacustrine mudstones (Yang & Xu, 2004). These three regional seals assembled with genetically different reservoirs that developed in different geologic times, forming four major reservoir and seal assemblages (He, Zhu, & Li, 2001).

Structural reservoirs include fault-block and fault-nose reservoirs (Figure 7.7). Nonstructural reservoirs include lithologic, stratum-overlap, and buried hill reservoirs (Figure 7.7). The Suizhong 36-1 oil field is one of the largest offshore oil fields in China. The structure is an anticline on

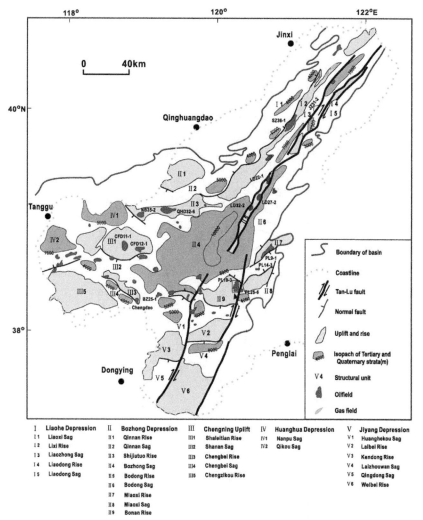

FIGURE 7.6 Cenozoic structural units and distribution of major oil/gas fields in the offshore Bohai Basin (Yang & Xu, 2004). The central faulted anticline belt along the Tanlu fault provides significant traps.

a faulted basement high and consists of Oligocene fluviodeltaic and lacustrine sandstones that contain heavy oil in a combination structural–stratigraphic trap (Figures 7.8 and 7.9; Gustavson & Gang, 1992). In fact, the structures of many of the large oil and gas fields in the Bohai Basin are anticlines overlying on buried basement highs (Zhu & Mi, 2010).

Stratigraphic traps are often related with unconformities. There are two first-order unconformities in the basin, namely, Neogene/Paleogene and

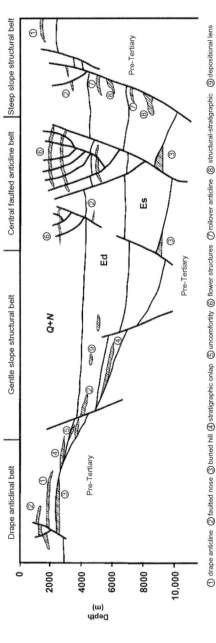

FIGURE 7.7 Schematic hydrocarbon accumulation model for the offshore Bohai Basin (Yang & Xu, 2004).

① drape anticline ② faulted nose ③ buried hill ④ stratigraphic onlap ⑤ uncomformity ⑥ flower structures ⑦ rollover anticline ⑧ structural-stratigraphic ⑨ depositional lens

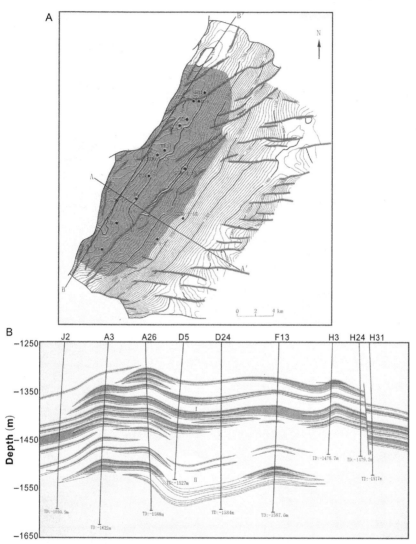

FIGURE 7.8   The Suizhong 36-1 oil field (Zhu & Mi, 2010). (A) Structural map of the oil-bearing Dongying Formation composed of lacustrine deltaic sandstones. (B) Profile of the Suizhong 36-1 oil field showing multiple layers of oil- and gas-bearing sandstones.

Paleogene/Mesozoic–Paleozoic unconformities. The former contains an over-lap trap reservoir in the steep slope belt and both an overlap trap and uncon-formable barrier reservoirs in the ramp belt (Jin, Yan, Cheng, & Su, 2009). The latter contains the buried hill reservoirs in swells and subswells. In the Paleogene, there are three second-order and ten third-order unconformities. Stratigraphic reservoirs can occur above and beneath the second-order

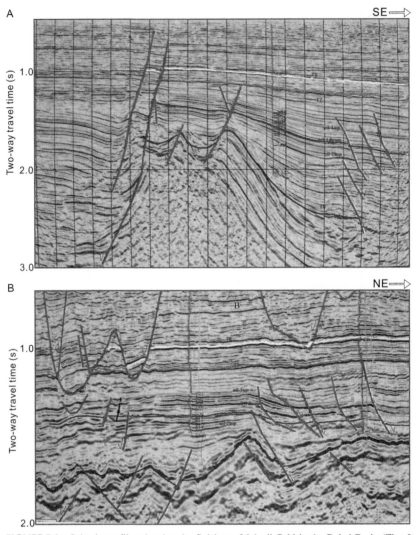

FIGURE 7.9 Seismic profiles showing the Suizhong 36-1 oil field in the Bohai Basin (Zhu & Mi, 2010).

unconformities, and those related to the third-order unconformities are in small scale but numerous (Jin et al., 2009).

Many structural traps formed by tectonic inversions in the last ~5 Ma. Several giant oil fields are discovered in the shallow Neogene reservoirs. A set of shallow faults and a central faulted anticlinal belt formed during the Neogene in the shallow sequences, under the influence of right-lateral strike-slip faulting of the Tanlu fault. These shallow faults networked with deep extensional faults, serving as hydrocarbon conduits. It is noticed that most petroliferous basins in China lie in the areas with intense neotectonism that has occurred since $\sim$5.1 Ma. Neotectonism is characterized by depocenter migration and angular unconformity, basalt extrusion and magma activities, and intensive later faults forming multitude anticlines.

The PL19-3 giant oil field (the largest one so far found offshore China), for example, is situated where faulting activities occurred violently during the late Neogene neotectonism (Zou, Gong, Teng, & Zhuang, 2011). In such places, recent, probably ongoing, petroleum charging into shallow Neogene reservoirs could result from active petroleum generation from multiple source rock intervals and the development of overpressure and the late-stage fault reactivation. These processes all lead to rapid petroleum accumulation (Hao, Zhou, Zou, Teng, & Yang, 2012). The oil accumulation rate can greatly exceed the rate of oil loss from in-reservoir biodegradation and vertical leakage through faults, and therefore, large oil fields can be preserved even within the active fault zones (Hao et al., 2012). Based on modeling of secondary petroleum migration pathways (PPMPs), Hao, Zou, Gong, and Deng (2007) suggested that most commercial petroleum accumulations are on the predicted preferential PPMPs formed by focusing of numerous "small petroleum streams" close to the kitchens, and most large fields have more than one PPMP to convey petroleum to the traps.

Different from that of the middle and shallow zones, deep zone reservoirs ($>$3500 m depth) in the Bohai Basin are generally distributed in ring form surrounding hydrocarbon generation center, showing an obvious source-controlling pattern (He, Wang, & Han, 1998). The disposition of structures associated with large high-density flow sand bodies is favorable for forming deep zone reservoir system; the existence of superhigh-pressure belt and abnormal porosity belt determines whether oil and gas can be gathered to have high production.

Geochemical analysis shows that shallow Neogene gas pools, in traps formed on top of the paleotopographic highs and along the margin of the secondary depressions, can be closely associated with heavy oils. They occur as gas caps or associated gases within the heavy oil pools or in separate gas pools above, or updip from, the heavy oil pools (Zhu et al., 2005). The stable carbon isotopes of methane in these gases are up to 1% more positive than those of the thermogenic gases in the deep Paleogene reservoirs. These signatures, together with the moderate to severe biodegradation of the associated heavy oils in the shallow Neogene strata, suggest that the formation of the shallow natural gases can result from the anaerobic degradation of accumulated oils in reservoir.

7.2.2 North Yellow Sea Basin

The North Yellow Sea Basin (NYSB) developed in the basement of the North China continental block and comprises four depressions and three uplifts (Figure 5.28; Cai, Dai, Chen, Li, & Sun, 2005; Li, Lu, Liu, & Xu, 2012). East–west- and northeast-trending faults mainly developed in the late Jurassic to the early Cretaceous. Minor Cretaceous to Paleogene NW-trending faults are found in the eastern depression. Large-scale NNE-trending normal–transcurrent faults in the central and western parts of the basin formed in the Eocene and may constitute boundaries of depressions (Li et al., 2012).

This basin has been so far not productive. The eastern depression has the biggest sediment thickness and is likely more prosperous in oil and gas resources. Oil–gas shows have been discovered in the depression from Jurassic, Cretaceous, and Paleogene units and from Upper Proterozoic and Lower Paleozoic carbonate rocks. Paraffinic oils recovered from Mesozoic and Paleogene sandstone intervals are thought to have been generated from different source beds (Massoud, Scott, & Killops, 1991, 1993). The Mesozoic and Cenozoic petroleum-bearing systems (source–reservoir–seal) are separated by a layer of mudstone at the base of the Eocene succession (Li et al., 2012).

The source rocks of the Mesozoic petroleum system consist of late Jurassic shallow to moderately deep lacustrine black shale (primary source) and dark early Cretaceous lacustrine shale (secondary source). Reservoir rocks are the late Jurassic–early Cretaceous alluvial fan, delta, turbidite fan, and shore facies sandstone and fractured and weathered pre-Mesozoic basement rocks. Fine-grained sandstone to siltstone intercalate with minor glutenite. The rock types include feldspar sandstone and feldspathic litharenite. The reservoir spaces are characterized by secondary solution openings and fissures, microporosity, and fine throat to microthroat, indicating the low-porosity and low-permeability reservoirs. The fissures, sedimentary facies, and diagenesis are believed to be the main controls on the reservoir quality of the sandstones in the study area. The fissure zones and secondary porosity zones are suggested to be the favorable areas for future exploration (Wang, Wang, & Zhang, 2010). The Eocene–Oligocene mudstone and the late Jurassic–early Cretaceous shallow to moderately deep lacustrine mudstone form cover rocks.

The primary source rocks of the Cenozoic petroleum system comprise the Eocene shallow–moderately deep lacustrine mudstone. The Eocene sandstone of fan delta, delta, turbidite fan, and shore facies forms the reservoir, with the Neogene and Oligocene basal transgressive mudstone as the cap rock in the NYSB (Li et al., 2012).

According to the drilling data of Korea, the kerogen types of hydrocarbon source rock (black shales) of Paleogene, Jurassic, and Cretaceous is III, I ,and II, respectively, and they have all become mature, with Ro between 0.4 and 1.32 (Cai et al., 2005; Massoud et al., 1993). In the eastern depression, the Eocene sequence contains organic-rich (TOC up to 7%) lacustrine mudstones with a mixture of type I and III kerogens; the Lower Cretaceous argillaceous sediments are organically lean (TOC 0.9% average) and contain gas-prone type III kerogen (Massoud et al., 1991).

Potential hydrocarbon traps include the stratigraphic overlap and underlying unconformity traps, truncated monoclinic traps, traps associated with stratigraphic overlap and buried hill, and traps associated with the faults and unconformity (Yuan et al., 2010).

In combination with regional geologic data and the proven oil–gas discoveries, anomalous hydrocarbon gases found at the lower atmosphere and microbial germs in the subsurface sediment in the NYSB indicate that there are three oil–gas prospects in the North Yellow Sea, which are located in the east, middle, and west of the area, respectively (Gong et al., 2007). The middle oil–gas prospect area shows homogeneously and continuously high-methane concentrations at the lower atmosphere and the strongest anomalies in microbial germs in the sediment. The west oil–gas prospect has occasionally high-methane concentrations at the lower atmosphere and relatively weak microbial anomalies in subsurface sediment. The east oil–gas prospect is between the middle and the west prospect in hydrocarbon potential based on the geochemical exploration (Gong et al., 2007).

7.2.3 South Yellow Sea Basin

The South Yellow Sea Basin consists of two parts, namely, the Northern South Yellow Sea Basin (SYSB-N) and the Southern South Yellow Sea Basin (SYSB-S). The eastern part of the SYSB-N is called Gunsan (Kunsan) Basin in Korea, and the SYSB-S is the eastern extension of the onshore Subei basin (Figure 7.10).

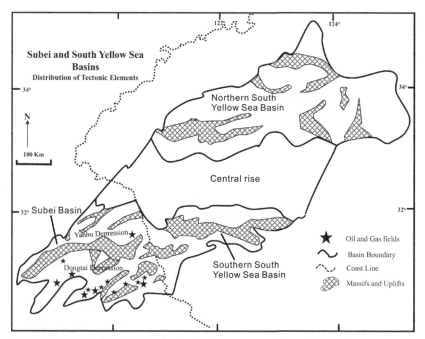

FIGURE 7.10 The South Yellow Sea Basin is composed of the northern and southern basins (Gong, Huang, & Chen, 2011).

As known in many Cenozoic rift basins in SE Asia, source rocks in continental rift basins include early synrift lacustrine, late synrift nonmarine and fluviodeltaic, and early postrift marine sediments (Doust & Sumner, 2007). Thus, lacustrine and nonmarine or fluviodeltaic sediments deposited during the rift phase are potentially the most important source rocks in the basin (Yoon et al., 2010).

However, aside from hydrocarbon display and small amount of oil occurrence in wells, no commercial oil and gas in the South Yellow Sea has been produced so far, and productive and favorably mature source rocks have yet to be discovered (Shin, 2013; Wu, Ni, & Cai, 2008; Yao et al., 2010). However, significant volumes of both gas and oil have been produced onshore from synrift reservoirs in the Subei basin.

The Cretaceous and Paleogene lacustrine organic-rich dark mudstones discovered by numerous wells are most probable source rocks (Figure 7.11; Tao, 1993; Yao et al., 2010; Zhang, Wei, Xu, Tao, & Chen, 1989). The known thick Mesozoic–Paleozoic marine facies deposited on the Yangtze craton are also taken to be important source rocks. In the Subei basin, the onshore extension of the SYSB-S, multiple groups of source rocks are indicated, but oil and gas found presently are primarily from middle to Upper Paleozoic marine source rocks. The known nonmarine source rocks in the Subei basin consist mainly of type III kerogen (TOC 1.08% average; Wu et al., 2008).

From observations of five wells drilled in the Kunsan Basin, the eastern part of the SYSB-N, reservoir rocks can be deposited in fluvial plain to shallow lacustrine and alluvial environments (Shin, 2013). The sediments are poorly sorted and are medium to fine or very fine-grained clastics. In all wells, porosities are low to moderate, ranging from as low as 2% to the highest 15%, and permeabilities are uniformly quite low due to possible early cementation with quartz or calcite or the presence of abundant clays in the pore throats (Shin, 2013). Adding to the poor potential for these sediments as reservoir rocks is the interbedding of volcanics in the Cretaceous portion, causing baking and alteration of clay fractions (Shin, 2013).

Thin mudstones are common in both the Cretaceous and Cenozoic sections and could become effective seals. Sealing capacity could be best developed

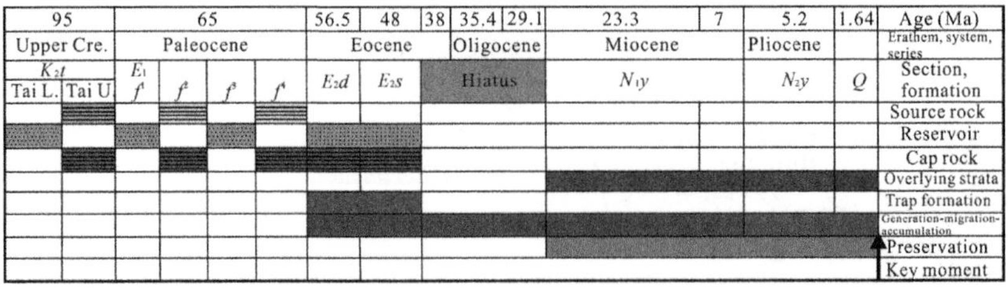

95	65		56.5	48	38	35.4	29.1	23.3	7	5.2	1.64	Age (Ma)
Upper Cre.	Paleocene		Eocene			Oligocene		Miocene		Pliocene		Erathem, system, series
K₂t Tai L.	Tai U.	E₁ f⁴	f³	f²	f¹	E₂d	E₂s	Hiatus	N₁y	N₂y	Q	Section, formation
												Source rock
												Reservoir
												Cap rock
												Overlying strata
												Trap formation
												Generation-migration-accumulation
												Preservation
												Key moment

FIGURE 7.11 Sketched diagram showing hydrocarbon reservoir formation in the Northern South Yellow Sea Basin (Yao et al., 2010). Cre., Cretaceous; Tai L., lower part of Taizhou Formation; Tai U., upper part of Taizhou Formation.

near the depocenter, but is expected to be reduced in the basin margins due to lateral lithofacies changes (Shin, 2013). Substantial but relatively young structural features (anticline, fault-nose, and fault-block) postdate hydrocarbon generation and migration in the basin and therefore may not form effective traps. Based on one-dimensional basin modeling using the basin subsidence and other geologic and geochemical data, Yoon et al. (2010) suggested that hydrocarbon expulsion peaked in the early Oligocene at the IIH-1Xa well, predating later tectonic inversions that created traps, and hydrocarbon generation at the dummy well began in the late Paleocene (Figure 7.12). Most source rocks in the depocenter passed the main expulsion phase except for the shallowest source rocks. Faulting that continued during the rift phase (~middle Miocene) of the subbasin probably acted as conduits for the escape of hydrocarbons. Thus, the anticlinal structure and associated faults may trap hydrocarbons that have been charged from the shallow source rocks in the depocenter since the middle Miocene (Yoon et al., 2010).

Recent research and exploration activities in the South Yellow Sea have focused on Paleozoic/early Mesozoic marine facies, because this area was flooded with the Paleo-Tethys and therefore should be of high hydrocarbon potential like other similar tectonic settings such as the Sichuan Basin. However, due to the intensive Cenozoic tectonic and magmatic events, promises from Paleozoic/early Mesozoic reservoirs remain unclear, at least not so optimistic as in the Sichuan Basin of the upper Yangtze region. The available reflection seismic data have not imaged the Paleozoic/early Mesozoic strata very well, making the interpretations of deep targets extremely difficult. Considering active Neogene faulting and magmatism in the basin, future exploration may focus on secondary traps and potential CO_2 fields.

7.2.4 ECS Basin

The early Cenozoic ECS basin is roughly NE–SW-oriented, nearly parallel to the East China shoreline and margin. It is bounded to the east by the Diaoyudao Uplift and the Okinawa Trough and to the west by the Zhemin uplift (Figures 3.12 and 7.13). Rifting started in the basin at the late Cretaceous/early Paleocene and continued to the late Eocene (~65 to ~40 Ma). Due to multiphased rifting and eastward migration of rifting events, the basin is characterized by a series of depressions and highs that are also roughly NE–SW-oriented (Figure 7.13; e.g., Hsu et al., 2001; Liu, 1992; Sun, 1981). The two largest oil- and gas-bearing depressions in the ECS basin are the Xihu Depression and the Lishui Depression, which will be further discussed next.

7.2.4.1 The Xihu Depression

The largest and most productive depression in the ECS basin is the Xihu Depression, which can be divided into different zones from the west to the east (Figure 7.14).

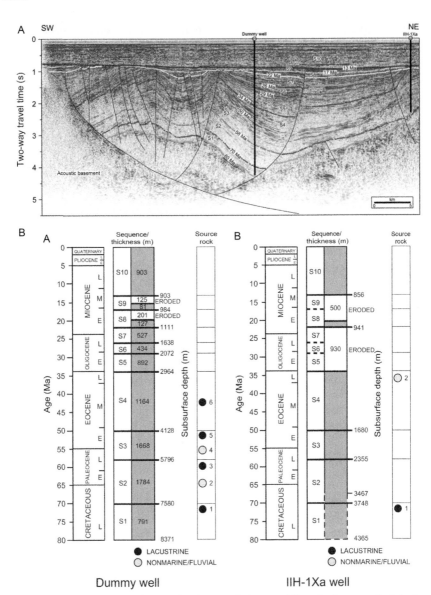

FIGURE 7.12 (A) Seismic line showing the curved, downward-flattening basement fault in the eastern part of the SYSB-N (Yoon et al., 2010). Rotation of the SW-dipping fault resulted in thicker sediment accumulation (S4–S6) in the hanging wall. Ages of the unconformities including the top of the acoustic basement were estimated from the correlation at the IIH-1Xa well with the unconformities reported by KIGAM (1997) and Ryu, Kim, Kwak, Kim, and Park (2000). (B) Stratigraphic column, thicknesses of sequences, and source rock locations of the dummy well (A). Stratigraphic column, thicknesses of sequences, and source rock locations of the IIH-1Xa well (B) (Yoon et al., 2010).

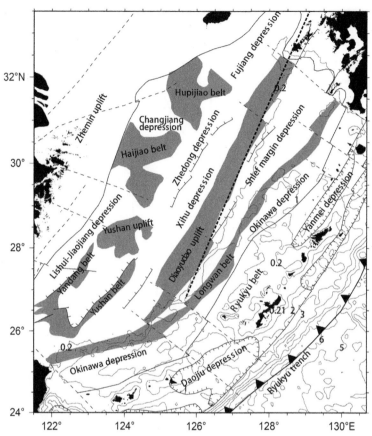

FIGURE 7.13   Depressions and uplifts in the East China Sea Basin. Dotted lines represent bathymetric contours. Dashed lines indicate possible strike-slip faults. Heavy dashed line indicates the possible dextral fault proposed by Le Pichon and Mazzotti (1997). *Modified from Sun (1981), Liu (1992), and Hsu, Sibuet, and Shyu (2001).*

Paleocene, Eocene, Oligocene, and Miocene strata in the ECS basin have hydrocarbon generation potential (Figure 7.15). Generally, in the northern part of the basin, these strata are mainly nonmarine sediments (in shallow lacustrine, fluvial, and swamp environments), with some marine–continental transitional facies and littoral and neritic sediments, but in the southern part, the Cenozoic strata are generally littoral–neritic sediments (Liu, Zhao, Chen, & Huo, 2005). Well Pinghu-1 in the Xihu Depression indicates that the middle and upper parts of the Eocene sunrift units are thin interbeds of dark gray mudstone with siltstone–fine sandstone containing interbedded layers of asphaltic coal and earbargillite, which appear to be the most important source rocks in the basin (Ye et al., 2007). Oligocene–Miocene postrift strata all coexist with coal seams formed in a swamp environment (Liu et al., 2005).

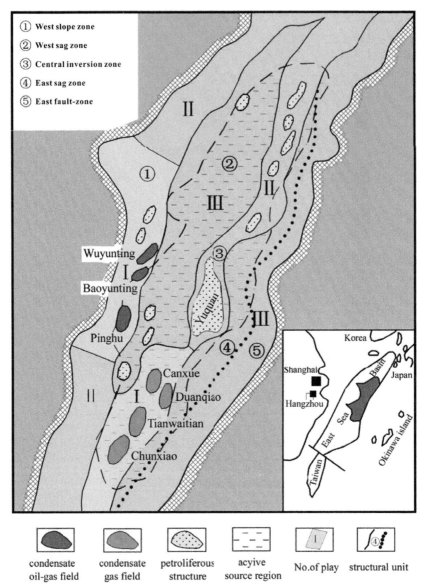

FIGURE 7.14 The distribution of gas fields and favorable gas exploration play in the Xihu Depression, ECS basin (Tao & Zou, 2005).

Experimental analyses revealed that gas and liquid can be generated and expelled from the coal structure of the Xihu Depression (Lin, Leon, Kevin, & Eric, 1997). In the southern part of the basin, these Oligocene–Miocene strata formed in closed shallow-marine–marine/terrestrial alternative environments (Wu & Xue, 1993).

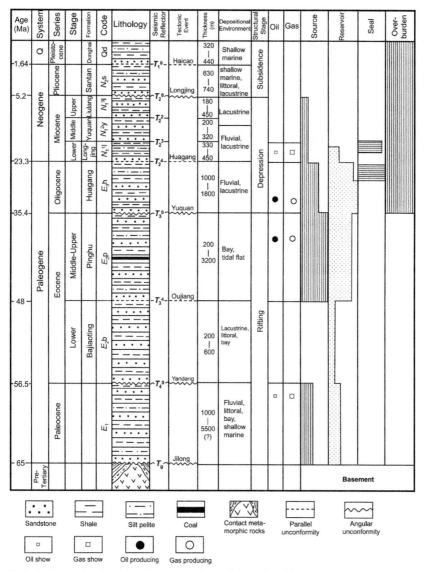

FIGURE 7.15 A schematic diagram showing major stratigraphic units, tectonic events, and characteristics of the petroleum system in the Xihu Depression (Ye, Qing, Bend, & Gu, 2007).

Most of these potential source rocks have entered the oil window and are believed to have provided the charge for the petroleum accumulations (Chen & Ge, 2003; Ye et al., 2007). Modeling shows that the source rocks have experienced three phases of hydrocarbon generation during Eocene–Oligocene, middle Miocene, and late Pliocene, respectively (Li & Li, 2003;

Yang, He, Jiang, & Liu, 2003; Zhang, Wu, & Zhou, 2005). Zhang et al. (2005) studied inclusions and diagenetic minerals and showed that oil and gas accumulations in the central inversion zone were mainly self-sourced and partly derived from the Eocene–Oligocene source rocks, with large-scale oil and gas charging from the late Miocene to the Pliocene.

Lee, Kim, Shin, and Sunwoo (2006) summarized that potential reservoir rocks are fluvial, lacustrine, deltaic, and shallow-marine sandstones and weathered and fractured basement rocks comprising the basement highs. In the Pinghu field, Eocene fluvial and littoral sandstones produce gas and condensate, and Oligocene lacustrine and fluvial sandstones produce oil (Silverman et al., 1996). In the southern Oujiang Depression, gas was discovered from Paleocene shallow-marine sandstones and lowstand deposits and oil from fractured basement rocks (Primeline, 2004). Clinoform seismic facies along the footwalls of the grabens and half-grabens in the southern ECS basin correlate with nonmarine to marginal-marine coarse-grained sediments with good reservoir potential (Armentrout et al., 1994).

In the Pinghu area, low-permeability shales are important local seals, whereas marine shales provide viable regional seals (Silverman et al., 1996). Many of the reservoirs in the Pinghu field are lenticular, and the intra-formational shales serve both as vertical seals and as elements in combinational structural–stratigraphic traps (Silverman et al., 1996).

Three condensate oil–gas fields and four condensate gas fields have been discovered in the Xihu Depression (Figure 7.14). Various types of hydrocarbon traps (e.g., faulted anticlines, overthrusts, rollover anticlines, faults, unconformity traps, combination structural–stratigraphic traps, weathered basement, and stratigraphic traps) have been recognized (Lee et al., 2006).

Inversion tectonics at the end of Miocene in the central part of the Xihu Depression has generated large-scale anticlines that are often bounded by deep-seated faults (Li, Zhou, Ge, & Mao, 2007; Figures 7.16 and 7.17). A large gas field in the ECS basin is mainly controlled by this inversion structure (Figure 7.14). The formation of inversion-related traps in the central inversion zone occurred at the end of the late Miocene, matching very well with the timing of oil and gas charge (Figure 7.18). According to the composition of gas, the relative content of methane is high, and it is a typical wet coal-derived gas (Dai et al., 2009). The carbon isotope of ethane is very heavy (Dai et al., 2009).

From seismic stratigraphy and sonic logging analysis, Li et al. (2007) found that wells in the northern part of the Xihu Depression have larger relative uplift and sharply increased erosions at the central inversion zone along the late Miocene unconformity. This may have resulted in less favorable preservation of traps and large-scale leakage of oil and gas (Figure 7.18; Yang et al., 2011). Furthermore, the deeper-water depths to the south in the initial rifting phase could also produce more favorable source rocks. Nevertheless, petroleum systems are dependent on many factors, and more favorable source

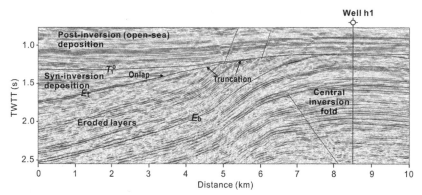

FIGURE 7.16 Seismic reflection profile showing late Miocene-faulted anticline and associated unconformity trap in the Xihu Depression (Li et al., 2007). T_1^0, inversion unconformity; E_t, top of eroded strata; E_b, base of eroded strata; TWTT, two-way travel time.

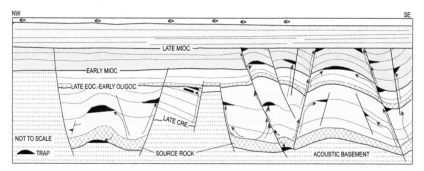

FIGURE 7.17 Schematic cross section of the northern ECS basin showing potential traps and possible petroleum migration pathways (Cukur, Horozal, Lee, Kim, & Han, 2012). Petroleum from the mature source rocks migrated preferentially toward the anticlinal structures and associated faults and may be accumulated within these traps.

and trap conditions to the south do not preclude potentially prosperous targets in the northern segment of the central inversion zone of the Xihu Depression.

7.2.4.2 The Lishui–Jiaojiang Depression

The Lishui–Jiaojiang Depression to the southwest of the Xihu Depression is another important area of hydrocarbon formation and accumulation in the ECS basin (Figure 7.19). The oil–gas discoveries made so far concentrate mainly in these two depressions.

In the Lishui Depression, there exist three suites of source rocks, namely, the Lower Paleocene lacustrine deposits, the Upper Paleocene marine deposits, and paralic coal-bearing deposits (Chen, Liang, & Zhang, 2005; Chen, Wu, & Wu, 2000; Jiang et al., 2004). The major source rock is the Lower Paleocene lacustrine deposit, which is about 2 km thick and rich in organic

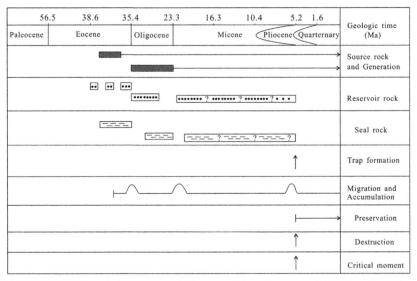

FIGURE 7.18 The petroleum system in the central inversion zone of the Xihu Depression (Yang, Xu, Zhao, & Sun, 2011).

matter (mainly type II kerogen). The Upper Paleocene marine deposit is over 5 km thick and moderately abundant in organic matter (dominated by type III kerogens; Chen et al., 2008).

Basin modeling showed that Lower Paleocene source rocks have been generally in oil windows except for those with shallower burial, and those at the central sag have become highly mature to overmature. Upper Paleocene source rocks have a slightly smaller area of maturity but those at the central sag have also been at peak generation. Upper Paleocene source rocks remain mostly at immature–low mature stage (Chen et al., 2008).

Organic inclusion analyses and basin modeling show that there are three periods of hydrocarbon migration and accumulation (Ge, Chen, Fang, & Shen, 2003; Jiang, Wang, & Jiang, 2005; Sun & Xi, 2003). The first petroleum migration occurred at the end of Paleocene, with hydrocarbon fluid generated mainly from the Lower Paleocene lacustrine deposits. The second migration happened during the middle–late Eocene, and hydrocarbon fluids were generated from, and nearly accumulated in, the Upper Paleogene suites. The third one charged during the later Oligocene to Miocene (Sun & Xi, 2003). Some of the oil–gas within the deeper reservoirs had remigrated upward along the faults and mixed with petroleum generated from the Upper Paleocene. The abiogenic CO_2- bearing fluid charged into the reservoir after the third stage (Sun & Xi, 2003).

Natural gases discovered in the Lishui Depression vary greatly in gaseous compositions. Nonhydrocarbon gases are dominated by abiogenic CO_2, and

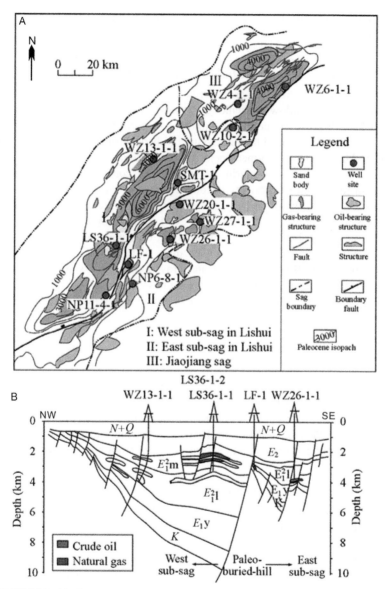

FIGURE 7.19   (A) Structural zoning of Lishui Sag (Chen, Ge, Chen, Deng, & Liang, 2008). (B) Sketched geologic cross section of Lishui Sag (Chen et al., 2008).

hydrocarbon gases are classified as wet gases produced by the mixed organic matter at peak generation (Chen et al., 2008). Currently, the LS36-1 oil–gas pool is the only commercial oil–gas reservoir in Lishui Sag, and natural gases in the pool are mainly derived from the Lower Paleocene lacustrine source rock rather than the Upper Paleocene marine or coal-bearing source rocks (Chen et al., 2008).

7.2.4.3 Late Quaternary Shallow Biogenic gas Reservoirs

In the coastal Hangzhou Bay area, northern Zhejiang Province, eastern China, late Quaternary shallow biogenic gas reservoirs buried at less than 100 m depth have been discovered and exploited (Figure 7.20; Lin et al., 2004). The river strongly incised the underlying old beds during a period of the Last Glacial Maximum and formed incised valleys filled with fluvial sediments and buried by postglacial marine sediments. All commercial gas fields are found in floodplain sand bodies surrounded by impermeable clays. The sand bodies are buried 30–60 m deep and are 3.0–7.0 m thick, with a maximum thickness of more than 10 m (Lin et al., 2004). Rapid deposition of overlying sublittoral-marine bay sediments supplied not only abundant gas sources but also good preservation conditions. The main hydrocarbon sources are dark gray clays of the floodplain facies and gray muds of the sublittoral-marine bay facies (Lin et al., 2004). Quaternary-incised valleys and floodplains in other coastal areas of eastern China remain promising targets of further exploration for shallow biogenic gas.

7.2.5 Northern SCS

There are numerous petroliferous basins around the SCS, located on wide continental shelves and slopes. Tectonics and sedimentary features of these basins have already been fully covered in Chapters 3 and 5. They are mostly related to the Paleogene rifting in the region, but their sedimentary environments and timings of rifting and subsidence vary from basin to basin.

7.2.5.1 Pearl River Mouth Basin

The major source rock in the Pearl River Mouth Basin (PRMB) is the Eocene synrift deep-lake deposition dominated by type I and II kerogens. The Oligocene synrift coal-bearing sequence dominated by mixed type II/III and type III kerogens is a minor source rock (e.g., Yu, Zou, & Gong, 2009). Lower Miocene postrift marine deposits also form source rocks. Measured vitrinite reflectance (Ro) increases regularly with burial depth (Chen et al., 2003), suggesting no suppression or retardation. Hydrocarbon generation in the Eocene source rock began at about 30 Ma and peaked between 23 and 18 Ma (Figure 7.21; Yu, Zou, et al., 2009), but this time window could change from place to place in the basin (Xiao et al., 2009).

The main reservoir rocks are widespread early Miocene sandstones of different depositional facies, including meandering stream, distributary channel, river mouth bar, longshore bar, and continental shelf (Figure 7.21).

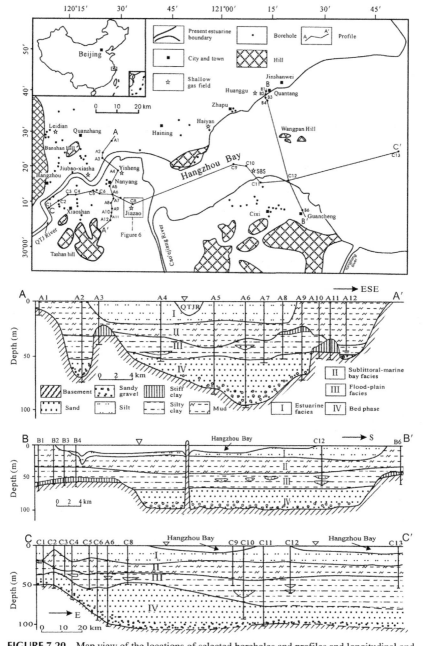

FIGURE 7.20 Map view of the locations of selected boreholes and profiles and longitudinal and cross sections of the late Quaternary strata in the Hangzhou Bay area (Lin, Gu, Li, Zhao, & Jiang, 2004). (A and B) The incised valley is entirely filled by fluvial sediments and buried by sublittoral-marine bay and estuary sediments. Fluvial-floor facies sediments are limited to the incised valley and are composed of sandy gravels and sands with normal upward grading (Lin, Huang, Zhu, Li, & Jiang, 1999). The floodplain facies occurs mainly in the valley and consists of gray silty clays. Several sand lenses are developed inside the floodplain facies. The thickness of the fluvial sediments reaches its maximum at the incised-valley thalweg. Both depth and width of the incised valley increase toward the east. (C) Late Quaternary sediments are separated from the underlying basement by a sinuous erosion surface in the incised valley, which increases in depth toward the sea. The thickness of the sediments and sand lenses of the floodplain facies also increases seaward.

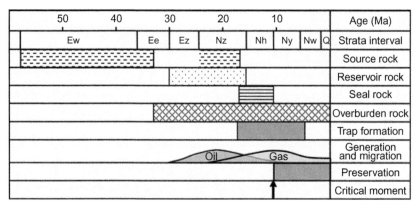

FIGURE 7.21 Event chart showing the major elements of the petroleum system and timing of petroleum generation and migration in the Pearl River Mouth Basin (Yu, Zhou, et al., 2009; Yu, Zou, et al., 2009). Ew., Wenchang Formation (Fm.); Ee., Enping Fm.; Ez., Zhuhai Fm.; Nz., Zhujiang Fm.; Nh., Hanjiang Fm.; Ny., Yuehai Fm.; Nw., Wanshan Fm.; Q., Quaternary.

Contemporary carbonate platforms developed near the Dongsha uplift are also major reservoir rocks (Yu, Zhou, et al., 2009; Yu, Zou, et al., 2009; Figure 7.21). The regional seal rocks are thick late Miocene shales directly overlying the reservoir rocks of the LH11-1 oil field, the largest one so far found in the PRMB (Figure 7.21). The shales are deposited in prodelta and shallow-marine environments during transgression.

Deepwater Plays

Recently, there are major exploration breakthroughs in the deepwater part of the northern SCS continental margin. Three sets of source rocks have been identified from the deepwater basins: (1) Eocene synrift lacustrine source rocks that are known to produce primarily oil in the shelfal basins, (2) Oligocene gas-generative coal-bearing source rocks developed in a late-rifting stage in a terrestrial–marine transitional environment, and (3) Miocene postrift marine source rocks (Zhu, Zhong, Li, Xu, & Fang, 2012). The Oligocene source rocks could be the primary source rock for the deepwater basins. The Miocene marine source rocks, first identified in the deepwater and ultra-deepwater basins south to the PRMB and Qiongdongnan basin, may have significant contribution to hydrocarbon generation. Previously, they have only been discovered in the Yinggehai basin, where it is the primary source rock (Zhu, Zhang, & Gao, 2008).

As already seen in Chapter 3, rifting and thermal subsidence of deepwater basins differ from that of the shelfal basins. This could have led to different mechanisms of hydrocarbon generation. Zhu et al. (2012) found that the source rocks are distributed in three zones. In the PRMB, the Eocene lacustrine source rocks dominate in the shallow shelf, the Oligocene transitional coal-bearing source rocks play a key role in the deepwater slope, and the Miocene marine suites are possible primary source rocks for the deepwater to ultra-deepwater areas south to the PRMB (Figure 7.22).

Simulations indicate that hydrocarbon distribution is primarily controlled by thermal evolution of the source rocks (Zhu et al., 2012). For example, the source rock in the Qiongdongnan basin is mature–overmature and mainly generated gas, whereas in the northern PRMB, the source rock is mature–highly mature and generated both gas and oil. In the Baiyun Sag, the source rock is found mainly in the oil-generation window because the sag is less evolved thermally than the main depocenter to the north that is in the gas generation stage. In the deepwater to ultra-deepwater areas (Figure 7.23), hydrocarbon generation is possible under high heat flow conditions (Zhu et al., 2012).

Oligocene and Miocene deepwater deltas, submarine fans, deep-sea channel fill, and mass flow complexes, sourced from the Pearl River and Red River, are primary deepwater exploration targets (Figure 7.23; Pang, Chen, & Zhu, 2006; Wu, Han, Ma, Dong, & Lü, 2009; Zhu et al., 2012). These sandstones have favorable porosity and permeability. The Pearl and Red Rivers have long been primary sources of sediment for the depositional

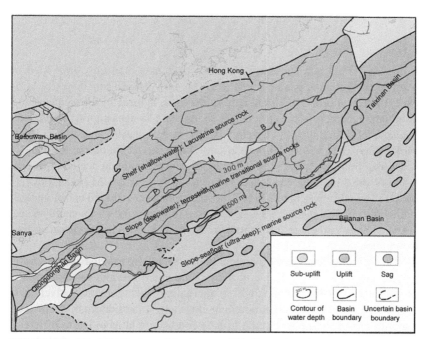

FIGURE 7.22   Zonal distributions of the source rocks in the northern South China Sea continental margin (Zhu et al., 2012). Taixinan Basin = Tainan Basin.

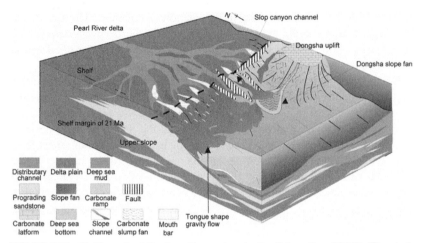

FIGURE 7.23 Shelf-margin delta deposition system in the Baiyun Sag, PRMB (Zhu et al., 2012).

system in the northern SCS (Chen et al., 2003; Gong et al., 1997; Pang, Chen, & Peng, 2007; Wang, Xu, et al., 2011; Zhu et al., 2008). Capping rocks on the top are mudstones with a huge thickness deposited in the postrifting stage.

Carbonate Plays

In the late Oligocene and early Miocene, carbonate platforms developed in many areas with the opening of the SCS (Figure 7.24). During the early Miocene, these platforms were gradually submerged, favored by climatic/eustatic optimum, except for some structurally highs (e.g., Dongsha uplift) where carbonate reefs continued to develop (Chen & Hu, 1989; Hutchison, 1989; Moldovanyi et al., 1995). From the late Oligocene to Holocene, the PRMB acted as a trap for clastic sediments, and this contributed to a prolonged isolation of the faulted basement and led to improved conditions for reef development on the basement highs of the Dongsha and Shenhu massifs (Fulthorpe & Schlanger, 1989; Tyrrell & Christian, 1992; Zampetti et al., 2005). These carbonate platforms are important targets for hydrocarbon exploration (Figure 7.25).

The Miocene carbonates at the Liuhua 11-1 field consist of six stratigraphic units (zones A to F; Figure 7.24). These units are zones with abundant porosity (F, D, and B with up to 50%) alternating with zones of relatively low porosity (E, C, and A) with a minimum close to 0% (Sattler et al., 2004). The majority of observed pores in the Liuhua cores are related to late leaching in the deep burial realm. Those pores are vugs with noncemented pore walls, which can be fabric-selective, or non-fabric-selective, truncating depositional

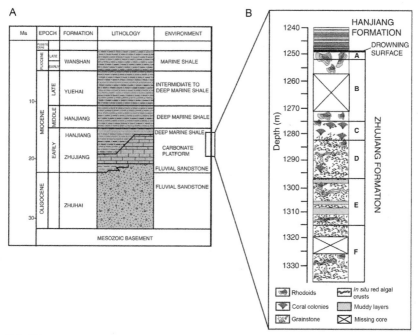

FIGURE 7.24 (A) Stratigraphic associations for sediments overlying the Dongsha massifs (Moldovanyi, Waal, & Yan, 1995; Zampetti, Sattler, & Braaksma, 2005). (B) Sedimentary log of well LH11-1-4 showing the five reservoir stratigraphic units A to F (Sattler, Zampetti, Schlager, & Immenhauser, 2004; Zampetti et al., 2005).

grains, and early diagenetic and burial cements. The vugs clearly postdate stylolitization. Two tight zones lack this porosity because of meteoric cementation and increased compaction and related calcite precipitation prior to late leaching (Sattler et al., 2004). Sattler et al. (2004) proposed that the following processes and factors influenced the spatial distribution of late-leached porosity: (i) depositional facies and early marine cementation, (ii) meteoric cementation, and (iii) compaction and related burial cementation. Migrating corrosive fluids were diverted along bedding-parallel flow barriers of the tight zones and leached the intervals in-between (the porous zones) thereby emphasizing the depositional pattern (Sattler et al., 2004).

7.2.5.2   Taixinan Basin

Like the Qiongdongnan basin, the formation of the Taixinan Basin (also called Tainan Basin) was closely related to the Oligocene opening of the SCS. Rifting and subsidence here did not occur until the late Oligocene. Therefore, promising Paleogene source rocks found in most other basins are not expected here. Potential source rocks are Lower Miocene, Upper Oligocene, and Cretaceous or even earlier units (Jin, 1989).

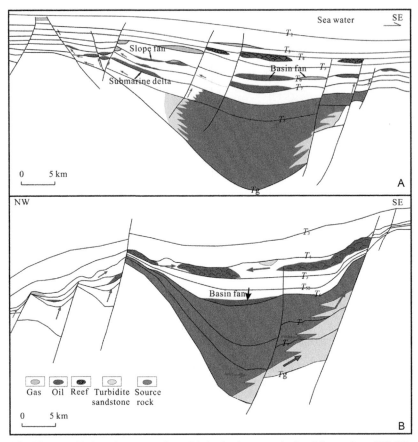

FIGURE 7.25 Hydrocarbon reservoir models in the deepwater basin of the northern SCS (Wu, Han, et al., 2009; Wu, Yuan, et al., 2009). (A) Siliciclastic reservoir section. Deep submarine fan sandstone reservoir is formed in the upper section, prograded delta front sandstone reservoir is formed in the middle section, and lacustrine subaqueous sandstone reservoir could be formed in the synrifting graben. (B) Reef carbonate reservoir section. Miocene carbonate reservoir could be formed in the upper section, and lacustrine subaqueous sandstone reservoir could be formed in the synrifting graben.

There are petroliferous structures in the central uplift zone, which are Oligocene sandstone reservoir and Cretaceous fractured reservoir (He, Xia, Chen, & Sun, 2006). The commercial oil and gas found so far are centralized within the central uplift zone.

Coral reefs can also be identified from seismic sections in the Taixinan Basin, and they developed mostly near the paleocontinental shelf break (Figure 7.26). They could form excellent reservoirs like in the PRMB. However, they have not been extensively explored, and their hydrocarbon potential remains unclear. Drilling efforts have focused on the central uplift of the

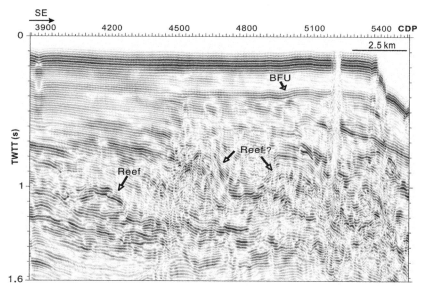

FIGURE 7.26   Seismic section showing coral reefs near the paleocontinental shelf break identified within the northern depression of the Taixinan Basin. BFU, foreland basin unconformity.

basin, because it is regionally situated in favorable petroleum migration area, with good petroleum accumulation conditions (He et al., 2006). Oil shows and gas are found in faulted anticlines and fractured basement highs from Lower Cretaceous, Oligocene, and Miocene units. Neogene sandstones in the area are regarded to have low porosity and limited areal distributions (Jin, 1989).

7.2.5.3   Yinggehai and Qiongdongnan Basins

These two basins are tectonically distinct but geographically connected basins (Figure 7.27). In petroleum exploration, these two basins are often taken as a single basin often called the Ying-Qiong Basin. Potential source rocks in the Yinggehai and Qiongdongnan basins include the early Oligocene synrift coal-bearing unit of marine–terrestrial alternative facies and early–middle Miocene and Pliocene postrift formations composed mainly of marine shales (Figure 7.28; Chen, 1990; Hao et al., 1998; Hao, Li, & Zhang, 1996; Hao, Sun, Li, & Zhang, 1995; Zhang, Li, & Huang, 1993). Potential source rocks in both basins contain higher plant-derived type III organic matter (Hao et al., 1995). Therefore, these rocks are gas-prone rather than oil-prone. The gases accumulate mainly in the Upper Oligocene, Miocene, Pliocene, and Quaternary strata.

The Yinggehai basin is characterized by thick Cenozoic sediments, rapid subsidence and sedimentation rates, and abnormal fluid pressures. All potential source rocks in the Yinggehai basin are dominated by higher plant-derived

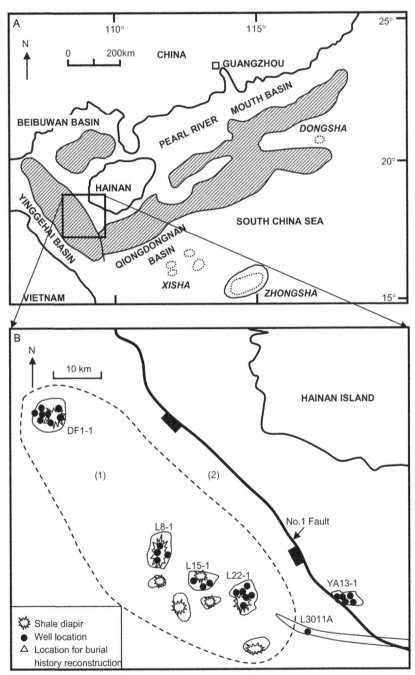

FIGURE 7.27 Maps showing the location of the DF1-1 gas field in the Yinggehai basin, SCS (Wang & Huang, 2008). (1) Diapir belt; (2) Yingdong slope.

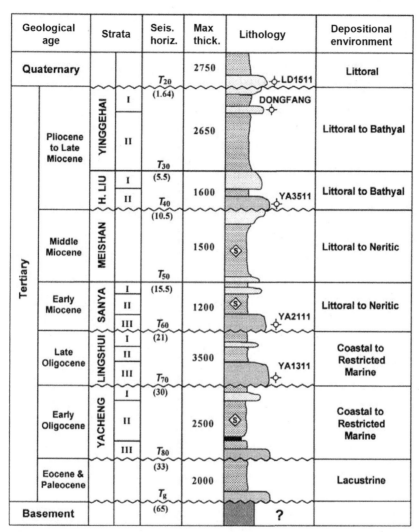

FIGURE 7.28 Generalized stratigraphy of the Yinggehai and Qiongdongnan basins (Hao, Li, Sun, & Zhang, 1998). SEIS. HORIZ., seismic horizons; MAX THICK., maximum thickness in meters; H. LIU, Huangliu Formation. Numbers in parentheses are ages of seismic horizons in million years. Potential source and reservoir rock intervals are marked.

type III kerogens with low organic richness. The strongly overpressured rocks have abnormally low vitrinite reflectance (Ro), caused probably by overpressure retardation that allows relatively low thermal stabilities to be preserved at higher temperatures than normal (Hao et al., 1996; Price & Wenger, 1992).

Rapid deposition, gas generation in the source rocks, and thermal expansion of pore fluids caused overpressuring in the Miocene shales in the center of the Yinggehai basin. This overpressuring together with active faulting formed diapir structures during 5–0.5 Ma (Huang, Xiao, & Dong, 2002). A number of commercial, thermogenic gas accumulations have been found in the area with shale diapirs (Figure 7.27). Hao et al. (1998) suggested that intense overpressuring in a closed fluid system triggered the development of diapirs under regional extensional stresses. These diapirs acted as conduits, through which fluids (both water and hydrocarbons) could rapidly migrate, leading to modification of thermal regime and enhancement of organic maturation and accumulation of commercial volumes of gas in a relatively short time interval (Hao et al., 1998).

The DF1-1 mud diapir field is the largest gas field so far discovered in the Yinggehai basin (Figure 7.29). Presently, it is also the largest offshore gas field. The gas pay zones are within the Pliocene fine sandstones and siltstones of the littoral and sandbar facies, with normal pressure regimes. The hydrocarbon gases (with the exception of biogenic methane) and nitrogen originated from the marine source rocks within the overpressured lower to middle Miocene strata, while CO_2 is most likely the product of thermal decomposition of the Cenozoic calcareous mudstones and pre-Paleogene carbonates through relatively long-distance vertical migration (Wang & Huang, 2008). He (1994), He, Zhang, and Chen (1995), and He, Li, and Chen (2001) concluded that there were volcanic activities in the basement of Yinggehai–Qiongdongnan basins during the spreading of the SCS. The ratio of R/Ra in Baodao reservoir is over 4.0 in the east, indicating the addition of mantle helium (Dai et al., 2009).

The field has received at least four late phases of natural gas charging, most likely after the Pliocene. Diapiric faults act as the main conduit for upward movement of fluids released from the deep high-pressure system, with the abnormally high pressure as the primary driving force for gas migration. Because of the episodic nature of the deep high-pressured fluid release and the diapir fault activity, this shallow diapir gas field is characterized by multiple gas sources and episodic fluid accumulation (Wang & Huang, 2008).

The combination of overpressure, high paleo- and present-day geothermal gradients, and diapirism has greatly influenced gas generation, migration, and accumulation in the basin. Natural gases found in the Yinggehai basin can be classified into nitrogen-, methane-, and CO_2-dominated gases, with nitrogen-dominated gases occurring only as gas shows (Hao et al., 1996). Hydrocarbons in all gas types should be of humic origin. Chemical and isotopic data indicate the presence of at least three genetic groups of gases in these basins: biogenic gas, thermogenic gas, and gases with mixed origin. Among them, thermogenic gases are dominant in the Yinggehai and Qiongdongnan basins (Huang, Xiao, & Li, 2003).

Drowned Miocene reef carbonate platform is also a promising reservoir unit, and two areas of reef carbonate reservoirs in deepwater areas (water depth >500 m) of the Qiongdongnan basin are identified (Figure 7.30; Wu, Han, et al., 2009; Wu, Yuan, et al., 2009). The seismic features of the reef

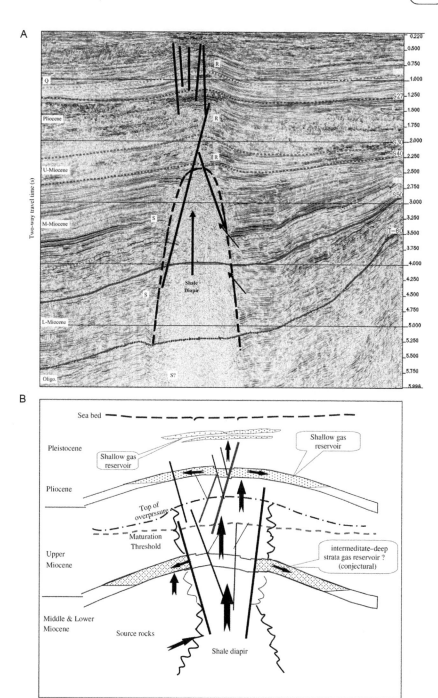

FIGURE 7.29   (A) A seismic cross section showing the key petroleum system elements in the DF1-1 gas field (Wang & Huang, 2008). R, reservoir; S, source rock. (B) Schematic gas migration and accumulation model of the DF1-1 diapiric structure (Wang & Huang, 2008).

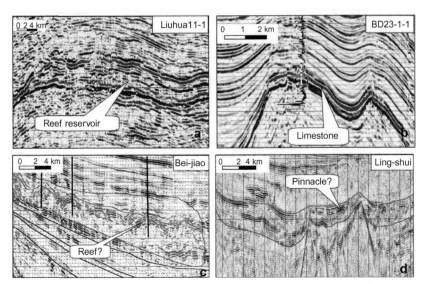

FIGURE 7.30 Seismic profiles of representative reef carbonates in different areas of the Qiong-dongnan basin (Wu, Han, et al., 2009; Wu, Yuan, et al., 2009). (A–D) are profiles across the LH11-1 field, the Beijiao Depression, well BD23-1-1, and the Ledong–Lingshui Depression, respectively.

carbonates in these two areas include chaotic bedding, intermittent internal reflections, chaotic or blank reflections, mounded reflections, and apparent amplitude anomalies (Figure 7.30). The mid-Miocene was dominated by off-shore and neritic environments. The input of terrigenous sediments was relatively minor and reef carbonates were probably widely distributed (Wu, Han, et al., 2009; Wu, Yuan, et al., 2009).

7.2.5.4 Beibuwan Basin

The Beibuwan Basin has good oil source conditions in the Cenozoic formations. In the Eocene late-rifting stage, thick organic-rich dark mudstones were deposited in a series of well-developed lakes, forming the most important hydrocarbon source rocks in this basin (Li et al., 2008). During the late Oligocene, the area was dominated by a shallow lake and swamps, leading to the deposition of organic-rich mudstones with thin coal interbeds, which is the second important source unit in the basin (Li et al., 2008). Depositional environments of the Beibuwan Basin were mainly semideep to deep lacustrine (Figure 7.31).

7.2.5.5 Mesozoic and Paleozoic hydrocarbon potential

Recently, Mesozoic and Paleozoic hydrocarbon potential has become another forefront of the next stage of exploration in the China Seas. There are two different types of hydrocarbon accumulations in the pre-Paleogene sediments: Hydrocarbons are generated in Cenozoic source beds and subsequently

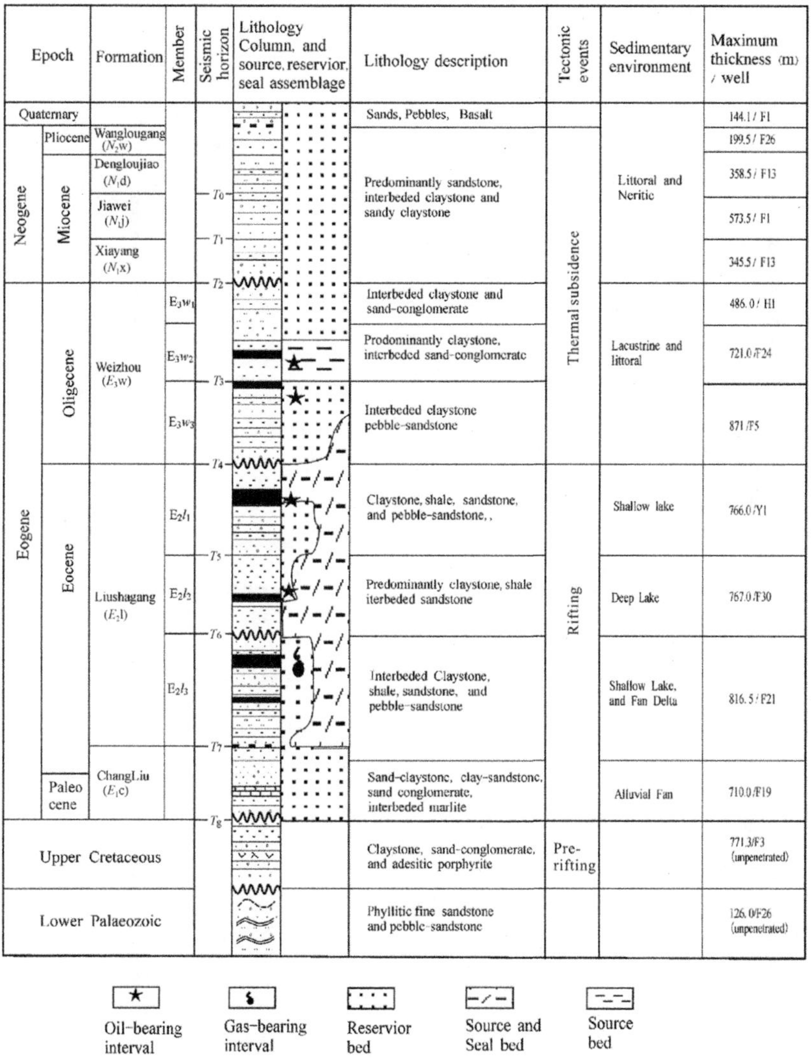

Epoch		Formation	Member	Seismic horizon	Lithology Column, and source, reservoir, seal assemblage	Lithology description	Tectonic events	Sedimentary environment	Maximum thickness (m) / well
Neogene		Quaternary				Sands, Pebbles, Basalt			144.1 / F1
		Wanglougang (N_2w)							199.5 / F26
	Pliocene	Dengloujiao (N_1d)		T_0		Predominantly sandstone, interbeded claystone and sandy claystone		Littoral and Neritic	358.5 / F13
	Miocene	Jiawei (N_1j)		T_1					573.5 / F1
		Xiayang (N_1x)		T_2					345.5 / F13
Eogene	Oligocene	Weizhou (E_3w)	E_3w_1			Interbeded claystone and sand-conglomerate	Thermal subsidence		486.0 / H1
			E_3w_2	T_3		Predominantly claystone, interbeded sand-conglomerate		Lacustrine and littoral	721.0 /F24
			E_3w_3	T_4		Interbeded claystone pebble-sandstone			871 /F5
	Eocene	Liushagang (E_2l)	E_2l_1	T_5		Claystone, shale, sandstone, and pebble-sandstone,,	Rifting	Shallow lake	766.0 /Y1
			E_2l_2	T_6		Predominantly claystone, shale iterbeded sandstone		Deep Lake	767.0 /F30
			E_2l_3	T_7		Interbeded Claystone, shale, sandstone, and pebble-sandstone		Shallow Lake, and Fan Delta	816.5 / F21
	Paleocene	ChangLiu (E_1c)		T_8		Sand-claystone, clay-sandstone, sand conglomerate, interbeded marlite		Alluvial Fan	710.0 /F19
Upper Cretaceous						Claystone, sand-conglomerate, and adesitic porphyrite	Pre-rifting		771.3/F3 (unpenetrated)
Lower Palaeozoic						Phyllitic fine sandstone and pebble-sandstone			126.0/F26 (unpenetrated)

Legend:
★ Oil-bearing interval | ⬧ Gas-bearing interval | [∴∴] Reservoir bed | [−/−] Source and Seal bed | [−−] Source bed

FIGURE 7.31 Generalized stratigraphy and tectonic events of the Fushan Depression, Beibuwan Basin (Li et al., 2008).

migrated into older rocks, or the source beds are attributable to pre-Paleogene rocks directly. Here, we are talking about the second type. Based on gravimetric, magnetic, seismic, and lithofacies data, Shi and Li (2012) had identified two large areas of Mesozoic marine sedimentation and three major phases of Mesozoic marine deposition (Figure 7.32).

The first phase was tied to the Paleozoic and early Triassic sequences within a mostly Tethyan affiliation. On the multichannel seismic section

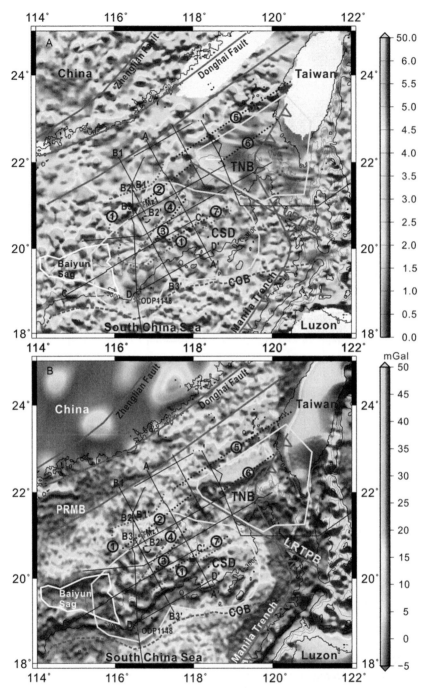

FIGURE 7.32 Detailed geologic structures revealed by magnetic and gravitational data on the northeastern SCS margin (Shi & Li, 2012). TNB, Taixinan Basin; CSD, Chaoshan depression; LRTPB, Luzon–Ryukyu transform plate boundary; COB, continent–ocean transition zone; (1), Circum Dongsha fault and sag (CDSFS); (2) and (5), offshore southeast China magnetic anomaly (SCMA); (3), belt of high gravity anomalies in the Chaoshan depression; (4), fault zone with low gravity anomalies; (6), belt of high gravity anomalies in the Taixinan Basin; (7), magmatic belt. Solid black lines labeled with AA′, B1B1′, B2B2′, B3B3′, CC′, and DD′ are locations of seismic sections shown in Figures 7.33 and 7.34.

(CC′ in Figures 7.33 and 7.34), a sequence of strong reflections at low frequencies is evident in the deep part, which differs sharply from other facies. Shi and Li (2012) interpreted it to be a carbonate formation, of most likely Permian or even older ages of the Tethyan domain. Early Permian and older carbonate rocks are found widespread along the coastal area of South China, but by the beginning of the late Permian, seawater retreated out of the coastal area due to tectonic uplift during the Triassic Indonesian orogeny and continent–continent collisions on a wide scale. The second phase was the late Triassic to early Jurassic paleo-Pacific transgression with coastal to shallow-marine clastic deposits, but this phase was then dismantled by a major southeastward regression synchronized with faulting and magmatism. The third phase of marine deposition, lasting from the middle Jurassic to the early Cretaceous, was limited to the forearc area, including the accretionary wedge and trench, of the paleo-Pacific subduction zone. Depositions from this phase, together with those from the previous two phases, were once again dismembered by pervasive late Mesozoic and Paleogene continental margin extension and rifting and by local uplift and erosion. Clearly, the sedimentary environments differed markedly from phase to phase. Mesozoic sedimentary basins as seen today are polycyclic superimposed basins that had undergone extensive tectonic compression, uplift, erosion, rifting, and magmatism (Shi & Li, 2012).

Nevertheless, the two large areas of Mesozoic sedimentation, one on each of the two conjugate margins of the SCS, have relatively thick pre-Paleogene marine deposition and weak magmatism and remain important potential areas for future hydrocarbon exploration. Before the opening of the SCS, these two regions would have been adjacent to each other, forming a large Mesozoic basin. Mesozoic strata are better preserved to the south of the Dongsha–Penghu Uplift, extending all the way to the continent–ocean boundary (Figure 7.32), and are correlative to Mesozoic strata on the southern SCS margin (Shi & Li, 2012).

7.2.6 Southern SCS

7.2.6.1 Palawan and Nansha (Dangerous Ground) area

Northeast Palawan and Reed Bank (together named as the Palawan continental block) was adjacent to the Chaoshan depression and Taixinan Basin located in the northern margin of SCS before opening of the SCS between about 32 and 16 Ma (Holloway, 1982; Shi & Li, 2012; Suzuki, Takemura, Yumul, David, & Asiedu, 2000; Taylor & Hayes, 1983). The Chaoshan depression and Taixinan Basin are also magnetically quiet and are underlain by thick Mesozoic sedimentary rocks (Shi & Li, 2012). It is found that both Mesozoic and Cenozoic magmatic activities generally subside southeastward, and the southern part had a longer period of deep Tethyan and Pacific

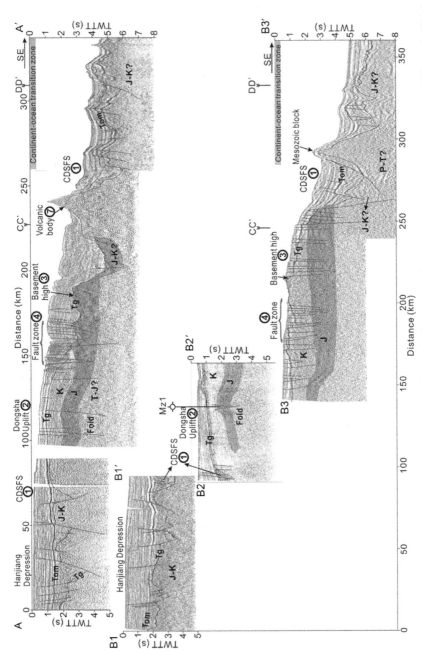

FIGURE 7.33 Multichannel northeast-striking seismic sections AA′ and B1B3′ showing major tectonic structures on the northeastern SCS margin (Shi & Li, 2012). T_g, Mesozoic/Cenozoic unconformity; T_{om}, Oligocene–Miocene unconformity; K, Cretaceous; J, Jurassic; T, Triassic; P, Permian. Subvertical line segments are interpretations of faults. See Figure 7.32 for locations of these seismic lines and further notations.

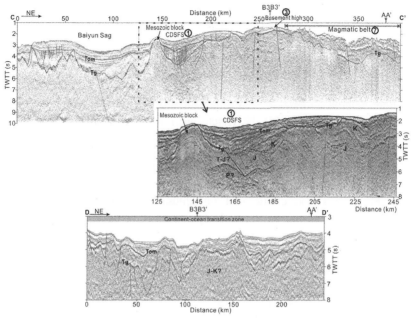

FIGURE 7.34   Southeast-striking multichannel seismic sections CC′ and DD′ showing major tectonic structures on the northeastern SCS margin (Shi & Li, 2012). T_g, Mesozoic/Cenozoic unconformity; T_{om}, Oligocene–Miocene unconformity; K, Cretaceous; J, Jurassic; T, Triassic; P, Permian. Subvertical line segments are interpretations of faults. Dashed line beneath the Baiyun Sag is the interpretation of a basement boundary bisecting an area of strong magmatism to the west and an area of little magmatism but thick Mesozoic sedimentary rocks to the east. See Figure 7.32 for locations of these seismic lines and further notations.

marine deposition (Shi & Li, 2012), making the hydrocarbon potential of Palawan even more favorable than in the Chaoshan depression and Taixinan Basin.

Magnetic data show that the Palawan continental block has very weak Mesozoic and Cenozoic volcanic activities and remained tectonically uniform during the Cenozoic (Shi & Li, 2012) and has moderate depths to Curie points and top of magnetic sources. It is interpreted that the Palawan continental block is underlain with thick economically important Mesozoic marine source rocks. Similar tectonic characteristics in the Mesozoic strata between the Chaoshan depression and Reed Bank are revealed by reflection seismic profiles (Sun, Sun, Zhou, & Liu, 2008; Yao et al., 2012). In the Taixinan Basin, commercial petroleum reservoirs, sourced from the Mesozoic strata, are found (He et al., 2006). Likewise, considerable total organic carbon content in the Lower Cretaceous sandstones is also found in the Sampaguita-1 well in the Reed Bank (e.g., Williams, 1997; Xia & Huang, 2000). Since the Chaoshan depression and Taixinan Basin show relatively high thermal maturity for the Mesozoic source rocks (He et al., 2006), the Mesozoic sequences in Palawan are also expected to be thermally mature for oil and gas generation. Thereby, Palawan and Reed Bank are a potentially favorable block for Mesozoic hydrocarbon explorations.

The Mesozoic strata in the prerift basement of the Dangerous Ground have a vitrinite reflectance of 1.0–2.5%, resulting from loading by the rock and the water column (Hutchison & Vijayan, 2010). This range is equivalent to medium volatile bituminous coal–anthracite and is generally overmature for oil—the oil window is 0.5–1.35 (Hutchison, 1983).

The offshore area of northwest Palawan, Philippines, also contains a number of proven Cenozoic exploration plays, including pinnacle reefs developed on Upper Oligocene to Lower Miocene Nido carbonate platforms (Figure 7.35), a seaward horst block reef fairway with large pinnacle reefs, early Miocene Galoc clastic unit turbidites, and four-way dip closures (Williams, 1997). Another potential exploration play is in deepwater Nido

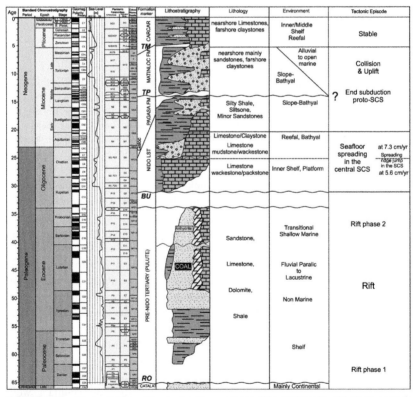

FIGURE 7.35 Chronostratigraphic chart with sequences, lithostratigraphy, lithology, tectonic events, and major unconformities interpreted on the NW Palawan shelf (Franke et al., 2011). RO, Rift onset unconformity; BU, Breakup unconformity; TP, top Pagasa unconformity; TM, top Matinloc formation.

Limestone turbidites that are found mainly in the deeper part of the grabens. Middle Miocene and Eocene sandstones are also of good reservoir potential (COMEXCO, 1994; Sales et al., 1997). The traditional and, to date, only economically productive play in northwest Palawan has been the Nido Limestone reefs, which grew on platforms and horst blocks separating the graben (Figure 7.36; Williams, 1997).

The most characteristic Nido carbonate sequence is made up of a shallow-marine carbonate platform, followed on top by limestones, wackestones, and packstones, deposited in an open marine, shelfal to bathyal setting (Franke et al., 2011). This sequence exhibits subparallel seismic reflections of high continuity and low frequencies. Above this unit, the shallow-water Nido carbonate reef complexes have developed (Figure 7.35). The platform carbonate formation started to establish in the early Oligocene (Grötsch & Mercadier, 1999). Limestone deposition may have ended at different times throughout the area, continuing on as reefal buildups almost up to the middle Miocene in some areas and elsewhere being terminated at the end of the late Oligocene (Williams, 1997) or in the early Miocene (Schlüter, Hinz, & Block, 1996). Therefore, the top Nido unconformity, though easily identifiable from seismic sections, is diachronous (Ding, Franke, Li, & Steuer, 2013; Franke et al., 2011; Hutchison & Vijayan, 2010).

Thermal maturity modeling shows that Nido Limestone prospects would have direct access to any hydrocarbons generated and migrating out of the underlying pre-Nido section. Economically, the drift sequence of Oligocene to early Miocene is of the greatest significance by far, as it contains almost

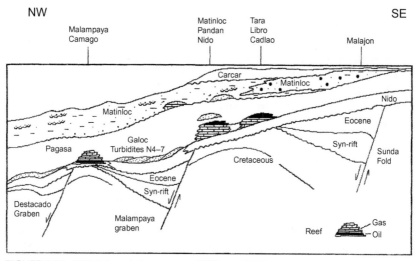

FIGURE 7.36 NW Palawan play concepts (Williams, 1997).

all the proven reservoirs, the primary seal unit, and is believed to contain the primary source rocks (Sales et al., 1997).

Probable source rocks are the Nido Limestone, the pre-Nido Paleogene synrift section, and pre-Tertiary sediments (Sales et al., 1997). The Lower Miocene to middle Miocene Pagasa Formation is organically lean with dominant kerogen of types III and IV, thus making it a gas-prone source. The early Miocene Nido Limestone platform is also generally lean, although parts of the section have been analyzed to contain significant amounts of algal debris in part and are likely to be oil-prone.

Sales et al. (1997) stated that the pre-Nido Eocene to Paleocene section as the most likely source of most of the oil and gas in northwest Palawan. The pre-Nido can be subdivided into two distinct sequences: an early (Paleocene to Eocene) synrift sequence in half-grabens separated unconformably from a marine turbidite sequence of late Eocene to possibly early Oligocene age (Figure 7.36). The half-graben sequences are believed to contain oil-prone source rocks, and all hydrocarbon discoveries to date are situated along the graben margin or overlying the graben (Williams, 1997). Drilling data show that the Eocene section consists primarily of type III and IV kerogen representing a gas-prone source, and a large percentage of middle to late Eocene shales, that is, those deposited within the rift graben environment, have good mixed gas-prone and oil-prone source potential (Sales et al., 1997).

The proven seal for the Oligocene and Miocene reservoirs is the Lower to middle Miocene claystone unit, which is regionally widespread in the northwest Palawan. For the Eocene reservoirs, intraformational claystones are anticipated to provide the seal (Sales et al., 1997).

Geochemical analyses of recovered oils in the northwest Palawan area indicate there are three major oil groups, suggesting at least three source rocks have contributed to these oils (Williams, 1997). Interpreted depositional environments for these source rocks include restricted marine carbonate, anoxic marine shale, fluviodeltaic, and lacustrine. The main phase of hydrocarbon maturation occurred about mid-Miocene time (Williams, 1997).

Like the northwest Palawan and the northern SCS continental margin, the Nansha (Dangerous Ground) area contains numerous Miocene carbonate buildups (Figure 7.37) and has proven to be a very significant gas province (Epting, 1980).

7.2.6.2 The Cuu Long and Wan'an (Nam Con Son) basins

The Cuu Long and Wan'an (Nam Con Son) Basins (Figure 3.27) in the southern Vietnam continental margin have proven active petroleum systems. The hydrocarbon occurrences are characterized by two distinct regions: (1) the oil-prone Cuu Long Basin where the oil is reserved mainly in the basement highs and (2) the gas-prone Wan'an (Nam Con Son) Basin where the gas is trapped in Miocene sands and late Miocene carbonates (Lee, Lee, & Watkins, 2001).

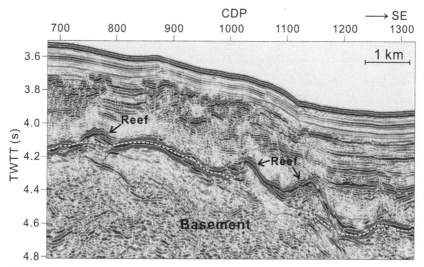

FIGURE 7.37   A seismic section showing the late Oligocene–early Miocene Nido reefs in the Nansha area. The dashed horizon is the interpreted base of the carbonate platform showing mostly strong seismic amplitudes.

These distinct trends may be attributed to differences in the timing of trap formation and the disruption of trap integrity caused by prolonged rifting and inversion in the Wan'an (Nam Con Son) Basin, an event that is naturally connected to the Oligocene–Miocene spreading of the SCS since the basin is located immediately to the west of the southwest subbasin (Figure 7.38). The prolonged rifting and inversion created shallow gas traps but adversely affected the integrity of deeper oil traps formed during the initial rifting phase (Lee et al., 2001) in the Wan'an (Nam Con Son) Basin. In the Cuu Long Basin, however, large traps are not likely in the postrift section because of the slow and quiet subsidence, and the lack of tectonic perturbations preserved the deeper traps without much oil remigration or leakage (Lee et al., 2001).

In the Cuu Long Basin, Eocene lacustrine deposits (Canh, Ha, Carstens, & Berstad, 1994) and Oligocene sediments (Areshev et al., 1992) are the known source rocks. Geochemical analysis of oil samples from the Wan'an (Nam Con Son) Basin has shown a dominance of kerogen from land plants, suggesting paralic carbonaceous mudstones and coastal plain sediments as the main source rocks (Matthews et al., 1997; Todd, Dunn, & Barwise, 1997). Coals and coaly shales are also the known source rocks for some plays in the Miocene reservoir in the Wan'an (Nam Con Son) Basin (Reid, 1997).

Miocene carbonate platforms have also been found widespread in the Wan'an (Nam Con Son) Basin, which initiated in the late early Miocene and reached their widest extent during the middle Miocene (Lü et al., 2013).

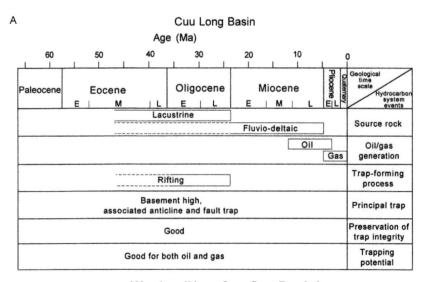

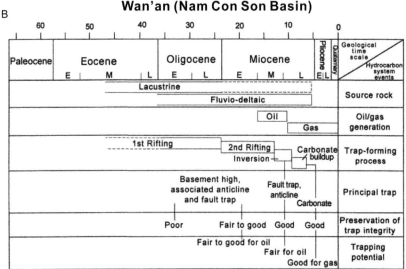

FIGURE 7.38 Source rocks, hydrocarbon generation, potential traps, and trap potential for (A) the Cuu Long and (B) Wan'an (Nam Con Son) Basins (Lee et al., 2001). In the Cuu Long Basin, large traps are associated with the older, synrift tectonics because oil formation began during the postrift phase. In the Wan'an (Nam Con Son) Basin, oil expulsion postdated the initial rifting and overlapped a second rifting and inversion but predated carbonate complexes. Thus, older and deeper traps in the Wan'an (Nam Con Son) Basin had potential to accumulate oil, whereas the carbonates are likely to contain only gas. Yet, no significant oil accumulations have been found in the deeper traps, probably due to the tectonic perturbations from prolonged rifting and inversion. Timing of hydrocarbon generation is from Areshev, Dong, San, and Shnip (1992) and Matthews, Fraser, Lowe, Todd, and Peel (1997).

7.2.6.3 Zengmu–Sabah Basin

Discussed in Chapter 5, the Zengmu–Sabah Basin developed offshore Sarawak and Sabah. The Zengmu Basin is also known as the Sarawak Basin. These two basins are characterized by high sedimentation rates and large sedimentary thickness and host many oil and gas fields. Most of the traps in these basins are anticlinal and associated with faults. The most common reservoir rock here is the Lower Miocene and Oligocene sandstone and middle Upper Miocene limestone or reef limestone. The biggest gas accumulations are found in limestone reservoirs in the Zengmu Basin (Chin, 2001).

The geochemistry of oils shows the generation of hydrocarbons from land plant-dominated source rocks, which are distributed favorably in the coastal fluvial, coastal, and lower coastal plain environments (Mat-Zin & Swarbrick, 1997). Oligocene lower coastal plain carbonaceous shale and Lower Miocene coaly/shale are the mature source rocks (Yao, Wu, Xia, & Wan, 2008). Maturity interval of the main oil-generating source rocks is in the range of 0.75–1.1 vitrinite reflectance (VR%; Gebregergis & Yukoff, 2010). The source rock containing type II and III kerogen with TOC in the range 0.5–2% generates waxy crude and gas found in oil fields in Zengmu and Sabah Basins (Abolins, 2007). The quantity and quality of the source rocks deteriorate with the distance from the coastline. Primarily related to difference in tectonic structures and sedimentation, the petroleum distribution in the Zengmu Basin is characterized by the occurrence of oil in the south and gas in the north (Yao et al., 2008).

7.3 MINERAL RESOURCES

7.3.1 Coastal Placers

Coastal placer deposits in the China Seas mainly exist in sandy and sandy-gravel coasts and formed primarily during the late Pleistocene and Holocene (Liu, 1989). Chen, Yang, et al. (2006) summarized distributions and divisions of placer deposits and ferromanganese nodules/crusts in the sea areas of China and outlined 3 prospecting zones and 24 divisions of the placer deposits (Figure 7.39).

The distribution of placer deposits is closely related to the coastal type. Minor gold placer deposits occur along the Chinese coast within river deltas that drain on-land areas of known gold occurrence. In Shandong Peninsula, placer deposits are mainly produced around lightly eroded coastal hills and bar lagoons, not in muddy tidal flat (Sun & Tan, 1986). Sun and Tan (1986) found that different kinds of placer deposits are accumulated in sediments of different grain sizes; zircon is commonly enriched in middle- and coarse-grained sediments, magnetite in fine- and middle-grained sand, placer gold in poorly sorted gravelly coarse sand with complicated composition, and quartz in fine-coarse sand.

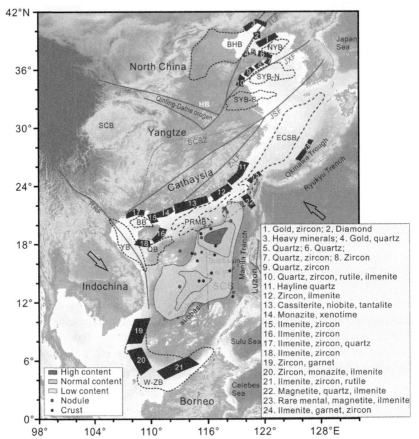

FIGURE 7.39 Prospecting zones and divisions of placer deposits in the sea areas of China. Also shown are site locations, content distributions, and zones of ferromanganese nodules/crusts in the South China Sea. BHB, Bohai Basin; NYB, North Yellow Sea Basin; SYB-N, Northern depression of the South Yellow Sea Basin; SYB-S, Southern depression of the South Yellow Sea Basin; ECSB, East China Sea Basin; TNB, Taixinan Basin; PRMB, Pearl River Mouth Basin; BB, Beibuwan Basin; QB, Qiongdongnan basin; YB, Yinggehai basin; W-ZB, Wan'an–Zengmu Basin; HB, Hefei basin; SCS, South China Sea; TLF, Tanlu fault; JXF, Jiaxiang fault; JSF, Jiangshao Fault; RRF, Red River Fault; Z-LF, Zhenghe–Lianhuashan Fault; ZNF, Zhongnan Fault; PF, Philippine Fault; SCSZ, South China suture zone; SCB, Sichuan Basin. *Modified from Qiu (1983), Chen, Yang, et al. (2006).*

Surrounding the ECS basin are titanium magnetite sands offshore north Taiwan and south Kyushu and placer deposits of ilmenite, zircon, and monazite on the west coast of Korea and east coast of China (Figure 7.40; Ludwig & Valencia, 1993). The former are mined largely for their iron content and the latter for their titanium, zircon, and rare earth elements (REEs; Clark & Li, 1991).

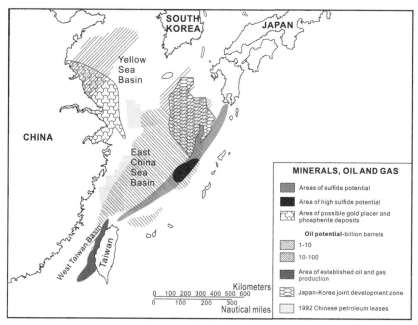

FIGURE 7.40 Petroleum and mineral resources of the East China Sea (Ludwig & Valencia, 1993).

In the SCS, the most intensively exploited mineral deposits are near-shore placer minerals of titaniferous magnetite, zircon, monazite, tin, gold, and chromite (Clark & Li, 1993). Seabed deposits of heavy minerals, such as gold, platinum, and chromite, are often a continuation of onshore placer deposits (Ritterbush, 1978). The SCS continental shelves are considered to be highly prospective for additional placer occurrence associated with such submarine features as submerged platforms and terraces, drowned rivers and sandbars, ancient beaches, and seafloors covered by relict sediments (Clark & Li, 1993).

7.3.2 Phosphorite

Phosphorite also occurs in lenses in sediments off the coasts of the Chinese mainland and the Korean Peninsula (Ludwig & Valencia, 1993). The Yellow River delta extends along more than 100 km of the coast, and river sediment can be traced all the way out to the edge of the continental shelf. Sediment wedges defining ancient deltas are prime prospects for the discovery of offshore placer and phosphorite deposits (Figure 7.40; Ludwig & Valencia, 1993).

In the SCS, phosphorite is also a major offshore mineral resource, a source of phosphate for chemicals and fertilizers. Significant deposits of phosphorite are found in the Xisha Islands (Ritterbush, 1978). Phosphorite is also formed in regions of oceanic upwelling where large volumes of phosphate-rich cold water rise from great depths to the surface. In the warmer surface waters, phosphate precipitates out of solution and then sinks to the seafloor forming deposits of phosphorite. There is a possibility that large reserves may be located off the east coast of Vietnam where a seasonal localized upwelling formed by the SCS's monsoon-driven current system generates the required oceanic conditions (Ritterbush, 1978).

7.3.3 Sulfide Deposits

Hydrothermal activities in the Okinawa Trough are introduced in Chapter 3. The formation of low- to medium-grade hydrothermal mineral deposits is possible in the Okinawa Trough from Japan all the way to Taiwan. The stronger and more consistent hydrothermal activity in the middle Okinawa Trough may produce larger and richer sulfide deposits. The middle Okinawa Trough is the area where the first sulfide deposits in the ECS were discovered. The heat flow within and around this area is anomalously high, implying unusually active hydrothermal flow and mineral deposition (Clark & Li, 1991; Halbach et al., 1989; Ludwig & Valencia, 1993; Uyeda, 1987). The known deposits contain either low-temperature sulfide of subeconomic grade (Ludwig & Valencia, 1993) or very high concentrations of gold (6 g/t), silver (900 g/t), lead (70/0), and zinc (11%), plus some copper and iron, making it the highest-grade polymetallic sulfide deposit yet discovered in the world's oceans (Clark & Li, 1991).

At least three types of polymetallic sulfide deposits may occur in the ECS: stratiform deposits, fault-associated deposits, and volcano-associated deposits (Clark & Li, 1991). The latter two are also known as massive sulfide deposits, since they generally form in the shape of a mound or dome. The first two types of deposits are possible throughout the Okinawa Trough. Both form in association with large-scale faults with horst and graben topographies (Ludwig & Valencia, 1993).

7.3.4 Ferromanganese Nodules/Crusts

Ferromanganese nodules were discovered in the Yellow Sea and ECS (Zhu & Wang, 1985), but they are of little commercial significance. Ferruginous nodules were formed when late Pleistocene sediments emerged from the water surface after sea regression, while others are of marine authigenic origin. The ferruginous nodules consist dominantly of goethite, with minor lepidocrocite.

Many samples of the manganese nodules and cobalt-rich crusts were collected in the SCS in pervious comprehensive surveys. Ferromanganese nodules/crusts, typically hydrogenetic in margin sea environment, are widely distributed in the northeastern continental slope and the central basin of the SCS (Figure 7.39; Lai, 1995; Lin, Ji, Zhang, Lin, & Shi, 2003). In 2013, the

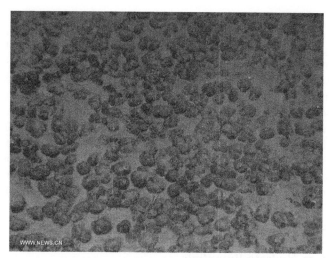

FIGURE 7.41 Photo taken by China's manned submersible Jiaolong on 3 July 2013 shows iron–manganese deposits in the SCS (Huaiyang Zhou, personal communication).

Chinese manned deep-sea submersible "Jiaolong" also discovered a deepwater caldera densely covered by iron–manganese nodules in the northeastern SCS (Figure 7.41). They are also even found in shallow-sea facies in the Beibu Gulf, with different chemical composition, texture, and origin from those of the deep-sea manganese nodules (Chen, 1984).

The nodules are usually concentrated at the lower edge of a gentle continental slope or in brown clay and siliceous ooze in the deep-sea basin in the SCS. They have girdle texture of rough laminated structure, colloid form girdle texture, and crypto-laminated crustal texture (Su & Wang, 1990). Generally, crustal minerals are manganite and MnO_2, and in the nucleus, dolomite, pyrite, and argillaceous detritus containing foraminifera are commonly seen (Su & Wang, 1990).

In the SCS, ferromanganese nodules/crusts have a high content of manganese oxides, though seldom exceeding 50% of the total content of iron and manganese oxides. Chemical components of the nodules are mainly Mn, Fe, Cu, Co, Ni, V, and Ti. The distribution is characterized by the dominance of iron oxides in the littoral area and an increase in manganese oxides in the deep sea, showing differentiations in iron and manganese during the transport, deposition, and diagenetic redistribution (Zhu & Wang, 1985). The growth of the nodules is affected by the environmental fluctuations and changing supply of the terrigenous sediments and Mn, Cu, Co, and Ni (Lin et al., 2003). Zhang et al. (2009) showed that, because of these effects, Fe, REE, and Si are enriched, but Mn and Ca are diluted.

REE contents of ferromanganese nodules/crusts occurring in the northern continental slope, sea mounts, and platforms are high, indicating a large potential prospect (Chen, Yan, et al., 2006; Chen, Yang, et al., 2006). The average concentration of REE in the SCS ferromanganese crusts and nodules is $1-2 \times$ higher than that in Pacific ferromanganese nodules, $5-6 \times$ higher than that in North Pacific sediments, and $10-20 \times$ higher than that in SCS sediments (Bao & Li, 1993). REEs in ferromanganese nodules and crusts are sourced from weathering of medium-acidic rocks in the SCS. Ferromanganese nodules and crusts consist of a relatively high content of Cu+Ni and low Co. The content of Ce is almost 50% of sum lanthanon (Yang, Chen, Yan, & Gu, 2002).

Fine ferromanganese particles were also found widespread in the central basin of the SCS (Li & Zhang, 1990). However, they are sourced primarily from the nearby continent, unlike ferromanganese nodules that are closely associated with basaltic seamounts (Liang, Yao, Liu, & Liang, 1991).

7.4　OTHER RESOURCES

7.4.1　Gas Hydrate Potentials

7.4.1.1　Geophysical Exploration in the Northern SCS

Since 1999, Guangzhou Marine Geological Survey (GMGS) has carried out extensive surveys for gas hydrate resource in the northern SCS (Zhang, Huang, Zhu, & Wu, 2002). Occurrence of gas hydrate in the SCS is often associated with diapiric structures, active faults, slumps, and gravity flows (Matsumoto et al., 2011; Wu, Zhang, Huang, Liang, & Wong, 2005). Gas hydrate is also associated with the occurrence of strong bottom-simulating reflectors (BSRs), which are widespread in the northern SCS continental slope (Chen, Huang, Yuan, & Cathles, 2005) and the accretionary prism offshore southern Taiwan (Deng, Yan, Liu, & Luo, 2006; Liu et al., 2006; Schnürle, Liu, Lin, & Lin, 2011; Shyu, Hsu, & Liu, 1998; Wang, Yang, Deng, Lee, & Liu, 2010; Figures 7.42 and 7.43). BSRs are found to be linked to cold seeps (Chen, Huang, et al., 2005). Normal faults, thrust faults, mud volcanoes, and mud diapirs within the accretionary prism provide good conduits for methane-charged fluids migrating from deeper sediments (Chiu, Tseng, & Liu, 2006; Liu et al., 2006). Very high concentrations of methane were found in sediments and in overlying bottom water (Chuang et al., 2006; Yang et al., 2006). Gas hydrate occurrence here is also marked by acoustic "pulldown" possibly related to the existence of free gas in the sediments beneath the BSRs (Wu et al., 2005).

BSRs have been identified in the Taixinan Basin (McDonnel, Max, Cherkis, & Czarneeki, 2000), the PRMB (Guo, Wu, Zhu, Fan, & Chen, 2004; Yang et al., 2008; Zhang et al., 2002), and the Xisha Trough and Qiongdongnan basin (Wang, Wu, Yuan, et al., 2010; Wu et al., 2005; Figure 7.44). However, the BSRs show different geophysical signatures and variable continuity (Wu, Wang, Wong, & Zhang, 2007). In June 2007, eight sites in the northern continental slope of the SCS were drilled. Gas hydrates were recovered in core samples at three sites (SH2, SH3, and SH7) at water depths of

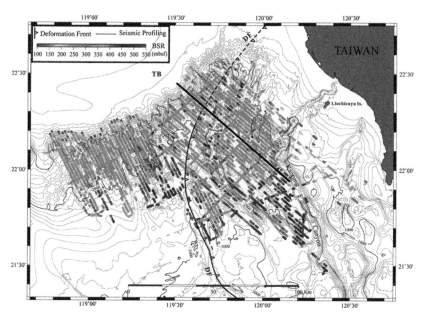

FIGURE 7.42   Map showing BSR distribution offshore SW Taiwan (Liu et al., 2006).

1105–1423 m in silt and silty clay sediments (Zhang et al., 2007). Gas hydrate-bearing sediments were identified at the depth range of 155–220 mbsf, and maximum gas hydrate saturations range from 25% to 47% of pore space (Wang, Hutchinson, Wu, Yang, & Guo, 2011; Yang et al., 2008). Despite these promises, Li, Moridis, Zhang, and Li (2010) estimated preliminarily the production potential of these hydrates as gas-producing resource but found that gas production is low and is burdened with significant water production, making the hydrate accumulations at the Shenhu area unattractive production targets with current technology. Wang et al. (2013) argued that BSRs tend to occur in the areas with a moderately high structural intensity, where faults frequently developed close to the seafloor possibly favor lateral migration of gases, and both BSR and the thermodynamic stability are necessary for hydrate occurrence in the subsurface.

Gas chimneys are also identified above or within a sequence of sediments with polygonal faults in the Qiongdongnan basin. Wang, Wu, Yuan, et al. (2010) argued that the high-amplitude reflections and slightly elevated velocities observed above gas chimneys indicate the presence of gas hydrate in the sediments. The presence of bright spots above or at the flanks of gas chimneys indicates lateral and vertical migration of gas/fluid in the Qiongdongnan basin.

Wu et al. (2005) summarized the mechanisms of gas hydrate formation on the continental slope of the northern SCS: biogenic gas generation followed by gas migration in the Xisha Trough, biogenic and thermogenic gas generation and migration southeast of the Dongsha Islands, *in situ* gas generation

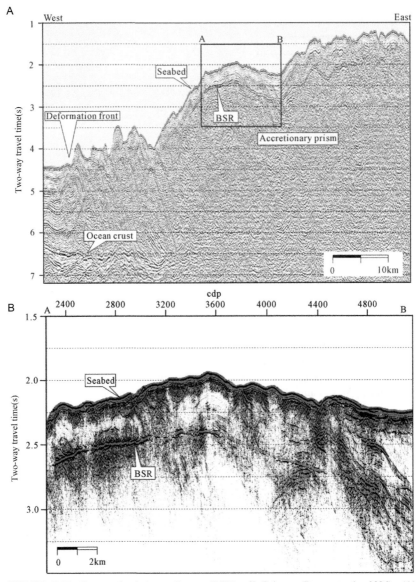

FIGURE 7.43 Bottom-simulating reflector (BSR) off Taiwan (Deng et al., 2006). (A) A migrated seismic profile in the Hengchun Ridge, offshore southern Taiwan. (B) The zoomed-in view of the box AB in (A).

with little migration on the Dongsha Rise, and thermogenic gas generation under a compressive regime and subsequent gas migration in the accretionary wedge of the Manila subduction zone. The conditions that must be satisfied include abundant gas source (deep thermogenic and shallow biogenic),

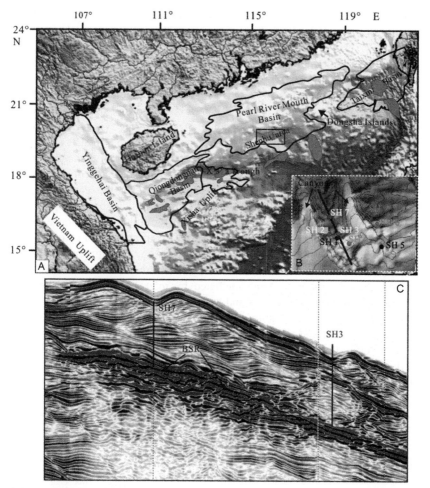

FIGURE 7.44   (A) Areas of gas hydrate exploration and the distribution of BSRs (gray zone) in the northern part of the SCS (Matsumoto et al., 2011). (B) Inset showing the drilling sites (gas hydrate samples obtained at SH2, SH3, and SH7; no gas hydrate samples obtained at SH1 and SH5; no samples were taken at SH4, SH6, and SH8). (C) Seismic line passing through drilling sites SH2 and SH3 (Matsumoto et al., 2011).

appropriate temperature and pressure conditions, and high sedimentation rates and tectonic activity (Wu et al., 2005).

7.4.1.2   Cold Seeps in the Northern SCS

Cold seeps are found to be linked to BSRs and gas hydrate in the northern SCS continental slope (Chen, Liang, et al., 2005). Carbonate precipitation at cold seeps is an important indication of seafloor gas venting and occurrence of shallow gas hydrates. Chemoherm carbonates, as well as numerous other types of methane seep carbonates, were discovered in 2004 along the northern passive margin of the SCS during the joint Chinese–German RV SONNE Cruise 177 (Han et al., 2005; Suess, 2005).

Studies of the mineralogy and oxygen and carbon stable isotopic compositions provide clues that the carbonates were probably precipitated from the cold seeps that originated from gas hydrate dissociation (Lu et al., 2005). Their moderate depleted ^{13}C reflects that carbon origin is possibly thermogenetic gas or mixed gas, which is evidence of gas ventings via microconduits/microchannels in the seafloor (Chen, Liang, et al., 2005; Chen, Yan, et al., 2006; Chen, Yang, et al., 2006). Massive worm tube fossils related to cold seeps were found on the surfaces of carbonate nodules. Microscopic and geochemical analyses of carbonates dredged suggest that the carbonates are hydrocarbon seep-related carbonates and their very light carbon isotopic compositions suggest that their carbon was derived from microbial methane oxidization (Chen, Liang, et al., 2005). From examination of the isotope characteristics, fabrics, abundant microbial structure, and mineralogies, Han et al. (2008) confirmed that these chimney-like carbonates are derived from anaerobic oxidation of methane mediated by microbes and established for the first time the exact locations and seafloor morphology where seep carbonates formed in the SCS (Figure 7.45). These characteristics suggest the chimney carbonates formed where gas vented from the seafloor in the SCS.

Tong et al. (2013) suggested from comprehensive geochemical studies of cold seep carbonates that gas hydrate dissociation during sea-level lowstands or falling stages in the SCS resulted from reduced hydrostatic pressures, which in turn enhanced the seep activities and promoted the formation of authigenic carbonates close to the seafloor.

7.4.1.3 Gas Hydrates in the Okinawa Trough

Gas hydrates have been identified in southern Okinawa Trough from seismic interpretation, velocity analysis, and impedance inversion (Xu et al., 2009). The gas hydrate resources often coexist with mud diapirs, and gas–fluid escape plays an important role in the formation of mud diapirs. Xu et al. (2009) identified an area of mud diapirs of about 11 km^2 in southern Okinawa Trough. In the outer shelf of the ECS, mud diapirs vary from tens to hundreds of meters in diameter and from a few meters up to 40 m in height (Figure 7.46; Yin et al., 2003). On seismic records, bright spots, phase inversions, and other acoustic anomalies indicate that gas and/or fluid escape formed these mounds (Yin et al., 2003). This implies that gas hydrate associated with mud diapirs may exist in a large area of the ECS.

7.4.2 CO$_2$ gas Fields

Abiogenic CO_2 gas fields also distribute widely in the China Seas (Figure 7.47), and they have CO_2 content more than 60% and $^{13}\delta C_{CO_2}$ heavier

FIGURE 7.45 (Left) Typical cold seep carbonates at the northern continental slope of the South China Sea. (A) Field of carbonate rubble; tubular carbonates are several centimeters in diameter and up to 40 cm long. (B) Protruding and broken carbonate chimneys; 4–10 cm in diameter. (C) Chemoherm buildup; diameter of image approx. 150 cm; yellow and white bacterial patches lining wall fractures and rock surfaces (Han et al., 2008). (Right) Typical morphologies of seep carbonates. (A) Tubular concretions: mold of bioturbation tube or fluid channel; Fe–Mn oxide coating; (B) Massive slab: abundant millimeter-sized holes with chitinous walls probably originating from tube worms. (C) Irregular tabular concretion: Fe–Mn coating, interconnected circuitous channels, and protuberances. (D) Chemoherm fragment: abundant cemented shell debris; aragonite void linings; elongated layered voids (1–6 × 1–2 cm) with fresh sediment fill. (E) Carbonate chimney: fresh appearance; central orifice 2–4 cm diameter; soft clay filling. (F) Carbonate breccia: network of aragonite-filled fractures; matrix highly brecciated and cemented (Han et al., 2008).

than $-28‰$ (Figure 7.48; Dai et al., 2009). Tectonically, these CO_2 fields mainly distribute in the rifting belt of eastern China and coincide well with Cenozoic magmatism (Dai et al., 2009). The volcanic activities are related to the reactivation of faults and triggered abiogenic alkane gas discharge and accumulation.

Gas pools with high CO_2 contents were observed previously in many basins of the SCS (e.g., Hao et al., 1996; He & Liu, 2004; Li et al., 2008). CO_2 in the Yinggehai basin appears to be inorganic in origin, formed from high-temperature thermal decomposition of calcareous shales with deep burial in the Cenozoic strata and pre-Cenozoic carbonates in the basement (Chen,

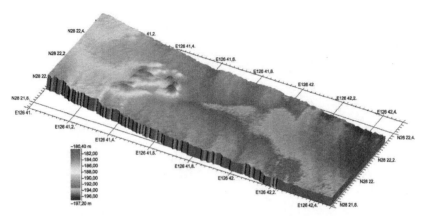

FIGURE 7.46 Bathymetric diagram showing mud diapirs in the outer shelf of the ECS (Yin et al., 2003).

He, & Yang, 2004; He, Wang, Liu, Zhang, & Zhong, 2004; Huang et al., 2002, Huang, Xiao, & Zhu, 2004; Liu, Xia, Liu, Zhang, & Chen, 2004). In contrast, CO_2 in the PRMB and Qiongdongnan basin is thought to have a mixed mantle and crustal origin (He et al., 2004; Lai, 1994). A dominant mantle source for the CO_2 in the Beibuwan Basin was determined from a simple binary mixing model using the $\delta^{13}C_{CO2}$ values and $^3He/^4He$ ratios of associated helium in these gases (Li et al., 2008).

7.5 SUMMARY

Paleogene lacustrine clastic sediments are known to be the main source rocks in the Cenozoic rift basins of Southeast Asia (Longman, 1993) and China (Watson, Hayward, Parkirkson, & Zhang, 1987). Paleogene is the period of strong regional extension and rapid initial subsidence in most rifting basins in the China Seas. Deepwater environments were favorable for deposition of organism-rich mudstones. Eocene lacustrine carbonate source rocks and reservoirs have been found in the Bohai Basin (Peng, 2011; Wang, Wang, & Zhang, 2010). Another secondary period of source rock deposition is the early Neogene, when the thermal subsidence just started with relatively fast subsidence rates.

Source rocks of marine facies are also important in the SCS and southern part of the ECS. The SCS area had two major episodes of marine deposition, in late Paleocene/early Eocene and late Oligocene–early Miocene, respectively (e.g., Jin, 1989; Shi & Li, 2012). Therefore, most basins around the SCS have both lacustrine and marine source rocks. While most source rocks were deposited during the early rifting phase of basin development before Miocene, Miocene source rocks exist in basins/depressions that are with higher geothermal gradients and normally close to the central oceanic basin

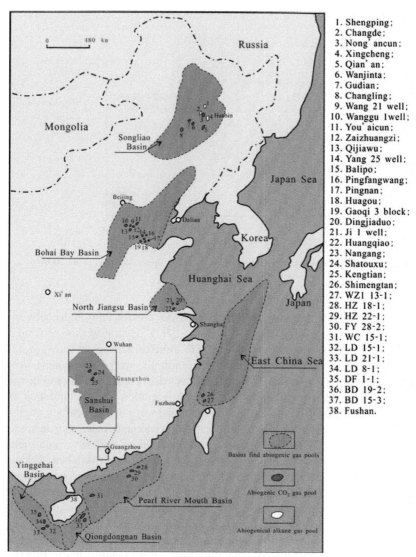

1. Shengping;
2. Changde;
3. Nong' ancun;
4. Xingcheng;
5. Qian' an;
6. Wanjinta;
7. Gudian;
8. Changling;
9. Wang 21 well;
10. Wanggu 1well;
11. You' aicun;
12. Zaizhuangzi;
13. Qijiawu;
14. Yang 25 well;
15. Balipo;
16. Pingfangwang;
17. Pingnan;
18. Huagou;
19. Gaoqi 3 block;
20. Dingjiaduo;
21. Ji 1 well;
22. Huangqiao;
23. Nangang;
24. Shatouxu;
25. Kengtian;
26. Shimengtan;
27. WZ1 13-1;
28. HZ 18-1;
29. HZ 22-1;
30. FY 28-2;
31. WC 15-1;
32. LD 15-1;
33. LD 21-1;
34. LD 8-1;
35. DF 1-1;
36. BD 19-2;
37. BD 15-3;
38. Fushan.

FIGURE 7.47 Petroliferous basins in the China Seas. Also shown are abiogenic alkane gas fields in eastern China (Dai et al., 2009). In this figure, Huanghai Sea, Yellow Sea; North Jiangsu Basin, Subei basin. Numbers 1–38 label the names of gas fields.

of the SCS. Paleogene source rocks became mature in the middle to late Miocene, slightly later than the formation of most traps (Jin, 1989).

Reef reservoirs become extremely important in the SCS. Late Oligocene and early Miocene marine carbonate platforms and coral reefs developed in the early spreading of the SCS are primary reservoirs. In the Miocene, they

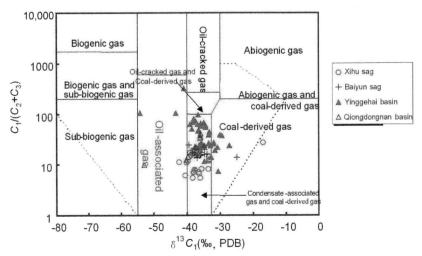

FIGURE 7.48   The classification diagram of gas from the China Seas (Dai et al., 2009).

occurred all around the SCS and often developed on structural highs between early rifted half-grabens (e.g., Williams, 1997). Hydrocarbons generated within the grabens can migrate upward into these carbonate structures.

Tectonics exerts critical roles during the entire life span of a petroleum system. Paleogene tectonics controls not only the basin configurations and distributions of source/reservoir rocks but also the migration and accumulation of hydrocarbons. Synrift faults can be important pathways of hydrocarbon migration and can form fault-related traps. Late Miocene and Quaternary tectonic activity offshore China also generated unconformities and new faults and formed new structural traps, migration routes, and source kitchens (Gong et al., 2011). Neotectonism can modify preexisting hydrocarbon accumulations and formed new large pools. For example, the strong neotectonism associated with large strike-slip faults in the Bohai Basin and Yinggehai basin is found to be closely related to large oil and gas pools there (Gong et al., 2011).

Large biogenic gas fields discovered so far in China are mainly distributed in the continental shelf basins—the Yinggehai–Qiongdongnan basin, PRMB, ECS basin (Chunxiao gas field), and Taixi Basin (Figure 7.1). Carbon isotope of ethane from these fields is heavier than −28‰, indicating that the gas is coal-derived (Dai et al., 2009). Thermogenic gases are also dominant in the Yinggehai and Qiongdongnan basins, where the geothermal gradients are high (Huang et al., 2003).

Mineral resources in the China Seas include coastal placers, phosphorite, sulfide deposits, and ferromanganese nodules/crusts. Recently, many research activities were carried out on gas hydrates, which are found in the SCS and Okinawa Trough.

REFERENCES

Abolins, P. (2007). Source rock depositional setting of Northwest Borneo: From deep peat swamps to ultra deep water. In *23rd International meeting of organic geochemistry conference, Torquay, England, 9-14 September 2007, poster session.*

Allen, M. B., MacDonald, D. I. M., Zhao, X., Vincent, S. J., & Brouet-Menzies, C. (1997). Early Cenozoic two-phase extension and late Cenozoic thermal subsidence and inversion of the Bohai Basin, northern China. *Marine and Petroleum Geology, 14*, 951–972.

Areshev, E. G., Dong, T. L., San, N. T., & Shnip, O. A. (1992). Reservoirs in fractured basement on the continental shelf of southern Vietnam. *Journal of Petroleum Geology, 15*, 451–464.

Armentrout, J. M., Prebish, M., Cunningham, A. C., Echols, R. J., Braithwaite, P., Sarg, J. F., et al. (1994). Cenozoic tectonostratigraphic sequences of the shelf rift basin, East China Sea. *AAPG Bulletin, 78*, 1169.

Bao, G.-D., & Li, Q.-X. (1993). Geochemistry of rare earth elements in ferromanganese nodules crusts of the South China Sea. *Oceanologia et Limnologia Sinica, 24*, 304–313, (in Chinese).

Cai, Q. (2005). *Oil and gas geology in China seas.* Beijing: China Ocean Press, 406 pp (in Chinese).

Cai, F., Dai, C.-S., Chen, J.-W., Li, G., & Sun, P. (2005). Hydrocarbon potential of Pre-Cenozoic strata in the North Yellow Sea Basin. *Marine Science Bulletin, 7*(1), 21–36.

Canh, T., Ha, D. V., Carstens, H., & Berstad, S. (1994). Vietnam-attractive plays in a new geological province. *Oil & Gas Journal, 92*, 78–83.

Chang, C. (1991). Geological characteristics and distribution patterns of hydrocarbon deposits in the Bohai Bay Basin, east China. *Marine and Petroleum Geology, 8*, 98–106.

Chen, J. (1984). Features of the ferromanganese nodules in the Beibu Gulf, the South China Sea. *Marine Science Bulletin, 3*(3), 46–50 (In Chinese).

Chen, W. H. (1990). *Natural gas resource prediction in major sedimentary basins in the north continental shelf of the South China Sea: CNOOC research report, 238 pp (in Chinese).*

Chen, Z. Y., & Ge, H. P. (2003). Inversion structures and hydrocarbon accumulation in Xihu Sag, East China Sea Basin. *China Offshore Oil and Gas (Geology), 17*(1), 20–24 (in Chinese with English abstract).

Chen, J., Ge, H., Chen, X., Deng, C., & Liang, D. (2008). Classification and origin of natural gases from Lishui Sag, the East China Sea Basin. *Science in China Series D: Earth Science, 51*(S1), 122–130.

Chen, C. P., He, J. X., & Yang, J. (2004). A possible lithochemical origin of shallow CO_2 in Yinggehai Basin. *China Offshore Oil and Gas (Geology), 16*(3), 170–171, (in Chinese).

Chen, S., & Hu, P. (1989). Tertiary reef complexes in the Zhujiangkou (Pearl River mouth) Basin and their significance for hydrocarbon exploration. *China Earth Sciences, 1*(1), 21–29.

Chen, D. F., Huang, Y. Y., Yuan, X. L., & Cathles, L. M., III. (2005). Seep carbonates and preserved methane oxidizing archaea and sulfate reducing bacteria fossils suggest recent gas venting on the seafloor in the Northeastern South China Sea. *Marine and Petroleum Geology, 22*, 613–621.

Chen, J. P., Liang, D. G., & Zhang, C. J. (2005). A study on characteristics and hydrocarbon-generative potential of source rocks in Lishui-Jiaojiang Depression. *Resource Report,* 1–208 (in Chinese).

Chen, C. M., Shi, H. S., & Xu, S. C. (2003). *Formation conditions of oil and gas reservoir in tertiary in the Eastern Pearl River Mouth Basin.* Beijing: Science Press, 266 pp (in Chinese).

Chen, Z. Y., Wu, P. K., & Wu, Z. X. (2000). Petroleum geology and exploration potential of Lishui Sag. *China Offshore Oil and Gas (Geology), 14*(6), 384–391 (in Chinese).

Chen, Z., Yan, W., Chen, M., Wang, S., Lu, J., Zhen, F., et al. (2006). Discovery of seep carbonate nodules as new evidence for gas venting on the northern continental slope of South China Sea. *Chinese Science Bulletin*, *51*(10), 1228–1237.

Chen, Z., Yang, H.-N., Yan, W., Wu, B.-H., Chen, M.-H., & Yang, H.-P. (2006). Distributions and divisions of mineral resources in the sea areas of china: Placer deposit and ferromanganese nodule/crust. *Marine Geology and Quaternary Geology*, *26*(5), 101–108 (In Chinese).

Chi, Y. (2001). Basic characteristics and hydrocarbon distribution in Cenozoic petroleum systems, Bohai Sea. *China Offshore Oil and Gas (Geology)*, *15*(1), 3–10.

Chin, K. L. C. (2001). Oil and gas. In O. J. Eong & G. W. Khoon (Eds.), *The seas. The encyclopedia of malaysia* (pp. 114–115). Kuala Lumpur: Archipelago Press.

Chiu, J. K., Tseng, W. H., & Liu, C. S. (2006). Distribution of gassy sediments and mud volcanoes offshore southwestern Taiwan. *Terrestrial, Atmospheric and Oceanic Sciences*, *17*, 703–722.

Chuang, P. C., Yang, T. F., Lin, S., Lee, H. F., Lan, T. F., Hong, W. L., et al. (2006). Extremely high methane concentration in bottom water and cored sediments from offshore southwestern Taiwan. *Terrestrial, Atmospheric and Oceanic Sciences*, *17*, 903–920.

Clark, A., & Li, C. (1991). Marine mineral resources of the East China Sea: Scientific and economic opportunities. *Marine Mining*, *10*, 117–144.

Clark, A., & Li, C. (1993). Marine mineral resources of the South China Sea. *Marine Georesources & Geotechnology*, *11*(1), 101–126.

COMEXCO. (1994). *Technical brochure-northwest Malampaya (Gsec 66) Philippines*. Manila, Philippines: The Philodrill Corporation, 22 pp.

Cukur, D., Horozal, S., Lee, G. H., Kim, D. C., & Han, H. C. (2012). Timing of trap formation and petroleum generation in the northern East China Sea Shelf Basin. *Marine and Petroleum Geology*, *36*, 154–163.

Dai, J., Hu, G., Ni, Y., Li, J., Luo, X., Yang, C., et al. (2009). Natural gas accumulation in eastern China. *Energy, Exploration & Exploitation*, *27*(4), 225–259.

Deng, H., Yan, P., Liu, H., & Luo, W. (2006). Seismic data processing and the characterization of a gas hydrate bearing zone offshore of southwestern Taiwan. *Terrestrial, Atmospheric and Oceanic Sciences*, *17*(4), 781–797.

Ding, W., Franke, D., Li, J., & Steuer, S. (2013). Seismic stratigraphy and tectonic structure from a composite multi-channel seismic profile across the entire Dangerous Grounds, South China Sea. *Tectonophysics*, *582*, 162–176.

Doust, H., & Sumner, H. S. (2007). Petroleum systems in rift basins-a collective approach in Southeast Asian basins. *Petroleum Geoscience*, *13*, 127–144.

Epting, M. (1980). Sedimentology of Miocene carbonate build-ups, Central Luconia, offshore Sarawak. *Geological Society of Malaysia Bulletin*, *12*, 17–30.

Franke, D., Barckhausen, U., Baristeas, N., Engels, M., Ladage, S., Lutz, R., et al. (2011). The continent-ocean transition at the southeastern margin of the South China Sea. *Marine and Petroleum Geology*, *28*, 1187–1204.

Fulthorpe, C. S., & Schlanger, O. S. (1989). Paleo-oceanographic and tectonic settings of early Miocene reefs and associated carbonates of offshore Southeast Asia. *AAPG Bulletin*, *73*(6), 729–756.

Ge, H., Chen, Z., Fang, L., & Shen, W. (2003). A discussion on hydrocarbon accumulation periods in Lishui sag, East China Sea basin. *China Offshore Oil and Gas (Geology)*, *17*(1), 44–50 (in Chinese).

Gebregergis, T. M., & Yukoff, W. I. W. (2010). Burial and thermal history model to evaluate source rock, in Tatau Province, offshore Sarawak Basin, Malaysia. In *Expanded abstract, AAPG international conference and exhibition, Calgary, Alberta, Canada, September 12-15*.

Gong, Z. S. (1997). *The major oil and gas fields of China offshore*. Beijing: Petroleum Industry Press, 223 pp.

Gong, Z. S., Huang, L. F., & Chen, P. H. (2011). Neotectonic controls on petroleum accumulations, offshore China. *Journal of Petroleum Geology, 34*(1), 5–28.

Gong, Z. S., Li, S. T., & Xie, T. J. (1997). *Continental margin basin analysis and hydrocarbon accumulation of the northern South China Sea*. (pp. 63–74). Beijing: Science Press (in Chinese).

Gong, Z. S., & Li, S. T. (2004). *Dynamic research of oil and gas accumulation in northern marginal basins of South China Sea*. Beijing: Science Press, 339 pp.

Gong, J., Lu, Z., Cao, Z., Chen, J., Yang, Y., Li, W., et al. (2007). Oil-gas geochemical exploration and its perspective prediction in the North Yellow Sea Basin. *Acta Geologica Sinica, 81* (10), 1416–1422.

Gong, Z.-S., & Wang, G.-C. (2001). Neotectonism and late hydrocarbon accumulation in Bohai Sea. *Acta Petrolei Sinica, 22*(2), 1–7.

Grötsch, J., & Mercadier, C. (1999). Integrated 3-D Reservoir modeling based on 3-D seismic: The Tertiary Malampaya and Camago buildups, offshore Palawan, Philippines. *AAPG Bulletin, 83*, 1703–1728.

Guo, T. M., Wu, B. H., Zhu, Y. H., Fan, S. S., & Chen, G. J. (2004). A review on the gas hydrate research in China. *Journal of Petroleum Science and Engineering, 41*(1/3), 11–20.

Gustavson, J. B., & Gang, X. H. (1992). The Suizhong 36-1 oil field, Bohai Gulf, offshore China: Reservoir delineation by geophysical methods. In M. T. Halbouty (Ed.), *AAPG Memoir*: *Vol. 54*. Giant oil and gas fields of the decade 1978–1988 (pp. 459–470).

Halbach, P., Nakamura, K., Wahsner, M., Lange, J., Sakai, H., Kaselitz, L., et al. (1989). Probable modern analogue of Kuroko-type massive sulphide deposits in the Okinawa Trough back-arc basin. *Nature, 338*, 496–499.

Han, X., Suess, E., Huang, Y., Wu, N., Bohrmann, G., Su, X., et al. (2008). Jiulong methane reef: Microbial mediation of seep carbonates in the South China Sea. *Marine Geology, 249*, 243–256.

Han, X., Suess, E., Huang, Y., Wu, N., Eisenhauer, A., Bohrmann, G., et al. (2005). Jiulong Methane reef: First direct evidence of methane seepage in the South China Sea. *Geophysical Research Abstracts, 7*, 04055, European Geosciences Union.

Hao, F., Li, S. T., Sun, Y. C., & Zhang, Q. (1998). Geology, compositional heterogeneities and geochemical origin of the Yacheng Gas Field in the Qiongdongnan Basin, South China Sea. *AAPG Bulletin, 82*, 1372–1384.

Hao, F., Li, S. T., & Zhang, Q. M. (1996). Characteristics and origin of the gas and condensate in the Yinggehai Basin, offshore South China Sea: Evidence for effects of overpressure on petroleum generation and migration. *Organic Geochemistry, 24*(3), 363–375.

Hao, F., Sun, Y. C., Li, S. T., & Zhang, Q. (1995). Overpressure retardation of organic-matter maturation and hydrocarbon generation: A case study from the Yinggehai and Qiongdongnan basins, offshore South China Sea. *AAPG Bulletin, 79*, 551–562.

Hao, F., Zhou, X., Zou, H., Teng, C., & Yang, Y. (2012). Petroleum charging and leakage in the BZ25-1 Field, Bohai Bay Basin. *Journal of Earth Science, 23*(3), 253–267.

Hao, F., Zou, H. Y., Gong, Z. S., & Deng, Y. (2007). Petroleum migration and accumulation in the Bozhong Sub-basin, Bohai Bay Basin, China: Significance of preferential petroleum migration pathways (PPMP) for the formation of large oilfields in lacustrine fault basins. *Marine and Petroleum Geology, 24*, 1–13.

He, J. X. (1994). Natural gas geology and geochemistry characteristics and origin of Dongfang 1-1 structure in Yinggehai Basin. *Natural Gas Geoscience, 5*(3), 1–8 (in Chinese).

He, J. X., Li, M. X., & Chen, W. H. (2001). Origin and source rock identification of CO_2 in Yinggehai-Southeast Hainan Basin. *Natural Gas Industry, 21*(3), 15–21 (in Chinese).

He, J. X., & Liu, Q. W. (2004). The analysis and discussion to the characteristics and generative cause, migration and distribution of CO_2 in Ying–Qiong Basins in the north of the South China Sea. *Natural Gas Geosciences, 15*(1), 12–19 (in Chinese).

He, J. X., Wang, Z. F., Liu, B. M., Zhang, S. L., & Zhong, Z. H. (2004). Main control factors of CO_2 reservoir formation in the marginal basins of the north part of South China Sea. *Natural Gas Industry, 24*(9), 19–22 (in Chinese).

He, H., Wang, Z., & Han, P. (1998). Deep zone reservoir type and oil gas distribution pattern in Bohai Gulf basin. *Petroleum Exploration and Development, 25*(3), 6–9.

He, J. X., Xia, B., Chen, G. Y., & Sun, D.-S. (2006). Petroleum Geology and Exploration Prospect of Mesozoic and Cenozoic in Taixinan Basin, Northern South China Sea. *Xinjiang Petroleum Geology, 27*, 398–402.

He, J. X., Zhang, W. H., & Chen, G. (1995). A preliminary study on CO_2 origin and migration and accumulation characteristics in Yinggehai Basin. *Petroleum Exploration and Development, 22*(6), 8–15 (in Chinese).

He, S., Zhu, W., & Li, L. (2001). Sedimentary evolution and Neogene reservoir—Seal assemblage analysis of Bozhong Depression. *Acta Petrolei Sinica, 22*(2), 38–43.

Holloway, N. (1982). North Palawan block, Philippines—Its relation to Asian mainland and role in evolution of South China Sea. *AAPG Bulletin, 66*, 1355–1383.

Hsu, S.-K., Sibuet, J.-C., & Shyu, C.-T. (2001). Magnetic inversion in the East China Sea and Okinawa Trough: Tectonic implications. *Tectonophysics, 333*, 111–122.

Huang, B. J., Xiao, X. M., & Dong, W. L. (2002). Multiphase natural gas migration and accumulation and its relationship to diapir structures in the DF1-1 gas field, South China Sea. *Marine and Petroleum Geology, 19*, 861–872.

Huang, B., Xiao, X., & Li, X. (2003). Geochemistry and origins of natural gases in the Yinggehai and Qiongdongnan basins, offshore South China Sea. *Organic Geochemistry, 34*, 1009–1025.

Huang, B. J., Xiao, X. M., & Zhu, W. L. (2004). Geochemistry, origin, and accumulation of CO_2 in natural gases of the Yinggehai Basin, offshore South China Sea. *AAPG Bulletin, 88*(9), 1277–1293.

Hutchison, C. S. (1983). *Economic deposits and their tectonic setting.* London: The MacMillan Press Ltd.

Hutchison, C. S. (1989). *Geological evolution of South East Asia Monographs on geology and geophysics.* Oxford: Oxford University Press.

Hutchison, C. S., & Vijayan, V. R. (2010). What are the Spratly Islands? *Journal of Asian Earth Sciences, 39*, 371–385.

Jia, J., & Gu, H. (2002). *Oil and gas bearing system and potential evaluation in the East China Sea.* Beijing: Geological Publishing House, 204 pp (in Chinese).

Jiang, L., Li, B. H., Zhong, S. L., Xiao, J. K., Zhang, Y. Y., & Wang, J. P. (2004). Biostratigraphy and paleoenvironment of the Yueguifeng Formation in the Taipei Depression of the continental shelf basin of the East China Sea. *Marine Geology and Quaternary Geology, 24*(1), 37–41 (in Chinese).

Jiang, Z.-L., Wang, Y.-B., & Jiang, L. (2005). Hydrocarbon accumulation process of Lishui sag. *Marine Geology Letters, 21*(1), 29–34 (in Chinese).

Jin, Q., Yan, Z.-M., Cheng, F.-Q., & Su, C.-G. (2009). Different orders of unconformities and their oil-gas accumulations in Bohai Bay Basin. *Xinjiang Petroleum Geology, 30*(4), 440–444.

Jin, Q. (1989). *Geology and hydrocarbon resources of the South China Sea.* Beijing: Geological Press, 417 pp (in Chinese).

KIGAM (1997). *Petroleum resources assessment: Korea Institute of Geoscience and Mineral Resources report KR-97(C)-17.* (pp. 71–190). (in Korean with English abstract).

Lai, W. Z. (1994). Origin of carbon dioxide in northern continental shelf, South China Sea. *China Offshore Oil and Gas (Geology), 8*(5), 319–327 (in Chinese).

Lai, L.-R. (1995). The evolving of formation of ferromanganese nodules in the South China Sea. *Journal of Mineral Resources and Geology, 94,* 293–298 (In Chinese).

Lee, G. H., Kim, B., Shin, K. S., & Sunwoo, D. (2006). Geologic evolution and aspects of the petroleum geology of the northern East China Sea shelf basin. *AAPG Bulletin, 90*(2), 237–260.

Lee, G. H., Lee, K., & Watkins, J. S. (2001). Geologic evolution of the Cuu Long and Nam Con Son basins, offshore southern Vietnam, South China Sea. *AAPG Bulletin, 85*(6), 1055–1082.

Le Pichon, X., & Mazzotti, S. (1997). A new model for the early opening of the Okinawa basin. In *Proceedings of Chinese Taipei ODP Consortium 1997 annual meeting and long-range plan for Chinese Taipei Ocean Drilling Program Workshop, Taipei, Taiwan, 6 November 1997,* 34 pp.

Li, S. Q., & Li, C. J. (2003). Distribution of oil and gas resources and analysis of the exploration potential for the Xihu Depression in the East China Sea Basin. *Petroleum Geology and Experiment, 25*(6), 721–728 (in Chinese with English abstract).

Li, W., Lu, W., Liu, Y., & Xu, J. (2012). Superimposed versus residual basin: The North Yellow Sea Basin. *Geoscience Frontiers, 3*(1), 33–39.

Li, G., Moridis, G. J., Zhang, K., & Li, X.-S. (2010). Evaluation of gas production potential from marine gas hydrate deposits in Shenhu Area of South China Sea. *Energy and Fuels, 24,* 6018–6033.

Li, S., Pang, X., Liu, K., & Jin, Z. (2006). Origin of the high waxy oils in Bohai Bay Basin, east China: Insight from geochemical and fluid inclusion analyses. *Journal of Geochemical Exploration, 89*(1–3), 218–221.

Li, M., Wang, T., Liu, J., Lu, H., Wu, W., & Gao, L. (2008). Occurrence and origin of carbon dioxide in the Fushan Depression, Beibuwan Basin, South China Sea. *Marine and Petroleum Geology, 25,* 500–513.

Li, Z., & Zhang, F. (1990). Geochemistry of elements in ferromanganese particles at depths of South Sea. *Marine Science Bulletin, 9*(6), 41–50 (In Chinese).

Li, C.-F., Zhou, Z., Ge, H. P., & Mao, Y. X. (2007). Correlations between erosions and relative uplifts from the central inversion zone of the Xihu depression, East China Sea Basin. *Terrestrial, Atmospheric and Oceanic Sciences, 18,* 757–776.

Liang, H.-F., Yao, D., Liu, X.-B., & Liang, D.-H. (1991). Geochemistry of polymetallic crust from Jianfeng Seamount, South China Sea. *Marine Geology and Quaternary Geology, 11* (4), 49–58 (In Chinese).

Lin, C. M., Gu, L. X., Li, G. Y., Zhao, Y. Y., & Jiang, W. S. (2004). Geology and formation mechanism of late Quaternary shallow biogenic gas reservoirs in the Hangzhou Bay area, eastern China. *AAPG Bulletin, 88*(5), 613–625.

Lin, C. M., Huang, Z. C., Zhu, S. Z., Li, C. X., & Jiang, W. S. (1999). Sedimentary characteristics and processes of late Quaternary in the Hangzhou Bay coastal plain. *Acta Geologica Sinica, 73*(2), 120–130 (in Chinese with English abstract).

Lin, Z.-H., Ji, F.-W., Zhang, F.-Y., Lin, X.-T., & Shi, Z.-B. (2003). Characteristics and origin of ferromanganese nodules from the northeastern continental slope of the South China Sea. *Marine Geology and Quaternary Geology, 23*, 7–12 (In Chinese).

Lin, R., Leon, D., Kevin, D., & Eric, J. (1997). *Far-east tertiary coal-related petroleum system. http://karl.nrcce.wvu.edu/esaapg/abstracts/97mtg-abstracts/lin1.htm (accessed July 2013).*

Liu, H. (1989). Enrichment laws of quaternary littoral placer deposits in china. *Marine Geology and Quaternary Geology, 9*(2), 41–49 (In Chinese).

Liu, G. (Ed.), (1992). *Geophysical series maps in China Seas and Adjacent Regions.* Beijing: Science Press.

Liu, S., & Li, S. (2001). *Geophysical exploration in the East China Sea.* Beijing: Geological Publishing House, 278 pp (in Chinese).

Liu, C. S., Schnurle, P., Wang, Y., Chung, S. H., Chen, S. C., & Hsiuan, T. H. (2006). Distribution and characters of gas hydrate offshore of southwestern Taiwan. *Terrestrial, Atmospheric and Oceanic Sciences, 17*, 615–644.

Liu, B. M., Xia, B., Liu, X. X., Zhang, M. Q., & Chen, Z. H. (2004). The genetic mechanism of CO_2 in east China and the south western China Sea. *Bulletin of Mineralogy, Petrology and Geochemistry, 23*(3), 207–212 (in Chinese).

Liu, L., Zhao, S., Chen, L., & Huo, H. (2005). Distribution of major hydrocarbon source rocks in the major oil-gas-bearing basins in China. *Chinese Journal of Geochemistry, 24*(2), 116–128.

Longman, M. W. (1993). Future bright for Tertiary carbonate reservoirs in Southeast Asia. *Oil and Gas Journal, 91*(51), 107–112.

Lu, H., Liu, J., Chen, F., Liao, Z., Sun, X., & Su, X. (2005). Mineralogy and stable isotope composition of authigenic carbonates in bottom sediments on the offshore area of southwest Taiwan, South China Sea: Evidence for gas hydrates occurrence. *Earth Science Frontiers, 12*, 268–276 (in Chinese with English abstract).

Lü, C., Wu, S., Yao, Y., & Fulthorpe, C. S. (2013). Development and controlling factors of Miocene carbonate platform in the Nam Con Son Basin, southwestern South China Sea. *Marine and Petroleum Geology, 45*, 55–68.

Ludwig, N. A., & Valencia, M. J. (1993). Oil and mineral resources of the East China Sea: Prospects in relation to maritime boundaries. *GeoJournal, 30*(4), 381–387.

Massoud, M. S., Scott, A. C., & Killops, S. D. (1991). Oil source rock potential of the lacustrine Jurassic Sim Uuju formation, West Korea Bay Basin. Part: Oil source rock correlation and environment of deposition. *Journal of Petroleum Geology, 14*(4), 365–386.

Massoud, M. S., Scott, A. C., & Killops, S. D. (1993). Oil source rock potential of the lacustrine Jurassic Sim Uuju formation, West Korea Bay Basin. Part II: Nature of the organic matter and hydrocarbon-generation history. *Journal of Petroleum Geology, 16*(3), 265–284.

Matsumoto, R., Ryu, B.-J., Lee, S.-R., Lin, S., Wu, S., Sain, K., et al. (2011). Occurrence and exploration of gas hydrate in the marginal seas and continental margin of the Asia and Oceania region. *Marine and Petroleum Geology, 28*, 1751–1767.

Matthews, S. J., Fraser, A. J., Lowe, S., Todd, S. P., & Peel, F. J. (1997). Structure, stratigraphy, and petroleum geology of the SE Nam Con Son Basin, offshore Vietnam. In A. J. Fraser, S. J. Matthews, & R. W. Murphy (Eds.), *Petroleum geology of Southeast Asia: Vol. 126.* (pp. 89–106). London: Geological Society Special Publication.

Mat-Zin, I. C., & Swarbrick, R. E. (1997). The tectonic evolution and associated sedimentation history of Sarawak Basin, eastern Malaysia: A guide for future hydrocarbon exploration. *Geological Society, London, Special Publications, 126*, 237–245.

McDonnel, S. L., Max, M. D., Cherkis, N. Z., & Czarneeki, M. F. (2000). Tectono-sedimentary controls on the likelihood of gas hydrate occurrence near Taiwan. *Marine and Petroleum Geology, 17,* 929–936.

Moldovanyi, E. P., Waal, F. M., & Yan, Z. J. (1995). Regional exposure events and platform evolution of Zhuijang Formation carbonates, Pearl river Mouth Basin: Evidence for primary and diagenetic seismic facies. In D. A. Budd, A. H. Saller, & P. M. Harris (Eds.), *American Association of Petroleum Geologists Memoir: Vol. 63. Unconformities and porosity in carbonate strata* (pp. 133–145).

Pang, X., Chen, C. M., & Peng, D. J. (2007). *The Pearl River deep-water fan system and petroleum in South China Sea.* Beijing: Science Press, (in Chinese).

Pang, X., Chen, C. M., & Zhu, M. (2006). A discussion about hydrocarbon accumulation conditions in the Baiyun deep-water area, northern continental slope, South China Sea (in Chinese). *China Offshore Oil & Gas, 18,* 145–149.

Peng, C. (2011). Distribution of favorable lacustrine carbonate reservoirs: A case from the Upper Es4 of Zhanhua Sag, Bohai Bay Basin. *Petroleum Exploration and Development, 38*(4), 435–443.

Price, L. C., & Wenger, L. M. (1992). The influence of pressure on petroleum generation and maturation as suggested by aqueous pyrolysis. *Organic Geochemistry, 19,* 141–159.

Primeline. (2004). *Corporate information. http:/www.primelineenergy.com/i/pdf/primeline.pdf* (accessed June 2013).

Qi, J., & Yang, Q. (2010). Cenozoic structural deformation and dynamic processes of the Bohai Bay basin province. *China Marine and Petroleum Geology, 27,* 757–771.

Qiu, C. (1983). A study on the characteristics and distribution of ferromanganese sediment and volcanic debris sediment in South China Sea. *Tropic Oceanology, 2*(4), 269–277 (In Chinese).

Reid, I. (1997). Petroleum exploration in Vietnam. *Petromin (Singapore), 23*(5), 32–34.

Ritterbush, S. W. (1978). *Marine resources and the potential for conflict in the South China Sea.* In *Fletcher Forum of World Affairs 1983-2000: Vol. 2.* (1) Winter, http://hdl.handle.net/10427/76180.

Ryu, I. C., Kim, B. Y., Kwak, W. J., Kim, G. H., & Park, S. J. (2000). Stratigraphic response to tectonic evolution of sedimentary basins in the Yellow Sea and adjustment area. *Korean Journal of Petroleum Geology, 8,* 19–26 (in Korean with English abstract).

Sales, A. O., Jacobsen, E. C., Morado, A. A., Benavidez, J. J., Navarro, F. A., & Lim, A. E. (1997). The petroleum potential of deep-water northwest Palawan Block GSEG 66. *Journal of Asian Earth Sciences, 15*(2–3), 217–240.

Sattler, U., Zampetti, V., Schlager, W., & Immenhauser, A. (2004). Late leaching under deep burial conditions: A case study from The Miocene Zhujiang Carbonate Reservoir, South China Sea. *Marine and Petroleum Geology, 21,* 977–992.

Schlüter, H. U., Hinz, K., & Block, M. (1996). Tectono-stratigraphic terranes and detachment faulting of the South China Sea and Sulu Sea. *Marine Geology, 130,* 39–51.

Schnürle, P., Liu, C.-S., Lin, A. T., & Lin, S. (2011). Structural controls on the formation of BSR over a diapiric anticline from a dense MCS survey offshore southwestern Taiwan. *Marine and Petroleum Geology, 28,* 1932–1942.

Shi, H., & Li, C.-F. (2012). Mesozoic and early Cenozoic tectonic convergence-to-rifting transition prior to opening of the South China Sea. *International Geology Review, 54*(15), 1801–1828.

Shin, K.-S. (2013). *Hydrocarbon potential of the Yellow Sea Kunsan basin western Korea offshore. http:/www.wilsoncenter.org/sites/default/files/Kook-sun_Shin.pdf.*

Shyu, C.-T., Hsu, S.-K., & Liu, C.-S. (1998). Heat flow off Southwest Taiwan: Measurements over mud diapirs and estimated from bottom simulating reflectors. *Terrestrial, Atmospheric and Oceanic Sciences, 9*(4), 795–812.

Silverman, M. R., Wang, Q., Byrd, B. L., Kun, Z. J., Ying, Z. H., Ming, C., et al. (1996). 3-D evaluation of the Ping Hu field, East China Sea. In P. Weimer & T. L. Davis (Eds.), *AAPG Studies in Geology*: Vol. 42. *Applications of 3-D seismic data to exploration and production* (pp. 91–101).

Su, G.-L., & Wang, T.-X. (1990). Fe-Mn nodules in the South China Sea. *Journal of Tropical Oceanography*, *9*, 29–36 (In Chinese).

Suess, E. (2005). *RV SONNE cruise report SO 177, Sino-German cooperative project, South China Sea Continental Margin: geological methane budget and environmental effects of methane emissions and gas hydrates: IFM-GEOMAR reports.* http://store.pangaea.de/documentation/Reports/SO177.pdf.

Sun, S.-C. (1981). The Tertiary basin offshore Taiwan. In *Proceedings of ASCOPE Conference and Exhibition, Manila, Philippines* (pp. 126–135).

Sun, L., Sun, Z., Zhou, D., & Liu, H. (2008). Stratigraphic and structural characteristics of Lile Basin in Nansha area. *Geotectonica et Metallogenia*, *32*, 151–158 (in Chinese).

Sun, Y., & Tan, Q. (1986). Relationship between formation and existance of coastal placer deposits and geomorphology in Shandong peninsula. *Marine Geology & Quaternary Geology*, *6*(3), 43–52 (in Chinese with English abstract).

Sun, Y.-M., & Xi, X.-Y. (2003). Petroleum reservoir filling history and oil source correlation in the Lishui Sag, East China Sea Basin. *Petroleum Exploration and Development*, *30*(6), 24–28.

Suzuki, S., Takemura, S., Yumul, G. P., David, S. D., & Asiedu, D. K. (2000). Composition and provenance of the Upper Cretaceous to Eocene sandstones in Central Palawan, Philippines: Constraints on the tectonic development of Palawan. *Island Arc*, *9*, 611–626.

Tao, R. (1993). Upper Taizhou Formation: The main hydrocarbon source rock in the northern sag of Northern Basin of South Yellow Sea. *China Offshore Oil & Gas Geology*, *7*(3), 10–14 (in Chinese).

Tao, S. Z., & Zou, C. N. (2005). Accumulation and distribution of natural gases in Xihu Sag, East China Sea Basin. *Petroleum Exploration and Development*, *32*(4), 103–110 (in Chinese).

Taylor, B., & Hayes, D. E. (1983). Origin and history of the South China Sea basin. *Geophysical Monograph Series*, *27*, 23–56.

Tian, K., & Yu, Z. (2000). *Deep lower Paleogene Petroleum Geology and Exploration in the Bohai Basin.* Beijing: Petroleum Industry Press, 292 pp (in Chinese).

Todd, S. P., Dunn, M. E., & Barwise, A. J. G. (1997). Characterizing petroleum charge systems in the Tertiary of SE Asia. In A. J. Fraser, S. J. Matthews, & R. W. Murphy (Eds.), *Geological Society Special Publication*: Vol. 126. *Petroleum geology of Southeast Asia* (pp. 25–47).

Tong, H., Feng, D., Cheng, H., Yang, S., Wang, H., Min, A. G., et al. (2013). Authigenic carbonates from seeps on the northern continental slope of the South China Sea: New insights into fluid sources and geochronology. *Marine and Petroleum Geology*, *43*, 260–271.

Tyrrell, W. W. J., & Christian, H. E. (1992). Exploration history of Liuhua 11-1 Field, Pearl River Mouth basin, China. *AAPG Bulletin*, *76*, 1209–1223.

Uyeda, S. (1987). Active hydrothermal mounds in the Okinawa Back-arc Trough. *Eos, Transactions American Geophysical Union*, *36*, 737.

Wang, Z., & Huang, B. (2008). Dongfang 1-1 gas field in the mud diapir belt of the Yinggehai Basin, South China Sea. *Marine and Petroleum Geology*, *25*, 445–455.

Wang, X. J., Hutchinson, D. R., Wu, S. G., Yang, S. X., & Guo, Y. Q. (2011). Elevated gas hydrate saturations within silt and silty-clay sediments in the Shenhu area, South China Sea. *Journal of Geophysical Research*, *116*. http://dx.doi.org/10.1029/2010JB007944.

Wang, G., Wang, T.-G., Simoneit, B. R. T., Zhang, L., & Zhang, X. (2010). Sulfur rich petroleum derived from lacustrine carbonate source rocks in Bohai Bay Basin, East China. *Organic Geochemistry*, *41*, 340–354.

Wang, Q., Wang, Y.-B., & Zhang, Y. (2010). Characteristics of the Mesozoic reservoirs in the eastern depression of the Northern Yellow Sea Basin. *Sedimentary Geology and Tethyan Geology, 30*(4), 97–103.

Wang, X., Wu, S., Yuan, S., Wang, D., Ma, Y., Yao, G., et al. (2010). Geophysical signatures associated with fluid flow and gas hydrate occurrence in a tectonically quiescent sequence, Qiongdongnan Basin, South China Sea. *Geofluids, 10*, 351–368.

Wang, Y. M., Xu, Q., Li, D., Han, J., Lu, M., Wang, Y., et al. (2011). Late Miocene Red River submarine fan, northwestern South China Sea. *Chinese Science Bulletin, 56*, 1488–1494.

Wang, T. K., Yang, B. J., Deng, J. M., Lee, C. S., & Liu, C. S. (2010). Seismic imaging of gas hydrates in the northernmost South China Sea. *Marine Geophysical Research, 31*, 59–76.

Wang, H. B., Yang, S. X., Wu, N. Y., Zhang, G. X., Liang, J. Q., & Chen, D. F. (2013). Controlling factors for gas hydrate occurrence in Shenhu area on the northern slope of the South China Sea. *Science China Earth Sciences, 56*, 513–520.

Watson, M. P., Hayward, A. B., Parkirkson, D. N., & Zhang, Z. M. (1987). Plate tectonics history, basin development and petroleum source rock deposition onshore China. *Marine and Petroleum Geology, 4*, 205–225.

Williams, H. H. (1997). Play concepts-northwest Palawan, Philippines. *Journal of Asian Earth Sciences, 15*, 251–273.

Wu, S., Han, Q., Ma, Y., Dong, D., & Lü, F. (2009). Petroleum system in deep water basins of the northern South China Sea. *Journal of Earth Science, 20*(1), 124–135.

Wu, S., Ni, X., & Cai, F. (2008). Petroleum geological framework and hydrocarbon potential in the Yellow Sea. *Chinese Journal of Oceanology and Limnology, 26*(1), 23–34.

Wu, S., Wang, X., Wong, H. K., & Zhang, G. (2007). Low-amplitude BSRs and gas hydrate concentration on the northern margin of the South China Sea. *Marine Geophysical Researches, 28*, 127–138.

Wu, C., & Xue, S. (1993). *Sedimentology of Oil-gas bearing basins in China.* (pp. 115–267). Beijing: Petroleum Industry Press (in Chinese).

Wu, S., Yuan, S., Zhang, G., Ma, Y., Mi, L., & Xu, N. (2009). Seismic characteristics of a reef carbonate reservoir and implications for hydrocarbon exploration in deep water of the Qiongdongnan Basin, northern South China Sea. *Marine and Petroleum Geology, 26*, 817–823.

Wu, S., Zhang, G., Huang, Y., Liang, J., & Wong, H. K. (2005). Gas hydrate occurrence on the continental slope of the northern South China Sea. *Marine and Petroleum Geology, 22*, 403–412.

Xia, K., & Huang, C. (2000). The discovery of Meso-Tethys sedimentary basins in the South China Sea and their oil and gas perspective. *Earth Science Frontiers (China University of Geosciences, Beijing), 7*, 227–238 (in China).

Xiao, X. M., Li, N. X., Gan, H. J., Jin, Y. B., Tian, H., Huang, B. J., et al. (2009). Tracing of deeply-buried source rock: A case study of the WC9-2 petroleum pool in the Pearl River Mouth Basin, South China Sea. *Marine and Petroleum Geology, 26*, 1365–1378.

Xu, N., Wu, S., Shi, B., Lu, B., Xue, L., Wang, X., et al. (2009). Gas hydrate associated with mud diapirs in southern Okinawa Trough. *Marine and Petroleum Geology, 26*, 1413–1418.

Yang, H., Chen, Z., Yan, W., & Gu, S. (2002). Solid mineral resource and its distribution in South China Sea area. *The papers on investigation of special economic zones and continental shelf of China* (pp. 102–109). Beijing: China Ocean Press.

Yang, T. F., Chuang, P. C., Lin, S., Chen, J. C., Wang, Y., & Chung, S. H. (2006). Methane venting in gas hydrate potential area offshore of SW Taiwan: Evidence of gas analysis of water column samples. *Terrestrial, Atmospheric and Oceanic Sciences, 17*, 933–950.

Yang, F. L., He, J. Q., Jiang, X. G., & Liu, J. N. (2003). Petroleum migration and accumulation in tertiary for Xihu Depression, East China Sea. *Offshore Oil, 23,* 25–29 (in Chinese with English abstract).

Yang, Y., & Xu, T. (2004). Hydrocarbon habitat of the offshore Bohai Basin, China. *Marine and Petroleum Geology, 21,* 691–708.

Yang, F.-L., Xu, X., Zhao, W.-F., & Sun, Z. (2011). Petroleum accumulations and inversion structures in the Xihu Depression, East China Sea Basin. *Journal of Petroleum Geology, 34*(4), 429–440.

Yang, S. X., Zhang, H. Q., Wu, N. Y., Su, X., Schultheiss, P., Holland, M., et al. (2008). High concentration hydrate in disseminated forms obtained in Shenhu Area, North Slope of South China Sea. In *Proceedings of the 6th international conference on gas hydrates (ICGH 2008), July 6–10. Vancouver, British Columbia, Canada.* (p. 10).

Yao, Y., Chen, C., Feng, Z., Zhang, S., Hao, T., & Wan, R. (2010). Tectonic evolution and hydrocarbon potential in northern area of the South Yellow Sea. *Journal of Earth Science, 21*(1), 71–82.

Yao, Y., Liu, H., Yang, C., Han, B., Tian, J., Yin, Z., et al. (2012). Characteristics and evolution of Cenozoic sediments in the Liyue Basin, SE South China Sea. *Journal of Asian Earth Sciences, 60,* 114–129.

Yao, Y.-J., Wu, N.-Y., Xia, B., & Wan, R.-S. (2008). Petroleum geology of the Zengmu basin in the southern South China Sea. *Geology in China, 35*(3), 503–513.

Ye, J., Qing, H., Bend, S. L., & Gu, H. (2007). Petroleum systems in the offshore Xihu Basin on the continental shelf of the East China Sea. *AAPG Bulletin, 91*(8), 1167–1188.

Yin, P., Berne, S., Vagner, P., Vagner, P., Loubrieu, B., & Liu, Z. (2003). Mud volcanoes at the shelf margin of the East China Sea. *Marine Geology, 194,* 135–149.

Yoon, Y., Lee, G. H., Han, S., Yoo, D. G., Han, H. C., Choi, K., et al. (2010). Cross-section restoration and one-dimensional basin modeling of the Central Subbasin in the southern Kunsan Basin, Yellow Sea. *Marine and Petroleum Geology, 27,* 1325–1339.

Yu, Y., Zhou, X., Tang, L., Peng, W., Lu, D., & Li, W. (2009). Salt structures in the Laizhouwan depression, offshore Bohai Bay basin, eastern China: New insights from 3D seismic data. *Marine and Petroleum Geology, 26*(8), 1600–1607.

Yu, J., Zou, H., & Gong, Z. (2009). Preferential petroleum migration pathways in eastern Pearl River Mouth basin, offshore South China Sea. *Journal of Earth Science, 20*(4), 720–730.

Yuan, S. K., Wang, Y. M., Liu, Z. H., Luo, H., Zhu, Y. H., Li, Y. Y., et al. (2010). Unconformity types and hydrocarbon accumulation models in the eastern depression of the North Yellow Sea basin. *Petroleum Exploration and Development, 37*(6), 663–667.

Zampetti, V., Sattler, U., & Braaksma, H. (2005). Well log and seismic character of Liuhua 11-1 Field, South China Sea: Relationship between diagenesis and seismic reflections. *Sedimentary Geology, 175,* 217–236.

Zhang, Z.-G., Fang, N.-Q., Du, Y.-S., Gao, L.-F., Yang, S.-X., Liu, J., et al. (2009). Geochemical characteristics and their causative mechanism of polymetallic nodules from the northwest continental margin of the South China Sea. *Earth Science—Journal of China University of Geosciences, 34*(6), 955–962 (In Chinese).

Zhang, G. X., Huang, Y. Y., Zhu, Y. H., & Wu, B. H. (2002). Prospect of gas hydrate resources in the South China Sea. *Marine Geology and Quaternary Geology, 22*(1), 75–81 (in Chinese).

Zhang, Q. X., Li, L., & Huang, B. J. (1993). Discrimination of hydrocarbon source rocks in the gas field Ya13-1 using GC-MS data. In Q. M. Zhang (Ed.), *A collection on the petroleum geology of the Yinggehai Basin, South China Sea* (pp. 37–43). Beijing: Seismic Press (in Chinese).

Zhang, X., Sui, F., & Lin, H. (2010). *Pre-paleogene petroleum geology and prospect evaluation in the Bohai Basin*. Beijing: Geological Publishing House, 446 pp (in Chinese).

Zhang, Y., Wei, Z., Xu, W., Tao, R., & Chen, R. (1989). The North Jiangsu–South Yellow Sea basin. In X. Zhu (Ed.), *Chinese sedimentary basin* (pp. 107–123). Amsterdam: Elsevier.

Zhang, Z. M., Wu, N. J., & Zhou, J. (2005). The hydrocarbon accumulation model of the central anticline zone for the Xihu Depression in the East China Sea Basin. *Gas Industry, 25*(10), 8–10 (in Chinese with English abstract).

Zhang, H. Q., Yang, S. X., Wu, N. Y., Xu, X., Holland, M., Schultheiss, P., et al. (2007). *Successful and surprising results for China's first gas hydrate drilling expedition*. (pp. 6–9). Fire in the Ice, Methane Hydrate Newsletter, National Energy Technology Laboratory, US Department of Energy, Fall issue.

Zhang, G.-C., Zhu, W.-L., & Shao, L. (2001). Pull-apart tectonics and hydrocarbon prospecting in Bohai Bay and its nearby area. *Acta Petrolei Sinica, 22*(2), 14–18.

Zhu, W. L. (2009). *Characteristics of paleolimnology and source rock in oil-bearing basins offshore China*. Beijing: Geological Publishing House, 239 pp (in Chinese).

Zhu, W. L. (2010). Advancements of oil and gas exploration in china offshore area: Overview and perspective. *China Engineering Sciences, 12*(5), 18–24 (in Chinese).

Zhu, W. L., & Ge, J.-D. (2001). Gas exploration potential in offshore Bohai Bay Basin. *Acta Petrolei Sinica, 22*(2), 8–13.

Zhu, W. L., & Mi, L. J. (2010). *Atlas of oil and gas Basins*. China Sea: Petroleum Industry Press, 316 pp.

Zhu, E.-L., & Wang, Q. (1985). Ferromanganese nodules in the offshore areas of China. *Geological Review, 31*(5), 404–409 (In Chinese).

Zhu, W. L., Zhang, G. C., & Gao, L. (2008). Geological characteristics and exploration objectives of hydrocarbons in the northern continental margin basin of South China Sea. *Acta Petrolei Sinica, 29*, 1–9 (in Chinese).

Zhu, G., Zhang, S., Jin, Q., Dai, J., Zhang, L., & Li, J. (2005). Origin of the Neogene shallow gas accumulations in the Jiyang Superdepression, Bohai Bay Basin. *Organic Geochemistry, 36*, 1650–1663.

Zhu, W. L., Zhong, K., Li, Y. C., Xu, Q., & Fang, D. Y. (2012). Characteristics of hydrocarbon accumulation and exploration potential of the northern South China Sea deepwater basins. *Chinese Science Bulletin, 57*, 3121–3129.

Zou, H. Y., Gong, Z. S., Teng, C. Y., & Zhuang, X. (2011). Late-stage rapid accumulation of the PL19-3 giant oilfield in an active fault zone during Neotectonism in the Bozhong depression, Bohai Bay. *Science China Earth Science, 54*, 388–398.

Chapter 8

History of the China Seas

8.1 INTRODUCTION

Numerous publications emerged over the last decades to outline certain aspects of the China Seas, yet the huge amounts of accumulated geologic data still lack in-depth summary on the seas' evolution. Some critical issues remain ambiguous: What is the "China Seas" in a geologic aspect? When was it formed in the geologic history? In the 1980s, the Chinese geologic community compiled all data then available to generate the "Atlas of Palaeogeography of China" covering the entire Phanerozoic, but the offshore was not included due to a lack of data (Wang, 1985a). Recently, many papers discussed the evolution history of the SCS, especially after ODP Leg 184 (e.g., Li et al., 2006; Wang et al., 2003), but the discussion has been focused on sediment records of the tectonic opening of the deep basin, without considering much on the shallow water parts.

The history of the China Seas has been a subject of debate since the 1970s, stimulated mainly by paleontological discoveries. According to onshore geology records, the seawater retreated from Southeast China after the early Mesozoic (Wang, 1985a). In the 1970s, some marine-style fossils were discovered in several areas of SE China in large distance from the present coast, which naturally led to a suspicion whether the seawater invaded the interior of East China in the late Mesozoic or early Cenozoic. A typical case was the discovery of foraminifera in the nonmarine Oligocene deposits in Hubei, nearly 800 km upstream from the Yangtze estuary (Wang & Lin, 1974). Marine transgression became a particularly important topic of discussion because of the dominantly nonmarine Cenozoic hydrocarbons found in China at the time. Now, the debate has been settled with a recognition that some marine-style fossils may occur at lacustrine settings with suitable water chemistry conditions (see Section 8.4).

This chapter appraises the new evidence for an attempt to answer the old question: What does the "China Seas" mean in geology? In order to minimize possible semantic misunderstanding, the term "China Seas" here is defined as a part of the West Pacific marginal basins. We start from examining in Section 8.2 the occurrences of marine deposits before the formation of the China Seas, although the late Mesozoic and early Paleogene seas might have a Tethys affiliation and should not be confused with the China Seas defined here. Section 8.3 discusses the history of the China Seas with the onset marked by the opening of their deep-sea basins. In the last part (Section 8.4), we will reconsider the debates on Cenozoic "marine transgressions" into the interior of China and pinpoint the unsolved problems.

8.2 OCCURRENCES OF PRE-OLIGOCENE MARINE DEPOSITS

As seen from the previous chapters, the opening of the SCS basin started in the Oligocene, and the Okinawa Trough has opened much later since the late Miocene. However, the occurrences of marine deposits in the South China Sea (SCS) and East China Sea (ECS) can be traced back to the late Mesozoic and Paleogene. However, these pre-Oligocene marine records are most probably related to the active margin stage of the western Pacific, when the East Asian continent was directly facing the subducting Pacific Plate.

8.2.1 Late Mesozoic Marine Deposits

Three groups of late Mesozoic marine deposits have been reported from onshore and offshore Eastern China: Heilongjiang in NE China, the coastal areas of the ECS, and the coastal areas of the SCS. In eastern Heilongjiang, the nonmarine coal-bearing formations are intercalated with marine layers, containing fossils such as ammonites, buchiid bivalves, and dinoflagellate cysts indicative of a late Jurassic–early Cretaceous age (Sha, Hirano, Yao, & Pan, 2008). More interesting is the recent discovery of benthic and planktonic foraminiferal fauna in the late Cretaceous deposits of the Songliao Basin, central Heilongjiang (Xi et al., 2011). All these are likely the results of marine transgression from the north and northeast, associated with the north-western Paleo-Pacific Ocean and the Arctic Ocean (Sha et al., 2008), and thus have no direct relation to the present discussion of the China Seas.

8.2.1.1 East China Sea

The question of late Mesozoic marine transgression to coastal southeastern China was introduced also with the discovery of possibly marine fossils. The "Shi-Pu limestone" in the coastal area of northern Zhejiang, for example, was reported as evidence for the late Jurassic–early Cretaceous transgression because the limestone contains stromatolites, calcareous warm tubes (serpulids), and "nannofossils" (Figure 8.1) (Xu & Zheng, 1989). Later, the limestone was ascribed to representing tidal flat facies on the basis of biomarker and isotope analyses (Hu et al., 2012). Similar late Mesozoic transgressions were also reported from other Cretaceous lacustrine basins in the Zhejiang and Fujian Provinces near the modern ECS coast, according to lithologic or

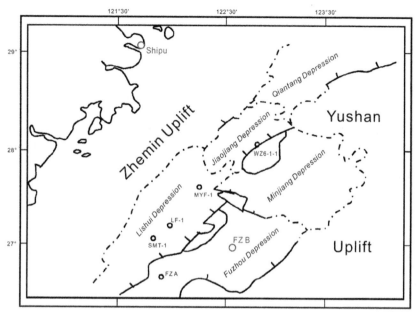

FIGURE 8.1   Locations of possibly marine deposits of late Mesozoic age as reported from the East China Sea region. *Modified from Wang et al. (2000).*

paleontological records (Hu et al., 2012), including ichthyofauna (Chang & Chow, 1978).

Late Mesozoic marine fossils also occur offshore in the ECS. An example is drill hole FZB in the inner shelf off Zhejiang coast where middle Jurassic calcareous nannofossils *Lotharingius velatus* and *Cyclagelosphaera margerelii* were found in mudstones of the Fuzhou Formation (Figure 8.1) (Wang et al., 2000).

Although the late Mesozoic is dominated by fluviolacustrine deposits in the modern ECS shelf and its coastal zone, the East Asian continent then was directly facing the subducting Pacific Plate from the east. Therefore, it is no surprise that marine transgressions had reached the modern shelf and coastal areas. After a recent compilation of the seismic and geologic data, it was found that the late Mesozoic deposits achieve a considerable thickness and are widely distributed in the southern ECS Shelf Basin. On this basis, a hypothetical "Great Mesozoic East China Sea" was proposed (Li, Gong, et al., 2012). Unfortunately, these data are insufficient to draw any conclusion about the marine nature of the late Mesozoic deposits in the shelf basins, and it is premature to reconstruct any meaningful paleogeographic pattern for a "Mesozoic ECS." Obviously, much more geologic data on a regional scope are needed for understanding the nature of the Mesozoic basin in the region.

8.2.1.2 South China Sea

Late Mesozoic marine deposits have been observed in the northern and southern SCS, as exemplified by the recent discovery of radiolarian siliceous rocks in Borehole MZ-1-1 in the northern slope (Figure 8.2). The siliceous rocks from burial depth of 1725–1997 m yielded radiolarians *Caneta hsui*, *Loopus primitivus*, and *Dibolachras* cf. *apletopora*, indicative of a late Jurassic–early Cretaceous age (Wu, Wang, Hao, & Shao, 2007).

Borehole MZ-1-1 was drilled in the Chaoshan Depression, northeastern SCS. The radiolarian-bearing rocks are underlain by mudstones with foraminifers and oolitic limestones, implying deepening of the sediment basin (Shao et al., 2007). The Chaoshan Depression is distinguished by its thick Mesozoic deposits, which might reach several thousand meters in thickness. The Mesozoic deposits are not restricted to the Chaoshan Depression, but distributed broadly in the NE SCS region, including the eastern part of the Pearl River Mouth (Zhujiangkou) Basin and the Taixinan Basin (Figure 8.2) (Hao, Lin, Yang, Xue, & Chen, 2001). Two sequences of marine depositions have been recognized: late Triassic to early Jurassic and early Cretaceous, both having significant prospects in hydrocarbon exploration (Yang, Lin, & Hao, 2002).

Marine Mesozoic deposits occur also in the southern SCS, with the distribution area extending from north of Palawan to the Nansha Islands (Dangerous Grounds) (Wei, Deng, & Chen, 2005). Although the early to middle Triassic and late Triassic to early Jurassic marine deposits were also found, the most extensive Mesozoic sea transgression in the southern SCS region occurred in the late Jurassic to early Cretaceous, resulting in the accumulation of shallow- and deepwater deposits there (Zhou, Chen, Sun, & Xu, 2005).

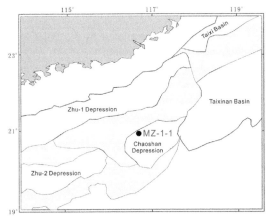

FIGURE 8.2 Location of Borehole MZ-1-1 in the Chaoshan Depression, NE South China Sea, where late Mesozoic marine deposits were discovered (Shao et al., 2007).

The northern SCS Mesozoic marine records have been ascribed to signifying the eastern extension of the Tethys Ocean (Zhou, Yan, Qiu, Chen, & Sun, 2003). The Tethys connection of the modern SCS region in the Mesozoic is discussed in detail in Chapter 3 (Section 3.1.2).

8.2.2 Paleogene Marine Deposits

Full marine deposits of Paleocene and Eocene age are now known from three regions: southern ECS, northern SCS, and southern SCS. Together with those found in western Taiwan, the Paleocene and Eocene marine deposits from southern ECS and northern SCS appear to have formed mainly in rifted (half) grabens. However, those found in the southern SCS appear to have accumulated mostly as carbonate platforms in the shallow "proto-SCS" (see Chapter 5). This section examines the age and occurrences of the Paleocene and Eocene marine deposits from offshore Mainland China and their bearings on the evolution of the early China Seas.

8.2.2.1 Marine Records

Three groups of marine biota are commonly employed for discriminating marine from lacustrine and other depositional settings: foraminifera, calcareous nanno-fossils, and dinoflagellates (see Chapter 4). Among them, planktonic foraminif-era, normal marine benthic foraminifera, and pelagic nannofossils are good indicators of full marine environments, while others due to their brackish nature may serve only as coastal marine proxies or proxies of distal marine influence. For example, there are reports of benthic foraminiferal assemblages character-ized by *Ammonia beccarii*, *Discorbis* spp., and *Elphidium* spp. and dinoflagellate assemblages from various Cenozoic strata of the ESC and SCS, indicating bay, lagoon, or salt lake settings (He et al., 1978; Hou, Li, Jin, & Wang, 1981; Yang, Cui, & Zhang, 2010; Yang, Sun, Li, & Liu, 1989b; Zhu, 2009).

East China Sea

Marine deposits containing Paleocene and Eocene planktonic foraminifera and pelagic nannofossils occur widely in the ESC, with about 40 well local-ities so far reported. Most of the records are from the Oujiang Depression (or Taipei Depression) in the southwestern ECS Shelf Basin, plus a couple of localities from other southern depressions (Li, 2012; Wu & Zhou, 2000; Yang et al., 2010, 1989b; Yang, Sun, Li, & Liu, 1989a; Zhong, Jiang, Li, & Wang, 2006). In wells Mingfeng-1, Mingyuefeng-1, and Shimentan-1, for example, planktonic foraminifera *Pseudohastigerina pseudomenardii*, *Acari-nina* spp., and *Morozovella* spp. and nannofossils *Heliolithus* spp. and *Dis-coaster* spp. indicate late Paleocene (mainly Lingfeng Formation) and early Eocene age (mainly Mingyuefeng Formation) (Table 8.1) (Su, Wu, Zhou, & Liu, 2000; Wu & Zhou, 2000; Yang et al., 2010, 1989a, 1989b). Middle to

TABLE 8.1 List of Paleocene–Eocene Marine Deposits with Planktonic Foraminifers and Nannofossils

Area	Locality/Well	Formation	Pl Foram Zone	Nannozone	Age	Reference
East China Sea						
East China Sea	Wenzhou-6-1 (Pinghu-1)	Pinghu/Wenzhou	P14–P16	NP17–NP19	M.–L. Eocene	Yang et al. (1989a, 1989b), Wu and Zhou (2000), Yang et al. (2010)
	Mingyuefeng-1	Oujiang	P8–P12	NP12–NP16	E.–M. Eocene	
	Shimentan-1	Mingyuefeng	P5–P7	NP10–NP12	E. Eocene	
	Lingfeng-1	Lingfeng	P3–P5	NP5–NP9	L. Paleocene	
W Taiwan	THS-1; WG-1			NP5–NP9	L. Paleocene	Ji (1984) (in Yang et al., 1989a)
South China Sea						
W Taiwan	Nantou outcrop	Chungliao	(*Discocyclina*)	NP14–NP15	M. Eocene	Huang et al. (2013)
Taixinan Basin	DP21-1-1		P6–P9, P10?		E.–M. Eocene	This study
NE Pearl River Mouth Basin	HJ15-1-1	Wenchang		NP15	M. Eocene	Huang (1997)
SW Pearl River Mouth Basin	BY7-1-1	Enping	P15–P16		Late Eocene	Wan, Hao, and Dong (1996), this study

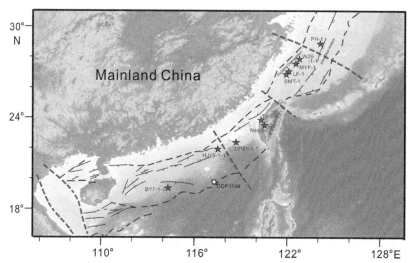

FIGURE 8.3   Location of sites with known Paleocene and Eocene marine deposits (data listed in Table 8.1), superimposed by major faults that control the distribution of basins (dash lines) or rifted paleolakes (short lines). *Modified from Zhu (2009).*

late Eocene marine microfossils, although sometimes discontinuous, also occur in these and neighboring wells. A group of wells represented by Pinghu-1 from the central part of the ESC Shelf Basin (or western Xihu Depression) probably mark the northern limit of normal marine influence during the Eocene (Figure 8.3). Due to sediment mixing in cutting samples, however, biostratigraphy of many sequences or formation in industrial wells from the ECS (and other localities alike) still requires efforts to resolve dating discrepancies between different authors (e.g., Su et al., 2000). Nevertheless, the ECS records listed in Table 8.1 represent the oldest Paleogene marine deposition at 60–62 Ma in the China Seas.

Taiwan

In wells THS-1 and WG-1 from the coastal western Taiwan, marine strata containing zones NP5–NP9 nannofossils indicating late Paleocene age were found by Wenrong Ji (probably in a well completion report presented in 1984, as quoted in Yang et al., 1989a). This age, if confirmed, will unite these strata with those found in the ECS to indicate the earliest marine deposition ~60 Ma (Table 8.1).

Indigenous larger benthic foraminifer *Discocyclina dispansa* and NP14–NP15 nannofossils were recently reported in the newly named Chungliao Formation from Nantou outcrop in the western foothills of Taiwan (Huang et al., 2013). This discovery provides solid evidence for an early to middle Eocene marine rift basin now exposed along the western foothill area, although whether similar rift sequences are exposed also in other parts of the Taiwan mountain belt is not clear.

Noteworthy is that a larger benthic foraminifer assemblage typified by *Discocyclina sowerbyi* and *Nummulites nuttali* has been reported from the Oujiang Formation in well Lingfeng-1 of the ECS (Figure 8.3) (Yang et al., 1989b). As the *Discocyclina* assemblages found in Taiwan and ECS both indicate early to middle Eocene age, systematic studies on their true taxonomic position may help to establish a possible link between the two regions and to categorize the marine influence during that time.

South China Sea

Compared to the ECS, the northern SCS records of Paleogene and Eocene marine deposits are sparse and young in age. Paleocene marine sequences are virtually absent, while Eocene marine deposits are now known only from three industrial wells: DP21-1-1, HJ15-1-1, and BY7-1-1 (Table 8.1; Figure 8.3).

Planktonic foraminifera typical of zones N6–N9 were found in well DP21-1-1 from the western Taixinan Basin in the northeastern SCS (Q. Li, unpublished report), indicating marine deposition during the early Eocene, at ~50–55 Ma. Eocene species also occur in the overlying sections (of probably middle Eocene) but their scarcity and poor preservation prevent precise age dating. This record may imply that the locality of DP21-1-1 and the exposed Nantou section in the western foothills of Taiwan could have been once under similar marine influence or even shared a single rifted basin of the early Eocene.

In the Hanjiang Depression of the Pearl River Mouth Basin, zone NP15 calcareous nannofossils were reported from a thin marine mudstone layer between 1547.85 and 1548.90 m in the Wenchang Formation of well HJ15-1-1 (Huang, 1997). Characterized by *Sphenolithus furcatolithoides* and *Helicosphaera seminulum*, the nannofossil assemblage indicates deposition in the middle Eocene, ~45 Ma.

Further west, in the deepwater Pearl River Mouth Basin, late Eocene microfossils, including planktonic foraminifera (small *Globigerina*, Wan et al., 1996; and small *Acarinina*, Q. Li, unpublished report) and dinoflagellates (*Homotryblium tenuispinosum* and *Hystrichosphaeridium tubiferum*, Mao & Lei, 1996), have been found to characterize the marine deposition in the Enping Formation during the late Eocene. Overall, however, large-scale full marine influence in the northern SCS did not start until the Oligocene–early Miocene.

8.2.2.2   Stepwise Southwesterly Advance of the Paleocene–Eocene sea

The marine deposition records presented in the preceding text provide some insight into the early evolution of the China Seas. As shown in Figure 8.4, marine environment first established about 60–62 Ma in the southern ECS–Taiwan region, leading to the deposition of marine sediment sequences with abundant calcareous plankton and other marine microfossils. During the early and middle Eocene, marine environment expanded in both directions, NE (into the central ECS) and SW (into the Taixinan Basin), before shrinking

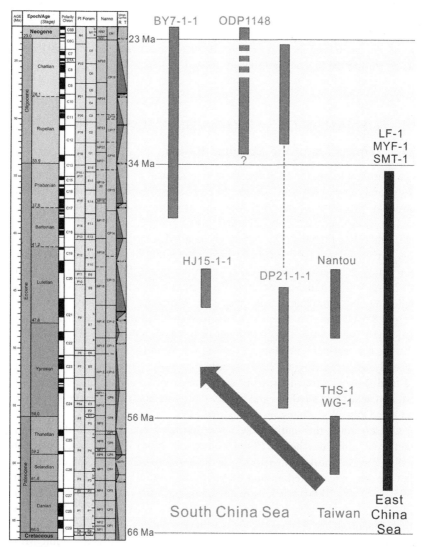

FIGURE 8.4 Stratigraphic distribution of marine deposits in the Paleogene China Seas according to the timescale and planktonic foraminiferal and nannofossil zonations given in Vandenberghe, Hilgen, and Speijer (2012). As indicated by the arrow bar, marine deposition started earlier in the NE (the ECS sector) than in the SW (the SCS sector). Refer to Figure 8.3 for site/well locations and Table 8.1 for data.

from the ECS and Taiwan areas in the later part of the middle Eocene and late Eocene. However, the southwesterly advance of the sea continued in the SCS in the middle and late Eocene.

Although these marine records reveal a generalized picture of marine influence, the lateral extent of the Paleocene–Eocene sea basins still awaits to be defined. The distribution of these marine records in the NE–SW direction along the main faults, however, may suggest that early sea basins were confined by rifted (half) grabens, NE-extending, and likely in a shape of elongated troughs. Presumably, the major NW–SE-trending fault zones (Figure 8.3) had been acting as the main passages connecting between these troughs and the Paleo-Pacific during the Paleocene–Eocene.

8.3   FORMATION OF MODERN CHINA SEAS

8.3.1   Formation of the Marginal Basins

As a marginal basin, the SCS was formed as a result of the continental margin rifting and subsequent seafloor spreading. As discussed in Section 3.3.1, the formation processes of the SCS went through continental margin rifting and final breakup, and seafloor spreading lasted roughly from ~32 to ~16 Ma (see Table 3.1; Figure 3.33). Although detailed paleogeographic reconstructions are prohibited by insufficient data, the tectonic reconstructions enabled us to draw general outlines for the formation of the SCS (Figure 8.5; Hall, 2012). As believed, there was an active margin between the East Asian continent and western Pacific in the late Mesozoic and early Paleogene (Figure 8.5A). The two deepwater basins of the SCS differ in their formation age: The eastern basin began to open in the early Oligocene (Figure 8.5B), whereas the seafloor spreading of the southwestern basin occurred later, from the early Miocene (Figure 8.5D). Seafloor spreading propagated from NE to SW and showed a transition from steady seafloor spreading in the northeast, to initial seafloor spreading, and finally to continental rifting in the southwest end (Li, Ding, Wu, Zhang, & Dong, 2012).

Consequently, the southern SCS was formed much later than its northern counterpart along the shelf of China (Hutchison, 2004). The difference can also be discerned in stratigraphic records: In the northern SCS, the tectonics-induced hiatuses are concentrated around 25 Ma in the late Oligocene (Li, Jian, & Su, 2005; Li et al., 2006), whereas major hiatuses in the southern SCS occurred around 19–16 Ma, the so-called mid-Miocene unconformity (Hutchison, 2010; Madon, Kim, & Wong, 2013). With the southwestern propagation, the SCS increased its size from the early Oligocene (Figure 8.6A) to middle Miocene (Figure 8.6B). The extent of the SCS further increased in the Pliocene with further subsidence of the Sundaland (Figure 8.6C) and reached the maximal size in the late Quaternary.

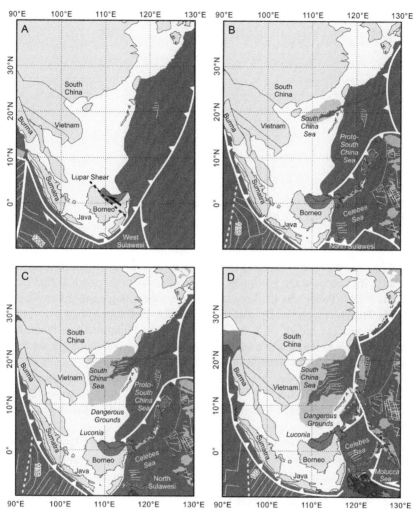

FIGURE 8.5   Plate reconstructions of the Southeast Asia. (A) 55 Ma (Eocene); (B) 30 Ma (early Oligocene); (C) 25 Ma (latest Oligocene); (D) 15 Ma (middle Miocene) (Hall, 2012).

The age of the modern ECS is much younger than the SCS. As mentioned in the preceding text and in Chapter 3, the middle and northern Okinawa Trough may be related to the opening of the Japan Sea with initial opening in the late Miocene, whereas the southern trough opened later, in the earliest Pleistocene. Since the ECS shelf, including the Yellow Sea and Bohai Gulf, could have been flooded only after the opening of the Okinawa Trough, the history of the ECS as a marginal basin is restricted to the post-Miocene. Accordingly, the Okinawa Trough first appeared only in the paleomap of the Pliocene (Figure 8.6C).

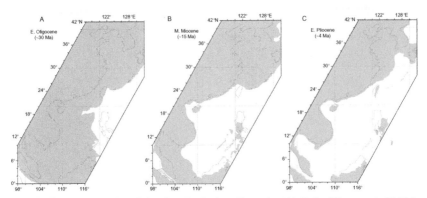

FIGURE 8.6   The growing China Seas in the late Cenozoic. (A) Early Oligocene (~30 Ma); (B) middle Miocene (~15 Ma); (C) early Pliocene (~4 Ma).

When the SCS basin opened, it must have had a free connection to the open Pacific, because the Philippine Islands were far in the south at that time (Figure 8.7A). Today, the SCS oceanic crust is subducting eastward along the Manila Trench (Figure 8.7D), and the subduction probably started before the spreading process ended. With the NW shift of the Philippine Sea Plate, the Manila Trench has been approaching the South China continent together with the accretionary prism and island arc resulted from the SCS subduction, subsequently reducing the size of the sea basin and restricting its connection to the open ocean (Figure 8.7B and C) (Sibuet et al., 2002), causing fundamental reorganization of deepwater circulations in the SCS basin (see Section 6.2.2.3).

In sum, the major events in the development history of the China Seas can be summarized in Table 8.2.

8.3.2   China Seas in Glacial Cycles

One of the outstanding features of the China Seas, as seen from the previous discussions, is the stepwise increase in area since their formation, with the maximal size occurring only in the late Quaternary. Due to the development of extensive shelves, the late Quaternary China Seas have been extremely sensitive to sea-level fluctuations. The best illustration is the Last Glacial Maximum (LGM). With the estimated 130 and 120 m sea-level drop in the ECS and SCS, respectively, the total area of the China Seas at the LGM was about 3,100,000 km^2, or two-thirds of its present size (4,700,000 km^2) (Table 8.3), with the emergence of all the extensive Sunda Shelf and the ECS shelf, including the entire Yellow Sea and Bohai Gulf. The sea areal reduction was accompanied by remarkable changes in regional geomorphology. As shown in Figure 8.8A, the ECS was then an elongated trough and the SCS

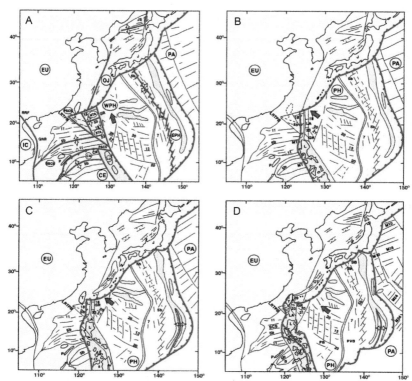

FIGURE 8.7   Plate kinematic reconstruction of East Asia since 15 Ma. (A) Early–middle Miocene (15 Ma); (B) late Miocene (8 Ma); (C) Pliocene (3 Ma); (D) present day. Thick red lines: active plate boundaries. Fracture zones and main magnetic lineations with their identifications are also shown (Sibuet et al., 2002).

was a semi-closed gulf (Wang, 1990; Wang & Sun, 1994). By contrast, during the high sea-level stand in interglacials, the marine transgression inundated extensive coastal plains, in particular along the Bohai and Yellow Sea coasts where large river deltas are located. This is best represented by the configuration of the China Seas during the mid-Holocene transgression shown in Figure 8.8B.

Extensive marine deposits marking late Quaternary transgressions were discovered largely in the 1970s and 1980s during hydro- and engineering geologic explorations in the coastal plains of East China. The recognition of marine or brackish water deposits was based on micropaleontological analyses supplemented later by geochemical approaches such as strontium isotopes or a combination of oxygen and carbon isotopes. Generally, three layers of marine transgressions have been identified: MIS 1 in the Holocene (~9–6 kyr BP; Figure 8.9A), MIS 3 (~39–26 kyr BP), and MIS 5 (around ~100 kyr BP) in the late Pleistocene (Figure 8.9B) (Wang, Min, Bian, &

TABLE 8.2 Major Tectonic and Sedimentologic Events of the China Seas

Age	Events
Oligocene, 32–25 Ma	Opening of the Eastern Subbasin, SCS
Latest Oligocene, 25–23 Ma	Hiatuses in northern SCS
Early Miocene, 22–16 Ma	Opening of the Southwest Subbasin, SCS
Early–middle Miocene	Formation of coral reefs and carbonate platforms
Middle Miocene, ~16 Ma	"Mid-Miocene unconformity" in southern SCS
Late Miocene–early Pliocene	Opening of the Okinawa Trough
Latest Miocene	Closure of the eastern side of the SCS
Late Pleistocene	Formation of the modern China Seas

TABLE 8.3 Estimations of the Area Reduction of the China Seas at the LGM

Sea		Present Area (10^3 km^2)	Area at LGM (10^3 km^2)
South China Sea		3600	2800
East China Sea (s.l.)	East China Sea (s.s.)	700	350
	Yellow Sea	380	0
	Bohai Gulf	77	0
	Total	1230	350
Total		4700	3150

From Wang (1990).

Cheng, 1985). The MIS 3 transgression seems somehow counterintuitive as it was within the glaciation, but many radiocarbon dating results indicated the same time interval about ~39–26 kyr BP, and the age was reconfirmed by many recent works (e.g., Wang, Zhang, Yuan, & Zhang, 2008; Wang, Zhao, Chen, & Li, 2008; Zhang et al., 2010).

A logical question follows: Why was East China subject to marine transgressions only since MIS 5, but not earlier? The answer lies in the subsidence of the coastal China continent, as the coastal areas in East China were too high for marine transgressions in the earlier interglacials. The interpretation is further supported by pollen data from the northern SCS. The pollen ratio

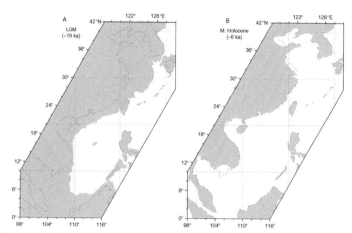

FIGURE 8.8   Changing extent of the China Seas in glacial cycles of the late Quaternary. (A) Last Glacial Maximum (~18 kyr BP); (B) middle Holocene (~6 kyr BP).

between shore plants (represented by herbaceous pollen, H) and upland ones (indicated by pine pollen, P) may indicate the relative distance of the studied site to the coast, with high ratio of H/P implying shorter distance and low ratio vice versa. Figure 8.10 shows the palynological diagrams of the last million years at Site 1144, northern SCS, revealing changes in the distance of the site from the coast due to eustatic sea-level changes in glacial cycles. Since a broader emerged shelf can also result in a higher H/P ratio, the small amplitudes in variations and low values of H/P before MIS 6 suggest a narrow and steep continental shelf, with only a limited area exposed at the lower sea-level stand (Figure 8.11), and the broadening of the continental shelf around MIS 6 was probably caused by neotectonic subsidence of East China. This also explains the occurrence of marine transgressions in the coastal areas of China only since MIS 5 (Sun et al., 2003). The conclusion is, therefore, that the China Seas reached the maximal size only in the late Quaternary interglacials.

8.4   DEBATES ON CENOZOIC MARINE TRANSGRESSIONS IN CHINA

In the 1970s and 1980s, the Chinese paleontological community was puzzled by the unexpected discoveries of marine-style microfossils in nonmarine Cenozoic basins far away from the known marine deposits in the country. These include (1) a benthic foraminiferal fauna featuring *Discorbis* sp. together with *Eucypris* and other nonmarine ostracods in the Oligocene lacustrine basin of Hubei Province, in the middle reach of the Yangtze River (Figure 8.12A; Wang & Lin, 1974); (2) a euryhaline ostracod and

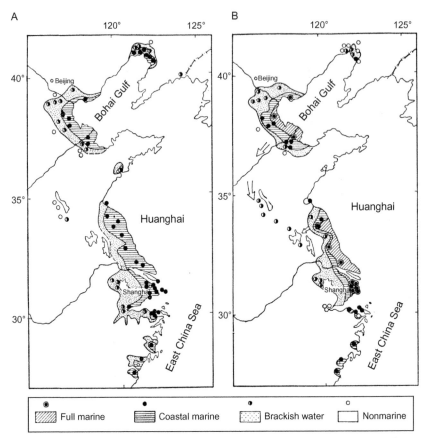

FIGURE 8.9   Areal distribution of the late Quaternary marine transgressions in East China.
(A) MIS 1 (Holocene); (B) MIS 3 and 5 (late Pleistocene). *From Wang, Min, Bian, and Cheng (1985).*

foraminiferal fauna, plus serpulid warm tubes in the Eocene lacustrine deposits of Jiangsu Province, on the western coast of the present Yellow Sea
(Figure 8.12B); (3) miliolid foraminifers with warm tubes and algal stromatolites in the Eocene carbonates from the northern Pearl River delta, to the north
of Guangzhou (Figure 8.12C); and (4) the separate occurrences of benthic foraminifers, serpulid tubes, and calcareous nannofossils in the Eocene and Oligocene lacustrine deposits in the Bohai Gulf basins (Figure 8.12D) (Wang,
1985b). Since these basins are hydrocarbon-bearing, the discoveries raised
some fundamental questions: Was seawater inundated inland China in the
Paleogene? Is the Paleogene oil in China really nonmarine in origin?

   The "marine transgression" question became even more challenging when
other foraminiferal faunas were found in late Pliocene to early Pleistocene
deposits in the middle reach of the Yellow River (Figure 8.13). However,

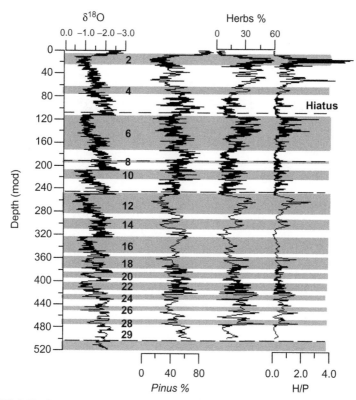

FIGURE 8.10 Downcore variations in pollen assemblages over the last million years at ODP Site 1144, northern South China Sea. (A) Planktonic foraminiferal $\delta^{18}O$, numbers denote marine isotope stages; (B) *Pinus* % (P); (C) herbs % (H); (D) H/P ratio. *From Sun, Luo, Huang, Tian, and Wang (2003).*

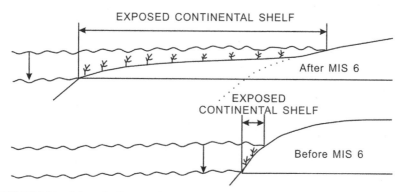

FIGURE 8.11 Schematic diagrams show changes in width of the northern continental shelf of the South China Sea, during glacial stages before and after MIS 6, as inferred from the pollen record. *From Sun et al. (2003).*

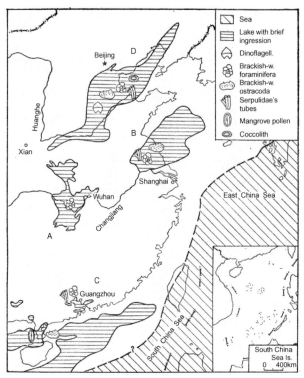

FIGURE 8.12 Sketch map showing distribution of Paleogene marine-style fossils in nonmarine basins in East China. (A) Qianjiang Basin, Hubei Province; (B) Subei Basin, Jiangsu Province; (C) Sanshui Basin, northern Pearl River delta; (D) Bohai basins. *From Wang (1985a).*

all these Paleogene and Pliocene–Pleistocene foraminifers belong to euryhaline benthic forms, with extremely low species diversity (mostly monospecific) and with abnormal test morphology, indicating unusual ecological conditions (Wang, Min, & Bian, 1985).

Nevertheless, the specific and assemblage features of these "marine-style" microfossils reveal their origin. In the modern world, nonmarine foraminifers occur in all continents but Antarctic, and some euryhaline benthic foraminifers can adapt to certain lacustrine environments with suitable water chemistry (Wang, Min, & Bian, 1985). In East China, the marine-style forms are mixed with typical nonmarine microfossils and occur in individual layers bounded by thick sequences with nonmarine fossils. An example is the Yongledian Group of late Pliocene–early Pleistocene age in Shaanxi and Shanxi Provinces, from the middle reach of the Yellow River (Figure 8.14), where monospecific foraminifers occur in thin layers intercalated within deposits with euryhaline ostracod *Cyprideis torosa* indicative of increased salinity. It

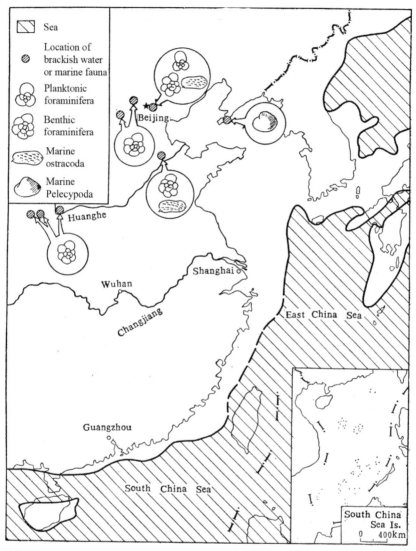

FIGURE 8.13 Sketch map showing distribution of marine-style fossils in late Pliocene–early Pleistocene deposits in East China. *From Wang (1985b).*

is the water chemistry that allowed the appearance of "marine-style" faunas in nonmarine basins. Consequently, the previously discussed discoveries cannot be used to indicate marine transgressions in paleoreconstructions of the China Seas.

However, the calcareous nannofossils found in the Paleogene deposits of Bohai basins appear to tell a different story from that of foraminifers

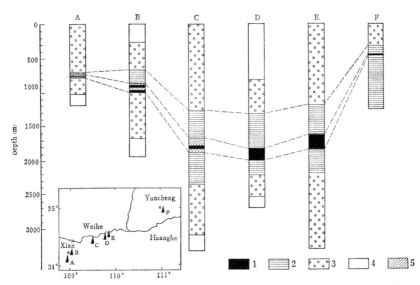

FIGURE 8.14 Stratigraphic correlation of the Yongledian Group of late Pliocene–early Pleistocene age in Shaanxi and Shanxi Provinces, China. Microfossil faunas: 1. foraminifers; 2. euryhaline ostracod *Cyprideis torosa*; 3. nonmarine ostracods; 4. barren of microfossils; 5. foraminifers and ostracod *Cytheromorpha* sp. *From Wang, Min, and Bian (1985).*

mentioned in the preceding text. The Eocene and Oligocene calcareous shales, which form important source rocks of the basins, are composed almost entirely of one species of coccolith *Reticulofenestra bohaiensis* (Zhong, Zhong, Fei, Nie, & Lin, 1988), yet the paleontological and geochemical features of the entire sequences indicate a nonmarine nature of the basins. In addition, the strontium isotope of the calcareous shale also excluded its connection with ocean water (Liu & Cheng, 1996), similar to the case of nonmarine foraminifer occurrence (Wand et al., 1996). Again, it was the lake water chemistry that gave rise to algal blooms, resulting in accumulation of laminated lacustrine source rocks (Liu & Wang, 2013).

The earlier discussion does not exhaust the decade-long debates on marine transgressions in China. Awaiting for final explanation is the mysterious occurrence of a full marine layer near Beijing. In the late 1970s, a highly diverse foraminiferal fauna with both planktonic and benthic species was found from a borehole NE of Beijing at a depth of 428.6 m below surface, in a layer paleomagnetically dated to 2.26 Ma (An, Wei, Lu, Wang, & Ding, 1979). The foraminiferal fauna was accompanied by rich coccolith nannofossils, and the finding was confirmed later in several other holes nearby (Wang & He, 1983). Unlike the previously discussed cases of "marine-style" faunas, this finding involves a full marine assemblage, which can be brought about only by sea transgression. However, the nearest occurrence of marine deposits at that age is about thousand kilometers away. Further research efforts and new discoveries are needed to solve the riddle, although solving the mystery might change our present perception about the history of the China Seas.

REFERENCES

An, Z., Wei, L., Lu, Y., Wang, N., & Ding, S. (1979). Magnetostratigraphy of the core S5 and the transgression in the Beijing area during the early Matuyama Epoch. *Geochimica, 4,* 343–346 (in Chinese with English abstract).

Chang, M. M., & Chow, C. C. (1978). On the fossil fishes in Mesozoic and Cenozoic oil-bearing strata from East China and their sedimentary environments. *Vertebrata PalAsiatica, 16*(4), 229–237, Beijing, (in Chinese, with English abstract).

Hall, R. (2012). Late Jurassic–Cenozoic reconstructions of the Indonesian region and the Indian Ocean. *Tectonophysics, 570–571,* 1–41.

Hao, H. J., Lin, H. M., Yang, M. X., Xue, H. Y., & Chen, J. (2001). The Mesozoic in Chaoshan Depression: A new domain of petroleum exploration. *China Offshore Oil and Gas (Geology), 15*(3), 157–163 (in Chinese, with English abstract).

He, Y., Hu, L., & Research Institute of Ministry of Petroleum and Chemical Industries and Nanjing Institute of Geology and Paleontology CAS. (1978). *Cenozoic Foraminifera from the coastal region of Bohai.* Beijing: China Science Press, 48 pp.+9 pls. (in Chinese).

Hou, Y., Li, Y., Jin, Q., & Wang, P. (Eds.), (1981). *Tertiary paleontology of the northern continental shelf of South China Sea.* Guangzhou: Guangdong Science and Technology Press, 274 pp.+108 pls (in Chinese).

Hu, G., Hu, W. X., Cao, J., Yao, S. P., Xia, X. M., Li, Y. X., et al. (2012). Deciphering the Early Cretaceous transgression in coastal southeastern China: Constraints based on petrography, paleontology and geochemistry. *Palaeogeography Palaeoclimatology Palaeoecology, 317–318,* 182–195.

Huang, L. S. (1997). Calcareous nannofossil biostratigraphy in the Pearl River Mouth Basin, South China Sea, and Neogene reticulofenestrid coccoliths size distribution pattern. *Marine Micropaleontology, 32,* 31–57.

Huang, C.-Y., Yen, Y., Liew, P.-M., He, D.-J., Chi, W.-R., Wu, M.-S., et al. (2013). Significance of indigenous Eocene larger foraminifera *Discocyclina dispansa* in Western Foothills, Central Taiwan: A Paleogene marine rift basin in Chinese continental margin. *Journal of Asian Earth Sciences, 62,* 425–437.

Hutchison, C. S. (2004). Marginal basin evolution: The southern South China Sea. *Marine and Petroleum Geology, 21,* 1129–1148.

Hutchison, C. S. (2010). What are the Spratly Islands? *Journal of Asian Earth Sciences, 39,* 371–385.

Li, J. B. (Ed.), (2012). *Regional oceanography of China Seas—Marine geology.* Beijing: China Ocean Press, 547 pp. (in Chinese).

Li, J. B., Ding, W. W., Wu, Z. Y., Zhang, J., & Dong, C. (2012). The propagation of seafloor spreading in the southwestern subbasin, South China Sea. *Chinese Science Bulletin, 57,* 3182–3191.

Li, G., Gong, J. M., Yang, C. Q., Yang, C. S., Wang, W. J., & Wang, H. R. (2012). Stratigraphic features of the Mesozoic "Great East China Sea"—A new exploration field. *Marine Geology and Quaternary Geology, 32*(3), 93–104 (in Chinese, with English abstract).

Li, Q., Jian, Z., & Su, X. (2005). Late Oligocene rapid transformations in the South China Sea. *Marine Micropaleontology, 54,* 5–25.

Li, Q., Wang, P., Zhao, Q., Shao, L., Zhong, G., Tian, J., et al. (2006). A 33 Ma lithostratigraphic record of tectonic and paleoceanographic evolution of the South China Sea. *Marine Geology*, *230*, 217–235.

Liu, C. L., & Cheng, X. R. (1996). Non-marine calcareous nannofossils in Paleogene of the Bohai Gulf Basin: Evidence from the $^{87}Sr/^{86}Sr$ ratio. *Chinese Science Bulletin*, *41*(10), 908–910.

Liu, C. L., & Wang, P. X. (2013). The role of algal blooms in the formation of lacustrine petroleum source rocks—Evidence from Jiyang depression, Bohai Gulf Rift Basin, eastern China. *Palaeogeography Palaeoclimatology Palaeoecology*, *388*, 15–22.

Madon, M., Kim, C. L., & Wong, R. (2013). The structure and stratigraphy of deepwater Sarawak, Malaysia: Implications for tectonic evolution. *Journal of Asian Earth Sciences*, *76*, 312–333.

Mao, S., & Lei, Z. (1996). Tertiary dinoflagellates and paleoenvironmental evolution of Pearl River Mouth Basin. In Y. Hao, Y. Xu, & S. Xu (Eds.), *Research on micropalaeontology and paleoceanography in Pearl River Mouth Basin, South China Sea* (pp. 52–66). Beijing: China University of Geoscience Press (in Chinese).

Sha, J. G., Hirano, H., Yao, X. G., & Pan, Y. H. (2008). Late Mesozoic transgressions of eastern Heilongjiang and their significance in tectonics, and coal and oil accumulation in northeast China. *Palaeogeography Palaeoclimatology Palaeoecology*, *263*, 119–130.

Shao, L., You, H. Q., Hao, H. J., Wu, G. X., Qiao, P. J., & Lei, Y. C. (2007). Petrology and depositional environments of Mesozoic strata in the northeastern South China Sea. *Geological Review*, *53*(2), 164–169 (in Chinese, with English abstract).

Sibuet, J. C., Hsu, S. K., Le Pichon, X., Le Formal, J.-P., Reed, D., Moore, G., et al. (2002). East Asia plate tectonics since 15 Ma: Constraints from the Taiwan region. *Tectonophysics*, *344*, 103–134.

Su, X., Wu, F. D., Zhou, P., & Liu, J. S. (2000). Cenozoic calcareous nannofossil biostratigraphy of sediments from Lingfeng Well in the East China Sea. *Journal of China University of Geosciences*, *11*(3), 259–263.

Sun, X., Luo, Y., Huang, F., Tian, J., & Wang, P. (2003). Deep-sea pollen from the South China Sea: Pleistocene indicators of East Asian monsoon. *Marine Geology*, *201*, 97–118.

Vandenberghe, N., Hilgen, F. J., & Speijer, R. P. (2012). The Paleogene period. In F. M. Gradstein, J. G. Ogg, M. Schmitz, & G. Ogg (Eds.), *The geologic time scale 2012* (pp. 855–921). Netherlands: Elsevier.

Wan, X., Hao, Y., & Dong, J. (1996). Tertiary planktonic foraminiferal zonations in Pearl River Mouth Basin. In Y. Hao, Y. Xu, & S. Xu (Eds.), *Research on micropalaeontology and paleoceanography in Pearl River Mouth Basin, South China Sea* (pp. 10–18). Beijing: China University of Geoscience Press (in Chinese).

Wand, S., Dong, L., Lin, W., Li, C., Wang, P., Zhao, Q., et al. (1996). Sr isotope of fossil foraminiferal assemblages in the Nihewan Group. *Chinese Science Bulletin*, *41*(7), 583–586.

Wang, H. Z. (Ed.). (1985a). *Atlas of the palaeogeography of China* (p. 143). Beijing: Cartographic Publishing House.

Wang, P. (1985b). Paleontological evidence for the formation of the seas neighbouring China. In P. X. Wang, et al. (Eds.), *Marine micropaleontology of China* (pp. 1–14): Beijing: China Ocean Press and Springer-Verlag.

Wang, P. (1990). The Ice-Age China Sea—Research results and problems. In *Proceedings of the first international conference on Asian marine geology, September 7–10, 1988, Shanghai* (pp. 181–197): Beijing: China Ocean Press.

Wang, N. W., & He, X. X. (1983). Preliminary study of the Quaternary calcareous nannofossils of Beijing Plain. *Chinese Science Bulletin*, *28*(9), 1238–1242.

Wang, P., Jian, Z., Zhao, Q., Li, Q., Wang, R., Liu, Z., et al. (2003). Evolution of the South China Sea and monsoon history revealed in deep-sea records. *Chinese Science Bulletin*, *48*(23), 2549–2561.

Wang, P., & Lin, J. (1974). Discovery of Paleogene brackish-water foraminifers in a certain basin, Central China, and its significance. *Acta Geologica Sinica*, *1979*(2), 175–183 (in Chinese, with English abstract).

Wang, P., Min, Q., & Bian, Y. (1985). On marine-continental transitional faunas in Cenozoic deposits of East China. In P. X. Wang, et al. (Ed.), *Marine micropaleontology of China* (pp. 15–33). Beijing: China Ocean Press and Springer-Verlag.

Wang, P., Min, Q., Bian, Y., & Cheng, X. (1985). On micropaleontology and stratigraphy of Quaternary marine transgressions in East China. In P. X. Wang, et al. (Ed.), *Marine micropaleontology of China* (pp. 265–284). Beijing: China Ocean Press and Springer-Verlag.

Wang, P., & Sun, X. (1994). Last glacial maximum in China: Comparison between land and sea. *Catena*, *23*(3/4), 341–353.

Wang, K. D., Wang, J. P., Xu, G. Q., Zhong, S. L., Zhang, Y. Y., & Yang, H. R. (2000). The discovery and division of the Mesozoic strata in the southwest of Donghai shelf basin. *Journal of Stratigraphy*, *24*(2), 129–131.

Wang, Q., Zhang, Y. F., Yuan, G. B., & Zhang, W. Q. (2008). Since MIS 3 stage the correlation between transgression and climatic changes in the North Huanghua Sea, Hebei. *Quaternary Sciences*, *28*(1), 79–95 (in Chinese, with English abstract).

Wang, Z. H., Zhao, B. C., Chen, J., & Li, X. (2008). Chronostratigraphy and two transgressions during the Late Quaternary in Changjiang delta area. *Journal of Palaeogeography*, *10*(1), 99–110 (in Chinese, with English abstract).

Wei, X., Deng, J. F., & Chen, Y. H. (2005). Distribution character and exploration potential of Mesozoic sea facies sedimentary strata in the South China Sea Basin. *Journal of Jilin University (Earth Science Edition)*, *35*(4), 456–461 (in Chinese, with English abstract).

Wu, G. X., Wang, R. J., Hao, H. J., & Shao, L. (2007). Microfossil evidence for development of marine Mesozoic in the North of South China Sea. *Marine Geology and Quaternary Geology*, *27*(1), 79–85 (in Chinese, with English abstract).

Wu, F., & Zhou, P. (Eds.). (2000). *Analyses of Tertiary sequence stratigraphy and depositional systems in the Xihu Depression of East China Sea Shelf Basin*. Beijing: Geological Publishing House, 179 pp. (in Chinese).

Xi, D. P., Wan, X. Q., Feng, Z. Q., Li, S., Feng, Z., Jia, J., et al. (2011). Discovery of Late Cretaceous foraminifera in the Songliao Basin: Evidence from SK-1 and implications for identifying seawater incursions. *Chinese Science Bulletin*, *56*, 253–256.

Xu, B. M., & Zheng, S. P. (1989). The age and characteristics of sedimentary facies of the "Shipu limestone" in Xiangshan, Zhejiang Province. *Geological Review*, *35*(3), 221–230 (In Chinese with English abstract).

Yang, W., Cui, Z., & Zhang, Y. (Eds.). (2010). *Geology and minerals of East China Sea* (p. 780). Beijing: China Ocean Press (in Chinese).

Yang, S. K., Lin, H. M., & Hao, H. J. (2002). Oil and gas exploration prospect of Mesozoic in the eastern part of the Pearl River Mouth Basin. *Acta Petrolei Sinica*, *23*(5), 28–33 (in Chinese, with English abstract).

Yang, J., Sun, M., Li, Z., & Liu, Z. (Eds.) (Research Party of Marine Geology, Ministry of Geology and Mineral Resources, and Geological Institute CAS), 1989a. Cenozoic paleobiota of the continental shelf of the East China Sea (Donghai) micropaleobotanical volume. Geological Publishing House, Beijing, 324 pp.+107 pls (in Chinese).

Yang, J., Sun, M., Li, Z., & Liu, Z. (Eds.) (Research Party of Marine Geology, Ministry of Geology and Mineral Resources, and Geological Institute CAS), 1989b. Cenozoic paleobiota of the continental shelf of the East China Sea (Donghai) paleozoological volume. Geological Publishing House, Beijing, 280 pp.+75 pls (in Chinese).

Zhang, Z. K., Xie, L., Zhang, Y. F., Xu, J., Li, S. H., & Wang, Y. (2010). Sedimentary records of the MIS 3 transgression event in the North Jiangsu Plain, China. *Quaternary Sciences*, *30*(5), 883–891 (in Chinese, with English abstract).

Zhong, S., Jiang, L., Li, B., & Wang, J. (2006). *High resolution Paleogene biostratigraphy and sequence stratigraphy of the Taipei depression in the East China Sea*. Beijing: Petroleum Industry Press, 119 pp.+39 pls (in Chinese).

Zhong, X. C., Zhong, S. L., Fei, X. D., Nie, Y. G., & Lin, T. (1988). Calcareous nannofossils from the Oligocene Shahejie I member in the Bohai basin and their sedimentary environment. *Acta Micropalaeontologica Sinica*, *5*(2), 145–151 (in Chinese, with English abstract).

Zhou, D., Chen, H. Z., Sun, Z., & Xu, H. H. (2005). Three Mesozoic sea basins in eastern and southern South China Sea and their relation to Tethys and Paleo-Pacific domains. *Journal of Tropical Oceanography*, *24*(2), 16–25 (in Chinese, with English abstract).

Zhou, D., Yan, J. X., Qiu, Y. X., Chen, H. Z., & Sun, Z. (2003). Route for the eastern extension of Meso-Tethys in the western environs of the South China Sea. *Earth Science Frontiers*, *10*(4), 469–477 (in Chinese, with English abstract).

Zhu, W. L. (2009). *Paleolimnology and source rock studies of Cenozoic Hydrocarbon-bearing offshore basins in China*. Beijing: Geological Publishing House, 239 pp. (in Chinese).

Chapter 9

Concluding Remarks

In the preceding chapters, we have examined various aspects of geology of the China Seas. Here, as a conclusion, we try to broaden the scope by viewing the China Seas as a whole in the aspect of the western Pacific marginal seas. In the modern global ocean, more than 75% of the marginal basins are concentrated in the western Pacific continental margin (Tamaki & Honza, 1991). Asia and the Pacific, the largest continent and the largest ocean, are separated by a series of marginal seas of which the China Seas are the major components (Figure 9.1). All the processes concerning the formation and development of the China Seas took place on the background of the interactions between the Asian continent and the Pacific Ocean. By summarizing the existing data, we recognize the following major features for the geological history of the China Seas.

9.1 RELATIVELY YOUNG AGE OF FORMATION

Geologically, the China Seas are among the youngest marginal seas in the low-latitude ocean. Being opened in the late Cenozoic, the series of western Pacific marginal basins comprise a relatively young feature in the global tectonics. As discussed in Chapter 8, the South China Sea (SCS) opened in the Oligocene and the East China Sea (ECS) in the late Miocene, but both stepwise increased their size until the late Pleistocene when the China Seas reached their full size. In other words, the broad shelves of the China Seas, including the entire Yellow Sea and Bohai Gulf, did not exist until the Pleistocene. This specific feature of the China Seas must have its profound effect on climate and environmental changes and should no more be ignored in paleoreconstructions.

9.2 EFFECT OF TOPOGRAPHIC REVERSAL OF EAST ASIA

One of the most radical changes in Asia-Pacific interactions was the reversal of west-tilting to east-tilting topography in China. From the late Mesozoic to early Paleogene, China displayed a general west-tilting relief (Figure 9.2A) contrasting with the present east-tilting topography (Figure 9.2B), and the reversal occurred sometime in the late Oligocene. Consequently, Paleogene

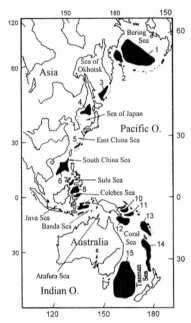

FIGURE 9.1 Western Pacific marginal seas with marginal basins (black areas). Marginal basins remote from continents (e.g., Philippine Basin, Mariana Basin, Lau Basin, and Fiji Basin) are not shown. Marginal basins: *1* Aleutian Basin; *2* Komandosky Basin; *3* Kuril Basin; *4* Japan Basin; *5* Okinawa Trough; *6* South China Sea Basin; *7* Sulu Basin; *8* Celebes Basin; *9* Banda Basin; *10* Soloman Basin; *11* Woodlark Basin; *12* Coral Basin; *13* N. Loyalty Basin; *14* New Caledonia Basin; *15* Tasman Basin (Wang, 1999).

sediments in China mostly accumulated in large lacustrine basins ponded in rift grabens, providing the main source of nonmarine hydrocarbon production in China today, whereas the Neogene fluvial–lacustrine deposits covered large areas as a result of general subsidence in eastern China (Figure 9.2D–E) (Wang, 2004). After the reversal, the large rivers such as the Yangtze started to flow eastward before the Oligocene/Miocene boundary (Zheng et al., 2013), but the hydrocarbon source rocks under the modern China Seas were formed mostly in the Paleogene (see Chapter 7).

9.3   RESTRICTED CONNECTION WITH THE OPEN OCEAN

The western Pacific marginal seas differ from each other in terms of their extent of connection with the open ocean due to their morphological features (Wang, 1999). The SCS belongs to the enclosed-basin type with the Bashi Strait as the only deep-water connection with the Pacific, and the geological evolution of the Bashi Strait largely determined the deep-water ventilation of the SCS (see Section 6.2.2.3). Although the ECS has an open connection with the Pacific, the inflow of the Kuroshio water into the Okinawa Trough

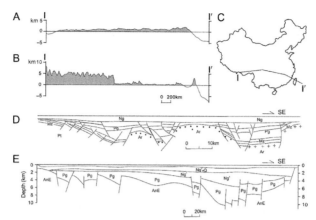

FIGURE 9.2   Mid-Cenozoic topographic reversal in China. Upper panels show a comparison of Paleogene and Neogene transects of South China, with gray areas indicating land above sea level. (A) A hypothetical transect of southern China (I–I′) during the Eocene about 50 Ma. (B) Modern transect (I–I′) showing east-tilting topography. (C) Location of the transect I–I′. Lower panels display two typical transects of Cenozoic basins onshore and offshore eastern China showing the different attitude of Paleogene and Neogene deposited: (D) the Liao River Mouth Basin, Northeast China, and (E) the East China Sea shelf basin. Stratigraphic units used in transects: Q, Quaternary; Ng, Neogene; Pg, Paleogene; Mz, Mesozoic; AnE, Pre-Paleogene; Pt, Proterozoic; Ar, Achaean (Wang, 2004).

varies significantly with sea-level fluctuations (see Section 6.3.1.2). In the present North Pacific, two western boundary currents, that is, the Kuroshio and the Oyashio, flow through several marginal seas to mix there with large river runoff from Asia (see Figure 2.1A and B in Chapter 2), providing vigorous water exchanges between land and sea and between the high and low latitudes (Wang, 2004). The system of exchanges is not stable, but vulnerable to sea-level fluctuations and local tectonic changes of the gateways, thereby exerting various climates and environmental impacts on the entire region (see Section 6.3).

9.4   EXTENSIVE CONTINENTAL SHELVES

The modern China Seas are characterized by the extensive continental shelves, yet it is a relatively young feature formed mainly in the late Quaternary. As marginal basins, the China Seas intercept the flux of terrigenous sediment supplied from the Asian continent. Although South and East Asia with their islands together provide over 70% of the terrigenous suspended material to the global ocean (Milliman & Meade, 1983), there is no large submarine deep-sea fan developed in the western Pacific comparable to Indian Ocean fans. Instead, most of large rivers in East Asia provide sediments into the ECS and SCS to build up large deltas and extensive shelves (see Figure 2.1), resulting in long-distance migration of coastline in glacial cycles.

Thus, during the last deglaciation, the ECS coastline had migrated ~1200 km landward, corresponding to coastline retreat of >0.4 m per day in average (see Section 6.3.1.3). With the broad continental shelves, the China Seas sensitively respond to all kinds of onshore and offshore changes, from variations in the Kuroshio flow and river runoff to sea-level fluctuations.

In a broader sense, the sea-land interactions involve all the Earth's surface systems from atmosphere to lithosphere. The formation and evolution of the China Seas are closely related to that of the Asian monsoon system, to the hydrology in the western Pacific, and to the plate dynamics between Asia and the Pacific. The top research field defined in the new science plan of the International Ocean Discovery Program (2013–2023) is "Earth connections" that examine the links between the surface, lithospheric, and deep Earth processes (IODP, 2011). This new, emerging direction in the Earth system science is particularly relevant to the western Pacific. As recently proposed, the development of numerous marginal basins in the western Pacific has its deep root. The mantle underlying the western Pacific has globally the largest amounts of subducting oceanic slabs and hence the highest water content, which may have stimulated the formation of microplates and marginal seas (Komiya & Maruyama, 2007). Likely, catastrophic collapse of the slabs to the core–mantle boundary between 30 and 20 Ma had triggered large-scale deformation in Asia and possibly the development of marginal seas (Maruyama, Santosh, & Zhao, 2007). If proved, the deep processes in the mantle have accounted at least partly for the major tectonic events observed in the China Seas.

Entering the twenty-first century, the China Seas become the target of active hydrocarbon explorations and basic geological researches, such as the major project of "South China Sea Deep" funded by the Chinese National Science Foundation (Wang, 2012) and IODP Expedition 349 to drill the oceanic crust under the SCS (Li, Wang, Franke, Lin, & Tian, 2012; Qiu, 2014). A new IODP proposal to drill the rifting stage of the SCS is under reviewing. It is expected that these endeavors will help reveal the life history of the China Seas and deepen our understanding of the development of global marginal basins.

REFERENCES

IODP (2011). *Illustrating earth's past, present, and future. The international ocean discovery program science plan, 2013-2023*. Washington, DC: IODP-MI, 84 pp.

Komiya, T., & Maruyama, S. (2007). A very hydrous mantle under the western Pacific region: Implications for formation of marginal basins and style of Archean plate tectonics. *Gondwana Research, 11*, 132–147.

Li, C. F., Wang, P. X., Franke, D., Lin, J., & Tian, J. (2012). Unlocking the opening processes of the South China Sea. *Scientific Drilling, 14*, 55–59.

Maruyama, S., Santosh, M., & Zhao, D. (2007). Superplume, supercontinent, and post-perovskite: Mantle dynamics and anti-plate tectonics on the Core–Mantle Boundary. *Gondwana Research, 11*, 7–37.

Milliman, J. D., & Meade, R. H. (1983). World-wide delivery of river sediment to the ocean. *Journal of Geology, 91*(1), 1–21.

Qiu, J. (2014). Sea drilling project launches. *Nature, 505*, 466–467.

Tamaki, K., & Honza, E. (1991). Global tectonics and formation of marginal basins: Role of the western Pacific. *Episodes, 14*(3), 224–230.

Wang, P. X. (1999). Response of Western Pacific marginal seas to glacial cycles: Paleoceanographic and sedimentological features. *Marine Geology, 156*, 5–39.

Wang, P. X. (2004). Cenozoic deformation and the history of sea-land interactions in Asia. In P. Clift, P. Wang, W. Kuhnt, & D. Hayes (Eds.), *AGU Geophysical Monograph Series: Vol. 149. Continent-ocean interactions in the East Asian marginal seas.* (pp. 1–22).

Wang, P. X. (2012). Tracing the life history of a marginal sea—On "The South China Sea Deep" research program. *Chinese Science Bulletin, 57*(24), 3093–3114.

Zheng, H. B., Clift, P. D., Wang, P., Tada, R., Jia, J. T., He, M. Y., et al. (2013). Pre-miocene birth of the Yangtze River. *Proceedings of the National Academy of Sciences of the United States of America, 110*(19), 7556–7561.

Note: Page numbers followed by "*f*" indicate figures, and "*t*" indicate tables.

A

A (anorogenic)-type granites, 144–145
Abiespollenites, 475, 477*f*
Accretionary prism, 128–129, 130–131, 133, 134–136, 363, 478, 622, 654
Adjacent waters and Pearl River estuary, 31–32, 32*f*
Aira-Tanzawa (AT) caldera, 310
Alkenone-based SST, 483*f*, 495, 522–523
Alkenone unsaturation index, 479–481
Ammonia beccarii, 283, 286*t*, 287, 647
Ammonia ketienziensis, 283, 287, 509–514, 513*f*
Annual mean sediment discharge, Pearl River, 32–33
Annual sediment load, Huayuankou and Lijin stations, 28–29, 29*f*, 38*f*
Annual water budget, SCS, 57–58, 58*t*
Arctic ice-sheet development, Pliocene, 470–471, 481, 490*f*
Asean Council on Petroleum (ASCOPE), 346–347, 356–357, 357*f*, 358–359, 372–373, 425–428, 430*f*, 432*f*, 433–434, 437–438
Asia, 4, 11–12, 23–26, 35–36
    Mekong River, 34–35
    nannofossil ooze and detrital clays, 226–228
    Pacific Plate, 116–117
    Yangtze River, 29
Asian monsoon system, 39*f*, 481, 522–523
Asian rivers, journals on, 24–26, 27*t*
Atlantic cycles, 271–272, 272*f*, 273, 274*f*
Axial Malay Fault Zone (AMFZ), 429–430

B

Bank erosion, 31, 35
Basalt and rhyolite, 145–146
Bashi Strait, 22–23, 40–41, 57–59, 60*f*, 61–62, 61*f*, 63–64, 63*f*, 66–67, 287–289, 293–295, 303, 477–478, 479*f*, 533, 668–669
Beibuwan Basin, 6*t*, 77*f*, 110–112, 342–343, 352–354, 356, 373, 377*t*, 416–419, 448, 449, 572*f*, 606, 607*f*, 618*f*, 627–628

basin-bounding fault systems, 416
Cenozoic basins, SCS, 352–354, 354*f*, 355*t*, 416
Cenozoic sediments, 416, 417*f*
lithostratigraphy, 416, 418*f*
secondary faults, 416
seismic reflectors, 419
spore–pollen assemblages, 419
Benham Plateau, 151–153
Benthic foraminifera, 213–215, 276–278, 368–369, 373, 376–378, 441*f*, 443, 445*f*, 446*f*, 478, 479*f*, 480*f*, 481–483, 513*f*, 524, 528, 647, 657–658
Benthic foraminiferal assemblages, 213–215, 285, 286*t*, 376–378, 379*f*, 647, 650
Benthic foraminifer *Discocyclina dispansa*, 649–650
"Biogenic bloom", 273, 378–380, 472–475
Biogenic sediments
    calcareous components, 280–293
    carbonate and coral reefs, 268–280
    organic carbon, 299–309
    siliceous components, 293–299
Biostratigraphic Framework, 341–342, 368–381
    benthic foraminiferal assemblages, 376–378
    fossil groups, 378–380
    microfossils, 368–369
    nannofossil and planktonic foraminiferal biostratigraphy, 369
    ostracod assemblages, 373–376
    quaternary stratigraphic events, 380–381
    spores and pollen assemblages, 369–373
Bohai and Yellow Seas, 1, 90–93, 103–104, 186, 196–197, 197*f*, 289–290, 295, 297, 342–343, 368–369
    Cenozoic basins, 349*f*, 350*t*
    Modern Sediment Distribution, 183–184
    Neogene postrift sequences, 349–352
    Paleozoic and Mesozoic sequence, 346
    summer SST gradient, 49
    synrift and postrift sediments, 349–352
    topographic maps, ECS, 17*f*
    water column, 65–66

Bohai Basin, 75, 77*f*, 89–91, 103–104, 142,
    143–144, 153–154, 162*f*, 341,
    342–343, 346, 349–352, 349*f*, 351*f*,
    368, 373, 374*t*, 376–378, 381–387,
    388*f*, 447, 448, 449, 571, 572–581,
    618*f*, 628, 630, 660*f*, 661–662
  Cenozoic basins, ECS, 349*f*
  depressions (subbasins) and uplifts,
    381–382, 382*f*
  large-scale seismic features, sequences,
    386–387, 388*f*
  lithofacies, temporospatial variations, 386
  postrift sequences, 387
  rifting, 382–383
  sedimentary sequences, 383, 385*f*
  synrift sequences, 383–386
  Tan-Lu fault, 342, 351*f*
Bohai Sea foraminifer assemblage, 283
Bohai Sea in China, 16, 19–20, 74–75, 75*f*,
    90–91, 92*f*, 183, 210–215, 213*f*,
    259–260, 261*f*, 282, 283, 287, 381,
    494–495, 572
Bølling–Allerød warming, 495–496
Borehole MZ-1-1, 646*f*
Borneo Deltas, 259
Bottom-simulating reflectors (BSRs), 107,
    130*f*, 622–623, 623*f*, 624*f*, 625–626,
    625*f*
Bottom temperature, ECS, 51*f*
Bouguer gravity map, 84*f*
Boundary currents, North Pacific, 668–669
BSRs. *See* Bottom-simulating reflectors
    (BSRs)
Bulimina alazanensis, 478

C

CaCO$_3$ production by reefs, 278–279, 279*f*
Calcareous *Cibicidoides*, 475
Calcareous components, 280–293
  foraminifers, 281–289
  nannoplankton, 289–291
  ostracods, 292–293
Calcareous nannoplankton, 186
  coccolith species, 289–290
  distribution, 290*f*
  F–EG ratio, 291
  $U_{37}^{K'}$-derived SST, 292*f*
Carbonate and coral reefs, 268–280, 629–630
Carbonate platforms, 21, 278, 280, 356–357,
    358–359, 360*f*, 378, 437–438, 448,
    598, 604–606, 615*f*, 647
Carbonate Plays, 598–599

Carbonates, 11–12, 14*f*, 21, 82, 85–86,
    108–109, 184, 201, 205, 208, 220,
    228, 230, 268–280, 304*f*, 307, 346,
    356–357, 358–359, 360*f*, 378,
    381–382, 426–428, 437–438, 439*f*,
    448, 450–451, 472–475, 474*f*,
    483–485, 491, 493–494, 515–516,
    528–530, 551–553, 573–575, 576,
    582, 594–596, 598–599, 604–606,
    613, 614, 615, 625–626, 627*f*,
    647–652, 657–658
Carbon cycling, 299–301, 479, 488–489
Carbon dioxide-rich fluids, 158
Carbon isotopic records, 489, 490*f*
Catchments, 23–37, 25*t*, 26*f*, 186, 258
Cedripites, 475, 477*f*
Cenozoic Manila subduction, 121
Cenozoic marine transgressions, China,
    657–663
Cenozoic petroleum-bearing system, 582
Cenozoic rifting, 76, 90–91, 102–115,
    119–120, 121, 164
Cenozoic sedimentary basins, 76–78, 77*f*, 341,
    344*t*, 397, 572*f*
Cenozoic sedimentary layers, 96–97
Cenozoic stratigraphy, 341–342, 347–368
  deep-water stratigraphy, 359–368
  shelf-slope basins, 347–368
  units (formations), 348*f*
Changjiang, 16–18, 23–24, 29–31, 237–245,
    249, 249*f*, 316–318, 318*f*, 399–402
Changjiang Delta, 237–245
Chaoshan Depression, 646, 646*f*
Cheju Island, 142, 510*f*, 514
Chemical index of alteration (CIA), 486–487,
    487*f*
China Sea Basins, 449
  Cenozoic Sedimentary Basins, 342, 343*f*,
    344*t*
  rifting, 343–346
China Seas, 1, 2, 2*t*, 3*t*, 4–5, 16*f*, 74–75, 75*f*.
    See also East China Sea (ECS);
    South China Sea (SCS)
  and Cenozoic sedimentary basins, 76–78,
    77*f*, 572*f*
Chinese economic reform, 1
Chinese National Science Foundation, 670
Chinese Sedimentary Basins, 341, 368
Chlorite, 189*t*, 203–205, 208, 318
CIA. *See* Chemical index of alteration (CIA)
Cibicidoides wuellerstorfi, 286*t*, 443, 478
Climatological mean precipitation rates,
    39–40, 39*f*

Climatology, 37–64, 472
Coastal ecosystems, 24–26
Coastal placers, 571, 617–619, 618f, 619f
Coastal upwelling, Vietnam, 55–56, 66
Coastline migration and sea-level changes, 501–504
COB and COT, 117–120
CO_2 gas fields, 626–628, 627f, 628f
Cold seeps, Northern SCS, 4, 625–626
Color spectral reflectance (CSR) record, 444–445
Complete Wilson cycle, 79
Continental shelfs, 11–12, 16–18, 21, 96–97, 194–198, 201–205, 366, 500–514, 594–596, 656–657, 669–670
Continental slope and deep basin, 205–210
  sand dunes, sediment waves and drifts, 208–210
  sediment distribution and mineral composition, 205–208
Continent–ocean boundary (COB), 88, 100, 117–118, 118f, 119f, 120f, 121–122, 133, 608f
Continent–ocean transition zone (COT), 80f, 99–100, 101f, 110, 110f, 113, 117–120, 121
Coral debris, 205
Coral reefs, 21, 600–601, 629–630
  $CaCO_3$ production, 278–279
  carbonate buildups and, 268–280
  modern distribution, 273–278, 276f, 277t
  SCS, 280
Core BC1, 213–215, 213f, 236–237
Core MD97-2151, southwestern SCS, 529f
Core Y127, 310
Cretaceous and Paleogene lacustrine, 584
Cultural Revolution, 1, 183
Curie depths, 100–102, 101f
Cuu Long Basin and Wan'an (Nam Con Son) Basin, 113–115, 114f, 115f, 116f, 420–423, 614–616
  active petroleum systems, 614, 615f
  Eocene lacustrine deposits, 615
  faults and Cenozoic seismic sequences, 420, 421f
  lithobiostratigraphy, 420, 422f
  Miocene carbonate platforms, 615
  Miocene synrift megasequence, 422–423
  NE–SW-trending, 420, 421f
  Oligocene-Miocene spreading, 615, 616f
  postrift megasequence, 422–423
  rifting phases, 420

D

Dam constructions, 24–26, 34, 35–36
3D analytic signal amplitudes (ASA), 89f, 142, 143–145, 143f, 144f, 148f
Dangerous Grounds, 21, 280, 356–357, 424, 425f, 437–438, 438f, 450–451, 609–614, 646
Dansgaard-Oeschger (D-O) events, 531–533
Deep crustal structures, 90–102
  Bohai and Yellow Sea, 90–93
  East China Sea (ECS), 93–96
  South China Sea (SCS), 96–102
Deep-dwelling nannoplankton *Florisphaera profunda*, 491
Deep reflection seismic surveys, 99–100
Deep regional unconformity (DRU), 356–357, 357f, 435–437
Deep Sea, 226–233
  Okinawa Trough, 230–233
  SCS, 226–230
Deepwater circulation, 57–64, 287–289, 288f, 654
The Deepwater oligocene, 475–477
Deepwater Plays, 596–598
Deep-water stratigraphy, 359–368
  lithostratigraphic characteristics, 359, 362f
  Oligocene section, 359–363
  preserved sediments, 363
  sediment, variations, 359, 361f
  turbidite deposits, 363
  volcanic ash and clay layers, 363–364
Delta growth, 548–551
Deltaic–estuarine–prodeltaic–shoreface facies series, 215
Deltaic sands and offshore muds, 233–259
DeltaMAP. *See* Deltas in the Monsoon Asia-Pacific Region (DeltaMAP)
Deltas in the Monsoon Asia-Pacific Region (DeltaMAP), 4
Dendritic drainage systems, 215, 216f
Depositional history
  Deep Sea, 226–233
  ECS, 210–215
  SCS, 215–225
Depositional mass
  average accumulation rates, 230, 232f
  distribution, deposit thickness, 230, 231f
  estimated accumulation rate, 229f
Detrital and biogenic carbonate, 268–270
DF1-1 gas field, 601, 602f, 605f
DF1-1 mud diapir field, 604

Diatoms, 256, 295–297, 298*f*, 302*f*, 307, 528–531

Discorbis spp., 647, 657–658

Domi Basin (Fujiang Depression), 399–402

Dongsha–Penghu uplift, 100

Downcore variations, planktonic foraminifers, 483–485, 485*f*

Drainage basin, Yangtze River, 24, 29, 30*f*

E

East Asian monsoon, 37–40, 41–42, 65–66, 267–268, 518, 537–539, 553

East China earthquakes, locations and depths, 155*f*, 156*f*

East China Sea (ECS), 1, 2, 11, 15*f*, 16–20, 17*f*, 42–45, 49, 73, 75*f*, 77*f*, 93–96, 104–106, 105*f*, 183–184, 187–189, 193–201, 197*f*, 210–215, 245, 247–251, 260–266, 265*f*, 284*f*, 342, 349–352, 353*f*, 364, 365*f*, 366*f*, 381–402, 400*f*, 469, 470–472, 494–514, 502*f*, 510*f*, 571, 572*f*, 585, 587*f*, 618*f*, 619*f*, 644–645, 647–649, 667, 668–670, 669*f*

  characteristics, 193–194

  continental shelf, 194–198

  depositional history, 210–233

  Okinawa Trough, 199–201

  Sand Ridges, 260–266

East China Sea Sand Ridges, 260–266

East China Sea (ECS) Shelf, 210–215, 397–402

East China Sea Shelf Basin (ECSSB), 342, 349*f*, 353*f*, 365*f*, 397–402, 398*f*, 400*f*, 669*f*

  basement rock types, 399

  Cenozoic basins, 349*f*

  Cenozoic Sedimentary Basins, 344*t*, 397

  depressions or subbasins, 397, 398*f*

  Donghai fault, 342

  evolution, stages, 397–399

  Mesozoic–Cenozoic strata, 399, 400*f*

  regional unconformities, megasequences, 399–402

  sediment facies, distribution, 399, 401*f*

  sequence stratigraphy, 402, 403*f*

  structural units, 397, 398*f*

  western depression belt, 397

ECS. *See* East China Sea (ECS)

ECSSB. *See* East China Sea Shelf Basin (ECSSB)

Elphidium spp., 283, 287, 647

Emerged shelves (solid areas), western Pacific marginal seas, 11–12, 15*f*

Environmental response, sea-floor spreading, 472–475

Eocene and Oligocene source rocks, 573–575, 589–590

ETOPO5 dataset, 46*f*

Eurasian Plate, subducted slab, 133, 134*f*

Extensive continental shelves, 22–23, 469, 669–670

F

"Female deltas", 24

Ferromanganese nodules/crusts, 205–208, 617, 620–622, 621*f*, 630

Flow velocity across Bashi Strait, 60*f*

Foraminiferal assemblage, planktonic species, 475

Foraminifers, 280, 286*t*, 292, 491–492, 657–658

  as indicators of depositional environments, 287–289

  modern distribution, 282–286

Forearc basin, 128–129, 363

G

Gas hydrate exploration, 4, 625*f*

Gas hydrate potentials, 622–626

  cold seeps, Northern SCS, 625–626

  geophysical exploration, Northern SCS, 622–625

  Okinawa Trough, 626

GDEM. *See* US Navy Generalized Digital Environment Model (GDEM)

Geographic and basin names, 5, 6*t*

Geology of the Bohai Sea, 494–495

Geology of the East China Sea, 494–495

Geomorphological features

  ECS, 16–18, 19*f*

  SCS, 21, 23*f*

Geomorphology, 12–23, 549–550, 550*f*, 654–655

Glacial cycles, 220, 273, 306–309, 516, 518–540, 654–657

Glacial periods, Okinawa trough, 222*f*, 306, 498*f*, 522–523, 539–540

Globigerinoides, 483–485

GMGS (Guangzhou Marine Geological Survey) scheme, 74, 422*f*, 428–429, 622

Goto Submarine Canyon, 500, 501*f*

Great Mesozoic East China Sea, 645
Green clay layers, 229, 363–364
Greenland ice-core and Chinese speleothem records, 522–523
Guangzhou Marine Geological Survey (GMGS), 74, 422f, 428–429, 622
Gulf of Thailand, 35, 203, 258, 429, 434–435, 450
Gunsan Basin, 392–396, 392f, 395f, 396f
  palynomorph assemblages, 395–396, 396f
  seismic units, 394–395, 395f

H

Hengchun Ridge, 128–129, 130–131, 624f
High-oxygen group (HOG), 478, 479f
Hòa Bình reservoirs, 34
Ho Basin, 364, 365f
HOG. See High-oxygen group (HOG)
Holocene brackish-water microfauna, 514
Holocene depocenter, 251
Holocene millennial-scale SST oscillations, 495–496
Holocene temperature changes, northeastern ECS, 515f
Holocene transgression, 501–503, 503f
HOV. See Human-operated vehicles (HOV)
Huanghe. See Yellow Sea
Huanghe Delta, 194–196, 235–237, 247–249
Huatung Basin, 121, 137
Human-operated vehicles (HOV), 2
Hydrocarbon reservoir formation, 584f
Hydrocarbon resources, 435–437, 572–617
  Bohai Basin, 572–581
  cenozoic sedimentary basins, oil and gas fields, 572f
  ECS Basin, 585–594
  Northern SCS (see Northern SCS)
  NYSB (see North Yellow Sea Basin (NYSB))
  Southern SCS (see Southern SCS)
  South Yellow Sea Basin, 583–585
Hydrologic system, marginal sea, 495

I

Iheya North Knoll, 157–158, 157f, 159f
IIH-1Xa well, 584–585, 586f
Illite, 187, 203–205, 208, 259, 318, 539
Imbrie-Kipp transfer function, 518–519
India–Eurasia continental collision, 78, 126f
Indian/South Asian monsoon, 37–39
Indosinian orogeny, 79, 85, 381–382

Infaunal detritus feeders, 528
Infaunal species, 478, 480f
Integrated Ocean Drilling Program (IODP), 74, 157–158, 312, 367–368, 670
International geological cruises, China Seas, 2, 3t
International Ocean Discovery Program (IODP), 157–158, 367–368, 367f, 670
Intertropical convergence zone (ITCZ), 40
IODP. See International Ocean Discovery Program (IODP)
Isotopic and astronomical stratigraphy
  $\delta^{18}O$ and $\delta^{13}C$ sequences, 438–440
  Neogene Isotopic Records, 443–447
  Pliocene–Pleistocene Isotopic Records, 440–443
  resolution records, 438–440
ITCZ. See Intertropical convergence zone (ITCZ)

J

Jeju Basin (Zhedong Depression), 399–402
Jiangshao Fault, 75–76, 75f, 77f, 91–93, 145–146, 155f, 618f
Jiangsu Offshore Sand Ridges, 267–268, 269f
Jiaxiang Fault, 93
Jilong Depression, 93–95
Jurassic, 82, 85–86, 87–89, 87f, 96–97, 121, 145–146, 146f, 346–347, 381–382, 581–582, 645, 646
Jurassic-early Cretaceous transgression, 644–645

K

Kaolinite, 187, 203–205, 208, 316
Kaoping Slope, 128–129, 130–131
Khmer Basin, 434–435
Kikai-Akahoya (K-Ah) caldera, 310
K-rich orogenic suites, 87–88
Kuroshio/Black Tide, 43
Kuroshio intrusion, 43–44, 57–58, 291, 470–471
Kuroshio migration, 496–499

L

Lancang, 34–35
Land climate, 534–540
Last Glacial Maximum (LGM), 2, 15f, 267–268, 287–289, 288f, 501–503, 519, 524–527, 528, 535, 542–543, 546–548, 547f, 548f, 594, 654–655, 656t

Late mesozoic marine deposits, 644–647
Late Pleistocene transgression, 501–503, 503*f*
Late quaternary paleoceanography,
     494–514
LGM littoral zone, 501–503, 502*f*
Lishui–Jiaojiang depression, 591–594, 593*f*
Lithostratigraphy
    basins and Pre-Cenozoic basements,
        342–347
    basins, basement types, 447
    biostratigraphic framework, 368–381
    Cenozoic stratigraphy, 347–368
    depositional successions, 449
    limestones and carbonate platforms, 448
    marine deposition, 449
    marine sequences, 448
    mudstones, layers, 448
    nonmarine deposits, 447
    synrift and postrift depositions, 449
Long-eccentricity cycles, 488, 489
Long-term trends in carbonate deposition, 273,
     275*f*
Luzon Arc, 78–79, 117, 128–129, 130*f*, 131,
     133, 134–136, 151–153, 165, 312
Luzon–Ryukyu Transform Plate Boundary
     (LRTPB), 89*f*, 119*f*, 121, 608*f*
Luzon Trough, 21, 128–129, 363
Luzon Volcanic Arc, 478

M

1.6 Ma Event, 490–491
Magmatism, 5, 76, 79, 80–81, 82, 85–86,
     87–88, 90, 99–100, 106, 116–117,
     122, 137, 585, 607–609, 611*f*,
     626–627
    hydrothermal activities, 156–160, 165, 620
    magmatic episodes, 145–153
    post-miocene continental margin
        neotectonics, 161–163
    regional seismicity (*see* Regional
        Seismicity)
Malay Basin
    Cenozoic basins, SCS, 352–354, 354*f*, 355*t*
    in Gulf of Thailand, distribution, 429–430,
        430*f*
    palynological assemblages, 372–373, 372*f*,
        433
    seismic sequences, 431, 431*f*
    stratigraphy, 431–432, 432*f*
    and Three Pagodas Fault, 429–430
    time rifting, 430
Male deltas, 24

Manila subduction zone, 121, 128–133,
     151–153, 156, 623–625
Manila Trench, 21, 79, 117, 128–129, 130,
     130*f*, 131, 132*f*, 133, 151–153, 156,
     363, 472, 478, 654
MAR. *See* Mass accumulation rate (MAR)
Marginal basins formation, 652–654
Marine geological researches, SCS, 2
Marine geology, 183
Marine records, 647–650
Marine-style fossils, 643
Mass accumulation rate (MAR), 524
MBE. *See* Mid-Brunhes Event (MBE)
Mekong Delta, 550*f*
Mekong Delta site localities, 257*f*, 320*f*,
     548–549
Mekong River Delta, 34–36, 36*f*, 256–258
Meltwater pulse (MWP), 546
Mesozoic and Paleozoic hydrocarbon
     potential, 606–609, 608*f*
Mesozoic and Paleozoic strata, 82
Mesozoic ECS, 645
Mesozoic northwestward subduction, 121
Mesozoic petroleum system, 582
Mesozoic tectonics, 76, 79–89, 102–103, 164
Mg/Ca-derived SST, 481–483, 484*f*
Mid-Brunhes Event (MBE), 493–494
Mid-Cenozoic topographic reversal in China,
     667–668, 669*f*
Middle Okinawa trough, 495–496
Mid-Miocene unconformity (MMU), 356–357
Mid-ocean ridge basalts (MORB), 150–151
Mid-Okinawa Trough, 2, 158
Mid-Pleistocene transition (MPT), 491–493,
     494
Mineral compositions
    average concentrations of characteristic
        elements, 187, 191*t*
    average percentage abundance, 187, 188*t*
    mean abundance, 187, 189*t*
    positioning, sediment sources, 187, 190*f*
    XRD spectral analyses, 187, 190*f*
Mineral provinces, 197, 199*t*
Mineral resources, 4–5, 73, 617–622, 630
    coastal placers, 617–619, 618*f*, 619*f*
    ferromanganese nodules/crusts,
        620–622, 621*f*
    phosphorite, 619–620
    sulfide deposits, 620
Miocene erosional event, 137–140
Miocene–Pliocene boundary, 107
Miocene sandstones, 470, 594–596, 596*f*
Mixed layer depth, SCS, 55*f*

Mixing mill, China Seas, 58, 66–67
Modeled annual mean flow patterns, Kuroshio, 498–499, 499*f*
Modern and quaternary carbonates
  detrital and biogenic carbonate distribution, 268–270
  long-term trends in carbonate deposition, 273
  quaternary carbonate cycles, 271–273
Modern sediment distribution
  East China Sea (ECS), 193–201
  patterns, 184–193
  sedimentologic features, 184
  South China Sea (SCS), 201–210
Moho depths, 90, 95–96, 96*f*, 100–102, 102*f*
Moho topography, 84*f*, 102*f*
Monographs and atlases, China Seas, 2, 2*t*
Monsoon, 4, 37–42, 256, 306, 469, 478–494, 516, 531–533, 534, 539–540, 551, 619–620
Monsoonal circulation system, 478–479
Monsoon climate and weathering rate, 486–488
Monsoon-driven biogeochemical processes, 489
Morphological features, marginal seas, 11, 13*t*, 668–669
MPT. *See* Mid-Pleistocene transition (MPT)
Mud areas, 508–514
Muddy and sandy sediments, 500
Multibeam morphotectonic analysis, 131, 132*f*
MWP. *See* Meltwater pulse (MWP)

N
Nannofossil analysis, 494
Nannofossil and planktonic foraminiferal biostratigraphy, 369, 370*f*, 371*f*
Nannoplankton, 280, 289–291, 492*f*
Nansha (NW Palawan) Trough, 354*f*, 435, 437–438
National Science Foundation of China (NSFC), 74, 670
Natural gamma ray (NGR) data, 444
N. dutertrei δ^{18}O, 292*f*
Neogene Isotopic Records
  benthic species, 443
  chronobiostratigraphic timescale, 443–444
  δ^{18}O and δ^{13}C events, 444, 445–447, 446*f*
  ETP records, 445
  NGR and CSR, 444–445
  tuned orbital age model, 444, 445

Neogene stratigraphy, Ryukyu Islands, 470, 470*t*
Neogene thermal subsidence, 103–104
Neogloboquadrina dutertrei, 483–485
NE-trending Xihu Depression, 93–95, 94*f*, 95*f*
NHG. *See* Northern Hemisphere glaciation (NHG)
Nodosariids, 478
Nonmarine Cenozoic hydrocarbons, 643
Northeast China Basin. *See* Bohai Basin
Northern Hemisphere glaciation (NHG), 481–483, 484*f*
Northern Okinawa trough, 500
Northern SCS, 1, 2, 4, 31, 76–78, 82, 87–89, 99–100, 110, 136–137, 161, 185–186, 187, 206*f*, 208, 209*f*, 215–220, 218*f*, 219*f*, 221*f*, 226–228, 278, 280, 285, 299–301, 324, 324*f*, 346–347, 352–354, 356, 356*f*, 371*f*, 373, 443, 450, 472, 482*f*, 486–487, 488*f*, 489*f*, 518–519, 521*f*, 523*f*, 529*f*, 532*f*, 534*f*, 535*f*, 537*f*, 546, 594–609, 622–626, 647, 650, 656–657
  Beibuwan Basin, 606, 607*f*
  Mesozoic and Paleozoic hydrocarbon potential, 606–609, 608*f*
  Pearl River Mouth Basin (PRMB), 594–599
  Taixinan Basin, 599–601
  Yinggehai and Qiongdongnan Basins, 601–606
Northern South China Sea shelf, 215–220, 659*f*
Northern South Yellow Sea Basins (NSYSB), 104, 151, 152*f*, 153*f*, 349*f*, 373, 374*t*, 376*f*, 387, 392–396, 583–585
  Cenozoic basins, 349*f*, 373, 376*f*
  eastern depression (Gunsan Basin), 394–395, 396*f*
  seismic units, 394–395, 395*f*
  sequence stratigraphy, 392–394, 393*f*
  structural characteristics, 392, 392*f*
  Subei Basin, 394
North Jiangsu Basin. *See* Subei Basin
North Korea's South Hwanghae Province, 18–19
The North Pacific Intermediate Water, 57–58, 66
North Sunda River valley, 548*f*
Northwest Subbasin, 79, 121–122, 123–124
North Yellow Sea Basin (NYSB), 18–19, 75*f*, 77*f*, 103–104, 153–154, 187–189, 297, 349–352, 349*f*, 352*f*, 387–391, 392–394, 572*f*, 581–583

North Yellow Sea Basin (NYSB) (*Continued*)
Cenozoic basins, ECS, 349*f*
seismic–stratigraphic mapping, 390, 391*f*
structural units, 387–389, 390*f*
synrift sequence sets, 389–390, 390*f*
NSFC. *See* National Science Foundation of
China (NSFC)
NW–SE-trending Red River Fault, 411, 412*f*

O

Ocean connection, 57–62, 472–478
Ocean connection and basin evolution,
472–478
Ocean Drilling Program (ODP), 122–123,
136–137, 273, 359, 362*f*, 364–366,
409*f*, 438–440, 441*f*, 443, 444, 445,
445*f*, 446*f*, 449, 472, 473*f*, 474*f*, 475,
476*f*, 477*f*, 478, 479*f*, 486–487, 487*f*
Oceanography, 37–66, 67, 469, 478–479, 494,
509–514, 531
ODP. *See* Ocean Drilling Program (ODP)
ODP Leg 184 Records, 226–229
ODP Site 1148, 122–123, 136–137, 273, 359,
362*f*, 364–366, 409*f*, 438–440, 441*f*,
443, 444, 445, 445*f*, 446*f*, 449, 472,
473*f*, 474*f*, 475, 476*f*, 477*f*, 478,
479*f*, 486–487, 487*f*
ODP Site 1202, 232–233, 291, 292*f*, 366, 367*f*
Offshore mud fields, 247–251
Oil exploration, Bohai Gulf, 1
Okinawa Trough, 2, 16–18, 22–23, 62, 76–78,
93–95, 115–140, 151, 156–157, 157*f*,
160*f*, 183–184, 186, 194, 195*f*,
199–201, 230–233, 272*f*, 292*f*, 309,
310–312, 364–368, 367*f*, 398*f*, 400*f*,
449–450, 470–472, 495–500, 497*f*,
501*f*, 620, 626, 644, 653, 668–669,
668*f*
Deep Sea, 230–233
East China Sea (ECS), 199–201
IODP Expedition 331, 367–368, 367*f*
Miocene erosional event, 137–140
Pleistocene and Holocene sediments,
dominance, 364–366, 366*f*
refraction data, 137
SCS basin, 116–137
sediment characteristics, 366, 367*f*
sediments, variations, 364, 365*f*
seismic reflections, 137, 138*f*, 139*f*
southern Ryukyu arc, 140, 141*f*
tectonic evolution, stages, 140
Okinawa Trough volcanic deposition, 310–312

downcore variations, grain size, volcanic
glass, and related minerals, 311*f*
drilling, 312
K-Ah caldera, 310
mainland-originated sediments, 201
ODP Site 1202, 232–233
sediment cores, 233*f*
silt-dominated southern trough, 200–201
silty clay, 199–200
^{87}Sr/^{86}Sr ratios and eNd values, 230–232, 234*f*
Old and new Baram Delta, 549*f*
Oligocene, 472
Oligocene basins, 435–437
Oligocene geochemistry and paleontology,
ODP Site 1148, 475, 476*f*
Oligocene–Miocene transition, 122, 136–137,
147–148, 147*f*
Onshore and offshore sectors, 573, 574*f*, 576
Optically stimulated luminescence (OSL)
dating, 213–215
Organic carbon, 299–309
particulate organic carbon (POC), 299–301
TOC
in glacial cycle, 306–309
in surface sediments, 301–306
Origin of sediments
abundance variations, detrital carbonate, 317*f*
clustering clay mineral assemblages, 319*f*
mineral and elemental properties, 316
REE variations and UCC-normalized REE
distribution, 317*f*
^{87}Sr/^{86}Sr ratios, carbonate-free fraction, 318*f*
strontium isotopic ratios, 316–318
terrigenous sediment, 316
Ostracod assemblages, 373–376, 377*t*
Ostracods, 201, 292–293, 347, 373–376, 377*t*,
475, 662*f*
Outer continental shelf, ECS, 506, 506*f*

P

Pacific cycles, 271–272, 272*f*, 273
Palawan and Nansha (Dangerous Ground)
area, 609–614
chronostratigraphic chart, 612–613, 612*f*
geochemical analyses, 614
magnetic data, 611–612
Miocene carbonate buildup, 614, 615*f*
Nido carbonate sequence, 612*f*, 613
Oligocene and Miocene reservoirs, 614
play concepts, 613*f*, 614
thermal maturity modeling, 613–614
Palawan Shelf Basin, 437–438

Paleoceanographic records
  ODP site, 491–492, 492*f*
  SCS, 481, 482*f*
Paleoceanography, 470–551
Paleocene and Eocene marine deposits, 649*f*
Paleocene-Eocene marine deposits, 648*t*
Paleocene-Eocene sea, 650–652
Paleocoastlines reconstructions, 547*f*
Paleogene marine deposits, 647–652
Paleogene postrift thermal subsidence, 90–91, 92*f*
Paleogene tectonic depocenter, 576, 577*f*
Paleolakes, rifted shelf-slope basins, 348*f*
Paleo-Pacific Plate
  Mesozoic Andes-type subduction, 80–81
  northwestward subduction, 79, 85, 128–129
Paleo-Pacific transgression, 85–86, 87*f*, 88–89, 607–609
Paleoreconstructions, 41–42, 660–661, 667
Paleoriver–paleodelta system, 215–218, 218*f*
Paleo-Sunda rivers, 21, 533, 540–542
Paleo-Tethys
  Mesozoic closure, 79
  and Paleo-Pacific interaction, 82–85
Paleozoic/early Mesozoic reservoirs, 585
Paraffinic oils, 582
Parker–Oldenburg algorithm, 100–102, 102*f*
Particulate and dissolved organic carbon (POC/DOC), 489
Pattani Basin and Khmer Basin, 434–435
Pearl River (Zhujiang), 31–33, 251–255, 405–408, 408*f*, 646
Pearl River Delta, 31–33, 251–255, 254*f*, 301, 303*f*, 551, 657–658, 660*f*
Pearl River estuary, 31–32, 32*f*, 53, 53*f*, 297, 298*f*
Pearl River Mouth Basin (PRMB), 77*f*, 78, 100, 101*f*, 107, 108–110, 111–112, 113, 136–137, 146, 161, 218*f*, 312–313, 342–346, 347, 352–354, 356, 369–371, 373, 403–404, 405–411, 447, 448, 449, 572*f*, 594–599, 618*f*, 622–623, 627–628, 630, 650
  Cenozoic basins, SCS, 352–354, 354*f*, 355*t*, 408
  dinoflagellate assemblages, 373
  fault-bounded blocks, 109*f*
  faults and sediment packages, 408, 409*f*
  lithology and sequence stratigraphy, 410–411, 410*f*
  palynological assemblages, 369–371, 372*f*
  postrift thermal subsidence, 108–109

  pre-Miocene synrifting sediments, 110, 110*f*
  rifting phase, 408–410
  sedimentation, 108–109
  sediment sequence, 410–411, 410*f*
  structural units, 405–408, 408*f*
Phase III paleodeltas, 215–218
Phase II paleodeltas, 215–218
Phase I paleodeltas, 215–218
Philippine Sea lithosphere, 128–129
Phosphorite, 619–620, 630
Phytoplankton/pollen ratios, 475, 477*f*
Piceaepollenites, 475
Planktonic foraminifera, 479–481
Planktonic foraminiferal deep-dwelling species, 485, 486*f*
Planktonic foraminiferal shells, 522
Planktonic foraminifers and nannofossils, 648*t*
Pleistocene paleoceanography, 490–494
Pleurostomellids, 478
Pliocene–Pleistocene Isotopic Records
  astronomical solution, 440
  ETP, 440–443, 442*f*
  marine isotopic stages (MIS 1 to MIS T1), 440, 441*f*
  tuned age model, 443
Plio–Pleistocene, 113
PL19-3 oil field, 581
Postglacial sea-level changes, SCS, 545*f*
Post-miocene continental margin neotectonics, 161–163
Postrift deposit, 342–343, 344*t*
Pre-Cenozoic basements
  of ECS Shelf Basin, 346
  NE–SW stretching Pearl River Mouth Basin, 347
  in northern SCS basins, 346–347
  in southern SCS basins, 346–347
Pre-oligocene marine deposits, 644–652
PRMB. *See* Pearl River Mouth Basin (PRMB)
Productivity, SCS, 524–531
Pulleniatina Minimum Event, 514
Pulleniatina obliquiloculata, 491–492, 496–497, 497*f*

Q

Qinling–Dabie–Sulu orogenic belt, 74–75, 75*f*, 142
Qinling–Dabie–Sulu orogeny, 77*f*, 81
Qiongdongnan Basin, 77*f*, 100, 110–113, 161, 342–343, 352–354, 356, 363, 411–416, 448, 572*f*, 596, 597, 599, 601–606, 618*f*, 623, 627–628, 630

Qiongdongnan Basin (*Continued*)
  basin-wide unconformities, 410*f*, 413–416
  Cenozoic basins, SCS, 352–354, 354*f*, 355*t*
  lithostratigraphic evidence, 413
  NW–SE-trending Red River Fault, 411, 412*f*
Quasiactive sand ridges, 504–506
Quaternary carbonate cycles
  Atlantic cycles, 271–272
  late quaternary carbonate fluctuations, 272*f*
  Pacific cycles, 271–272
  proxies variations, 274*f*
Quaternary deposits, 183–184, 218–220,
    312–313, 394, 504
Quaternary Shallow Biogenic gas reservoirs,
    594, 595*f*
Quaternary stratigraphic events, 380–381, 380*t*
QuikSCAT surface winds over SCS, 40–41,
    42*f*

R

Radiating sand ridge field, South Yellow Sea,
    506, 507*f*
Radiolarian-bearing rocks, 646
Radiolarians, 297–299, 300*f*, 302*f*
Red River Delta, 255–256, 551
Red River Fault, 75*f*, 77*f*, 78, 79, 89–90, 100,
    111, 124–128, 161, 164, 342,
    411–413, 618*f*
Red River system (Sông Hồng), 33–34
Reed Bank Basin, 354*f*, 435
Regional deglacial warming, 495–496
Regional lithostratigraphy, characteristics,
    447–449
Regional seismicity, 153–156
  China Earthquake Networks Center,
    153–154
  earthquakes, South China Sea area, 156*f*
  East China earthquakes, locations and
    depths, 155*f*
  integrated geologic model, 155*f*
  magnitudes, stations and earthquakes,
    153–154, 154*f*
  offshore intraplate earthquakes, 153–154
  Sumatra–Andaman earthquake, 153
Regional stratigraphy and sequence
    stratigraphy
  Bohai Basin, 381–387
  ECSSB, 397–402
  Yellow Sea basins, 387–396
Regional topographic map, Manila subduction
    zone, 117, 129*f*
Relict sands and sand ridges

depositional sequences, outer continental
    shelf, 265*f*
East China Sea Sand Ridges, 260–266
evolution of sand ridges, 266*f*
Jiangsu Offshore Sand Ridges, 267–268
linear sand ridge, 263*f*
major sand ridge fields, 261*f*, 262*t*
seismic profile, 263*f*
Remotely operated vehicles (ROV), 2
Rhyolites, 144–146, 312–313, 314–315
Rifting basin. *See* Okinawa Trough
River deltas, 23–37, 183–184, 233–235, 545,
    617
River runoff, 500
River system, 540–544
ROV. *See* Remotely operated vehicles (ROV)
Ryukyu arc, 76–79, 137–140, 141*f*, 151–153,
    364, 365*f*, 471–472, 497–498
Ryukyu Archipelago, 143–144
Ryukyu-Taiwan land bridge, 497–498

S

Sabah Basin, 423–424, 435–437, 617
Sand dunes, sediment waves and drifts
  echosounder data, time series composite,
    209*f*
  mesoscale internal waves, 208
  slump and turbidite deposition, 210
  wave fields, 210, 211*f*
Sand ridges, 19–20, 185–186, 194, 202–203,
    215, 259–268, 504–508
Sarawak and East Natuna Basins. *See* Zengmu
    Basin
Satellite and modeled chlorophyll distribution,
    SCS, 527*f*
SCS. *See* South China Sea (SCS)
SCS basin, 53, 59, 66, 78, 89*f*, 96, 100–103,
    113, 116–137, 148–150, 153–154,
    158, 165, 226–228, 273, 293,
    299–301, 346–347, 355*t*, 356–357,
    357*f*, 358, 368–369, 372*f*, 378,
    433–434, 447, 448, 449–451,
    472–475, 478, 518, 602*f*, 644, 654
SCS coral reef history, 280, 281*f*
SCSMEX. *See* South China Sea Monsoon
    Experiment (SCSMEX)
Seafloor spreading, 78–79, 90, 110, 113–115,
    117, 122, 147–150, 164–165, 228,
    312, 444, 472–475, 652
  COB, 117–118
  complete Wilson cycle, 79
  Miocene, 88

Oligocene–Miocene transition, 136–137
opening sequences, 123
postspreading volcanisms, 158
in SCS, 78–79, 99–100, 113
Sea–land interactions, 11, 469, 509, 514, 552, 670
Sea-level and environmental changes, China, 500
Seasonality and monsoon, 65–66, 553
Seasonal thermocline, 49, 53, 55*t*
Sea surface salinity (SSS), 49–53, 50*f*, 52*f*, 481–483
Sea surface temperature (SST), 49–51, 50*f*, 52*f*, 61–62, 66, 67, 291, 292*f*, 479–485, 482*f*, 493–494, 495–496, 518–524, 521*f*
Sediment cores and boreholes, 511*t*
Sediment distribution patterns
  chemical compositions, 187–193
  mineral compositions, 187
  sedimentologic features, 184
  sediment types and distribution, 184–186
Sediment-hosted CO_2 lakes, southern Okinawa Trough, 2
Sedimentologic features, 184
Sedimentologic response, SCS, 472, 473*f*
Sedimentology
  biogenic sediments, 268–309
  depositional history, 210–233
  modern sediment distribution, 184–210
  origin of sediments, 316–318
  sediment transport dynamics, 319–324
  terrigenous sediments, 233–268
  volcanic sediments, 310–315
Sediments and tectonic pattern, Middle Triassic distribution, 80*f*
Seismic profile, 221*f*
Seismic reflections, 73, 75–76, 82, 83*f*, 90, 93–95, 96*f*, 99–100, 103, 117, 120*f*, 122, 130–131, 137, 138*f*, 139*f*, 151, 158, 580–581, 585, 611–612, 613
Sensitivity to eustatic changes, 22–23
Sequence stratigraphy, 381–438, 449
Shallow seismic profiles, 503
Shelf-slope basins, 373, 423
  ECS sector, 349–352, 349*f*, 350*t*, 351*f*, 352*f*
  lithostratigraphy, 347
  paleolakes in rifted shelf-slope basins, 347, 348*f*
  SCS sector, 352–358, 354*f*, 355*t*, 356*f*, 357*f*
  synrift and postrift sequences, 347
Shelf–slope seismic profile, 215, 217*f*
The Shi-Pu limestone, 644–645

Siliceous components
  diatoms, 295–297
  radiolarians, 297–299
  total opal data and radiolarian data, 293–295, 296*f*
SIMMAX-28 transfer function, 519
Slab geometry and deformation patterns, 131
Small mountainous rivers, 37, 548
Socotra Basin (Changjiang Depression), 399–402
Sông Hồng, 33–34
South Asian subsystem, 37–39
South Asian summer monsoon, 37–39
South China Sea (SCS), 53, 59, 66, 78, 89*f*, 96–103, 113, 116–137, 148–150, 153–154, 158, 165, 201–210, 226–228, 273, 293, 299–301, 346–347, 355*t*, 356–357, 357*f*, 358, 368–369, 372*f*, 378, 433–434, 447, 448, 449–451, 472–478, 515–551, 602*f*, 644, 646–647, 650, 654, 667, 668–670
  cenozoic rifting, 102–115
  characteristics, 201
  continental shelf, 201–205
  continental slope and deep basin, 205–210
  Deep Sea, 226–230
  depositional history, 215–225
  heat flow measurement, 73
  joint geophysical surveys, 74
  and Okinawa trough, 115–140
  Pearl River Mouth Basin, 78
  seafloor spreading, 78–79
  subbasins, 79
  volcanic deposition, 312–315
South China Sea Deep (SCSD), 74
South China Sea Monsoon Experiment (SCSMEX), 41–42
South China Sea Shelf, 215–225
South China Sea volcanic deposition, 312–315
Southern Okinawa Trough, 2, 137, 140, 186, 291, 297, 316, 364, 496–497, 500, 626
Southern Ryukyu arc, 140, 141*f*
Southern SCS, 11–12, 37, 53, 61–62, 82–85, 121, 226–228, 230, 259, 278, 280, 302*f*, 314–315, 346–347, 356–357, 356*f*, 357*f*, 358–359, 372–373, 372*f*, 420, 440, 448, 449–450, 478, 481, 485, 488*f*, 490–491, 519, 522, 523*f*, 529*f*, 533, 533*f*, 535*f*, 541*f*, 545, 609–617, 646, 647, 652

Southern SCS (*Continued*)
  Cuu Long and Wan'an (Nam Con Son)
    basins, 614–616
  Palawan and Nansha (Dangerous Ground)
    area, 609–614
  Zengmu–Sabah Basin, 617
Southern South Yellow Sea Basins (SSYSB),
    104, 349*f*, 387, 392–396, 583, 583*f*,
    584
  Cenozoic basins, 349*f*
  eastern depression (Gunsan Basin),
    394–395, 396*f*
  seismic units, 394–395, 395*f*
  sequence stratigraphy, 392–394, 393*f*
  structural characteristics, 392, 392*f*
South Yellow Sea Basin, 89–90, 104, 106, 151,
    152*f*, 342–343, 349–352, 387, 447,
    448, 449, 572*f*, 583–585,
    584*f*, 618*f*
Spores and pollen assemblages, 419, 475, 659*f*
  dinoflagellate assemblages, 373
  in ECS, 373, 374*t*
  palynological assemblages or zones, SCS,
    369–371, 372–373
SSS. *See* Sea surface salinity (SSS)
SST. *See* Sea surface temperature (SST)
Stilostomellids, 478
Stratigraphical Lexicon of China, 341
Stratigraphy and evolution, China Seas
  lacustrine, accumulation, 450–451
  marine influence, 450
  nonmarine deposition, 450
  synrift and postrift sequences, 449–450
Subei Basin, 390*f*, 392–394
Suizhong 36-1 oil field, 576–577, 579*f*, 580*f*
Sulfide deposits, 2, 620, 630
Sulu–Dabie orogen, 143–144
Sulu orogen, 75, 81–82, 90, 93, 93*f*
Summer southeast Vietnam offshore current,
    47–48
Sun Basin. *See* North Yellow Sea Basin
Sunda Shelf, 21, 22–23, 56, 59, 193, 205,
    220–225, 224*t*, 316, 516–517, 524,
    540–551, 544*f*, 654–655
  interpreted environmental dynamics, 226*f*
  sediment-acoustic (parasound) profiles,
    220, 222*f*
  stratigraphic units, 220, 224*t*
  west–east profile in cores along Mekong
    incised-valley system, 227*f*
Surface circulation
  ECS, 42–48, 45*f*, 518
  SCS, 47*f*, 48*f*

Surface pressure and wind field, China Seas,
    40–41, 41*f*
Surface sediments
  TOC, 301–306
  western Pacific marginal seas, 11–12, 14*f*
Synrift deposit, 344*t*, 356–357, 378, 449–450
  ECS Basins, 349–352, 350*t*
  SCS Basins, 349–352, 355*t*
  stratigraphic evidence, 449–450
SYSB-N, 104, 583, 584, 586*f*

T

Taiwan Orogen, 133–137, 136*f*
Taiwan rivers, 37, 38*f*, 194, 323*f*
Taiwan–Tsushima–Tsugaru Warm Current
    System, 43–44
Taixinan Basin, 77*f*, 89*f*, 96–97, 99, 100, 107,
    113, 121, 342–343, 352–354, 363,
    369, 403–408, 406*f*, 407*f*, 448, 450,
    572*f*, 597*f*, 599–601, 608*f*, 609–612,
    618*f*, 622–623, 646, 650–652
  bottom-simulating reflectors (BSRs), 107
  Cenozoic basins, SCS, 352–354, 354*f*, 355*t*,
    404–405
  coral reefs, 600–601, 601*f*
  faults and structural units, 403–404, 404*f*
  high-angle normal faults, 107
  Miocene–Pliocene boundary, 107
  NE–SW-trending basins, 403–404
  Oligocene and younger deposits, 405
  Oligocene transgressive sandstones, 107
  petroliferous structures, 600
  postrift sediments, 107
  rifting and subsidence, 599
  rocks source, 599
  stratigraphic framework, 405, 407*f*
Tan-Lu fault, 75–76, 75*f*, 77*f*, 81–82, 83*f*, 84*f*,
    89–93, 92*f*, 142, 155*f*, 161, 162*f*,
    164, 342, 383, 387, 392
Taxodiaceaepollenites, 475
Tectonic and sedimentologic events, China
    sea, 656*t*
Tectonic events, Southeast China, 86*f*
Tectonic evolution, stages, 140, 141*f*
Tectonic framework, 74–79, 135*f*, 164, 343*f*
  Bohai Basin, Cenozoic evolution, 75
  cenozoic continental margin rifting and
    subsidence, 89–115
  China Seas, 74–75, 75*f*
  China Seas and major Cenozoic sedimentary
    basins, 76–78, 77*f*
  faulting and volcanisms, 76–78

India–Eurasia continental collision, 78
Jiangshao Fault, 76
NE strikes, 75–76
orogenic belts and large-scale faults, 75
Qinling–Dabie–Sulu orogenic belt, 74–75, 75f
Red River Fault zone, extrusion process, 78
Sulu orogen, 75
western Philippine Sea Plate, 76–78
Yangtze Block, 76
Tectonic inversion and denudation, 104
Temperature and salinity, 49–56, 481–483
Tephra layers, 310
Terrigenous sediments
deltaic sands and offshore muds, 233–259
relict sands and sand ridges, 260–266, 267–268
Tethys Ocean, 647
TEX$_{86}$-based SST, 495
TF. See Transfer function (TF)
TGD. See Three Gorges Dam (TGD)
Thai Basin Group, 342–343
in Gulf of Thailand, distribution, 430f
lithostratigraphy, 436f
The International Conference on Asian Marine Geology, 4, 5t
Thermocline, 479–485
Three Gorges Dam (TGD), 30f, 31
Tibetan Plateau, 24, 26–27, 29, 34–35, 37–39, 116–117
Tidal sand ridges, ECS shelf, 504, 505f
TOC. See Trend of organic carbon (TOC)
Topographic effect, East Asia, 667–668
Topographic maps
ECS, 16, 17f, 117, 128–129, 129f
Yellow Sea, 18–19, 20f
Topography
Bohai Gulf, 19–20, 21f, 102f, 117, 346, 397–399, 545, 667–668
SCS, 20–21, 22f
Total organic carbon (TOC)
in glacial cycle, 306–309, 307f, 308f
in surface sediments, 301–306, 304f, 305f, 306f
Transfer function (TF) method, 479–481
Trend of organic carbon (TOC), 491, 500
Triassic carbonate sequences, 82
Triassic tectonic convergence, 80–81
Tsugaepollenites, 475
Tsushima Warm Current west of Kyushu, 43
Turbidite-associated agglutinated foraminifers, 475
Types and distribution, sediments, 184–186

U

$U^{K'}_{37}$, 479–481, 514, 520–521, 521f, 525t
$U^{K'}_{37}$-derived and Mg/Ca-derived SST records, 525
$U^{K'}_{37}$-derived SST, 292f
Ulreung-Oki (U-Oki) caldera, 310
UNESCO/IOC projects, 4
Upper waters and monsoon climate, 478–494
Upper water structure, 531–533
US Navy Generalized Digital Environment Model (GDEM), 63

V

Volcanic arc, 128–129, 142, 143–144, 151–153, 397–399, 478
Volcanic deposition, SCS
igneous rocks in and surrounding SCS, 312–313, 313f
large-scale eruptions, Toba volcano, 315
positions, thickness, and estimated ages, ash layers, 315f
tephra layers of Pleistocene age, 314
volcanic ash, 313
Volcanic sediments
Okinawa Trough volcanic deposition, 310–312
SCS volcanic deposition, 312–315
Volcanisms, 76–78, 87–88, 90–91, 100, 103, 128, 145, 147–148, 147f, 150–151, 156–157, 158, 165, 230, 289–290, 297–299, 310, 312, 363–364, 367–368

W

Wan'an (Nam Con Son) Basin, 113–115, 342–343, 352–354, 420–424, 447, 448
carbonate platforms, 358–359
Cenozoic basins, SCS, 352–354, 354f, 355t
faults and Cenozoic seismic sequences, 420, 421f
lithobiostratigraphy, 420, 422f
Miocene synrift megasequence, 422–423
NE–SW-trending, 420, 421f
Water column and surface sediments, 196
Water exchange with oceans, 57–62
Weathering rates, northern SCS records, 486–487, 487f
West Bay Basin. See North Yellow Sea Basin
Western Hinge Fault (WHF), 429–430

Western Pacific marginal seas, 11, 12f, 14f,
    15f, 116–117, 183, 519, 667,
    668–669, 668f
Western Pacific warm pool (WPWP),
    37–39, 485
Western Philippine Sea Plate, 76–78, 133
Western Philippine Sea water, 59
West Natuna Basin, 372–373, 423–424
    ASCOPE, 433–434
    Barat Formation, 434
    Cenozoic sediment succession, 433–434
    early Oligocene Gabus Formation, 434
    in Gulf of Thailand, distribution, 430f, 433
    Muda Formation, 434
    Sunda Folds, 433
West Pacific marginal basins, 643–644
WPWP. See Western Pacific warm pool
    (WPWP)

X

Xialiao Basin, 381, 383–387
Xihu Depression, 93–95, 94f, 95f, 104–105,
    105f, 106, 106f, 140, 142, 147–148,
    147f, 151, 162f, 369, 397–399, 402,
    585–591, 588f, 589f, 591f, 592f,
    647–649
    ECS Basin, 142, 151, 162f, 585, 587–588,
    588f, 590, 591, 593f

Y

Yangtze Block, 75–76, 81, 82, 91–93
Yangtze clays, 508
Yangtze River (Changjiang), 16–18, 29–31,
    41–42, 44–45, 49, 65–66, 237–245,
    249, 249f, 318f
Yangtze River Delta, 237–245, 241f, 243f,
    244f
Yangtze River Distal Subaqueous Delta,
    245–247, 246f
Yangtze upstream and downstream,
    1984–2008, 30–31, 30f
Yanshanian event, 145–146, 146f
Yellow and East China Seas, 247–251
Yellow River Delta, 235–237, 236f, 238f
Yellow River Distal Subaqueous Delta, 237,
    240f
The Yellow River drainage basin, 26–27, 28f
Yellow Sea (Huanghe), 12f, 15f, 16, 17f,
    18–20, 22–23, 28–29, 43–45, 45f, 49,
    57, 65–66, 235–237, 316, 317f
Yellow Sea basins, 1, 2, 15f, 16, 17f, 18–19,
    20f, 22–23, 43, 49, 57, 65–66, 73,

    90–93, 103–104, 142, 183, 186, 193f,
    194–196, 197f, 236–237, 247–249,
    283, 287, 297, 347, 368–369, 387,
    389f, 450–451, 494–495, 500,
    509–514, 509f, 513f, 515f, 571, 620,
    653, 654–655, 657–658, 667
    Cenozoic basins, ECS, 349f
    Jiaxiang fault, 342
Yellow Sea Warm Current (YSWC), 43–44,
    44f, 45, 45f, 46f, 49, 194–196,
    320–322, 509–514
Yinggehai and Qiongdongnan Basins,
    111–113, 161, 342–343, 352–354,
    356, 411–416, 448, 572f, 601–606,
    618f, 623, 630
    depocenters migration, 112f
    DF1-1 gas field, 601, 602f, 605f
    DF1-1 mud diapir field, 604
    Diapiric faults, 604
    Drowned Miocene reef carbonate platform,
    604–606
    gas generation, 603–604
    natural gases, 604
    NS-trending uplift, 111
    one-dimensional backstripping analysis,
    113f
    one-dimensional tectonic subsidence
    analysis, 111
    petroleum system elements, 605f
    Plio–Pleistocene, 113
    Red River Fault, 111
    reef carbonates representative, 606f
    stratigraphy, 603f
    thermal cooling, 111–112
Yinggehai Basins, 77f, 89–90, 100, 111–113,
    112f, 161, 163f, 342–343, 352–354,
    411–413, 414f, 447, 572f, 596,
    601–604, 602f, 618f, 627–628, 630
    basin-wide unconformities, 410f, 413–416
    Cenozoic basins, SCS, 352–354, 354f, 355t
    Cenozoic deposits, 413
    evolution, stages, 410f, 413–416
    extensional tectonic processes, 411–413
    mass transport deposits (MTD), 413
    NW–SE-trending Red River Fault, 411, 412f
YSWC. See Yellow Sea Warm Current
    (YSWC)

Z

Zengmu Basin, 342–343, 617
    ASCOPE, 426–428, 427f
    Cenozoic basins, SCS, 352–354, 354f, 355t

evolution, tectonic model, 424, 425*f*
faults and structural–stratigraphic provinces,
    423–424, 424*f*
Rajang Fold–Thrust Belt, 423–424
Sarawak Basin, 428–429, 435
SCS sector, 352–354
sedimentary cycle concept, 426, 427*f*
sediment succession, 428–429
Shell's cycle concept, limitations, 428
Terumbu Formation L-structure, 358–359

Zengmu–Sabah Basin, 617
Zhenghe–Lianhuashan Fault, 77*f*, 82, 87*f*, 135*f*,
    146*f*
Zhongnan Fault, 77*f*, 89*f*, 119*f*, 121–122,
    123–124, 161–163, 164*f*, 618*f*
Zhongnan ridge, 119*f*, 161–163, 164*f*
Zhujiang, 23–24, 31–33, 251–255, 405–408,
    408*f*, 472, 646
Zhujiang Delta, 251–255
Zoophycos, 359, 475

Printed and bound by CPI Group (UK) Ltd, Croydon, CR0 4YY

08/05/2025

01864822-0001